T0301025

# PR🔴BLEMS
### 4th Edition
## AND
# SOLUTIONS
## IN
## QUANTUM
## COMPUTING
## AND
## QUANTUM
## INFORMATION

# Willi-Hans Steeb
*University of Johannesburg, South Africa*

# Yorick Hardy
*University of the Witwatersrand, South Africa*

# PR**O**BLEMS
**4th Edition**
## AND
# SOLUTIONS
## IN
## QUANTUM
## COMPUTING
## AND
## QUANTUM
## INFORMATION

**World Scientific**

NEW JERSEY · LONDON · SINGAPORE · BEIJING · SHANGHAI · HONG KONG · TAIPEI · CHENNAI

*Published by*

World Scientific Publishing Co. Pte. Ltd.

5 Toh Tuck Link, Singapore 596224

*USA office:* 27 Warren Street, Suite 401-402, Hackensack, NJ 07601

*UK office:* 57 Shelton Street, Covent Garden, London WC2H 9HE

**Library of Congress Cataloging-in-Publication Data**

Names: Steeb, W.-H., author. | Hardy, Yorick, 1976–   author.

Title: Problems and solutions in quantum computing and quantum information /
    Willi-Hans Steeb and Yorick Hardy, University of Johannesburg, South Africa.

Description: 4th edition. | Singapore ; Hackensack, NJ : World Scientific, [2018] |
    Includes bibliographical references and index.

Identifiers: LCCN 2018001917| ISBN 9789813238404 (hardcover ; alk. paper) |
    ISBN 9813238402 (hardcover ; alk. paper) | ISBN 9789813239289 (pbk ; alk. paper) |
    ISBN 981323928X (pbk ; alk. paper)

Subjects: LCSH: Quantum computers. | Quantum theory.

Classification: LCC QA76.889 .S74 2018 | DDC 006.3/843--dc23

LC record available at https://lccn.loc.gov/2018001917

**British Library Cataloguing-in-Publication Data**

A catalogue record for this book is available from the British Library.

For any available supplementary material, please visit
http://www.worldscientific.com/worldscibooks/10.1142/10943#t=suppl

Printed in Singapore

# Preface

The purpose of this book is to supply a collection of problems in quantum computing and quantum information together with their detailed solutions which will prove to be valuable to graduate students as well as to research workers in these fields. All the important concepts and topics such as quantum gates and quantum circuits, quantum channels, entanglement, teleportation, Bell states, Bell inequality, Schmidt decomposition, quantum Fourier transform, magic gate, von Neumann entropy, quantum channels, quantum cryptography, quantum error correction, coherent states, coherent Bell states, squeezed states, POVM measurement, beam splitter, homodyne detection and Kerr Hamilton operator are included. The topics range in difficulty from elementary to advanced. Almost all problems are solved in detail and most of the problems are self-contained. All relevant definitions are given. Students can learn important principles and strategies required for problem solving. Teachers will also find this text useful as a supplement, since important concepts and techniques are developed in the problems. The book can also be used as a text or a supplement for linear and multilinear algebra or matrix theory. Each chapter also includes supplementary problems.

Most chapters also include programming problems in Maxima and SymbolicC++.

The material was tested in our lectures given around the world.

Any useful suggestions and comments are welcome.

The International School for Scientific Computing (ISSC) provides certificate courses for this subject. Please contact the first author if you want to do this course. More quantum computing exercises can be found on the web page given below.

e-mail addresses of the authors:

whsteeb@uj.ac.za
steebwilli@gmail.com
yorick.hardy@wits.ac.za
yorickhardy@gmail.com

Home page of the authors: http://issc.uj.ac.za

# Contents

# Notation

| | |
|---|---|
| $\emptyset$ | empty set |
| $\mathbb{N}$ | natural numbers |
| $\mathbb{N}_0$ | natural numbers including 0 |
| $\mathbb{Z}$ | integers |
| $\mathbb{Q}$ | rational numbers |
| $\mathbb{R}$ | real numbers |
| $\mathbb{R}^+$ | nonnegative real numbers |
| $\mathbb{C}$ | complex numbers |
| $\mathbb{R}^n$ | $n$-dimensional Euclidean space |
| $\mathbb{C}^n$ | $n$-dimensional complex linear space |
| $\mathcal{H}$ | Hilbert space |
| $i$ | $\sqrt{-1}$ |
| $\omega_n := \exp(2\pi i/n)$ | $n$-th root of unity $n \in \mathbb{N}$ |
| $\Re(z)$ | real part of the complex number $z$ |
| $\Im(z)$ | imaginary part of the complex number $z$ |
| $A \subset B$ | subset $A$ of set $B$ |
| $A \cap B$ | the intersection of the sets $A$ and $B$ |
| $A \cup B$ | the union of the sets $A$ and $B$ |
| $f \circ g$ | composition of two mappings $(f \circ g)(x) = f(g(x))$ |
| $\mathbf{x}$ | column vector in $\mathbb{C}^n$ |
| $\mathbf{x}^T$ | transpose of $\mathbf{x}$ (row vector) |
| $\times$ | vector product in $\mathbb{R}^3$ |
| $\|\cdot\|$ | norm |
| $\mathbf{x} \cdot \mathbf{y} \equiv \mathbf{x}^*\mathbf{y}$ | scalar product (inner product) in $\mathbb{C}^n$ |
| $\langle . | . \rangle$ | scalar product in Hilbert space |
| $\mathbf{x} \times \mathbf{y}$ | vector product |
| $A \oplus B$ | direct sum of matrices $A$ and $B$ |
| $A \otimes B$ | Kronecker product of matrices $A$ and $B$ |
| $f \otimes g$ | tensor product of elements $f$ and $g$ of Hilbert spaces |
| $\det(A)$ | determinant of a square matrix $A$ |
| $\mathrm{tr}(A)$ | trace of a square matrix $A$ |
| $\mathrm{rank}(A)$ | rank of matrix $A$ |
| $A^T$ | transpose of matrix $A$ |
| $\overline{A}$ | conjugate of matrix $A$ |
| $A^*$ | conjugate transpose of matrix $A$ |
| $A^\dagger$ | conjugate transpose of matrix $A$ (notation used in physics) |

| | |
|---|---|
| $I_n$ | $n \times n$ unit matrix |
| $I$ | unit operator |
| $U$ | unitary operator, unitary matrix |
| $\Pi$ | projection operator, projection matrix |
| $P$ | permutation matrix |
| $\sigma_1, \sigma_2, \sigma_3$ | Pauli spin matrices |
| $S_1, S_2, S_3$ | spin matrices for spin $s = 1/2, 1, 3/2, 2, \ldots$ |
| $\rho$ | density operator, density matrix |
| $[A, B] := AB - BA$ | commutator for square matrices $A$ and $B$ |
| $[A, B]_+ := AB + BA$ | anticommutator for square matrices $A$ and $B$ |
| $\delta_{jk}$ | Kronecker delta with $\delta_{jk} = 1$ for $j = k$ |
| | and $\delta_{jk} = 0$ for $j \neq k$ |
| $E_{jk}$ | elementary matrices with 1 at $jk$ and 0 otherwise |
| $\lambda$ | eigenvalue |
| $\epsilon$ | real parameter |
| $H$ | Hamilton function |
| $\hat{H}$ | Hamilton operator |
| $\{ |0\rangle, |1\rangle, \ldots, |n-1\rangle \}$ | arbitrary orthonormal basis for $\mathbb{C}^n$ |
| $|\Phi^+\rangle, |\Phi^-\rangle, |\Psi^+\rangle, |\Psi^-\rangle$ | Bell states in $\mathbb{C}^4$ |
| $\hbar$ | $h/2\pi$ with $h$ the Planck constant |
| $t$ | time |
| $\omega$ | frequency |
| $\mathbf{k}$ | wave vector |
| $\mathbf{r}$ | space coordinates |
| $\mathbf{p}$ | momentum |
| $b, b^\dagger$ | Bose annihilation and creation operators |
| $|n\rangle$ | number states, Fock states $n = 0, 1, 2, \ldots$ |
| $|\beta\rangle$ | coherent state, $\beta \in \mathbb{C}$ |
| $|\zeta\rangle$ | squeezed state, $\zeta \in \mathbb{C}$ |
| $D(\beta)$ | displacement operator, $\beta \in \mathbb{C}$ |
| $S(\zeta)$ | one-mode squeezing operator, $\zeta \in \mathbb{C}$ |
| $s$ | squeezing parameter |
| $c, c^\dagger$ | Fermi annihilation and creation operators |
| $\mathbf{E}$ | electric field |
| $\mathbf{B}$ | magnetic induction |
| $\mathbf{P}$ | electric polarization |

The *Pauli spin matrices* are used extensively in the book. They are given by

$$\sigma_1 := \begin{pmatrix} 0 & 1 \\ 1 & 0 \end{pmatrix}, \quad \sigma_2 := \begin{pmatrix} 0 & -i \\ i & 0 \end{pmatrix}, \quad \sigma_3 := \begin{pmatrix} 1 & 0 \\ 0 & -1 \end{pmatrix}.$$

The spin-$\frac{1}{2}$ matrix $S_1, S_2, S_3$ are given by $S_1 = \frac{1}{2}\sigma_1$, $S_2 = \frac{1}{2}\sigma_2$, $S_3 = \frac{1}{2}\sigma_3$.

The Dirac *gamma matrices* are given by

$$\gamma_1 = -\sigma_2 \otimes \sigma_1, \quad \gamma_2 = \sigma_2 \otimes \sigma_2, \quad \gamma_3 = \sigma_2 \otimes \sigma_3, \quad \gamma_4 = \sigma_3 \otimes I_2.$$

The concept of a Hilbert space will be used throughout the book. A *Hilbert space* is a set, $\mathcal{H}$ of elements, or vectors, $(f, g, h, \ldots)$ which satisfies the following conditions (1) -- (5).

(1) If $f$ and $g$ belong to $\mathcal{H}$, then there is a unique element of $\mathcal{H}$, denoted by $f + g$, the operation of addition (+) being invertible, commutative and associative.

(2) If $c$ is a complex number, then for any $f$ in $\mathcal{H}$, there is an element $cf$ of $\mathcal{H}$; and the multiplication of vectors by complex numbers thereby defined satisfies the distributive conditions

$$c(f + g) = cf + cg, \qquad (c_1 + c_2)f = c_1 f + c_2 f.$$

(3) Hilbert spaces $\mathcal{H}$ possess a zero element, 0, characterized by the property that $0 + f = f$ for all vectors $f$ in $\mathcal{H}$.

(4) For each pair of vectors $f$, $g$ in $\mathcal{H}$, there is a complex number $\langle f|g \rangle$, termed the inner product or scalar product, of $f$ with $g$, such that

$$\langle f|g \rangle = \overline{\langle g|f \rangle}$$

$$\langle f|g + h \rangle = \langle f|g \rangle + \langle f|h \rangle$$

$$\langle f|cg \rangle = c\langle f|g \rangle$$

and

$$\langle f|f \rangle \geq 0.$$

Equality in the last formula occurs only if $f = 0$. The scalar product defines the norm $\|f\| = \langle f|f \rangle^{1/2}$.

(5) If $\{ f_n \}$ is a sequence in $\mathcal{H}$ satisfying the Cauchy condition that

$$\|f_m - f_n\| \to 0$$

as $m$ and $n$ tend independently to infinity, then there is a unique element $f$ of $\mathcal{H}$ such that $\|f_n - f\| \to 0$ as $n \to \infty$.

Let $B = \{ \phi_n : n \in I \}$ be an orthonormal basis in the Hilbert space $\mathcal{H}$. $I$ is the countable index set. Then

$$(1) \qquad \langle \phi_j|\phi_k \rangle = \delta_{jk}$$

$$(2) \quad \bigwedge_{f \in \mathcal{H}} \quad f = \sum_{j \in I} \langle f | \phi_j \rangle \phi_j$$

$$(3) \quad \bigwedge_{f, g \in \mathcal{H}} \quad \langle f | g \rangle = \sum_{j \in I} \overline{\langle f | \phi_j \rangle} \langle g | \phi_j \rangle$$

$$(4) \quad \left( \bigwedge_{\phi_j \in B} \langle f, \phi_j \rangle = 0 \right) \Rightarrow f = 0$$

$$(5) \quad \bigwedge_{f \in \mathcal{H}} \| f \|^2 = \sum_{j \in I} |\langle f, \phi_j \rangle|^2 .$$

Let $f, g \in \mathcal{H}$. Then we have the inequalities

$$|\langle f, g \rangle| \leq \|f\| \cdot \|g\|$$
$$\|f + g\| \leq \|f\| + \|g\|$$

and the equality

$$\|f + g\|^2 + \|f - g\|^2 = 2(\|f\|^2 + \|g\|^2).$$

We will also use the so-called Dirac notation. Let $\mathcal{H}$ be a Hilbert space and $\mathcal{H}_*$ be the dual space endowed with a multiplication law of the form

$$(c, \phi) = \bar{c}\phi$$

where $c \in \mathbb{C}$ and $\phi \in \mathcal{H}$. The inner product can be viewed as a bilinear form (duality)

$$\langle \cdot | \cdot \rangle \; : \; \mathcal{H}_* \times \mathcal{H} \to \mathbb{C}$$

such that the linear maps

$$\langle \phi | \; : \; \psi \to \langle \phi | \psi \rangle, \quad \langle \cdot | \; : \; \mathcal{H}_* \to \mathcal{H}'$$

$$|\psi\rangle \; : \; \phi \to \langle \phi | \psi \rangle, \quad |\cdot\rangle \; : \; \mathcal{H} \to \mathcal{H}'_*$$

where prime denotes the space of linear continuous functionals on the corresponding space, are monomorphisms. The vectors $\langle \phi |$ and $|\psi\rangle$ are called bra and ket vectors, respectively. The ket vector $|\phi\rangle$ is uniquely determined by a vector $\phi \in \mathcal{H}$, therefore we can write $|\phi\rangle \in \mathcal{H}$.

# Part I

# Finite-Dimensional
# Hilbert Spaces

# Chapter 1

# Qubits

## 1.1  Introduction

A single *qubit* is a two-state system, such as a two-level atom. The states (kets) $|h\rangle$ and $|v\rangle$ of the horizontal and vertical polarization of a photon can also be considered as a two-state system. Another example is the relative phase and intensity of a single photon in two arms of an interferometer. The underlying Hilbert space for the qubit is $\mathbb{C}^2$. An arbitrary orthonormal basis for $\mathbb{C}^2$ is denoted by $\{\,|0\rangle,\,|1\rangle\,\}$, where (scalar product)

$$\langle 0|0\rangle = \langle 1|1\rangle = 1, \qquad \langle 0|1\rangle = \langle 1|0\rangle = 0.$$

Any pure quantum state $|\psi\rangle$ (qubit) of this system can be written, up to a phase, as a *superposition* (linear combination of the states)

$$|\psi\rangle = \alpha|0\rangle + \beta|1\rangle, \quad |\alpha|^2 + |\beta|^2 = 1, \quad \alpha, \beta \in \mathbb{C}.$$

The classical boolean states, 0 and 1, can be represented by a fixed pair of orthonormal states of the qubit. The *standard basis* in $\mathbb{C}^2$ is given by

$$|0\rangle = \begin{pmatrix} 1 \\ 0 \end{pmatrix}, \quad |1\rangle = \begin{pmatrix} 0 \\ 1 \end{pmatrix}$$

and the *Hadamard basis* in $\mathbb{C}^2$ is given by

$$|0\rangle = \frac{1}{\sqrt{2}}\begin{pmatrix} 1 \\ 1 \end{pmatrix}, \quad |1\rangle = \frac{1}{\sqrt{2}}\begin{pmatrix} 1 \\ -1 \end{pmatrix}.$$

Up to an overall phase an arbitrary normalized state in $\mathbb{C}^2$ can be written as

$$|\psi\rangle = \begin{pmatrix} e^{i\phi}\cos(\theta) \\ \sin(\theta) \end{pmatrix}.$$

For any orthonormal basis $\{|0\rangle, |1\rangle\}$ in $\mathbb{C}^2$ we have

$$|0\rangle\langle 0| + |1\rangle\langle 1| = I_2$$

where $I_2$ is the $2 \times 2$ identity matrix. The $2 \times 2$ matrices

$$|0\rangle\langle 0|, \qquad |1\rangle\langle 1|$$

are projection matrices with

$$|0\rangle\langle 0|1\rangle\langle 1| = 0_2.$$

Furthermore

$$(|0\rangle\langle 1| + |1\rangle\langle 0|)^2 = |1\rangle\langle 1| + |0\rangle\langle 0| = I_2.$$

Given two normalized states $|\psi\rangle$, $|\phi\rangle$ in $\mathbb{C}^2$, then $0 \leq |\langle\psi|\phi\rangle|^2 \leq 1$ provides a probability. Let $|\psi\rangle \in \mathbb{C}^2$ and normalized. Then

$$\rho = |\psi\rangle\langle\psi|$$

is a density matrix (*pure state*). We have

$$\rho^2 = |\psi\rangle\langle\psi|\psi\rangle\langle\psi| = |\psi\rangle\langle\psi| = \rho.$$

If the qubit represents a *mixed state* one uses a two-dimensional *density matrix* $\rho$ for its representation. We therefore express one qubit as

$$\rho = \frac{1}{2}(I_2 + \mathbf{n} \cdot \boldsymbol{\sigma}) \equiv \frac{1}{2}(I_2 + n_1\sigma_1 + n_2\sigma_2 + n_3\sigma_3)$$

where $\mathbf{n} \in \mathbb{R}^3$,

$$\mathbf{n} \cdot \mathbf{n} \equiv n_1^2 + n_2^2 + n_3^2 \leq 1$$

and $\boldsymbol{\sigma} = (\sigma_1, \sigma_2, \sigma_3)$ denote the *Pauli spin matrices*

$$\sigma_1 = \begin{pmatrix} 0 & 1 \\ 1 & 0 \end{pmatrix}, \quad \sigma_2 = \begin{pmatrix} 0 & -i \\ i & 0 \end{pmatrix}, \quad \sigma_3 = \begin{pmatrix} 1 & 0 \\ 0 & -1 \end{pmatrix}.$$

For pure states we have $\mathbf{n} \cdot \mathbf{n} = 1$ and $\rho = |\psi\rangle\langle\psi|$. The Pauli spin matrices are hermitian and unitary and admit the eigenvalues $+1$ and $-1$.

## 1.2   Solved Problems

**Problem 1.**   Any normalized state (qubit) in $\mathbb{C}^2$ can be written as

$$\begin{pmatrix} \alpha \\ \beta \end{pmatrix}, \qquad \alpha, \beta \in \mathbb{C}, \quad |\alpha|^2 + |\beta|^2 = 1.$$

Find a parameter representation (i) if the underlying field is the set of real numbers (ii) if the underlying field is the set of complex numbers.

**Solution 1.**   (i) Using $\alpha = \cos(\theta)$, $\beta = \sin(\theta)$ and the identity $\cos^2(\theta) + \sin^2(\theta) \equiv 1$ for all $\theta \in \mathbb{R}$ we have

$$\begin{pmatrix} \cos(\theta) \\ \sin(\theta) \end{pmatrix}.$$

With the map $\theta \to \theta + \pi/2$ we can construct the orthonormal basis

$$\begin{pmatrix} \cos(\theta) \\ \sin(\theta) \end{pmatrix}, \quad \begin{pmatrix} \cos(\theta + \pi/2) \\ \sin(\theta + \pi/2) \end{pmatrix} = \begin{pmatrix} -\sin(\theta) \\ \cos(\theta) \end{pmatrix}.$$

(ii) We have as a representation

$$\begin{pmatrix} e^{i\phi} \cos(\theta) \\ \sin(\theta) \end{pmatrix}$$

where $\theta, \phi \in \mathbb{R}$ and $e^{i\phi} e^{-i\phi} = 1$.

**Problem 2.**   Consider the normalized states in $\mathbb{C}^2$ ($\theta_1, \theta_2 \in [0, 2\pi)$)

$$\begin{pmatrix} \cos(\theta_1) \\ \sin(\theta_1) \end{pmatrix}, \quad \begin{pmatrix} \cos(\theta_2) \\ \sin(\theta_2) \end{pmatrix}.$$

Find the condition on $\theta_1$ and $\theta_2$ such that the vector

$$\begin{pmatrix} \cos(\theta_1) \\ \sin(\theta_1) \end{pmatrix} + \begin{pmatrix} \cos(\theta_2) \\ \sin(\theta_2) \end{pmatrix}$$

is normalized.

**Solution 2.**   From the condition that the vector

$$\begin{pmatrix} \cos(\theta_1) + \cos(\theta_2) \\ \sin(\theta_1) + \sin(\theta_2) \end{pmatrix}$$

is normalized we obtain $(\sin(\theta_1) + \sin(\theta_2))^2 + (\cos(\theta_1) + \cos(\theta_2))^2 = 1$. Thus

$$\sin(\theta_1) \sin(\theta_2) + \cos(\theta_1) \cos(\theta_2) = -\frac{1}{2} \ \Rightarrow \ \cos(\theta_1 - \theta_2) = -\frac{1}{2}.$$

Therefore, $\theta_1 - \theta_2 = 2\pi/3$ or $\theta_1 - \theta_2 = 4\pi/3$.

**Problem 3.**   Let $\{\,|0\rangle, |1\rangle\,\}$ be an orthonormal basis in the Hilbert space $\mathbb{R}^2$ and $A := |0\rangle\langle 0| + |1\rangle\langle 1|$. Consider the three cases

$$\text{(i)}\ |0\rangle := \begin{pmatrix} 1 \\ 0 \end{pmatrix}, \qquad |1\rangle := \begin{pmatrix} 0 \\ 1 \end{pmatrix}$$

$$\text{(ii)}\ |0\rangle := \frac{1}{\sqrt{2}} \begin{pmatrix} 1 \\ 1 \end{pmatrix}, \qquad |1\rangle := \frac{1}{\sqrt{2}} \begin{pmatrix} 1 \\ -1 \end{pmatrix}$$

$$\text{(iii)}\ |0\rangle := \begin{pmatrix} \cos(\theta) \\ \sin(\theta) \end{pmatrix}, \qquad |1\rangle := \begin{pmatrix} \sin(\theta) \\ -\cos(\theta) \end{pmatrix}.$$

Find the matrix representation of $A$ in these bases.

**Solution 3.**   We find

$$\text{(i)}\quad A = \begin{pmatrix} 1 & 0 \\ 0 & 0 \end{pmatrix} + \begin{pmatrix} 0 & 0 \\ 0 & 1 \end{pmatrix} = \begin{pmatrix} 1 & 0 \\ 0 & 1 \end{pmatrix}$$

$$\text{(ii)}\quad A = \frac{1}{2} \begin{pmatrix} 1 & 1 \\ 1 & 1 \end{pmatrix} + \frac{1}{2} \begin{pmatrix} 1 & -1 \\ -1 & 1 \end{pmatrix} = \begin{pmatrix} 1 & 0 \\ 0 & 1 \end{pmatrix}$$

$$\text{(iii)}\quad A = \begin{pmatrix} \cos^2(\theta) & \cos(\theta)\sin(\theta) \\ \cos(\theta)\sin(\theta) & \sin^2(\theta) \end{pmatrix} + \begin{pmatrix} \sin^2(\theta) & -\cos(\theta)\sin(\theta) \\ -\cos(\theta)\sin(\theta) & \cos^2(\theta) \end{pmatrix}$$

$$= \begin{pmatrix} 1 & 0 \\ 0 & 1 \end{pmatrix}.$$

For all three cases $A = I_2$, where $I_2$ is the $2 \times 2$ unit matrix. Obviously, the third case contains the first two as special cases. This is the *completeness relation*.

**Problem 4.**   Let $\{\,|0\rangle, |1\rangle\,\}$ be an orthonormal basis in the Hilbert space $\mathbb{C}^2$. The *NOT operation* (unitary operator) is defined as

$$|0\rangle \to |1\rangle, \qquad |1\rangle \to |0\rangle.$$

(i) Find the unitary operator $U_{NOT}$ which implements the NOT operation with respect to the basis $\{\,|0\rangle, |1\rangle\,\}$.

(ii) Consider the *standard basis*

$$|0\rangle = \begin{pmatrix} 1 \\ 0 \end{pmatrix}, \qquad |1\rangle = \begin{pmatrix} 0 \\ 1 \end{pmatrix}.$$

Find the matrix representation of $U_{NOT}$ for this basis.

(iii) Consider the *Hadamard basis*

$$|0\rangle = \frac{1}{\sqrt{2}} \begin{pmatrix} 1 \\ 1 \end{pmatrix}, \qquad |1\rangle = \frac{1}{\sqrt{2}} \begin{pmatrix} 1 \\ -1 \end{pmatrix}.$$

Find the matrix representation of $U_{NOT}$ for this basis.

**Solution 4.** (i) Obviously, $U_{NOT} = |0\rangle\langle 1| + |1\rangle\langle 0|$ since $\langle 0|0\rangle = \langle 1|1\rangle = 1$ and $\langle 0|1\rangle = \langle 1|0\rangle = 0$.
(ii) For the standard basis we find

$$U_{NOT} = \begin{pmatrix} 0 & 1 \\ 1 & 0 \end{pmatrix}.$$

(iii) For the Hadamard basis we find

$$U_{NOT} = \begin{pmatrix} 1 & 0 \\ 0 & -1 \end{pmatrix}.$$

Thus we see that the respective matrix representations for the two bases are different.

**Problem 5.** Let $|0\rangle$, $|1\rangle$ be an orthonormal basis in $\mathbb{C}^2$. The *Walsh-Hadamard transform* is a 1-qubit operation, denoted by $H$, and performs the linear transform

$$|0\rangle \to \frac{1}{\sqrt{2}}(|0\rangle + |1\rangle), \qquad |1\rangle \to \frac{1}{\sqrt{2}}(|0\rangle - |1\rangle).$$

(i) Find the unitary operator $U_H$ which implements $H$ with respect to the basis $\{\,|0\rangle, |1\rangle\,\}$.
(ii) Find the inverse of this operator.
(iii) Consider the *standard basis*

$$|0\rangle = \begin{pmatrix} 1 \\ 0 \end{pmatrix}, \qquad |1\rangle = \begin{pmatrix} 0 \\ 1 \end{pmatrix}$$

in $\mathbb{C}^2$. Find the matrix representation of $U_H$ for this basis.
(iv) Consider the *Hadamard basis*

$$|0\rangle = \frac{1}{\sqrt{2}} \begin{pmatrix} 1 \\ 1 \end{pmatrix}, \qquad |1\rangle = \frac{1}{\sqrt{2}} \begin{pmatrix} 1 \\ -1 \end{pmatrix}$$

in $\mathbb{C}^2$. Find the matrix representation of $U_H$ for this basis.

**Solution 5.** (i) Obviously,

$$U_H = \frac{1}{\sqrt{2}}(|0\rangle + |1\rangle)\langle 0| + \frac{1}{\sqrt{2}}(|0\rangle - |1\rangle)\langle 1|$$

$$= \frac{1}{\sqrt{2}}|0\rangle(\langle 0| + \langle 1|) + \frac{1}{\sqrt{2}}|1\rangle(\langle 0| - \langle 1|).$$

(ii) The operator $U_H$ is unitary and the inverse is given by $U_H^{-1} = U_H^* = U_H$, where $*$ denotes the adjoint.

(iii) For the standard basis we find

$$U_H = \frac{1}{\sqrt{2}} \begin{pmatrix} 1 & 1 \\ 1 & -1 \end{pmatrix}.$$

(iv) For the Hadamard basis we find

$$U_H = \frac{1}{\sqrt{2}} \begin{pmatrix} 1 & 1 \\ 1 & -1 \end{pmatrix}.$$

We see that the matrix representations for each of the two bases are the same.

**Problem 6.**   The *Hadamard operator* on one qubit can be written as

$$U_H = \frac{1}{\sqrt{2}}((|0\rangle + |1\rangle)\langle 0| + (|0\rangle - |1\rangle)\langle 1|).$$

Calculate the states $U_H|0\rangle$ and $U_H|1\rangle$. Calculate $U_H U_H$.

**Solution 6.**   We obtain the normalized states

$$U_H|0\rangle = \frac{1}{\sqrt{2}}(|0\rangle + |1\rangle), \quad U_H|1\rangle = \frac{1}{\sqrt{2}}(|0\rangle - |1\rangle).$$

Since $\langle 0|0\rangle = \langle 1|1\rangle = 1$ and $\langle 0|1\rangle = \langle 1|0\rangle = 0$ we obtain

$$U_H U_H = |0\rangle\langle 0| + |1\rangle\langle 1| = I$$

where $I$ is the identity operator ($2 \times 2$ unit matrix).

**Problem 7.**   Let $\sigma_1$, $\sigma_2$, $\sigma_3$ be the Pauli spin matrices and $I_2$ be the $2 \times 2$ unit matrix. Consider the Hilbert space $\mathbb{C}^2$ and the linear operator ($2 \times 2$ matrix)

$$\Pi(\mathbf{n}) := \frac{1}{2}\left(I_2 + \sum_{j=1}^{3} n_j \sigma_j\right)$$

where $\mathbf{n} := (n_1, n_2, n_3)$ ($n_j \in \mathbb{R}$) is a unit vector, i.e. $n_1^2 + n_2^2 + n_3^2 = 1$.

(i) Describe the properties of $\Pi(\mathbf{n})$, i.e. find $\Pi^\dagger(\mathbf{n})$, $\text{tr}(\Pi(\mathbf{n}))$ and $\Pi^2(\mathbf{n})$.

(ii) Find the vector ($\phi, \theta \in \mathbb{R}$)

$$\Pi(\mathbf{n}) \begin{pmatrix} e^{i\phi}\cos(\theta) \\ \sin(\theta) \end{pmatrix}.$$

**Solution 7.** (i) For the Pauli matrices we have $\sigma_1^\dagger = \sigma_1$, $\sigma_2^\dagger = \sigma_2$, $\sigma_3^\dagger = \sigma_3$. Thus $\Pi(\mathbf{n}) = \Pi^\dagger(\mathbf{n})$. Since $\mathrm{tr}(\sigma_1) = \mathrm{tr}(\sigma_2) = \mathrm{tr}(\sigma_3) = 0$, $\mathrm{tr}(I_2) = 2$, and the trace operation is linear, we obtain $\mathrm{tr}(\Pi(\mathbf{n})) = 1$. Since $\sigma_1^2 = \sigma_2^2 = \sigma_3^2 = I_2$ and

$$[\sigma_1, \sigma_2]_+ = 0_2, \quad [\sigma_2, \sigma_3]_+ = 0_2, \quad [\sigma_3, \sigma_1]_+ = 0_2$$

where $[A, B]_+ := AB + BA$ denotes the *anticommutator*, the expression

$$\Pi^2(\mathbf{n}) = \frac{1}{4}\left(I_2 + \sum_{j=1}^3 n_j \sigma_j\right)^2 = \frac{1}{4}I_2 + \frac{1}{2}\sum_{j=1}^3 n_j \sigma_j + \frac{1}{4}\sum_{j=1}^3\sum_{k=1}^3 n_j n_k \sigma_j \sigma_k$$

simplifies to

$$\Pi^2(\mathbf{n}) = \frac{1}{4}I_2 + \frac{1}{2}\sum_{j=1}^3 n_j \sigma_j + \frac{1}{4}\sum_{j=1}^3 n_j^2 I_2.$$

Using $n_1^2 + n_2^2 + n_3^2 = 1$ we obtain $\Pi^2(\mathbf{n}) = \Pi(\mathbf{n})$. Thus $\Pi(\mathbf{n})$ is a projection matrix.

(ii) We find

$$\Pi(\mathbf{n})\begin{pmatrix} e^{i\phi}\cos(\theta) \\ \sin(\theta) \end{pmatrix} = \frac{1}{2}\begin{pmatrix} (1 + n_3)e^{i\phi}\cos(\theta) + (n_1 - in_2)\sin(\theta) \\ (n_1 + in_2)e^{i\phi}\cos(\theta) + (1 - n_3)\sin(\theta) \end{pmatrix}.$$

**Problem 8.** The *qubit trine* is defined by the following states

$$|\psi_0\rangle = |0\rangle, \quad |\psi_1\rangle = -\frac{1}{2}|0\rangle - \frac{\sqrt{3}}{2}|1\rangle, \quad |\psi_2\rangle = -\frac{1}{2}|0\rangle + \frac{\sqrt{3}}{2}|1\rangle$$

where $\{\,|0\rangle, |1\rangle\,\}$ is an orthonormal basis. Find the probabilities

$$|\langle\psi_0|\psi_1\rangle|^2, \quad |\langle\psi_1|\psi_2\rangle|^2, \quad |\langle\psi_2|\psi_0\rangle|^2.$$

**Solution 8.** Using $\langle 0|0\rangle = 1$, $\langle 1|1\rangle = 1$ and $\langle 0|1\rangle = 0$ we find

$$|\langle\psi_0|\psi_1\rangle|^2 = \frac{1}{4}, \quad |\langle\psi_1|\psi_2\rangle|^2 = \frac{1}{4}, \quad |\langle\psi_2|\psi_0\rangle|^2 = \frac{1}{4}.$$

**Problem 9.** The kets $|h\rangle$ and $|v\rangle$ are states of horizontal and vertical polarization, respectively. Consider the normalized states

$$|\psi_1\rangle = -\frac{1}{2}(|h\rangle + \sqrt{3}|v\rangle)$$

$$|\psi_2\rangle = -\frac{1}{2}(|h\rangle - \sqrt{3}|v\rangle)$$

$$|\psi_3\rangle = |h\rangle$$

$$|\phi_1\rangle = \frac{1}{\sqrt{3}}(-|h\rangle + \sqrt{2}e^{-2\pi i/3}|v\rangle)$$

$$|\phi_2\rangle = \frac{1}{\sqrt{3}}(-|h\rangle + \sqrt{2}e^{+2\pi i/3}|v\rangle)$$

$$|\phi_3\rangle = \frac{1}{\sqrt{3}}(-|h\rangle + \sqrt{2}|v\rangle).$$

Give an interpretation of these states.

**Solution 9.**   Since $\langle h|h\rangle = \langle v|v\rangle = 1$ and $\langle v|h\rangle = \langle h|v\rangle = 0$ we find

$$\langle \psi_1|\psi_2\rangle = -\frac{1}{2}, \qquad \langle \psi_1|\psi_3\rangle = -\frac{1}{2}, \qquad \langle \psi_2|\psi_3\rangle = -\frac{1}{2}.$$

Since the solution to $\cos(\alpha) = -1/2$ is given by $\alpha = 120^\circ$ $(2\pi/3)$ or $\alpha = 240^\circ$ $(4\pi/3)$ we find that the first three states $|\psi_1\rangle$, $|\psi_2\rangle$, $|\psi_3\rangle$ correspond to states of linear polarization separated by $120^\circ$ $(2\pi/3)$. We find the scalar product

$$\langle \phi_1|\phi_2\rangle = -\frac{i}{\sqrt{3}}.$$

The states $|\phi_1\rangle$ and $|\phi_2\rangle$ correspond to elliptic polarization and the third state $|\phi_3\rangle$ corresponds to linear polarization.

**Problem 10.**   Let

$$|\psi\rangle = \begin{pmatrix} e^{i\phi}\cos(\theta) \\ \sin(\theta) \end{pmatrix}$$

be a normalized state in the Hilbert space $\mathbb{C}^2$, where $\phi, \theta \in \mathbb{R}$. Find the density matrix $\rho := |\psi\rangle\langle\psi|$, $\mathrm{tr}(\rho)$ and $\rho^2$.

**Solution 10.**   Since

$$\langle\psi| = (e^{-i\phi}\cos(\theta), \sin(\theta))$$

we obtain the $2 \times 2$ density matrix

$$\rho = |\psi\rangle\langle\psi| = \begin{pmatrix} \cos^2(\theta) & e^{i\phi}\sin(\theta)\cos(\theta) \\ e^{-i\phi}\sin(\theta)\cos(\theta) & \sin^2(\theta) \end{pmatrix}.$$

Since $\cos^2(\theta) + \sin^2(\theta) = 1$ we obtain that $\mathrm{tr}(\rho) = 1$. With $\langle\psi|\psi\rangle = 1$ we obtain $\rho^2 = (|\psi\rangle\langle\psi|)^2 = |\psi\rangle\langle\psi|\psi\rangle\langle\psi| = |\psi\rangle\langle\psi| = \rho$. Thus we have a pure state.

**Problem 11.**   Given the Hamilton operator $\hat{H} = \hbar\omega\sigma_1$.

(i) Find the solution

$$|\psi(t)\rangle = e^{-i\hat{H}t/\hbar}|\psi(t=0)\rangle$$

of the *Schrödinger equation*

$$i\hbar\frac{d}{dt}|\psi\rangle = \hat{H}|\psi\rangle$$

with the initial conditions

$$|\psi(t=0)\rangle = \begin{pmatrix} 1 \\ 0 \end{pmatrix}.$$

(ii) Find the probability $|\langle\psi(t=0)|\psi(t)\rangle|^2$.
(iii) The solution of the *Heisenberg equation of motion*

$$i\hbar\frac{d\sigma_3}{dt} = [\sigma_3, \hat{H}](t)$$

is given by

$$\sigma_3(t) = e^{i\hat{H}t/\hbar}\sigma_3 e^{-i\hat{H}t/\hbar}$$

where $\sigma_3(t=0) = \sigma_3$. Calculate $\sigma_3(t)$.
(iv) Show that $\langle\psi(t=0)|\sigma_3(t)|\psi(t=0)\rangle = \langle\psi(t)|\sigma_3|\psi(t)\rangle$.

**Solution 11.** (i) The solution of the Schrödinger equation is given by

$$|\psi(t)\rangle = \exp(-i\hat{H}t/\hbar)|\psi(t=0)\rangle.$$

Since $\sigma_1^2 = I_2$ we find the unitary matrix

$$\exp(-i\hat{H}t/\hbar) \equiv U(t) = \begin{pmatrix} \cos(\omega t) & -i\sin(\omega t) \\ -i\sin(\omega t) & \cos(\omega t) \end{pmatrix}.$$

Thus the normalized state at time $t$ is

$$|\psi(t)\rangle = U(t)\begin{pmatrix} 1 \\ 0 \end{pmatrix} = \begin{pmatrix} \cos(\omega t) \\ -i\sin(\omega t) \end{pmatrix}.$$

(ii) We find the probability $|\langle\psi(t=0)|\psi(t)\rangle|^2 = \cos^2(\omega t)$.
(iii) Since the commutators are given by

$$[\sigma_3, \hat{H}] = \hbar\omega[\sigma_3, \sigma_1] = 2i\hbar\omega\sigma_2, \qquad [\sigma_2, \hat{H}] = \hbar\omega[\sigma_2, \sigma_1] = -2i\hbar\omega\sigma_3$$

we obtain the linear system of matrix-valued differential equations

$$\frac{d\sigma_3}{dt} = 2\omega\sigma_2(t), \qquad \frac{d\sigma_2}{dt} = -2\omega\sigma_3(t)$$

with the initial conditions $\sigma_3(t = 0) = \sigma_3$ and $\sigma_2(t = 0) = \sigma_2$. Here we used the Heisenberg equation of motion for $\sigma_2$ to obtain the second differential equation. The solution of this system of matrix-valued linear differential equations is given by

$$\sigma_3(t) = \sigma_3 \cos(2\omega t) + \sigma_2 \sin(2\omega t)$$

$$\sigma_2(t) = \sigma_2 \cos(2\omega t) - \sigma_3 \sin(2\omega t).$$

(iv) We find
$$\langle \psi(t = 0)|\sigma_3(t)|\psi(t = 0)\rangle = \cos(2\omega t)$$

and
$$\langle \psi(t)|\sigma_3|\psi(t)\rangle = \cos^2(\omega t) - \sin^2(\omega t) \equiv \cos(2\omega t).$$

**Problem 12.**   Consider a *Mach-Zehnder interferometer* in which the beam pair spans a two-dimensional Hilbert space with orthonormal basis $\{\,|0\rangle,\ |1\rangle\,\}$. The state vectors $|0\rangle$ and $|1\rangle$ can be considered as orthonormal wave packets that move in two given directions defined by the geometry of the interferometer. We may represent mirrors, beam splitters and relative $U_P$ phase shifts by the unitary matrices

$$U_M = \begin{pmatrix} 0 & 1 \\ 1 & 0 \end{pmatrix}, \quad U_B = \frac{1}{\sqrt{2}} \begin{pmatrix} 1 & 1 \\ 1 & -1 \end{pmatrix}, \quad U_P = \begin{pmatrix} e^{i\chi} & 0 \\ 0 & 1 \end{pmatrix}$$

respectively. Consider the density matrix

$$\rho_{in} = |0\rangle\langle 0|$$

where $\{\,|0\rangle,\ |1\rangle\,\}$ denotes the standard basis. Using this basis find

$$\rho_{out} = U_B U_M U_P U_B \rho_{in} U_B^\dagger U_P^\dagger U_M^\dagger U_B^\dagger.$$

Give an interpretation of the result.

**Solution 12.**   Since

$$\rho_{in} = |0\rangle\langle 0| = \begin{pmatrix} 1 \\ 0 \end{pmatrix} (1\ 0) = \begin{pmatrix} 1 & 0 \\ 0 & 0 \end{pmatrix}$$

and
$$U_B U_M U_P U_B = \frac{1}{2} \begin{pmatrix} e^{i\chi} + 1 & e^{i\chi} - 1 \\ -e^{i\chi} + 1 & -e^{i\chi} - 1 \end{pmatrix}$$

we obtain
$$\rho_{out} = \frac{1}{2} \begin{pmatrix} 1 + \cos(\chi) & i\sin(\chi) \\ -i\sin(\chi) & 1 - \cos(\chi) \end{pmatrix}.$$

This yields the intensity along $|0\rangle$ as $I \propto 1 + \cos(\chi)$. Thus the relative $U_P$ phase $\chi$ could be observed in the output signal of the interferometer.

**Problem 13.**  Let $\{\,|0\rangle, |1\rangle\,\}$ be an orthonormal basis in $\mathbb{C}^2$.
(i) Find the commutator $[\,|0\rangle\langle 1|, |1\rangle\langle 0|\,]$.
(ii) Find the operators $\exp(t|0\rangle\langle 1|)$ and $\exp(t|1\rangle\langle 0|)$.
(iii) Find the operator $\exp(t|0\rangle\langle 1|)\exp(t|1\rangle\langle 0|)$.
(iv) Find the operator $\exp(t(|0\rangle\langle 1| + |1\rangle\langle 0|))$.
(v) Is $\exp(t(|0\rangle\langle 1| + |1\rangle\langle 0|)) = \exp(t|0\rangle\langle 1|)\exp(t|1\rangle\langle 0|)$?

**Solution 13.**  (i) We have

$$\left[\,|0\rangle\langle 1|, |1\rangle\langle 0|\,\right] = |0\rangle\langle 0| - |1\rangle\langle 1|$$

since $\langle 0|0\rangle = \langle 1|1\rangle = 1$ and $\langle 0|1\rangle = \langle 1|0\rangle = 0$. We see that the commutator is nonzero.
(ii) Since $\langle 0|1\rangle = \langle 1|0\rangle = 0$ we find

$$\exp(t|0\rangle\langle 1|) = \sum_{j=0}^{\infty} \frac{t^j}{j!}(|0\rangle\langle 1|)^j = I_2 + t|0\rangle\langle 1|.$$

Analogously

$$\exp(t|1\rangle\langle 0|) = \sum_{j=0}^{\infty} \frac{t^j}{j!}(|1\rangle\langle 0|)^j = I_2 + t|1\rangle\langle 0|.$$

(iii) Multiplying the results found above we obtain

$$\exp(t|0\rangle\langle 1|)\exp(t|1\rangle\langle 0|) = I_2 + t(|0\rangle\langle 1| + |1\rangle\langle 0|) + t^2|0\rangle\langle 0|.$$

(iv) Since $(|0\rangle\langle 1| + |1\rangle\langle 0|)^2 = I_2$ we obtain

$$\exp(t|0\rangle\langle 1| + t|1\rangle\langle 0|) = \sum_{j=0}^{\infty} \frac{t^{2j}}{(2j)!}I_2 + \sum_{j=0}^{\infty} \frac{t^{2j+1}}{(2j+1)!}(|0\rangle\langle 1| + |1\rangle\langle 0|)$$
$$= \cosh(t)I_2 + \sinh(t)(|0\rangle\langle 1| + |1\rangle\langle 0|).$$

(v) Clearly when $t \neq 0$ we have

$$\exp(t(|0\rangle\langle 1| + |1\rangle\langle 0|)) \neq \exp(t|0\rangle\langle 1|)\exp(t|1\rangle\langle 0|).$$

**Problem 14.**  Consider the unitary matrix for the NOT gate

$$U_{NOT} = \begin{pmatrix} 0 & 1 \\ 1 & 0 \end{pmatrix}.$$

Show that we can find a unitary matrix $V$ such that $V^2 = U_{NOT}$. Thus $V$ would be the square root NOT gate. What are the eigenvalues of $V$?

**Solution 14.**   We find the unitary matrix

$$V = \frac{1}{2}\begin{pmatrix} 1+i & 1-i \\ 1-i & 1+i \end{pmatrix}.$$

Obviously $-V$ is also a square root. The eigenvalues of $V$ are $1$ and $i$. The eigenvalues of $-V$ are $-1$ and $-i$. Note that the eigenvalues of $U_{NOT}$ are $1$ and $-1$ and $\sqrt{1} = \pm 1$, $\sqrt{-1} = \pm i$.

**Problem 15.**   Let $\sigma_1, \sigma_2, \sigma_3$ be the Pauli spin matrices. Let $\mathbf{n}$ be a unit vector in $\mathbb{R}^3$. We define the operator

$$\Sigma := \mathbf{n} \cdot \boldsymbol{\sigma} \equiv n_1\sigma_1 + n_2\sigma_2 + n_3\sigma_3.$$

(i) Calculate the matrix $\Sigma^2$. From this result and the fact that $\Sigma$ is hermitian show that $\Sigma$ is unitary. Find the eigenvalues of $\Sigma$.
(ii) Let

$$|\psi\rangle = \begin{pmatrix} 1 \\ 0 \end{pmatrix}.$$

Calculate the state $\Sigma|\psi\rangle$ and the probability $|\langle\psi|\Sigma|\psi\rangle|^2$.

**Solution 15.**   (i) Using $n_1^2 + n_2^2 + n_3^2 = 1$, $\sigma_1^2 = \sigma_2^2 = \sigma_3^2 = I_2$ and

$$\sigma_1\sigma_2 + \sigma_2\sigma_1 = 0_2, \quad \sigma_1\sigma_3 + \sigma_3\sigma_1 = 0_2, \quad \sigma_2\sigma_3 + \sigma_3\sigma_2 = 0_2$$

we obtain

$$\begin{aligned}
\Sigma^2 &= (n_1\sigma_1 + n_2\sigma_2 + n_3\sigma_3)^2 \\
&= (n_1^2 + n_2^2 + n_3^2)I_2 \\
&\quad + n_1n_2(\sigma_1\sigma_2 + \sigma_2\sigma_1) + n_1n_3(\sigma_3\sigma_1 + \sigma_1\sigma_3) + n_2n_3(\sigma_2\sigma_3 + \sigma_3\sigma_2) \\
&= I_2.
\end{aligned}$$

Since $\Sigma$ is hermitian, i.e. $\Sigma = \Sigma^*$ and $\Sigma^2 = I_2$ we find that $\Sigma$ is a unitary matrix with $\Sigma = \Sigma^{-1}$. Since $\Sigma$ is hermitian and unitary the eigenvalues $\lambda_1, \lambda_2$ can only be $\pm 1$. Since $\mathrm{tr}(\Sigma) = 0 = \lambda_1 + \lambda_2$ we obtain that the eigenvalues are $+1$ and $-1$.
(ii) We find

$$\Sigma|\psi\rangle = n_1\begin{pmatrix} 0 \\ 1 \end{pmatrix} + n_2\begin{pmatrix} 0 \\ i \end{pmatrix} + n_3\begin{pmatrix} 1 \\ 0 \end{pmatrix}.$$

It follows that $|\langle\psi|\Sigma|\psi\rangle|^2 = n_3^2$.

**Problem 16.** Let **n** be a unit vector in $\mathbb{R}^3$, $\boldsymbol{\sigma} = (\sigma_1, \sigma_2, \sigma_3)$ and

$$\mathbf{n} \cdot \boldsymbol{\sigma} := n_1 \sigma_1 + n_2 \sigma_2 + n_3 \sigma_3.$$

(i) Find the unitary matrix $\exp(i\theta \mathbf{n} \cdot \boldsymbol{\sigma})$, where $\theta \in \mathbb{R}$.
(ii) Find the normalized state

$$\exp(i\theta \mathbf{n} \cdot \boldsymbol{\sigma}) \begin{pmatrix} 1 \\ 0 \end{pmatrix}.$$

**Solution 16.** (i) Since

$$\sigma_j \sigma_k = \delta_{jk} I_2 + i \sum_{\ell=1}^{3} \epsilon_{jk\ell} \sigma_\ell$$

where $\epsilon_{123} = \epsilon_{231} = \epsilon_{312} = 1$, $\epsilon_{321} = \epsilon_{213} = \epsilon_{132} = -1$ and 0 otherwise, we obtain

$$\exp(i\theta \mathbf{n} \cdot \boldsymbol{\sigma}) = I_2 \cos(\theta) + i(\mathbf{n} \cdot \boldsymbol{\sigma}) \sin(\theta)$$
$$= \begin{pmatrix} \cos(\theta) + in_3 \sin(\theta) & i(n_1 - in_2) \sin(\theta) \\ i(n_1 + in_2) \sin(\theta) & \cos(\theta) - in_3 \sin(\theta) \end{pmatrix}.$$

Note that we could also use $(\mathbf{n} \cdot \boldsymbol{\sigma})^2 = I_2$ to find the result.
(ii) Using (i) we find the state

$$\exp(i\theta \mathbf{n} \cdot \boldsymbol{\sigma}) \begin{pmatrix} 1 \\ 0 \end{pmatrix} = \begin{pmatrix} \cos(\theta) + in_3 \sin(\theta) \\ i(n_1 + in_2) \sin(\theta) \end{pmatrix}.$$

**Problem 17.** Consider the Hamilton operator

$$\hat{H} = \hbar\omega \begin{pmatrix} 0 & \alpha \\ \alpha & 1 \end{pmatrix}$$

where $\alpha \geq 0$. Find $\alpha$ where the energy gap between the two energy levels is the smallest.

**Solution 17.** From the eigenvalue equation we find $E^2 - \hbar\omega E = \hbar^2 \omega^2 \alpha^2$. Consequently

$$E_0(\alpha) = \hbar\omega(1 - \sqrt{1 + \alpha^2}), \qquad E_1(\alpha) = \hbar\omega(1 + \sqrt{1 + \alpha^2}).$$

Thus

$$E_1(\alpha) - E_0(\alpha) = 2\hbar\omega\sqrt{1 + \alpha^2}.$$

Therefore the shortest energy gap is for $\alpha = 0$.

**Problem 18.**   Consider the Hamilton operator

$$\hat{H} = \hbar\omega\sigma_3 + \Delta\sigma_1 = \begin{pmatrix} \hbar\omega & \Delta \\ \Delta & -\hbar\omega \end{pmatrix}$$

where $\Delta \geq 0$.
(i) Find the eigenvalues and the normalized eigenvectors of $\hat{H}$.
(ii) Use the *Cayley-Hamilton theorem* to calculate $\exp(-i\hat{H}t/\hbar)$.

**Solution 18.**    (i) From $\det(\hat{H} - EI_2) = 0$ we find the two eigenvalues

$$E_{\pm} = \pm\sqrt{\hbar^2\omega^2 + \Delta^2}.$$

We set $E := \sqrt{\hbar^2\omega^2 + \Delta^2}$. Then from the eigenvalue equation

$$\begin{pmatrix} \hbar\omega & \Delta \\ \Delta & -\hbar\omega \end{pmatrix} \begin{pmatrix} u_1 \\ u_2 \end{pmatrix} = E_+ \begin{pmatrix} u_1 \\ u_2 \end{pmatrix}$$

for the eigenvalue $E_+ = E$ we find $\Delta u_2 = (E - \hbar\omega)u_1$. Thus the eigenvector is given by

$$\begin{pmatrix} \Delta \\ E - \hbar\omega \end{pmatrix}.$$

After normalization we have

$$\frac{1}{\sqrt{\Delta^2 + (E - \hbar\omega)^2}} \begin{pmatrix} \Delta \\ E - \hbar\omega \end{pmatrix}.$$

Analogously we find for the eigenvalue $E_- = -E$ the normalized eigenvector

$$\frac{1}{\sqrt{\Delta^2 + (E + \hbar\omega)^2}} \begin{pmatrix} \Delta \\ -E - \hbar\omega \end{pmatrix}.$$

(ii) Since $E_+ \neq E_-$ and $E_+ = E$, $E_- = -E$ we have to solve the system of equations

$$e^{-iEt/\hbar} = c_0 + c_1 E, \qquad e^{iEt/\hbar} = c_0 - c_1 E$$

for $c_0$ and $c_1$. Then

$$e^{-i\hat{H}t/\hbar} = c_0 I_2 + c_1 \hat{H} = \begin{pmatrix} c_0 + c_1\hbar\omega & c_1\Delta \\ c_1\Delta & c_0 - c_1\hbar\omega \end{pmatrix}.$$

The solution of the system of equations is given by

$$c_0 = \cos(Et/\hbar), \quad c_1 = \frac{e^{-iEt/\hbar} - e^{iEt/\hbar}}{2E} = \frac{-i\sin(Et/\hbar)}{E}.$$

Thus

$$e^{-i\hat{H}t/\hbar} = \begin{pmatrix} \cos(Et/\hbar) - i\sin(Et/\hbar)\hbar\omega/E & -i\sin(Et/\hbar)\Delta/E \\ -i\sin(Et/\hbar)\Delta/E & \cos(Et/\hbar) + i\sin(Et/\hbar)\hbar\omega/E \end{pmatrix}.$$

Obviously, $\exp(-i\hat{H}t/\hbar)$ is a unitary matrix.

**Problem 19.** Consider the Pauli spin matrices $\sigma_1$, $\sigma_2$, $\sigma_3$. Can one find an $\alpha \in \mathbb{R}$ such that $\exp(i\alpha\sigma_3)\sigma_1\exp(-i\alpha\sigma_3) = \sigma_2$?

**Solution 19.** We have

$$\exp(i\alpha\sigma_3)\sigma_1\exp(-i\alpha\sigma_3) = \begin{pmatrix} 0 & e^{2i\alpha} \\ e^{-2i\alpha} & 0 \end{pmatrix}.$$

Thus we have to solve the equations $\exp(2i\alpha) = -i$, $\exp(-2i\alpha) = i$. For $\alpha \in [0, 2\pi)$ we obtain $\alpha = 3\pi/4$.

**Problem 20.** Let $\mathbf{n}$ and $\mathbf{m}$ be a unit vectors in $\mathbb{R}^3$, $\boldsymbol{\sigma} = (\sigma_1, \sigma_2, \sigma_3)$ and

$$\mathbf{n} \cdot \boldsymbol{\sigma} := n_1\sigma_1 + n_2\sigma_2 + n_3\sigma_3.$$

Calculate the commutator $[\mathbf{n} \cdot \boldsymbol{\sigma}, \mathbf{m} \cdot \boldsymbol{\sigma}]$.

**Solution 20.** We find

$$\begin{aligned}[\mathbf{n} \cdot \boldsymbol{\sigma}, \mathbf{m} \cdot \boldsymbol{\sigma}] = {} & 2i((n_2m_3 - m_2n_3)\sigma_1 + (n_3m_1 - m_3n_1)\sigma_2 \\ & + (n_1m_2 - m_1n_2)\sigma_3) \\ = {} & 2i(\mathbf{n} \times \mathbf{m}) \cdot \boldsymbol{\sigma} \end{aligned}$$

where $\times$ denotes the *vector product*. The vector $\mathbf{n} \times \mathbf{m}$ is perpendicular to the plane spanned by the vectors $\mathbf{n}$ and $\mathbf{m}$.

**Problem 21.** Let $|\psi_1\rangle$ and $|\psi_2\rangle$ be two normalized states in a Hilbert space $\mathcal{H}$. A distance $d$ with $0 \le d \le \pi/2$ can be defined as

$$\cos^2(d) := |\langle\psi_1|\psi_2\rangle|^2.$$

Let $\mathcal{H} = \mathbb{C}^2$ and consider the normalized states

$$|\psi_1\rangle = \frac{1}{\sqrt{2}}\begin{pmatrix} 1 \\ 1 \end{pmatrix}, \qquad |\psi_2\rangle = \frac{1}{\sqrt{2}}\begin{pmatrix} 1 \\ -1 \end{pmatrix}.$$

Find $d$.

**Solution 21.** Since $\langle\psi_1|\psi_2\rangle = 0$ we have $\cos^2(d) = 0$ and therefore $d = \pi/2$.

**Problem 22.**    Let $\rho_1$ and $\rho_2$ be density matrices in the same Hilbert space. The *Bures distance* between the two density matrices is defined as

$$D_B(\rho_1, \rho_2) := \sqrt{2(1 - \mathrm{tr}((\rho_1^{1/2} \rho_2 \rho_1^{1/2})^{1/2}))}.$$

Consider the density matrices

$$\rho_1 = \begin{pmatrix} 1 & 0 \\ 0 & 0 \end{pmatrix}, \qquad \rho_2 = \begin{pmatrix} 1/2 & 0 \\ 0 & 1/2 \end{pmatrix}$$

acting in the Hilbert space $\mathbb{C}^2$. Find the Bures distance.

**Solution 22.**    Since

$$\rho_1^{1/2} = \rho_1 = \begin{pmatrix} 1 & 0 \\ 0 & 0 \end{pmatrix}$$

we obtain

$$\rho_1^{1/2} \rho_2 \rho_1^{1/2} = \begin{pmatrix} 1/2 & 0 \\ 0 & 0 \end{pmatrix}.$$

Thus $D_B(\rho_1, \rho_2) = \sqrt{2(1 - 1/\sqrt{2})}$.

**Problem 23.**    (i) Consider the Hilbert space $\mathbb{C}^2$. Show that

$$\Pi_S = \frac{1}{2} \begin{pmatrix} 1 & -i \\ i & 1 \end{pmatrix}, \qquad \Pi_A = \frac{1}{2} \begin{pmatrix} 1 & i \\ -i & 1 \end{pmatrix}$$

are *projection matrices*.
(ii) Decompose the Hilbert space into sub-Hilbert spaces using the result from (i).

**Solution 23.**    (i) We have

$$\Pi_S = \Pi_S^*, \quad \Pi_S^2 = \Pi_S, \quad \Pi_A = \Pi_A^*, \quad \Pi_A^2 = \Pi_A$$

and $\Pi_S + \Pi_A = I_2$, $\Pi_S \Pi_A = 0_2$.
(ii) Consider the normalized state

$$|\psi\rangle = \begin{pmatrix} e^{i\phi} \sin(\theta) \\ \cos(\theta) \end{pmatrix}.$$

Then

$$\Pi_S |\psi\rangle = \begin{pmatrix} e^{i\phi} \sin(\theta) - i \cos(\theta) \\ ie^{i\phi} \sin(\theta) + \cos(\theta) \end{pmatrix}, \qquad \Pi_A |\psi\rangle = \begin{pmatrix} e^{i\phi} \sin(\theta) + i \cos(\theta) \\ -ie^{i\phi} \sin(\theta) + \cos(\theta) \end{pmatrix}$$

with $\langle \psi | \Pi_A \Pi_S | \psi \rangle = 0$.

**Problem 24.** Let $\sigma_1$, $\sigma_2$, $\sigma_3$ be the Pauli spin matrices. Consider the normalized state in $\mathbb{C}^2$

$$\psi = \begin{pmatrix} e^{i\phi}\cos(\theta) \\ \sin(\theta) \end{pmatrix} \Rightarrow \psi^* = \left( e^{-i\phi}\cos(\theta) \quad \sin(\theta) \right).$$

Find the vector $\mathbf{v} = (v_1\ v_2\ v_3)^T$ in $\mathbb{R}^3$ with

$$v_1 = \psi^*\sigma_1\psi, \quad v_2 = \psi^*\sigma_2\psi, \quad v_3 = \psi^*\sigma_3\psi.$$

Is the vector $\mathbf{v}$ normalized?

**Solution 24.** Utilizing that

$$e^{i\phi} + e^{-i\phi} \equiv 2\cos(\phi), \quad ie^{i\phi} - ie^{-i\phi} \equiv -2\sin(\phi),$$

$$\cos(\theta)\sin(\theta) \equiv \frac{1}{2}\sin(2\theta), \quad \cos^2(\theta) - \sin^2(\theta) \equiv \cos(2\theta)$$

we obtain

$$v_1 = \psi^*\sigma_1\psi = \cos(\phi)\sin(2\theta)$$
$$v_2 = \psi^*\sigma_2\psi = -\sin(\phi)\sin(2\theta)$$
$$v_3 = \psi^*\sigma_3\psi = \cos(2\theta).$$

The vector

$$\mathbf{v} = \begin{pmatrix} v_1 \\ v_2 \\ v_3 \end{pmatrix} = \begin{pmatrix} \cos(\phi)\sin(2\theta) \\ -\sin(\phi)\sin(2\theta) \\ \cos(2\theta) \end{pmatrix}$$

is normalized, i.e. $\|\mathbf{v}\|^2 = 1$.

**Problem 25.** (i) Consider the symmetric matrix over $\mathbb{R}$

$$H = \begin{pmatrix} h_{11} & h_{12} \\ h_{12} & h_{22} \end{pmatrix}$$

and the normalized state

$$|\psi\rangle = \begin{pmatrix} \cos(\theta) \\ \sin(\theta) \end{pmatrix}.$$

Calculate the *variance* $V_H(|\psi\rangle) := \langle\psi|H^2|\psi\rangle - (\langle\psi|H|\psi\rangle)^2$.
(ii) Consider the Hadamard matrix

$$H = \frac{1}{\sqrt{2}} \begin{pmatrix} 1 & 1 \\ 1 & -1 \end{pmatrix}$$

and the normalized state

$$|\psi\rangle = \begin{pmatrix} \cos(\theta) \\ \sin(\theta) \end{pmatrix}.$$

Calculate the variance $V_H(\psi) := \langle\psi|H^2|\psi\rangle - (\langle\psi|H|\psi\rangle)^2$ and discuss the dependence on $\theta$.

**Solution 25.**   (i) We find

$$\langle\psi|H^2|\psi\rangle = h_{11}^2\cos^2(\theta) + h_{22}^2\sin^2(\theta) + h_{12}^2 + 2h_{12}(h_{11} + h_{22})\cos(\theta)\sin(\theta)$$

and

$$(\langle\psi|H|\psi\rangle)^2 = h_{11}^2\cos^4(\theta) + h_{22}^2\sin^4(\theta) + 2h_{11}h_{22}\cos^2(\theta)\sin^2(\theta)$$
$$+4h_{11}h_{12}\cos^3(\theta)\sin(\theta) + 4h_{22}h_{12}\cos(\theta)\sin^3(\theta).$$

Thus

$$V_H(|\psi\rangle) = (h_{11}^2 + h_{22}^2 - 2h_{11}h_{22})\sin^2(\theta)\cos^2(\theta) + h_{12}^2(1 - 4\cos^2(\theta)\sin^2(\theta))$$
$$+2h_{12}h_{11}\sin(\theta)\cos(\theta)(1 - 2\cos^2(\theta))$$
$$+2h_{12}h_{22}\sin(\theta)\cos(\theta)(1 - 2\sin^2(\theta)).$$

(ii) Using that $H^2 = I_2$ we have

$$V_H(\psi) = \langle\psi|I_2|\psi\rangle - (\langle\psi|H|\psi\rangle)^2 = 1 - \frac{1}{2}(\cos(2\theta) + \sin(2\theta))^2$$
$$= \frac{1}{2}(1 - \sin(4\theta)).$$

For $\theta = 0$ we have $V_H(\psi) = 1/2$. The minimum value is 0, for example for $\theta = \pi/8$. The maximum value is 1, for example for $\theta = 3\pi/8$.

**Problem 26.**   Let $c^\dagger$, $c$ be Fermi creation and annihilation operators with

$$[c^\dagger, c]_+ = c^\dagger c + cc^\dagger = I$$

and $(c^\dagger) = 0$, $c^2 = 0$, where $I$ is the identity operator and 0 the zero operator. Consider the Hamilton operator

$$\hat{H} = \hbar\omega_1(e^{i\phi}c^\dagger + e^{-i\phi}c) + \hbar\omega_2 c^\dagger c$$

and the basis $|0\rangle$, $c^\dagger|0\rangle$. The dual basis is $\langle 0|$, $\langle 0|c$. Find the matrix representation of $\hat{H}$ and the eigenvalues of the matrix.

**Solution 26.**   We obtain

$$\hat{H}|0\rangle = \hbar\omega_1 e^{i\phi}c^\dagger|0\rangle$$

$$\hat{H}c^\dagger|0\rangle = \hbar\omega_1 e^{-i\phi}|0\rangle + \hbar\omega_2 c^\dagger|0\rangle.$$

Hence we obtain the matrix representation of $\hat{H}$

$$\begin{pmatrix} 0 & \hbar\omega_1 e^{-i\phi} \\ \hbar\omega_1 e^{i\phi} & \hbar\omega_2 \end{pmatrix}.$$

The eigenvalues are given by

$$E_\pm = \frac{\hbar\omega_2}{2} \pm \frac{1}{2}\sqrt{4\hbar^2\omega_1^2 + \hbar^2\omega_2^2}.$$

**Problem 27.** Let $\sigma_1$, $\sigma_2$, $\sigma_3$ be the Pauli spin matrices and $I_2$ the $2 \times 2$ identity matrix. Find the eigenvalues and normalized eigenvectors of the Hamilton operator

$$\hat{H} = \varepsilon_0 I_2 + \hbar\omega\sigma_3 + \Delta_1\sigma_1 + \Delta_2\sigma_2$$

where $\varepsilon_0 > 0$. Are the normalized eigenvectors orthonormal to each other?

**Solution 27.** In matrix form we have the (hermitian) $2 \times 2$ matrix

$$\hat{H} = \begin{pmatrix} \varepsilon_0 + \hbar\omega & \Delta_1 - i\Delta_2 \\ \Delta_1 + i\Delta_2 & \varepsilon_0 - \hbar\omega \end{pmatrix}.$$

From

$$\det(\hat{H} - EI_2) = (\varepsilon + \hbar\omega - E)(\varepsilon - \hbar\omega - E) - (\Delta_1 + i\Delta_2)(\Delta_1 - i\Delta_2) = 0$$

we obtain the characteristic equation

$$E^2 - 2\varepsilon_0 E + \varepsilon_0^2 = \hbar^2\omega^2 + \Delta_1^2 + \Delta_2^2.$$

Thus the two eigenvalues $E_+$, $E_-$ are

$$E_\pm = \varepsilon_0 \pm \sqrt{\hbar^2\omega^2 + \Delta_1^2 + \Delta_2^2}.$$

Let $S := \sqrt{\hbar^2\omega^2 + \Delta_1^2 + \Delta_2^2}$. For the eigenvector of $E_+$ we have to solve

$$\begin{pmatrix} \varepsilon_0 + \hbar\omega & \Delta_1 - i\Delta_2 \\ \Delta_1 + i\Delta_2 & \varepsilon_0 - \hbar\omega \end{pmatrix} \begin{pmatrix} v_1 \\ v_2 \end{pmatrix} = E_+ \begin{pmatrix} v_1 \\ v_2 \end{pmatrix}$$

or

$$(\varepsilon_0 + \hbar\omega)v_1 + (\Delta_1 - i\Delta_2)v_2 = E_+ v_1 = (\varepsilon_0 + S)v_1$$
$$(\Delta_1 + i\Delta_2)v_1 + (\varepsilon_0 - \hbar\omega)v_2 = (\varepsilon_0 + S)v_2.$$

We can set $v_1 = S + \hbar\omega$. Thus $v_2 = \Delta_1 + i\Delta_2$. After normalizing we have the eigenvector

$$\frac{1}{\sqrt{(S + \hbar\omega)^2 + \Delta_1^2 + \Delta_2^2}} \begin{pmatrix} S + \hbar\omega \\ \Delta_1 + i\Delta_2 \end{pmatrix}.$$

Analogously for $E_-$ we obtain the normalized eigenvector

$$\frac{1}{\sqrt{(S + \hbar\omega)^2 + \Delta_1^2 + \Delta_2^2}} \begin{pmatrix} \Delta_1 - i\Delta_2 \\ -S - \hbar\omega \end{pmatrix}.$$

Obviously the two eigenvectors are orthonormal to each other, i.e. the scalar product vanishes. The eigenvectors do not depend on $\varepsilon_0$.

**Problem 28.** Let $\hat{A}$ and $\hat{B}$ be $n \times n$ hermitian matrices. Let $|\psi\rangle$ be a normalized state in the Hilbert space $\mathbb{C}^n$. Then we have the inequality

$$(\Delta\hat{A})(\Delta\hat{B}) \geq \frac{1}{2}|\langle[\hat{A}, \hat{B}]\rangle|$$

where

$$\Delta\hat{A} := \sqrt{\langle\hat{A}^2\rangle - \langle\hat{A}\rangle^2}, \qquad \Delta\hat{B} := \sqrt{\langle\hat{B}^2\rangle - \langle\hat{B}\rangle^2}$$

and

$$\langle\hat{A}\rangle := \langle\psi|\hat{A}|\psi\rangle, \qquad \langle\hat{B}\rangle := \langle\psi|\hat{B}|\psi\rangle.$$

Consider the hermitian spin-$\frac{1}{2}$ matrices

$$S_1 = \frac{1}{2}\begin{pmatrix} 0 & 1 \\ 1 & 0 \end{pmatrix}, \qquad S_2 = \frac{1}{2}\begin{pmatrix} 0 & -i \\ i & 0 \end{pmatrix}, \qquad S_3 = \frac{1}{2}\begin{pmatrix} 1 & 0 \\ 0 & -1 \end{pmatrix}.$$

Let $\hat{A} = S_1$ and $\hat{B} = S_2$. Find states $|\psi\rangle$ such that

$$(\Delta\hat{A})(\Delta\hat{B}) = \frac{1}{2}|\langle[\hat{A}, \hat{B}]\rangle|$$

i.e. the inequality given above should be an equality.

**Solution 28.** For the commutator we find $[S_1, S_2] = iS_3$. Now

$$S_1^2 = S_2^2 = S_3^2 = \frac{1}{4}I_2.$$

We set

$$|\psi\rangle = \begin{pmatrix} c_1 \\ c_2 \end{pmatrix}, \qquad \langle\psi| = (c_1^* \quad c_2^*)$$

with $c_1 c_1^* + c_2 c_2^* = 1$ (normalization). Thus we have for the right-hand side of the equality

$$\frac{1}{2}|\langle[\hat{A},\hat{B}]\rangle| = \frac{1}{2}|\langle\psi|iS_3|\psi\rangle| = \frac{1}{4}\left|(c_1^* \;\; c_2^*)\begin{pmatrix} 1 & 0 \\ 0 & -1 \end{pmatrix}\begin{pmatrix} c_1 \\ c_2 \end{pmatrix}\right| = \frac{1}{4}|r_1^2 - r_2^2|$$

where we set $c_1 = r_1 e^{i\phi_1}$, $c_2 = r_2 e^{i\phi_2}$. Now we have

$$\Delta S_1 = \sqrt{\langle\frac{1}{4}I_2\rangle - \langle S_1\rangle^2} = \frac{1}{2}\sqrt{1 - (c_1^* c_2 + c_1 c_2^*)^2}$$

$$\Delta S_2 = \sqrt{\langle\frac{1}{4}I_2\rangle - \langle S_2\rangle^2} = \frac{1}{2}\sqrt{1 - (c_1 c_2^* - c_1^* c_2)^2}.$$

Thus for the left-hand side we find

$$(\Delta S_1)(\Delta S_2) = \frac{1}{2\sqrt{2}}\sqrt{1 - (c_1 c_2^*)^2 - (c_1^* c_2)^2} = \frac{1}{2\sqrt{2}}\sqrt{1 - 2r_1^2 r_2^2 \cos(2(\phi_1 - \phi_2))}.$$

Thus the condition from the equality is

$$\frac{1}{2\sqrt{2}}\sqrt{1 - 2r_1^2 r_2^2 \cos(2(\phi_1 - \phi_2))} = \frac{1}{4}|r_1^2 - r_2^2|.$$

**Problem 29.** Consider a $d$-dimensional Hilbert space with two orthonormal bases

$$|b_{11}\rangle, \quad |b_{12}\rangle, \quad \dots \quad |b_{1d}\rangle \in \mathcal{B}_1$$

$$|b_{21}\rangle, \quad |b_{22}\rangle, \quad \dots \quad |b_{2d}\rangle \in \mathcal{B}_2.$$

The two bases are said to be *mutually unbiased bases* if

$$|\langle b_{2j}|b_{1k}\rangle| = \frac{1}{\sqrt{d}}$$

for all $j, k = 1, \dots, d$ and $\langle|\rangle$ denotes the scalar product in the Hilbert space. Consider the Hilbert space $M_2(\mathbb{C})$ of $2 \times 2$ matrices over $\mathbb{C}$, where the scalar product is defined as

$$\langle A|B\rangle = \text{tr}(AB^*), \qquad A, B \in M_2(\mathbb{C}).$$

Thus $d = \dim(M_2(\mathbb{C})) = 4$. The standard basis in this Hilbert space is given by

$$E_{11} = \begin{pmatrix} 1 & 0 \\ 0 & 0 \end{pmatrix}, \quad E_{12} = \begin{pmatrix} 0 & 1 \\ 0 & 0 \end{pmatrix}, \quad E_{21} = \begin{pmatrix} 0 & 0 \\ 1 & 0 \end{pmatrix}, \quad E_{22} = \begin{pmatrix} 0 & 0 \\ 0 & 1 \end{pmatrix}.$$

Let $U_H$ be the Hadamard matrix

$$U_H = \frac{1}{\sqrt{2}} \begin{pmatrix} 1 & 1 \\ 1 & -1 \end{pmatrix}, \qquad U_H^* = U_H.$$

Show that the matrices $\widetilde{E}_{jk}$ $(j, k = 1, 2)$

$$\widetilde{E}_{jk} = U_H E_{jk} U_H^*, \quad j, k = 1, 2$$

and the standard basis form mutually unbiased bases.

**Solution 29.**   Straightforward calculations yield

$$\widetilde{E}_{11} = U_H E_{11} U_H^* = \frac{1}{2} \begin{pmatrix} 1 & 1 \\ 1 & 1 \end{pmatrix}$$

$$\widetilde{E}_{12} = U_H E_{12} U_H^* = \frac{1}{2} \begin{pmatrix} 1 & -1 \\ 1 & -1 \end{pmatrix}$$

$$\widetilde{E}_{21} = U_H E_{21} U_H^* = \frac{1}{2} \begin{pmatrix} 1 & 1 \\ -1 & -1 \end{pmatrix}$$

$$\widetilde{E}_{22} = U_H E_{22} U_H^* = \frac{1}{2} \begin{pmatrix} 1 & -1 \\ -1 & 1 \end{pmatrix}.$$

It follows that

$$|\text{tr}(E_{1j} \widetilde{E}_{2k})| = \frac{1}{2} \quad \text{for all } j, k = 1, 2.$$

Thus we have mutually unbiased bases. Apply the vec-operator to the matrices $E_{jk}$ and $\widetilde{E}_{jk}$ $(j, k = 1, 2)$ to find mutually unbiased bases in the Hilbert space $\mathbb{C}^4$.

**Problem 30.**   Let $x_0 = ct$. We define a linear *bijection*, $h$, between $\mathbb{R}^4$ and $\mathbb{H}(2)$, the set of complex $2 \times 2$ hermitian matrices, by

$$(x_0, x_1, x_2, x_3) \rightarrow \begin{pmatrix} x_0 + x_1 & x_2 - ix_3 \\ x_2 + ix_3 & x_0 - x_1 \end{pmatrix}.$$

We denote the matrix on the right hand side by $H$.
(i) Show that the matrix can be written as a linear combination of the Pauli spin matrices and the identity matrix $I_2$.
(ii) Find the inverse map.
(iii) Calculate the determinant of the $2 \times 2$ hermitian matrix $H$. Discuss.

**Solution 30.**   (i) We have $H = x_0 I_2 + x_1 \sigma_3 + x_2 \sigma_1 + x_3 \sigma_2$.
(ii) Consider $(a, b \in \mathbb{R})$

$$\begin{pmatrix} a & c \\ c^* & b \end{pmatrix} = \begin{pmatrix} x_0 + x_1 & x_2 - ix_3 \\ x_2 + ix_3 & x_0 - x_1 \end{pmatrix}.$$

Comparing the entries we obtain

$$x_0 = \frac{a+b}{2}, \quad x_1 = \frac{a-b}{2}, \quad x_2 = \frac{c+c^*}{2}, \quad x_3 = \frac{c^*-c}{2i}.$$

(iii) We obtain

$$\det(H) = x_0^2 - x_1^2 - x_2^2 - x_3^2.$$

This is the *Lorentz metric*. Let $U$ be a unitary $2 \times 2$ matrix. Then $\det(UHU^*) = \det(H)$.


## Programming Problems

**Problem 1.** Consider the *unary gates* ($2 \times 2$ unitary matrices)

$$N = \begin{pmatrix} 0 & 1 \\ 1 & 0 \end{pmatrix}, \qquad H = \frac{1}{\sqrt{2}} \begin{pmatrix} 1 & 1 \\ 1 & -1 \end{pmatrix},$$

$$V = \begin{pmatrix} 1 & 0 \\ 0 & e^{i\pi/2} \end{pmatrix}, \qquad W = \begin{pmatrix} 1 & 0 \\ 0 & e^{i\pi/4} \end{pmatrix}$$

and the normalized state

$$|\psi\rangle = \frac{1}{\sqrt{2}} \begin{pmatrix} 1 \\ 1 \end{pmatrix}.$$

Calculate the state $NHVW|\psi\rangle$ and the expectation value $\langle\psi|NHVW|\psi\rangle$.

**Solution 1.** Applying the Maxima program

```
/* unary.mac */
N: matrix([0,1],[1,0]);
H: matrix([1/sqrt(2),1/sqrt(2)],[1/sqrt(2),-1/\sqrt(2)]);
V: matrix([1,0],[0,exp(%i*%pi/2)]);
W: matrix([1,0],[0,exp(%i*%pi/4)]);
psi: matrix([1/sqrt(2)],[1/sqrt(2)]);
psiT: matrix([1/sqrt(2),1/sqrt(2)]);
R1: N . H . V . W; R1: ratsimp(R1);
R2: R1 . psi; R2: ratsimp(R2);
R3: psiT . R2; R3: ratsimp(R3);
```

we find the unitary matrix

$$NHVW = \frac{1}{\sqrt{2}} \begin{pmatrix} 1 & -e^{i3\pi/4} \\ 1 & e^{i3\pi/4} \end{pmatrix}$$

the normalized state

$$NHVW|\psi\rangle = \frac{1}{2} \begin{pmatrix} 1 - e^{i3\pi/4} \\ 1 + e^{i3\pi/4} \end{pmatrix}$$

and
$$\langle\psi|NHVW|\psi\rangle = \frac{1}{\sqrt{2}}.$$

**Problem 2.**    Consider the Pauli spin matrix $\sigma_1$ and the normalized state in $\mathbb{C}^2$
$$|\psi\rangle = \begin{pmatrix} \cos(\theta) \\ \sin(\theta) \end{pmatrix}.$$
Calculate the *variance*
$$V_{\sigma_1}(\psi) := \langle\psi|\sigma_1^2|\psi\rangle - (\langle\psi|\sigma_1|\psi\rangle)^2$$
and discuss the dependence on $\theta$

**Solution 2.**    The following Maxima program

```
/* Variancesig1 */
sig1: matrix([0,1],[1,0]);
sig2: matrix([0,-%i],[%i,0]);
sig3: matrix([1,0],[0,-1]);
psi: matrix([cos(theta)],[sin(theta)]);
psiT: transpose(psi);
Vs1: psiT . (sig1 . sig1) . psi - (psiT . sig1 . psi)^2;
Vs1: trigsimp(Vs1);
D1: diff(Vs1,theta);
list: solve(D1=0,theta);
theta1: rhs(part(list,1));
theta2: rhs(part(list,2));
theta3: rhs(part(list,3));
theta4: rhs(part(list,4));
r11: subst(theta1,theta,Vs1); r12: subst(theta2,theta,Vs1);
r13: subst(theta3,theta,Vs1); r14: subst(theta4,theta,Vs1);
```

provides
$$V_{\sigma_1}(\psi) := \langle\psi|\sigma_1^2|\psi\rangle - (\langle\psi|\sigma_1|\psi\rangle)^2 = 1 + 4(\cos^4(\theta) - \cos^2(\theta)).$$

Differentiation with respect to $\theta$ and solving the resulting equation provides that the variance is 1 for $\theta = 0$ and $\theta = \pi/2$. The variance is 0 for $\theta = \pi/4$ and $\theta = 3\pi/4$.

**Problem 3.**    Find the eigenvalues and normalized eigenvectors of the Hamilton operator
$$\hat{K} = \frac{\hat{H}}{\hbar\omega} = \frac{1}{\sqrt{2}} \begin{pmatrix} 1 & 1 \\ 1 & -1 \end{pmatrix}.$$

**Solution 3.**    The Maxima program

```
/* Hamilton1.mac */
H: matrix([1,1],[1,-1])/sqrt(2);
list: eigenvectors(H);
p1: part(list,1);
p11: part(p1,1);
lam1: part(p11,1);
lam2: part(p11,2);
p2: part(list,2);
v1: part(p2,1); v1: part(v1,1);
v2: part(p2,2); v2: part(v2,1);
v2T: transpose(v2);
scalar: v1 . v2T;
scalar: ratsimp(scalar);
```

provides the eigenvalues $\lambda_1 = -1$, $\lambda_2 = 1$ with the corresponding (nonnormalized) eigenvectors

$$\mathbf{v}_1 = \begin{pmatrix} 1 \\ -\sqrt{2} - 1 \end{pmatrix}, \quad \mathbf{v}_2 = \begin{pmatrix} 1 \\ \sqrt{2} - 1 \end{pmatrix}.$$

The two eigenvectors are orthogonal to each other, i.e. `scalar=0`.

## 1.3  Supplementary Problems

**Problem 1.**  Consider the map $\mathbf{f} : \mathbb{C}^2 \to \mathbb{R}^3$ defined by

$$\mathbf{f} \; : \; \begin{pmatrix} \cos(\theta) \\ e^{i\phi} \sin(\theta) \end{pmatrix} \mapsto \begin{pmatrix} \sin(2\theta)\cos(\phi) \\ \sin(2\theta)\sin(\phi) \\ \cos(2\theta) \end{pmatrix}.$$

Are the vectors in $\mathbb{C}^2$ and $\mathbb{R}^3$ normalized? Consider the four normalized vectors in $\mathbb{C}^2$

$$\frac{1}{\sqrt{2}}\begin{pmatrix} 1 \\ 1 \end{pmatrix}, \quad \frac{1}{\sqrt{2}}\begin{pmatrix} 1 \\ -1 \end{pmatrix}, \quad \frac{1}{\sqrt{2}}\begin{pmatrix} 1 \\ i \end{pmatrix}, \quad \frac{1}{\sqrt{2}}\begin{pmatrix} 1 \\ -i \end{pmatrix}.$$

Find the vectors in $\mathbb{R}^3$.

**Problem 2.**  (i) Let $x_1, x_2, x_3 \in \mathbb{R}$ and $\sigma_1$, $\sigma_2$, $\sigma_3$ be the Pauli spin matrices. Show that

$$e^{i(x_1\sigma_1 + x_2\sigma_2 + x_3\sigma_3)} = \cos(r)I_2 + i\frac{\sin(r)}{r}(x_1\sigma_1 + x_2\sigma_2 + x_3\sigma_3)$$

$$= \begin{pmatrix} \cos(r) + ix_3\sin(r)/r & i(x_1 - ix_2)\sin(r)/r \\ i(x_1 + ix_2)\sin(r)/r & \cos(r) - ix_3\sin(r)/r \end{pmatrix}$$

where $r := \sqrt{x_1^2 + x_2^2 + x_3^2}$.

(ii) Let $y_1, y_2, y_3 \in \mathbb{R}$ and

$$X := x_1\sigma_1 + x_2\sigma_2 + x_3\sigma_3, \qquad Y := y_1\sigma_1 + y_2\sigma_2 + y_3\sigma_3.$$

Consider the maps

$$X \leftrightarrow \mathbf{x} = \begin{pmatrix} x_1 \\ x_2 \\ x_3 \end{pmatrix}, \qquad Y \leftrightarrow \mathbf{y} = \begin{pmatrix} y_1 \\ y_2 \\ y_3 \end{pmatrix}.$$

Let $\mathbf{x} \cdot \mathbf{y} := x_1y_1 + x_2y_2 + x_3y_3$ (scalar product). Show that

$$\mathbf{x} \cdot \mathbf{y} = \frac{1}{2}\mathrm{tr}(XY).$$

(iii) Show that

$$-\frac{i}{2}[X,Y] \leftrightarrow \mathbf{x} \times \mathbf{y} = \begin{pmatrix} x_2y_3 - x_3y_2 \\ x_3y_1 - x_1y_3 \\ x_1y_2 - x_2y_1 \end{pmatrix}.$$

**Problem 3.** Let $\hat{H}$ be a $2 \times 2$ hermitian matrix. Consider the normalized state

$$|\psi\rangle = \begin{pmatrix} e^{i\phi}\cos(\theta) \\ \sin(\theta) \end{pmatrix}$$

in the Hilbert space $\mathbb{C}^2$. Assume that

$$\langle\psi|\hat{H}|\psi\rangle = \hbar\omega\cos(\phi)\sin(2\theta), \qquad \langle\psi|\hat{H}^2|\psi\rangle = \hbar^2\omega^2.$$

Reconstruct the hermitian matrix $\hat{H}$ from these three assumptions. Note that

$$\cos(\theta)\sin(\theta) \equiv \frac{1}{2}\sin(2\theta), \quad e^{i\phi} = \cos(\phi)+i\sin(\phi), \quad e^{-i\phi} = \cos(\phi)-i\sin(\phi).$$

Show that $H = \hbar\omega\sigma_1$.

**Problem 4.** Let $A$, $B$ be $n \times n$ matrices over $\mathbb{C}$. Let $\mathbf{v}$ be a normalized (column) vector in $\mathbb{C}^n$. Let $\langle A\rangle := \mathbf{v}^*A\mathbf{v}$ and $\langle B\rangle := \mathbf{v}^*B\mathbf{v}$. We have the identity

$$AB \equiv (A - \langle A\rangle I_n)(B - \langle B\rangle I_n) + A\langle B\rangle + B\langle A\rangle - \langle A\rangle\langle B\rangle I_n.$$

We approximate $AB$ as $AB \approx A\langle B\rangle + B\langle A\rangle - \langle A\rangle\langle B\rangle I_n$.

(i) Let $n = 2$ and

$$A = \sigma_1, \quad B = \sigma_2, \quad \mathbf{v} = \frac{1}{\sqrt{2}} \begin{pmatrix} 1 \\ 1 \end{pmatrix}.$$

Find $AB$ and $A\langle B \rangle + B\langle A \rangle - \langle A \rangle \langle B \rangle I_2$ and the distance (Frobenius norm) between the two matrices.

(ii) Let $n = 2$ and

$$A = \sigma_1, \quad B = \sigma_2, \quad \mathbf{v} = \frac{1}{\sqrt{2}} \begin{pmatrix} 1 \\ -1 \end{pmatrix}.$$

Find $AB$ and $A\langle B \rangle + B\langle A \rangle - \langle A \rangle \langle B \rangle I_2$ and the distance (Frobenius norm) between the two matrices.

(iii) Consider the case

$$A = \sigma_1, \quad B = \sigma_2, \quad \mathbf{v} = \begin{pmatrix} \cos(\theta) \\ \sin(\theta) \end{pmatrix}.$$

Find $AB$ and $A\langle B \rangle + B\langle A \rangle - \langle A \rangle \langle B \rangle I_2$ and the distance (Frobenius norm) between the two matrices.

**Problem 5.**   Let $\alpha \in \mathbb{R}$. Show that the vectors

$$\mathbf{v}_1 = \frac{1}{\sqrt{1 + 2\sinh^2(\alpha)}} \begin{pmatrix} \cosh(\alpha) \\ \sinh(\alpha) \end{pmatrix}, \quad \mathbf{v}_2 = \frac{1}{\sqrt{1 + 2\sinh^2(\alpha)}} \begin{pmatrix} \sinh(\alpha) \\ -\cosh(\alpha) \end{pmatrix}$$

form an orthonormal basis in $\mathbb{C}^2$. Find $\mathbf{v}_1 \mathbf{v}_1^T$, $\mathbf{v}_2 \mathbf{v}_2^T$, $\mathbf{v}_1 \mathbf{v}_2^T$.

**Problem 6.**   Let $\sigma_1, \sigma_2, \sigma_3$ be the Pauli spin matrices. Show that

$$\Pi_+ = \frac{1}{2}(I_2 + \sigma_j), \quad \Pi_- = \frac{1}{2}(I_2 - \sigma_j)$$

$(j = 1, 2, 3)$ are projection matrices. Find the vectors

$$\Pi_+ \frac{1}{\sqrt{2}} \begin{pmatrix} 1 \\ 1 \end{pmatrix}, \quad \Pi_- \frac{1}{\sqrt{2}} \begin{pmatrix} 1 \\ 1 \end{pmatrix}.$$

Are the vectors normalized?

**Problem 7.**   Given two arbitrary normalized states $|\psi\rangle$ and $|\phi\rangle$ in $\mathbb{C}^2$. Find a $2 \times 2$ unitary matrix $U$ such that $|\psi\rangle = U|\phi\rangle$, i.e. $U$ must be expressed in terms of the components of the states $|\psi\rangle$ and $\phi\rangle$. Since $U$ is unitary we have $U^{-1} = U^*$.

**Problem 8.**   Let $A$, $B$ be $2 \times 2$ hermitian matrices and

$$|\psi\rangle = \begin{pmatrix} \cos(\theta) \\ \sin(\theta) \end{pmatrix}.$$

Find the minima of the function

$$f(\theta) = \|AB - A\langle\psi|B|\psi\rangle - \langle\psi|A|\psi\rangle B + \langle\psi|A|\psi\rangle\langle\psi|B|\psi\rangle I_2\|$$

where $\|.\|$ denotes the norm.

**Problem 9.**   Let $|0\rangle$, $|1\rangle$ be an orthonormal basis in $\mathbb{C}^2$ and $z_{00}$, $z_{01}$, $z_{10}$, $z_{11}$ be complex numbers. Calculate

$$\exp(z_{00}|0\rangle\langle0| + z_{01}|0\rangle\langle1| + z_{10}|1\rangle\langle0| + z_{11}|1\rangle\langle1|).$$

Then set

$$z_{00} = -i\hbar\omega_1, \quad z_{01} = -z_{10} = -i\hbar\omega_2, \quad z_{11} = -i\hbar\omega_3.$$

# Chapter 2

# Kronecker Product and Tensor Product

## 2.1 Introduction

Let $\mathcal{H}_1$ and $\mathcal{H}_2$ be two Hilbert spaces and $\mathcal{H}$ be a third Hilbert space defined in terms of $\mathcal{H}_1$ and $\mathcal{H}_2$ with the following specifications. For each pair of vectors $f_1$, $f_2$ in $\mathcal{H}_1$, $\mathcal{H}_2$, respectively, there are vectors in $\mathcal{H}$ denoted by $f_1 \otimes f_2$ and $g_1 \otimes g_2$, respectively such that

$$\langle f_1 \otimes f_2 | g_1 \otimes g_2 \rangle = \langle f_1 | g_1 \rangle_{\mathcal{H}_1} \langle f_2 | g_2 \rangle_{\mathcal{H}_2}$$

where $\langle f_1 | g_1 \rangle$ is the scalar product in the Hilbert space $\mathcal{H}_1$. The vector $f_1 \otimes f_2$ is called the tensor product of the vectors $f_1$ and $f_2$. The Hilbert space $\mathcal{H}$ consists of the linear combinations of the vectors $f_1 \otimes f_2$ together with the strong limits of their Cauchy sequences. We term $\mathcal{H}$ the *tensor product* of $\mathcal{H}_1$ and $\mathcal{H}_2$ and denote it by $\mathcal{H}_1 \otimes \mathcal{H}_2$. Given a basis $\{ |\phi_i\rangle : i \in I \}$ in the Hilbert $\mathcal{H}_1$ and a basis $\{ |\psi_j\rangle : j \in J \}$ in the Hilbert space $\mathcal{H}_2$ we can construct a basis

$$\{ |\phi_i\rangle \otimes |\psi_j\rangle : i \in I, j \in J \}$$

in the product Hilbert space. The tensor product is associative and distributive. If $\hat{A}_1$ and $\hat{A}_2$ are linear operators in $\mathcal{H}_1$ and $\mathcal{H}_2$, respectively, we define the operator $\hat{A}_1 \otimes \hat{A}_2$ in $\mathcal{H}_1 \otimes \mathcal{H}_2$ by the formula

$$(\hat{A}_1 \otimes \hat{A}_2)(f_1 \otimes f_2) = (\hat{A}_1 f_1) \otimes (\hat{A}_2 f_2).$$

$\hat{A}_1 \otimes \hat{A}_2$ is called the tensor product of $\hat{A}_1$ and $\hat{A}_2$. Similarly we can define the tensor product of $n$ Hilbert spaces.

For the finite-dimensional Hilbert spaces $\mathbb{C}^n$ and $\mathbb{R}^n$ the tensor product is realized by the Kronecker product.

Let $A$ be an $m \times n$ matrix and $B$ be an $r \times s$ matrix. The *Kronecker product* of $A$ and $B$ is defined as the $(m \cdot r) \times (n \cdot s)$ matrix

$$
A \otimes B := \begin{pmatrix} a_{11}B & a_{12}B & \cdots & a_{1n}B \\ a_{21}B & a_{22}B & \cdots & a_{2n}B \\ \vdots & \vdots & \ddots & \vdots \\ a_{m1}B & a_{m2}B & \cdots & a_{mn}B \end{pmatrix}.
$$

We have the following properties.
1) If $\mathbf{v}_j$ $(j = 1, \ldots, n)$ is an orthonormal basis in $\mathbb{C}^n$, then

$$
\mathbf{v}_j \otimes \mathbf{v}_k, \quad (j, k = 1, \ldots, n)
$$

is an orthonormal basis in $\mathbb{C}^{n^2}$.
2) If $A$, $B$ are normal matrices, then $A \otimes B$ is a normal matrix.
3) If $A$, $B$ are hermitian matrices, then $A \otimes B$ is a hermitian matrix.
4) If $A$, $B$ are unitary matrices, then $A \otimes B$ is a unitary matrix.
5) If $A$, $B$ are projection matrices, then $A \otimes B$ is a projection matrix.
6) If $A$, $B$ are nilpotent matrices, then $A \otimes B$ is a nilpotent matrix.
7) If $P_1$, $P_2$ are permutation matrices, then $P_1 \otimes P_2$ is a permutation matrix.
8) If $A$ and $B$ are invertible, then $A \otimes B$ is invertible with

$$
(A \otimes B)^{-1} \equiv A^{-1} \otimes B^{-1}.
$$

Let $A$, $B$, $C$, $D$ be matrices and assume that the matrix products $AC$ and $BD$ exist. Then

$$
(A \otimes B)(C \otimes B) = (AC) \otimes (BD).
$$

Let $A$ be an $m \times m$ matrix and $B$ be an $n \times n$ matrix. The underlying field is $\mathbb{C}$. Let $I_m$, $I_n$ be the $m \times m$ and $n \times n$ unit matrix, respectively. Then

$$
\mathrm{tr}(A \otimes B) = \mathrm{tr}(A)\mathrm{tr}(B)
$$

and

$$
\mathrm{tr}(A \otimes I_n + I_m \otimes B) = n\,\mathrm{tr}(A) + m\,\mathrm{tr}(B).
$$

## 2.2   Solved Problems

**Problem 1.**   (i) Let

$$|\phi_1\rangle := \begin{pmatrix} 1 \\ 0 \end{pmatrix}, \qquad |\phi_2\rangle := \begin{pmatrix} 0 \\ 1 \end{pmatrix}.$$

Thus the set $\{ |\phi_1\rangle, |\phi_2\rangle \}$ forms a basis in $\mathbb{C}^2$ (the standard basis). Calculate the vectors in $\mathbb{C}^4$

$$|\phi_1\rangle \otimes |\phi_1\rangle, \quad |\phi_1\rangle \otimes |\phi_2\rangle, \quad |\phi_2\rangle \otimes |\phi_1\rangle, \quad |\phi_2\rangle \otimes |\phi_2\rangle$$

and interpret the result.
(ii) Consider the Pauli matrices $\sigma_1$ and $\sigma_3$. Find $\sigma_1 \otimes \sigma_3$ and $\sigma_3 \otimes \sigma_1$ and discuss. Both $\sigma_1$ and $\sigma_3$ are hermitian. Are $\sigma_1 \otimes \sigma_3$ and $\sigma_3 \otimes \sigma_1$ hermitian? Both $\sigma_1$ and $\sigma_3$ are unitary. Is $\sigma_1 \otimes \sigma_3$ and $\sigma_3 \otimes \sigma_1$ unitary?

**Solution 1.**   (i) We obtain

$$|\phi_1\rangle \otimes |\phi_1\rangle = \begin{pmatrix} 1 \\ 0 \\ 0 \\ 0 \end{pmatrix}, \qquad |\phi_1\rangle \otimes |\phi_2\rangle = \begin{pmatrix} 0 \\ 1 \\ 0 \\ 0 \end{pmatrix},$$

$$|\phi_2\rangle \otimes |\phi_1\rangle = \begin{pmatrix} 0 \\ 0 \\ 1 \\ 0 \end{pmatrix}, \qquad |\phi_2\rangle \otimes |\phi_2\rangle = \begin{pmatrix} 0 \\ 0 \\ 0 \\ 1 \end{pmatrix}.$$

Thus we find the standard basis in $\mathbb{C}^4$ from the standard basis in $\mathbb{C}^2$.
(ii) We obtain

$$\sigma_1 \otimes \sigma_3 = \begin{pmatrix} 0 & 0 & 1 & 0 \\ 0 & 0 & 0 & -1 \\ 1 & 0 & 0 & 0 \\ 0 & -1 & 0 & 0 \end{pmatrix}$$

and

$$\sigma_3 \otimes \sigma_1 = \begin{pmatrix} 0 & 1 & 0 & 0 \\ 1 & 0 & 0 & 0 \\ 0 & 0 & 0 & -1 \\ 0 & 0 & -1 & 0 \end{pmatrix}.$$

We note that $\sigma_1 \otimes \sigma_3 \neq \sigma_3 \otimes \sigma_1$. $\sigma_1 \otimes \sigma_3$ and $\sigma_3 \otimes \sigma_1$ are hermitian. $\sigma_1 \otimes \sigma_3$ and $\sigma_3 \otimes \sigma_1$ are unitary.

**Problem 2.**   Given the orthonormal basis

$$|\psi_1\rangle = \begin{pmatrix} e^{i\phi} \cos(\theta) \\ \sin(\theta) \end{pmatrix}, \qquad |\psi_2\rangle = \begin{pmatrix} -\sin(\theta) \\ e^{-i\phi} \cos(\theta) \end{pmatrix}$$

in the Hilbert space $\mathbb{C}^2$. Use this basis to find a basis in $\mathbb{C}^4$.

**Solution 2.**   A basis in $\mathbb{C}^4$ is given by

$$\{\,|\psi_1\rangle \otimes |\psi_1\rangle, \quad |\psi_1\rangle \otimes |\psi_2\rangle, \quad |\psi_2\rangle \otimes |\psi_1\rangle, \quad |\psi_2\rangle \otimes |\psi_2\rangle\,\}$$

since

$$((\langle\psi_j| \otimes \langle\psi_k|)(|\psi_m\rangle \otimes |\psi_n\rangle)) = \delta_{jm}\delta_{kn}$$

where $j, k, m, n = 1, 2$.

**Problem 3.**   A system of $n$-qubits can be represented by a finite-dimensional Hilbert space over the complex numbers of dimension $2^n$. A state $|\psi\rangle$ of the system is a superposition of the basic states

$$|\psi\rangle = \sum_{j_1,j_2,\dots,j_n=0}^{1} c_{j_1 j_2 \dots j_n} |j_1\rangle \otimes |j_2\rangle \otimes \cdots \otimes |j_n\rangle.$$

In a short cut notation this state is written as

$$|\psi\rangle = \sum_{j_1,j_2,\dots,j_n=0}^{1} c_{j_1 j_2 \dots j_n} |j_1 j_2 \dots j_n\rangle.$$

Consider as a special case the state

$$|\psi\rangle = \frac{1}{2}(|0\rangle\otimes|0\rangle+|0\rangle\otimes|1\rangle+|1\rangle\otimes|0\rangle+|1\rangle\otimes|1\rangle) \equiv \frac{1}{2}(|00\rangle+|01\rangle+|10\rangle+|11\rangle)$$

in the Hilbert space $\mathcal{H} = \mathbb{C}^4$ ($n = 2$). Can this state be written as a product state?

**Solution 3.**   Yes, the state can be written as product state. We have

$$\frac{1}{\sqrt{2}}(|0\rangle + |1\rangle) \otimes \frac{1}{\sqrt{2}}(|0\rangle + |1\rangle).$$

**Problem 4.**   The single-bit *Walsh-Hadamard transform* is the unitary map $W$ given by

$$W|0\rangle = \frac{1}{\sqrt{2}}(|0\rangle + |1\rangle), \qquad W|1\rangle = \frac{1}{\sqrt{2}}(|0\rangle - |1\rangle).$$

The $n$-bit Walsh-Hadamard transform $W_n$ is defined as

$$W_n := W \otimes W \otimes \cdots \otimes W \qquad (n - \text{times}).$$

Consider $n = 2$. Find the normalized state $W_2(|0\rangle \otimes |0\rangle)$.

**Solution 4.**   We have

$$W_2(|0\rangle \otimes |0\rangle) = (W \otimes W)(|0\rangle \otimes |0\rangle) = W|0\rangle \otimes W|0\rangle.$$

Thus

$$W_2(|0\rangle \otimes |0\rangle) = \frac{1}{2}((|0\rangle + |1\rangle) \otimes (|0\rangle + |1\rangle)).$$

Finally

$$W_2(|0\rangle \otimes |0\rangle) = \frac{1}{2}(|0\rangle \otimes |0\rangle + |0\rangle \otimes |1\rangle + |1\rangle \otimes |0\rangle + |1\rangle \otimes |1\rangle).$$

$W_2$ generates a linear combination of all states. This also applies to $W_n$.

**Problem 5.**   Consider the spin matrix $S_1$ for spin-$\frac{1}{2}$

$$S_1 = \frac{1}{2}\sigma_1 = \frac{1}{2}\begin{pmatrix} 0 & 1 \\ 1 & 0 \end{pmatrix}$$

with the eigenvalues $1/2$ and $-1/2$ and the corresponding normalized eigenvectors

$$\mathbf{e}_{1/2} = \frac{1}{\sqrt{2}}\begin{pmatrix} 1 \\ 1 \end{pmatrix}, \qquad \mathbf{e}_{-1/2} = \frac{1}{\sqrt{2}}\begin{pmatrix} 1 \\ -1 \end{pmatrix}.$$

Do the four vectors

$$\frac{1}{\sqrt{2}}(\mathbf{e}_{1/2} \otimes \mathbf{e}_{1/2} + \mathbf{e}_{-1/2} \otimes \mathbf{e}_{-1/2}), \qquad \frac{1}{\sqrt{2}}(\mathbf{e}_{1/2} \otimes \mathbf{e}_{1/2} - \mathbf{e}_{-1/2} \otimes \mathbf{e}_{-1/2}),$$

$$\frac{1}{\sqrt{2}}(\mathbf{e}_{1/2} \otimes \mathbf{e}_{-1/2} + \mathbf{e}_{-1/2} \otimes \mathbf{e}_{1/2}), \qquad \frac{1}{\sqrt{2}}(\mathbf{e}_{1/2} \otimes \mathbf{e}_{-1/2} - \mathbf{e}_{-1/2} \otimes \mathbf{e}_{1/2})$$

form a basis in $\mathbb{C}^4$? Prove or disprove.

**Solution 5.**   We obtain the *Bell basis*

$$\mathbf{v}_1 = \frac{1}{\sqrt{2}}\begin{pmatrix} 1 \\ 0 \\ 0 \\ 1 \end{pmatrix}, \quad \mathbf{v}_2 = \frac{1}{\sqrt{2}}\begin{pmatrix} 0 \\ 1 \\ 1 \\ 0 \end{pmatrix}, \quad \mathbf{v}_3 = \frac{1}{\sqrt{2}}\begin{pmatrix} 1 \\ 0 \\ 0 \\ -1 \end{pmatrix}, \quad \mathbf{v}_4 = \frac{1}{\sqrt{2}}\begin{pmatrix} 0 \\ -1 \\ 1 \\ 0 \end{pmatrix}.$$

which forms an orthonormal basis in $\mathbb{C}^4$

**Problem 6.**   Let $A$ be an arbitrary $n \times n$ matrix over $\mathbb{C}$. Show that

$$\exp(A \otimes I_n) \equiv \exp(A) \otimes I_n. \tag{1}$$

**Solution 6.** Using the expansion

$$\exp(A \otimes I_n) = \sum_{k=0}^{\infty} \frac{(A \otimes I_n)^k}{k!}$$

$$= I_n \otimes I_n + \frac{1}{1!}(A \otimes I_n) + \frac{1}{2!}(A \otimes I_n)^2 + \frac{1}{3!}(A \otimes I_n)^3 + \cdots$$

and $(A \otimes I_n)^k = A^k \otimes I_n$, $k \in \mathbb{N}$ we find identity (1).

**Problem 7.** Let $A$, $B$ be arbitrary $n \times n$ matrices over $\mathbb{C}$. Let $I_n$ be the $n \times n$ unit matrix. Show that

$$\exp(A \otimes I_n + I_n \otimes B) \equiv \exp(A) \otimes \exp(B).$$

**Solution 7.** The proof of this identity relies on $[A \otimes I_n, I_n \otimes B] = 0_{n^2}$ and

$$(A \otimes I_n)^r (I_n \otimes B)^s \equiv (A^r \otimes I_n)(I_n \otimes B^s) \equiv A^r \otimes B^s, \qquad r, s \in \mathbb{N}.$$

Thus

$$\exp(A \otimes I_n + I_n \otimes B) = \sum_{j=0}^{\infty} \frac{(A \otimes I_n + I_n \otimes B)^j}{j!}$$

$$= \sum_{j=0}^{\infty} \sum_{k=0}^{j} \frac{1}{j!} \binom{j}{k} (A \otimes I_n)^k (I_n \otimes B)^{j-k}$$

$$= \sum_{j=0}^{\infty} \sum_{k=0}^{j} \frac{1}{j!} \binom{j}{k} (A^k \otimes B^{j-k})$$

$$= \left( \sum_{j=0}^{\infty} \frac{A^j}{j!} \right) \otimes \left( \sum_{k=0}^{\infty} \frac{B^k}{k!} \right)$$

$$= \exp(A) \otimes \exp(B).$$

**Problem 8.** Let $A$ and $B$ be arbitrary $n \times n$ matrices over $\mathbb{C}$. Prove or disprove the equation

$$e^{A \otimes B} = e^A \otimes e^B.$$

**Solution 8.** Obviously this is not true in general. For example, let $A = B = I_n$. Then

$$e^{A \otimes B} = e^{I_{n^2}}$$

and

$$e^A \otimes e^B = e^{I_n} \otimes e^{I_n} \neq e^{I_{n^2}}.$$

**Problem 9.** Let $A$ be an $m \times m$ normal matrix and $B$ be an $n \times n$ normal matrix. The underlying field is $\mathbb{C}$. The eigenvalues and eigenvectors of $A$ are given by $\lambda_1, \lambda_2, \ldots, \lambda_m$ and $\mathbf{u}_1, \mathbf{u}_2, \ldots, \mathbf{u}_m$. The eigenvalues and eigenvectors of $B$ are given by $\mu_1, \mu_2, \ldots, \mu_n$ and $\mathbf{v}_1, \mathbf{v}_2, \ldots, \mathbf{v}_n$. Let $\epsilon_1$, $\epsilon_2$ and $\epsilon_3$ be real parameters. Find the eigenvalues and eigenvectors of the matrix

$$\epsilon_1 A \otimes B + \epsilon_2 A \otimes I_n + \epsilon_3 I_m \otimes B.$$

**Solution 9.** Let $\mathbf{x} \in \mathbb{C}^m$ and $\mathbf{y} \in \mathbb{C}^n$. Then we have

$$(A \otimes B)(\mathbf{x} \otimes \mathbf{y}) = (A\mathbf{x}) \otimes (B\mathbf{y}),$$

$$(A \otimes I_n)(\mathbf{x} \otimes \mathbf{y}) = (A\mathbf{x}) \otimes \mathbf{y}, \qquad (I_m \otimes B)(\mathbf{x} \otimes \mathbf{y}) = \mathbf{x} \otimes (B\mathbf{y}).$$

Thus the eigenvectors of the matrix are

$$\mathbf{u}_j \otimes \mathbf{v}_k, \qquad j = 1, 2, \ldots, m \quad k = 1, 2, \ldots, n.$$

The corresponding eigenvalues are given by $\epsilon_1 \lambda_j \mu_k + \epsilon_2 \lambda_j + \epsilon_3 \mu_k$.

**Problem 10.** Let $A$, $B$ be $n \times n$ matrices over $\mathbb{C}$. The $n \times n$ matrices form a vector space. A *scalar product* can be defined as

$$\langle A, B \rangle := \mathrm{tr}(AB^*).$$

This provides a Hilbert space. The scalar product implies a *norm*

$$\|A\|^2 = \langle A, A \rangle = \mathrm{tr}(AA^*).$$

This norm is called the *Hilbert-Schmidt norm*.
(i) Consider the *Dirac matrices*

$$\gamma_0 := \begin{pmatrix} 1 & 0 & 0 & 0 \\ 0 & 1 & 0 & 0 \\ 0 & 0 & -1 & 0 \\ 0 & 0 & 0 & -1 \end{pmatrix}, \qquad \gamma_1 := \begin{pmatrix} 0 & 0 & 0 & 1 \\ 0 & 0 & 1 & 0 \\ 0 & -1 & 0 & 0 \\ -1 & 0 & 0 & 0 \end{pmatrix}.$$

Calculate the scalar product $\langle \gamma_0, \gamma_1 \rangle$. Discuss.
(ii) Let $U$ be a unitary $n \times n$ matrix. Find the scalar product $\langle UA, UB \rangle$.
(iii) Let $C$, $D$ be $m \times m$ matrices over $\mathbb{C}$. Find $\langle A \otimes C, B \otimes D \rangle$.

**Solution 10.** (i) We find $\langle \gamma_0, \gamma_1 \rangle = \mathrm{tr}(\gamma_0 \gamma_1^*) = 0$. Thus $\gamma_0$ and $\gamma_1$ are orthogonal to each other.

(ii) Since

$$\text{tr}(UA(UB)^*) = \text{tr}(UAB^*U^*) = \text{tr}(U^*UAB^*) = \text{tr}(AB^*)$$

where we used the *cyclic invariance* for matrices, we find that

$$\langle UA, UB \rangle = \langle A, B \rangle.$$

Thus the scalar product is invariant under unitary transformations.
(iii) Since

$$\text{tr}((A \otimes C)(B \otimes D)^*) = \text{tr}((A \otimes C)(B^* \otimes D^*)) = \text{tr}((AB^*) \otimes (CD^*))$$
$$= \text{tr}(AB^*)\text{tr}(CD^*)$$

we find $\langle A \otimes C, B \otimes D \rangle = \langle A, B \rangle \langle C, D \rangle$.

**Problem 11.**   Let $T$ be the $4 \times 4$ matrix

$$T := \left( I_2 \otimes I_2 + \sum_{j=1}^{3} t_j \sigma_j \otimes \sigma_j \right)$$

where $\sigma_j$, $j = 1, 2, 3$ are the Pauli spin matrices and $-1 \leq t_j \leq +1$, $j = 1, 2, 3$. Find the matrix $T^2$.

**Solution 11.**   We have

$$T^2 = I_2 \otimes I_2 + 2 \sum_{j=1}^{3} t_j \sigma_j \otimes \sigma_j + \sum_{j=1}^{3} \sum_{k=1}^{3} t_j t_k \sigma_j \sigma_k \otimes \sigma_j \sigma_k.$$

Since

$$\sigma_1 \sigma_2 = i\sigma_3, \quad \sigma_2 \sigma_1 = -i\sigma_3, \quad \sigma_2 \sigma_3 = i\sigma_1,$$
$$\sigma_3 \sigma_2 = -i\sigma_1, \quad \sigma_3 \sigma_1 = i\sigma_2, \quad \sigma_1 \sigma_3 = -i\sigma_2$$

and $\sigma_1^2 = I_2$, $\sigma_2^2 = I_2$, $\sigma_3^2 = I_2$, we find

$$\sum_{j,k=1}^{3} t_j t_k \sigma_j \sigma_k \otimes \sigma_j \sigma_k \equiv I_2 \otimes I_2 \sum_{j=1}^{3} t_j^2 - 2(t_1 t_2 \sigma_3 \otimes \sigma_3 + t_2 t_3 \sigma_1 \otimes \sigma_1 + t_3 t_1 \sigma_2 \otimes \sigma_2).$$

Therefore

$$T^2 = (I_2 \otimes I_2) \left( 1 + \sum_{j=1}^{3} t_j^2 \right)$$
$$+ 2(t_1 - t_2 t_3)\sigma_1 \otimes \sigma_1 + 2(t_2 - t_3 t_1)\sigma_2 \otimes \sigma_2 + 2(t_3 - t_1 t_2)\sigma_3 \otimes \sigma_3.$$

**Problem 12.**   Let $\{\,|0\rangle, |1\rangle, \ldots, |n-1\rangle\,\}$ be an orthonormal basis in the Hilbert space $\mathbb{C}^n$. Is

$$|\psi\rangle = \frac{1}{\sqrt{n}} \left( \sum_{j=0}^{n-2} |j\rangle \otimes |j+1\rangle + |n-1\rangle \otimes |0\rangle \right)$$

independent of the chosen orthonormal basis? Prove or disprove.

**Solution 12.**   Consider the special case $\mathbb{R}^2$. Let

$$|0\rangle = \begin{pmatrix} 1 \\ 0 \end{pmatrix}, \qquad |1\rangle = \begin{pmatrix} 0 \\ 1 \end{pmatrix}.$$

Thus

$$|\psi\rangle = \frac{1}{\sqrt{2}} \begin{pmatrix} 1 \\ 0 \end{pmatrix} \otimes \begin{pmatrix} 0 \\ 1 \end{pmatrix} + \frac{1}{\sqrt{2}} \begin{pmatrix} 0 \\ 1 \end{pmatrix} \otimes \begin{pmatrix} 1 \\ 0 \end{pmatrix} = \frac{1}{\sqrt{2}} \begin{pmatrix} 0 \\ 1 \\ 1 \\ 0 \end{pmatrix}.$$

Now let

$$|0\rangle = \frac{1}{\sqrt{2}} \begin{pmatrix} 1 \\ 1 \end{pmatrix}, \qquad |1\rangle = \frac{1}{\sqrt{2}} \begin{pmatrix} 1 \\ -1 \end{pmatrix}.$$

Then

$$|\psi\rangle = \frac{1}{\sqrt{2}} \frac{1}{2} \begin{pmatrix} 1 \\ 1 \end{pmatrix} \otimes \begin{pmatrix} 1 \\ -1 \end{pmatrix} + \frac{1}{\sqrt{2}} \frac{1}{2} \begin{pmatrix} 1 \\ -1 \end{pmatrix} \otimes \begin{pmatrix} 1 \\ 1 \end{pmatrix} = \frac{1}{\sqrt{2}} \begin{pmatrix} 1 \\ 0 \\ 0 \\ -1 \end{pmatrix}.$$

Thus, $|\psi\rangle$ depends on the chosen basis.

**Problem 13.**   In the product Hilbert space $\mathbb{C}^4 \cong \mathbb{C}^2 \otimes \mathbb{C}^2$ the *Bell states* are given by

$$|\Phi^+\rangle = \frac{1}{\sqrt{2}}(|0\rangle \otimes |0\rangle + |1\rangle \otimes |1\rangle), \qquad |\Phi^-\rangle = \frac{1}{\sqrt{2}}(|0\rangle \otimes |0\rangle - |1\rangle \otimes |1\rangle),$$

$$|\Psi^+\rangle = \frac{1}{\sqrt{2}}(|0\rangle \otimes |1\rangle + |1\rangle \otimes |0\rangle), \qquad |\Psi^-\rangle = \frac{1}{\sqrt{2}}(|0\rangle \otimes |1\rangle - |1\rangle \otimes |0\rangle)$$

and form an orthonormal basis in $\mathbb{C}^4$. Here, $\{\,|0\rangle, |1\rangle\,\}$ is an arbitrary orthonormal basis in the Hilbert space $\mathbb{C}^2$. Let

$$|0\rangle = \begin{pmatrix} e^{i\phi}\cos(\theta) \\ \sin(\theta) \end{pmatrix}, \qquad |1\rangle = \begin{pmatrix} -e^{i\phi}\sin(\theta) \\ \cos(\theta) \end{pmatrix}.$$

(i) Find $|\Phi^+\rangle$, $|\Phi^-\rangle$, $|\Psi^+\rangle$, $|\Psi^-\rangle$ for this basis.
(ii) Consider the special case when $\phi = 0$ and $\theta = 0$.

**Solution 13.**  (i) We obtain

$$
|\Phi^+\rangle = \frac{1}{\sqrt{2}}\begin{pmatrix} e^{2i\phi} \\ 0 \\ 0 \\ 1 \end{pmatrix}, \qquad
|\Phi^-\rangle = \frac{1}{\sqrt{2}}\begin{pmatrix} e^{2i\phi}\cos(2\theta) \\ e^{i\phi}\sin(2\theta) \\ e^{i\phi}\sin(2\theta) \\ -\cos(2\theta) \end{pmatrix},
$$

$$
|\Psi^+\rangle = \frac{1}{\sqrt{2}}\begin{pmatrix} -e^{2i\phi}\sin(2\theta) \\ e^{i\phi}\cos(2\theta) \\ e^{i\phi}\cos(2\theta) \\ \sin(2\theta) \end{pmatrix}, \qquad
|\Psi^-\rangle = \frac{1}{\sqrt{2}}\begin{pmatrix} 0 \\ e^{i\phi} \\ -e^{i\phi} \\ 0 \end{pmatrix}.
$$

(ii) If we choose $\phi = 0$ and $\theta = 0$ which simply means we choose the standard basis for $|0\rangle$ and $|1\rangle$ (i.e. $|0\rangle = (1\ 0)^T$ and $|1\rangle = (0\ 1)^T$), we find that the Bell states take the form

$$
|\Phi^+\rangle = \frac{1}{\sqrt{2}}\begin{pmatrix} 1 \\ 0 \\ 0 \\ 1 \end{pmatrix}, \qquad
|\Phi^-\rangle = \frac{1}{\sqrt{2}}\begin{pmatrix} 1 \\ 0 \\ 0 \\ -1 \end{pmatrix},
$$

$$
|\Psi^+\rangle = \frac{1}{\sqrt{2}}\begin{pmatrix} 0 \\ 1 \\ 1 \\ 0 \end{pmatrix}, \qquad
|\Psi^-\rangle = \frac{1}{\sqrt{2}}\begin{pmatrix} 0 \\ 1 \\ -1 \\ 0 \end{pmatrix}.
$$

**Problem 14.**  Let $\mathcal{H}_A$ and $\mathcal{H}_B$ be two $p$-dimensional Hilbert spaces over $\mathbb{C}$, where $p$ is a prime number. Let

$$
\{\,|0\rangle_A, |1\rangle_A, \dots, |(p-1)\rangle_A\,\}, \qquad \{\,|0\rangle_B, |1\rangle_B, \dots, |(p-1)\rangle_B\,\}
$$

be orthonormal bases in these Hilbert spaces. We define the states

$$
|\psi(a,b)\rangle := (I_p \otimes X^a Z^b)\frac{1}{\sqrt{p}}\sum_{j=0}^{p-1}|j\rangle_A \otimes |j\rangle_B
$$

in the Hilbert space $\mathcal{H}_A \otimes \mathcal{H}_B$, where $a, b \in \{0, 1, \dots, p-1\}$. The $p \times p$ matrices $X$ and $Z$ are defined as

$$
X|j\rangle = |j+1 \bmod p\rangle, \qquad Z|j\rangle = \omega^j |j\rangle, \quad j = 0, 1, \dots, p
$$

with a complex primitive $p$th root $\omega$ of 1 and $\{\,|0\rangle, |1\rangle, \dots, |p-1\rangle\,\}$ is the orthonormal basis given above for the Hilbert space $\mathcal{H}_B$. Calculate the states $|\psi(0,0)\rangle$ and $|\psi(1,1)\rangle$.

**Solution 14.** Since

$$X^0 = Z^0 = I_p, \quad I_p|j\rangle_A = |j\rangle_A, \quad I_p|j\rangle_B = |j\rangle_B$$

we obtain

$$|\psi(0,0)\rangle = \frac{1}{\sqrt{p}} \sum_{j=0}^{p-1} |j\rangle_A \otimes |j\rangle_B.$$

Using

$$
\begin{aligned}
(I_p \otimes XZ)(|j\rangle_A \otimes |j\rangle_B &= |j\rangle_A \otimes (XZ|j\rangle_B) = |j\rangle_A \otimes \omega^j X|j\rangle_B \\
&= |j\rangle_A \otimes \omega^j |j_B + 1 \bmod p\rangle \\
&= \omega^j |j\rangle_A \otimes |j_B + 1 \bmod p\rangle
\end{aligned}
$$

we find

$$|\psi(1,1)\rangle = \frac{1}{\sqrt{p}} \sum_{j=0}^{p-1} \omega^j |j\rangle_A \otimes |j_B + 1 \bmod p\rangle.$$

The states $|\psi(a,b)\rangle$ are maximally entangled states in the Hilbert space $\mathcal{H}_A \otimes \mathcal{H}_B$.

**Problem 15.** Consider the Pauli matrices $\sigma_1$ and $\sigma_2$ and the *GHZ state*

$$|\psi\rangle = \frac{1}{\sqrt{2}} \left( \begin{pmatrix} 1 \\ 0 \end{pmatrix} \otimes \begin{pmatrix} 1 \\ 0 \end{pmatrix} \otimes \begin{pmatrix} 1 \\ 0 \end{pmatrix} + \begin{pmatrix} 0 \\ 1 \end{pmatrix} \otimes \begin{pmatrix} 0 \\ 1 \end{pmatrix} \otimes \begin{pmatrix} 0 \\ 1 \end{pmatrix} \right).$$

(i) Show that $|\psi\rangle$ is an eigenvector of the operator $\sigma_2 \otimes \sigma_2 \otimes \sigma_1$. What is the eigenvalue?
(ii) Is

$$|\phi\rangle = \frac{1}{\sqrt{2}} \left( \begin{pmatrix} 1 \\ 0 \end{pmatrix} \otimes \begin{pmatrix} 1 \\ 0 \end{pmatrix} \otimes \begin{pmatrix} 1 \\ 0 \end{pmatrix} + e^{i\phi} \begin{pmatrix} 0 \\ 1 \end{pmatrix} \otimes \begin{pmatrix} 0 \\ 1 \end{pmatrix} \otimes \begin{pmatrix} 0 \\ 1 \end{pmatrix} \right)$$

an eigenvector of $\sigma_2 \otimes \sigma_2 \otimes \sigma_1$?

**Solution 15.** (i) We have

$$\sigma_1 \begin{pmatrix} 1 \\ 0 \end{pmatrix} = \begin{pmatrix} 0 \\ 1 \end{pmatrix}, \quad \sigma_1 \begin{pmatrix} 0 \\ 1 \end{pmatrix} = \begin{pmatrix} 1 \\ 0 \end{pmatrix}, \quad \sigma_2 \begin{pmatrix} 1 \\ 0 \end{pmatrix} = \begin{pmatrix} 0 \\ i \end{pmatrix}, \quad \sigma_2 \begin{pmatrix} 0 \\ 1 \end{pmatrix} = \begin{pmatrix} -i \\ 0 \end{pmatrix}.$$

Thus

$$
\begin{aligned}
(\sigma_2 \otimes \sigma_2 \otimes \sigma_1)|\psi\rangle = \frac{1}{\sqrt{2}} \Bigg( &\sigma_2 \begin{pmatrix} 1 \\ 0 \end{pmatrix} \otimes \sigma_2 \begin{pmatrix} 1 \\ 0 \end{pmatrix} \otimes \sigma_1 \begin{pmatrix} 1 \\ 0 \end{pmatrix} \\
+ &\sigma_2 \begin{pmatrix} 0 \\ 1 \end{pmatrix} \otimes \sigma_2 \begin{pmatrix} 0 \\ 1 \end{pmatrix} \otimes \sigma_1 \begin{pmatrix} 0 \\ 1 \end{pmatrix} \Bigg)
\end{aligned}
$$

$$= \frac{1}{\sqrt{2}} \left( \begin{pmatrix} 0 \\ i \end{pmatrix} \otimes \begin{pmatrix} 0 \\ i \end{pmatrix} \otimes \begin{pmatrix} 0 \\ 1 \end{pmatrix} + \begin{pmatrix} -i \\ 0 \end{pmatrix} \otimes \begin{pmatrix} -i \\ 0 \end{pmatrix} \otimes \begin{pmatrix} 1 \\ 0 \end{pmatrix} \right)$$

$$= -\frac{1}{\sqrt{2}} \left( \begin{pmatrix} 0 \\ 1 \end{pmatrix} \otimes \begin{pmatrix} 0 \\ 1 \end{pmatrix} \otimes \begin{pmatrix} 0 \\ 1 \end{pmatrix} + \begin{pmatrix} 1 \\ 0 \end{pmatrix} \otimes \begin{pmatrix} 1 \\ 0 \end{pmatrix} \otimes \begin{pmatrix} 1 \\ 0 \end{pmatrix} \right)$$

$$= -|\psi\rangle.$$

Consequently, the eigenvalue is $-1$.
(ii) We find using the calculation from (i)

$$(\sigma_2 \otimes \sigma_2 \otimes \sigma_1)|\phi\rangle = -\frac{1}{\sqrt{2}} \left( e^{i\phi} \begin{pmatrix} 1 \\ 0 \end{pmatrix} \otimes \begin{pmatrix} 1 \\ 0 \end{pmatrix} \otimes \begin{pmatrix} 1 \\ 0 \end{pmatrix} + \begin{pmatrix} 0 \\ 1 \end{pmatrix} \otimes \begin{pmatrix} 0 \\ 1 \end{pmatrix} \otimes \begin{pmatrix} 0 \\ 1 \end{pmatrix} \right).$$

Thus, in general $|\phi\rangle$ is not an eigenvector of $\sigma_2 \otimes \sigma_2 \otimes \sigma_1$, for example if $\phi = \pi/4$.

**Problem 16.**   Consider the three-qubit *GHZ state*

$$|\psi\rangle = \frac{1}{\sqrt{2}} (|0\rangle \otimes |0\rangle \otimes |0\rangle + |1\rangle \otimes |1\rangle \otimes |1\rangle)$$

with the standard basis

$$|0\rangle = \begin{pmatrix} 1 \\ 0 \end{pmatrix}, \qquad |1\rangle = \begin{pmatrix} 0 \\ 1 \end{pmatrix}.$$

Let $\sigma_1$, $\sigma_2$ be the Pauli spin matrices.
(i) Calculate the expectation values

$$\langle\psi|(\sigma_2 \otimes \sigma_2 \otimes \sigma_2)|\psi\rangle, \quad \langle\psi|(\sigma_1 \otimes \sigma_1 \otimes \sigma_2)|\psi\rangle,$$

$$\langle\psi|(\sigma_1 \otimes \sigma_2 \otimes \sigma_1)|\psi\rangle, \quad \langle\psi|(\sigma_2 \otimes \sigma_1 \otimes \sigma_1)|\psi\rangle.$$

(ii) Calculate the expectation values

$$\langle\psi|(\sigma_1 \otimes \sigma_2 \otimes \sigma_2)|\psi\rangle, \quad \langle\psi|(\sigma_2 \otimes \sigma_1 \otimes \sigma_2)|\psi\rangle,$$

$$\langle\psi|(\sigma_2 \otimes \sigma_2 \otimes \sigma_1)|\psi\rangle, \quad \langle\psi|(\sigma_1 \otimes \sigma_1 \otimes \sigma_1)|\psi\rangle.$$

**Solution 16.**   (i) We find

$$\langle\psi|(\sigma_2 \otimes \sigma_2 \otimes \sigma_2)|\psi\rangle = 0, \quad \langle\psi|(\sigma_1 \otimes \sigma_1 \otimes \sigma_2)|\psi\rangle = 0,$$

$$\langle\psi|(\sigma_1 \otimes \sigma_2 \otimes \sigma_1)|\psi\rangle = 0, \quad \langle\psi|(\sigma_2 \otimes \sigma_1 \otimes \sigma_1)|\psi\rangle = 0.$$

(ii) We obtain

$$\langle\psi|(\sigma_1 \otimes \sigma_2 \otimes \sigma_2)|\psi\rangle = -1, \quad \langle\psi|(\sigma_2 \otimes \sigma_1 \otimes \sigma_2)|\psi\rangle = -1,$$

$$\langle\psi|(\sigma_2 \otimes \sigma_2 \otimes \sigma_1)|\psi\rangle = -1, \quad \langle\psi|(\sigma_1 \otimes \sigma_1 \otimes \sigma_1)|\psi\rangle = 1.$$

**Problem 17.** Consider the *Bell state*

$$|\Psi^-\rangle = \frac{1}{\sqrt{2}}\left(\begin{pmatrix}1\\0\end{pmatrix}\otimes\begin{pmatrix}0\\1\end{pmatrix} - \begin{pmatrix}0\\1\end{pmatrix}\otimes\begin{pmatrix}1\\0\end{pmatrix}\right).$$

Let $\mathbf{n}$, $\mathbf{m}$ be unit vectors in $\mathbb{R}^3$. Calculate the expectation value

$$E(\mathbf{n},\mathbf{m}) = \langle\Psi^-|(\mathbf{n}\cdot\boldsymbol{\sigma})\otimes(\mathbf{m}\cdot\boldsymbol{\sigma})|\Psi^-\rangle.$$

**Solution 17.** Straightforward calculation yields

$$E(\mathbf{n},\mathbf{m}) = -n_1m_1-n_2m_2-n_3m_3 = -\mathbf{n}\cdot\mathbf{m} = -|\mathbf{n}|\cdot|\mathbf{m}|\cos(\theta) = -\cos(\theta_{\mathbf{m},\mathbf{n}})$$

since $|\mathbf{m}| = |\mathbf{n}| = 1$. We write $\cos(\theta_{\mathbf{m},\mathbf{n}})$ instead of $\cos(\theta)$ in order to indicate that $\theta_{\mathbf{n},\mathbf{m}}$ is the angle between the quantization directions $\mathbf{m}$ and $\mathbf{n}$.

**Problem 18.** Let $A_1$, $A_2$, ..., $A_m$ and $B_1$, $B_2$, ..., $B_n$ be two sets of $4 \times 4$ matrices over $\mathbb{C}$. Assume that $A_jB_k = B_kA_j$ for all $j$, $k$ with $j = 1,2,\ldots,m$ and $k = 1,2,\ldots,n$. Find two such sets of matrices using the Kronecker product of $2 \times 2$ matrices.

**Solution 18.** If $A_j = C_j \otimes I_2$, $B_k = I_2 \otimes D_k$, where $C_j$, $D_k$ are arbitrary $2 \times 2$ matrices we have

$$[A_j, B_k] = [C_j \otimes I_2, I_2 \otimes D_k] = C_j \otimes D_k - C_j \otimes D_k = 0_4$$

for all $j$, $k$.

**Problem 19.** Let $A$, $B$ be $n \times n$ hermitian matrices over $\mathbb{C}$ and

$$H = \hbar\omega(A \otimes B)$$

be a Hamilton operator, where $\hbar$ is the Planck constant and $\omega$ the frequency. The *Heisenberg equation of motion* for the operator $B \otimes A$ is given by

$$i\hbar\frac{d(B \otimes A)}{dt} = [B \otimes A, H](t).$$

(i) Assume that $[A, B] = 0_n$. Simplify the Heisenberg equation of motion using this condition.

(ii) Assume that $[A, B]_+ = 0_n$ Simplify the Heisenberg equation of motion using this condition. Give an example for such matrices.

(iii) Assume that $[A, B]_+ = I_n$. Simplify the Heisenberg equation of motion using this condition. Give an example for such matrices.

**Solution 19.** (i) We have

$$[B \otimes A, \hat{H}](t) = \hbar\omega[B \otimes A, A \otimes B](t)$$
$$= \hbar\omega((B \otimes A)(A \otimes B) - (A \otimes B)(B \otimes A))(t)$$
$$= \hbar\omega((BA) \otimes (AB) - (AB) \otimes (BA))(t).$$

In general, $(AB) \otimes (BA) \neq (BA) \otimes (AB)$. If $[A, B] = 0_n$ we find

$$i\hbar\frac{d(B \otimes A)}{dt} = 0_{n^2}.$$

Thus $B \otimes A$ is a constant of motion.
(ii) From $[A, B]_+ = 0_n$ we have $AB = -BA$. Thus we also have

$$i\hbar\frac{d(B \otimes A)}{dt} = 0_{n^2}$$

in this case. An example are the Pauli spin matrices. For example, let $A = \sigma_1$ and $B = \sigma_2$. Then $[A, B]_+ = 0_2$. Another example is given by Fermi operators. They have the matrix representation $(j = 1, 2, \ldots, N)$

$$c_j^\dagger = \sigma_3 \otimes \cdots \otimes \sigma_3 \otimes \left(\frac{1}{2}\sigma_+\right) \otimes I_2 \otimes \cdots \otimes I_2$$

$$c_j = \sigma_3 \otimes \cdots \otimes \sigma_3 \otimes \left(\frac{1}{2}\sigma_-\right) \otimes I_2 \cdots \otimes I_2$$

where $\sigma_+$ and $\sigma_-$ appears in the $j$-th place and where $\sigma_+ := \sigma_1 + i\sigma_2$, $\sigma_- := \sigma_1 - i\sigma_2$. We have $[c_k^\dagger, c_j]_+ = \delta_{jk}I$ and $[c_k^\dagger, c_j^\dagger]_+ = [c_k, c_j]_+ = 0$.
(iii) Since $BA = -AB + I_n$ we find

$$i\hbar\frac{d(B \otimes A)}{dt} = \hbar\omega(I_n \otimes (AB) - (AB) \otimes I_n)(t).$$

**Problem 20.** The four *Bell states* with spin $J$ ($J = 1/2, 1, 3/2, 2, \ldots$) are given by

$$|B_1\rangle = \frac{1}{\sqrt{2J+1}} \sum_{k=0}^{2J} |k\rangle \otimes |k\rangle$$

$$|B_2\rangle = \frac{1}{\sqrt{2J+1}} \sum_{k=0}^{2J} (-1)^k |k\rangle \otimes |k\rangle$$

$$|B_3\rangle = \frac{1}{\sqrt{2J+1}} \sum_{k=0}^{2J} |k\rangle \otimes |2J - k\rangle$$

$$|B_4\rangle = \frac{1}{\sqrt{2J+1}} \sum_{k=0}^{2J} (-1)^k |k\rangle \otimes |2J - k\rangle.$$

For $J = 1/2$ we obtain the standard Bell basis in $\mathbb{C}^4$. Note that for $J = 1/2$ the four Bell states form an orthonormal basis in $\mathbb{C}^4$. Show that for spin $J = 1$ the Bell states are linearly dependent.

**Solution 20.** For $J = 1$ the Bell states are

$$|B_1\rangle = \frac{1}{\sqrt{3}}(|0\rangle \otimes |0\rangle + |1\rangle \otimes |1\rangle + |2\rangle \otimes |2\rangle)$$

$$|B_2\rangle = \frac{1}{\sqrt{3}}(|0\rangle \otimes |0\rangle - |1\rangle \otimes |1\rangle + |2\rangle \otimes |2\rangle)$$

$$|B_3\rangle = \frac{1}{\sqrt{3}}(|0\rangle \otimes |2\rangle + |1\rangle \otimes |1\rangle + |2\rangle \otimes |0\rangle)$$

$$|B_4\rangle = \frac{1}{\sqrt{3}}(|0\rangle \otimes |2\rangle - |1\rangle \otimes |1\rangle + |2\rangle \otimes |0\rangle).$$

Let $c_1, c_2, c_3, c_4 \in \mathbb{C}$. Then the equation

$$c_1|B_1\rangle + c_2|B_2\rangle + c_3|B_3\rangle + c_4|B_4\rangle = \mathbf{0}$$

provides the solution $c_2 = -c_1$, $c_3 = -c_1$, $c_4 = c_1$ with $c_1$ arbitrary. Thus the four states are not linearly independent. We also have

$$\langle B_1|B_2\rangle = \frac{1}{3}, \qquad \langle B_1|B_3\rangle = \frac{1}{3}, \qquad \langle B_1|B_4\rangle = -\frac{1}{3},$$

$$\langle B_2|B_3\rangle = -\frac{1}{3}, \qquad \langle B_2|B_4\rangle = \frac{1}{3}, \qquad \langle B_3|B_4\rangle = \frac{1}{3}.$$

**Problem 21.** Consider the state in $\mathbb{C}^4$

$$|\psi\rangle = \sum_{j_1, j_2 = 0}^{1} c_{j_1 j_2} |j_1\rangle \otimes |j_2\rangle$$

where

$$|0\rangle = \begin{pmatrix} 1 \\ 0 \end{pmatrix}, \qquad |1\rangle = \begin{pmatrix} 0 \\ 1 \end{pmatrix}.$$

Let $S, T$ be $2 \times 2$ matrices over $\mathbb{C}$

$$S = \begin{pmatrix} s_{11} & s_{12} \\ s_{21} & s_{22} \end{pmatrix}, \qquad T = \begin{pmatrix} t_{11} & t_{12} \\ t_{21} & t_{22} \end{pmatrix}$$

where $\det(S) = 1$, $\det(T) = 1$. This means that $S$ and $T$ are elements of the Lie group $SL(2, \mathbb{C})$. Let

$$\sum_{j_1, j_2 = 0}^{1} d_{j_1 j_2} |j_1\rangle \otimes |j_2\rangle = (S \otimes T) \sum_{j_1, j_2 = 0}^{1} c_{j_1 j_2} |j_1\rangle \otimes |j_2\rangle.$$

Show that

$$d_{00} d_{11} - d_{01} d_{10} = c_{00} c_{11} - c_{01} c_{10}. \tag{1}$$

Owing to (1) the quantity $c_{00} c_{11} - c_{01} c_{10}$ is called an *invariant*.

**Solution 21.**   Since

$$S \otimes T = \begin{pmatrix} s_{11} t_{11} & s_{11} t_{12} & s_{12} t_{11} & s_{12} t_{12} \\ s_{11} t_{21} & s_{11} t_{22} & s_{12} t_{21} & s_{12} t_{22} \\ s_{21} t_{11} & s_{21} t_{12} & s_{22} t_{11} & s_{22} t_{12} \\ s_{21} t_{21} & s_{21} t_{22} & s_{22} t_{21} & s_{22} t_{22} \end{pmatrix}$$

we have

$$(S \otimes T)|\psi\rangle = c_{00} \begin{pmatrix} s_{11} t_{11} \\ s_{11} t_{21} \\ s_{21} t_{11} \\ s_{21} t_{21} \end{pmatrix} + c_{01} \begin{pmatrix} s_{11} t_{12} \\ s_{11} t_{22} \\ s_{21} t_{12} \\ s_{21} t_{22} \end{pmatrix} + c_{10} \begin{pmatrix} s_{12} t_{11} \\ s_{12} t_{21} \\ s_{22} t_{11} \\ s_{22} t_{21} \end{pmatrix} + c_{11} \begin{pmatrix} s_{12} t_{12} \\ s_{12} t_{22} \\ s_{22} t_{12} \\ s_{22} t_{22} \end{pmatrix}.$$

Thus

$$d_{00} = c_{00} s_{11} t_{11} + c_{01} s_{11} t_{12} + c_{10} s_{12} t_{11} + c_{11} s_{12} t_{12}$$
$$d_{01} = c_{00} s_{11} t_{21} + c_{01} s_{11} t_{22} + c_{10} s_{12} t_{21} + c_{11} s_{12} t_{22}$$
$$d_{10} = c_{00} s_{21} t_{11} + c_{01} s_{21} t_{12} + c_{10} s_{22} t_{11} + c_{11} s_{22} t_{12}$$
$$d_{11} = c_{00} s_{21} t_{21} + c_{01} s_{21} t_{22} + c_{10} s_{22} t_{21} + c_{11} s_{22} t_{22}.$$

From $\det(S) = 1$ and $\det(T) = 1$ it follows that $s_{11} s_{22} - s_{12} s_{21} = 1$ and $t_{11} t_{22} - t_{12} t_{21} = 1$. Thus

$$s_{11} s_{22} t_{11} t_{22} + s_{12} s_{21} t_{12} t_{21} - s_{11} s_{22} t_{12} t_{21} - s_{12} s_{21} t_{11} t_{22} = 1.$$

Using this result we obtain (1).

**Problem 22.**   Let $I_2$ be the $2 \times 2$ identity matrix and let $\sigma_1$, $\sigma_2$ and $\sigma_3$ be the Pauli spin matrices.
(i) Let $A$ be an $m \times m$ matrix over $\mathbb{C}$ and $B$ be an $n \times n$ matrix over $\mathbb{C}$. Find all solutions for $m$, $n$, $A$ and $B$ satisfying

$$\frac{1}{2} I_2 \otimes I_2 + \frac{1}{2} \sigma_1 \otimes \sigma_1 + \frac{1}{2} \sigma_2 \otimes \sigma_2 + \frac{1}{2} \sigma_3 \otimes \sigma_3 = A \otimes B.$$

(ii) Let $C$, $D$, $E$, and $F$ be $2 \times 2$ matrices over $\mathbb{C}$. Find all solutions for $C$, $D$, $E$ and $F$ satisfying

$$\frac{1}{2}I_2 \otimes I_2 + \frac{1}{2}\sigma_1 \otimes \sigma_1 + \frac{1}{2}\sigma_2 \otimes \sigma_2 + \frac{1}{2}\sigma_3 \otimes \sigma_3 = C \otimes D + E \otimes F.$$

(iii) Find the set *stabilized* by $S = \{\, \sigma_1 \otimes I_2, \, I_2 \otimes \sigma_1 \,\}$ i.e. find

$$\{\, \mathbf{u} \in \mathbb{C}^4 \mid \forall A \in S : \; A\mathbf{u} = \mathbf{u} \,\}.$$

**Solution 22.** (i) Straightforward calculation yields

$$\frac{1}{2}I_2 \otimes I_2 + \frac{1}{2}\sigma_1 \otimes \sigma_1 + \frac{1}{2}\sigma_2 \otimes \sigma_2 + \frac{1}{2}\sigma_3 \otimes \sigma_3 = \begin{pmatrix} 1 & 0 & 0 & 0 \\ 0 & 0 & 1 & 0 \\ 0 & 1 & 0 & 0 \\ 0 & 0 & 0 & 1 \end{pmatrix} = A \otimes B.$$

Since $A \otimes B$ is an $mn \times mn$ matrix we have $mn = 4$. For $m = n = 2$ we consider

$$A = \begin{pmatrix} a_1 & a_2 \\ a_3 & a_4 \end{pmatrix}, \qquad B = \begin{pmatrix} b_1 & b_2 \\ b_3 & b_4 \end{pmatrix}.$$

Thus we have $a_1 b_1 = 1$, $a_1 b_4 = 0$ and $a_4 b_4 = 1$. Consequently $a_1 \neq 0$ and $b_4 \neq 0$, so that $a_1 b_4 \neq 0$. Thus $m = n = 2$ does not yield a solution.

For $m = 1$ and $n = 4$ we find the solution

$$A = (a), \qquad B = \frac{1}{a}\begin{pmatrix} 1 & 0 & 0 & 0 \\ 0 & 0 & 1 & 0 \\ 0 & 1 & 0 & 0 \\ 0 & 0 & 0 & 1 \end{pmatrix}, \qquad a \in \mathbb{C}/\{0\}.$$

For $m = 4$ and $n = 1$ we find the solution

$$A = \frac{1}{b}\begin{pmatrix} 1 & 0 & 0 & 0 \\ 0 & 0 & 1 & 0 \\ 0 & 1 & 0 & 0 \\ 0 & 0 & 0 & 1 \end{pmatrix}, \qquad B = (b), \qquad b \in \mathbb{C}/\{0\}.$$

(ii) Clearly $C$ and $E$, and $D$ and $F$ must be linearly independent, otherwise we would have the case $m = n = 2$ discussed in (a) which has no solution. Let

$$C = \begin{pmatrix} c_1 & c_2 \\ c_3 & c_4 \end{pmatrix}, \qquad E = \begin{pmatrix} e_1 & e_2 \\ e_3 & e_4 \end{pmatrix}.$$

We have to satisfy the equations

$$c_1 D + e_1 F = \begin{pmatrix} 1 & 0 \\ 0 & 0 \end{pmatrix}, \qquad c_2 D + e_2 F = \begin{pmatrix} 0 & 0 \\ 1 & 0 \end{pmatrix},$$

$$c_3 D + e_3 F = \begin{pmatrix} 0 & 1 \\ 0 & 0 \end{pmatrix}, \qquad c_4 D + e_4 F = \begin{pmatrix} 0 & 0 \\ 0 & 1 \end{pmatrix}.$$

It follows that $\{D, F\}$ should be a basis for the $2 \times 2$ matrices over $\mathbb{C}$, however the span of $\{D, F\}$ is a two dimensional space whereas the $2 \times 2$ matrices over $\mathbb{C}$ forms a four dimensional space. Thus we have a contradiction. There are no solutions.

(iii) Thus the vector $\mathbf{u}$ must simultaneously be an eigenvector of $\sigma_1 \otimes I_2$ and $I_2 \otimes \sigma_1$ with eigenvalue 1. The eigenspace corresponding to the eigenvalue 1 from $\sigma_1 \otimes I_2$ yields

$$\mathbf{u} = \alpha \begin{pmatrix} 1 \\ 1 \end{pmatrix} \otimes \begin{pmatrix} 1 \\ 0 \end{pmatrix} + \beta \begin{pmatrix} 1 \\ 1 \end{pmatrix} \otimes \begin{pmatrix} 0 \\ 1 \end{pmatrix}$$

where $\alpha, \beta \in \mathbb{C}$. We must now satisfy

$$(I_2 \otimes \sigma_1)\mathbf{u} = \alpha(I_2 \otimes \sigma_1) \begin{pmatrix} 1 \\ 1 \end{pmatrix} \otimes \begin{pmatrix} 1 \\ 0 \end{pmatrix} + \beta(I_2 \otimes \sigma_1) \begin{pmatrix} 1 \\ 1 \end{pmatrix} \otimes \begin{pmatrix} 0 \\ 1 \end{pmatrix}$$

$$= \alpha \begin{pmatrix} 1 \\ 1 \end{pmatrix} \otimes \begin{pmatrix} 0 \\ 1 \end{pmatrix} + \beta \begin{pmatrix} 1 \\ 1 \end{pmatrix} \otimes \begin{pmatrix} 1 \\ 0 \end{pmatrix}$$

$$= \mathbf{u}$$

so that $\alpha = \beta$. Thus

$$\mathbf{u} = \alpha \begin{pmatrix} 1 \\ 1 \end{pmatrix} \otimes \begin{pmatrix} 1 \\ 0 \end{pmatrix} + \alpha \begin{pmatrix} 1 \\ 1 \end{pmatrix} \otimes \begin{pmatrix} 0 \\ 1 \end{pmatrix}, \qquad \alpha \in \mathbb{C}.$$

**Problem 23.** Let $N \geq 1$. Consider the Hilbert space $\mathbb{C}^{2^N}$. The $(N+1)$ *Dicke states* are defined by

$$|N/2, \ell - N/2\rangle := \frac{1}{\sqrt{{}^N C_\ell}} (\underbrace{|0\rangle \otimes \cdots \otimes |0\rangle}_{\ell} \otimes \underbrace{|1\rangle \otimes \cdots \otimes |1\rangle}_{N-\ell} + \text{permutations})$$

where $\ell = 0, 1, \ldots, N$ and

$$^N C_\ell = N!/(\ell!(N - \ell)!).$$

Write down the Dicke states for $N = 2$ and $N = 3$. Which of the states are entangled?

**Solution 23.** For $N = 2$ we have the three states in the Hilbert space $\mathbb{C}^4$

$$|1, -1\rangle = |1\rangle \otimes |1\rangle, \quad |1, 0\rangle = \frac{1}{\sqrt{2}}(|0\rangle \otimes |1\rangle + |1\rangle \otimes |0\rangle), \quad |1, 1\rangle = |0\rangle \otimes |0\rangle.$$

The first and the last states are product states. The second state is a Bell state and fully entangled. For $N = 3$ we find the four states in the Hilbert space $\mathbb{C}^8$

$$\left|\frac{3}{2}, -\frac{3}{2}\right\rangle = |1\rangle \otimes |1\rangle \otimes |1\rangle$$

$$\left|\frac{3}{2}, -\frac{1}{2}\right\rangle = \frac{1}{\sqrt{3}}(|0\rangle \otimes |1\rangle \otimes |1\rangle + |1\rangle \otimes |0\rangle \otimes |1\rangle + |1\rangle \otimes |1\rangle \otimes |0\rangle)$$

$$\left|\frac{3}{2}, \frac{1}{2}\right\rangle = \frac{1}{\sqrt{3}}(|0\rangle \otimes |0\rangle \otimes |1\rangle + |0\rangle \otimes |1\rangle \otimes |0\rangle + |1\rangle \otimes |0\rangle \otimes |0\rangle)$$

$$\left|\frac{3}{2}, \frac{3}{2}\right\rangle = |0\rangle \otimes |0\rangle \otimes |0\rangle.$$

Obviously the first and last states are product states. The other two states are entangled.

**Problem 24.** Can we find $2 \times 2$ matrices $A$, $B$, $C$ with $\det(A) = 1$, $\det(B) = 1$ and $\det(C) = 1$ such that

$$\frac{1}{\sqrt{3}}\begin{pmatrix} 0 \\ 1 \\ 1 \\ 0 \\ 1 \\ 0 \\ 0 \\ 0 \end{pmatrix} = (A \otimes B \otimes C)\frac{1}{\sqrt{2}}\begin{pmatrix} 1 \\ 0 \\ 0 \\ 0 \\ 0 \\ 0 \\ 0 \\ 1 \end{pmatrix} ?$$

On the left-hand side we have the $W$ *state* and on the right-hand side we have the *GHZ state*.

**Solution 24.** Let

$$A = \begin{pmatrix} a_{11} & a_{12} \\ a_{21} & a_{22} \end{pmatrix}, \qquad \det(A) = a_{11}a_{22} - a_{12}a_{21} = 1$$

etc. Thus the condition yields 8 equations and we have the three constraints

$$a_{11}a_{22} - a_{12}a_{21} = 1, \quad b_{11}b_{22} - b_{12}b_{21} = 1, \quad c_{11}c_{22} - c_{12}c_{21} = 1.$$

There is no solution for this system.

**Problem 25.** Consider the vector

$$\mathbf{v} = \frac{1}{2}(1 \quad 0 \quad 1 \quad 1 \quad 0 \quad 1)^T \in \mathbb{C}^6.$$

Find a *Schmidt decomposition* of $\mathbf{v}$ over $\mathbb{C}^6 = \mathbb{C}^2 \otimes \mathbb{C}^3$ and over $\mathbb{C}^6 = \mathbb{C}^3 \otimes \mathbb{C}^2$.

**Solution 25.**   Over $\mathbb{C}^2 \otimes \mathbb{C}^3$ we have

$$
\mathbf{v} = \frac{1}{2}\begin{pmatrix} 1 \\ 0 \end{pmatrix} \otimes \begin{pmatrix} 1 \\ 0 \\ 1 \end{pmatrix} + \frac{1}{2}\begin{pmatrix} 0 \\ 1 \end{pmatrix} \otimes \begin{pmatrix} 1 \\ 0 \\ 1 \end{pmatrix} = \frac{1}{\sqrt{2}}\begin{pmatrix} 1 \\ 1 \end{pmatrix} \otimes \frac{1}{\sqrt{2}}\begin{pmatrix} 1 \\ 0 \\ 1 \end{pmatrix}
$$

which is trivially a Schmidt decomposition of $\mathbf{v}$. Thus, over $\mathbb{C}^2 \otimes \mathbb{C}^3$, $\mathbf{v}$ has Schmidt rank 1. Over $\mathbb{C}^3 \otimes \mathbb{C}^2$ we have

$$
\mathbf{v} = \frac{1}{2}\begin{pmatrix} 1 \\ 0 \\ 0 \end{pmatrix} \otimes \begin{pmatrix} 1 \\ 0 \end{pmatrix} + \frac{1}{2}\begin{pmatrix} 0 \\ 1 \\ 0 \end{pmatrix} \otimes \begin{pmatrix} 1 \\ 1 \end{pmatrix} + \frac{1}{2}\begin{pmatrix} 0 \\ 0 \\ 1 \end{pmatrix} \otimes \begin{pmatrix} 0 \\ 1 \end{pmatrix}.
$$

Identifying $\mathbb{C}^3$ with the columns of a matrix and $\mathbb{C}^2$ with the rows of a matrix, we rearrange $\mathbf{v}$ to yield the matrix

$$
A = \frac{1}{2}\begin{pmatrix} 1 & 0 \\ 1 & 1 \\ 0 & 1 \end{pmatrix}
$$

which has the *singular value decomposition*

$$
A = \begin{pmatrix} \frac{1}{\sqrt{6}} & \frac{1}{\sqrt{2}} & \frac{1}{\sqrt{3}} \\ \frac{2}{\sqrt{6}} & 0 & -\frac{1}{\sqrt{3}} \\ \frac{1}{\sqrt{6}} & -\frac{1}{\sqrt{2}} & \frac{1}{\sqrt{3}} \end{pmatrix}\begin{pmatrix} \frac{\sqrt{3}}{2} & 0 \\ 0 & \frac{1}{2} \\ 0 & 0 \end{pmatrix}\begin{pmatrix} \frac{1}{\sqrt{2}} & \frac{1}{\sqrt{2}} \\ \frac{1}{\sqrt{2}} & -\frac{1}{\sqrt{2}} \end{pmatrix}^{*}.
$$

From the singular value decomposition we obtain the Schmidt decomposition

$$
\mathbf{v} = \frac{\sqrt{3}}{2} \cdot \frac{1}{\sqrt{6}}\begin{pmatrix} 1 \\ 2 \\ 1 \end{pmatrix} \otimes \frac{1}{\sqrt{2}}\begin{pmatrix} 1 \\ 1 \end{pmatrix} + \frac{1}{2} \cdot \frac{1}{\sqrt{2}}\begin{pmatrix} 1 \\ 0 \\ -1 \end{pmatrix} \otimes \frac{1}{\sqrt{2}}\begin{pmatrix} 1 \\ -1 \end{pmatrix}.
$$

Thus, over $\mathbb{C}^3 \otimes \mathbb{C}^2$, $\mathbf{v}$ has Schmidt rank 2.

**Problem 26.**   Let $A$ be an $m \times m$ hermitian matrix and let $B$ be an $n \times n$ hermitian matrix. Then $A \otimes B$, $A \otimes I_n$, $I_m \otimes B$ are also hermitian matrices, where $I_m$ is the $m \times m$ identity matrix. Let $\epsilon_1$, $\epsilon_2$ and $\epsilon_3$ be real parameters. Consider the Hamilton operator

$$
H = \hbar\omega(\epsilon_1 A \otimes B + \epsilon_2 A \otimes I_n + \epsilon_3 I_m \otimes B).
$$

The *partition function* $Z(\beta)$ is given by $Z(\beta) = \text{tr}(\exp(-\beta H))$, where $H$ is the (hermitian) Hamilton operator and tr denotes the trace. From the partition function we obtain the Helmholtz free energy, entropy and specific heat.

(i) Calculate $Z(\beta)$ for the Hamilton operator given above.

(ii) Consider the special case that $n = m = 2$ and $A$, $B$ are any of the Pauli spin matrices $\sigma_1$, $\sigma_2$, $\sigma_3$.

**Solution 26.** Since $A$ is an $m \times m$ matrix we can find a $m \times m$ unitary matrix $U_A$ such that $\widetilde{A} = U_A^* A U_A$ is a diagonal matrix. We set $\mathrm{diag}(\widetilde{A}) = (\lambda_1, \lambda_2, \dots, \lambda_m)$. Analogously for the $n \times n$ hermitian matrix $B$ we find a $n \times n$ hermitian matrix $U_B$ such that $\widetilde{B} = U_B^* B U_B$ is a diagonal matrix. We set $\mathrm{diag}(\widetilde{B}) = (\mu_1, \mu_2, \dots, \mu_n)$. Since $A$ and $B$ are hermitian the diagonal elements of $\widetilde{A}$ and $\widetilde{B}$ are real. Since $U_A$ and $U_B$ are unitary matrices we find that $U_A \otimes U_B$ is also a unitary matrix and $(U_A \otimes U_B)^* = U_A^* \otimes U_B^*$. Now we find

$$
\begin{aligned}
\mathrm{tr}(e^{-\beta H}) &= \mathrm{tr}((U_A^* \otimes U_B^*) e^{-\beta H} (U_A \otimes U_B)) \\
&= \mathrm{tr}\, e^{-\beta(U_A^* \otimes U_B^*) H (U_A \otimes U_B)} \\
&= \mathrm{tr}\, e^{-\beta \hbar \omega (\epsilon_1 (U_A^* A U_A) \otimes (U_B^* B U_B) + \epsilon_2 (U_A^* A U_A) \otimes I_n + \epsilon_3 (I_m \otimes (U_B^* B U_B))} \\
&= \mathrm{tr}\, e^{-\beta \hbar \omega (\epsilon_1 \widetilde{A} \otimes \widetilde{B} + \epsilon_2 \widetilde{A} \otimes I_n + \epsilon_3 I_m \otimes \widetilde{B})} \\
&= \sum_{j=1}^{m} \sum_{k=1}^{n} e^{-\beta \hbar \omega (\epsilon_1 \lambda_j \mu_k + \epsilon_2 \lambda_j + \epsilon_3 \mu_k)}.
\end{aligned}
$$

This calculation can be extended straightforward to the matrix

$$
A_1 \otimes A_2 \otimes A_3 + A_1 \otimes I_{n_2} \otimes I_{n_3} + I_{n_1} \otimes A_2 \otimes I_{n_3} + I_{n_1} \otimes I_{n_2} \otimes A_3
$$

and so on, where $A_1$, $A_2$, $A_3$ are $n_1 \times n_1$, $n_2 \times n_2$, $n_3 \times n_3$ matrices, respectively.

(ii) The eigenvalues of any Pauli spin matrix are $+1$ and $-1$. Thus for any combination $\sigma_j \otimes \sigma_k$ for $A \otimes B$ we find

$$
Z(\beta) = e^{-\beta \hbar \omega \epsilon_1} 2 \cosh(\beta \hbar \omega (\epsilon_2 + \epsilon_3)) + e^{\beta \hbar \omega \epsilon_1} 2 \cosh(\beta \hbar \omega (\epsilon_2 - \epsilon_3)).
$$

**Problem 27.** (i) Let $\{\, |0\rangle, |1\rangle, \dots, |n-1\rangle \,\}$ be an orthonormal basis in $\mathbb{C}^n$. Is

$$
|\psi\rangle := \frac{1}{\sqrt{n}} \sum_{j=0}^{n-1} |j\rangle \otimes |j\rangle
$$

independent of the chosen orthonormal basis?

(ii) Find the density matrix $|\psi\rangle\langle\psi|$.

(iii) Consider the linear operator

$$
\Pi := \frac{1}{4} \sum_{j=0}^{n-1} \sum_{\substack{k=0 \\ k \neq j}}^{n-1} (|jk\rangle - |kj\rangle)(\langle jk| - \langle kj|)
$$

where we used the short-cut notation $|jk\rangle \equiv |j\rangle \otimes |k\rangle$. Calculate $\Pi^*$ and $\Pi^2$. What is the use of this operator?

**Solution 27.**   (i) Let $\{ |\phi_0\rangle, |\phi_1\rangle, \ldots, |\phi_{n-1}\rangle \}$ be an orthonormal basis in $\mathbb{C}^n$. Then we have the expansion for the state $|j\rangle$

$$|j\rangle = \sum_{k=0}^{n-1} \langle j|\phi_k\rangle |\phi_k\rangle.$$

Thus $|\psi\rangle$ can be written as

$$|\psi\rangle = \frac{1}{\sqrt{n}} \sum_{j=0}^{n-1} \left( \left( \sum_{k=0}^{n-1} \langle j|\phi_k\rangle |\phi_k\rangle \right) \otimes \left( \sum_{l=0}^{n-1} \langle j|\phi_l\rangle |\phi_l\rangle \right) \right)$$

$$= \frac{1}{\sqrt{n}} \sum_{j=0}^{n-1} \sum_{k=0}^{n-1} \sum_{l=0}^{n-1} \overline{\langle \phi_k|j\rangle} \langle j|\phi_l\rangle |\phi_k\rangle \otimes |\phi_l\rangle$$

$$= \frac{1}{\sqrt{n}} \sum_{k=0}^{n-1} \sum_{l=0}^{n-1} \left( \sum_{j=0}^{n-1} \overline{\langle \phi_k|j\rangle} \langle j|\phi_l\rangle \right) |\phi_k\rangle \otimes |\phi_l\rangle$$

where we used $\langle j|\phi_k\rangle = \overline{\langle \phi_k|j\rangle}$. Note that for the sum

$$\sum_{j=0}^{n-1} \overline{\langle \phi_k|j\rangle} \langle j|\phi_l\rangle$$

we cannot apply *Parseval's relation*. Parseval's relation would apply to

$$\sum_{j=0}^{n-1} \overline{\langle \phi_k|j\rangle} \langle \phi_l|j\rangle = \langle \phi_k|\phi_l\rangle = \delta_{kl}.$$

Thus the Bell state $|\psi\rangle$ is dependent on the chosen basis. However, if all scalar products $\langle j|\phi_k\rangle$ are real numbers then $|\psi\rangle$ is independent of the chosen basis.

(ii) We have

$$|\psi\rangle\langle\psi| = \frac{1}{n} \sum_{j=0}^{n-1} \sum_{k=0}^{n-1} (|j\rangle \otimes |j\rangle)(\langle k| \otimes \langle k|) = \frac{1}{n} \sum_{j=0}^{n-1} \sum_{k=0}^{n-1} |j\rangle\langle k| \otimes |j\rangle\langle k|.$$

(iii) Clearly, $\Pi^* = \Pi$. Furthermore

$$\Pi^2 = \frac{1}{16} \sum_{j=0}^{n-1} \sum_{j\neq k} \sum_{l=0}^{n-1} \sum_{l\neq m} (|jk\rangle - |kj\rangle)(\langle jk| - \langle kj|)(|lm\rangle - |ml\rangle)(\langle lm| - \langle ml|)$$

$$= \frac{1}{16} \sum_{j=0}^{n-1} \sum_{j \neq k} \sum_{l=0}^{n-1} \sum_{l \neq m} (|jk\rangle - |kj\rangle)(2\delta_{jl}\delta_{km} - 2\delta_{jm}\delta_{lk})(\langle lm| - \langle ml|)$$

$$= \Pi.$$

Thus $\Pi$ is a projection matrix. It projects onto the space spanned by

$$\left\{ \frac{1}{\sqrt{2}}(|jk\rangle - |kj\rangle) \ : \ j, k \in \{0, 1, \ldots, n-1\}, \ k > j \right\}.$$

**Problem 28.** Let $\mathcal{H}_A$ and $\mathcal{H}_B$ be two finite-dimensional Hilbert spaces. The *Schmidt rank* of a linear operator $L : \mathcal{H}_A \otimes \mathcal{H}_B \to \mathcal{H}_A \otimes \mathcal{H}_B$ over $\mathcal{H}_A \otimes \mathcal{H}_B$ is the smallest non-negative integer $\mathrm{Sch}(L, \mathcal{H}_A, \mathcal{H}_B)$ such that $L$ can be written as

$$L = \sum_{j=1}^{\mathrm{Sch}(L, \mathcal{H}_A, \mathcal{H}_B)} L_{j,A} \otimes L_{j,B}$$

where $L_{j,A} : \mathcal{H}_A \to \mathcal{H}_A$ and $L_{j,B} : \mathcal{H}_B \to \mathcal{H}_B$ are linear operators.

Let $\{|0\rangle, |1\rangle\}$ denote an orthonormal basis in $\mathbb{C}^2$. Find the Schmidt rank $\mathrm{Sch}(U_{CNOT}, \mathbb{C}^2, \mathbb{C}^2)$ and $\mathrm{Sch}(U_{SWAP}, \mathbb{C}^2, \mathbb{C}^2)$ where

$$U_{CNOT} = |00\rangle\langle 00| + |01\rangle\langle 01| + |11\rangle\langle 10| + |10\rangle\langle 11|$$
$$U_{SWAP} = |00\rangle\langle 00| + |10\rangle\langle 01| + |01\rangle\langle 10| + |11\rangle\langle 11|.$$

**Solution 28.** We note that

$$U_{CNOT} = |0\rangle\langle 0| \otimes I_2 + |1\rangle\langle 1| \otimes U_{NOT}$$

where $U_{NOT} := |0\rangle\langle 1| + |1\rangle\langle 0|$. In other words

$$0 < \mathrm{Sch}(U_{CNOT}, \mathbb{C}^2, \mathbb{C}^2) \leq 2.$$

Now suppose $U_{CNOT}$ can be written as the product $A \otimes B$ where

$$A := a_0|0\rangle\langle 0| + a_1|0\rangle\langle 1| + a_2|1\rangle\langle 0| + a_3|1\rangle\langle 1|$$
$$B := b_0|0\rangle\langle 0| + b_1|0\rangle\langle 1| + b_2|1\rangle\langle 0| + b_3|1\rangle\langle 1|.$$

This yields the conditions $a_0 b_0 = 1$, $a_0 b_1 = 0$ and $a_3 b_1 = 1$. These equations are inconsistent, i.e.

$$\mathrm{Sch}(U_{CNOT}, \mathbb{C}^2, \mathbb{C}^2) \neq 1.$$

Thus

$$\mathrm{Sch}(U_{CNOT}, \mathbb{C}^2, \mathbb{C}^2) = 2.$$

The operator $U_{SWAP}$ has the eigenvalue 1 (three times) with corresponding orthonormal eigenvectors

$$\left\{ |00\rangle, |11\rangle, \frac{1}{\sqrt{2}}(|01\rangle + |10\rangle) \right\}$$

and the eigenvalue $-1$ with corresponding eigenvector $\frac{1}{\sqrt{2}}(|01\rangle - |10\rangle)$. Defining

$$|\phi_1\rangle := \frac{1}{\sqrt{2}}(|01\rangle + |10\rangle), \qquad |\phi_2\rangle := \frac{1}{\sqrt{2}}(|01\rangle - |10\rangle)$$

we find that

$$U_{SWAP} := |00\rangle\langle 00| + |\phi_1\rangle\langle\phi_1| - |\phi_2\rangle\langle\phi_2| + |11\rangle\langle 11|$$

where $\{ |00\rangle, |\phi_1\rangle, |\phi_2\rangle, |11\rangle \}$ forms an orthonormal basis in $\mathbb{C}^4$. In this basis $U_{SWAP}$ is the diagonal matrix

$$U_{SWAP} = \begin{pmatrix} 1 & 0 & 0 & 0 \\ 0 & 1 & 0 & 0 \\ 0 & 0 & -1 & 0 \\ 0 & 0 & 0 & 1 \end{pmatrix}.$$

Clearly, the matrices

$$|00\rangle\langle 00|, \quad |11\rangle\langle 11|, \quad |\phi_1\rangle\langle\phi_1| \quad \text{and} \quad |\phi_2\rangle\langle\phi_2|$$

are linearly independent. Thus $\text{Sch}(U_{SWAP}, \mathbb{C}^2, \mathbb{C}^2) = 4$.

**Problem 29.** The *operator-Schmidt decomposition* of a linear operator $Q$ acting in the product Hilbert space $\mathcal{H} = \mathcal{H}_1 \otimes \mathcal{H}_2$ of two finite-dimensional Hilbert spaces $(\dim(\mathcal{H}_1) = m, \dim(\mathcal{H}_2) = n)$ with $\mathcal{H}_1 = \mathbb{C}^m$ and $\mathcal{H}_2 = \mathbb{C}^n$ can be constructed as follows. Let $X$, $Y$ be $d \times d$ matrices over $\mathbb{C}$. Then we can define a scalar product or inner product $\langle X, Y \rangle := \text{tr}(XY^*)$. Using this inner product we can define an orthonormal set of $d \times d$ matrices $\{ X_j : j = 1, 2, \ldots, d^2 \}$ which satisfies the condition

$$\langle X_j, X_k \rangle = \text{tr}(X_j X_k^*) = \delta_{jk}.$$

Thus we can write the matrix $Q$ as

$$Q = \sum_{j=1}^{m^2} \sum_{k=1}^{n^2} c_{jk} A_j \otimes B_k$$

where $\{ A_j : j = 1, 2, \ldots, m^2 \}$ and $\{ B_k : k = 1, 2, \ldots, n^2 \}$ are fixed orthonormal bases of $m \times m$ and $n \times n$ matrices in the Hilbert spaces $\mathbb{C}^m$

and $\mathbb{C}^n$ respectively, and $c_{jk}$ are complex coefficients. Thus $C = (c_{jk})$, with $j = 1, 2, \ldots, m^2$ and $k = 1, 2, \ldots, n^2$ is an $m^2 \times n^2$ matrix. The singular value decomposition theorem states that the matrix $C$ can be written as

$$C = U\Sigma V^*$$

where $U$ is an $m^2 \times m^2$ unitary matrix, $V$ is an $n^2 \times n^2$ unitary matrix and $\Sigma$ is an $m^2 \times n^2$ diagonal matrix. The matrix $\Sigma$ is of the form

$$\Sigma = \begin{pmatrix} s_1 & \cdots & 0 \\ \vdots & \ddots & \vdots \\ 0 & \cdots & s_{n^2} \\ 0 & \cdots & 0 \\ \vdots & \ddots & \vdots \\ 0 & \cdots & 0 \end{pmatrix}.$$

It is assumed that $C$, $U$ and $V$ are calculated in orthonormal bases, for example the standard basis. Thus we obtain

$$Q = \sum_{j=1}^{m^2} \sum_{k=1}^{n^2} \sum_{\ell=1}^{n^2} U_{j\ell} s_\ell V_{\ell k} A_j \otimes B_k$$

where $s_\ell$ is the $\ell$-th diagonal entry of the $m^2 \times n^2$ diagonal matrix $\Sigma$. Defining

$$H_\ell := \sum_{j=1}^{m^2} U_{j\ell} A_j, \qquad K_\ell := \sum_{k=1}^{n^2} V_{\ell k} B_k$$

where $\ell = 1, 2, \ldots, n^2$ we find the operator-Schmidt decomposition

$$Q = \sum_{\ell=1}^{n^2} s_\ell H_\ell \otimes K_\ell.$$

(i) Consider the *CNOT gate*

$$U_{CNOT} = \begin{pmatrix} 1 & 0 & 0 & 0 \\ 0 & 1 & 0 & 0 \\ 0 & 0 & 0 & 1 \\ 0 & 0 & 1 & 0 \end{pmatrix}.$$

Find the operator-Schmidt decomposition of $U_{CNOT}$.
(ii) Consider the SWAP operator

$$U_{SWAP} = \begin{pmatrix} 1 & 0 & 0 & 0 \\ 0 & 0 & 1 & 0 \\ 0 & 1 & 0 & 0 \\ 0 & 0 & 0 & 1 \end{pmatrix}.$$

Find the operator-Schmidt decomposition of $U_{SWAP}$.

(iii) Let

$$Z := \left( \sqrt{1-p} I_2 \otimes I_2 + i\sqrt{p}\sigma_1 \otimes \sigma_1 \right) \left( \sqrt{1-p} I_2 \otimes I_2 + i\sqrt{p}\sigma_3 \otimes \sigma_3 \right)$$

where $\sigma_1$, $\sigma_2$ and $\sigma_3$ are the Pauli spin matrices. Find the operator-Schmidt decomposition of $Z$.

**Solution 29.**    (i) We have

$$U_{CNOT} = \begin{pmatrix} 1 & 0 \\ 0 & 0 \end{pmatrix} \otimes \begin{pmatrix} 1 & 0 \\ 0 & 1 \end{pmatrix} + \begin{pmatrix} 0 & 0 \\ 0 & 1 \end{pmatrix} \otimes \begin{pmatrix} 0 & 1 \\ 1 & 0 \end{pmatrix}$$

$$= \begin{pmatrix} 1 & 0 \\ 0 & 0 \end{pmatrix} \otimes I_2 + \begin{pmatrix} 0 & 0 \\ 0 & 1 \end{pmatrix} \otimes \sigma_1.$$

(ii) We have

$$U_{SWAP} = \frac{1}{2}(I_2 \otimes I_2 + \sigma_1 \otimes \sigma_1 + \sigma_2 \otimes \sigma_2 + \sigma_3 \otimes \sigma_3).$$

(iii) We have

$$Z = (1-p)I_2 \otimes I_2 + p\sigma_2 \otimes \sigma_2 + \sqrt{p(1-p)} \left[ \left( e^{i\pi/4}\sigma_1 \right) \otimes \sigma_1 + \left( e^{i\pi/4}\sigma_3 \right) \otimes \sigma_3 \right].$$

**Programming Problems**

**Problem 1.**    Consider the Hadamard basis in $\mathbb{C}^2$

$$v1 = \frac{1}{\sqrt{2}} \begin{pmatrix} 1 \\ 1 \end{pmatrix}, \quad v2 = \frac{1}{\sqrt{2}} \begin{pmatrix} 1 \\ -1 \end{pmatrix}.$$

Apply the Kronecker product to find a basis in $\mathbb{C}^4$.

**Solution 1.**    The following Maxima program

```
/* hadamardbasis.mac */
v1: matrix([1/sqrt(2)],[1/sqrt(2)]);
v1T: transpose(v1);
v2: matrix([1/sqrt(2)],[-1/sqrt(2)]);
v2T: transpose(v2);
r1: v1T . v1;
r2: v1T . v2;
r3: v2T . v2;
```

```
v1v1: kronecker_product(v1,v1);
v1v2: kronecker_product(v1,v2);
v2v1: kronecker_product(v2,v1);
v2v2: kronecker_product(v2,v2);
v1v1T: transpose(v1v1);
v1v2T: transpose(v1v2);
v2v1T: transpose(v2v1);
v2v2T: transpose(v2v2);
r4: v1v1T . v2v2T;
```

provides the orthonormal basis

$$\frac{1}{2}\begin{pmatrix}1\\1\\1\\1\end{pmatrix}, \quad \frac{1}{2}\begin{pmatrix}1\\-1\\1\\-1\end{pmatrix}, \quad \frac{1}{2}\begin{pmatrix}1\\1\\-1\\-1\end{pmatrix}, \quad \frac{1}{2}\begin{pmatrix}1\\-1\\-1\\1\end{pmatrix}.$$

**Problem 2.** Let $S_1$, $S_2$, $S_3$ be the spin matrices for spin-$\frac{1}{2}$

$$S_1 = \frac{1}{2}\begin{pmatrix}0&1\\1&0\end{pmatrix}, \quad S_2 = \frac{1}{2}\begin{pmatrix}0&-i\\i&0\end{pmatrix}, \quad S_3 = \frac{1}{2}\begin{pmatrix}1&0\\0&-1\end{pmatrix}.$$

Consider the Hamilton operators

$$\widetilde{H} = \frac{\hat{H}}{\hbar\omega} = S_1 \otimes S_1 + S_2 \otimes S_2 + S_3 \otimes S_3$$

$$\widetilde{K} = \frac{\hat{H}}{\hbar\omega} = S_1 \otimes S_2 + S_2 \otimes S_3 + S_3 \otimes S_1.$$

Find the eigenvalues and eigenvectors of $\widetilde{H}$ and $\widetilde{K}$.

**Solution 2.** Applying the Maxima program

```
/* eigenS1S2S3.mac */
I2: matrix([1,0],[0,1]);
S1: matrix([0,1/2],[1/2,0]);
S2: matrix([0,-%i/2],[%i/2,0]);
S3: matrix([1/2,0],[0,-1/2]);
T1: kronecker_product(S1,S1);
T2: kronecker_product(S2,S2);
T3: kronecker_product(S3,S3);
H: T1 + T2 + T3;
EH: eigenvectors(H);
X1: kronecker_product(S1,S2);
X2: kronecker_product(S2,S3);
X3: kronecker_product(S3,S1);
K: X1 + X2 + X3;
EK: eigenvectors(K);
```

we find for $\widetilde{H}$ the eigenvalues

$$-\frac{3}{4}\,(1\times),\quad \frac{1}{4}\,(3\times)$$

with the corresponding eigenvectors

$$\frac{1}{\sqrt{2}}\begin{pmatrix}0\\1\\-1\\0\end{pmatrix},\quad \begin{pmatrix}1\\0\\0\\0\end{pmatrix},\quad \frac{1}{\sqrt{2}}\begin{pmatrix}0\\1\\1\\0\end{pmatrix},\quad \begin{pmatrix}0\\0\\0\\1\end{pmatrix}$$

and for $\widetilde{K}$ the eigenvalues

$$-\frac{3}{4}\,(1\times),\quad \frac{1}{4}\,(3\times)$$

with the corresponding eigenvectors

$$\frac{1}{2}\begin{pmatrix}1\\-1\\-i\\i\end{pmatrix},\quad \frac{1}{\sqrt{2}}\begin{pmatrix}1\\0\\0\\i\end{pmatrix},\quad \frac{1}{\sqrt{2}}\begin{pmatrix}0\\1\\0\\-i\end{pmatrix},\quad \frac{1}{\sqrt{2}}\begin{pmatrix}0\\0\\1\\-1\end{pmatrix}.$$

**Problem 3.**    Let $M$ be a $2 \times 2$ matrix and $M^T$ the transpose. Let $|0\rangle, |1\rangle$ be the standard basis in $\mathbb{C}^2$. Show that

$$(M \otimes I_2)(|0\rangle \otimes |0\rangle + |1\rangle \otimes |1\rangle) = (I_2 \otimes M^T)(|0\rangle \otimes |0\rangle + |1\rangle \otimes |1\rangle).$$

**Solution 3.**    The following Maxima program will do the job

```
/* M22.mac */
b1: matrix([1],[0]); b2: matrix([0],[1]);
b1b1: kronecker_product(b1,b1); b2b2: kronecker_product(b2,b2);
I2: matrix([1,0],[0,1]);
M: matrix([m11,m12],[m21,m22]);
MT: matrix([m11,m21],[m12,m22]);
T1: kronecker_product(M,I2);
T2: kronecker_product(I2,MT);
R1: T1 . b1b1 + T1 . b2b2;
R2: T2 . b1b1 + T2 . b2b2;
F: R1 - R2;
```

Do we find the same result if we select the orthonormal basis

$$\begin{pmatrix}\cos(\theta)\\\sin(\theta)\end{pmatrix},\quad \begin{pmatrix}\sin(\theta)\\-\cos(\theta)\end{pmatrix}?$$

## 2.3   Supplementary Problems

**Problem 1.**   Consider the standard basis in the vector space of $2 \times 2$ matrices

$$E_{00} = \begin{pmatrix} 1 & 0 \\ 0 & 0 \end{pmatrix}, \quad E_{01} = \begin{pmatrix} 0 & 1 \\ 0 & 0 \end{pmatrix}, \quad E_{10} = \begin{pmatrix} 0 & 0 \\ 1 & 0 \end{pmatrix}, \quad E_{11} = \begin{pmatrix} 0 & 0 \\ 0 & 1 \end{pmatrix}$$

and the mutually unbiased basis

$$\mu_0 = \frac{1}{\sqrt{2}} \begin{pmatrix} 1 & 0 \\ 0 & 1 \end{pmatrix}, \quad \mu_1 = \frac{1}{\sqrt{2}} \begin{pmatrix} 0 & 1 \\ 1 & 0 \end{pmatrix},$$

$$\mu_2 = \frac{1}{\sqrt{2}} \begin{pmatrix} 0 & -i \\ i & 0 \end{pmatrix}, \quad \mu_3 = \frac{1}{\sqrt{2}} \begin{pmatrix} 1 & 0 \\ 0 & -1 \end{pmatrix}.$$

Express the *Bell matrix*

$$B = \frac{1}{\sqrt{2}} \begin{pmatrix} 1 & 0 & 0 & 1 \\ 0 & 1 & 1 & 0 \\ 0 & 1 & -1 & 0 \\ 1 & 0 & 0 & -1 \end{pmatrix}$$

with the basis given by $\mu_j \otimes \mu_k$ $(j, k = 0, 1, 2, 3)$.

**Problem 2.**   The following states form an orthonormal basis in the Hilbert space $\mathbb{C}^3$

$$|\pi^+\rangle = \frac{1}{\sqrt{2}} \begin{pmatrix} 1 \\ 0 \\ 1 \end{pmatrix}, \quad |\pi^0\rangle = \begin{pmatrix} 0 \\ 1 \\ 0 \end{pmatrix}, \quad |\pi^-\rangle = \frac{1}{\sqrt{2}} \begin{pmatrix} 1 \\ 0 \\ -1 \end{pmatrix}.$$

These states play a role for the $\pi$-mesons. Show that the states

$$|\pi^+\rangle \otimes |\pi^+\rangle, \qquad |\pi^-\rangle \otimes |\pi^-\rangle$$

$$\frac{1}{\sqrt{2}}(|\pi^+\rangle \otimes |\pi^0\rangle + |\pi^0\rangle \otimes |\pi^+\rangle), \quad \frac{1}{\sqrt{2}}(|\pi^0\rangle \otimes |\pi^-\rangle + |\pi^-\rangle \otimes |\pi^0\rangle)$$

$$\frac{1}{\sqrt{2}}(|\pi^+\rangle \otimes |\pi^0\rangle - |\pi^0\rangle \otimes |\pi^+\rangle), \quad \frac{1}{\sqrt{2}}(|\pi^+\rangle \otimes |\pi^-\rangle - |\pi^-\rangle \otimes |\pi^+\rangle),$$

$$\frac{1}{\sqrt{2}}(|\pi^0\rangle \otimes |\pi^-\rangle - |\pi^-\rangle \otimes |\pi^0\rangle)$$

$$\frac{1}{\sqrt{6}}(2|\pi^0\rangle \otimes |\pi^0\rangle + |\pi^+\rangle \otimes |\pi^-\rangle + |\pi^-\rangle \otimes |\pi^+\rangle),$$

$$\frac{1}{\sqrt{3}}(|\pi^+\rangle \otimes |\pi^-\rangle + |\pi^-\rangle \otimes |\pi^+\rangle - |\pi^0\rangle \otimes |\pi^0\rangle)$$

form an orthonormal basis in the Hilbert space $\mathbb{C}^9$. Which of these states are entangled?

**Problem 3.**   Let $\mathbf{v}_1$, $\mathbf{v}_2$, $\mathbf{v}_3$ be elements of $\mathbb{C}^2$. Find the conditions on $\mathbf{v}_1$, $\mathbf{v}_2$, $\mathbf{v}_3$ such that

$$\mathbf{v}_1 \otimes \mathbf{v}_2 \otimes \mathbf{v}_3 = \mathbf{v}_3 \otimes \mathbf{v}_2 \otimes \mathbf{v}_1.$$

**Problem 4.**   Let $\sigma_1$, $\sigma_2$, $\sigma_3$ be the Pauli spin matrices. Consider the $8 \times 8$ matrices

$$K = \sigma_1 \otimes \sigma_2 \otimes \sigma_3, \qquad S = \sigma_1 \otimes \sigma_1 \otimes \sigma_1.$$

Note that the matrices $K$ and $S$ are unitary and hermitian. Show that $[K, S] = 0_8$. Show that

$$\Pi_1 = \frac{1}{2}(I_8 + S), \quad \Pi_2 = \frac{1}{2}(I_8 - S)$$

are projection matrices and $\Pi_1 \Pi_2 = 0_8$. Show that

$$\Pi_3 = \frac{1}{2}(I_8 - K), \quad \Pi_4 = \frac{1}{2}(I_8 + K)$$

are projection matrices.

**Problem 5.**   Let $\sigma_1$, $\sigma_2$, $\sigma_3$ be the Pauli spin matrices. We have

$$\sigma_1\sigma_2 = i\sigma_3, \quad \sigma_2\sigma_3 = i\sigma_1, \quad \sigma_3\sigma_1 = i\sigma_2$$

and

$$\sigma_2\sigma_1 = -i\sigma_3, \quad \sigma_3\sigma_2 = -i\sigma_1, \quad \sigma_1\sigma_3 = -i\sigma_2.$$

Show that

$$[\sigma_1 \otimes \sigma_1, \sigma_2 \otimes \sigma_2] = 0_4$$

and

$$[\sigma_1 \otimes \sigma_1, \sigma_3 \otimes \sigma_3] = 0_4.$$

**Problem 6.**   Can the $\mathbb{Z}_4$ Fourier matrix

$$\begin{pmatrix} 1 & 1 & 1 & 1 \\ 1 & i & -1 & -i \\ 1 & -1 & 1 & -1 \\ 1 & -i & -1 & i \end{pmatrix}$$

be written as the Kronecker product of two $2 \times 2$ unitary matrices?

**Problem 7.** Let $|a\rangle$, $|b\rangle$ be normalized states in $\mathbb{C}^n$ and $X$, $Y$ be $n \times n$ matrices over $\mathbb{C}$. Show that

$$((\langle a| \otimes \langle b|)((X \otimes I_n)(I_n \otimes Y))(|a\rangle \otimes |b\rangle) = \langle a|X|a\rangle\langle b|Y|b\rangle.$$

Note that

$$(|a\rangle \otimes |b\rangle)^* = \langle a| \otimes \langle b|.$$

**Problem 8.** (i) Consider the standard basis

$$|0\rangle = \begin{pmatrix} 1 \\ 0 \end{pmatrix}, \quad |1\rangle = \begin{pmatrix} 0 \\ 1 \end{pmatrix}$$

in $\mathbb{C}^2$. Calculate

$$|0\rangle\langle 0| \otimes I_2 + |1\rangle\langle 1| \otimes \begin{pmatrix} 0 & 1 \\ 1 & 0 \end{pmatrix}.$$

(ii) Consider the orthonormal basis

$$|0\rangle = \begin{pmatrix} \cos(\theta) \\ \sin(\theta) \end{pmatrix}, \quad |1\rangle = \begin{pmatrix} \sin(\theta) \\ -\cos(\theta) \end{pmatrix}$$

in $\mathbb{C}^2$. Calculate

$$|0\rangle\langle 0| \otimes I_2 + |1\rangle\langle 1| \otimes \begin{pmatrix} 0 & 1 \\ 1 & 0 \end{pmatrix}.$$

Discuss.

**Problem 9.** Consider the states in $\mathbb{C}^2$

$$|\psi_1\rangle = \begin{pmatrix} \cos(\theta) \\ \sin(\theta) \end{pmatrix}, \quad |\psi_2\rangle = \begin{pmatrix} \sin(\theta) \\ -\cos(\theta) \end{pmatrix}$$

which form an orthonormal basis in $\mathbb{C}^2$. Find

$$((\langle\psi_1| \otimes \langle\psi_2|)(\sigma_j \otimes I_2 + I_2 \otimes \sigma_j)(|\psi_1\rangle \otimes |\psi_2\rangle)), \quad j = 1, 2, 3$$

and discuss the dependence on $\theta$.

**Problem 10.** Consider the four Bell states

$$|\psi_1\rangle = \frac{1}{\sqrt{2}}(|0\rangle \otimes |0\rangle + |1\rangle \otimes |1\rangle)$$

$$|\psi_2\rangle = \frac{1}{\sqrt{2}}(|0\rangle \otimes |1\rangle + |1\rangle \otimes |0\rangle)$$

$$|\psi_3\rangle = \frac{1}{\sqrt{2}}(|0\rangle \otimes |1\rangle - |1\rangle \otimes |0\rangle)$$

$$|\psi_4\rangle = \frac{1}{\sqrt{2}}(|0\rangle \otimes |0\rangle - |1\rangle \otimes |1\rangle)$$

and the three $4 \times 4$ matrices

$$T_j := \sigma_j \otimes I_2 + I_2 \otimes \sigma_j, \quad j = 1, 2, 3.$$

Calculate

$$T_j |\psi_k\rangle, \quad j = 1, 2, 3 \quad k = 1, 2, 3, 4.$$

Which of these expressions is an eigenvalue equation? Calculate

$$\langle \psi_k | T_j | \psi_k \rangle.$$

**Problem 11.**   Let $A$ be an $m \times m$ and $B$ be an $n \times n$ matrix. Study the conditions on $A$ and $B$ such that

$$(A \otimes I_n + I_m \otimes B)\mathbf{v} = 0 \cdot \mathbf{0} = \mathbf{0}$$

where $\mathbf{v} \neq \mathbf{0}$, i.e. we have an eigenvalue problem with eigenvalue 0. As a consequence we have

$$\det(A \otimes I_n + I_m \otimes B) = 0.$$

Note that

$$\operatorname{tr}(A \otimes I_n + I_m \otimes B) = n\operatorname{tr}(A) + m\operatorname{tr}(B).$$

Study first the case with $m = n = 2$ and $A$ and $B$ the Pauli spin matrices.

**Problem 12.**   Given the hermitian matrices of the three dipole operators

$$L_1 = \frac{1}{\sqrt{2}} \begin{pmatrix} 0 & 1 & 0 \\ 1 & 0 & 1 \\ 0 & 1 & 0 \end{pmatrix}, \quad L_2 = \frac{1}{\sqrt{2}} \begin{pmatrix} 0 & -i & 0 \\ i & 0 & -i \\ 0 & i & 0 \end{pmatrix}, \quad L_3 = \begin{pmatrix} 1 & 0 & 0 \\ 0 & 0 & 0 \\ 0 & 0 & -1 \end{pmatrix}$$

and the hermitian matrices of five quadrupole operators

$$W_1 = \begin{pmatrix} 0 & 0 & 1 \\ 0 & 0 & 0 \\ 1 & 0 & 0 \end{pmatrix}, \quad W_2 = \begin{pmatrix} 0 & 0 & -i \\ 0 & 0 & 0 \\ i & 0 & 0 \end{pmatrix},$$

$$V_1 = \frac{1}{\sqrt{2}} \begin{pmatrix} 0 & 1 & 0 \\ 1 & 0 & -1 \\ 0 & -1 & 0 \end{pmatrix}, \quad V_2 = \frac{1}{\sqrt{2}} \begin{pmatrix} 0 & -i & 0 \\ i & 0 & i \\ 0 & -i & 0 \end{pmatrix},$$

$$Q_0 = \frac{1}{\sqrt{3}} \begin{pmatrix} 1 & 0 & 0 \\ 0 & -2 & 0 \\ 0 & 0 & 1 \end{pmatrix}.$$

Show that multiplying these eight hermitian matrices by $i$ we obtain a basis for the semi-simple Lie algebra $su(3)$. Consider the Hamilton operator

$$\hat{H} = \kappa_0 Q_0 \otimes Q_0 + \kappa_1 (V_1 \otimes V_1 + V_2 \otimes V_2) + \kappa_2 (W_1 \otimes W_1 + W_2 \otimes W_2).$$

Find the eigenvalues and eigenvectors of $\hat{H}$.

**Problem 13.** Let

$$\mathbf{w}_1 = \begin{pmatrix} 1 \\ 0 \end{pmatrix}, \quad \mathbf{w}_2 = \begin{pmatrix} 0 \\ 1 \end{pmatrix}$$

and

$$\mathbf{u} = \sum_{j_1, j_2, j_3 = 1}^{2} t_{j_1 j_2 j_3} \mathbf{w}_{j_1} \otimes \mathbf{w}_{j_2} \otimes \mathbf{w}_{j_3} = \mathbf{w}_1 \otimes \mathbf{w}_2 \otimes \mathbf{w}_2 + \mathbf{w}_2 \otimes \mathbf{w}_1 \otimes \mathbf{w}_1$$

$$= \begin{pmatrix} 0 & 0 & 0 & 1 & 1 & 0 & 0 & 0 \end{pmatrix}^T$$

i.e. $t_{122} = 1$, $t_{211} = 1$ and all other coefficients are equal to 0. Can the vector $\mathbf{u} \in \mathbb{C}^8$ be written as the Kronecker product of a vector in $\mathbb{C}^2$ and a vector in $\mathbb{C}^4$? Consider both cases $\mathbb{C}^2 \otimes \mathbb{C}^4$ and $\mathbb{C}^4 \otimes \mathbb{C}^2$.

**Problem 14.** Let $\sigma_1$, $\sigma_2$, $\sigma_3$ be the Pauli spin matrices. Consider the Hamilton operator

$$\hat{H} = \hbar \omega_1 \sigma_1 \otimes \sigma_1 + \hbar \omega_2 \sigma_2 \otimes \sigma_2 + \hbar \omega_3 \sigma_3 \otimes \sigma_3$$

acting in the Hilbert space $\mathbb{C}^4$. Show that

$$\begin{aligned}
e^{i\hat{H}t/\hbar} (\sigma_1 \otimes I_2) e^{-i\hat{H}t/\hbar} &= (\sigma_1 \otimes I_2) \cos(\omega_2 t) \cos(\omega_3 t) \\
&+ (I_2 \otimes \sigma_1) \sin(\omega_2 t) \sin(\omega_3 t) \\
&- (\sigma_2 \otimes \sigma_3) \cos(\omega_2 t) \sin(\omega_3 t) \\
&+ (\sigma_3 \otimes \sigma_2) \sin(\omega_2 t) \cos(\omega_3 t)
\end{aligned}$$

$$\begin{aligned}
e^{i\hat{H}t/\hbar} (\sigma_2 \otimes I_2) e^{-i\hat{H}t/\hbar} &= (\sigma_2 \otimes I_2) \cos(\omega_3 t) \cos(\omega_1 t) \\
&+ (I_2 \otimes \sigma_2) \sin(\omega_3 t) \sin(\omega_1 t) \\
&- (\sigma_3 \otimes \sigma_1) \cos(\omega_3 t) \sin(\omega_1 t) \\
&+ (\sigma_1 \otimes \sigma_3) \sin(\omega_3 t) \cos(\omega_1 t)
\end{aligned}$$

$$\begin{aligned}
e^{i\hat{H}t/\hbar} (\sigma_3 \otimes I_2) e^{-i\hat{H}t/\hbar} &= (\sigma_3 \otimes I_2) \cos(\omega_1 t) \cos(\omega_2 t) \\
&+ (I_2 \otimes \sigma_3) \sin(\omega_1 t) \sin(\omega_2 t) \\
&- (\sigma_1 \otimes \sigma_2) \cos(\omega_1 t) \sin(\omega_2 t) \\
&+ (\sigma_2 \otimes \sigma_1) \sin(\omega_1 t) \cos(\omega_2 t).
\end{aligned}$$

**Problem 15.**   Let $S_1$, $S_2$, $S_3$ be the spin-$\frac{1}{2}$ matrices

$$S_1 = \frac{1}{2}\begin{pmatrix} 0 & 1 \\ 1 & 0 \end{pmatrix}, \quad S_2 = \frac{1}{2}\begin{pmatrix} 0 & -i \\ i & 0 \end{pmatrix}, \quad S_3 = \frac{1}{2}\begin{pmatrix} 1 & 0 \\ 0 & -1 \end{pmatrix}.$$

Solve the eigenvalue problem for the Hamilton operator

$$\hat{H} = \hbar\omega_1(S_1 \otimes S_2 \otimes S_3 + S_3 \otimes S_1 \otimes S_2 + S_2 \otimes S_3 \otimes S_1)$$
$$+ \hbar\omega_2(S_3 \otimes I_2 \otimes I_2 + I_2 \otimes S_3 \otimes I_2 + I_2 \otimes I_2 \otimes S_3).$$

# Chapter 3

# Matrix Properties

## 3.1 Introduction

For finite-dimensional quantum systems finding the norm, eigenvalues, eigen-vectors, Schmidt rank and inverse (if it exists) of square matrices is impor-tant. Let $A$ be an $n \times n$ matrix over $\mathbb{C}$. Then we can define the sup-norm

$$\|A\| := \sup_{\|\mathbf{x}\|=1} \|A\mathbf{x}\|$$

where $\|A\mathbf{x}\|$ denotes the Euclidean norm in $\mathbb{C}^n$. The *Hilbert-Schmidt norm* of a square matrix $A$ is defined as

$$\|A\| := (\operatorname{tr}(AA^*))^{1/2}$$

where tr denotes the trace.

Let $A$ be an $n \times n$ matrix over $\mathbb{C}$. Then the *eigenvalue equation* is defined as

$$A\mathbf{x} = \lambda\mathbf{x}$$

where $\lambda \in \mathbb{C}$ is the eigenvalue and $\mathbf{x} \in \mathbb{C}^n$ with $\mathbf{x} \neq \mathbf{0}$ is a corresponding eigenvector. It follows that $\mathbf{x}^*A^* = \bar{\lambda}\mathbf{x}^*$.

The most important function in quantum computing is the exponential function of a square matrix $A$ defined by

$$\exp(A) := \sum_{j=0}^{\infty} \frac{A^j}{j!} = \lim_{k \to \infty} \left(I + \frac{A}{k}\right)^k.$$

65

An $n \times n$ matrix over $\mathbb{C}$ is called *normal* if

$$A^* A = AA^*.$$

Let $A$ be an $n \times n$ normal matrix over $\mathbb{C}$ with eigenvalues $\lambda_1, \ldots, \lambda_n$ and corresponding pairwise orthonormal eigenvectors $\mathbf{v}_j$ ($j = 1, \ldots, n$). Then the matrix $A$ can be written as (*spectral decomposition*)

$$A = \sum_{j=1}^{n} \lambda_j \mathbf{v}_j \mathbf{v}_j^*.$$

Note that

$$\mathbf{v}_j \mathbf{v}_j^*, \quad j = 1, \ldots, n$$

are projection matrices, i.e. $(\mathbf{v}_j \mathbf{v}_j^*)(\mathbf{v}_j \mathbf{v}_j^*) = \mathbf{v}_j \mathbf{v}_j^*$, $(\mathbf{v}_j \mathbf{v}_j^*)^* = \mathbf{v}_j \mathbf{v}_j^*$ and $\mathbf{v}_j^* \mathbf{v}_j = 1$ for $j = k$ and $\mathbf{v}_j^* \mathbf{v}_k = 0$ for $j \neq k$.

Consider an $n \times n$ matrix $A$ over $\mathbb{C}$ and the polynomial

$$p(\lambda) = \det(A - \lambda I_n)$$

with the characteristic equation

$$p(\lambda) = 0.$$

The *Cayley-Hamilton theorem* states that substituting the matrix $A$ in the characteristic polynomial results in the $n \times n$ zero matrix, i.e.

$$p(A) = 0_n.$$

Decompositions of square matrices such as the singular value decomposition, spectral decomposition, polar decomposition and Schur decomposition are necessary in quantum computing. Any unitary $2^n \times 2^n$ matrix $U$ can be decomposed as

$$U = \begin{pmatrix} U_1 & 0 \\ 0 & U_2 \end{pmatrix} \begin{pmatrix} C & S \\ -S & C \end{pmatrix} \begin{pmatrix} U_3 & 0 \\ 0 & U_4 \end{pmatrix}$$

where $U_1, U_2, U_3, U_4$ are $2^{n-1} \times 2^{n-1}$ unitary matrices and $C$ and $S$ are the $2^{n-1} \times 2^{n-1}$ diagonal matrices

$$C = \text{diag}(\cos(\alpha_1), \cos(\alpha_2), \ldots, \cos(\alpha_{2^n}/2)),$$
$$S = \text{diag}(\sin(\alpha_1), \sin(\alpha_2), \ldots, \sin(\alpha_{2^n}/2))$$

where $\alpha_j \in \mathbb{R}$. This decomposition is called *cosine-sine decomposition*.

# 3.2 Solved Problems

**Problem 1.** Consider the hermitian $4 \times 4$ matrix (Hamilton operator)

$$\hat{H} = \frac{\hbar\omega}{2}(\sigma_1 \otimes \sigma_1 - \sigma_2 \otimes \sigma_2)$$

where $\omega$ is the frequency. Find the *norm* of $\hat{H}$, i.e.

$$\|\hat{H}\| := \sup_{\|\mathbf{x}\|=1} \|\hat{H}\mathbf{x}\|, \qquad \mathbf{x} \in \mathbb{C}^4.$$

**Solution 1.** There are two methods to find the norm of $\hat{H}$. In the first method we use the *Lagrange multiplier method* where the constraint $\|\mathbf{x}\| = 1$ can be written as $x_1^2 + x_2^2 + x_3^2 + x_4^2 = 1$. Since

$$\sigma_1 \otimes \sigma_1 = \begin{pmatrix} 0 & 0 & 0 & 1 \\ 0 & 0 & 1 & 0 \\ 0 & 1 & 0 & 0 \\ 1 & 0 & 0 & 0 \end{pmatrix}, \quad \sigma_2 \otimes \sigma_2 = \begin{pmatrix} 0 & 0 & 0 & -1 \\ 0 & 0 & 1 & 0 \\ 0 & 1 & 0 & 0 \\ -1 & 0 & 0 & 0 \end{pmatrix}$$

we have

$$\hat{H} = \hbar\omega \begin{pmatrix} 0 & 0 & 0 & 1 \\ 0 & 0 & 0 & 0 \\ 0 & 0 & 0 & 0 \\ 1 & 0 & 0 & 0 \end{pmatrix}.$$

Let $\mathbf{x} = (x_1, x_2, x_3, x_4)^T \in \mathbb{C}^4$. We maximize

$$f(\mathbf{x}) := \|\hat{H}\mathbf{x}\|^2 - \lambda(x_1^2 + x_2^2 + x_3^2 + x_4^2 - 1)$$

where $\lambda$ is the Lagrange multiplier. To find the extrema we solve the four equations

$$\frac{\partial f}{\partial x_1} = 2\hbar^2\omega^2 x_1 - 2\lambda x_1 = 0$$

$$\frac{\partial f}{\partial x_2} = -2\lambda x_2 = 0$$

$$\frac{\partial f}{\partial x_3} = -2\lambda x_3 = 0$$

$$\frac{\partial f}{\partial x_4} = 2\hbar^2\omega^2 x_4 - 2\lambda x_4 = 0$$

together with the constraint $x_1^2 + x_2^2 + x_3^2 + x_4^2 = 1$. The four equations can be written in the matrix form

$$\begin{pmatrix} \hbar^2\omega^2 - \lambda & 0 & 0 & 0 \\ 0 & -\lambda & 0 & 0 \\ 0 & 0 & -\lambda & 0 \\ 0 & 0 & 0 & \hbar^2\omega^2 - \lambda \end{pmatrix} \begin{pmatrix} x_1 \\ x_2 \\ x_3 \\ x_4 \end{pmatrix} = \begin{pmatrix} 0 \\ 0 \\ 0 \\ 0 \end{pmatrix}.$$

If $\lambda = 0$ then $x_1 = x_4 = 0$ and $\|\hat{H}\mathbf{x}\| = 0$, which is a minimum. If $\lambda \neq 0$ then $x_2 = x_3 = 0$ and $x_1^2 + x_4^2 = 1$ so that $\|\hat{H}\mathbf{x}\| = \hbar\omega$, which is the maximum. Thus we find $\|\hat{H}\| = \hbar\omega$. In the second method we calculate the positive definite matrix $\hat{H}^*\hat{H}$ and find the square root of the largest eigenvalue of $\hat{H}^*\hat{H}$. Since $\hat{H}^* = \hat{H}$ we find the positive semi-definite

$$\hat{H}^*\hat{H} = \hbar^2\omega^2 \begin{pmatrix} 1 & 0 & 0 & 0 \\ 0 & 0 & 0 & 0 \\ 0 & 0 & 0 & 0 \\ 0 & 0 & 0 & 1 \end{pmatrix}.$$

Thus the maximum eigenvalue is $\hbar^2\omega^2$ (twice degenerate) and $\|\hat{H}\| = \hbar\omega$.

**Problem 2.**   Let $\hat{H}$ be a hermitian $n \times n$ matrix (Hamilton operator) with eigenvalues $E_0$, $E_1$, $\ldots$, $E_{n-1}$ with corresponding normalized eigenvectors $|\psi_0\rangle$, $\psi_1\rangle$, $\ldots$, $|\psi_{n-1}\rangle$. The *quantum correlation function* of two $n \times n$ hermitian matrices $A$ and $B$ is given by

$$Q_k(t) := \frac{1}{2}\langle\psi_k|(A(t)B - AB(t) + BA(t) - B(t)A)|\psi_k\rangle, \quad k = 0, 1, \ldots, n-1$$

where

$$A(t) = e^{i\hat{H}t/\hbar}Ae^{-i\hat{H}t/\hbar}, \qquad B(t) = e^{i\hat{H}t/\hbar}Be^{-i\hat{H}t/\hbar}.$$

Note that $Q_k(t)$ is real valued. Find $Q_k(t)$ using the properties

$$e^{i\hat{H}t/\hbar}|\psi_k\rangle = e^{iE_kt/\hbar}|\psi_k\rangle, \qquad \sum_{j=0}^{n-1}|\psi_j\rangle\langle\psi_j| = I_n.$$

**Solution 2.**   Since

$$\langle\psi_k|A(t)B|\psi_k\rangle = \sum_{j=0}^{n-1} e^{i(E_k - E_j)t/\hbar}\langle\psi_k|A|\psi_j\rangle\langle\psi_j|B|\psi_k\rangle$$

$$\langle\psi_k|AB(t)|\psi_k\rangle = \sum_{j=0}^{n-1} e^{i(E_j - E_k)t/\hbar}\langle\psi_k|A|\psi_j\rangle\langle\psi_j|B|\psi_k\rangle$$

$$\langle\psi_k|BA(t)|\psi_k\rangle = \sum_{j=0}^{n-1} e^{i(E_j - E_k)t/\hbar}\langle\psi_k|B|\psi_j\rangle\langle\psi_j|A|\psi_k\rangle$$

$$\langle\psi_k|B(t)A|\psi_k\rangle = \sum_{j=0}^{n-1} e^{i(E_k - E_j)t/\hbar}\langle\psi_k|B|\psi_j\rangle\langle\psi_j|A|\psi_k\rangle$$

and utilizing the identity $e^{i\alpha} - e^{-i\alpha} \equiv 2i\sin(\alpha)$ yields

$$Q_k(t) = i\sum_{j=0}^{n-1} \sin((E_j - E_k)t/\hbar)\left(\langle\psi_j|A|\psi_k\rangle\langle\psi_k|B|\psi_j\rangle - \langle\psi_k|A|\psi_j\rangle\langle\psi_j|B|\psi_k\rangle\right).$$

**Problem 3.** (i) Let $A$ and $B$ be two $n \times n$ matrices over $\mathbb{C}$. If there exists a non-singular $n \times n$ matrix $X$ such that $A = XBX^{-1}$, then $A$ and $B$ are said to be *similar matrices*. Show that the spectra (eigenvalues) of two similar matrices are equal.

(ii) Let $A$ and $B$ be $n \times n$ matrices over $\mathbb{C}$. Show that the matrices $AB$ and $BA$ have the same set of eigenvalues.

**Solution 3.** (i) We have

$$\det(A - \lambda I_n) = \det(XBX^{-1} - X\lambda I_n X^{-1}) = \det(X(B - \lambda I_n)X^{-1})$$
$$= \det(X)\det(B - \lambda I_n)\det(X^{-1}) = \det(B - \lambda I_n).$$

(ii) Consider first the case that $A$ is invertible. Then we have

$$AB = A(BA)A^{-1}.$$

Thus $AB$ and $BA$ are similar and therefore have the same set of eigenvalues. If $A$ is singular we apply the *continuity argument*: Consider the matrix $A + \epsilon I_n$. We choose $\delta > 0$ such that $A + \epsilon I_n$ is invertible for all $\epsilon$, $0 < \epsilon < \delta$. Thus $(A + \epsilon I_n)B$ and $B(A + \epsilon I_n)$ have the same set of eigenvalues for every $\epsilon \in (0, \delta)$. We equate their characteristic polynomials to obtain

$$\det(\lambda I_n - (A + \epsilon I_n)B) = \det(\lambda I_n - B(A + \epsilon I_n)), \quad 0 < \epsilon < \delta.$$

Since both sides are analytic functions of $\epsilon$ we find by letting $\epsilon \to 0^+$ that

$$\det(\lambda I_n - AB) = \det(\lambda I_n - BA).$$

**Problem 4.** Consider a square non-singular matrix $A$ over $\mathbb{C}$. The *polar decomposition theorem* states that $A$ can be written as

$$A = UP$$

where $U$ is a unitary matrix and $P$ is a positive definite matrix. Thus $P$ is hermitian. Show that $A$ has a unique polar decomposition.

**Solution 4.** Since $A$ is invertible, so are $A^*$ and $A^*A$. The positive square root $P$ of $A^*A$ is also invertible. Set $U := AP^{-1}$. Then $U$ is invertible and

$$U^*U = P^{-1}A^*AP^{-1} = P^{-1}P^2P^{-1} = I$$

so that $U$ is unitary. Since $P$ is invertible, it is obvious that $AP^{-1}$ is the only possible choice for $U$.

**Problem 5.**  Let $A$ and $B$ be $n \times n$ hermitian matrices. Suppose that

$$A^2 = I_n, \qquad B^2 = I_n \qquad (1)$$

and

$$[A, B]_+ \equiv AB + BA = 0_n \qquad (2)$$

where $0_n$ is the $n \times n$ zero matrix. Let $\mathbf{x} \in \mathbb{C}^n$ be normalized, i.e. $\|\mathbf{x}\| = 1$. Here $\mathbf{x}$ is considered as a column vector.
(i) Show that

$$(\mathbf{x}^* A \mathbf{x})^2 + (\mathbf{x}^* B \mathbf{x})^2 \le 1. \qquad (3)$$

(ii) Give an example for the matrices $A$ and $B$.

**Solution 5.**  (i) Let $a, b \in \mathbb{R}$ and let $r^2 := a^2 + b^2$. The matrix

$$C = aA + bB$$

is again hermitian. Then

$$C^2 = a^2 A^2 + ab AB + ba BA + b^2 B^2.$$

Using the properties (1) and (2) we find

$$C^2 = a^2 I_n + b^2 I_n = r^2 I_n.$$

Therefore $(\mathbf{x}^* C^2 \mathbf{x}) = r^2$ and $-r \le a(\mathbf{x}^* A \mathbf{x}) + b(\mathbf{x}^* B \mathbf{x}) \le r$. Let

$$a = \mathbf{x}^* A \mathbf{x}, \qquad b = \mathbf{x}^* B \mathbf{x}$$

then $a^2 + b^2 \le r$ or $r^2 \le r$. This implies $r \le 1$ and $r^2 \le 1$ from which (3) follows.
(ii) An example is $A = \sigma_1$ and $B = \sigma_2$ since $\sigma_1^2 = I_2$, $\sigma_2^2 = I_2$ and $\sigma_1 \sigma_2 + \sigma_2 \sigma_1 = 0_2$.

**Problem 6.**  Let $K$ be an $n \times n$ *skew-hermitian matrix* $K = -K^*$ with eigenvalues $\mu_1, \ldots, \mu_n$ (counted according to multiplicity) and the corresponding normalized eigenvectors $\mathbf{u}_1, \ldots, \mathbf{u}_n$, where $\mathbf{u}_j^* \mathbf{u}_k = 0$ for $k \ne j$. Then $K$ can be written as

$$K = \sum_{j=1}^{n} \mu_j \mathbf{u}_j \mathbf{u}_j^*$$

and $\mathbf{u}_j \mathbf{u}_j^* \mathbf{u}_k \mathbf{u}_k^* = 0$ for $k \neq j$ and $j,k = 1, 2, \dots, n$. Note that the $n \times n$ matrices $\mathbf{u}_j \mathbf{u}_j^*$ are projection matrices and

$$\sum_{j=1}^n \mathbf{u}_j \mathbf{u}_j^* = I_n.$$

(i) Calculate $\exp(K)$.
(ii) Every $n \times n$ unitary matrix can be written as $U = \exp(K)$, where $K$ is a skew-hermitian matrix. Find $U$ from a given $K$.
(iii) Use the result from (ii) to find for a given $U$ a possible $K$.
(iv) Apply the result from (ii) and (iii) to the unitary $2 \times 2$ matrix

$$U(\theta) = \begin{pmatrix} \cos(\theta) & \sin(\theta) \\ -\sin(\theta) & \cos(\theta) \end{pmatrix}.$$

(v) Apply the result from (ii) and (iii) to the $2 \times 2$ unitary matrix

$$V(\theta, \phi) = \begin{pmatrix} \cos(\theta) & -e^{i\phi}\sin(\theta) \\ e^{-i\phi}\sin(\theta) & \cos(\theta) \end{pmatrix}.$$

(vi) Every hermitian matrix $H$ can be written as $H = iK$, where $K$ is a skew-hermitian matrix. Find $H$ for the examples given above.

**Solution 6.** (i) Using the properties of the $n \times n$ matrix $\mathbf{u}_j \mathbf{u}_j^*$ we find

$$\exp(K) = \exp\left(\sum_{j=1}^n \mu_j \mathbf{u}_j \mathbf{u}_j^*\right) = \sum_{j=1}^n e^{\mu_j} \mathbf{u}_j \mathbf{u}_j^*.$$

(ii) From $U = \exp(K)$ we find

$$U = \sum_{j=1}^n e^{\mu_j} \mathbf{u}_j \mathbf{u}_j^*$$

where $\mathbf{u}_j$ $(j = 1, 2, \dots, n)$ are the normalized eigenvectors of $U$.
(iii) The matrix $K$ is given by

$$K = \sum_{j=1}^n \ln(\lambda_j) \mathbf{u}_j \mathbf{u}_j^*$$

where $\lambda_j$ $(j = 1, 2, \dots, n)$ are the eigenvalues of $U$ and $\mathbf{u}_j$ are the normalized eigenvectors of $U$. Note that the eigenvalues of $U$ are of the form $\exp(i\alpha)$ with $\alpha \in \mathbb{R}$. Thus we have $\ln(e^{i\alpha}) = i\alpha$.

(iv) The eigenvalues of the matrix $U(\theta)$ are $e^{i\theta}$ and $e^{-i\theta}$ with the corresponding normalized eigenvectors

$$\mathbf{u}_1 = \frac{1}{\sqrt{2}} \begin{pmatrix} 1 \\ i \end{pmatrix}, \qquad \mathbf{u}_2 = \frac{1}{\sqrt{2}} \begin{pmatrix} 1 \\ -i \end{pmatrix}.$$

Thus

$$K(\theta) = \ln(e^{i\theta})\mathbf{u}_1\mathbf{u}_1^* + \ln(e^{i\theta})\mathbf{u}_2\mathbf{u}_2 = \frac{i\theta}{2} \begin{pmatrix} 1 & -i \\ i & 1 \end{pmatrix} - \frac{i\theta}{2} \begin{pmatrix} 1 & i \\ -i & 1 \end{pmatrix} = \begin{pmatrix} 0 & \theta \\ -\theta & 0 \end{pmatrix}.$$

(v) For the matrix $V(\theta, \phi)$ the eigenvalues are $e^{-i\theta}$ and $e^{i\theta}$ with the corresponding normalized eigenvectors

$$\frac{1}{\sqrt{2}} \begin{pmatrix} 1 \\ ie^{-i\phi} \end{pmatrix}, \qquad \frac{1}{\sqrt{2}} \begin{pmatrix} 1 \\ -ie^{-i\phi} \end{pmatrix}.$$

Thus

$$K(\theta, \phi) = \ln(e^{-i\theta})\mathbf{u}_1\mathbf{u}_1^* + \ln(e^{i\theta})\mathbf{u}_2\mathbf{u}_2^* = \begin{pmatrix} 0 & -\theta e^{i\phi} \\ \theta e^{-i\phi} & 0 \end{pmatrix}.$$

(vi) For $U(\theta)$ we find the matrix

$$i\theta \begin{pmatrix} 0 & 1 \\ -1 & 0 \end{pmatrix}.$$

For $V(\theta, \phi)$ we find the matrix

$$\begin{pmatrix} 0 & -i\theta e^{i\phi} \\ i\theta e^{-i\phi} & 0 \end{pmatrix}.$$

**Problem 7.**   Let $A$ and $B$ be $n \times n$ hermitian matrices. Suppose that

$$A^2 = A, \qquad B^2 = B \tag{1}$$

and

$$[A, B]_+ \equiv AB + BA = 0_n \tag{2}$$

where $0_n$ is the $n \times n$ zero matrix. Let $\mathbf{x} \in \mathbb{C}^n$ be normalized, i.e. $\|\mathbf{x}\| = 1$. Here $\mathbf{x}$ is considered as a column vector. Show that

$$(\mathbf{x}^* A\mathbf{x})^2 + (\mathbf{x}^* B\mathbf{x})^2 \leq 1. \tag{3}$$

**Solution 7.**   For an arbitrary $n \times n$ hermitian matrix $M$ we have

$$0 \leq (\mathbf{x}^*(M - (\mathbf{x}^* M\mathbf{x})I_n)^2\mathbf{x}) = (\mathbf{x}^*(M^2 - 2(\mathbf{x}^* M\mathbf{x})M + (\mathbf{x}^* M\mathbf{x})^2 I_n)\mathbf{x})$$
$$= (\mathbf{x}^* M^2\mathbf{x}) - 2(\mathbf{x}^* M\mathbf{x})^2 + (\mathbf{x}^* M\mathbf{x})^2 = (\mathbf{x}^* M^2\mathbf{x}) - (\mathbf{x}^* M\mathbf{x})^2.$$

Thus
$$0 \le (\mathbf{x}^* M^2 \mathbf{x}) - (\mathbf{x}^* M \mathbf{x})^2 \quad \text{or} \quad (\mathbf{x}^* M \mathbf{x})^2 \le (\mathbf{x}^* M^2 \mathbf{x}).$$

Thus for $A = M$ we have using (1)

$$(\mathbf{x}^* A \mathbf{x})^2 \le \mathbf{x}^* A \mathbf{x}$$

and therefore $0 \le (\mathbf{x}^* A \mathbf{x}) \le 1$. Similarly $0 \le (\mathbf{x}^* B \mathbf{x}) \le 1$. Let $a, b \in \mathbb{R}$, $r^2 := a^2 + b^2$ and $C := aA + bB$. Then

$$C^2 = a^2 A^2 + b^2 B^2 + abAB + baBA.$$

Using (1) and (2) we arrive at $C^2 = a^2 A + b^2 B$. Thus

$$(\mathbf{x}^* C \mathbf{x})^2 \le (\mathbf{x}^* C^2 \mathbf{x}) \le a^2 + b^2.$$

Let $a := (\mathbf{x}^* A \mathbf{x})$, $b := (\mathbf{x}^* B \mathbf{x})$ then $(\mathbf{x}^* C \mathbf{x}) = a^2 + b^2 = r^2$ and therefore $(r^2)^2 \le r^2$ which implies that $r^2 \le 1$ and thus (3) follows.

**Problem 8.** Let $A, B$ be $n \times n$ matrices over $\mathbb{C}$. Assume that

$$[A, [A, B]] = [B, [A, B]] = 0_n. \tag{1}$$

Show that

$$e^{A+B} = e^A e^B e^{-\frac{1}{2}[A,B]} \tag{2a}$$

$$e^{A+B} = e^B e^A e^{+\frac{1}{2}[A,B]}. \tag{2b}$$

Hint. Use the *technique of parameter differentiation*. Consider the matrix-valued function

$$f(\epsilon) = e^{\epsilon A} e^{\epsilon B}$$

where $\epsilon$ is a real parameter and calculate the derivative $df/d\epsilon$.

**Solution 8.** If we differentiate $f(\epsilon)$ with respect to $\epsilon$ we find

$$\frac{df}{d\epsilon} = Ae^{\epsilon A} e^{\epsilon B} + e^{\epsilon A} e^{\epsilon B} B = (A + e^{\epsilon A} B e^{-\epsilon A}) f(\epsilon)$$

since $e^{\epsilon A} e^{-\epsilon A} = I_n$. Owing to (1) we have

$$e^{\epsilon A} B e^{-\epsilon A} = B + \epsilon [A, B].$$

Thus we obtain the linear matrix-valued differential equation

$$\frac{df}{d\epsilon} = ((A + B) + \epsilon [A, B]) f(\epsilon).$$

Since the matrix $A + B$ commutes with $[A, B]$ we may treat $A + B$ and $[A, B]$ as ordinary commuting variables and integrate this linear differential equation with the initial conditions $f(0) = I_n$. We find

$$f(\epsilon) = e^{\epsilon(A+B)+(\epsilon^2/2)[A,B]} = e^{\epsilon(A+B)}e^{(\epsilon^2/2)[A,B]}$$

since $A + B$ commutes with $[A, B]$. If we set $\epsilon = 1$ and multiply both sides by $e^{-[A,B]/2}$ then (2a) follows. Likewise we can prove the second form of the identity (2b).

**Problem 9.**    Let $A$ be an $n \times n$ matrix. Assume that the inverse matrix of $A$ exists. The inverse matrix can be calculated as follows (*Csanky's algorithm*). Let

$$p(x) := \det(xI_n - A) \qquad (1)$$

where $I_n$ is the $n \times n$ unit matrix. The roots are, by definition, the eigenvalues $\lambda_1, \lambda_2, \ldots, \lambda_n$ of $A$. We write

$$p(x) = x^n + c_1 x^{n-1} + \cdots + c_{n-1}x + c_n \qquad (2)$$

where $c_n = (-1)^n \det(A)$. Since $A$ is nonsingular we have $c_n \neq 0$ and vice versa. The *Cayley-Hamilton theorem* states that

$$p(A) = A^n + c_1 A^{n-1} + \cdots + c_{n-1}A + c_n I_n = 0_n. \qquad (3)$$

Multiplying this equation with $A^{-1}$ we obtain

$$A^{-1} = \frac{1}{-c_n}(A^{n-1} + c_1 A^{n-2} + \cdots + c_{n-1}I_n). \qquad (4)$$

If we have the coefficients $c_j$ we can calculate the inverse matrix $A$. Let

$$s_k := \sum_{j=1}^{n} \lambda_j^k.$$

Then the $s_j$ and $c_j$ satisfy the following $n \times n$ lower triangular system of linear equations

$$\begin{pmatrix} 1 & 0 & 0 & \cdots & 0 \\ s_1 & 2 & 0 & \cdots & 0 \\ s_2 & s_1 & 3 & \cdots & 0 \\ \vdots & \vdots & \vdots & \ddots & \vdots \\ s_{n-1} & s_{n-2} & \cdots & s_1 & n \end{pmatrix} \begin{pmatrix} c_1 \\ c_2 \\ c_3 \\ \vdots \\ c_n \end{pmatrix} = \begin{pmatrix} -s_1 \\ -s_2 \\ -s_3 \\ \vdots \\ -s_n \end{pmatrix}.$$

Since

$$\mathrm{tr}(A^k) = \lambda_1^k + \lambda_2^k + \cdots + \lambda_n^k = s_k$$

we find $s_k$ for $k = 1, 2, \ldots, n$. Thus we can solve the linear equation for $c_j$. Finally, using (4) we obtain the inverse matrix of $A$. Apply Csanky's algorithm to the $4 \times 4$ permutation matrix

$$U = \begin{pmatrix} 0 & 1 & 0 & 0 \\ 0 & 0 & 1 & 0 \\ 0 & 0 & 0 & 1 \\ 1 & 0 & 0 & 0 \end{pmatrix}.$$

**Solution 9.**  Since

$$U^2 = \begin{pmatrix} 0 & 0 & 1 & 0 \\ 0 & 0 & 0 & 1 \\ 1 & 0 & 0 & 0 \\ 0 & 1 & 0 & 0 \end{pmatrix}, \quad U^3 = \begin{pmatrix} 0 & 0 & 0 & 1 \\ 1 & 0 & 0 & 0 \\ 0 & 1 & 0 & 0 \\ 0 & 0 & 1 & 0 \end{pmatrix}$$

and $U^4 = I_4$ we find

$$\operatorname{tr}(U) = 0 = s_1, \quad \operatorname{tr}(U^2) = 0 = s_2, \quad \operatorname{tr}(U^3) = 0 = s_3, \quad \operatorname{tr}(U^4) = 4 = s_4.$$

We obtain the system of linear equations

$$\begin{pmatrix} 1 & 0 & 0 & 0 \\ 0 & 2 & 0 & 0 \\ 0 & 0 & 3 & 0 \\ 0 & 0 & 0 & 4 \end{pmatrix} \begin{pmatrix} c_1 \\ c_2 \\ c_3 \\ c_4 \end{pmatrix} = \begin{pmatrix} 0 \\ 0 \\ 0 \\ -4 \end{pmatrix}$$

with the solution $c_1 = 0$, $c_2 = 0$, $c_3 = 0$, $c_4 = -1$. Thus the inverse matrix of $U$ is given by

$$U^{-1} = U^3 = \begin{pmatrix} 0 & 0 & 0 & 1 \\ 1 & 0 & 0 & 0 \\ 0 & 1 & 0 & 0 \\ 0 & 0 & 1 & 0 \end{pmatrix}.$$

**Problem 10.**  Let

$$J^+ := \begin{pmatrix} 0 & 1 \\ 0 & 0 \end{pmatrix}, \qquad J^- := \begin{pmatrix} 0 & 0 \\ 1 & 0 \end{pmatrix}, \qquad J_3 := \frac{1}{2} \begin{pmatrix} 1 & 0 \\ 0 & -1 \end{pmatrix}.$$

(i) Let $\epsilon \in \mathbb{R}$. Find the matrices $e^{\epsilon J^+}$, $e^{\epsilon J^-}$, $e^{\epsilon (J^+ + J^-)}$.

(ii) Let $r \in \mathbb{R}$. Show that

$$e^{r(J^+ + J^-)} \equiv e^{J^- \tanh(r)} e^{2 J_3 \ln(\cosh(r))} e^{J^+ \tanh(r)}.$$

**Solution 10.**   (i) Using the expansion for an $n \times n$ matrix $A$

$$\exp(\epsilon A) = \sum_{j=0}^{\infty} \frac{\epsilon^j A^j}{j!}$$

we find

$$e^{\epsilon J^+} = \begin{pmatrix} 1 & 0 \\ 0 & 1 \end{pmatrix} + \epsilon \begin{pmatrix} 0 & 1 \\ 0 & 0 \end{pmatrix}, \qquad e^{\epsilon J^-} = \begin{pmatrix} 1 & 0 \\ 0 & 1 \end{pmatrix} + \epsilon \begin{pmatrix} 0 & 0 \\ 1 & 0 \end{pmatrix}$$

and

$$e^{\epsilon(J^+ + J^-)} = \begin{pmatrix} 1 & 0 \\ 0 & 1 \end{pmatrix} \cosh(\epsilon) + \begin{pmatrix} 0 & 1 \\ 1 & 0 \end{pmatrix} \sinh(\epsilon).$$

(ii) Since

$$e^{2J_3 \ln(\cosh(r))} = \begin{pmatrix} \cosh(r) & 0 \\ 0 & 1/\cosh(r) \end{pmatrix},$$

$1/\cosh(r) + \tanh(r) \sinh(r) \equiv \cosh(r)$ and using the results from (i) we find the identity.

**Problem 11.**   The *Heisenberg commutation relation* can be written as

$$[\hat{p}, \hat{q}] = -i\hbar I$$

where $\hat{p} := -i\hbar \partial/\partial q$ and $I$ is the identity operator. Let $\alpha, \beta \in \mathbb{R}$ and

$$U(\alpha) = \exp(i\alpha\hat{p}), \qquad V(\beta) = \exp(i\beta\hat{q}).$$

Then using the *Baker-Campbell-Hausdorff formula* we find

$$U(\alpha)V(\beta) = \exp(i\alpha\beta)V(\beta)U(\alpha).$$

This is called the *Weyl representation* of Heisenberg's commutation relation. Can we find finite-dimensional $n \times n$ unitary matrices $U$ ($U \neq I_n$) and $V$ ($V \neq I_n$) such that

$$UV = \omega VU$$

with $\omega \in \mathbb{C}$, $\omega^n = 1$ ?

**Solution 11.**   Such matrices can be found, namely the *permutation matrix*

$$U := \begin{pmatrix} 0 & 1 & 0 & \cdots & 0 \\ 0 & 0 & 1 & \cdots & 0 \\ \vdots & \vdots & \vdots & \ddots & \vdots \\ 0 & 0 & 0 & \cdots & 1 \\ 1 & 0 & 0 & \cdots & 0 \end{pmatrix}$$

and the diagonal matrix

$$V := \begin{pmatrix} 1 & 0 & 0 & \cdots & 0 \\ 0 & \omega & 0 & \cdots & 0 \\ 0 & 0 & \omega^2 & \cdots & 0 \\ \vdots & \vdots & \vdots & \ddots & \vdots \\ 0 & 0 & 0 & \cdots & \omega^{n-1} \end{pmatrix}.$$

**Problem 12.** Let $U$ be the $n \times n$ unitary matrix

$$U := \begin{pmatrix} 0 & 1 & 0 & \cdots & 0 \\ 0 & 0 & 1 & \cdots & 0 \\ \vdots & \vdots & \vdots & \ddots & \vdots \\ 0 & 0 & 0 & \cdots & 1 \\ 1 & 0 & 0 & \cdots & 0 \end{pmatrix}$$

and $V$ be the $n \times n$ unitary diagonal matrix ($\omega \in \mathbb{C}$)

$$V := \begin{pmatrix} 1 & 0 & 0 & \cdots & 0 \\ 0 & \omega & 0 & \cdots & 0 \\ 0 & 0 & \omega^2 & \cdots & 0 \\ \vdots & \vdots & \vdots & \ddots & \vdots \\ 0 & 0 & 0 & \cdots & \omega^{n-1} \end{pmatrix}$$

where $\omega^n = 1$ ($\omega \neq 1$). Then the set of matrices

$$\{ U^j V^k \ : \ j, k = 0, 1, 2, \ldots, n - 1 \}$$

provide a basis in the Hilbert space for all $n \times n$ matrices with the *scalar product*

$$\langle A, B \rangle := \frac{1}{n} \text{tr}(AB^*)$$

for $n \times n$ matrices $A$ and $B$. Write down the basis for $n = 2$.

**Solution 12.** For $n = 2$ we have the combinations

$$(j, k) \in \{ (0, 0), (0, 1), (1, 0), (1, 1) \}.$$

This yields the orthonormal basis (where $\omega = -1$)

$$I_2 = \begin{pmatrix} 1 & 0 \\ 0 & 1 \end{pmatrix}, \quad \sigma_1 = \begin{pmatrix} 0 & 1 \\ 1 & 0 \end{pmatrix}, \quad \sigma_3 = \begin{pmatrix} 1 & 0 \\ 0 & -1 \end{pmatrix}, \quad -i\sigma_2 = \begin{pmatrix} 0 & -1 \\ 1 & 0 \end{pmatrix}.$$

**Problem 13.**   An $n \times n$ *circulant matrix* $C$ is given by

$$C := \begin{pmatrix} c_0 & c_1 & c_2 & \cdots & c_{n-1} \\ c_{n-1} & c_0 & c_1 & \cdots & c_{n-2} \\ c_{n-2} & c_{n-1} & c_0 & \cdots & c_{n-3} \\ \vdots & \vdots & \vdots & \ddots & \vdots \\ c_1 & c_2 & c_3 & \cdots & c_0 \end{pmatrix}.$$

For example, the permutation matrix

$$P := \begin{pmatrix} 0 & 1 & 0 & \cdots & 0 \\ 0 & 0 & 1 & \cdots & 0 \\ \vdots & \vdots & \vdots & \ddots & \vdots \\ 0 & 0 & 0 & \cdots & 1 \\ 1 & 0 & 0 & \cdots & 0 \end{pmatrix}$$

is a circulant matrix. It is also called the $n \times n$ *primary permutation matrix*.
(i) Let $C$ and $P$ be the matrices given above. Let

$$f(\lambda) = c_0 + c_1 \lambda + \cdots + c_{n-1}\lambda^{n-1}.$$

Show that $C = f(P)$.
(ii) Show that $C$ is a *normal matrix*, that is $C^*C = CC^*$.
(iii) Show that the eigenvalues of $C$ are $f(\omega^k)$, $k = 0, 1, \ldots, n-1$, where $\omega$ is the $n$th primitive root of unity.
(iv) Show that

$$\det(C) = f(\omega^0)f(\omega^1)\cdots f(\omega^{n-1}).$$

(v) Show that $F^*CF$ is a diagonal matrix, where $F$ is the unitary matrix with $(j,k)$-entry equal to

$$\frac{1}{\sqrt{n}}\omega^{(j-1)(k-1)}, \quad j,k = 1, \ldots, n.$$

**Solution 13.**   (i) Direct calculation of

$$f(P) = c_0 I_n + c_1 P + c_2 P^2 + \cdots + c_{n-1}P^{n-1}$$

yields the matrix $C$, where $I_n$ is the $n \times n$ unit matrix. Notice that $P^2$, $P^3$, ..., $P^{n-1}$ are permutation matrices.
(ii) We have $PP^* = P^*P$. If two $n \times n$ matrices $A$ and $B$ commute, then $g(A)$ and $h(B)$ commute, where $g$ and $h$ are polynomials. Thus $C$ is a normal matrix.
(iii) The characteristic polynomial of $P$ is

$$\det(\lambda I_n - P) = \lambda^n - 1 = \prod_{k=0}^{n-1}(\lambda - \omega^k).$$

Thus the eigenvalues of $P$ and $P^j$ are, respectively, $\omega^k$ and $\omega^{jk}$, where $k = 0, 1, \ldots, n-1$. It follows that the eigenvalues of $C = f(P)$ are $f(\omega^k)$, $k = 0, 1, \ldots, n-1$.

(iv) Using the result from (iii) we find

$$\det(C) = \prod_{k=0}^{n-1} f(\omega^k).$$

(v) For each $k = 0, 1, \ldots, n-1$, let

$$\mathbf{x}_k = (1, \omega^k, \omega^{2k}, \ldots, \omega^{(n-1)k})^T$$

where $T$ denotes the transpose. If follows that

$$P\mathbf{x}_k = (\omega^k, \omega^{2k}, \ldots, \omega^{(n-1)k}, 1)^T = \omega^k \mathbf{x}_k$$

and

$$C\mathbf{x}_k = f(P)\mathbf{x}_k = f(\omega^k)\mathbf{x}_k.$$

Thus the vectors $\mathbf{x}_k$ are the eigenvectors of $P$ and $C$ corresponding to the respective eigenvalues $\omega^k$ and $f(\omega^k)$, $k = 0, 1, \ldots, n-1$. Since

$$\langle \mathbf{x}_j, \mathbf{x}_k \rangle \equiv \mathbf{x}_j^* \mathbf{x}_k = \sum_{\ell=0}^{n-1} \overline{\omega^{k\ell}} \omega^{j\ell} = \sum_{\ell=0}^{n-1} \omega^{(j-k)\ell} = \begin{cases} 0 & j \neq k \\ n & j = k \end{cases}$$

we find that

$$\left\{ \frac{1}{\sqrt{n}} \mathbf{x}_0, \ \frac{1}{\sqrt{n}} \mathbf{x}_1, \ \ldots, \ \frac{1}{\sqrt{n}} \mathbf{x}_{n-1} \right\}$$

is an orthonormal basis in the Hilbert space $\mathbb{C}^n$. Thus we obtain the unitary matrix

$$F = \frac{1}{\sqrt{n}} \begin{pmatrix} 1 & 1 & 1 & \cdots & 1 \\ 1 & \omega & \omega^2 & \cdots & \omega^{n-1} \\ 1 & \omega^2 & \omega^4 & \cdots & \omega^{2(n-1)} \\ \vdots & \vdots & \vdots & \ddots & \vdots \\ 1 & \omega^{n-1} & \omega^{2(n-1)} & \cdots & \omega^{(n-1)(n-1)} \end{pmatrix}$$

such that

$$F^* C F = \mathrm{diag}(f(\omega^0), f(\omega^1), \ldots, f(\omega^{n-1})).$$

The matrix $F$ is unitary and is called the *Fourier matrix*.

**Problem 14.** An $n \times n$ matrix $A$ is called a *Hadamard matrix* if each entry of $A$ is 1 or $-1$ and if the rows or columns of $A$ are orthogonal, i.e.

$$AA^T = nI_n \quad \text{or} \quad A^T A = nI_n.$$

Note that $AA^T = nI_n$ and $A^T A = nI_n$ are equivalent. Hadamard matrices $H_n$ of order $2^n$ can be generated recursively by defining

$$H_1 = \begin{pmatrix} 1 & 1 \\ 1 & -1 \end{pmatrix}, \qquad H_n = \begin{pmatrix} H_{n-1} & H_{n-1} \\ H_{n-1} & -H_{n-1} \end{pmatrix}$$

for $n \geq 2$. Show that the eigenvalues of $H_n$ are given by $+2^{n/2}$ and $-2^{n/2}$ each of multiplicity $2^{n-1}$.

**Solution 14.**   We use induction on $n$. The case $n = 1$ is obvious. Now for $n \geq 2$ we have

$$\det(\lambda I - H_n) = \begin{vmatrix} \lambda I - H_{n-1} & -H_{n-1} \\ -H_{n-1} & \lambda I + H_{n-1} \end{vmatrix}$$
$$= \det((\lambda I - H_{n-1})(\lambda I + H_{n-1}) - H_{n-1}^2).$$

Thus

$$\det(\lambda I - H_n) = \det(\lambda^2 I - 2H_{n-1}^2)$$
$$= \det(\lambda I - \sqrt{2}H_{n-1}) \det(\lambda I + \sqrt{2}H_{n-1}).$$

This shows that each eigenvalue $\mu$ of $H_{n-1}$ generates two eigenvalues $\pm\sqrt{2}\mu$ of $H_n$. The assertion then follows by the induction hypothesis, for $H_{n-1}$ has eigenvalues $+2^{(n-1)/2}$ and $-2^{(n-1)/2}$ each of multiplicity $2^{n-2}$.

**Problem 15.**   Let $U$ be an $n \times n$ unitary matrix. Then $U$ can be written as

$$U = V\mathrm{diag}(\lambda_1, \lambda_2, \ldots, \lambda_n)V^*$$

where $\lambda_1, \lambda_2, \ldots, \lambda_n$ are the eigenvalues of $U$ and $V$ is an $n \times n$ unitary matrix. Let

$$U = \begin{pmatrix} 0 & 1 \\ 1 & 0 \end{pmatrix}.$$

Find the decomposition for $U$ given above.

**Solution 15.**   The eigenvalues of $U$ are $+1$ and $-1$. Thus we have

$$U = V\mathrm{diag}(1, -1)V^*$$

with

$$V = \frac{1}{\sqrt{2}} \begin{pmatrix} 1 & 1 \\ 1 & -1 \end{pmatrix}.$$

Therefore $V = V^*$. The columns of $V$ are the eigenvectors of $U$.

**Problem 16.**   An $n \times n$ hermitian matrix $A$ over the complex numbers is called *positive semidefinite* (written as $A \geq 0$), if

$$\mathbf{x}^* A \mathbf{x} \geq 0 \quad \text{for all } \mathbf{x} \in \mathbb{C}^n.$$

Show that for every $A \geq 0$, there exists a unique $B \geq 0$ so that $B^2 = A$.

**Solution 16.**   Let $A = U^* \text{diag}(\lambda_1, \ldots, \lambda_n) U$, where $U$ is unitary. We take

$$B = U^* \text{diag}(\lambda_1^{1/2}, \ldots, \lambda_n^{1/2}) U.$$

Then the matrix $B$ is positive semidefinite and $B^2 = A$ since $U^* U = I_n$. To show the uniqueness, suppose that $C$ is an $n \times n$ positive semidefinite matrix satisfying $C^2 = A$. Since the eigenvalues of $C$ are the nonnegative square roots of the eigenvalues of $A$, we can write

$$C = V \text{diag}(\lambda_1^{1/2}, \ldots, \lambda_n^{1/2}) V^*$$

for some unitary matrix $V$. Then the identity $C^2 = A = B^2$ yields

$$T \text{diag}(\lambda_1, \ldots, \lambda_n) = \text{diag}(\lambda_1, \ldots, \lambda_n) T$$

where $T = UV$. This yields $t_{jk} \lambda_k = \lambda_j t_{jk}$. Thus

$$t_{jk} \lambda_k^{1/2} = \lambda_j^{1/2} t_{jk}.$$

Hence

$$T \text{diag}(\lambda_1^{1/2}, \ldots, \lambda_n^{1/2}) = \text{diag}(\lambda_1^{1/2}, \ldots, \lambda_n^{1/2}) T.$$

Since $T = UV$ it follows that $B = C$.

**Problem 17.**   An $n \times n$ matrix $A$ over the complex numbers is said to be *normal* if it commutes with its conjugate transpose $A^* A = AA^*$. The matrix $A$ can be written

$$A = \sum_{j=1}^{n} \lambda_j E_j$$

where $\lambda_j \in \mathbb{C}$ are the eigenvalues of $A$ and $E_j$ are $n \times n$ matrices satisfying

$$E_j^2 = E_j = E_j^*, \qquad E_j E_k = 0_n \text{ if } j \neq k, \qquad \sum_{j=1}^{n} E_j = I_n.$$

Let $n = 2$. Consider the *NOT gate*

$$A = \begin{pmatrix} 0 & 1 \\ 1 & 0 \end{pmatrix}.$$

Find the decomposition of $A$ given above.

**Solution 17.**    The eigenvalues of $A$ are given by $\lambda_1 = +1$, $\lambda_2 = -1$. The matrices $E_j$ are constructed from the normalized eigenvectors of $A$. The normalized eigenvectors of $A$ are given by

$$\mathbf{v}_1 = \frac{1}{\sqrt{2}} \begin{pmatrix} 1 \\ 1 \end{pmatrix}, \qquad \mathbf{v}_2 = \frac{1}{\sqrt{2}} \begin{pmatrix} 1 \\ -1 \end{pmatrix}.$$

Thus

$$E_1 = \mathbf{v}_1 \mathbf{v}_1^* = \frac{1}{2} \begin{pmatrix} 1 & 1 \\ 1 & 1 \end{pmatrix}, \qquad E_2 = \mathbf{v}_2 \mathbf{v}_2^* = \frac{1}{2} \begin{pmatrix} 1 & -1 \\ -1 & 1 \end{pmatrix}.$$

**Problem 18.**    Let $X$ be an $n \times n$ matrix over $\mathbb{C}$ with $X^2 = I_n$, where $I_n$ is the $n \times n$ identity matrix. Let $z \in \mathbb{C}$.
(i) Show that $e^{zX} = I_n \cosh(z) + X \sinh(z)$.
(ii) Let $z = i\alpha$, where $\alpha \in \mathbb{R}$. Simplify the result from (i).

**Solution 18.**    (i) We have

$$e^{zX} := \sum_{k=0}^{\infty} \frac{(zX)^k}{k!} = \sum_{k=0}^{\infty} \frac{z^k X^k}{k!} = I_n + zX + \frac{1}{2!} z^2 X^2 + \frac{1}{3!} z^3 X^3 + \cdots .$$

Since $X^2 = I_n$ we have

$$e^{zX} = I_n \left( 1 + \frac{1}{2!} z^2 + \frac{1}{4!} z^4 + \cdots \right) + X \left( z + \frac{1}{3!} z^3 + \frac{1}{5!} z^5 + \cdots \right)$$
$$= I_n \cosh(z) + X \sinh(z).$$

(ii) Since $\cosh(i\alpha) = \cos(\alpha)$, $\sinh(i\alpha) = i \sin(\alpha)$ we obtain

$$e^{i\alpha X} = I_n \cos(\alpha) + iX \sin(\alpha).$$

**Problem 19.**    Let $\sigma_1$, $\sigma_2$ be the Pauli spin matrices.
(i) Consider

$$V = \exp(i(\pi/4)\sigma_1), \qquad W = \exp(i(\pi/4)\sigma_2).$$

Show that

$$V = \frac{1}{\sqrt{2}} \begin{pmatrix} 1 & i \\ i & 1 \end{pmatrix}, \qquad W = \frac{1}{\sqrt{2}} \begin{pmatrix} 1 & 1 \\ -1 & 1 \end{pmatrix}.$$

(ii) Let

$$V = \exp(i(\pi/4)\sigma_1), \qquad W = \exp(i(\pi/4)\sigma_2)$$

be two unitary $2 \times 2$ matrices. Calculate the matrices $V^* \sigma_3 V$, $W^* \sigma_3 W$.

**Solution 19.** (i) Using $\sigma_1^2 = I_2$, $\sigma_2^2 = I_2$ we find

$$V = I_n \cos(\pi/4) + i\sigma_1 \sin(\pi/4) = \frac{1}{\sqrt{2}}(I_2 + i\sigma_1) = \frac{1}{\sqrt{2}} \begin{pmatrix} 1 & i \\ i & 1 \end{pmatrix}$$

and

$$W = I_n \cos(\pi/4) + i\sigma_2 \sin(\pi/4) = \frac{1}{\sqrt{2}}(I_2 + i\sigma_2) = \frac{1}{\sqrt{2}} \begin{pmatrix} 1 & 1 \\ -1 & 1 \end{pmatrix}.$$

(ii) We have

$$\begin{aligned} V^* \sigma_3 V &= \left( \frac{1}{\sqrt{2}}I_2 - \frac{i}{\sqrt{2}}\sigma_1 \right) \sigma_3 \left( \frac{1}{\sqrt{2}}I_2 + \frac{i}{\sqrt{2}}\sigma_1 \right) \\ &= \frac{1}{2}\sigma_3 + \frac{i}{2}\sigma_3\sigma_1 - \frac{i}{2}\sigma_1\sigma_3 + \frac{1}{2}\sigma_1\sigma_3\sigma_1 \\ &= -\sigma_2 \end{aligned}$$

and

$$\begin{aligned} W^* \sigma_3 W &= \left( \frac{1}{\sqrt{2}}I_2 - \frac{i}{\sqrt{2}}\sigma_2 \right) \sigma_3 \left( \frac{1}{\sqrt{2}}I_2 + \frac{i}{\sqrt{2}}\sigma_2 \right) \\ &= \frac{1}{2}\sigma_3 + \frac{i}{2}\sigma_3\sigma_2 - \frac{i}{2}\sigma_2\sigma_3 + \frac{1}{2}\sigma_2\sigma_3\sigma_2 \\ &= \sigma_1. \end{aligned}$$

**Problem 20.** Find the matrices

$$e^{(i\pi/4)\sigma_2} \sigma_1 e^{-(i\pi/4)\sigma_2}, \qquad e^{(i\pi/4)\sigma_2} \sigma_3 e^{-(i\pi/4)\sigma_2}.$$

Use the technique of *parameter differentiation*

$$f(\epsilon) = e^{\epsilon\sigma_2} \sigma_1 e^{-\epsilon\sigma_2}$$

with $f(\epsilon = 0) = \sigma_1$.

**Solution 20.** Differentiation of $f$ with respect to $\epsilon$ and using $[\sigma_2, \sigma_1] = -2i\sigma_3$ yields

$$\frac{df}{d\epsilon} = -2ie^{\epsilon\sigma_2} \sigma_3 e^{-\epsilon\sigma_2}$$

with $df(\epsilon = 0)/d\epsilon = -2i\sigma_3$. The second derivative and using $[\sigma_2, \sigma_3] = 2i\sigma_1$ yields

$$\frac{d^2 f}{d\epsilon^2} = 4f(\epsilon).$$

The solution of this second order linear differential equation with constant coefficients is

$$f(\epsilon) = C_1 \cosh(2\epsilon) + C_2 \sinh(2\epsilon).$$

Inserting the initial values provides

$$f(\epsilon) = \cosh(2\epsilon)\sigma_1 - i\sinh(2\epsilon)\sigma_3.$$

Since $\epsilon = i\pi/4$ we have $\cosh(i\pi/2) = \cos(\pi/2) = 0$ and $\sinh(i\pi/2) = i\sin(\pi/2) = i$. Thus $f(\epsilon = i\pi/4) = \sigma_3$ or

$$e^{(i\pi/4)\sigma_2}\sigma_1 e^{-(i\pi/4)\sigma_2} = \sigma_3.$$

Analogously we find $e^{(i\pi/4)\sigma_2}\sigma_3 e^{-(i\pi/4)\sigma_2} = \sigma_1$.

**Problem 21.**   Let

$$V = \exp(i(\pi/4)\sigma_1) \otimes \exp(i(\pi/4)\sigma_1), \quad W = \exp(i(\pi/4)\sigma_2) \otimes \exp(i(\pi/4)\sigma_2).$$

Calculate $V^*(\sigma_3 \otimes \sigma_3)V$ and $W^*(\sigma_3 \otimes \sigma_3)W$.

**Solution 21.**   We have

$$\begin{aligned}
V^*(\sigma_3 \otimes \sigma_3)V &= (\exp(-i(\pi/4)\sigma_1)\sigma_3 \exp(i(\pi/4)\sigma_1)) \\
&\quad \otimes(\exp(-i(\pi/4)\sigma_1)\sigma_3 \exp(i(\pi/4)\sigma_1)) \\
&= \sigma_2 \otimes \sigma_2.
\end{aligned}$$

Analogously $W^*(\sigma_3 \otimes \sigma_3)W = \sigma_1 \otimes \sigma_1$.

**Problem 22.**   Let $X$ be an $n \times n$ matrix over $\mathbb{C}$. Assume that $X^2 = I_n$. Let $Y$ be an arbitrary $n \times n$ matrix over $\mathbb{C}$. Let $z \in \mathbb{C}$.
(i) Calculate $\exp(zX)Y\exp(-zX)$ using the *Baker-Campbell-Hausdorff formula*

$$e^{zX}Ye^{-zX} = Y + z[X,Y] + \frac{z^2}{2!}[X,[X,Y]] + \frac{z^3}{3!}[X,[X,[X,Y]]] + \cdots.$$

(ii) Calculate $\exp(zX)Y\exp(-zX)$ by first calculating $\exp(zX)$ and $\exp(-zX)$ and then doing the matrix multiplication. Compare the two methods.

**Solution 22.**   (i) Using $X^2 = I_n$ we find for the first three commutators

$$\begin{aligned}
[X,[X,Y]] &= [X, XY - YX] = 2(Y - XYX) \\
[X,[X,[X,Y]]] &= 2^2[X,Y] \\
[X,[X,[X,[X,Y]]]] &= 2^3(Y - XYX).
\end{aligned}$$

If the number of $X$'s in the commutator is even (say $m$, $m \geq 2$) we have

$$[X, [X, \ldots [X, Y] \ldots]] = 2^{m-1}(Y - XYX).$$

If the number of $X$'s in the commutator is odd (say $m$, $m \geq 3$) we have

$$[X, [X, \ldots [X, Y] \ldots]] = 2^{m-1}[X, Y].$$

Thus

$$e^{zX} Y e^{-zX} = Y \left(1 + \frac{2^1 z^2}{2!} + \frac{2^3 z^4}{4!} + \cdots\right) + [X, Y] \left(z + \frac{2^2 z^3}{3!} + \cdots\right)$$
$$- XYX \left(\frac{2^1 z^2}{2!} + \frac{2^3 z^4}{4!} + \cdots\right).$$

Consequently

$$e^{zX} Y e^{-zX} = Y \cosh^2(z) + [X, Y] \sinh(z) \cosh(z) - XYX \sinh^2(z).$$

(ii) Using the expansion

$$e^{zX} = \sum_{j=0}^{\infty} \frac{(zX)^j}{j!}$$

and $X^2 = I_n$ we have

$$e^{zX} = I_n \cosh(z) + X \sinh(z), \qquad e^{-zX} = I_n \cosh(z) - X \sinh(z).$$

Matrix multiplication yields

$$e^{zX} Y e^{-zX} = Y \cosh^2(z) + [X, Y] \sinh(z) \cosh(z) - XYX \sinh^2(z).$$

**Problem 23.** The definition of the Lie group $SU(2)$ is

$$SU(2) := \{ A : A \text{ a } 2 \times 2 \text{ complex matrix}, \ \det(A) = 1, \ AA^* = A^*A = I_2 \}.$$

In the name $SU(2)$, the $S$ stands for special and refers to the condition $\det(A) = 1$ and the $U$ stands for unitary and refers to the conditions $AA^* = A^*A = I_2$. Show that $SU(2)$ can also be defined as

$$SU(2) := \{ x_0 I_2 + i\mathbf{x}^T \boldsymbol{\sigma} : (x_0, \mathbf{x})^T \in \mathbb{R}^4, \ x_0^2 + \|\mathbf{x}\|^2 = 1 \}$$

where $\boldsymbol{\sigma} = (\sigma_1, \sigma_2, \sigma_3)^T$ are the Pauli spin matrices. Here $\mathbf{x}^T \boldsymbol{\sigma} := x_1 \sigma_1 + x_2 \sigma_2 + x_3 \sigma_3$.

**Solution 23.** Let $A$ be any $2 \times 2$ complex matrix. Then $A$ can be written as

$$A = a_0 I_2 + i\mathbf{a}^T \boldsymbol{\sigma}$$

with $\mathbf{a} = (a_1, a_2, a_3)^T \in \mathbb{C}^3$. Thus we have

$$AA^* = (a_0 I_2 + i\mathbf{a}^T \boldsymbol{\sigma})(\overline{a_0} I_2 - i\overline{\mathbf{a}}^T \boldsymbol{\sigma})$$
$$= |a_0|^2 I_2 + i\overline{a_0}\mathbf{a}^T \boldsymbol{\sigma} - ia_0\overline{\mathbf{a}}^T \boldsymbol{\sigma} + \mathbf{a}^T \overline{\mathbf{a}} I_2 + i(\mathbf{a} \times \overline{\mathbf{a}})^T \boldsymbol{\sigma}$$
$$= (|a_0|^2 + \|\mathbf{a}\|^2) I_2 + i(\overline{a_0}\mathbf{a} - a_0\overline{\mathbf{a}} + \mathbf{a} \times \overline{\mathbf{a}})^T \boldsymbol{\sigma}$$

where we used that $\sigma_j^* = \sigma_j$ for $j = 1, 2, 3$. It follows that

$$AA^* = I_2 \Leftrightarrow |a_0|^2 + \|\mathbf{a}\|^2 = 1, \quad \overline{a_0}\mathbf{a} - a_0\overline{\mathbf{a}} + \mathbf{a} \times \overline{\mathbf{a}} = \mathbf{0}.$$

First, suppose that $\mathbf{a} \neq \mathbf{0}$. Since $\mathbf{a} \times \overline{\mathbf{a}}$ is orthogonal to both $\mathbf{a}$ and $\overline{\mathbf{a}}$, the equation $\overline{a_0}\mathbf{a} - a_0\overline{\mathbf{a}} + \mathbf{a} \times \overline{\mathbf{a}} = \mathbf{0}$ can only be satisfied if $\mathbf{a} \times \overline{\mathbf{a}} = \mathbf{0}$. That is, only if $\mathbf{a}$ and $\overline{\mathbf{a}}$ are parallel. Since $\mathbf{a}$ and $\overline{\mathbf{a}}$ have the same length, this is the case only if $\overline{\mathbf{a}} = e^{-2i\theta}\mathbf{a}$ for some real number $\theta$. This can be written as

$$\overline{e^{-i\theta}\mathbf{a}} = e^{-i\theta}\mathbf{a}$$

which says that $\mathbf{x} = e^{-i\theta}\mathbf{a}$ is real. Substituting $\mathbf{a} = e^{i\theta}\mathbf{x}$ into $\overline{a_0}\mathbf{a} - a_0\overline{\mathbf{a}} + \mathbf{a} \times \overline{\mathbf{a}} = \mathbf{0}$ gives

$$e^{i\theta}\overline{a_0}\mathbf{x} - e^{-i\theta}a_0\mathbf{x} = \mathbf{0}.$$

This forces $a_0 = e^{i\theta}x_0$ for some real $x_0$. If $\mathbf{a} = \mathbf{0}$, we may still choose $\theta$ so that $a_0 = e^{i\theta}x_0$. We have shown that

$$AA^* = I_2 \Leftrightarrow A = e^{i\theta}(x_0 I_2 + i\mathbf{x}^T \boldsymbol{\sigma})$$

for some $(x_0, \mathbf{x})^T \in \mathbb{R}^4$ with $|x_0|^2 + \|\mathbf{x}\|^2 = 1$ and some $\theta \in \mathbb{R}$. Since

$$\det(A) = \det(e^{i\theta}(x_0 I_2 + i\mathbf{x}^T \cdot \boldsymbol{\sigma}))$$
$$= e^{2i\theta}(x_0^2 + x_1^2 + x_2^2 + x_3^2)$$
$$= e^{2i\theta}$$

we have that $\det(A) = 1$ if and only if $e^{i\theta} = \pm 1$. If $e^{i\theta} = -1$, we can absorb the $-1$ into the vector $(x_0, \mathbf{x})^T$.

**Problem 24.** Let $A$, $B$ be two $n \times n$ matrices over $\mathbb{C}$. We introduce the scalar product

$$\langle A, B \rangle := \frac{\text{tr}(AB^*)}{\text{tr}(I_n)} = \frac{1}{n}\text{tr}(AB^*).$$

The Lie group $SU(N)$ is defined by the complex $n \times n$ matrices $U$

$$SU(N) := \{ U : U^*U = UU^* = I_n, \ \det(U) = 1 \}.$$

The dimension is $N^2 - 1$. The semi-simple Lie algebra $su(N)$ is defined by the $n \times n$ matrices $X$

$$su(N) := \{ X : X^* = -X, \ \text{tr}(X) = 0 \}.$$

(i) Let $U \in SU(N)$. Calculate the scalar product $\langle U, U \rangle$.
(ii) Let $A$ be an arbitrary complex $n \times n$ matrix. Let $U \in SU(N)$. Calculate the scalar product $\langle UA, UA \rangle$.
(iii) Consider the Lie algebra $su(2)$. Provide a basis. The elements of the basis should be orthogonal to each other with respect to the scalar product given above. Calculate the commutator of these matrices.

**Solution 24.** (i) We have

$$\langle U, U \rangle = \frac{1}{n} \text{tr}(UU^*) = 1$$

where we used that $U^* = U^{-1}$.
(ii) We obtain

$$\langle UA, UA \rangle = \frac{1}{n} \text{tr}(UA(UA)^*) = \frac{1}{n} \text{tr}(UAA^*U^*) = \frac{1}{n} \text{tr}(AA^*) = \langle A, A \rangle$$

where we used the *cyclic invariance* of the trace.
(iii) We are looking for three linear independent $2 \times 2$ matrices which are traceless and skew-hermitian matrices. A choice is

$$\tau_1 = \begin{pmatrix} i & 0 \\ 0 & -i \end{pmatrix}, \quad \tau_2 = \begin{pmatrix} 0 & -1 \\ 1 & 0 \end{pmatrix}, \quad \tau_3 = \begin{pmatrix} 0 & i \\ i & 0 \end{pmatrix}.$$

The matrices are also orthogonal to each other using the scalar product given above. For the commutators we find

$$[\tau_1, \tau_2] = -2\tau_3, \quad [\tau_3, \tau_1] = -2\tau_2, \quad [\tau_2, \tau_3] = -2\tau_1.$$

**Problem 25.** Let $H$ be an $n \times n$ matrix which depends on $n$ real parameters $\epsilon_1, \epsilon_2, \ldots, \epsilon_n$, where we assume that we can differentiate $H$ with respect to all $\epsilon$'s. Let $\beta > 0$ and $Z(\beta) := \text{tr}(\exp(-\beta H))$.
(i) Show that

$$\frac{\partial}{\partial \epsilon_j} e^{-\beta H} \equiv - \int_0^\beta d\tau e^{(\tau - \beta)H} \frac{\partial H}{\partial \epsilon_j} e^{-\tau H}. \tag{1}$$

(ii) Show that

$$\frac{\partial}{\partial \epsilon_j} Z = -\beta \text{tr} \left( \frac{\partial}{\partial \epsilon_j} H e^{-\beta H} \right).$$

**Solution 25.** (i) We set

$$f(\beta, \epsilon_1, \ldots, \epsilon_n) := e^{\beta H} \frac{\partial}{\partial \epsilon_j} e^{-\beta H}.$$

It follows that

$$\frac{\partial f}{\partial \beta} = H e^{\beta H} \frac{\partial}{\partial \epsilon_j} e^{-\beta H} + e^{\beta H} \frac{\partial}{\partial \beta} \left( \frac{\partial}{\partial \epsilon_j} e^{-\beta H} \right).$$

Since

$$\frac{\partial}{\partial \beta} \left( \frac{\partial}{\partial \epsilon_j} e^{-\beta H} \right) = \frac{\partial}{\partial \epsilon_j} \left( \frac{\partial}{\partial \beta} e^{-\beta H} \right) = -\frac{\partial}{\partial \epsilon_j} H e^{-\beta H}$$

$$= -\frac{\partial H}{\partial \epsilon_j} e^{-\beta H} - H \frac{\partial}{\partial \epsilon_j} e^{-\beta H}$$

we obtain

$$\frac{\partial f}{\partial \beta} = -e^{\beta H} \frac{\partial H}{\partial \epsilon_j} e^{-\beta H}$$

with the initial value $f(0, \epsilon_1, \dots, \epsilon_n) = 0$. Integrating provides

$$f(\beta, \epsilon_1, \dots, \epsilon_n) = -\int_0^\beta e^{\tau H} \frac{\partial H}{\partial \epsilon_j} e^{-\tau H} d\tau.$$

Multiplying by $\exp(-\beta H)$ yields identity (1).

(ii) We have

$$\frac{\partial Z}{\partial \epsilon_j} = \frac{\partial}{\partial \epsilon_j} \text{tr}(\exp(-\beta H)) = \text{tr}\left( \frac{\partial}{\partial \epsilon_j} \exp(-\beta H) \right)$$

$$= -\text{tr}\left( \int_0^\beta d\tau\, e^{(\tau-\beta)H} \frac{\partial H}{\partial \epsilon_j} e^{-\tau H} \right)$$

$$= -\beta \text{tr}\left( e^{-\beta H} \frac{\partial H}{\partial \epsilon_j} \right)$$

where we used the result from (i) and the cyclic invariance of the trace.

**Problem 26.**   To calculate $\exp(A)$ we can also use the *Cayley-Hamilton theorem* and the *Putzer method*. Using the Cayley-Hamilton theorem we can write

$$f(A) = a_{n-1} A^{n-1} + a_{n-2} A^{n-2} + \cdots + a_2 A^2 + a_1 A + a_0 I_n \qquad (1)$$

where the complex numbers $a_0, a_1, \dots, a_{n-1}$ are determined as follows: Let

$$r(\lambda) := a_{n-1} \lambda^{n-1} + a_{n-2} \lambda^{n-2} + \cdots + a_2 \lambda^2 + a_1 \lambda + a_0$$

which is the right-hand side of (1) with $A^j$ replaced by $\lambda^j$ ($j = 0, 1, \dots, n-1$). For each distinct eigenvalue $\lambda_j$ of the matrix $A$, we consider the equation

$$f(\lambda_j) = r(\lambda_j). \qquad (2)$$

If $\lambda_j$ is an eigenvalue of multiplicity $k$, for $k > 1$, then we consider also the following equations

$$f'(\lambda)|_{\lambda=\lambda_j} = r'(\lambda)|_{\lambda=\lambda_j}, \quad \cdots \quad , f^{(k-1)}(\lambda)\Big|_{\lambda=\lambda_j} = r^{(k-1)}(\lambda)\Big|_{\lambda=\lambda_j}.$$

Any unitary matrix $U$ can be written as $U = \exp(iK)$, where $K$ is hermitian. Apply this method to find $K$ for the *Hadamard gate*

$$U_H = \frac{1}{\sqrt{2}} \begin{pmatrix} 1 & 1 \\ 1 & -1 \end{pmatrix}.$$

**Solution 26.** The hermitian $2 \times 2$ matrix $K$ is given by

$$K = \begin{pmatrix} a & b \\ \overline{b} & c \end{pmatrix}, \qquad a, c \in \mathbb{R}, \quad b \in \mathbb{C}.$$

Then we find the condition on $a$, $b$ and $c$ such that $e^{iK} = U_H$. The eigenvalues of $iK$ are given by

$$\lambda_{1,2} = \frac{i(a+c)}{2} \pm \frac{1}{2}\sqrt{2ac - a^2 - c^2 - 4b\overline{b}}.$$

We set in the following

$$\Delta := \lambda_1 - \lambda_2 = \sqrt{2ac - a^2 - c^2 - 4b\overline{b}}.$$

To apply the method given above we have

$$r(\lambda) = \alpha_1\lambda + \alpha_0 = f(\lambda) = e^{\lambda}.$$

Thus we obtain the two equations $e^{\lambda_1} = \alpha_1\lambda_1 + \alpha_0$, $e^{\lambda_2} = \alpha_1\lambda_2 + \alpha_0$. It follows that

$$\alpha_1 = \frac{e^{\lambda_1} - e^{\lambda_2}}{\lambda_1 - \lambda_2}, \qquad \alpha_0 = \frac{e^{\lambda_2}\lambda_1 - e^{\lambda_1}\lambda_2}{\lambda_1 - \lambda_2}.$$

We have the condition

$$e^{iK} = \alpha_1 iK + \alpha_0 I_2 = \begin{pmatrix} i\alpha_1 a + \alpha_0 & i\alpha_1 b \\ i\alpha_1\overline{b} & i\alpha_1 c + \alpha_0 \end{pmatrix} = \frac{1}{\sqrt{2}} \begin{pmatrix} 1 & 1 \\ 1 & -1 \end{pmatrix}.$$

We obtain the four equations

$$i\alpha_1 a + \alpha_0 = \frac{1}{\sqrt{2}}, \quad i\alpha_1 c + \alpha_0 = -\frac{1}{\sqrt{2}}, \quad i\alpha_1 b = \frac{1}{\sqrt{2}}, \quad i\alpha_1\overline{b} = \frac{1}{\sqrt{2}}.$$

From the last two equations we find that $\overline{b} = b$, i.e. $b$ is real. From the first two equations we find $\alpha_0 = -i\alpha_1(a+c)/2$ and therefore, using the last two equations, $c = a - 2b$. Thus

$$\begin{pmatrix} i\alpha_1 a + \alpha_0 & i\alpha_1 b \\ i\alpha_1\overline{b} & i\alpha_1 c + \alpha_0 \end{pmatrix} = \begin{pmatrix} i\alpha_1 b & i\alpha_1 b \\ i\alpha_1 b & -i\alpha_1 b \end{pmatrix}.$$

From the eigenvalues of $e^{iK}$ we find $e^{\lambda_1} - e^{\lambda_2} = 2$ and

$$\Delta = \sqrt{2ac - a^2 - c^2 - 4b^2} = 2\sqrt{2}ib.$$

Furthermore $\lambda_1 = i(a - b) + \sqrt{2}ib$, $\lambda_2 = i(a - b) - \sqrt{2}ib$. Thus we arrive at the equation

$$e^{i(a-b)+\sqrt{2}ib} - e^{i(a-b)-\sqrt{2}ib} = 2.$$

It follows that $ie^{i(a-b)}\sin(\sqrt{2}b) = 1$ and therefore

$$i\cos(a - b)\sin(\sqrt{2}b) - \sin(a - b)\sin(\sqrt{2}b) = 1$$

with a solution

$$b = \frac{\pi}{2\sqrt{2}}, \qquad a = \frac{\pi}{2}\left(3 + \frac{1}{\sqrt{2}}\right), \qquad c = a - 2b = \frac{\pi}{2}\left(3 - \frac{1}{\sqrt{2}}\right).$$

Then the matrix $K$ is given by

$$K = \frac{\pi}{2}\begin{pmatrix} 3 + 1/\sqrt{2} & 1/\sqrt{2} \\ 1/\sqrt{2} & 3 - 1/\sqrt{2} \end{pmatrix} = \frac{3\pi}{2}\begin{pmatrix} 1 & 0 \\ 0 & 1 \end{pmatrix} + \frac{\pi}{2}\cdot\frac{1}{\sqrt{2}}\begin{pmatrix} 1 & 1 \\ 1 & -1 \end{pmatrix}.$$

We note that the second matrix on the right-hand side is the Hadamard gate again.

**Problem 27.** Let $A$, $B$, $C_2$, ..., $C_m$, ... be $n \times n$ matrices over $\mathbb{C}$. The *Zassenhaus formula* is given by

$$\exp(A + B) = \exp(A)\exp(B)\exp(C_2)\cdots\exp(C_m)\cdots$$

The left-hand side is called the *disentangled form* and the right hand side is called the *undisentangled form*. Find $C_2$, $C_3$, ..., using the *comparison method*. In the comparison method the disentangled and undisentangled form are expanded in terms of an ordering scalar $\alpha$ and matrix coefficients of equal powers of $\alpha$ are compared. From

$$\exp(\alpha(A + B)) = \exp(\alpha A)\exp(\alpha B)\exp(\alpha^2 C_2)\exp(\alpha^3 C_3)\cdots$$

we obtain

$$\sum_{k=0}^{\infty}\frac{\alpha^k}{k!}(A + B)^k = \sum_{r_0,r_1,r_2,r_3,\ldots=0}^{\infty}\frac{\alpha^{r_0+r_1+2r_2+3r_3+\ldots}}{r_0!r_1!r_2!r_3!\cdots}A^{r_0}B^{r_1}C_2^{r_2}C_3^{r_3}\cdots$$

(i) Find the matrices $C_2$ and $C_3$.
(ii) Assume that $[A, [A, B]] = 0_n$ and $[B, [A, B]] = 0_n$. What conclusion can we draw for the Zassenhaus formula?

**Solution 27.** (i) For $\alpha^2$ we have the decompositions $(r_0, r_1, r_2) = (2,0,0)$, $(1,1,0)$, $(0,2,0)$, $(0,0,1)$. Thus we obtain

$$(A+B)^2 = A^2 + 2AB + B^2 + 2C_2.$$

Thus it follows that

$$C_2 = -\frac{1}{2}[A, B].$$

For $\alpha^3$ we obtain

$$(A+B)^3 = A^3 + 3A^2B + 3AB^2 + B^3 + 6AC_2 + 6BC_2 + 6C_3.$$

Using $C_2$ given above we obtain

$$C_3 = \frac{1}{3}[B, [A, B]] + \frac{1}{6}[A, [A, B]].$$

(ii) Since $[B, [A, B]] = 0_n$ and $[A, [A, B]] = 0_n$ we find that $C_3 = C_4 = \cdots = 0$. Thus

$$\exp(\alpha(A+B)) = \exp(\alpha A)\exp(\alpha B)\exp(-\alpha^2[A, B]/2).$$

**Problem 28.** Let $H$ be a hermitian $n \times n$ matrix. Show that $\exp(H)$ is a positive definite matrix.

**Solution 28.** If $H$ is hermitian then $H^2$, $H^3$ etc are hermitian and also $\exp(H)$. Let $\lambda_j$ $(j = 1, 2, \ldots, n)$ be the real eigenvalues of $H$ since $H$ is hermitian. Then $e^{\lambda_j}$ $(j = 1, 2, \ldots, n)$ are the real eigenvalues of $\exp(H)$ and obviously $e^{\lambda_j} > 0$ for $(j = 1, 2, \ldots, n)$. Thus $\exp(H)$ is a positive definite matrix.

**Problem 29.** Let $\sigma_1$, $\sigma_2$, $\sigma_3$ be the Pauli spin matrices. Does the set of $4 \times 4$ matrices $\{I_2 \otimes I_2, \sigma_1 \otimes \sigma_1, -\sigma_2 \otimes \sigma_2, \sigma_3 \otimes \sigma_3\}$ form a *group* under matrix multiplication?

**Solution 29.** We have

$$\sigma_1\sigma_2 = i\sigma_3, \quad \sigma_2\sigma_1 = -i\sigma_3, \quad \sigma_2\sigma_3 = i\sigma_1,$$

$$\sigma_3\sigma_2 = i\sigma_1, \quad \sigma_3\sigma_1 = i\sigma_2, \quad \sigma_1\sigma_3 = -i\sigma_2.$$

Thus

$$(\sigma_1 \otimes \sigma_1)(-\sigma_2 \otimes \sigma_2) = -(\sigma_1\sigma_2) \otimes (\sigma_1\sigma_2) = \sigma_3 \otimes \sigma_3$$
$$(-\sigma_2 \otimes \sigma_2)(\sigma_3 \otimes \sigma_3) = -(\sigma_2\sigma_3) \otimes (\sigma_2\sigma_3) = \sigma_1 \otimes \sigma_1$$
$$(\sigma_3 \otimes \sigma_3)(\sigma_1 \otimes \sigma_1) = (\sigma_3\sigma_1) \otimes (\sigma_3\sigma_1) = -\sigma_2 \otimes \sigma_2.$$

The neutral element is $I_2 \otimes I_2$. Each element is its own inverse. Thus the set forms a group under matrix multiplication.

**Problem 30.** The spin matrices for spin-2 particles (for example *graviton*) are given by

$$
J_1 = \frac{1}{2}
\begin{pmatrix}
0 & 2 & 0 & 0 & 0 \\
2 & 0 & \sqrt{6} & 0 & 0 \\
0 & \sqrt{6} & 0 & \sqrt{6} & 0 \\
0 & 0 & \sqrt{6} & 0 & 2 \\
0 & 0 & 0 & 2 & 0
\end{pmatrix},
$$

$$
J_2 = \frac{i}{2}
\begin{pmatrix}
0 & -2 & 0 & 0 & 0 \\
2 & 0 & -\sqrt{6} & 0 & 0 \\
0 & \sqrt{6} & 0 & -\sqrt{6} & 0 \\
0 & 0 & \sqrt{6} & 0 & -2 \\
0 & 0 & 0 & 2 & 0
\end{pmatrix},
$$

$$
J_3 =
\begin{pmatrix}
2 & 0 & 0 & 0 & 0 \\
0 & 1 & 0 & 0 & 0 \\
0 & 0 & 0 & 0 & 0 \\
0 & 0 & 0 & -1 & 0 \\
0 & 0 & 0 & 0 & -2
\end{pmatrix}.
$$

(i) Show that the matrices are hermitian.
(ii) Find the eigenvalues and eigenvectors of these matrices.
(iii) Calculate the commutation relations.
(iv) Are the matrices unitary?

**Solution 30.** (i) Obviously the matrices are hermitian, i.e.

$$
J_1^* = J_1, \qquad J_2^* = J_2, \qquad J_3^* = J_3.
$$

(ii) The eigenvalues of $J_1$ are $-2, 2, -1, 1, 0$ with the corresponding normalized eigenvectors

$$
\mathbf{u}_1 = \frac{1}{4}
\begin{pmatrix}
1 \\ -2 \\ \sqrt{6} \\ 2 \\ 1
\end{pmatrix},
\quad
\mathbf{u}_2 = \frac{1}{4}
\begin{pmatrix}
1 \\ 2 \\ \sqrt{6} \\ 2 \\ 1
\end{pmatrix},
\quad
\mathbf{u}_3 = \frac{1}{2}
\begin{pmatrix}
1 \\ -1 \\ 0 \\ 1 \\ -1
\end{pmatrix},
$$

$$
\mathbf{u}_4 = \frac{1}{2}
\begin{pmatrix}
1 \\ 1 \\ 0 \\ -1 \\ -1
\end{pmatrix},
\quad
\mathbf{u}_5 = \frac{\sqrt{3}}{\sqrt{8}}
\begin{pmatrix}
1 \\ 0 \\ -\sqrt{2}/\sqrt{3} \\ 0 \\ 1
\end{pmatrix}.
$$

The eigenvectors form an orthonormal basis in $\mathbb{C}^5$. The eigenvalues of $J_2$ are 2, 1, 0, $-1$, $-2$ with the corresponding normalized eigenvectors

$$\mathbf{u}_1 = \frac{1}{4}\begin{pmatrix} 1 \\ -2i \\ 1 \\ -\sqrt{6} \\ 2i \\ 1 \end{pmatrix}, \quad \mathbf{u}_2 = \frac{1}{4}\begin{pmatrix} 1 \\ 2i \\ -\sqrt{6} \\ -2i \\ 1 \\ 1 \end{pmatrix}, \quad \mathbf{u}_3 = \frac{1}{2}\begin{pmatrix} 1 \\ -i \\ 0 \\ -i \\ -1 \end{pmatrix},$$

$$\mathbf{u}_4 = \frac{1}{2}\begin{pmatrix} 1 \\ i \\ 0 \\ i \\ -1 \end{pmatrix}, \quad \mathbf{u}_5 = \frac{\sqrt{3}}{\sqrt{8}}\begin{pmatrix} 1 \\ 0 \\ \sqrt{2}/\sqrt{3} \\ 0 \\ 1 \end{pmatrix}.$$

The eigenvalues of $J_3$ are 2, 1, 0, $-1$, $-2$ with the corresponding eigenvectors (standard basis)

$$\mathbf{u}_1 = \begin{pmatrix} 1 \\ 0 \\ 0 \\ 0 \\ 0 \end{pmatrix}, \quad \mathbf{u}_2 = \begin{pmatrix} 0 \\ 1 \\ 0 \\ 0 \\ 0 \end{pmatrix}, \quad \mathbf{u}_3 = \begin{pmatrix} 0 \\ 0 \\ 1 \\ 0 \\ 0 \end{pmatrix}, \quad \mathbf{u}_4 = \begin{pmatrix} 0 \\ 0 \\ 0 \\ 1 \\ 0 \end{pmatrix}, \quad \mathbf{u}_5 = \begin{pmatrix} 0 \\ 0 \\ 0 \\ 0 \\ 1 \end{pmatrix}.$$

(iii) The commutation relations are $[J_1, J_2] = iJ_3$, $[J_2, J_3] = iJ_1$, $[J_3, J_1] = iJ_2$.

(iv) No the matrices are not unitary. Note that

$$\det(J_1) = \det(J_2) = \det(J_3) = 0$$

owing to the eigenvalue 0.

**Problem 31.** Two orthonormal bases in an $n$-dimensional complex Hilbert space

$$\{\, |\mathbf{u}_j\rangle \,:\, j = 1, 2, \dots, n \,\}, \qquad \{\, |\mathbf{v}_j\rangle \,:\, j = 1, 2, \dots, n \,\}$$

are called *mutually unbiased* if the inner products (scalar products) between all possible pairs of vectors taken from distinct bases have the same magnitude $1/\sqrt{n}$, i.e.

$$|\langle \mathbf{u}_j | \mathbf{v}_k \rangle| = \frac{1}{\sqrt{n}} \qquad \text{for all} \quad j, k \in \{1, 2, \dots, n\}.$$

(i) Find such bases for the Hilbert space $\mathbb{C}^2$.

(ii) Find such bases for the Hilbert space $\mathbb{C}^3$.

**Solution 31.** (i) As the first bases we select the standard basis

$$\left\{ \begin{pmatrix} 1 \\ 0 \end{pmatrix}, \ \begin{pmatrix} 0 \\ 1 \end{pmatrix} \right\}.$$

For the second basis we could select

$$\left\{ \frac{1}{\sqrt{2}} \begin{pmatrix} 1 \\ 1 \end{pmatrix}, \ \frac{1}{\sqrt{2}} \begin{pmatrix} 1 \\ -1 \end{pmatrix} \right\}$$

or

$$\left\{ \frac{1}{\sqrt{2}} \begin{pmatrix} 1 \\ i \end{pmatrix}, \ \frac{1}{\sqrt{2}} \begin{pmatrix} 1 \\ -i \end{pmatrix} \right\}.$$

Applying a unitary matrix to these two sets provide other such sets.
(ii) An example is the standard basis

$$\left\{ \begin{pmatrix} 1 \\ 0 \\ 0 \end{pmatrix}, \ \begin{pmatrix} 0 \\ 1 \\ 0 \end{pmatrix}, \ \begin{pmatrix} 0 \\ 0 \\ 1 \end{pmatrix} \right\}$$

and

$$\left\{ \frac{1}{\sqrt{3}} \begin{pmatrix} 1 \\ 1 \\ 1 \end{pmatrix}, \ \frac{1}{\sqrt{3}} \begin{pmatrix} -1 \\ (1+\sqrt{3})/2 \\ (1-\sqrt{3})/2 \end{pmatrix}, \ \frac{1}{\sqrt{3}} \begin{pmatrix} (-1+i\sqrt{3})/2 \\ (-1-i\sqrt{3})/2 \\ 1 \end{pmatrix} \right\}.$$

Applying a unitary matrix to these two sets provide other such sets.

**Problem 32.** Find a $4 \times 4$ matrix $A$ such that $-A = A^{-1} = A^T = A^*$.

**Solution 32.** Let $0_2$ be the $2 \times 2$ zero matrix. We find

$$A = \begin{pmatrix} 0_2 & -i\sigma_2 \\ -i\sigma_2 & 0_2 \end{pmatrix} \equiv \begin{pmatrix} 0 & 0 & 0 & -1 \\ 0 & 0 & 1 & 0 \\ 0 & -1 & 0 & 0 \\ 1 & 0 & 0 & 0 \end{pmatrix}.$$

This matrix plays a role for the *charge conjugation* in the Dirac equation.

**Problem 33.** Consider the spin matrices for a spin-1 particle

$$S_1 = \frac{1}{\sqrt{2}} \begin{pmatrix} 0 & 1 & 0 \\ 1 & 0 & 1 \\ 0 & 1 & 0 \end{pmatrix}, \quad S_2 = \frac{1}{\sqrt{2}} \begin{pmatrix} 0 & -i & 0 \\ i & 0 & -i \\ 0 & i & 0 \end{pmatrix}, \quad S_3 = \begin{pmatrix} 1 & 0 & 0 \\ 0 & 0 & 0 \\ 0 & 0 & -1 \end{pmatrix}$$

and the unit vector $\mathbf{n} = (\ \sin(\theta)\cos(\phi) \quad \sin(\theta)\sin(\phi) \quad \cos(\theta)\ )$. We define
the scalar product

$$\mathbf{n} \cdot \mathbf{S} := n_1 S_1 + n_2 S_2 + n_3 S_3.$$

Consider the Hamilton operator

$$\hat{H} = \hbar\omega(\mathbf{n} \cdot \mathbf{S}).$$

(i) Calculate the Hamilton operator $\hat{H}$. Is the Hamilton operator $\hat{H}$ hermitian?
(ii) Calculate the trace of $\hat{H}$.
(iii) Find the eigenvalues and normalized eigenvectors of $\hat{H}$.
(iv) Do the eigenvectors form a basis in the Hilbert space $\mathbb{C}^3$?

**Solution 33.**   (i) The Hamilton operator is given by

$$\hat{H} = \hbar\omega \begin{pmatrix} \cos(\theta) & \sin(\theta)e^{-i\phi}/\sqrt{2} & 0 \\ \sin(\theta)e^{i\phi}/\sqrt{2} & 0 & \sin(\theta)e^{-i\phi}/\sqrt{2} \\ 0 & \sin(\theta)e^{i\phi}/\sqrt{2} & -\cos(\theta) \end{pmatrix}.$$

Since $\overline{\exp(i\phi)} = \exp(-i\phi)$ the Hamilton operator is hermitian. Thus the eigenvalues must be real.
(ii) The trace of $\hat{H}$ is 0. Thus the sum of the three eigenvalues of $\hat{H}$ must be 0.
(iii) The eigenvalues of $\hat{H}$ are $\hbar\omega$, $0$, $-\hbar\omega$. The corresponding normalized eigenvectors are

$$\begin{pmatrix} (1+\cos(\theta))e^{-i\phi}/2 \\ \sin(\theta)/\sqrt{2} \\ (1-\cos(\theta))e^{i\phi}/2 \end{pmatrix}, \quad \begin{pmatrix} -\sin(\theta)e^{-i\phi}/\sqrt{2} \\ \cos(\theta) \\ \sin(\theta)e^{i\phi}/\sqrt{2} \end{pmatrix}, \quad \begin{pmatrix} (1-\cos(\theta))e^{-i\phi}/2 \\ -\sin(\theta)/\sqrt{2} \\ (1+\cos(\theta))e^{i\phi}/2 \end{pmatrix}.$$

(iv) The Hamilton operator is hermitian and the three eigenvalues are different. Thus the normalized eigenvectors form an orthonormal basis in the Hilbert space $\mathbb{C}^3$.

**Problem 34.**   Consider a complex Hilbert space $\mathcal{H}$ and $|\phi_1\rangle, |\phi_2\rangle \in \mathcal{H}$. Let $c_1, c_2 \in \mathbb{C}$. An *antilinear operator* $K$ in this Hilbert space $\mathcal{H}$ is characterized by

$$K(c_1|\phi_1\rangle + c_2|\phi_2\rangle) = c_1^* K|\phi_1\rangle + c_2^* K|\phi_2\rangle.$$

A *comb* is an antilinear operator $K$ with zero expectation value for all states $|\psi\rangle$ of a certain complex Hilbert space $\mathcal{H}$. This means

$$\langle\psi|K|\psi\rangle = \langle\psi|LC|\psi\rangle = \langle\psi|L|\psi^*\rangle = 0$$

for all states $|\psi\rangle \in \mathcal{H}$, where $L$ is a linear operator and $C$ is the complex conjugation.
(i) Consider the two-dimensional Hilbert space $\mathcal{H} = \mathbb{C}^2$. Find a unitary $2 \times 2$ matrix such that $\langle\psi|UC|\psi\rangle = 0$.

(ii) Consider the Pauli spin matrices with $\sigma_0 = I_2$, $\sigma_1$, $\sigma_2$, $\sigma_3$. Find

$$\sum_{\mu=0}^{3}\sum_{\nu=0}^{3}\langle\psi|\sigma_\mu C|\psi\rangle g^{\mu,\nu}\langle\psi|\sigma_\nu C|\psi\rangle$$

where $g^{\mu,\nu} = \operatorname{diag}(-1,1,0,1)$.

**Solution 34.**   (i) We find $U = \sigma_2$ since

$$\langle\psi|\sigma_2 C|\psi\rangle = \langle\psi|\sigma_2|\psi^*\rangle = (\,\psi_1^* \quad \psi_2^*\,)\begin{pmatrix} 0 & -i \\ i & 0 \end{pmatrix}\begin{pmatrix} \psi_1^* \\ \psi_2^* \end{pmatrix} = 0.$$

(ii) We have

$$\sum_{\mu=0}^{3}\sum_{\nu=0}^{3}\langle\psi|\sigma_\mu C|\psi\rangle g^{\mu,\nu}\langle\psi|\sigma_\nu C|\psi\rangle = -\langle\psi|\sigma_0|\psi^*\rangle^2 + \langle\psi|\sigma_1|\psi^*\rangle^2 + \langle\psi|\sigma_3|\psi^*\rangle^2$$

$$= 0.$$

**Problem 35.**   Consider the Hilbert space $\mathbb{C}^d$. Let $|j\rangle$ $(j = 1,\ldots,d)$ be an orthonormal basis in $\mathbb{C}^d$. Then a $d \times d$ matrix $A$ acting in $\mathbb{C}^d$ can be written as

$$A = \sum_{j,k=1}^{d} a_{jk}|j\rangle\langle k|$$

with $a_{jk} \in \mathbb{C}$. Obviously $A$ depends on the underlying orthonormal basis. If we have the standard basis, then $A$ reduces to the matrix $A = (a_{jk})$. We can associate a vector $|\psi_A\rangle$ in the Hilbert space $\mathbb{C}^{d^2}$ with the matrix $A$ via

$$|\psi_A\rangle = \sum_{j,k=1}^{d} a_{jk}|j\rangle \otimes |k\rangle.$$

(i) Let $d = 2$ and consider the standard basis

$$|1\rangle = \begin{pmatrix} 1 \\ 0 \end{pmatrix}, \qquad |2\rangle = \begin{pmatrix} 0 \\ 1 \end{pmatrix}.$$

Find $A$ and $|\psi_A\rangle$.

(ii) Let $d = 2$ and consider the Hadamard basis

$$|1\rangle = \frac{1}{\sqrt{2}}\begin{pmatrix} 1 \\ 1 \end{pmatrix}, \qquad |2\rangle = \frac{1}{\sqrt{2}}\begin{pmatrix} 1 \\ -1 \end{pmatrix}.$$

Find $A$ and $|\psi_A\rangle$.

**Solution 35.** (i) We have

$$A = a_{11}|1\rangle\langle 1| + a_{12}|1\rangle\langle 2| + a_{21}|2\rangle\langle 1| + a_{22}|2\rangle\langle 2|$$
$$= \begin{pmatrix} a_{11} & 0 \\ 0 & 0 \end{pmatrix} + \begin{pmatrix} 0 & a_{12} \\ 0 & 0 \end{pmatrix} + \begin{pmatrix} 0 & 0 \\ a_{21} & 0 \end{pmatrix} + \begin{pmatrix} 0 & 0 \\ 0 & a_{22} \end{pmatrix} = \begin{pmatrix} a_{11} & a_{12} \\ a_{21} & a_{22} \end{pmatrix}$$

and

$$|\psi_A\rangle = a_{11}|1\rangle \otimes |1\rangle + a_{12}|1\rangle \otimes |2\rangle + a_{21}|2\rangle \otimes |1\rangle + a_{22}|2\rangle \otimes |2\rangle$$
$$= \begin{pmatrix} a_{11} \\ 0 \\ 0 \\ 0 \end{pmatrix} + \begin{pmatrix} 0 \\ a_{12} \\ 0 \\ 0 \end{pmatrix} + \begin{pmatrix} 0 \\ 0 \\ a_{21} \\ 0 \end{pmatrix} + \begin{pmatrix} 0 \\ 0 \\ 0 \\ a_{22} \end{pmatrix} = \begin{pmatrix} a_{11} \\ a_{12} \\ a_{21} \\ a_{22} \end{pmatrix}.$$

(ii) We have

$$A = a_{11}|1\rangle\langle 1| + a_{12}|1\rangle\langle 2| + a_{21}|2\rangle\langle 1| + a_{22}|2\rangle\langle 2|$$
$$= \frac{1}{2}a_{11}\begin{pmatrix} 1 & 1 \\ 1 & 1 \end{pmatrix} + \frac{1}{2}a_{12}\begin{pmatrix} 1 & -1 \\ 1 & -1 \end{pmatrix} + \frac{1}{2}a_{21}\begin{pmatrix} 1 & 1 \\ -1 & -1 \end{pmatrix} + \frac{1}{2}a_{22}\begin{pmatrix} 1 & -1 \\ -1 & 1 \end{pmatrix}$$
$$= \frac{1}{2}\begin{pmatrix} a_{11} + a_{12} + a_{21} + a_{22} & a_{11} - a_{12} + a_{21} - a_{22} \\ a_{11} + a_{12} - a_{21} - a_{22} & a_{11} - a_{12} - a_{21} + a_{22} \end{pmatrix}$$

and

$$|\psi_A\rangle = a_{11}|1\rangle \otimes |1\rangle + a_{12}|1\rangle \otimes |2\rangle + a_{21}|2\rangle \otimes |1\rangle + a_{22}|2\rangle \otimes |2\rangle$$
$$= \frac{1}{2}a_{11}\begin{pmatrix} 1 \\ 1 \\ 1 \\ 1 \end{pmatrix} + \frac{1}{2}a_{12}\begin{pmatrix} 1 \\ -1 \\ 1 \\ -1 \end{pmatrix} + \frac{1}{2}a_{21}\begin{pmatrix} 1 \\ 1 \\ -1 \\ -1 \end{pmatrix} + \frac{1}{2}a_{22}\begin{pmatrix} 1 \\ -1 \\ -1 \\ 1 \end{pmatrix}$$
$$= \frac{1}{2}\begin{pmatrix} a_{11} + a_{12} + a_{21} + a_{22} \\ a_{11} - a_{12} + a_{21} - a_{22} \\ a_{11} + a_{12} - a_{21} - a_{22} \\ a_{11} - a_{12} - a_{21} + a_{22} \end{pmatrix}.$$

Extend to $d = 3$ and consider the orthonormal basis

$$|1\rangle = \frac{1}{\sqrt{2}}\begin{pmatrix} 1 \\ 0 \\ 1 \end{pmatrix}, \quad |2\rangle = \begin{pmatrix} 0 \\ 1 \\ 0 \end{pmatrix}, \quad |3\rangle = \frac{1}{\sqrt{2}}\begin{pmatrix} 1 \\ 0 \\ -1 \end{pmatrix}.$$

Find $A$ and $|\psi_A\rangle$. Describe the connection of the map $A \mapsto |\psi_A\rangle$ with the vec-operator.

**Problem 36.** Let **s** be a spin with a fixed total angular momentum quantum number

$$s \in \{1/2, 1, 3/2, 2, \ldots\}.$$

The (normalized) eigenstates of $x_3$-angular momentum $|s, m\rangle$ form a ladder with

$$m = -s, -s+1, \ldots, s-1, s.$$

The eigenstates $|s, m\rangle$ form an orthonormal basis in a $2s + 1$ dimensional Hilbert space. For example if $s = 1/2$ we have the two states $|1/2, -1/2\rangle$, $|1/2, 1/2\rangle$ and can identify

$$|1/2, 1/2\rangle \mapsto \begin{pmatrix} 1 \\ 0 \end{pmatrix}, \quad |1/2, -1/2\rangle \mapsto \begin{pmatrix} 0 \\ 1 \end{pmatrix}.$$

Thus we have the Hilbert space $\mathbb{C}^2$. For $s = 1$ we have the three states $|1, -1\rangle, |1, 0\rangle, |1, 1\rangle$ and can identify

$$|1, -1\rangle \mapsto \begin{pmatrix} 0 \\ 0 \\ 1 \end{pmatrix}, \quad |1, 0\rangle \mapsto \begin{pmatrix} 0 \\ 1 \\ 0 \end{pmatrix}, \quad |1, 1\rangle \mapsto \begin{pmatrix} 1 \\ 0 \\ 0 \end{pmatrix}.$$

A *spin coherent state* $|s, \theta, \phi\rangle$ for $s = 1/2, 1, 3/2, \ldots$ can be given by

$$|s, \theta, \phi\rangle = \sum_{m=-s}^{m=s} \sqrt{\frac{(2s)!}{(s+m)!(s-m)!}} (\cos(\theta/2))^{s+m} (\sin(\theta/2))^{s-m} e^{-im\phi} |s, m\rangle.$$

(i) Find $|1/2, \theta, \phi\rangle$ and write it as a vector in $\mathbb{C}^2$.
(ii) Find $|1, \theta, \phi\rangle$ and write it as a vector in $\mathbb{C}^3$.
(iii) For a given $s$ find the scalar product $\langle s, m | s, \theta, \phi \rangle$.

**Solution 36.**  (i) We obtain

$$|1/2, \theta, \phi\rangle = \sin(\theta/2)e^{i\phi/2}|1/2, -1/2\rangle + \cos(\theta/2)e^{-i\phi/2}|1/2, 1/2\rangle.$$

Thus

$$|1/2, \theta, \phi\rangle \mapsto \begin{pmatrix} \cos(\theta/2)e^{-i\phi/2} \\ \sin(\theta/2)e^{i\phi/2} \end{pmatrix}.$$

(ii) For $s = 1$ we obtain

$$\sin^2(\theta/2)e^{i\phi}|1, -1\rangle + \sqrt{2}\cos(\theta/2)\sin(\theta/2)|1, 0\rangle + \cos^2(\theta/2)e^{-i\phi}|1, 1\rangle.$$

Thus we the state in $\mathbb{C}^3$

$$|1, \theta, \phi\rangle \mapsto \begin{pmatrix} \cos^2(\theta/2)e^{-i\phi} \\ \sqrt{2}\cos(\theta/2)\sin(\theta/2) \\ \sin^2(\theta/2)e^{i\phi} \end{pmatrix}.$$

(iii) Since $\langle s, m' | s, m \rangle = \delta_{m', m}$ we obtain

$$\langle s, m | s, \theta, \phi \rangle = \sqrt{\frac{(2s)!}{(s+m)!(s-m)!}} (\cos(\theta/2))^{s+m} (\sin(\theta/2))^{s-m} e^{-im\phi}.$$

**Problem 37.** Let $n \geq 1$ and $\{|0\rangle\rangle, |1\rangle, \ldots, |n\rangle\}$ be an orthonormal basis in $\mathbb{C}^{n+1}$. Consider the linear operators $((n+1) \times (n+1)$ matrices)

$$a_n = \sum_{j=1}^{n} \sqrt{j}|j-1\rangle\langle j|, \qquad a_n^{\dagger} = \sum_{k=1}^{n} \sqrt{k}|k\rangle\langle k-1|.$$

Find the commutator $[a_n, a_n^{\dagger}]$. Note that

$$\sum_{\ell=0}^{n} |\ell\rangle\langle \ell| = I_{n+1}.$$

**Solution 37.** We have

$$a_n a_n^{\dagger} = \sum_{j=1}^{n}\sum_{k=1}^{n} \sqrt{j}\sqrt{k}|j-1\rangle\langle j|k\rangle\langle k-1| = \sum_{k=1}^{n} k|k-1\rangle\langle k-1|$$

$$a_n^{\dagger} a_n = \sum_{k=1}^{n}\sum_{j=1}^{n} \sqrt{k}\sqrt{j}|k\rangle\langle k-1|j-1\rangle\langle j| = \sum_{j=1}^{n} j|j\rangle\langle j|.$$

Thus $[a_n, a_n^{\dagger}] = a_n a_n^{\dagger} - a_n^{\dagger} a_n = I_{n+1} - (n+1)|n\rangle\langle n|$.

**Problem 38.** Let $z \in \mathbb{C}$. Consider the spin-1 matrix

$$S_2 = \frac{1}{\sqrt{2}} \begin{pmatrix} 0 & -i & 0 \\ i & 0 & -i \\ 0 & i & 0 \end{pmatrix}.$$

Calculate $\exp(zS_2)$. Then substitute $z = -i\omega t$.

**Solution 38.** Since $S_2^3 = S_2$, $S_2^4 = S_2^2$ etc we obtain

$$\exp(zS_2) = I_3 + S_2\left(z + \frac{z^3}{3!} + \frac{z^5}{5!} + \cdots\right) + S_2^2\left(\frac{z^2}{2!} + \frac{z^4}{4!} + \cdots\right)$$

$$= I_3 + S_2\sinh(z) + S_2^2(\cosh(z) - 1).$$

With $z = -i\omega t$ we obtain

$$\exp(-i\omega t S_2) = I_3 - i\sin(\omega t)S_2 + (\cos(\omega t) - 1)S_2^2.$$

Since

$$S_2^2 = \frac{1}{2} \begin{pmatrix} 1 & 0 & -1 \\ 0 & 2 & 0 \\ -1 & 0 & 1 \end{pmatrix}$$

we end up with

$$\exp(-i\omega t S_2) = \begin{pmatrix} 1 + (\cos(\omega t) - 1)/2 & -\sin(\omega t)/\sqrt{2} & (-\cos(\omega t) + 1)/2 \\ -\sin(\omega t)/\sqrt{2} & \cos(\omega t) & -\sin(\omega t)/\sqrt{2} \\ (-\cos(\omega t) + 1)/2 & \sin(\omega t)/\sqrt{2} & 1 + (\cos(\omega t) - 1)/2 \end{pmatrix}.$$

**Problem 39.** Consider the Hilbert space $M_2(\mathbb{C})$ of all $2 \times 2$ matrices over $\mathbb{C}$ with scalar product

$$\langle A, B \rangle := \text{tr}(AB^*), \quad A, B \in M_2(\mathbb{C}).$$

The standard basis is

$$E_{11} = \begin{pmatrix} 1 & 0 \\ 0 & 0 \end{pmatrix}, \quad E_{12} = \begin{pmatrix} 0 & 1 \\ 0 & 0 \end{pmatrix}, \quad E_{21} = \begin{pmatrix} 0 & 0 \\ 1 & 0 \end{pmatrix}, \quad E_{22} = \begin{pmatrix} 0 & 0 \\ 0 & 1 \end{pmatrix}.$$

A *mutually unbiased basis* is

$$\mu_0 = \frac{1}{\sqrt{2}}\sigma_0 = \frac{1}{\sqrt{2}}\begin{pmatrix} 1 & 0 \\ 0 & 1 \end{pmatrix}, \quad \mu_1 = \frac{1}{\sqrt{2}}\sigma_1 = \frac{1}{\sqrt{2}}\begin{pmatrix} 0 & 1 \\ 1 & 0 \end{pmatrix},$$

$$\mu_2 = \frac{1}{\sqrt{2}}\sigma_2 = \frac{1}{\sqrt{2}}\begin{pmatrix} 0 & -i \\ i & 0 \end{pmatrix}, \quad \mu_3 = \frac{1}{\sqrt{2}}\sigma_3 = \frac{1}{\sqrt{2}}\begin{pmatrix} 1 & 0 \\ 0 & -1 \end{pmatrix}.$$

(i) Express the Hadamard matrix

$$A = \frac{1}{\sqrt{2}}\begin{pmatrix} 1 & 1 \\ 1 & -1 \end{pmatrix}$$

with this mutually unbiased basis.
(ii) Express the Bell matrix

$$B = \frac{1}{\sqrt{2}}\begin{pmatrix} 1 & 0 & 0 & 1 \\ 0 & 1 & 1 & 0 \\ 0 & 1 & -1 & 0 \\ 1 & 0 & 0 & -1 \end{pmatrix}$$

with the basis (sixteen dimensional) given by $\mu_j \otimes \mu_k$, $(j, k = 0, 1, 2, 3)$.

**Solution 39.**    (i) We have the expansion

$$A = \sum_{j=0}^{3} \langle A, \mu_j \rangle \mu_j$$

with

$$\langle A, \mu_0 \rangle = 0, \quad \langle A, \mu_1 \rangle = 1, \quad \langle A, \mu_2 \rangle = 0, \quad \langle A, \mu_3 \rangle = 1.$$

Hence $A = \mu_1 + \mu_3$.

(ii) We have the expansion

$$B = \sum_{j=0}^{3} \sum_{k=0}^{3} \langle B, \mu_j \otimes \mu_k \rangle (\mu_j \otimes \mu_k).$$

The only nonzero expansion coefficients are

$$\langle B, \mu_3 \otimes \mu_0 \rangle = \sqrt{2}, \quad \langle B, \mu_1 \otimes \mu_1 \rangle = \sqrt{2}.$$

Hence

$$B = \sqrt{2}\mu_3 \otimes \mu_0 + \sqrt{2}\mu_1 \otimes \mu_1.$$

**Problem 40.** Two orthonormal bases in an $n$-dimensional complex Hilbert space

$$\{ |\mathbf{u}_j\rangle : j = 1, 2, \ldots, n \}, \qquad \{ |\mathbf{v}_j\rangle : j = 1, 2, \ldots, n \}$$

are called *mutually unbiased* if inner products (scalar products) between all possible pairs of vectors taken from distinct bases have the same magnitude $1/\sqrt{n}$, i.e.

$$|\langle \mathbf{u}_j | \mathbf{v}_k \rangle| = \frac{1}{\sqrt{n}} \qquad \text{for all} \quad j, k \in \{1, 2, \ldots, n\}.$$

(i) Find such bases for the Hilbert space $\mathbb{C}^2$. Start of with the standard basis

$$\mathbf{u}_1 = \begin{pmatrix} 1 \\ 0 \end{pmatrix}, \qquad \mathbf{u}_2 = \begin{pmatrix} 0 \\ 1 \end{pmatrix}.$$

(ii) Find such bases for the Hilbert space $\mathbb{C}^3$. Start of with the standard basis

$$\mathbf{u}_1 = \begin{pmatrix} 1 \\ 0 \\ 0 \end{pmatrix}, \qquad \mathbf{u}_2 = \begin{pmatrix} 0 \\ 1 \\ 0 \end{pmatrix}, \qquad \mathbf{u}_3 = \begin{pmatrix} 0 \\ 0 \\ 1 \end{pmatrix}.$$

(iii) Find such bases for the Hilbert space $\mathbb{C}^4$ using the result from $\mathbb{C}^2$ and the Kronecker product.

**Solution 40.** (i) For the second basis we could select

$$\mathbf{v}_1 = \frac{1}{\sqrt{2}} \begin{pmatrix} 1 \\ 1 \end{pmatrix}, \qquad \mathbf{v}_2 = \frac{1}{\sqrt{2}} \begin{pmatrix} 1 \\ -1 \end{pmatrix}.$$

Another selection would be

$$\mathbf{v}_1 = \frac{1}{\sqrt{2}} \begin{pmatrix} 1 \\ i \end{pmatrix}, \qquad \mathbf{v}_2 = \frac{1}{\sqrt{2}} \begin{pmatrix} 1 \\ -i \end{pmatrix}.$$

(ii) A possible solution is

$$\mathbf{v_1} = \frac{1}{\sqrt{3}}\begin{pmatrix} 1 \\ 1 \\ 1 \end{pmatrix}, \quad \mathbf{v_2} = \begin{pmatrix} 1/\sqrt{3} \\ -i/2 - 1/(2\sqrt{3}) \\ i/2 - 1/(2\sqrt{3}) \end{pmatrix}, \quad \mathbf{v_3} = \begin{pmatrix} 1/\sqrt{3} \\ i/2 - 1/(2\sqrt{3}) \\ -i/2 - 1/(2\sqrt{3}) \end{pmatrix}.$$

**Problem 41.**    (i) Let $A$, $B$ be $n \times n$ matrices over $\mathbb{C}$ such that $A^2 = I_n$ and $B^2 = I_n$. Furthermore assume that

$$[A, B]_+ \equiv AB + BA = 0_n$$

i.e. the anticommutator vanishes. Let $\alpha, \beta \in \mathbb{C}$. Calculate $e^{\alpha A + \beta B}$ using

$$e^{\alpha A + \beta B} = \sum_{j=0}^{\infty} \frac{(\alpha A + \beta B)^j}{j!}.$$

(ii) Consider the case that $n = 2$ and

$$\alpha = -i\omega t, \quad A = \sigma_3 = \begin{pmatrix} 1 & 0 \\ 0 & -1 \end{pmatrix}$$

$$\beta = -i\Delta t/\hbar, \quad B = \sigma_1 = \begin{pmatrix} 0 & 1 \\ 1 & 0 \end{pmatrix}.$$

(iii) Consider the case that $n = 8$ and

$$\alpha = -i\omega t, \quad A = \sigma_3 \otimes \sigma_3 \otimes \sigma_3$$

$$\beta = -i\Delta t/\hbar, \quad B = \sigma_1 \otimes \sigma_1 \otimes \sigma_1.$$

**Solution 41.**    (i) Since $BA = -AB$ we have

$$(\alpha A + \beta B)^2 = (\alpha^2 + \beta)^2 I_n, \quad (\alpha A + \beta B)^3 = (\alpha^2 + \beta^2)(\alpha A + \beta B).$$

Thus in general we have for positive $n$

$$(\alpha A + \beta B)^n = (\alpha^2 + \beta^2)^{n/2} I_n \quad \text{for} \quad n \text{ even}$$

and

$$(\alpha A + \beta B)^n = (\alpha^2 + \beta^2)^{n/2-1} \quad \text{for} \quad n \text{ odd}$$

Thus we have the expansion

$$e^{\alpha A + \beta B} = I_n (1 + \frac{1}{2!}(\alpha^2 + \beta^2) + \frac{1}{4!}(\alpha^2 + \beta^2)^2 + \frac{1}{6!}(\alpha^2 + \beta^2)^3 + \cdots)$$

$$+ (\alpha A + \beta B)(1 + \frac{1}{3!}(\alpha^2 + \beta^2) + \frac{1}{5!}(\alpha^2 + \beta^2)^2 + \cdots).$$

This can be summed up to

$$e^{\alpha A + \beta B} = I_n \cosh(\sqrt{\alpha^2 + \beta^2}) + \frac{\alpha A + \beta B}{\sqrt{\alpha^2 + \beta^2}} \sinh(\sqrt{\alpha^2 + \beta^2}).$$

(ii) We have

$$\sqrt{\alpha^2 + \beta^2} = \sqrt{-\omega^2 t^2 - \Delta^2 t^2/\hbar^2} = \frac{it}{\hbar}\sqrt{\hbar^2\omega^2 + \Delta^2}.$$

We set $E := \sqrt{\hbar^2\omega^2 + \Delta^2}$. It follows that

$$\frac{\alpha A + \beta B}{\alpha^2 + \beta^2} = -\frac{\hbar\omega A + \Delta B}{E} = \begin{pmatrix} -\hbar\omega/E & -\Delta/E \\ -\Delta/E & \hbar\omega \end{pmatrix}.$$

Thus

$$\frac{\alpha A + \beta B}{\sqrt{\alpha^2 + \beta^2}} \sinh(\sqrt{\alpha^2 + \beta^2}) = \begin{pmatrix} -i\sin(Et/\hbar)\hbar\omega & -i\sin(Et/\hbar)\Delta/E \\ -i\sin(Et/\hbar)\Delta/E & i\sin(Et/\hbar)\hbar\omega \end{pmatrix}$$

and

$$e^{\alpha A + \beta B} = \begin{pmatrix} \cos(Et/\hbar) - i\sin(Et/\hbar)\hbar\omega/E & -i\sin(Et/\hbar)\Delta/E \\ -i\sin(Et/\hbar)\Delta/E & \cos(Et/\hbar) + i\sin(Et/\hbar)\hbar\omega/E \end{pmatrix}.$$

**Problem 42.** Let $H$ be an $n \times n$ hermitian matrix and $\lambda_1, \ldots, \lambda_n$ be the eigenvalues with the pairwise orthogonal normalized eigenvectors $\mathbf{v}_1, \ldots,$ $\mathbf{v}_n$. Then we can write

$$H = \sum_{\ell=1}^{n} \lambda_\ell \mathbf{v}_\ell \mathbf{v}_\ell^*.$$

Let

$$P = I_n - \mathbf{v}_j \mathbf{v}_j^* - \mathbf{v}_k \mathbf{v}_k^* + \mathbf{v}_j \mathbf{v}_k^* + \mathbf{v}_k \mathbf{v}_j^*, \qquad j \neq k.$$

(i) What is the condition on the eigenvalues of $H$ such that $PHP^* = H$.
(ii) Find $P^2$.

**Solution 42.**   (i) Note that $P$ is hermitian. Utilizing $\mathbf{v}_\ell^* \mathbf{v}_j = \delta_{\ell j}$ we find by straightforward calculation

$$PHP^* = \sum_{\ell=1}^{n} \lambda_\ell \mathbf{v}_\ell \mathbf{v}_\ell^* + (\lambda_k - \lambda_j)\mathbf{v}_j \mathbf{v}_j^* + (\lambda_j - \lambda_k)\mathbf{v}_k \mathbf{v}_k^*.$$

Thus $\lambda_j = \lambda_k$.
(ii) We obtain $P^2 = I_n$.

**Problem 43.** Consider the orthogonal group $O(n, \mathbb{R}) \subset \mathbb{R}^{n \times n}$ with the linear product ($v, w \in O(n, \mathbb{R})$)

$$\langle v, w \rangle := \text{tr}(v^T w).$$

The orthogonal projection $\Pi(g) : \mathbb{R}^{n \times n} \to T_g O(n)$ is given by

$$\Pi(g)v := \frac{1}{2}(v - gv^T g).$$

(i) Let $n = 2$ and

$$g = \begin{pmatrix} 0 & 1 \\ 1 & 0 \end{pmatrix}, \quad v = \begin{pmatrix} a & b \\ c & d \end{pmatrix}.$$

Find the orthogonal projection.

(ii) Let $n = 2$ and

$$g = \begin{pmatrix} 0 & 1 \\ 1 & 0 \end{pmatrix}, \quad v = \begin{pmatrix} a & b \\ c & d \end{pmatrix}.$$

Find the orthogonal projection.

**Solution 43.** (i) We have

$$\Pi(g)v = \frac{1}{2}\left( \begin{pmatrix} a & b \\ c & d \end{pmatrix} - \begin{pmatrix} 0 & 1 \\ 1 & 0 \end{pmatrix} \begin{pmatrix} a & c \\ b & d \end{pmatrix} \begin{pmatrix} 0 & 1 \\ 1 & 0 \end{pmatrix} \right) = \frac{1}{2}\begin{pmatrix} a - d & 0 \\ 0 & d - a \end{pmatrix}.$$

(ii) We have

$$\Pi(g)v = \frac{1}{2}\left( \begin{pmatrix} a & b \\ c & d \end{pmatrix} - \begin{pmatrix} 0 & 1 \\ -1 & 0 \end{pmatrix} \begin{pmatrix} a & c \\ b & d \end{pmatrix} \begin{pmatrix} 0 & 1 \\ -1 & 0 \end{pmatrix} \right)$$

$$= \frac{1}{2}\begin{pmatrix} a + d & 0 \\ 0 & a + d \end{pmatrix}.$$

Note that

$$\text{tr}\left( \begin{pmatrix} a - d & 0 \\ 0 & d - a \end{pmatrix} \begin{pmatrix} a + d & 0 \\ 0 & a + d \end{pmatrix} \right) = 0.$$

Consider the case that

$$g = \frac{1}{\sqrt{2}}\begin{pmatrix} 1 & 1 \\ 1 & -1 \end{pmatrix}.$$

**Problem 44.** Let $n \geq 1$ and $m \geq 1$. Consider the $T = (t_{j_1 \dots j_m})$ order-$m$ tensor of size $(n \times \cdots \times n)$ ($m$-times), ($j_1, \dots, j_m = 1, \dots, n$). One defines the operator on $\mathbf{v} \in \mathbb{C}^n$ written as

$$(T\mathbf{v}^{m-1})_k := \sum_{j_2=1}^{n} \cdots \sum_{j_m=1}^{n} t_{kj_2 \dots j_m} v_{j_2} \cdots v_{j_m}, \quad k = 1, \dots, n.$$

The $(E-)$ eigenvector of $T$ are the fixed points (up to scaling) of this operator

$$T\mathbf{v}^{m-1} = \lambda\mathbf{v} \quad \text{where} \quad \mathbf{v} \neq \mathbf{0}.$$

Let $m = 3$, $n = 2$ with $t_{122} = 1$, $t_{211} = 1$ and all other entries are 0. Solve the eigenvalue problem.

**Solution 44.** We obtain

$$(T\mathbf{v}^2)_1 = a_{122}v_2^2 = v_2^2 = \lambda v_1, \quad (T\mathbf{v}^2)_2 = a_{211}v_1^2 = v_1^2 = \lambda v_2$$

Now $\lambda = 0$ is not a solution, since $v_1^2 = v_2^2 = 0$ implies $v_1 = v_2 = 0$. If $\lambda \neq 0$, then $v_1 \neq 0$ and $v_2 \neq 0$. We also have $v_1^3 = v_2^3$.

**Problem 45.** Starting from *Maxwell's equations* in vacuum

$$\text{curl}(\mathbf{B}) = \frac{1}{c^2}\frac{\partial \mathbf{E}}{\partial t}, \quad \text{curl}(\mathbf{E}) = -\frac{\partial \mathbf{B}}{\partial t}, \quad \text{div}(\mathbf{E}) = 0, \quad \text{div}(\mathbf{B}) = 0$$

and *Kramer's vector* $\mathbf{F} := \mathbf{E} + ic\mathbf{B}$, $\mathbf{F}^* := \mathbf{E} - ic\mathbf{B}$ show that the *photon* is a spin-1 particle.

**Solution 45.** Using Kramer's vector we can write

$$\text{curl}(\mathbf{F}) = \frac{i}{c}\frac{\partial \mathbf{F}}{\partial t}, \quad \text{curl}(\mathbf{F}^*) = -\frac{i}{c}\frac{\partial \mathbf{F}^*}{\partial t}, \quad \text{div}(\mathbf{F}) = 0, \quad \text{div}(\mathbf{F}^*) = 0.$$

Let

$$\epsilon_{jk\ell} = \begin{cases} +1 & \text{if } jk\ell \text{ are an even permutation of the integers 123} \\ -1 & \text{if } jk\ell \text{ are an odd permutation of the integers 123} \\ 0 & \text{otherwise} \end{cases}.$$

Since

$$(\text{curl}\mathbf{F})_j = \sum_{k=1}^{3}\sum_{\ell=1}^{3} \epsilon_{jk\ell}\frac{\partial}{\partial x_k}F_\ell = -\sum_{k=1}^{3}\sum_{\ell=1}^{3}\frac{\partial}{\partial x_k}\epsilon_{kj\ell}F_\ell$$

we can write

$$\sum_{k=1}^{3}\sum_{\ell=1}^{3}\left(-i\frac{\partial}{\partial x_k}\right)(-i\epsilon_{kj\ell})F_\ell = \frac{i}{c}\frac{\partial F_j}{\partial t}.$$

Introducing the differential operator

$$\hat{p}_k := -i\frac{\partial}{\partial x_k}$$

we find

$$-\sum_{j=1}^{3}\sum_{k=1}^{3}\hat{p}_k i\epsilon_{kj\ell}F_\ell = \frac{i}{c}\frac{\partial F_\ell}{\partial t}.$$

For fixed $k$, $-i\epsilon_{kj\ell}$ is a $3 \times 3$ matrix, $S_{k(j,\ell)}$. The equation for **F** then takes the form

$$\left(\sum_{k=1}^{3} \hat{p}_k S_k\right) \mathbf{F} = (\hat{\mathbf{p}} \cdot \mathbf{S})\mathbf{F} = \frac{i}{c}\frac{\partial \mathbf{F}}{\partial t}.$$

Using the definition of $\epsilon_{jk\ell}$, we obtain the representation for the $3 \times 3$ matrices

$$S_1 = i\begin{pmatrix} 0 & 0 & 0 \\ 0 & 0 & -1 \\ 0 & 1 & 0 \end{pmatrix}, \quad S_2 = i\begin{pmatrix} 0 & 0 & 1 \\ 0 & 0 & 0 \\ -1 & 0 & 0 \end{pmatrix}, \quad S_3 = i\begin{pmatrix} 0 & -1 & 0 \\ 1 & 0 & 0 \\ 0 & 0 & 0 \end{pmatrix}.$$

The commutators are $[S_1, S_2] = iS_3$, $[S_2, S_3] = iS_1$, $[S_3, S_1] = iS_2$. We have $\mathbf{S} \times \mathbf{S} = i\mathbf{S}$ and

$$S_1^2 + S_2^2 + S_3^2 = 2I_3$$

where $I_3$ is the $3 \times 3$ identity matrix. Thus Maxwell's equations describe a particle of spin-1.

## Programming Problems

**Problem 1.**   Let $\sigma_1$, $\sigma_2$, $\sigma_3$ be the Pauli spin matrices. Find the eigenvalues and eigenvectors of the $8 \times 8$ hermitian matrix

$$H = (\sigma_1 \otimes \sigma_1 + \sigma_2 \otimes \sigma_2 + \sigma_3 \otimes \sigma_3) \otimes \sigma_1.$$

The $8 \times 8$ hermitian matrix can be written as a *direct sum*

$$H = \begin{pmatrix} 0 & 1 \\ 1 & 0 \end{pmatrix} \oplus \begin{pmatrix} 0 & -1 & 0 & 2 \\ -1 & 0 & 2 & 0 \\ 0 & 2 & 0 & -1 \\ 2 & 0 & -1 & 0 \end{pmatrix} \oplus \begin{pmatrix} 0 & 1 \\ 1 & 0 \end{pmatrix}.$$

**Solution 1.**   Thus the eigenvalues can be calculated from the two $2 \times 2$ matrices and the $4 \times 4$ matrix. Applying the Maxima program

```
/* directsum.mac */
sig1: matrix([0,1],[1,0]);
sig2: matrix([0,-%i],[%i,0]);
sig3: matrix([1,0],[0,-1]);
sig11: kronecker_product(sig1,sig1);
sig22: kronecker_product(sig2,sig2);
sig33: kronecker_product(sig3,sig3);
S: sig11 + sig22 + sig33;
H: kronecker_product(S,sig1);
```

```
eigenvectors(H);
D: matrix([0,-1,0,2],[-1,0,2,0],[0,2,0,-1],[2,0,-1,0]);
eigenvectors(D);
```

we find the eigenvalues $-1$ (3 times), $+1$ (3 times), $+3$ (1 times), $-3$ (1 times) of the matrix $H$. The eigenvalues of the $4 \times 4$ matrix are $-1, 1, -3, 3$ and the eigenvectors are

$$\begin{pmatrix} 1 \\ -1 \\ 1 \\ -1 \end{pmatrix}, \quad \begin{pmatrix} 1 \\ 1 \\ 1 \\ 1 \end{pmatrix}, \quad \begin{pmatrix} 1 \\ 1 \\ -1 \\ -1 \end{pmatrix}, \quad \begin{pmatrix} 1 \\ -1 \\ -1 \\ 1 \end{pmatrix}.$$

**Problem 2.** Let $I_n$ be the $n \times n$ identity matrix. An invertible matrix $X \in \mathbb{C}^{n^2 \times n^2}$ satisfies the *Yang-Baxter equation* if

$$(X \otimes I_n)(I_n \otimes X)(X \otimes I_n) = (I_n \otimes X)(X \otimes I_n)(I_n \otimes X).$$

If $X$ satisfies the Yang-Baxter equation, then $X^*$ satisfies the Yang-Baxter equation. If $X$ satisfies the Yang-Baxter equation, then $X^{-1}$ satisfies the Yang-Baxter equation. If $X$ satisfies the Yang-Baxter equation and $Q \in \mathbb{C}^{n \times n}$ is an arbitrary invertible matrix. Then

$$\tilde{X} = (Q \otimes Q)X(Q \otimes Q)^{-1}$$

also satisfies the Yang-Baxter equation. Show that

$$X = \frac{(1+i)}{2} \begin{pmatrix} 1 & 0 & 0 & 1 \\ 0 & 1 & 1 & 0 \\ 0 & -1 & 1 & 0 \\ -1 & 0 & 0 & 1 \end{pmatrix}$$

satisfies the Yang-Baxter equation with $n = 2$.

**Solution 2.** The following Maxima program provides the proof

```
/* YBBell.mac */
I2: matrix([1,0],[0,1]);
X: ((1+%i)/2)*matrix([1,0,0,1],[0,1,1,0],[0,-1,1,0],[-1,0,0,1]);
T1: kronecker_product(X,I2);
T2: kronecker_product(I2,X);
F: (T1 . T2 . T1) - (T2 . T1 . T2);
```

## 3.3    Supplementary Problems

**Problem 1.**    Let $\sigma_1$, $\sigma_2$, $\sigma_3$ be the Pauli spin matrices. Consider the sixteen $4 \times 4$ matrices

$$I_{16}, \quad \Gamma_1 = \begin{pmatrix} \sigma_1 & 0_2 \\ 0_2 & \sigma_1 \end{pmatrix}, \quad \Gamma_2 = \begin{pmatrix} \sigma_2 & 0_2 \\ 0_2 & \sigma_2 \end{pmatrix},$$

$$\Gamma_3 = \begin{pmatrix} 0_2 & \sigma_3 \\ \sigma_3 & 0_2 \end{pmatrix}, \quad \Gamma_4 = \begin{pmatrix} 0_2 & -i\sigma_3 \\ i\sigma_3 & 0_2 \end{pmatrix}, \quad \Gamma_5 = \begin{pmatrix} -\sigma_3 & 0_2 \\ 0_2 & \sigma_3 \end{pmatrix},$$

$$\Gamma_{[\mu,\nu]} = \frac{1}{2}i(\Gamma_\mu\Gamma_\nu - \Gamma_\nu\Gamma_\mu), \quad \mu, \nu = 1, 2, 3, 4, 5.$$

Show that

$$\Gamma_\nu\Gamma_\mu + \Gamma_\mu\Gamma_\nu = 2\delta_{\nu\mu}I_4$$

$$[\Gamma_\nu, \Gamma_{[\lambda,\mu]}] = 2i\delta_{\nu\lambda}\Gamma_\mu - 2i\delta_{\nu\mu}\Gamma_\lambda$$

$$[\Gamma_{[\nu,\mu]}, \Gamma_{[\lambda,\sigma]}] = 2i\delta_{\mu\lambda}\Gamma_{[\nu,\sigma]} - 2i\delta_{\nu\lambda}\Gamma_{[\mu,\sigma]} + 2i\delta_{\nu\sigma}\Gamma_{[\nu,\lambda]} - 2i\delta_{\mu\sigma}\Gamma_{[\nu,\lambda]}.$$

Do the 16 matrices form an orthonormal basis in the Hilbert space of the $4 \times 4$ matrices?

**Problem 2.**    Consider the $2^5 \times 2^5$ matrices

$$A = I_2 \otimes I_2 \otimes \sigma_1 \otimes \sigma_3 \otimes \sigma_3$$
$$B = I_2 \otimes I_2 \otimes \sigma_2 \otimes \sigma_3 \otimes \sigma_3$$
$$C = \sigma_3 \otimes \sigma_3 \otimes \sigma_3 \otimes \sigma_3 \otimes \sigma_3.$$

Find $A^2$, $B^2$, $C^2$, $[A, B]$, $[B, C]$, $[C, A]$, $[A, B]_+$, $[B, C]_+$, $[C, A]_+$.

**Problem 3.**    Let $A$ be an $n \times n$ matrix over $\mathbb{C}$ and $f : \mathbb{C} \to \mathbb{C}$ be analytic in a region $D$ containing the spectrum of $A$. Then the matrix $f(A)$ can be defined as the *Cauchy integral formula*

$$f(A) = \frac{1}{2\pi i} \int_{\partial D} (zI_n - A)^{-1}f(z)dz.$$

Let

$$A = \begin{pmatrix} 0 & 0 & 1 \\ 0 & 1 & 0 \\ 1 & 0 & 0 \end{pmatrix}$$

with the spectrum $+1$ (twice) and $-1$. Find $\exp(A)$ applying the Cauchy integral formula.

**Problem 4.** Let $\alpha > 0$ and $A$ be an $n \times n$ matrix over $\mathbb{C}$. Show that

$$e^{-\alpha A} = \frac{i}{2\pi} \int_C e^{-\alpha\lambda}(A - \lambda I_n)^{-1} d\lambda$$

where $C$ is the contour in the complex $\lambda$ plane which encloses all eigenvalues of the matrix $A$. Let

$$A = \begin{pmatrix} 0 & 1 \\ 0 & 0 \end{pmatrix}.$$

Calculate the right-hand side.

**Problem 5.** Show that the vector is normalized in $\mathbb{R}^4$

$$\begin{pmatrix} \cos(\theta_1) \\ \sin(\theta_1)\cos(\theta_2) \\ \sin(\theta_1)\sin(\theta_2)\cos(\theta_3) \\ \sin(\theta_1)\sin(\theta_2)\sin(\theta_3) \end{pmatrix}.$$

**Problem 6.** Show that the equation of a *hyperplane* passing through the points $x_1, x_2, \ldots, x_n$ in $\mathbb{R}^n$ can be given in the form

$$\det \begin{pmatrix} 1 & 1 & 1 & \cdots & 1 \\ x & x_1 & x_2 & \cdots & x_n \end{pmatrix} = 0.$$

Apply it to $n = 4$ with (Bell basis)

$$x_1 = \frac{1}{\sqrt{2}}\begin{pmatrix} 1 \\ 0 \\ 0 \\ 1 \end{pmatrix}, \quad x_2 = \frac{1}{\sqrt{2}}\begin{pmatrix} 1 \\ 0 \\ 0 \\ -1 \end{pmatrix}, \quad x_3 = \frac{1}{\sqrt{2}}\begin{pmatrix} 0 \\ 1 \\ 1 \\ 0 \end{pmatrix}, \quad x_4 = \frac{1}{\sqrt{2}}\begin{pmatrix} 0 \\ 1 \\ -1 \\ 0 \end{pmatrix}.$$

**Problem 7.** Let $\sigma_1, \sigma_2, \sigma_3$ be the Pauli spin matrices.
(i) Show that the matrices

$$\frac{1}{2}(I_2 + \sigma_j), \quad \frac{1}{2}(I_2 - \sigma_j), \quad j = 1, 2, 3$$

are projection matrices.
(ii) Show that the matrices

$$\frac{1}{2}(I_4 + \sigma_j \otimes \sigma_k), \quad \frac{1}{2}(I_4 - \sigma_j \otimes \sigma_k), \quad j, k = 1, 2, 3$$

are projection matrices.

(iii) Show that the matrices

$$\frac{1}{2}(I_8 + \sigma_j \otimes \sigma_k \otimes \sigma_\ell), \quad \frac{1}{2}(I_8 - \sigma_j \otimes \sigma_k \otimes \sigma_\ell), \quad j, k, \ell = 1, 2, 3$$

are projection matrices.

(iv) Let $U$ be an $n \times n$ unitary and hermitian matrix. Show that

$$\Pi = \frac{1}{2}(I_n - U)$$

is a projection matrix. Show that $\Pi \otimes \Pi$ is a projection matrix.

**Problem 8.**    Consider the matrices

$$A = \begin{pmatrix} 0 & 1 & 0 \\ 1 & 0 & 1 \\ 0 & 1 & 0 \end{pmatrix}, \quad B = \begin{pmatrix} 1 & 0 & 1 \\ 0 & 1 & 0 \\ 1 & 0 & 1 \end{pmatrix}.$$

Write down all six $3 \times 3$ permutation matrices with

$$P_0 = \begin{pmatrix} 1 & 0 & 0 \\ 0 & 1 & 0 \\ 0 & 0 & 1 \end{pmatrix}, \quad P_5 = \begin{pmatrix} 0 & 0 & 1 \\ 0 & 1 & 0 \\ 1 & 0 & 0 \end{pmatrix}.$$

Find the permutation matrices in this set such that $P_j A P_j^T = A$. Find the permutation matrices in this set such that $P_j B P_j^T = B$.

**Problem 9.**    Consider the $4 \times 4$ matrix

$$A = \begin{pmatrix} -1/2 & 1/2 & 1/2 & 1/2 \\ 1/2 & -1/2 & 1/2 & 1/2 \\ 1/2 & 1/2 & -1/2 & 1/2 \\ 1/2 & 1/2 & 1/2 & -1/2 \end{pmatrix}.$$

Find the eigenvalues of $A$ without calculating the eigenvalues. Utilize the information from $A^2$, $\mathrm{tr}(A)$ and that the matrix $A$ is symmetric over $\mathbb{R}$. Then find the eigenvalues of $A \otimes A$ and $A \otimes I_4 + I_4 \otimes A$.

**Problem 10.**    Find the eigenvalues and normalized eigenvectors of the Hamilton operator ($16 \times 16$ hermitian matrix)

$$\hat{H} = \hbar\omega_1(\sigma_3 \otimes \sigma_3 \otimes I_2 \otimes I_2 + I_2 \otimes I_2 \otimes \sigma_3 \otimes \sigma_3) + \hbar\omega_2(\sigma_1 \otimes \sigma_1 \otimes \sigma_1 \otimes \sigma_1).$$

**Problem 11.**    Consider the Hamilton operator

$$\hat{H} = \hbar\omega_1(\sigma_3 \otimes I_2 + I_2 \otimes \sigma_3) + \hbar\omega_2\sigma_1 \otimes \sigma_1 + \hbar\omega_3\sigma_2 \otimes \sigma_2.$$

Find the eigenvalues and normalized eigenvectors of $\hat{H}$.

**Problem 12.** Let $n \geq 2$ and $\omega = e^{2\pi i/n}$. Consider the $n \times n$ matrices

$$D = \text{diag}(1 \ \omega \ \cdots \ \omega^{n-1}), \quad \Gamma = \begin{pmatrix} 0 & 0 & \cdots & 1 \\ 1 & 0 & \cdots & 0 \\ \vdots & \ddots & \vdots & 0 \\ 0 & \cdots & 1 & 0 \end{pmatrix}.$$

So $\Gamma$ is a permutation matrix with $\Gamma^n = I_n$. Furthermore $D^n = I_n$. Find the commutator $[D, \Gamma]$.

**Problem 13.** Let $\mathbf{v}_0$, $\mathbf{v}_1$, $\mathbf{v}_2$, $\mathbf{v}_3$ be an orthonormal basis in the Hilbert space $\mathbb{C}^4$. Show that the vectors

$$\mathbf{u}_0 = \frac{1}{2}(\mathbf{v}_0 + \mathbf{v}_1 + \mathbf{v}_2 + \mathbf{v}_3), \quad \mathbf{u}_1 = \frac{1}{2}(\mathbf{v}_0 - \mathbf{v}_1 + \mathbf{v}_2 - \mathbf{v}_3),$$

$$\mathbf{u}_2 = \frac{1}{2}(\mathbf{v}_0 + \mathbf{v}_1 - \mathbf{v}_2 - \mathbf{v}_3), \quad \mathbf{u}_3 = \frac{1}{2}(\mathbf{v}_0 - \mathbf{v}_1 - \mathbf{v}_2 + \mathbf{v}_3)$$

also form an orthonormal basis in $\mathbb{C}^4$.

**Problem 14.** Let $x \in \mathbb{R}$. Are the vectors

$$\mathbf{v}_1 = \begin{pmatrix} 1 \\ x \\ x^2 \\ x^3 \end{pmatrix}, \quad \mathbf{v}_2 = \begin{pmatrix} 0 \\ 1 \\ 2x \\ 3x^2 \end{pmatrix}, \quad \mathbf{v}_3 = \begin{pmatrix} 0 \\ 0 \\ 2 \\ 6x \end{pmatrix}, \quad \mathbf{v}_4 = \begin{pmatrix} 0 \\ 0 \\ 0 \\ 6 \end{pmatrix}$$

linearly independent? Find *Gram's matrix*

$$G = (\mathbf{v}_j^T \mathbf{v}_k), \quad j, k = 1, 2, 3, 4$$

and its determinant.

**Problem 15.** Let

$$\mathbf{v}(\theta) = \begin{pmatrix} \cos(\theta) \\ \sin(\theta) \end{pmatrix}.$$

Find all $2 \times 2$ matrices $A$, $B$ such that $\mathbf{v}^* A B \mathbf{v} = (\mathbf{v}^* A \mathbf{v})(\mathbf{v}^* B \mathbf{v})$.

**Problem 16.** Consider the symmetric binary matrices

$$A = \begin{pmatrix} 0 & 1 & 0 \\ 1 & 0 & 1 \\ 0 & 1 & 0 \end{pmatrix}, \quad B = \begin{pmatrix} 1 & 0 & 1 \\ 0 & 1 & 0 \\ 1 & 0 & 1 \end{pmatrix}.$$

Find the eigenvalues and eigenvectors of $A$ and $B$. Find the eigenvalues and eigenvectors of the anti-commutator $[A, B]_+$. Discuss.

**Problem 17.**    Find all $n \times n$ matrices $A$ and $B$ such that

$$e^A \otimes e^B = e^{A \otimes B}.$$

**Problem 18.**    Consider the $2 \times 2$ elementary matrices

$$E_{11} = \begin{pmatrix} 1 & 0 \\ 0 & 0 \end{pmatrix}, \quad E_{22} = \begin{pmatrix} 0 & 0 \\ 0 & 1 \end{pmatrix}, \quad E_{12} = \begin{pmatrix} 0 & 1 \\ 0 & 0 \end{pmatrix}, \quad E_{21} = \begin{pmatrix} 0 & 0 \\ 1 & 0 \end{pmatrix}$$

with the commutator $[E_{12}, E_{21}] = E_{11} - E_{22}$. Let $\theta \in \mathbb{R}$. Show that

$$\exp\left(-\theta(E_{12} + E_{21})\right) =$$

$$\exp(-\tanh(\theta)E_{12}) \exp(\ln(\cosh(\theta))(E_{22} - E_{11})) \exp(-\tanh(\theta)E_{21}).$$

**Problem 19.**    Let $\mu \in \mathbb{C}$, $S$ the spin $(S = 0, 1/2, 1, 3/2, 2, \ldots)$ and $|0\rangle$, $|1\rangle$, ..., $|2S\rangle$ be the standard basis in $\mathbb{C}^{2S+1}$. The *Bloch coherent states* $|\mu\rangle$ are defined by

$$|\mu\rangle = \frac{1}{(1 + |\mu|^2)^S} \sum_{p=0}^{2S} \left( \frac{(2S)!}{p!(2S - p)!} \right)^{1/2} \mu^p |p\rangle.$$

Show that

$$\langle \mu | = \frac{1}{(1 + |\mu|^2)^S} \sum_{p=0}^{2S} \left( \frac{(2S)!}{p!(2S - p)!} \right)^{1/2} (\mu^*)^p \langle p|.$$

Show that (*completeness relation*)

$$\frac{1 + 2S}{\pi} \int_{\mathbb{C}} \frac{d^2\mu}{(1 + |\mu|^2)^2} |\mu\rangle\langle \mu| = I_{2S+1}$$

where $d^2\mu = d(\Re(\mu))(d(\Im(\mu))$. Show that the scalar product of two Bloch coherent states is given as

$$\langle \nu | \mu \rangle = \frac{(1 + \nu^*\mu)^{2S}}{(1 + |\nu|^2)^S(1 + |\mu|^2)^S}.$$

Let $S = 1/2$. Show that

$$|\mu\rangle = \frac{1}{(1 + |\mu|^2)^{1/2}}(|0\rangle + \mu|1\rangle).$$

**Problem 20.** (i) Let $A$, $B$ be $n \times n$ matrices over $\mathbb{C}$ and $C = A + B$. Show that (*Trotter formula*)

$$e^{-\tau C} = \lim_{n \to \infty} \left( e^{-\tau A/n} e^{-\tau B/n} \right)^n, \qquad \tau \geq 0.$$

(ii) Let $A$, $B$ be $n \times n$ matrices. Show that using exponential theory such that

$$\exp(\tau(A + B)) = \exp(\tau A/2) \exp(\tau B) \exp(\tau A/2) + O(\tau^3).$$

**Problem 21.** Let $s = 1/2, 1, 3/2, 2, \ldots$ be the spin and let $m = -s, -s + 1, \ldots, s - 1, s$. The simple Lie algebra $su(2)$ has generators $\{S_3, S_+, S_-\}$ with commutation relations

$$[S_+, S_-] = 2S_3, \qquad [S_3, S_\pm] = \pm S_\pm.$$

Given a finite-dimensional module with highest weight $s$ the action of $S_3$, $S_-$, $S_+$ on the weight basis is

$$S_3|s, m\rangle = m|s, m\rangle \text{ eigenvalue equation}$$
$$S_+|s, m\rangle = \sqrt{(s - m)(s + m + 1)}|s, m + 1\rangle$$
$$S_-|s, m\rangle = \sqrt{(s + m)(s - m + 1)}|s, m - 1\rangle.$$

Study the eigenvalue problem of the Hamilton operators

$$H_1 = \hbar\omega(S_1 \otimes S_1 + S_2 \otimes S_2 + S_3 \otimes S_3)$$
$$H_2 = \hbar\omega(S_1 \otimes S_2 + S_2 \otimes S_3 + S_3 \otimes S_1)$$

with the basis

$$\{ |s, m_1\rangle \otimes |s, m_2\rangle : m_j = -s, -s + 1, \ldots, s - 1, s; j = 1, 2 \}.$$

**Problem 22.** Consider $\mathbb{C}^8 \cong \mathbb{C}^2 \otimes \mathbb{C}^2 \otimes \mathbb{C}^2$ and

$$|\psi\rangle = \frac{1}{\sqrt{8}} \begin{pmatrix} 1 & -1 & 1 & -1 & 1 & -1 & 1 & -1 \end{pmatrix}^T$$

Find the Schmidt decomposition.

**Problem 23.** Consider the four Bell states

$$|\psi_1\rangle = \frac{1}{\sqrt{2}} \begin{pmatrix} 1 \\ 0 \\ 0 \\ 1 \end{pmatrix}, \qquad |\psi_2\rangle = \frac{1}{\sqrt{2}} \begin{pmatrix} 0 \\ 1 \\ 1 \\ 0 \end{pmatrix},$$

$$|\psi_3\rangle = \frac{1}{\sqrt{2}}\begin{pmatrix} 0 \\ 1 \\ -1 \\ 0 \end{pmatrix}, \qquad |\psi_4\rangle = \frac{1}{\sqrt{2}}\begin{pmatrix} 1 \\ 0 \\ 0 \\ -1 \end{pmatrix}$$

and the Pauli spin matrices $\sigma_1$, $\sigma_2$, $\sigma_3$. Find

$$\langle \psi_j | \sigma_k \otimes \sigma_\ell | \psi_j \rangle$$

for $j = 1, 2, 3, 4$ and $k, \ell = 1, 2, 3$. Apply Computer Algebra.

**Problem 24.**   Consider the normalized (column) vector $\mathbf{v} \in \mathbb{C}^n$ and the $n \times n$ matrix $M = I_n - 2\mathbf{v}\mathbf{v}^*$, where $I_n$ is the $n \times n$ identity matrix. Show that the matrix is hermitian and unitary.

**Problem 25.**   Let $c_1^\dagger$, $c_2^\dagger$, $c_1$, $c_2$ be Fermi creation and annihilation operators, respectively. Let $\alpha \in \mathbb{R}$. Show that

$$U(\alpha) = \exp(i\alpha(c_1^\dagger c_2 + c_2^\dagger c_1))$$
$$= I + i\sin(\alpha)(c_1^\dagger c_2 + c_2^\dagger c_1) + (\cos(\alpha) - 1)(\hat{N}_1 + \hat{N}_2 - 2\hat{N}_1\hat{N}_2)$$

where $\hat{N}_1 = c_1^\dagger c_1$, $\hat{N}_2 = c_2^\dagger c_2$. First show that

$$(c_1^\dagger c_2 + c_2^\dagger c_1)^2 = \hat{N}_1 + \hat{N}_2 - 2\hat{N}_1\hat{N}_2$$
$$(c_1^\dagger c_2 + c_2^\dagger c_1)^3 = c_1^\dagger c_2 + c_2^\dagger c_1$$
$$(c_1^\dagger c_2 + c_2^\dagger c_1)^4 = \hat{N}_1 + \hat{N}_2 - 2\hat{N}_1\hat{N}_2.$$

Show that $\hat{N}_1 + \hat{N}_2 - 2\hat{N}_1\hat{N}_2$ is a projection operator.

# Chapter 4

# Density Operators

## 4.1 Introduction

A *density operator* $\rho$ or *density matrix* is a positive semidefinite operator on a Hilbert space with unit trace. An operator is *positive semidefinite* if it is hermitian and none of its (necessarily real) eigenvalues are less than zero. The state of a quantum-mechanical system is characterized by a density operator $\rho$ with $\operatorname{tr}(\rho) = 1$. The *expectation value* of an observable $\hat{A}$ (self-adjoint operator, hermitian matrix), determined in an experiment as the average value $\langle \hat{A} \rangle$ is given by

$$\langle \hat{A} \rangle := \operatorname{tr}(\hat{A}\rho).$$

A density matrix or density operator is used in quantum theory to describe the statistical state of a quantum system. If we have a *pure state* $|\psi\rangle$ in a Hilbert space then

$$\rho = |\psi\rangle\langle\psi|$$

defines a density matrix with ($\langle\psi|\psi\rangle = 1$)

$$\rho^2 = |\psi\rangle\langle\psi|\psi\rangle\langle\psi| = |\psi\rangle\langle\psi| = \rho.$$

For a *mixed state* we have the spectral representation

$$\rho = \sum_{j=1}^{n} p_j |\psi_j\rangle\langle\psi_j|$$

where $p_j \geq 0$ for $j = 1, 2, \ldots, n$,

$$\sum_{j=1}^{n} p_j = 1 \quad \text{and} \quad \langle \psi_j | \psi_k \rangle = \delta_{jk}.$$

Then for the expectation value of a hermitian operator $\hat{A}$ in the Hilbert space we have

$$\text{tr}(\rho \hat{A}) = \sum_{j=1}^{n} p_j \langle \psi_j | \hat{A} | \psi_j \rangle.$$

Let $M$ be an arbitrary nonzero $n \times n$ matrix over $\mathbb{C}$. Then

$$\rho = \frac{MM^*}{\text{tr}(MM^*)}$$

is a density matrix.

If $\rho_1$ and $\rho_2$ are density matrices, then $\rho_1 \otimes \rho_2$, $\rho_1 \oplus \rho_2$ are density matrices. If $\rho_1$, $\rho_2$ are pure states, then $\rho_1 \otimes \rho_2$ and $\rho_1 \oplus \rho_2$ are pure states. If $\rho$ is an $n \times n$ density matrix and $U$ is an $n \times n$ unitary matrix ($U^{-1} = U^*$), then $U\rho U^*$ is a density matrix.

The eigenvalues of an $n \times n$ density matrix which is a pure state are 1 and 0 ($n - 1$ times), since $\rho^2 = \rho$, $\text{tr}(\rho) = 1$ and $\rho^* = \rho$.

Consider a quantum system of spin-1/2 particles. The density matrix describing the spin degree of freedom is a $2 \times 2$ matrix which can be written as

$$\rho(\mathbf{n}) = \frac{1}{2}(I_2 + \mathbf{n} \cdot \boldsymbol{\sigma}) \equiv \frac{1}{2}(I_2 + n_1\sigma_1 + n_2\sigma_2 + n_3\sigma_3)$$

where $\sigma_1$, $\sigma_2$, $\sigma_3$ denote the Pauli spin matrices and $|\mathbf{n}| \leq 1$. For $|\mathbf{n}| = 1$ the density matrix describes a pure state, whereas for $|\mathbf{n}| < 1$ one has a mixed state. The density matrix $\rho$ is thus uniquely determined by a point of the unit sphere $|\mathbf{n}| \leq 1$.

The *variance* of an observable $\hat{A}$ and a density operator $\rho$ in a Hilbert space $\mathcal{H}$ is defined as

$$V(\rho, \hat{A}) := \text{tr}(\rho \hat{A}^2) - (\text{tr}(\rho \hat{A}))^2.$$

Let $|\psi\rangle$ be a normalized state in the Hilbert space $\mathcal{H}$. If $\rho = |\psi\rangle\langle\psi|$ (pure state) we obtain

$$V(|\psi\rangle\langle\psi|, \hat{A}) = \langle\psi|\hat{A}^2|\psi\rangle - \langle\psi|\hat{A}|\psi\rangle^2.$$

# 4.2   Solved Problems

**Problem 1.**   Is the $2 \times 2$ matrix

$$\rho = \begin{pmatrix} 1/2 & -1/2 \\ -1/2 & 1/2 \end{pmatrix}$$

a density matrix? If so is it a pure state? If so find the normalized state in $\mathbb{C}^2$ that provides this density matrix.

**Solution 1.**   We find that $\operatorname{tr}(\rho) = 1$ and the matrix is hermitian over $\mathbb{C}$. The eigenvalues of $\rho$ are 0 and 1. Thus $\rho$ is a density matrix and a pure state. The state

$$|\psi\rangle = \frac{1}{\sqrt{2}} \begin{pmatrix} 1 \\ -1 \end{pmatrix}$$

is a normalized state that provides the density matrix $\rho = |\psi\rangle\langle\psi|$.

**Problem 2.**   Let

$$|\psi\rangle = \begin{pmatrix} \cos(\theta) \\ e^{i\phi}\sin(\theta) \end{pmatrix}, \qquad \theta, \phi \in \mathbb{R}.$$

be a normalized state in $\mathbb{C}^2$. Does $\rho := |\psi\rangle\langle\psi|$ define a density matrix?

**Solution 2.**   We find the $2 \times 2$ matrix for $\rho$

$$\rho = |\psi\rangle\langle\psi| = \begin{pmatrix} \cos^2(\theta) & e^{-i\phi}\cos(\theta)\sin(\theta) \\ e^{i\phi}\cos(\theta)\sin(\theta) & \sin^2(\theta) \end{pmatrix}.$$

Now $\rho = \rho^*$, $\operatorname{tr}(\rho) = \cos^2(\theta) + \sin^2(\theta) = 1$ and $\rho^2 = \rho$. Hence we have a density matrix (pure state).

**Problem 3.**   Let $r \geq 0$. Is the $2 \times 2$ matrix

$$\rho = \frac{1}{2} \begin{pmatrix} 1 + r\cos(\theta) & r\sin(\theta)e^{-i\phi} \\ r\sin(\theta)e^{i\phi} & 1 - r\cos(\theta) \end{pmatrix}$$

a density matrix? What are the conditions on $r$, $\theta$, $\phi$?

**Solution 3.**   We have that $\operatorname{tr}(\rho) = 1$ and the matrix is hermitian. Thus the eigenvalues are real. The eigenvalues are given by

$$\lambda_{1,2} = \frac{1}{2} \pm \frac{1}{2}r.$$

Thus the condition that $\rho$ is density matrix is $r \leq 1$. There is no condition on $\phi$ and $\theta$.

**Problem 4.**   Let $\sigma_1$, $\sigma_2$ and $\sigma_3$ be the Pauli spin matrices. Let $\boldsymbol{\sigma} = (\sigma_1, \sigma_2, \sigma_3)$ and $\mathbf{r} \in \mathbb{R}^3$ with $\mathbf{r}^2 \leq 1$. Consider the $2 \times 2$ matrix (density matrix)

$$\rho := \frac{1}{2}(I_2 + \mathbf{r} \cdot \boldsymbol{\sigma})$$

where $\mathbf{r} \cdot \boldsymbol{\sigma} := r_1 \sigma_1 + r_2 \sigma_2 + r_3 \sigma_3$. Let $\mathbf{n}$ be an arbitrary unit length vector in $\mathbb{R}^3$, i.e. $\mathbf{n}^2 \equiv n_1^2 + n_2^2 + n_3^2 = 1$. Calculate $\mathrm{tr}((\mathbf{n} \cdot \boldsymbol{\sigma})\rho)$ i.e. we calculate the expectation value of $\mathbf{n} \cdot \boldsymbol{\sigma}$. Give an interpretation of the result.

**Solution 4.**   We have

$$\mathrm{tr}((\mathbf{n} \cdot \boldsymbol{\sigma})\rho) = \mathrm{tr}\left((\mathbf{n} \cdot \boldsymbol{\sigma})\frac{1}{2}(I_2 + \mathbf{r} \cdot \boldsymbol{\sigma})\right) = \frac{1}{2}\mathrm{tr}(\mathbf{n} \cdot \boldsymbol{\sigma} + (\mathbf{n} \cdot \boldsymbol{\sigma})(\mathbf{r} \cdot \boldsymbol{\sigma}))$$

$$= \frac{1}{2}\mathrm{tr}((\mathbf{n} \cdot \boldsymbol{\sigma})(\mathbf{r} \cdot \boldsymbol{\sigma})) = \frac{1}{2}\mathrm{tr}\left(\sum_{i,j=1}^{3} n_i r_j \sigma_i \sigma_j\right)$$

$$= \frac{1}{2}\mathrm{tr}\left(\sum_{i=1}^{3} n_i r_i I_2\right) = \sum_{i=1}^{3} n_i r_i = \mathbf{n} \cdot \mathbf{r}.$$

The vector $\mathbf{r}$ can be thought of as an expectation value of spin polarization, and it can be obtained by measuring $\mathbf{n} \cdot \boldsymbol{\sigma}$ along each direction $\mathbf{e}_1$, $\mathbf{e}_2$ and $\mathbf{e}_3$.

**Problem 5.**   Let $A$ be a nonzero $n \times n$ matrix over $\mathbb{C}$. Then $\mathrm{tr}(AA^*) > 0$. Consider the map

$$A \to \rho = \frac{AA^*}{\mathrm{tr}(AA^*)}.$$

(i) Show that $\rho$ is a density matrix.
(ii) Show that $\rho$ is invariant under the map $A \to AU$, where $U$ is an $n \times n$ unitary matrix.
(iii) Is $AA^* = A^*A$ in general?   A matrix is called a *normal matrix* if $AA^* = A^*A$.
(iv) Consider the map

$$A \to \sigma = \frac{A^*A}{\mathrm{tr}(A^*A)}.$$

Is $\sigma = \rho$?

**Solution 5.**   (i) Since $A$ is a nonzero matrix we find that $AA^*$ is nonzero and $\mathrm{tr}(AA^*) \neq 0$. The matrix $\rho$ is positive-semidefinite and $\mathrm{tr}(\rho) = 1$.
(ii) We set $A' = AU$. Thus we have

$$\rho \to \rho' = \frac{A'A'^*}{\mathrm{tr}(A'A'^*)} = \frac{(AU)(AU)^*}{\mathrm{tr}((AU)(AU)^*)} = \frac{AUU^*A^*}{\mathrm{tr}(AUU^*A^*)} = \frac{AA^*}{\mathrm{tr}(AA^*)}$$

$$= \rho$$

where we used that $UU^* = I$.

(iii) In general we have $AA^* \neq A^*A$. For example, let

$$A = \begin{pmatrix} 0 & 1 \\ 0 & 0 \end{pmatrix}, \qquad A^* = \begin{pmatrix} 0 & 0 \\ 1 & 0 \end{pmatrix}.$$

Then

$$AA^* = \begin{pmatrix} 1 & 0 \\ 0 & 0 \end{pmatrix}, \qquad A^*A = \begin{pmatrix} 0 & 0 \\ 0 & 1 \end{pmatrix}.$$

Thus $AA^* \neq A^*A$. However, we have $\mathrm{tr}(AA^*) = \mathrm{tr}(A^*A)$.

(iv) From (iii) it follows that in general we have $\rho \neq \sigma$.

**Problem 6.** Find a normalized state $|\phi\rangle$ in the Hilbert space $\mathbb{C}^2$ such that we have the density matrix

$$|\phi\rangle\langle\phi| = \frac{1}{2}\left(I_2 + \frac{1}{\sqrt{2}}(\sigma_1 + \sigma_3)\right).$$

**Solution 6.** We have to solve

$$|\phi\rangle\langle\phi| = \frac{1}{2}\begin{pmatrix} 1 + 1/\sqrt{2} & 1/\sqrt{2} \\ 1/\sqrt{2} & 1 - 1/\sqrt{2} \end{pmatrix}.$$

We obtain the normalized state

$$|\phi\rangle = \begin{pmatrix} \sqrt{(\sqrt{2}+1)/(2\sqrt{2})} \\ \sqrt{(\sqrt{2}-1)/(2\sqrt{2})} \end{pmatrix}.$$

**Problem 7.** Consider the $2 \times 2$ matrix

$$\rho = \begin{pmatrix} 3/4 & \sqrt{2}e^{-i\phi}/4 \\ \sqrt{2}e^{i\phi}/4 & 1/4 \end{pmatrix}.$$

(i) Is the matrix a density matrix?
(ii) If so do we have a pure state or a mixed state?
(iii) Find the eigenvalues of $\rho$.
(iv) Find $\mathrm{tr}(\sigma_1\rho)$, where $\sigma_1$ is the first Pauli spin matrix.

**Solution 7.** (i) We have $\mathrm{tr}(\rho) = 1$ and the matrix is hermitian. Furthermore the eigenvalues are nonnegative. Thus we have a density matrix.
(ii) Since $\rho^2 \neq \rho$ we have a mixed state.
(iii) The eigenvalues are $\lambda_1 = (2 + \sqrt{3})/4$, $\lambda_2 = (2 - \sqrt{3})/4$ which also indicate the state is mixed.

(iv) We obtain

$$\mathrm{tr}(\sigma_1 \rho) = \frac{1}{\sqrt{2}} \cos(\phi).$$

**Problem 8.**  (i) The *Hilbert-Schmidt distance* between any two density operators $\rho_1$ and $\rho_2$ is given by the Frobenius-Hilbert-Schmidt norm of their differences

$$D_{HS}(\rho_1, \rho_2) := \sqrt{\mathrm{tr}((\rho_1 - \rho_2)^2)}.$$

Let

$$\rho_1 = \begin{pmatrix} 1 & 0 \\ 0 & 0 \end{pmatrix}, \qquad \rho_2 = \begin{pmatrix} 0 & 0 \\ 0 & 1 \end{pmatrix}.$$

Calculate $D_{HS}(\rho_1, \rho_2)$.
(ii) The *Bures distance* in the space of mixed quantum states described by the density matrices $\rho_1$ and $\rho_2$ is defined as

$$D_B(\rho_1, \rho_2) := \sqrt{2(1 - \mathrm{tr}((\rho_1^{1/2} \rho_2 \rho_1^{1/2})^{1/2}))}.$$

Let

$$\rho_1 = \begin{pmatrix} 1/2 & 0 \\ 0 & 1/2 \end{pmatrix}, \qquad \rho_2 = \begin{pmatrix} 3/4 & 0 \\ 0 & 1/4 \end{pmatrix}.$$

Calculate the Bures distance $D_B(\rho_1, \rho_2)$.
(iii) Let $\rho$, $\sigma$ be two density operators acting in the same finite dimensional Hilbert space. The *trace distance* between $\rho$ and $\sigma$ is defined as

$$D(\rho, \sigma) := \frac{1}{2} \mathrm{tr}(\sqrt{(\rho^* - \sigma^*)(\rho - \sigma)}).$$

Let

$$\rho = \begin{pmatrix} 1/2 & 0 & 0 & 1/2 \\ 0 & 0 & 0 & 0 \\ 0 & 0 & 0 & 0 \\ 1/2 & 0 & 0 & 1/2 \end{pmatrix}, \qquad \sigma = \begin{pmatrix} 1/2 & 0 & 0 & 0 \\ 0 & 0 & 0 & 0 \\ 0 & 0 & 0 & 0 \\ 0 & 0 & 0 & 1/2 \end{pmatrix}.$$

Find the trace distance.

**Solution 8.**  (i) Since

$$\rho_1 - \rho_2 = \begin{pmatrix} -1/4 & 0 \\ 0 & 1/4 \end{pmatrix}$$

we find $D_{HS}(\rho_1, \rho_2) = 1/\sqrt{8}$.
(ii) Since

$$\rho_1^{1/2} \rho_2 \rho_1^{1/2} = \begin{pmatrix} 3/8 & 0 \\ 0 & 1/8 \end{pmatrix}$$

we find

$$D_B(\rho_1, \rho_2) = \sqrt{\frac{2\sqrt{2} - \sqrt{3} - 1}{\sqrt{2}}}.$$

(iii) We have

$$\rho - \sigma = \begin{pmatrix} 0 & 0 & 0 & 1/2 \\ 0 & 0 & 0 & 0 \\ 0 & 0 & 0 & 0 \\ 1/2 & 0 & 0 & 0 \end{pmatrix}.$$

Thus

$$(\rho^* - \sigma^*)(\rho - \sigma) = \begin{pmatrix} 1/4 & 0 & 0 & 0 \\ 0 & 0 & 0 & 0 \\ 0 & 0 & 0 & 0 \\ 0 & 0 & 0 & 1/4 \end{pmatrix}.$$

It follows that $D(\rho, \sigma) = 1/2$.

**Problem 9.** Consider the linear operator ($4 \times 4$ matrix) in the Hilbert space $\mathbb{C}^4$

$$\rho = \frac{1}{4}(1 - \epsilon)I_4 + \epsilon(|0\rangle \otimes |0\rangle)(\langle 0| \otimes \langle 0|)$$

where $\epsilon$ is a real parameter with $\epsilon \in [0, 1]$ and the state

$$|0\rangle = \begin{pmatrix} 1 \\ 0 \end{pmatrix}.$$

Does $\rho$ define a density matrix?

**Solution 9.** We find the diagonal matrix for $\rho$

$$\rho = \begin{pmatrix} (1-\epsilon)/4 + \epsilon & 0 & 0 & 0 \\ 0 & (1-\epsilon)/4 & 0 & 0 \\ 0 & 0 & (1-\epsilon)/4 & 0 \\ 0 & 0 & 0 & (1-\epsilon)/4 \end{pmatrix}.$$

Thus $\rho = \rho^*$, $\text{tr}(\rho) = 1$, and $\langle \mathbf{x}|\rho|\mathbf{x}\rangle \geq 0$, for all $\mathbf{x} \in \mathbb{C}^4$. The last property follows since all entries on the diagonal are non-negative. Thus $\rho$ defines a density matrix.

**Problem 10.** A mixed state is a statistical mixture of pure states, i.e. the state is described by pairs of probabilities and pure states. Given a mixture $\{ (p_1, |\psi_1\rangle), \ldots, (p_n, |\psi_n\rangle) \}$ we define its *density matrix* to be the positive hermitian matrix

$$\rho = \sum_{j=1}^{n} p_j |\psi_j\rangle\langle\psi_j|$$

where the pure states $|\psi_j\rangle$ are normalized (i.e. $\langle\psi_j|\psi_j\rangle = 1$), and $p_j \geq 0$ for $j = 1, 2, \ldots, n$ with $p_1 + p_2 + \cdots + p_n = 1$.
(i) Find the probability that measurement in the orthonormal basis

$$\{\, |k_1\rangle, \ldots, |k_n\rangle \,\}$$

will yield $|k_j\rangle$.
(ii) Find the density matrix $\rho_U$ when the mixture is transformed according to the unitary matrix $U$.

**Solution 10.**    (i) From the probability distribution of states in the mixture we have for the probability $P(k_j)$ of measuring the state $|k_j\rangle$ $(j = 1, 2, \ldots, n)$

$$P(k_j) = \sum_{l=1}^{n} p_l |\langle k_j | \psi_l \rangle|^2 = \sum_{l=1}^{n} p_l \langle k_j | \psi_l \rangle \langle \psi_l | k_j \rangle = \langle k_j | \rho | k_j \rangle.$$

(ii) After applying the transform $U$ to the states in the mixture we have the new mixture $\{\, (p_1, U|\psi_1\rangle), \ldots, (p_n, U|\psi_n\rangle) \,\}$, with the density matrix

$$\rho_U = \sum_{j=1}^{n} p_j U |\psi_j\rangle\langle\psi_j| U^* = U \left( \sum_{j=1}^{n} p_j |\psi_j\rangle\langle\psi_j| \right) U^* = U\rho U^*.$$

**Problem 11.**    (i) The Bell state

$$|\psi\rangle = \frac{1}{\sqrt{2}}(|0\rangle \otimes |0\rangle + |1\rangle \otimes |1\rangle)$$

has the density matrix

$$\rho = \frac{1}{2}\begin{pmatrix} 1 & 0 & 0 & 1 \\ 0 & 0 & 0 & 0 \\ 0 & 0 & 0 & 0 \\ 1 & 0 & 0 & 1 \end{pmatrix}.$$

Show that $\rho$ can be written as linear combination of the matrices $\Lambda_{00} = \frac{1}{2}(I_2 \otimes I_2)$, $\Lambda_{11} = \frac{1}{2}(\sigma_1 \otimes \sigma_1)$, $\Lambda_{22} = \frac{1}{2}(\sigma_2 \otimes \sigma_2)$ and $\Lambda_{33} = \frac{1}{2}(\sigma_3 \otimes \sigma_3)$.
(ii) The *Werner state* is described by the density matrix

$$\rho_W = \begin{pmatrix} (1-x)/4 & 0 & 0 & 0 \\ 0 & (1+x)/4 & -x/2 & 0 \\ 0 & -x/2 & (1+x)/4 & 0 \\ 0 & 0 & 0 & (1-x)/4 \end{pmatrix}$$

where $x \in [0,1]$. Show that $\rho_W$ can also be written as linear combination of the operators given in (i).

**Solution 11.** (i) We find the linear combination

$$\rho = \frac{1}{2}\Lambda_{00} + \frac{1}{2}\Lambda_{11} - \frac{1}{2}\Lambda_{22} + \frac{1}{2}\Lambda_{33}.$$

(ii) We obtain the linear combination

$$\rho_W = \frac{1}{2}\Lambda_{00} - \frac{x}{2}\Lambda_{11} - \frac{x}{2}\Lambda_{22} - \frac{x}{2}\Lambda_{33}.$$

**Problem 12.** Suppose we expand a density matrix for $N$ qubits in terms of Kronecker products of Pauli spin matrices

$$\rho = \frac{1}{2^N} \sum_{j_0=0}^{3} \sum_{j_1=0}^{3} \cdots \sum_{j_{N-1}=0}^{3} c_{j_0 j_1 \ldots j_{N-1}} \sigma_{j_0} \otimes \sigma_{j_1} \otimes \cdots \otimes \sigma_{j_{N-1}}$$

where $\sigma_0 = I_2$.
(i) What is condition on the expansion coefficients if we impose $\rho^* = \rho$?
(ii) What is the condition on the expansion coefficients if we impose $\mathrm{tr}(\rho) = 1$?
(iii) Calculate $\mathrm{tr}(\rho \sigma_{k_0} \otimes \sigma_{k_1} \otimes \cdots \otimes \sigma_{k_{N-1}})$.

**Solution 12.** (i) Since $\sigma_1 = \sigma_1^*$, $\sigma_2 = \sigma_2^*$, $\sigma_3 = \sigma_3^*$ and $I_2 = I_2^*$ we find that the expansion coefficients are real.
(ii) Since $\mathrm{tr}(A \otimes B) = \mathrm{tr}(A)\mathrm{tr}(B)$ for square matrices $A$ and $B$ and

$$\mathrm{tr}(\sigma_1) = \mathrm{tr}(\sigma_2) = \mathrm{tr}(\sigma_3) = 0, \qquad \mathrm{tr}(I_2) = 2$$

we find $c_{00\ldots 0} = 1$.
(iii) Since $\mathrm{tr}(\sigma_1 \sigma_2) = 0$, $\mathrm{tr}(\sigma_2 \sigma_3) = 0$, $\mathrm{tr}(\sigma_3 \sigma_1) = 0$ we find

$$\mathrm{tr}(\rho \sigma_{k_0} \otimes \sigma_{k_1} \otimes \cdots \otimes \sigma_{k_{N-1}}) = c_{k_0 k_1 \ldots k_{N-1}}.$$

**Problem 13.** Let $A$ and $B$ be a pair of qubits and let the density matrix of the pair be $\rho_{AB}$, which may be pure or mixed. We define the *spin flipped density matrix* to be

$$\tilde{\rho}_{AB} := (\sigma_2 \otimes \sigma_2)\rho_{AB}^*(\sigma_2 \otimes \sigma_2)$$

where the asterisk denotes complex conjugation and transpose in the standard basis

$$\{\, |0\rangle \otimes |0\rangle, \quad |0\rangle \otimes |1\rangle, \quad |1\rangle \otimes |0\rangle, \quad |1\rangle \otimes |1\rangle \,\}$$

and

$$\sigma_2 = \begin{pmatrix} 0 & -i \\ i & 0 \end{pmatrix}.$$

Since both $\rho_{AB}$ and $\tilde{\rho}_{AB}$ are positive operators, it follows that the product $\rho_{AB}\tilde{\rho}_{AB}$, though non-hermitian, also has only real and non-negative eigenvalues. Consider the Bell state

$$|\psi\rangle := \frac{1}{\sqrt{2}}(|0\rangle \otimes |0\rangle + |1\rangle \otimes |1\rangle)$$

and $\rho := |\psi\rangle\langle\psi|$. Find the eigenvalues of $\rho_{AB}\tilde{\rho}_{AB}$.

**Solution 13.**    Since

$$\rho = |\psi\rangle\langle\psi| = \frac{1}{2}\begin{pmatrix} 1 & 0 & 0 & 1 \\ 0 & 0 & 0 & 0 \\ 0 & 0 & 0 & 0 \\ 1 & 0 & 0 & 1 \end{pmatrix}$$

we have $\rho^* = \rho$. Furthermore

$$\sigma_2 \otimes \sigma_2 = \begin{pmatrix} 0 & 0 & 0 & -1 \\ 0 & 0 & 1 & 0 \\ 0 & 1 & 0 & 0 \\ -1 & 0 & 0 & 0 \end{pmatrix}.$$

Thus $\tilde{\rho} = \rho$ and $\rho\tilde{\rho} = \rho$ with eigenvalues $1, 0, 0, 0$. The *tangle* of the density matrix $\rho_{AB}$ is defined as

$$\tau_{AB} := [\max\{\mu_1 - \mu_2 - \mu_3 - \mu_4, 0\}]^2$$

where $\mu_j$ are the square root of the eigenvalues of $\rho_{AB}\tilde{\rho}_{AB}$ ordered in decreasing order. For the special case in which the state of $AB$ is pure, the matrix $\rho_{AB}\tilde{\rho}_{AB}$ has only one non-zero eigenvalue. One can show that

$$\tau_{AB} = 4\det(\rho_A)$$

where $\rho_A$ is the density matrix of qubit $A$, that is, the trace of $\rho_{AB}$ over qubit $B$.

**Problem 14.**    Consider the density matrix

$$\rho = \frac{1}{2}(I_2 + \mathbf{r} \cdot \boldsymbol{\sigma})$$

where $\mathbf{r} \cdot \boldsymbol{\sigma} := r_1\sigma_1 + r_2\sigma_2 + r_3\sigma_3$ and $\mathbf{r}^2 \leq 1$. Consider the four normalized vectors $\mathbf{a}_1, \mathbf{a}_2, \mathbf{a}_3, \mathbf{a}_4$ in $\mathbb{R}^3$

$$\mathbf{a}_1 = \frac{1}{\sqrt{3}}\begin{pmatrix} 1 \\ 1 \\ 1 \end{pmatrix}, \quad \mathbf{a}_2 = \frac{1}{\sqrt{3}}\begin{pmatrix} 1 \\ -1 \\ -1 \end{pmatrix}, \quad \mathbf{a}_3 = \frac{1}{\sqrt{3}}\begin{pmatrix} -1 \\ 1 \\ -1 \end{pmatrix}, \quad \mathbf{a}_4 = \frac{1}{\sqrt{3}}\begin{pmatrix} -1 \\ -1 \\ 1 \end{pmatrix}$$

such that
$$\mathbf{a}_j^T \mathbf{a}_k = \frac{4}{3}\delta_{jk} - \frac{1}{3} = \begin{cases} 1 & \text{for } j = k \\ -1/3 & \text{for } j \neq k \end{cases}.$$

We have
$$\sum_{j=1}^{4} \mathbf{a}_j = \mathbf{0}, \qquad \frac{3}{4}\sum_{j=1}^{4} \mathbf{a}_j\mathbf{a}_j^T = I_3.$$

Such a quartet of vectors consists of the vectors pointing from the center of a cube to nonadjacent corners. These four vectors can be viewed as the normal vectors for the faces of the *tetrahedron* that is defined by the other four corners of the cube. Owing to the conditions the four vectors are normalized. Each such quartets of $\mathbf{a}_j$'s defines a positive operator-valued measure for minimal four-state *tomography* owing to

$$\sum_{j=1}^{4} P_j = I_2, \qquad P_j := \frac{1}{4}(I_2 + \mathbf{a}_j \cdot \boldsymbol{\sigma}).$$

(i) Show that $p_j := \langle P_j \rangle = \text{tr}(P_j\rho) = \frac{1}{4}(1 + \mathbf{a}_j \cdot \mathbf{r})$.
(ii) Given $p_j$ for $j = 1, 2, 3, 4$ find the density matrix $\rho$.

**Solution 14.** (i) Since $\text{tr}(\sigma_1) = \text{tr}(\sigma_2) = \text{tr}(\sigma_3) = 0$ we have

$$\langle P_j \rangle = \text{tr}(P_j\rho) = \frac{1}{8}\text{tr}((I_2 + \mathbf{a}_j \cdot \boldsymbol{\sigma})(I_2 + \mathbf{r} \cdot \boldsymbol{\sigma}))$$
$$= \frac{1}{8}\text{tr}(I_2 + \mathbf{a}_j \cdot \boldsymbol{\sigma} + \mathbf{r} \cdot \boldsymbol{\sigma} + (\mathbf{a}_j \cdot \boldsymbol{\sigma})(\mathbf{r} \cdot \boldsymbol{\sigma})) = \frac{1}{8}\text{tr}(I_2 + (\mathbf{a}_j \cdot \boldsymbol{\sigma})(\mathbf{r} \cdot \boldsymbol{\sigma}))$$
$$= \frac{1}{8}\text{tr}(I_2 + (\mathbf{a}_j \cdot \mathbf{r})I_2)$$
$$= \frac{1}{4}(1 + \mathbf{a}_j \cdot \mathbf{r}).$$

(ii) Since
$$I_3\mathbf{r} = \left(\frac{3}{4}\sum_{j=1}^{4} \mathbf{a}_j\mathbf{a}_j^T\right)\mathbf{r}$$

the vector $\mathbf{r}$ is obtained as

$$\mathbf{r} = \frac{3}{4}\sum_{j=1}^{4}(\mathbf{a}_j \cdot \mathbf{r})\mathbf{a}_j = 3\sum_{j=1}^{4}\frac{1}{4}(1 + \mathbf{a}_j \cdot \mathbf{r})\mathbf{a}_j = 3\sum_{j=1}^{4} p_j\mathbf{a}_j$$

where we used $\sum_{j=1}^{4} \mathbf{a}_j = \mathbf{0}$. From $\mathbf{a}_j \cdot \boldsymbol{\sigma} = 4P_j - I_2$ and substituting $\mathbf{r}$ from above yields the density matrix

$$\rho = 6\sum_{j=1}^{4} p_j P_j - I_2 = \sum_{j=1}^{4}\langle P_j\rangle(6P_j - I_2) = \frac{1}{2}(I_2 + \mathbf{r} \cdot \boldsymbol{\sigma}).$$

It follows that $p_j$ is restricted to the range $0 \le p_j \le 1/2$ and the probabilities $p_j$ obey the inequalities

$$\frac{1}{4} \le \sum_{j=1}^{4} p_j^2 = \frac{3 + \mathbf{r}^2}{12} \le \frac{1}{3}.$$

The upper bound is reached by all pure states, $\rho = \rho^2$ and $\mathbf{r}^2 = 1$. The lower bound is reached for the completely mixed state, $\rho = \frac{1}{3} I_2$ and $\mathbf{r}^2 = 0$.

**Problem 15.** Let $|0\rangle$, $|1\rangle$ be the standard basis in $\mathbb{C}^2$. Consider the mixed states

$$\frac{1}{2} \left\{ \frac{3}{5} |0\rangle + \frac{4}{5} |1\rangle \right\} + \frac{1}{2} \left\{ \frac{3}{5} |0\rangle - \frac{4}{5} |1\rangle \right\}$$

and

$$\frac{9}{25} \{ |0\rangle \} + \frac{16}{25} \{ |1\rangle \}.$$

Find the density matrices. Discuss.

**Solution 15.** In the first case we have

$$\rho_1 = \frac{1}{2} \begin{pmatrix} 9/25 & 12/25 \\ 12/25 & 16/25 \end{pmatrix} + \frac{1}{2} \begin{pmatrix} 9/25 & -12/25 \\ -12/25 & 16/25 \end{pmatrix} = \begin{pmatrix} 9/25 & 0 \\ 0 & 16/25 \end{pmatrix}.$$

In the second case we have

$$\rho_2 = \frac{9}{25} \begin{pmatrix} 1 & 0 \\ 0 & 0 \end{pmatrix} + \frac{16}{25} \begin{pmatrix} 0 & 0 \\ 0 & 1 \end{pmatrix} = \begin{pmatrix} 9/25 & 0 \\ 0 & 16/25 \end{pmatrix}.$$

Thus these two different mixed states correspond to the same density matrix and thus they are indistinguishable.

**Problem 16.** Let $\rho_1$ and $\rho_2$ be $n \times n$ density matrices. Let $\lambda_j$ denote the eigenvalues of $\rho_1 - \rho_2$ with corresponding orthonormal eigenvectors $|\phi_j\rangle$ where $j = 1, 2, \ldots, n$.
(i) Find the difference $|D_1 - D_2|$ between the probability distributions $D_1$ and $D_2$ for the measurement of the mixtures $\rho_1$ and $\rho_2$ in the basis $\{ |\phi_1\rangle, \ldots, |\phi_n\rangle \}$.
(ii) Show that measurement in the basis $\{ |\phi_1\rangle, \ldots, |\phi_n\rangle \}$ maximizes the difference $|D_1 - D_2|$.
Hint. Use *Schur's theorem*. For any hermitian matrix $A$, let

$$a_1 \ge a_2 \ge \cdots \ge a_n$$

be the non increasing diagonal entries of $A$ and

$$\mu_1 \ge \mu_2 \ge \cdots \ge \mu_n$$

the non increasing eigenvalues of $A$. Then for $1 \leq k \leq n$

$$\sum_{j=1}^{k} \mu_j \geq \sum_{j=1}^{k} a_j$$

where equality holds for $k = n$.

**Solution 16.** (i) We write $\rho_1$ and $\rho_2$ in the basis $\{\,|\phi_1\rangle, \ldots, |\phi_n\rangle\,\}$. In this basis we have

$$|D_1 - D_2| = \sum_{j=1}^{n} |\langle \phi_j|\rho_1|\phi_j\rangle - \langle \phi_j|\rho_2|\phi_j\rangle| = \sum_{j=1}^{n} |\langle \phi_j|(\rho_1 - \rho_2)|\phi_j\rangle| = \sum_{j=1}^{n} |\lambda_j|.$$

(ii) Let $U$ be an arbitrary unitary transform (change of basis). We define $P := U\rho_1 U^*$ and $Q := U\rho_2 U^*$. The matrix $P - Q$ is hermitian. Let

$$q_1 \geq q_2 \geq \cdots \geq q_n$$

denote the non decreasing diagonal entries of $P - Q$ in the $\{\,|\phi_1\rangle, \ldots, |\phi_n\rangle\,\}$ basis and

$$\nu_1 \geq \nu_2 \geq \cdots \geq \nu_n$$

be the non decreasing eigenvalues (i.e. $\lambda_j$) of $P - Q$. Consider the difference $|D'_1 - D'_2|$ between the probability distributions $D'_1$ and $D'_2$ for the measurement of the mixtures $\rho_1$ and $\rho_2$ in the basis $\{\,U|\phi_1\rangle, \ldots, U|\phi_n\rangle\,\}$

$$|D'_1 - D'_2| = \sum_{j=1}^{n} |\langle \phi_j|U^*\rho_1 U|\phi_j\rangle - \langle \phi_j|U^*\rho_2 U|\phi_j\rangle|$$

$$= \sum_{j=1}^{n} |\langle \phi_j|P|\phi_j\rangle - \langle \phi_j|Q|\phi_j\rangle| = \sum_{j=1}^{n} |q_j|.$$

Since $\operatorname{tr}(P - Q) = \operatorname{tr}(P) - \operatorname{tr}(Q) = 1 - 1 = 0$ and

$$\operatorname{tr}(P) - \operatorname{tr}(Q) = \sum_{j=1}^{n} (\langle \phi_j|P|\phi_j\rangle - \langle \phi_j|Q|\phi_j\rangle) = \sum_{j=1}^{n} q_j$$

we have for all $1 \leq k \leq n$

$$\left| \sum_{j=1}^{k} q_j \right| = \left| \sum_{j=k+1}^{n} q_j \right|.$$

We conclude from the *triangle inequality* that

$$\sum_{j=1}^{n} |q_j| \geq 2 \left| \sum_{j=1}^{k} q_j \right|$$

where equality holds for some $1 \leq k_0 \leq n$. Similarly

$$\sum_{j=1}^{n} |\nu_j| \geq 2 \left| \sum_{j=1}^{k} \nu_j \right|.$$

From Schur's theorem we have

$$\sum_{j=1}^{n} |\nu_j| \geq \sum_{j=1}^{k_0} \nu_j \geq \sum_{j=1}^{k_0} q_j = \sum_{j=1}^{n} |q_j|.$$

Thus

$$\sum_{j=1}^{n} |\nu_j| = |D_1 - D_2| \geq |D_1' - D_2'| = \sum_{j=1}^{n} |q_j|.$$

**Problem 17.** Let $\sigma_1$, $\sigma_2$, $\sigma_3$ be the Pauli spin matrices.
(i) Is the $4 \times 4$ matrix $\rho = \frac{1}{4}(I_2 \otimes I_2 + \sigma_1 \otimes \sigma_1)$ a density matrix?
(ii) Is the $4 \times 4$ matrix $\rho = \frac{1}{4}(I_4 - \sigma_1 \otimes \sigma_1 - \sigma_2 \otimes \sigma_2 - \sigma_3 \otimes \sigma_3)$ a density matrix?

**Solution 17.** (i) We have $\mathrm{tr}(\rho) = 1$ and $\rho$ is hermitian. The eigenvalues of $\rho$ are $1/2, 1/2, 0, 0$. Thus the matrix $\rho$ is a density matrix (mixed state).
(ii) Obviously we have $\rho^* = \rho$, $\mathrm{tr}(\rho) = 1$. Furthermore the matrix is positive semidefinite. We have $\rho^2 = \rho$. Thus $\rho$ is a density matrix. We have a pure state. The density matrix is given by

$$\rho = \begin{pmatrix} 0 & 0 & 0 & 0 \\ 0 & 1/2 & -1/2 & 0 \\ 0 & -1/2 & 1/2 & 0 \\ 0 & 0 & 0 & 0 \end{pmatrix}.$$

**Problem 18.** Consider the eight *Gell-Mann matrices*

$$G_1 = \begin{pmatrix} 0 & 1 & 0 \\ 1 & 0 & 0 \\ 0 & 0 & 0 \end{pmatrix}, \quad G_2 = \begin{pmatrix} 0 & -i & 0 \\ i & 0 & 0 \\ 0 & 0 & 0 \end{pmatrix}, \quad G_3 = \begin{pmatrix} 1 & 0 & 0 \\ 0 & -1 & 0 \\ 0 & 0 & 0 \end{pmatrix},$$

$$G_4 = \begin{pmatrix} 0 & 0 & 1 \\ 0 & 0 & 0 \\ 1 & 0 & 0 \end{pmatrix}, \quad G_5 = \begin{pmatrix} 0 & 0 & -i \\ 0 & 0 & 0 \\ i & 0 & 0 \end{pmatrix}, \quad G_6 = \begin{pmatrix} 0 & 0 & 0 \\ 0 & 0 & 1 \\ 0 & 1 & 0 \end{pmatrix},$$

$$G_7 = \begin{pmatrix} 0 & 0 & 0 \\ 0 & 0 & -i \\ 0 & i & 0 \end{pmatrix}, \quad G_8 = \frac{1}{\sqrt{3}} \begin{pmatrix} 1 & 0 & 0 \\ 0 & 1 & 0 \\ 0 & 0 & -2 \end{pmatrix}.$$

They all have trace 0 and they are hermitian.
(i) Find the anticommutation relations for these matrices.
(ii) Consider the matrix

$$\rho = \frac{1}{3}\left(I_3 + \sqrt{3}\sum_{j=1}^{8} n_j G_j\right)$$

where $n_j \in \mathbb{R}$. What is the condition on the vector $\mathbf{n} = (n_1\ n_2\ \dots\ n_8)^T$ such that $\rho$ is a density matrix for a pure state?

**Solution 18.** (i) We obtain for the anticommutators

$$[G_j, G_k]_+ = \frac{4}{3}\delta_{jk}I_3 + 2\sum_{\ell=1}^{8} d_{jk\ell}G_\ell$$

where the nonzero components of the completely symmetric tensor $d_{jk\ell}$ are

$$d_{118} = d_{228} = d_{338} = -d_{888} = \frac{1}{\sqrt{3}}, \quad d_{448} = d_{558} = d_{668} = d_{778} = -\frac{1}{2\sqrt{3}}$$

$$d_{146} = d_{157} = -d_{247} = d_{256} = d_{344} = d_{355} = -d_{366} = -d_{377} = \frac{1}{2}.$$

(ii) The conditions for a *density matrix* $\rho$ of a *pure state* are

$$\rho^* = \rho, \quad \rho^2 = \rho, \quad \mathrm{tr}(\rho) = 1.$$

Imposing these conditions we obtain for the vector $\mathbf{n}$ that

$$\bar{\mathbf{n}} = \mathbf{n}, \quad \mathbf{n}^T\mathbf{n} = 1, \quad \mathbf{n} \star \mathbf{n} = \mathbf{n}$$

where

$$(\mathbf{a} \star \mathbf{b})_j := \sqrt{3}\sum_{k,\ell=1}^{8} d_{jk\ell}a_k b_\ell.$$

**Problem 19.** Let $\rho$ denote the density matrix (mixed state)

$$\rho := \frac{1}{2}\begin{pmatrix} 1 & 0 \\ 0 & 1 \end{pmatrix}$$

in $\mathbb{C}^2$. Find a pure state $|\Psi\rangle \in \mathbb{C}^2 \otimes \mathbb{C}^2$ such that the reduced density matrix found by taking the partial trace over the second system ($\mathbb{C}^2$) is $\rho$. In other words purify the density matrix $\rho$ to obtain a pure state $|\Psi\rangle$.

**Solution 19.** We begin with the *Schmidt decomposition* of $|\Psi\rangle$ over the Hilbert space $\mathbb{C}^2 \otimes \mathbb{C}^2$

$$|\Psi\rangle = \sum_{j=1}^{\mathrm{Sch}(|\Psi\rangle,\mathbb{C}^2,\mathbb{C}^2)} \sqrt{\lambda_j}|\psi_j\rangle \otimes |\phi_j\rangle$$

where $\lambda_1$ and $\lambda_2$ are the eigenvalues of $\rho$ and $|\psi_1\rangle$ and $|\psi_2\rangle$ are the corresponding orthonormal eigenvectors of $\rho$. The states $|\phi_1\rangle$ and $|\phi_2\rangle$ in $\mathbb{C}^2$ are also orthonormal. The eigenvalues and eigenvectors of $\rho$ are given by $\lambda_1 = \lambda_2 = 1/2$ and

$$|\psi_1\rangle = \begin{pmatrix} 1 \\ 0 \end{pmatrix}, \qquad |\psi_2\rangle = \begin{pmatrix} 0 \\ 1 \end{pmatrix}.$$

Thus the spectral decomposition of $\rho$ is given by

$$\rho = \frac{1}{2}\begin{pmatrix} 1 \\ 0 \end{pmatrix}(1 \quad 0) + \frac{1}{2}\begin{pmatrix} 0 \\ 1 \end{pmatrix}(0 \quad 1).$$

Hence

$$|\Psi\rangle = \frac{1}{\sqrt{2}}\begin{pmatrix} 1 \\ 0 \end{pmatrix} \otimes |\phi_1\rangle + \frac{1}{\sqrt{2}}\begin{pmatrix} 0 \\ 1 \end{pmatrix} \otimes |\phi_2\rangle$$

where $\langle\phi_1|\phi_1\rangle = \langle\phi_2|\phi_2\rangle = 1$ and $\langle\phi_1|\phi_2\rangle = \langle\phi_2|\phi_1\rangle = 0$. Thus we could take $|\Psi\rangle$ as one of the *Bell states*

$$\frac{1}{\sqrt{2}}\begin{pmatrix} 1 \\ 0 \\ 0 \\ 1 \end{pmatrix}, \quad \frac{1}{\sqrt{2}}\begin{pmatrix} 1 \\ 0 \\ 0 \\ -1 \end{pmatrix}, \quad \frac{1}{\sqrt{2}}\begin{pmatrix} 0 \\ 1 \\ 1 \\ 0 \end{pmatrix}, \quad \frac{1}{\sqrt{2}}\begin{pmatrix} 0 \\ 1 \\ -1 \\ 0 \end{pmatrix}$$

but not a product state.

**Problem 20.** Let $r \in [0, 1]$. Consider the density matrix

$$\rho = r|\Phi^+\rangle\langle\Phi^+| + (1 - r)|00\rangle\langle00|$$

where $|\Phi^+\rangle$ is the Bell state

$$|\Phi^+\rangle = \frac{1}{\sqrt{2}}(|00\rangle + |11\rangle).$$

Calculate the eigenvalues of $\rho$.

**Solution 20.** The matrix representation of $\rho$ is

$$\rho = \begin{pmatrix} 1 - r/2 & 0 & 0 & r/2 \\ 0 & 0 & 0 & 0 \\ 0 & 0 & 0 & 0 \\ r/2 & 0 & 0 & r/2 \end{pmatrix}.$$

The characteristic equation is

$$\left(\left(1-\frac{r}{2}\right)-\lambda\right)(-\lambda)(-\lambda)\left(\frac{r}{2}-\lambda\right)-\frac{r}{2}(-\lambda)(-\lambda)\frac{r}{2}=0.$$

Thus two eigenvalues are 0 ($\lambda_2 = 0$, $\lambda_3 = 0$) with the corresponding eigenvectors

$$\mathbf{u}_2 = \begin{pmatrix} 0 \\ 1 \\ 0 \\ 0 \end{pmatrix}, \qquad \mathbf{u}_3 = \begin{pmatrix} 0 \\ 0 \\ 1 \\ 0 \end{pmatrix}.$$

The characteristic equation reduces to

$$\left(\left(1-\frac{r}{2}\right)-\lambda\right)\left(\frac{r}{2}-\lambda\right)-\frac{r^2}{4}=0$$

with the eigenvalues

$$\lambda_{1,4} = \frac{1}{2} \pm \frac{1}{2}\sqrt{1+2r(r-1)}.$$

If $r = 0$ the eigenvalues reduce to 1 and 0. If $r = 1$ the eigenvalues also reduce to 1 and 0.

**Problem 21.** Let $\rho_j^{(1)}$ ($j = 1, 2, \ldots, n$) be density matrices in a finite-dimensional Hilbert space $\mathcal{H}_1$. Let $\rho_j^{(2)}$ ($j = 1, 2, \ldots, n$) be density matrices in a finite-dimensional Hilbert space $\mathcal{H}_2$. Show that the *convex combination*

$$\rho = \sum_{j=1}^n \lambda_j \rho_j^{(1)} \otimes \rho_j^{(2)}, \quad \lambda_j \geq 0, \quad \sum_{j=1}^n \lambda_j = 1$$

is also a density matrix.

**Solution 21.** We have

$$\mathrm{tr}(\rho) = \mathrm{tr}\left(\sum_{j=1}^n \lambda_j \rho_j^{(1)} \otimes \rho_j^{(2)}\right) = \sum_{j=1}^n \lambda_j \mathrm{tr}(\rho_j^{(1)} \otimes \rho_j^{(2)})$$

$$= \sum_{j=1}^n \lambda_j \mathrm{tr}(\rho_j^{(1)})\mathrm{tr}(\rho_j^{(2)}) = \sum_{j=1}^n \lambda_j$$

$$= 1.$$

Obviously $\rho \geq 0$, since $\rho_j^{(1)} \geq 0$, $\rho_j^{(2)} \geq 0$ and $\lambda_j \geq 0$ for $j = 1, 2, \ldots, n$.

**Problem 22.** Given the *Schrödinger equation*

$$i\hbar\frac{\partial}{\partial t}|\psi\rangle = \hat{H}|\psi\rangle.$$

Find the time-evolution of the density matrix

$$\rho(t) := \sum_{j=1}^{n} |\psi^{(j)}(t)\rangle\langle\psi^{(j)}(t)|.$$

**Solution 22.**    From the Schrödinger equation we find

$$-i\hbar\frac{\partial}{\partial t}\langle\psi^{(j)}(t)| = \langle\psi^{(j)}(t)|\hat{H}.$$

Thus

$$
\begin{aligned}
\frac{\partial\rho}{\partial t} &= \sum_{j=1}^{n}\left(\left(\frac{\partial}{\partial t}|\psi^{(j)}(t)\rangle\right)\langle\psi^{(j)}(t)| + |\psi^{(j)}(t)\rangle\left(\frac{\partial}{\partial t}\langle\psi^{(j)}(t)|\right)\right) \\
&= \frac{1}{i\hbar}\sum_{j=1}^{n}\left(\left(\hat{H}|\psi^{(j)}(t)\rangle\right)\langle\psi^{(j)}(t)| - |\psi^{(j)}(t)\rangle\left(\langle\psi^{(j)}(t)|\hat{H}\right)\right) \\
&= \frac{1}{i\hbar}(\hat{H}\rho(t) - \rho(t)\hat{H}) \\
&= \frac{1}{i\hbar}[\hat{H}, \rho(t)].
\end{aligned}
$$

Note that the equation of motion for $\rho(t)$ differs from the Heisenberg equation of motion by a minus sign. Since $\rho(t)$ is constructed from state vectors it is not an observable like other hermitian operators, so there is no reason to expect that its time-evolution will be the same. The solution to the equation of motion is given by

$$\rho(t) = e^{-i\hat{H}t/\hbar}\rho(0)e^{i\hat{H}t/\hbar}.$$

**Problem 23.**    Consider the Hamilton operator

$$\hat{H}(t) = -\frac{\gamma}{2}\boldsymbol{\sigma}\cdot\mathbf{B}(t) \equiv -\frac{\gamma}{2}(\sigma_1 B_1(t) + \sigma_2 B_2(t) + \sigma_3 B_3(t))$$

where $\gamma$ denotes the gyromagnetic ratio and $\mathbf{B}(t)$ denotes the time-dependent magnetic induction. The time-evolution of the density matrix $\rho(t)$ obeys the *von Neumann equation*

$$i\hbar\frac{d\rho(t)}{dt} = [\hat{H}(t), \rho(t)]$$

and the time-dependent expectation value of the spin vector is given by

$$\langle\boldsymbol{\sigma}(t)\rangle := \text{tr}(\boldsymbol{\sigma}\rho(t))$$

or, written in components

$$\langle\sigma_1(t)\rangle = \mathrm{tr}(\sigma_1\rho(t)), \quad \langle\sigma_2(t)\rangle = \mathrm{tr}(\sigma_2\rho(t)), \quad \langle\sigma_3(t)\rangle = \mathrm{tr}(\sigma_3\rho(t)).$$

It follows that the *Bloch vector* $\mathbf{n}(t)$ pertaining to $\rho(t)$ is related to the spin vector as follows

$$\mathbf{n}(t) = \langle\boldsymbol{\sigma}(t)\rangle$$

or, written in components

$$n_1(t) = \langle\sigma_1(t)\rangle, \quad n_2(t) = \langle\sigma_2(t)\rangle, \quad n_3(t) = \langle\sigma_3(t)\rangle.$$

Find the time-evolution of $\mathbf{n}(t)$.

**Solution 23.** We have

$$\frac{dn_j}{dt} = \left\langle \frac{d\sigma_j(t)}{dt} \right\rangle = \mathrm{tr}\left(\sigma_j\frac{d\rho(t)}{dt}\right)$$

where $j = 1, 2, 3$. Inserting the right-hand side of the von Neumann equation, using the *cyclic invariance* of the trace

$$\mathrm{tr}(XYZ) = \mathrm{tr}(ZXY) = \mathrm{tr}(YZX)$$

and the properties $\sigma_1\sigma_2 = i\sigma_3$, $\sigma_2\sigma_3 = i\sigma_1$, $\sigma_3\sigma_1 = i\sigma_2$, we obtain

$$\frac{d}{dt}\mathbf{n}(t) = \frac{\gamma}{\hbar}\mathbf{n}(t) \times \mathbf{B}(t)$$

where $\times$ denotes the vector product.

**Problem 24.** Consider the state

$$|\psi\rangle = \begin{pmatrix} \cos(\theta) \\ e^{i\phi}\sin(\theta) \end{pmatrix}$$

in the Hilbert space $\mathbb{C}^2$, where $\phi, \theta \in \mathbb{R}$. Let $\rho(t = 0) = \rho(0) = |\psi\rangle\langle\psi|$ be a density matrix at time $t = 0$. Given the Hamilton operator $\hat{H} = \hbar\omega\sigma_1$. Solve the von Neumann equation to find $\rho(t)$.

**Solution 24.** We obtain for the density matrix at time $t = 0$

$$\rho(0) = |\psi\rangle\langle\psi| = \begin{pmatrix} \cos^2(\theta) & e^{-i\phi}\cos(\theta)\sin(\theta) \\ e^{i\phi}\cos(\theta)\sin(\theta) & \sin^2(\theta) \end{pmatrix}.$$

Now

$$\rho(t) = e^{-i\hat{H}t/\hbar}\rho(0)e^{i\hat{H}t/\hbar} = e^{-i\omega t\sigma_1}\rho(0)e^{i\omega t\sigma_1}.$$

Since

$$e^{i\hat{H}t/\hbar} = e^{i\omega t\sigma_1} = \begin{pmatrix} \cos(\omega t) & i\sin(\omega t) \\ i\sin(\omega t) & \cos(\omega t) \end{pmatrix}$$

it follows that

$$\rho(t) = U^*(t) \begin{pmatrix} \cos^2(\theta) & e^{-i\phi}\cos(\theta)\sin(\theta) \\ e^{i\phi}\cos(\theta)\sin(\theta) & \sin^2(\theta) \end{pmatrix} U(t)$$

where

$$U(t) = \begin{pmatrix} \cos(\omega t) & i\sin(\omega t) \\ i\sin(\omega t) & \cos(\omega t) \end{pmatrix}.$$

**Problem 25.** Consider the Hilbert space $\mathbb{C}^n$. Let $\rho$ be a density matrix, i.e. $\rho \geq 0$ and $\text{tr}(\rho) = 1$. The mean value of an observable $A$ (hermitian $n \times n$ matrix) is given by

$$\langle A \rangle = \text{tr}(\rho A).$$

If the density $\rho$ is unknown, then it may be determined using $n^2$ mean values $\langle A^{(k)} \rangle$ $(k = 1, 2, \ldots, n^2)$ obtained from measurement if the set $\{A^{(k)}\}$ is a basis in the space of all hermitian $n \times n$ matrices.
(i) Let $n = 2$,

$$A = \sigma_2 = \begin{pmatrix} 0 & -i \\ i & 0 \end{pmatrix}$$

and $\text{tr}(\rho A) = 0$, $\text{tr}(\rho A^2) = 1$, $\text{tr}(\rho A^3) = 0$, $\text{tr}(\rho A^4) = 1$. Find the density matrix.
(ii) Let $n = 2$ and $\text{tr}(\rho I_2) = 1$, $\text{tr}(\rho \sigma_1) = -1$, $\text{tr}(\rho \sigma_2) = 0$, $\text{tr}(\rho \sigma_3) = 0$. Find $\rho$.

**Solution 25.** (i) Note that $\sigma_2^2 = I_2$, $\sigma_2^3 = \sigma_2$, $\sigma_2^4 = I_2$. The density matrix is

$$\rho = \begin{pmatrix} 1 - \epsilon & 0 \\ 0 & \epsilon \end{pmatrix}, \qquad \epsilon \in [0, 1].$$

(ii) The $2 \times 2$ matrices $I_2, \sigma_1, \sigma_2, \sigma_3$ form an orthogonal basis in the Hilbert space of the $2 \times 2$ matrices with scalar product $\langle X, Y \rangle = \text{tr}(XY^*)$. The density matrix is

$$\rho = \begin{pmatrix} 1/2 & -1/2 \\ -1/2 & 1/2 \end{pmatrix}.$$

**Problem 26.** Let $|0\rangle$, $|1\rangle$ be the standard basis in $\mathbb{C}^2$. Consider the entangled state

$$|\psi\rangle = \frac{1}{\sqrt{2}}(|0\rangle \otimes |1\rangle - |1\rangle \otimes |0\rangle)$$

with the density matrix $\rho = |\psi\rangle\langle\psi|$. Find the reduced density matrix $\rho_1$. Discuss.

**Solution 26.**   We obtain

$$\rho_1 = \frac{1}{2}(|0\rangle\langle0| + |1\rangle\langle1|) = \frac{1}{2}\begin{pmatrix} 1 & 0 \\ 0 & 1 \end{pmatrix}.$$

Thus we have mixed state.

**Problem 27.**   Consider a mixture of 25% of the pure state $(1,0)^T$, 25% of the pure state $(0,1)^T$ and 50% of the pure state $\frac{1}{\sqrt{2}}(1,1)^T$ described by the density matrix

$$\rho = \frac{1}{4}\begin{pmatrix} 1 \\ 0 \end{pmatrix}(1 \quad 0) + \frac{1}{4}\begin{pmatrix} 0 \\ 1 \end{pmatrix}(0 \quad 1) + \frac{1}{2}\frac{1}{\sqrt{2}}\begin{pmatrix} 1 \\ 1 \end{pmatrix}\frac{1}{\sqrt{2}}(1 \quad 1).$$

Find the *spectral representation* of $\rho$. Use the spectral representation of $\rho$ to find another mixture of pure states with the same (measurement) statistical properties as $\rho$.

**Solution 27.**   The density matrix takes the form

$$\rho = \frac{1}{4}\begin{pmatrix} 2 & 1 \\ 1 & 2 \end{pmatrix}$$

with the eigenvalues $\lambda_1 = 3/4$ and $\lambda_2 = 1/4$ with the corresponding normalized eigenvectors

$$\mathbf{v}_1 = \frac{1}{\sqrt{2}}\begin{pmatrix} 1 \\ 1 \end{pmatrix}, \qquad \mathbf{v}_2 = \frac{1}{\sqrt{2}}\begin{pmatrix} 1 \\ -1 \end{pmatrix}.$$

Applying the spectral theorem $\rho$ can be written as

$$\rho = \frac{3}{4}\frac{1}{\sqrt{2}}\begin{pmatrix} 1 \\ 1 \end{pmatrix}\frac{1}{\sqrt{2}}(1 \quad 1) + \frac{1}{4}\frac{1}{\sqrt{2}}\begin{pmatrix} 1 \\ -1 \end{pmatrix}\frac{1}{\sqrt{2}}(1 \quad -1).$$

Consequently $\rho$ can be realized by a mixture of 75% of the state $(1,1)^T/\sqrt{2}$ and 25% of the state $(1,-1)^T/\sqrt{2}$.

**Problem 28.**   Consider the Hilbert space $\mathbb{C}^n$. Let $\rho$ be a density matrix in this Hilbert space and $H$ and $K$ be two hermitian $n \times n$ matrices. One defines $\langle H \rangle := \mathrm{tr}(\rho H)$, $\langle H^2 \rangle := \mathrm{tr}(\rho H^2)$ and analogously for $K$. Let

$$\Delta H := \sqrt{\langle H^2 \rangle - \langle H \rangle^2}, \qquad \Delta K := \sqrt{\langle K^2 \rangle - \langle K \rangle^2}.$$

Then we have the *uncertainty relation*

$$(\Delta H)(\Delta K) \geq \frac{1}{2} |\langle i[H, K] \rangle| \, .$$

Let

$$\rho = \frac{1}{2} \begin{pmatrix} 1 & 0 & 0 \\ 0 & 0 & 0 \\ 0 & 0 & 1 \end{pmatrix}, \quad H = \begin{pmatrix} 0 & 1 & 0 \\ 1 & 2 & 0 \\ 0 & 0 & 0 \end{pmatrix}, \quad K = \begin{pmatrix} 0 & i & 0 \\ -i & 0 & 0 \\ 0 & 0 & 0 \end{pmatrix}.$$

Show that the uncertainty relation becomes an equality for the given $\rho$, $H$ and $K$.

**Solution 28.**   Note that $\rho$ is a mixed state since $\rho^2 \neq \rho$. Straightforward calculations yield

$$\langle H \rangle = 0, \quad \langle K \rangle = 0, \quad \langle H^2 \rangle = \frac{1}{2}, \quad \langle K^2 \rangle = \frac{1}{2}$$

and the commutator of $H$ and $K$ is given by

$$[H, K] = \begin{pmatrix} -2i & 2i & 0 \\ 2i & 2i & 0 \\ 0 & 0 & 0 \end{pmatrix}.$$

Thus $|\langle i[H, K] \rangle| = 1/2$ and the equality follows.

## Programming Problems

**Problem 1.**   Consider the matrix

$$\rho = \begin{pmatrix} 3/4 & \sqrt{2}e^{-i\phi}/4 \\ \sqrt{2}e^{i\phi}/4 & 1/4 \end{pmatrix}.$$

Check that the matrix is a density matrix. Is it a pure or mixed state? Apply computer algebra. Find $\mathrm{tr}(\sigma_1 \rho)$.

**Solution 1.**   The following Maxima program will do the job

```
/* densitycheck.mac */
load("nchrpl");
rho: matrix([3/4,sqrt(2)*exp(-%i*phi)/4],[sqrt(2)*exp(%i*phi)/4,1/4]);
rhoT: transpose(rho); rhoTC: conjugate(rhoT);
tr: mattrace(rho);
E: eigenvalues(rho);
e1: first(first(E)); e2: second(first(E));
```

```
if(rho=rhoTC and tr=1 and e1>0 and e2>0)
then print("matrix is density matrix")
else print("matrix is not a density matrix");
rho2: rho . rho;
if(rho2=rho and tr=1 and rho=rhoTC) then print("pure state")
else print("not a pure state");
sig1: matrix([0,1],[1,0]);
trsig1: mattrace(rho . sig1);
```

The matrix is a density matrix and a mixed state. The two eigenvalues are $(2 + \sqrt{3})/4$, $(2 - \sqrt{3})/4$. We find $\text{tr}(\sigma_1 \rho) = \frac{1}{\sqrt{2}} \cos(\phi)$.

**Problem 2.** Consider the density matrix

$$\rho = \frac{1}{2} \begin{pmatrix} 1 & -1 \\ -1 & 1 \end{pmatrix}$$

and let $A$ be an $2 \times 2$ real symmetric matrix. Assume that $\text{tr}(\rho A) = -1$, $\text{tr}(\rho A^2) = 1$. Reconstruct the matrix from this information.

**Solution 2.** Utilizing the Maxima program

```
/* density1.mac */
load("nchrpl");
rho: matrix([1/2,-1/2],[-1/2,1/2]);
A: matrix([a11,a12],[a12,a22]);
A2: A . A;
r1: mattrace(rho . A);  r1: ratsimp(r1);
r2: mattrace(rho . A2); r2: ratsimp(r2);
solve([r1+1=0,r2-1=0],[a11,a12,a22]);
```

we obtain

$$A = \begin{pmatrix} r & 1+r \\ 1+r & r \end{pmatrix}$$

with $r$ an arbitrary real constant.

**Problem 3.** Let $S$ be the set of unit vectors in the Hilbert space $\mathbb{C}^n$. Let $\mathbf{u} \in S$. A function $\mu(\mathbf{u})$ from $S$ to $\mathbb{R}$ is called a generalized probability measure if the following two conditions hold: (i) for $\mathbf{u} \in S$, $0 \leq \mu(\mathbf{u}) \leq 1$, (ii) if $\mathbf{u}_1, \ldots, \mathbf{u}_n$ form an orthonormal basis in the Hilbert space $\mathbb{C}^n$, then $\sum_{j=1}^n \mu(\mathbf{u}_j) = 1$.
Let $n \geq 3$. Then any generalized probability measure $\mu$ on $\mathbb{C}^n$ has the form

$$\mu(\rho) = \text{tr}(\rho \mathbf{u} \mathbf{u}^*)$$

for a uniquely defined density matrix $\rho$ (Gleason 1957).

(i) Consider the Hilbert space $\mathbb{C}^3$, the orthonormal basis

$$\mathbf{u}_1 = \frac{1}{\sqrt{2}} \begin{pmatrix} 1 \\ 0 \\ 1 \end{pmatrix}, \quad \mathbf{u}_2 = \begin{pmatrix} 0 \\ 1 \\ 0 \end{pmatrix}, \quad \mathbf{u}_3 = \frac{1}{\sqrt{2}} \begin{pmatrix} 1 \\ 0 \\ -1 \end{pmatrix}$$

and the density matrix

$$\rho = \frac{1}{3} \begin{pmatrix} 1 & 1 & 1 \\ 1 & 1 & 1 \\ 1 & 1 & 1 \end{pmatrix}.$$

Find $\mu(\mathbf{u}_1)$, $\mu(\mathbf{u}_2)$, $\mu(\mathbf{u}_3)$.

(ii) Consider the Hilbert space $\mathbb{C}^4$, the orthonormal basis

$$\mathbf{u}_1 = \frac{1}{\sqrt{2}} \begin{pmatrix} e^{i\phi} \\ 0 \\ 0 \\ e^{i\phi} \end{pmatrix}, \quad \mathbf{u}_2 = \frac{1}{\sqrt{2}} \begin{pmatrix} e^{i\phi} \\ 0 \\ 0 \\ -e^{i\phi} \end{pmatrix},$$

$$\mathbf{u}_3 = \frac{1}{\sqrt{2}} \begin{pmatrix} 0 \\ e^{i\phi} \\ e^{i\phi} \\ 0 \end{pmatrix}, \quad \mathbf{u}_4 = \frac{1}{\sqrt{2}} \begin{pmatrix} 0 \\ e^{i\phi} \\ -e^{i\phi} \\ 0 \end{pmatrix}$$

and the density matrix

$$\rho = \frac{1}{4} \begin{pmatrix} 1 & 1 & 1 & 1 \\ 1 & 1 & 1 & 1 \\ 1 & 1 & 1 & 1 \\ 1 & 1 & 1 & 1 \end{pmatrix}.$$

Find $\mu(\mathbf{u}_1)$, $\mu(\mathbf{u}_2)$, $\mu(\mathbf{u}_3)$, $\mu(\mathbf{u}_4)$.

**Solution 3.** Applying the Maxima program

```
/* Gleason.mac */
load("nchrpl");
u1: matrix([1],[0],[1])/sqrt(2); u1T: transpose(u1);
u2: matrix([0],[1],[0]); u2T: transpose(u2);
u3: matrix([1],[0],[-1])/sqrt(2); u3T: transpose(u3);
rho: matrix([1,1,1],[1,1,1],[1,1,1])/3;
muu1: mattrace(rho . u1 . u1T);
muu2: mattrace(rho . u2 . u2T);
muu3: mattrace(rho . u3 . u3T);
v1: matrix([exp(%i*phi)],[0],[0],[exp(%i*phi)])/sqrt(2);
v1T: transpose(v1); v1TC: conjugate(v1T);
v2: matrix([exp(%i*phi)],[0],[0],[exp(%i*phi)])/sqrt(2);
```

```
v2T: transpose(v2); v2TC: conjugate(v2T);
v3: matrix([0],[exp(%i*phi)],[exp(%i*phi)],[0])/sqrt(2);
v3T: transpose(v3); v3TC: conjugate(v3T);
v4: matrix([0],[exp(%i*phi)],[exp(%i*phi)],[0])/sqrt(2);
v4T: transpose(v4); v4TC: conjugate(v4T);
rho4: matrix([1,1,1,1],[1,1,1,1],[1,1,1,1],[1,1,1,1])/4;
muv1: mattrace(rho4 . v1 . v1TC); muv2: mattrace(rho4 . v2 . v2TC);
muv3: mattrace(rho4 . v3 . v3TC); muv4: mattrace(rho4 . v4 . v4TC);
```

we find for (i) $\mu(\mathbf{u}_1) = 2/3$, $\mu(\mathbf{u}_2) = 1/3$, $\mu(\mathbf{u}_3) = 0$ and for (ii) we find $\mu(\mathbf{u}_1) = 1/2$, $\mu(\mathbf{u}_2) = 1/2$, $\mu(\mathbf{u}_3) = 1/2$, $\mu(\mathbf{u}_4) = 1/2$.

## 4.3 Supplementary Problems

**Problem 1.** Consider a spin-1 system. Any pure state can be parametrized, with a suitable choice of its phase as

$$|\psi\rangle = \begin{pmatrix} e^{i\alpha} \sin(\theta) \cos(\phi) \\ e^{i\beta} \sin(\theta) \sin(\phi) \\ \cos(\theta) \end{pmatrix}$$

where $0 \le \theta, \phi \le \pi/2$ and $0 \le \alpha, \beta < 2\pi$. Find the density matrix $\rho = |\psi\rangle\langle\psi|$ and $\text{tr}(\rho S_1)$, where $S_1$ is the spin-1 matrix

$$S_1 = \frac{1}{\sqrt{2}} \begin{pmatrix} 1 & 0 & 1 \\ 0 & 1 & 0 \\ 1 & 0 & 1 \end{pmatrix}.$$

Note that the density matrix depends on all four parameters.

**Problem 2.** Let $\sigma_0, \sigma_1, \sigma_2, \sigma_3$ be the Pauli spin matrices, where $\sigma_0 = I_2$ is the $2 \times 2$ unit matrix and let

$$\mathbf{v} = \begin{pmatrix} v_1 & v_2 & v_3 \end{pmatrix}^T$$

be a vector in $\mathbb{R}^3$ with $\|\mathbf{v}\| \le 1$.
(i) Show that $\rho_{\mathbf{v}} = \frac{1}{2}(\sigma_0 + v_1\sigma_1 + v_2\sigma_2 + v_3\sigma_3)$ is a density matrix.
(ii) Is $\rho = \frac{1}{4}(\sigma_0 \otimes \sigma_0 + \sum_{j=1}^{3} v_j\sigma_j \otimes \sigma_j)$ a density matrix?
(iii) Is

$$\rho = \frac{1}{2^3}(\sigma_0 \otimes \sigma_0 \otimes \sigma_0 + \sum_{j=1}^{3} v_j\sigma_j \otimes \sigma_j \otimes \sigma_j)$$

a density matrix? Extend the result to $n$ Kronecker products.

**Problem 3.**   Let $|0\rangle$, $|1\rangle$, $\ldots$, $|n\rangle$ be an orthonormal basis in $\mathbb{C}^{n+1}$. Are the states

$$|\psi_0\rangle = \frac{1}{\sqrt{2}}|0\rangle \otimes |0\rangle + \frac{1}{\sqrt{2n}}\sum_{j=1}^{n}|j\rangle \otimes |j\rangle$$

$$|\psi_1\rangle = \frac{1}{\sqrt{2}}|0\rangle \otimes |0\rangle - \frac{1}{\sqrt{2n}}\sum_{j=1}^{n}|j\rangle \otimes |j\rangle$$

normalized? Are the state orthogonal to each other? Is

$$\rho = (|\psi_0\rangle\langle\psi_0|) \otimes (|\psi_1\rangle\langle\psi_1|)$$

a density matrix?

**Problem 4.**   Consider the Pauli spin matrices $\sigma_1$ and $\sigma_2$. Let $\rho$ be a $2 \times 2$ density matrix with $\text{tr}(\rho\sigma_1) = 1$, $\text{tr}(\rho\sigma_2) = 0$. Reconstruct the density matrix $\rho$ from this information. Show that

$$\rho = \frac{1}{2}\begin{pmatrix} 1 & 1 \\ 1 & 1 \end{pmatrix}.$$

**Problem 5.**   Let $\mathcal{H}_1$ and $\mathcal{H}_2$ be two Hilbert spaces and $\mathcal{H}_1 \otimes \mathcal{H}_2$ be the product Hilbert space. Let $\rho$ be a density operators of the Hilbert space $\mathcal{H}_1 \otimes \mathcal{H}_2$. Show that if one of the reduced density operators $\text{tr}_{\mathcal{H}_2}(\rho) = \rho_1$ or $\text{tr}_{\mathcal{H}_1}(\rho) = \rho_2$ is pure, then $\rho = \rho_1 \otimes \rho_2$. If both $\rho_1$ and $\rho_2$ are pure, then $\rho$ is pure too.

**Problem 6.**   Find a normalized state $|\psi\rangle$ in the Hilbert space $\mathbb{C}^2$ such that we have the density matrix

$$|\psi\rangle\langle\psi| = \frac{1}{2}\left(I_2 + \frac{1}{\sqrt{3}}(\sigma_1 + \sigma_2 + \sigma_3)\right).$$

# Chapter 5

# Trace and Partial Trace

## 5.1 Introduction

Let $\mathcal{H}$ be the finite dimensional Hilbert space $\mathbb{C}^n$ with an orthonormal basis $\{\, |\phi_j\rangle : j = 1, 2, \ldots, n \,\}$. Let $A$ be a linear operator ($n \times n$ matrix) acting in this Hilbert space. Then the *trace* of $A$ is defined as

$$\mathrm{tr}(A) := \sum_{j=1}^{n} \langle \phi_j | A | \phi_j \rangle .$$

The trace is independent of the chosen orthonormal basis. For the trace we have cyclic invariance. Let $A$, $B$, $C$ be $n \times n$ matrices over $\mathbb{C}$. Then

$$\mathrm{tr}(AB) = \mathrm{tr}(BA)$$

and (*cyclic invariance*)

$$\mathrm{tr}(ABC) = \mathrm{tr}(CAB) = \mathrm{tr}(BCA).$$

The trace of an $n \times n$ matrix $A$ is the sum of the eigenvalues counting multiplicities. The eigenvalues of $A$ can be reconstructed from

$$\mathrm{tr}(A) = \lambda_1 + \lambda_2 + \cdots + \lambda_n$$
$$\mathrm{tr}(A^2) = \lambda_1^2 + \lambda_2^2 + \cdots + \lambda_n^2$$
$$\vdots$$
$$\mathrm{tr}(A^n) = \lambda_1^n + \lambda_2^n + \cdots + \lambda_n^n.$$

If $|\psi\rangle$ is a normalized state in $\mathbb{C}^n$, then

$$\text{tr}(|\psi\rangle\langle\psi|) = 1.$$

For any $n \times n$ matrix over $\mathbb{C}$ we have the identity

$$\det(\exp(A)) = \exp(\text{tr}(A)).$$

Let $A$ be an $n \times n$ matrix over $\mathbb{C}$ and $B$ be an $m \times m$ matrix over $\mathbb{C}$. Then

$$\text{tr}(A \otimes B) = \text{tr}(A)\text{tr}(B).$$

The calculation of the *partial trace* plays a central role in quantum computing. Suppose a finite dimensional quantum system $S_{AB}$ is a system composed of two subsystems $S_A$ and $S_B$. The finite dimensional Hilbert space $\mathcal{H}$ of $S_{AB}$ is given by the tensor product of the individual Hilbert spaces $\mathcal{H}_A \otimes \mathcal{H}_B$. Let $N_A := \dim(\mathcal{H}_A)$ and $N_B := \dim(\mathcal{H}_B)$. Let $\rho_{AB}$ be the density matrix of $S_{AB}$. Using the partial trace we can define the density operators $\rho_A$ and $\rho_B$ in the subspaces $\mathcal{H}_A$ and $\mathcal{H}_B$ as follows

$$\rho_A := \text{tr}_B(\rho_{AB}) \equiv \sum_{j=1}^{N_B}(I_A \otimes \langle\phi_j|)\rho_{AB}(I_A \otimes |\phi_j\rangle)$$

and

$$\rho_B := \text{tr}_A(\rho_{AB}) \equiv \sum_{j=1}^{N_A}(\langle\psi_j| \otimes I_B)\rho_{AB}(|\psi_j\rangle \otimes I_B))$$

where $I_A$ is the identity operator in $\mathcal{H}_A$, $I_B$ is the identity operator in $\mathcal{H}_B$ and

$$|\phi_j\rangle, \quad (j = 1, 2, \ldots, N_B)$$

is an orthonormal basis in $\mathcal{H}_B$ and

$$|\psi_j\rangle, \quad (j = 1, 2, \ldots, N_A)$$

is an orthonormal basis in $\mathcal{H}_A$. For example we could select the standard bases in the two finite dimensional Hilbert spaces $\mathcal{H}_A$ and $\mathcal{H}_B$.

The partial trace can also be calculated as follows. Consider a bipartite state

$$|\psi\rangle = \sum_{j=0}^{n-1}\sum_{k=0}^{n-1} c_{jk}|jk\rangle \equiv \sum_{j=0}^{n-1}\sum_{k=0}^{n-1} c_{jk}|j\rangle \otimes |k\rangle, \qquad \sum_{j=0}^{n-1}\sum_{k=0}^{n-1} c_{jk}c_{jk}^* = 1$$

in the finite-dimensional Hilbert space $\mathcal{H} = \mathbb{C}^n \otimes \mathbb{C}^n$. We can define the $n \times n$ matrix

$$\Lambda_{jk} := c_{jk}, \qquad j, k = 0, 1, \ldots, n - 1.$$

Then we have (prove it)

$$\rho_A = \text{tr}_B(\rho) = \text{tr}_B(|\psi\rangle\langle\psi|) = \Lambda\Lambda^\dagger.$$

## 5.2   Solved Problems

**Problem 1.**   Consider the Pauli spin matrix $\sigma_1$. Calculate the trace of $\sigma_1$ with respect to the standard basis. Calculate the trace of $\sigma_1$ with respect to the *Hadamard basis*

$$\left\{ \frac{1}{\sqrt{2}} \begin{pmatrix} 1 \\ 1 \end{pmatrix}, \ \frac{1}{\sqrt{2}} \begin{pmatrix} 1 \\ -1 \end{pmatrix} \right\}.$$

**Solution 1.**   Obviously for the standard basis we find $\text{tr}(\sigma_1) = 0$. Since

$$\frac{1}{\sqrt{2}}(1 \ 1) \begin{pmatrix} 0 & 1 \\ 1 & 0 \end{pmatrix} \frac{1}{\sqrt{2}} \begin{pmatrix} 1 \\ 1 \end{pmatrix} = 1, \quad \frac{1}{\sqrt{2}}(1 \ -1) \begin{pmatrix} 0 & 1 \\ 1 & 0 \end{pmatrix} \frac{1}{\sqrt{2}} \begin{pmatrix} 1 \\ -1 \end{pmatrix} = -1$$

we also obtain (as expected since the trace is independent of the chosen orthonormal basis) for the Hadamard basis the result 0. In the Hadamard basis we have the Pauli spin matrix $\sigma_3$.

**Problem 2.**   Consider the hermitian matrix

$$H = \frac{1}{\sqrt{2}} \begin{pmatrix} 1 & 1 \\ 1 & -1 \end{pmatrix}.$$

Find $\text{tr}(H)$ and $\text{tr}(H^2)$ and then the eigenvalues from this information.

**Solution 2.**   Since $\text{tr}(H) = 0$, $\text{tr}(H^2) = 2$ and $\lambda_1 + \lambda_2 = 0$, $\lambda_1^2 + \lambda_2^2 = 2$ we obtain the eigenvalues $\lambda_1 = 1$, $\lambda_2 = -1$.

**Problem 3.**   Consider the entangled state in $\mathbb{C}^4$

$$|\psi\rangle = \frac{1}{\sqrt{2}} \left( \begin{pmatrix} 1 \\ 0 \end{pmatrix} \otimes \begin{pmatrix} 0 \\ 1 \end{pmatrix} - \begin{pmatrix} 0 \\ 1 \end{pmatrix} \otimes \begin{pmatrix} 1 \\ 0 \end{pmatrix} \right) = \frac{1}{\sqrt{2}} \begin{pmatrix} 0 \\ 1 \\ -1 \\ 0 \end{pmatrix}$$

and the density matrix $\rho = |\psi\rangle\langle\psi|$. Find $\rho_1 = \text{tr}_2(\rho)$, $\rho_2 = \text{tr}_1(\rho)$ i.e. calculate the partial trace.

**Solution 3.**   We have

$$\langle\psi| = \frac{1}{\sqrt{2}}(0 \ \ 1 \ \ -1 \ \ 0).$$

Thus

$$\rho = \frac{1}{2} \begin{pmatrix} 0 & 0 & 0 & 0 \\ 0 & 1 & -1 & 0 \\ 0 & -1 & 1 & 0 \\ 0 & 0 & 0 & 0 \end{pmatrix}.$$

Using the basis

$$\begin{pmatrix} 1 \\ 0 \end{pmatrix} \otimes \begin{pmatrix} 1 & 0 \\ 0 & 1 \end{pmatrix}, \qquad \begin{pmatrix} 0 \\ 1 \end{pmatrix} \otimes \begin{pmatrix} 1 & 0 \\ 0 & 1 \end{pmatrix}$$

and the basis

$$\begin{pmatrix} 1 & 0 \\ 0 & 1 \end{pmatrix} \otimes \begin{pmatrix} 1 \\ 0 \end{pmatrix}, \qquad \begin{pmatrix} 1 & 0 \\ 0 & 1 \end{pmatrix} \otimes \begin{pmatrix} 0 \\ 1 \end{pmatrix}$$

we find the density matrix (mixed state)

$$\rho_1 = \rho_2 = \frac{1}{2} \begin{pmatrix} 1 & 0 \\ 0 & 1 \end{pmatrix}.$$

**Problem 4.** Consider the $4 \times 4$ matrix (*density matrix*)

$$|\mathbf{u}\rangle\langle\mathbf{u}| = (u_j \bar{u}_k), \qquad j, k = 1, \ldots, 4$$

in the product Hilbert space $\mathcal{H}_A \otimes \mathcal{H}_B \equiv \mathbb{C}^4$, where $\mathcal{H}_A = \mathcal{H}_B = \mathbb{C}^2$.
(i) Calculate $\mathrm{tr}_A(|\mathbf{u}\rangle\langle\mathbf{u}|)$, where the basis is given by

$$\begin{pmatrix} 1 \\ 0 \end{pmatrix} \otimes I_2, \qquad \begin{pmatrix} 0 \\ 1 \end{pmatrix} \otimes I_2$$

and $I_2$ denotes the $2 \times 2$ unit matrix.
(ii) Find the partial trace $\mathrm{tr}_B(|\mathbf{u}\rangle\langle\mathbf{u}|)$, where the basis is given by

$$I_2 \otimes \begin{pmatrix} 1 \\ 0 \end{pmatrix}, \qquad I_2 \otimes \begin{pmatrix} 0 \\ 1 \end{pmatrix}.$$

**Solution 4.**   (i) Since

$$\begin{pmatrix} 1 \\ 0 \end{pmatrix} \otimes I_2 = \begin{pmatrix} 1 & 0 \\ 0 & 1 \\ 0 & 0 \\ 0 & 0 \end{pmatrix}, \qquad \begin{pmatrix} 0 \\ 1 \end{pmatrix} \otimes I_2 = \begin{pmatrix} 0 & 0 \\ 0 & 0 \\ 1 & 0 \\ 0 & 1 \end{pmatrix}$$

we find, using the transpose of these matrices on the left-hand side of $|\mathbf{u}\rangle\langle\mathbf{u}|$, that

$$\mathrm{tr}_A(|\mathbf{u}\rangle\langle\mathbf{u}|) = \begin{pmatrix} 1 & 0 & 0 & 0 \\ 0 & 1 & 0 & 0 \end{pmatrix} \begin{pmatrix} u_1\bar{u}_1 & u_1\bar{u}_2 & u_1\bar{u}_3 & u_1\bar{u}_4 \\ u_2\bar{u}_1 & u_2\bar{u}_2 & u_2\bar{u}_3 & u_2\bar{u}_4 \\ u_3\bar{u}_1 & u_3\bar{u}_2 & u_3\bar{u}_3 & u_3\bar{u}_4 \\ u_4\bar{u}_1 & u_4\bar{u}_2 & u_4\bar{u}_3 & u_4\bar{u}_4 \end{pmatrix} \begin{pmatrix} 1 & 0 \\ 0 & 1 \\ 0 & 0 \\ 0 & 0 \end{pmatrix}$$

$$+ \begin{pmatrix} 0 & 0 & 1 & 0 \\ 0 & 0 & 0 & 1 \end{pmatrix} \begin{pmatrix} u_1\bar{u}_1 & u_1\bar{u}_2 & u_1\bar{u}_3 & u_1\bar{u}_4 \\ u_2\bar{u}_1 & u_2\bar{u}_2 & u_2\bar{u}_3 & u_2\bar{u}_4 \\ u_3\bar{u}_1 & u_3\bar{u}_2 & u_3\bar{u}_3 & u_3\bar{u}_4 \\ u_4\bar{u}_1 & u_4\bar{u}_2 & u_4\bar{u}_3 & u_4\bar{u}_4 \end{pmatrix} \begin{pmatrix} 0 & 0 \\ 0 & 0 \\ 1 & 0 \\ 0 & 1 \end{pmatrix}.$$

Using matrix multiplication and matrix addition we obtain

$$\text{tr}_A(|\mathbf{u}\rangle\langle\mathbf{u}|) = \begin{pmatrix} u_1\bar{u}_1 + u_3\bar{u}_3 & u_1\bar{u}_2 + u_3\bar{u}_4 \\ u_2\bar{u}_1 + u_4\bar{u}_3 & u_2\bar{u}_2 + u_4\bar{u}_4 \end{pmatrix}.$$

(ii) Since

$$I_2 \otimes \begin{pmatrix} 1 \\ 0 \end{pmatrix} = \begin{pmatrix} 1 & 0 \\ 0 & 0 \\ 0 & 1 \\ 0 & 0 \end{pmatrix}, \qquad I_2 \otimes \begin{pmatrix} 0 \\ 1 \end{pmatrix} = \begin{pmatrix} 0 & 0 \\ 1 & 0 \\ 0 & 0 \\ 0 & 1 \end{pmatrix}$$

we find

$$\text{tr}_B(|\mathbf{u}\rangle\langle\mathbf{u}|) = \begin{pmatrix} 1 & 0 & 0 & 0 \\ 0 & 0 & 1 & 0 \end{pmatrix} \begin{pmatrix} u_1\bar{u}_1 & u_1\bar{u}_2 & u_1\bar{u}_3 & u_1\bar{u}_4 \\ u_2\bar{u}_1 & u_2\bar{u}_2 & u_2\bar{u}_3 & u_2\bar{u}_4 \\ u_3\bar{u}_1 & u_3\bar{u}_2 & u_3\bar{u}_3 & u_3\bar{u}_4 \\ u_4\bar{u}_1 & u_4\bar{u}_2 & u_4\bar{u}_3 & u_4\bar{u}_4 \end{pmatrix} \begin{pmatrix} 1 & 0 \\ 0 & 0 \\ 0 & 1 \\ 0 & 0 \end{pmatrix}$$

$$+ \begin{pmatrix} 0 & 1 & 0 & 0 \\ 0 & 0 & 0 & 1 \end{pmatrix} \begin{pmatrix} u_1\bar{u}_1 & u_1\bar{u}_2 & u_1\bar{u}_3 & u_1\bar{u}_4 \\ u_2\bar{u}_1 & u_2\bar{u}_2 & u_2\bar{u}_3 & u_2\bar{u}_4 \\ u_3\bar{u}_1 & u_3\bar{u}_2 & u_3\bar{u}_3 & u_3\bar{u}_4 \\ u_4\bar{u}_1 & u_4\bar{u}_2 & u_4\bar{u}_3 & u_4\bar{u}_4 \end{pmatrix} \begin{pmatrix} 0 & 0 \\ 1 & 0 \\ 0 & 0 \\ 0 & 1 \end{pmatrix}.$$

Using matrix multiplication and matrix addition yields

$$\text{tr}_B(|\mathbf{u}\rangle\langle\mathbf{u}|) = \begin{pmatrix} u_1\bar{u}_1 + u_2\bar{u}_2 & u_1\bar{u}_3 + u_2\bar{u}_4 \\ u_3\bar{u}_1 + u_4\bar{u}_2 & u_3\bar{u}_3 + u_4\bar{u}_4 \end{pmatrix}.$$

We see that $\text{tr}_A(|\mathbf{u}\rangle\langle\mathbf{u}|) \neq \text{tr}_B(|\mathbf{u}\rangle\langle\mathbf{u}|)$. However

$$\text{tr}\,(\text{tr}_A(|\mathbf{u}\rangle\langle\mathbf{u}|)) = \text{tr}\,(\text{tr}_B(|\mathbf{u}\rangle\langle\mathbf{u}|)), \quad \det\,(\text{tr}_A(|\mathbf{u}\rangle\langle\mathbf{u}|)) = \det\,(\text{tr}_B(|\mathbf{u}\rangle\langle\mathbf{u}|)).$$

**Problem 5.** Consider the $9 \times 9$ matrix (density matrix)

$$|\mathbf{u}\rangle\langle\mathbf{u}| = (u_j\bar{u}_k), \qquad j,k = 1,\ldots,9.$$

Find the partial trace $\text{tr}_{\mathbb{C}^3}(|\mathbf{u}\rangle\langle\mathbf{u}|)$, where the basis is given by

$$\begin{pmatrix} 1 \\ 0 \\ 0 \end{pmatrix} \otimes I_3, \qquad \begin{pmatrix} 0 \\ 1 \\ 0 \end{pmatrix} \otimes I_3, \qquad \begin{pmatrix} 0 \\ 0 \\ 1 \end{pmatrix} \otimes I_3$$

and $I_3$ denotes the $3 \times 3$ unit matrix.

**Solution 5.**   We have

$$
\begin{pmatrix} 1 \\ 0 \\ 0 \end{pmatrix} \otimes I_3 = \begin{pmatrix} 1 & 0 & 0 \\ 0 & 1 & 0 \\ 0 & 0 & 1 \\ 0 & 0 & 0 \\ 0 & 0 & 0 \\ 0 & 0 & 0 \\ 0 & 0 & 0 \\ 0 & 0 & 0 \\ 0 & 0 & 0 \end{pmatrix}, \qquad \begin{pmatrix} 0 \\ 1 \\ 0 \end{pmatrix} \otimes I_3 = \begin{pmatrix} 0 & 0 & 0 \\ 0 & 0 & 0 \\ 0 & 0 & 0 \\ 1 & 0 & 0 \\ 0 & 1 & 0 \\ 0 & 0 & 1 \\ 0 & 0 & 0 \\ 0 & 0 & 0 \\ 0 & 0 & 0 \end{pmatrix}
$$

and

$$
\begin{pmatrix} 0 \\ 0 \\ 1 \end{pmatrix} \otimes I_3 = \begin{pmatrix} 0 & 0 & 0 \\ 0 & 0 & 0 \\ 0 & 0 & 0 \\ 0 & 0 & 0 \\ 0 & 0 & 0 \\ 0 & 0 & 0 \\ 1 & 0 & 0 \\ 0 & 1 & 0 \\ 0 & 0 & 1 \end{pmatrix}.
$$

The respective transposes of the above matrices are given by

$$
(1\ 0\ 0) \otimes I_3 = \begin{pmatrix} 1 & 0 & 0 & 0 & 0 & 0 & 0 & 0 & 0 \\ 0 & 1 & 0 & 0 & 0 & 0 & 0 & 0 & 0 \\ 0 & 0 & 1 & 0 & 0 & 0 & 0 & 0 & 0 \end{pmatrix}
$$

$$
(0\ 1\ 0) \otimes I_3 = \begin{pmatrix} 0 & 0 & 0 & 1 & 0 & 0 & 0 & 0 & 0 \\ 0 & 0 & 0 & 0 & 1 & 0 & 0 & 0 & 0 \\ 0 & 0 & 0 & 0 & 0 & 1 & 0 & 0 & 0 \end{pmatrix}
$$

$$
(0\ 0\ 1) \otimes I_3 = \begin{pmatrix} 0 & 0 & 0 & 0 & 0 & 0 & 1 & 0 & 0 \\ 0 & 0 & 0 & 0 & 0 & 0 & 0 & 1 & 0 \\ 0 & 0 & 0 & 0 & 0 & 0 & 0 & 0 & 1 \end{pmatrix}.
$$

Taking this basis we find

$$
\text{tr}_A(|\mathbf{u}\rangle\langle\mathbf{u}|) =
$$

$$
\begin{pmatrix} u_1\bar{u}_1 + u_4\bar{u}_4 + u_7\bar{u}_7 & u_1\bar{u}_2 + u_4\bar{u}_5 + u_7\bar{u}_8 & u_1\bar{u}_3 + u_4\bar{u}_6 + u_7\bar{u}_9 \\ u_2\bar{u}_1 + u_5\bar{u}_4 + u_8\bar{u}_7 & u_2\bar{u}_2 + u_5\bar{u}_5 + u_8\bar{u}_8 & u_2\bar{u}_3 + u_5\bar{u}_6 + u_8\bar{u}_9 \\ u_3\bar{u}_1 + u_6\bar{u}_4 + u_9\bar{u}_7 & u_3\bar{u}_2 + u_6\bar{u}_5 + u_9\bar{u}_8 & u_3\bar{u}_3 + u_6\bar{u}_6 + u_9\bar{u}_9 \end{pmatrix}.
$$

**Problem 6.**   (i) Consider the Bell state

$$
|\psi\rangle = \frac{1}{\sqrt{2}}(|00\rangle + |11\rangle).
$$

Hence $c_{00} = c_{11} = 1/\sqrt{2}$, $c_{01} = c_{10} = 0$. Find $\rho_A$.

(ii) Under a unitary transformation $U$, $V$ ($U$ and $V$ are $n \times n$ unitary matrices) the matrix $\Lambda$ is changed to $\Lambda \to U^T \Lambda V$, where $^T$ denotes the transpose. Apply the transformation to $\Lambda\Lambda^\dagger$. Calculate the Wigner function $\text{tr}(\Lambda\Lambda^\dagger)^2$.

**Solution 6.** (i) Since $c_{00} = c_{11} = 1/\sqrt{2}$, $c_{01} = c_{10} = 0$, we find the matrix

$$\Lambda = \begin{pmatrix} \frac{1}{\sqrt{2}} & 0 \\ 0 & \frac{1}{\sqrt{2}} \end{pmatrix}.$$

Thus we obtain the density matrix (mixed state)

$$\rho_A = \Lambda\Lambda^\dagger = \begin{pmatrix} \frac{1}{2} & 0 \\ 0 & \frac{1}{2} \end{pmatrix}.$$

For the other three Bell states we find the same result.

(ii) We have

$$\Lambda\Lambda^\dagger \to (U^T \Lambda V)(V^\dagger \Lambda^\dagger U^{T\dagger}) = U^\dagger \Lambda\Lambda^\dagger U^{T\dagger}$$

since $V^\dagger V = VV^\dagger = I_n$. Furthermore, $\text{tr}(\Lambda\Lambda^\dagger)^2$ stays invariant under the transformation since $U^T U^{T\dagger} = (U^\dagger U)^T = I_n$.

**Problem 7.** Let $\{\, |0\rangle, |1\rangle, \ldots, |d-1\rangle \,\}$ be an orthonormal basis in the Hilbert space $\mathbb{C}^d$. The discrete *Wigner operator* is defined as

$$\hat{A}(q,p) := \sum_{r=0}^{d-1}\sum_{s=0}^{d-1} \delta_{2q,r+s} \exp\left(i\frac{2\pi}{d}p(r-s)\right) |r\rangle\langle s|$$

where $q$ and $p$ take integer values from 0 to $d-1$ and $\delta_{m,u}$ denotes the Kronecker delta. The arithmetic in the subscript is modulo $N$ arithmetic, i.e. $2q \bmod d$ and $(r+s) \bmod d$. The $(p,q)$ pairs constitute the discrete phase space. For a state described by the density matrix $\rho$ the discrete *Wigner function* is defined as

$$W(p,q) := \frac{1}{d}\text{tr}(\rho\hat{A}).$$

Let $\rho = |0\rangle\langle 0|$. Calculate $W(p,q)$.

**Solution 7.** Since $\langle 0|r\rangle = \delta_{0r}$ we obtain

$$W(p,q) = \frac{1}{d}\text{tr}\left( |0\rangle \sum_{s=0}^{d-1} \delta_{2q,s} \exp\left(-i\frac{2\pi}{N}ps\right) \langle s| \right).$$

To calculate the trace we have

$$W(p,q) = \frac{1}{d} \sum_{k=0}^{d-1} \left( \langle k|0 \rangle \sum_{s=0}^{d-1} \delta_{2q,s} \exp\left( -i\frac{2\pi}{N}ps \right) \langle s|k \rangle \right).$$

Using $\langle k|0 \rangle = \delta_{k0}$ and $\langle s|k \rangle = \delta_{sk}$ we arrive at

$$W(p,q) = \frac{1}{d}\delta_{2q,0}.$$

**Problem 8.**  For a bipartite state with subsystems 1 and 2 described by the joint density matrix the joint *Wigner function* is given by

$$W(q_1,q_2,p_1,p_2) := \frac{1}{d^2}\mathrm{tr}(\rho^{(12)}(\hat{A}_1(q_1,p_1) \otimes \hat{A}_2(q_2,p_2)))$$

where the Wigner operators are given by

$$\hat{A}_1(q_1,p_1) := \sum_{r=0}^{d-1}\sum_{s=0}^{d-1} \delta_{2q_1,r+s} \exp\left( i\frac{2\pi}{d}p_1(r-s) \right) |r\rangle\langle s|$$

and

$$\hat{A}_2(q_2,p_2) := \sum_{r=0}^{d-1}\sum_{s=0}^{d-1} \delta_{2q_2,r+s} \exp\left( i\frac{2\pi}{d}p_2(r-s) \right) |r\rangle\langle s|.$$

Wigner functions describing a subsystem are obtained by summing the joint Wigner functions in the corresponding set of the respective variables, e.g.

$$W(q_1,p_1) = \sum_{q_2=0}^{d-1}\sum_{p_2=0}^{d-1} W(q_1,p_1,q_2,p_2)$$

$$W(q_2,p_2) = \sum_{q_1=0}^{d-1}\sum_{p_1=0}^{d-1} W(q_1,p_1,q_2,p_2).$$

Consider the *EPR state*

$$|\psi\rangle = \frac{1}{\sqrt{d}} \sum_{k=0}^{d-1} |k\rangle \otimes |k\rangle.$$

Let $\rho = |\psi\rangle\langle\psi|$. Find $W(q_1,q_2,p_1,p_2)$. Discuss.

**Solution 8.**  Straightforward calculation yields the Wigner function

$$W(q_1,q_2,p_1,p_2) = \frac{1}{d^2}\delta_{q_1,q_2}\delta_{p_1,-p_2}.$$

The Wigner function given above shows the connection with the EPR state for continuous-variable teleportation

$$\delta(q_1 - q_2) \otimes \delta(p_1 + p_2)$$

where $\delta$ denotes the Dirac delta function.

**Problem 9.** The trace of an $m \times m$ matrix $A$ over $\mathbb{C}$ is defined as

$$\mathrm{tr}_m(A) = \sum_{j=1}^{m} (\mathbf{b}_{m,j}^* A \, \mathbf{b}_{m,j})$$

where $\{ \mathbf{b}_{m,1}, \mathbf{b}_{m,2}, \ldots, \mathbf{b}_{m,m} \}$ is an orthonormal basis for $\mathbb{C}^m$. The partial traces of an $mn \times mn$ matrix $C$ over $\mathbb{C}$ are defined as

$$\mathrm{tr}_{m,n}^1(C) = \sum_{j=1}^{m} \left( \mathbf{b}_{m,j}^* \otimes I_n \right) C \left( \mathbf{b}_{m,j} \otimes I_n \right)$$

$$\mathrm{tr}_{m,n}^2(C) = \sum_{k=1}^{n} \left( I_m \otimes \mathbf{b}_{n,k}^* \right) C \left( I_m \otimes \mathbf{b}_{n,k} \right)$$

where $\{ \mathbf{b}_{n,1}, \mathbf{b}_{n,2}, \ldots, \mathbf{b}_{n,n} \}$ is an orthonormal basis for $\mathbb{C}^n$ and $I_m$ and $I_n$ denote the $m \times m$ and $n \times n$ identity matrices respectively. Let $B$ be an $n \times n$ matrix over $\mathbb{C}$.
(i) Show that the above definition of $\mathrm{tr}_m(A)$ is independent of the choice of orthonormal basis $\{ \mathbf{b}_{m,1}, \mathbf{b}_{m,2}, \ldots, \mathbf{b}_{m,m} \}$.
(ii) Show that

$$\mathrm{tr}_{mn}(C) = \mathrm{tr}_n \left( \mathrm{tr}_{m,n}^1(C) \right) \quad \text{and} \quad \mathrm{tr}_{mn}(C) = \mathrm{tr}_m \left( \mathrm{tr}_{m,n}^2(C) \right).$$

(iii) Use your favourite orthonormal basis for $\mathbb{C}^2$ to calculate

$$\mathrm{tr}_{2,2}^1 \left( \begin{pmatrix} 0 & 1 \\ 1 & 0 \end{pmatrix} \otimes \begin{pmatrix} 1 & 2 \\ 3 & 4 \end{pmatrix} \right) \quad \text{and} \quad \mathrm{tr}_{2,2}^2 \begin{pmatrix} 0 & 1 & 1 & 0 \\ 1 & 0 & 0 & 1 \\ 1 & 0 & 0 & -1 \\ 0 & 1 & -1 & 0 \end{pmatrix}.$$

**Solution 9.** (i) Let $\{ \phi_1, \phi_2, \ldots, \phi_m \}$ be an orthonormal basis for $\mathbb{C}^n$. Expanding each of the $\mathbf{b}_{m,k}$ in terms of the $\phi_j$ and vice versa yields

$$\mathbf{b}_{m,j} = \sum_{k=1}^{m} \phi_k^* \mathbf{b}_{m,j} \phi_k, \qquad \phi_j = \sum_{k=1}^{m} \mathbf{b}_{m,k}^* \phi_j \mathbf{b}_{m,k}$$

where we require

$$\mathbf{b}_{m,j}^*\mathbf{b}_{m,k} = \sum_{u,v=1}^{m} \mathbf{b}_{m,j}^*\phi_u\phi_u^*\phi_v^*\mathbf{b}_{m,k}\phi_v = \sum_{u,v=1}^{m} \mathbf{b}_{m,j}^*\phi_u\phi_v^*\mathbf{b}_{m,k}(\phi_u^*\phi_v)$$

$$= \sum_{u=1}^{m} \phi_u^*\mathbf{b}_{m,k}\mathbf{b}_{m,j}^*\phi_u = \delta_{j,k}$$

so that orthonormality holds. It follows that

$$\sum_{k=1}^{m} \phi_k^*A\phi_k = \sum_{k=1}^{m}\sum_{j=1}^{m}\sum_{l=1}^{m} \phi_k^*\mathbf{b}_{m,j}\mathbf{b}_{m,j}^*A\mathbf{b}_{m,l}\phi_k\mathbf{b}_{m,l}$$

$$= \sum_{j=1}^{m}\sum_{l=1}^{m} \left( \sum_{k=1}^{m} \phi_k^*\mathbf{b}_{m,j}\mathbf{b}_{m,l}^*\phi_k \right) \mathbf{b}_{m,j}^*A\mathbf{b}_{m,l}$$

$$= \sum_{j=1}^{m}\sum_{l=1}^{m} \delta_{j,l}\mathbf{b}_{m,j}^*A\mathbf{b}_{m,l}$$

$$= \sum_{j=1}^{m} \mathbf{b}_{m,j}^*A\mathbf{b}_{m,j} = \operatorname{tr}_m(A).$$

(ii) We have

$$\operatorname{tr}_n\left(\operatorname{tr}_{m,n}^1(C)\right) = \sum_{j=1}^{n} \mathbf{b}_{n,j}^* \left( \sum_{k=1}^{m}(\mathbf{b}_{m,k}^* \otimes I_n)C(\mathbf{b}_{m,k} \otimes I_n) \right) \mathbf{b}_{n,j}$$

$$= \sum_{j=1}^{n}\sum_{k=1}^{m}(I_1 \otimes \mathbf{b}_{n,j}^*)(\mathbf{b}_{m,k}^* \otimes I_n)C(\mathbf{b}_{m,k} \otimes I_n)(I_1 \otimes \mathbf{b}_{n,j})$$

$$= \sum_{j=1}^{n}\sum_{k=1}^{m}(\mathbf{b}_{m,k}^* \otimes \mathbf{b}_{n,j}^*)C(\mathbf{b}_{m,k} \otimes \mathbf{b}_{n,j})$$

$$= \operatorname{tr}_{mn}(C)$$

where the last equivalence follows from taking the trace of $C$ using the basis

$$\{\, \mathbf{b}_{m,j} \otimes \mathbf{b}_{n,k} \ : \ j \in \{1,2,\dots,m\}\ k \in \{1,2,\dots,n\}\,\}.$$

The result $\operatorname{tr}_{mn}(C) = \operatorname{tr}_m\left(\operatorname{tr}_{m,n}^2(C)\right)$ follows similarly.
(iii) Using any basis, for example the standard basis

$$\left\{ \begin{pmatrix} 1 \\ 0 \end{pmatrix}, \begin{pmatrix} 0 \\ 1 \end{pmatrix} \right\}$$

we find

$$\operatorname{tr}_{2,2}^1\left( \begin{pmatrix} 0 & 1 \\ 1 & 0 \end{pmatrix} \otimes \begin{pmatrix} 1 & 2 \\ 3 & 4 \end{pmatrix} \right) = \begin{pmatrix} 0 & 0 \\ 0 & 0 \end{pmatrix}$$

and

$$\text{tr}_{2,2}^2 \begin{pmatrix} 0 & 1 & 1 & 0 \\ 1 & 0 & 0 & 1 \\ 1 & 0 & 0 & -1 \\ 0 & 1 & -1 & 0 \end{pmatrix} = \begin{pmatrix} 0 & 2 \\ 2 & 0 \end{pmatrix}.$$

**Problem 10.** Consider the GHZ state in the Hilbert space $\mathbb{C}^8$ ($\mathbb{C}^8 \cong \mathbb{C}^2 \otimes \mathbb{C}^2 \otimes \mathbb{C}^2$)

$$|GHZ\rangle = \frac{1}{\sqrt{2}} \left( \begin{pmatrix} 1 \\ 0 \end{pmatrix} \otimes \begin{pmatrix} 1 \\ 0 \end{pmatrix} \otimes \begin{pmatrix} 1 \\ 0 \end{pmatrix} + \begin{pmatrix} 0 \\ 1 \end{pmatrix} \otimes \begin{pmatrix} 0 \\ 1 \end{pmatrix} \otimes \begin{pmatrix} 0 \\ 1 \end{pmatrix} \right).$$

Then the density matrix is given by the $8 \times 8$ matrix

$$\rho = |GHZ\rangle\langle GHZ| = \frac{1}{2} \begin{pmatrix} 1 & 0 & 0 & 0 & 0 & 0 & 0 & 1 \\ 0 & 0 & 0 & 0 & 0 & 0 & 0 & 0 \\ 0 & 0 & 0 & 0 & 0 & 0 & 0 & 0 \\ 0 & 0 & 0 & 0 & 0 & 0 & 0 & 0 \\ 0 & 0 & 0 & 0 & 0 & 0 & 0 & 0 \\ 0 & 0 & 0 & 0 & 0 & 0 & 0 & 0 \\ 0 & 0 & 0 & 0 & 0 & 0 & 0 & 0 \\ 1 & 0 & 0 & 0 & 0 & 0 & 0 & 1 \end{pmatrix}.$$

(i) Calculate the partial trace $\rho_{AB} = \text{tr}_C(\rho)$ with the basis

$$I_4 \otimes \begin{pmatrix} 1 \\ 0 \end{pmatrix}, \qquad I_4 \otimes \begin{pmatrix} 0 \\ 1 \end{pmatrix}.$$

(ii) Calculate the partial trace $\rho_A = \text{tr}_B(\rho_{AB})$ with the basis

$$I_2 \otimes \begin{pmatrix} 1 \\ 0 \end{pmatrix}, \qquad I_2 \otimes \begin{pmatrix} 0 \\ 1 \end{pmatrix}.$$

**Solution 10.** (i) We have

$$\text{tr}_C(\rho) = \left( I_4 \otimes \begin{pmatrix} 1 \\ 0 \end{pmatrix} \right)^* \rho \left( I_4 \otimes \begin{pmatrix} 1 \\ 0 \end{pmatrix} \right) + \left( I_4 \otimes \begin{pmatrix} 0 \\ 1 \end{pmatrix} \right)^* \rho \left( I_4 \otimes \begin{pmatrix} 0 \\ 1 \end{pmatrix} \right)$$

$$= \frac{1}{2} \begin{pmatrix} 1 & 0 & 0 & 0 \\ 0 & 0 & 0 & 0 \\ 0 & 0 & 0 & 0 \\ 0 & 0 & 0 & 0 \end{pmatrix} + \frac{1}{2} \begin{pmatrix} 0 & 0 & 0 & 0 \\ 0 & 0 & 0 & 0 \\ 0 & 0 & 0 & 0 \\ 0 & 0 & 0 & 1 \end{pmatrix}$$

$$= \frac{1}{2} \begin{pmatrix} 1 & 0 & 0 & 0 \\ 0 & 0 & 0 & 0 \\ 0 & 0 & 0 & 0 \\ 0 & 0 & 0 & 1 \end{pmatrix}.$$

(ii) We find

$$\rho_A = \text{tr}_B(\rho_{AB})$$

$$= \left(I_2 \otimes \begin{pmatrix} 1 \\ 0 \end{pmatrix}\right)^* \rho_{AB} \left(I_2 \otimes \begin{pmatrix} 1 \\ 0 \end{pmatrix}\right) + \left(I_2 \otimes \begin{pmatrix} 0 \\ 1 \end{pmatrix}\right)^* \rho_{AB} \left(I_2 \otimes \begin{pmatrix} 0 \\ 1 \end{pmatrix}\right)$$

$$= \frac{1}{2}\begin{pmatrix} 1 & 0 \\ 0 & 0 \end{pmatrix} + \frac{1}{2}\begin{pmatrix} 0 & 0 \\ 0 & 1 \end{pmatrix}$$

$$= \frac{1}{2}\begin{pmatrix} 1 & 0 \\ 0 & 1 \end{pmatrix}.$$

## Programming Problems

**Problem 1.** Consider the hermitian matrix

$$A = \begin{pmatrix} 0 & 0 & i \\ 0 & 0 & 0 \\ -i & 0 & 0 \end{pmatrix}.$$

Find the eigenvalues of $A$ utilizing

$$\text{tr}(A) = \lambda_1 + \lambda_2 + \lambda_3, \quad \text{tr}(A^2) = \lambda_1^2 + \lambda_2^2 + \lambda_3^2, \quad \text{tr}(A^3) = \lambda_1^3 + \lambda_2^3 + \lambda_3^3.$$

**Solution 1.** The following Maxima program will do the job

```
/* traceeigen.mac */
load("nchrpl");
A: matrix([0,0,%i],[0,0,0],[-%i,0,0]);
r1: mattrace(A); r2: mattrace(A . A); r3: mattrace(A . A . A);
solve([x1+x2+x3-r1=0,x1*x1+x2*x2+x3*x3-r2=0,
       x1*x1*x1+x2*x2*x2+x3*x3*x3-r3=0], [x1,x2,x3]);
```

The eigenvalues are $\lambda_1 = -1$, $\lambda_2 = 0$, $\lambda_3 = +1$.

**Problem 2.** Consider the $6 \times 6$ matrix $A = (a_{jk})$. Calculate the partial trace with the basis

$$\begin{pmatrix} 1 \\ 0 \end{pmatrix} \otimes I_3, \quad \begin{pmatrix} 0 \\ 1 \end{pmatrix} \otimes I_3.$$

Calculate the partial trace with the basis

$$\begin{pmatrix} 1 \\ 0 \\ 0 \end{pmatrix} \otimes I_2, \quad \begin{pmatrix} 0 \\ 1 \\ 0 \end{pmatrix} \otimes I_2, \quad \begin{pmatrix} 0 \\ 0 \\ 1 \end{pmatrix} \otimes I_2.$$

Apply computer algebra.

**Solution 2.** The following Maxima program will do the job.

```
/* partialtrace.mac */
load("nchrpl");
A: matrix([a11,a12,a13,a14,a15,a16],[a21,a22,a23,a24,a25,a26],
          [a31,a32,a33,a34,a35,a36],[a41,a42,a43,a44,a45,a46],
          [a51,a52,a53,a54,a55,a46],[a61,a62,a63,a64,a65,a66]);
I2: matrix([1,0],[0,1]);
I3: matrix([1,0,0],[0,1,0],[0,0,1]);
v1: matrix([1],[0]); v2: matrix([0],[1]);
u1: matrix([1],[0],[0]); u2: matrix([0],[1],[0]);
u3: matrix([0],[0],[1]);
b1: kronecker_product(v1,I3); b1T: transpose(b1);
b2: kronecker_product(v2,I3); b2T: transpose(b2);
ptrA1: b1T . A . b1 + b2T . A . b2;
c1: kronecker_product(u1,I2); c1T: transpose(c1);
c2: kronecker_product(u2,I2); c2T: transpose(c2);
c3: kronecker_product(u3,I2); c3T: transpose(c3);
ptrA2: c1T . A . c1 + c2T . A. c2 + c3T . A. c3;
```

The output is the $3 \times 3$ matrix for the first basis

```
[ a44 + a11   a45 + a12   a46 + a13 ]
[ a54 + a21   a55 + a22   a46 + a23 ]
[ a64 + a31   a65 + a32   a66 + a33 ]
```

for the first two-dimensional basis and the $2 \times 2$ matrix

```
[ a55 + a33 + a11   a46 + a34 + a12 ]
[ a65 + a43 + a21   a66 + a44 + a22 ]
```

for the second three dimensional basis.

## 5.3   Supplementary Problems

**Problem 1.** Let $c_1, c_2, c_3 \in \mathbb{C}$ and $|c_1|^2 + |c_2|^2 + |c_3|^2 = 1$. Show that

$$
\rho = \begin{pmatrix}
|c_1|^2 & c_1 c_2^*/\sqrt{2} & c_1 c_2^*/\sqrt{2} & c_1 c_3^* \\
c_1^* c_2/\sqrt{2} & |c_2|^2/2 & |c_2|^2/2 & c_2 c_3^*/\sqrt{2} \\
c_1^* c_2/\sqrt{2} & |c_2|^2/2 & |c_2|^2/2 & c_2 c_3^*/\sqrt{2} \\
c_1^* c_3 & c_2^* c_3/\sqrt{2} & c_2^* c_3/\sqrt{2} & |c_3|^2
\end{pmatrix}
$$

is a density matrix. Taking the partial trace we obtain

$$\rho_p = \begin{pmatrix} |c_1|^2 + |c_2|^2/2 & (c_1 c_2^* + c_2 c_3^*)/\sqrt{2} \\ (c_1^* c_2 + c_2^* c_3)/\sqrt{2} & |c_3|^2 + |c_2|^2/2 \end{pmatrix}.$$

Show that $\rho_p$ is a density matrix. Discuss.

**Problem 2.**    Consider the finite dimensional Hilbert spaces $\mathcal{H}_1 = \mathbb{C}^{n_1}$ and $\mathcal{H}_2 = \mathbb{C}^{n_2}$. Let $\hat{Q}_1$ be an $n_1 \times n_1$ matrix and $\hat{Q}_2$ be an $n_2 \times n_2$ matrix. Let $\rho_1$ be a density matrix in $\mathbb{C}^{n_1}$ and $\rho_2$ be a density matrix. Show that

$$\text{tr}((\hat{Q}_1 \otimes \hat{Q}_2)(\rho_1 \otimes \rho_2)) = \text{tr}(\hat{Q}_1 \rho_1)\text{tr}(Q_2 \rho_2).$$

**Problem 3.**    Consider the finite-dimensional Hilbert spaces $\mathcal{H}_1 = \mathbb{C}^{n_1}$ and $\mathcal{H}_2 = \mathbb{C}^{n_2}$. Let $\mathcal{H}_1 \otimes \mathcal{H}_2$ be the product Hilbert space. Let $|\psi\rangle$ and $|\phi\rangle$ be states in the product Hilbert space $\mathcal{H}_1 \otimes \mathcal{H}_2$. Show that if

$$\text{tr}_{\mathcal{H}_2}(|\psi\rangle\langle\psi|) = \text{tr}_{\mathcal{H}_2}(|\phi\rangle\langle\phi|)$$

then there exists a unitary matrix $U$ acting in the Hilbert space $\mathcal{H}_2$ such that

$$|\psi\rangle = (I_{n_1} \otimes U)|\phi\rangle$$

where $I_{n_1}$ is the identity matrix in the Hilbert space $\mathcal{H}_1$.

**Problem 4.**    Consider the finite dimensional Hilbert spaces $\mathcal{H}_1 = \mathbb{C}^{n_1}$ and $\mathcal{H}_2 = \mathbb{C}^{n_2}$. Let $\mathbf{v}$, $\mathbf{u}$ be normalized vectors in the product Hilbert space $\mathbb{C}^{n_1} \otimes \mathbb{C}^{n_2}$. Show that if

$$\text{tr}_1(\mathbf{v}\mathbf{v}^*) = \text{tr}_1(\mathbf{u}\mathbf{u}^*)$$

then there exists a $n_2 \times n_2$ unitary matrix $U$ such that

$$\mathbf{v} = (I_{n_1} \otimes U)\mathbf{u}.$$

**Problem 5.**    Consider the product Hilbert space $\ell_2(\mathbb{N}_0) \otimes \mathbb{C}^{2s+1}$, where $s = 1/2, 1, 3/2, 2, \dots$ is the spin. Find the partial trace over $\mathbb{C}^{2s+1}$.

# Chapter 6

# Boolean Functions and Quantum Gates

## 6.1 Introduction

A *truth table* (or function table) is a tabular description of a combinational circuit (such as an AND gate, OR gate, NAND gate) listing all possible states of the input variables together with a statement of the output variable(s) for each of those possible states. The truth table for the AND gate, OR gate, XOR gate and NOT gate are

| AND | | | OR | | | XOR | | | NOT | |
|---|---|---|---|---|---|---|---|---|---|---|
| 0 | 0 | 0 | 0 | 0 | 0 | 0 | 0 | 0 | 0 | 1 |
| 0 | 1 | 0 | 0 | 1 | 1 | 0 | 1 | 1 | 1 | 0 |
| 1 | 0 | 0 | 1 | 0 | 1 | 1 | 0 | 1 | | |
| 1 | 1 | 1 | 1 | 1 | 1 | 1 | 1 | 0 | | |

The NAND gate is an AND gate followed by a NOT gate. The NOR gate is an OR gate followed by a NOT gate. Both are *universal gates*, i.e. all other gates can be built from these gates.

A boolean function $f$ on $n$ variables is a mapping $\{0,1\}^n$ into $\{0,1\}$. Let $x_j \in \{0,1\}$ for $j = 1, \ldots, n$. We set $\mathbf{x} = (x_1, x_2, \ldots, x_n)$. In the following $\cdot$ denotes the AND operation, $+$ denotes the OR operation, $\oplus$ the XOR operation and $^-$ is the NOT operation. For $n = 1$ we have the four boolean functions

$$f_1(x) = 0, \quad f_2(x) = 1, \quad f_3(x) = x, \quad f_4(x) = \bar{x}.$$

The last two are of course reversible.

Let $X = \{0, 1\}$ and $x_j \in X$. A boolean function $\mathbf{f}$ with $n$ input variables, $x_1, \ldots, x_n$ and $n$ output variables, $y_1, \ldots, y_n$ is a function $\mathbf{f} : X^n \to X^n$ obeying

$$\mathbf{f}(x_1, \ldots, x_n) \mapsto (y_1, \ldots, y_n).$$

Here $(x_1, \ldots, x_n) \in X^n$ is called the input vector and $(y_1, \ldots, y_n) \in X^n$ is called the output vector. An $n$-input and $n$-output boolean function $\mathbf{f}$ is reversible if it maps each input vector to a unique output vector, i.e. the map is a bijection.

Quantum gates are described by unitary operators. In the finite dimensional Hilbert space $\mathbb{C}^d$ we have $d \times d$ unitary matrices. We describe how $2^{n+1} \times 2^{n+1}$ unitary matrices can be associated with a non-reversible boolean function $f$ and how $2^n \times 2^n$ unitary matrices can be associated with reversible boolean functions $\mathbf{f}$. Finally the associated Hamilton operator has to be constructed.

Reversible gates are gates that function in both directions. CMOS implementations of such gates have been designed. A special pass transistor logic family has been applied: reversible MOS. Many different reversible logic gates are candidates as universal building blocks. The Feynman gate, the controlled NOT gate, the Fredkin gate can be implemented. They dissipate very little energy. Owing to their use of reversible truth tables, they are even candidates for zero-power computing. Circuit synthesis takes advantage of mathematical group theory. Algorithms have been developed for the synthesis of arbitrary reversible circuits. A reversible logic gate has a corresponding quantum version, whose properties are completely defined by the truth table for the classical version.

Reversible circuits are applicable to nanotechnolgy, quantum and optical computing as well as reducing power in CMOS implementations. In adiabatic circuits, current is restricted to flow across devices with low voltage drop and the energy stored on their capacitors is recycled. One uses reversible energy recovery gates capable of realizing the functions $\{AND, OR\}$ or $\{NAND, NOR\}$.

# 6.2 Solved Problems

**Problem 1.** The *Feynman gate* is a 2 input/2 output gate given by

$$x_1' = x_1, \quad x_2' = x_1 \oplus x_2.$$

(i) Give the truth table for the Feynman gate.
(ii) Show that copying can be implemented using the Feynman gate.
(iii) Show that the complement can be implemented using the Feynman gate.
(iv) Is the Feynman gate invertible?

**Solution 1.** (i) The truth table is

| $x_1$ | $x_2$ | $x_1'$ | $x_2'$ |
|---|---|---|---|
| 0 | 0 | 0 | 0 |
| 0 | 1 | 0 | 1 |
| 1 | 0 | 1 | 1 |
| 1 | 1 | 1 | 0 |

(ii) Setting $x_2 = 0$, we have $x_2' = x_1 \oplus 0 = x_1$. Thus we have a copy.
(iii) Setting $x_2 = 1$, we have $x_2' = x_1 \oplus 1 = \bar{x}_1$. Thus we generated the complement.
(iv) From the truth table we see that the transformation is invertible. The inverse transformation can be found as follows. Since $x_1 \oplus x_1 = 0$ we have

$$x_1' \oplus x_2' = x_1 \oplus x_1 \oplus x_2 = 0 \oplus x_2 = x_2.$$

Thus $x_1 = x_1'$, $x_2 = x_1' \oplus x_2'$.

**Problem 2.** Consider the 3-input/3-output gate given by

$$x_1' = x_1, \quad x_2' = x_1 \oplus x_2, \quad x_3' = x_1 \oplus x_2 \oplus x_3.$$

Give the truth table. Is the transformation invertible?

**Solution 2.** The truth table is given by

| $x_1$ | $x_2$ | $x_3$ | $x_1'$ | $x_2'$ | $x_3'$ |
|---|---|---|---|---|---|
| 0 | 0 | 0 | 0 | 0 | 0 |
| 0 | 0 | 1 | 0 | 0 | 1 |
| 0 | 1 | 0 | 0 | 1 | 1 |
| 0 | 1 | 1 | 0 | 1 | 0 |
| 1 | 0 | 0 | 1 | 1 | 1 |
| 1 | 0 | 1 | 1 | 1 | 0 |
| 1 | 1 | 0 | 1 | 0 | 0 |
| 1 | 1 | 1 | 1 | 0 | 1 |

From the truth table we see that the transformation is invertible, i.e. we have a $1-1$ map. The inverse transformation is given by

$$x_1 = x_1', \quad x_2 = x_1' \oplus x_2', \quad x_3 = x_1' \oplus x_2' \oplus x_3'.$$

**Problem 3.** Consider the 3-input/3-output gate given by

$$x_1' = x_1, \quad x_2' = x_1 \oplus x_2, \quad x_3' = x_3 \oplus (x_1 \cdot x_2).$$

Give the truth table. Is the gate invertible?

**Solution 3.** The truth table is given by

| $x_1$ | $x_2$ | $x_3$ | $x_1'$ | $x_2'$ | $x_3'$ |
|-------|-------|-------|--------|--------|--------|
| 0 | 0 | 0 | 0 | 0 | 0 |
| 0 | 0 | 1 | 0 | 0 | 1 |
| 0 | 1 | 0 | 0 | 1 | 0 |
| 0 | 1 | 1 | 0 | 1 | 1 |
| 1 | 0 | 0 | 1 | 1 | 0 |
| 1 | 0 | 1 | 1 | 1 | 1 |
| 1 | 1 | 0 | 1 | 0 | 1 |
| 1 | 1 | 1 | 1 | 0 | 0 |

From the truth table we see that the map is invertible, i.e. we have a 1-1 map. The inverse transformation is given by

$$x_1 = x_1'$$
$$x_2 = x_1' \oplus x_2'$$
$$x_3 = x_3' \oplus (x_1' \cdot (x_1' \oplus x_2'))$$

where we used that $x \oplus x = 0$.

**Problem 4.** For reversible gates the following boolean expression plays an important role

$$(a_{11} \cdot a_{22}) \oplus (a_{12} \cdot a_{21})$$

where $a_{11}, a_{12}, a_{21}, a_{22} \in \{0, 1\}$. It could be considered as the *determinant* of the $2 \times 2$ binary matrix

$$\begin{pmatrix} a_{11} & a_{12} \\ a_{21} & a_{22} \end{pmatrix}.$$

Discuss. Find the inverse of the matrix when it exists.

**Solution 4.** The inverse exists iff $(a_{11} \cdot a_{22}) \oplus (a_{12} \cdot a_{21}) = 1$. The inverse is given by

$$\begin{pmatrix} a_{22} & a_{12} \\ a_{21} & a_{11} \end{pmatrix}$$

since

$$\begin{pmatrix} a_{11} & a_{12} \\ a_{21} & a_{22} \end{pmatrix} \begin{pmatrix} a_{22} & a_{12} \\ a_{21} & a_{11} \end{pmatrix} =$$

$$\begin{pmatrix} (a_{11} \cdot a_{22}) \oplus (a_{12} \cdot a_{21}) & (a_{11} \cdot a_{12}) \oplus (a_{12} \cdot a_{11}) \\ (a_{21} \cdot a_{22}) \oplus (a_{22} \cdot a_{21}) & (a_{21} \cdot a_{12}) \oplus (a_{11} \cdot a_{22}) \end{pmatrix}.$$

**Problem 5.** Consider a two input gate $(x, y)$ / two output gate $(x', y')$ given by

$$x' = a \cdot x \oplus b \cdot y \oplus c, \qquad y' = a' \cdot x \oplus b' \cdot y \oplus c'$$

where $a, b, a', b', c, c' \in \{0, 1\}$.
(i) Let $a = 0, b = 1, a' = 1, b' = 0$ and $c = c' = 0$. Find the output $(x', y')$ for all possible inputs $(x, y)$. Is the transformation invertible?
(ii) Let $a = 1, b = 1, a' = 1, b' = 1$ and $c = c' = 0$. Find the output $(x', y')$ for all possible inputs $(x, y)$. Is the transformation invertible?

**Solution 5.** (i) We have

$$x' = 0 \cdot x \oplus 1 \cdot y \oplus 0, \qquad y' = 1 \cdot x \oplus 0 \cdot y \oplus 0.$$

Thus

$$x' = 0 \oplus y \oplus 0 = y, \qquad y' = x \oplus 0 \oplus 0 = x.$$

The truth table follows as

| $x$ | $y$ | $x'$ | $y'$ |
|---|---|---|---|
| 0 | 0 | 0 | 0 |
| 0 | 1 | 1 | 0 |
| 1 | 0 | 0 | 1 |
| 1 | 1 | 1 | 1 |

Therefore the transformation is invertible.
(ii) We have

$$x' = 1 \cdot x \oplus 1 \cdot y \oplus 0, \qquad y' = 1 \cdot x \oplus 1 \cdot y \oplus 0.$$

Thus

$$x' = x \oplus y \oplus 0 = x \oplus y, \qquad y' = x \oplus y \oplus 0 = x \oplus y.$$

The truth table follows as

| $x$ | $y$ | $x'$ | $y'$ |
|-----|-----|------|------|
| 0 | 0 | 0 | 0 |
| 0 | 1 | 1 | 1 |
| 1 | 0 | 1 | 1 |
| 1 | 1 | 0 | 0 |

Therefore the transformation is not invertible.

**Problem 6.** Consider the *Toffoli gate*

$$T : \{0,1\}^3 \to \{0,1\}^3, \qquad T(a,b,c) := (a,b,(a \cdot b) \oplus c)$$

and the *Fredkin gate*

$$F : \{0,1\}^3 \to \{0,1\}^3, \qquad F(a,b,c) := (a, a \cdot b + \bar{a} \cdot c, a \cdot c + \bar{a} \cdot b)$$

where $\bar{a}$ is the NOT operation, $+$ is the OR operation, $\cdot$ is the AND operation and $\oplus$ is the XOR operation.

1. Express $NOT(a)$ exclusively in terms of the TOFFOLI gate.

2. Express $NOT(a)$ exclusively in terms of the FREDKIN gate.

3. Express $AND(a,b)$ exclusively in terms of the TOFFOLI gate.

4. Express $AND(a,b)$ exclusively in terms of the FREDKIN gate.

5. Express $OR(a,b)$ exclusively in terms of the TOFFOLI gate.

6. Express $OR(a,b)$ exclusively in terms of the FREDKIN gate.

7. Show that the TOFFOLI gate is invertible.

8. Show that the FREDKIN gate is invertible.

Thus the TOFFOLI and FREDKIN gates are each universal and reversible (invertible).

**Solution 6.** (1) We have $NOT(a) = BIT3(T(a,a,1))$, where $BIT3(a,b,c) = c$. This follows from $T(a,a,1) = (a,a,a \oplus 1)$.

(2) $NOT(a) = BIT3(F(a,1,0))$.

(3) $AND(a,b) = BIT3(T(a,b,0))$.

(4) $AND(a,b) = BIT3(F(a,0,b))$.

(5) We can use

$$OR(a, b) = NOT(AND(NOT(a), NOT(b)))$$

and simply expand *NOT* and *AND* in terms of TOFFOLI gate. Noting that

$$T(\bar{a}, \bar{b}, 1) = (\bar{a}, \bar{b}, (\bar{a} \cdot \bar{b}) \oplus 1) = (\bar{a}, \bar{b}, \overline{\bar{a} \cdot \bar{b}}) = (\bar{a}, \bar{b}, a + b)$$

we implement the *OR* operation as

$$OR(a, b) = BIT3(T(NOT(a), NOT(b), 1))$$
$$= BIT3(T(BIT3(T(a, a, 1)), BIT3(T(b, b, 1)), 1)).$$

(6) We can use

$$OR(a, b) = NOT(AND(NOT(a), NOT(b)))$$

and simply expand *NOT* and *AND* in terms of the FREDKIN gate. Noting that

$$F(a, b, 1) = (a, a \cdot b + \bar{a} \cdot 1, a \cdot 1 + \bar{a} \cdot b) = (a, b + \bar{a}, a + b)$$

we implement the *OR* operation as

$$OR(a, b) = BIT3(F(a, b, 1)).$$

(7) We have $T^{-1}(a, b, c) = T(a, b, c)$ since the XOR ($\oplus$) is invertible if we remember at least one of the arguments. In other words

$$T(T(a, b, c)) = T(a, b, (a \cdot b) \oplus c)$$
$$= (a, b, (a \cdot b) \oplus ((a \cdot b) \oplus c))$$
$$= (a, b, ((a \cdot b) \oplus (a \cdot b)) \oplus c)$$
$$= (a, b, c).$$

(8) We have $F^{-1}(a, b, c) = F(a, b, c)$ since the swap operation on two bits is its own inverse, and the FREDKIN gate swaps the last two bits whenever the first argument is 0. In other words

$$F(F(a, b, c)) = F(a, a \cdot b + \bar{a} \cdot c, a \cdot c + \bar{a} \cdot b)$$
$$= (a, a \cdot (a \cdot b + \bar{a} \cdot c) + \bar{a} \cdot (a \cdot c + \bar{a} \cdot b), a \cdot (a \cdot c + \bar{a} \cdot b)$$
$$+ \bar{a} \cdot (a \cdot b + \bar{a} \cdot c))$$
$$= (a, a \cdot b + \bar{a} \cdot b, a \cdot c + \bar{a} \cdot c)$$
$$= (a, b, c).$$

**Problem 7.** A *generalized Toffoli gate* $T(x_1, x_2, \ldots, x_n, x_{n+1})$ is a gate that maps a boolean pattern $(x_1, x_2, \ldots, x_n, x_{n+1})$ to

$$(x_1, x_2, \ldots, x_n, x_{n+1} \oplus (x_1 \cdot x_2 \cdot \ldots \cdot x_n))$$

where $\oplus$ is the XOR operation and $\cdot$ the AND operation. Show that the generalized Toffoli gate includes the NOT gate, CNOT gate and the original Toffoli gate.

**Solution 7.** The NOT gate is given by $\mathrm{T}(1,a)$, where $a \in \{0,1\}$. The CNOT-gate is given by $\mathrm{T}(a,b)$, where $a,b \in \{0,1\}$. The original Toffoli gate is given by $\mathrm{T}(a,b,c)$.

**Problem 8.** The *Fredkin gate* $\mathrm{F}(x_1,x_2,x_3)$ has 3 inputs $(x_1,x_2,x_3)$ and three outputs $(y_1,y_2,y_3)$. It maps boolean patterns

$$(x_1,x_2,x_3) \to (x_1,x_3,x_2)$$

if and only if $x_1 = 1$, otherwise it passes the boolean pattern unchanged. Give the truth table.

**Solution 8.** We have

| $x_1$ | $x_2$ | $x_3$ | $y_1$ | $y_2$ | $y_3$ |
|---|---|---|---|---|---|
| 0 | 0 | 0 | 0 | 0 | 0 |
| 0 | 0 | 1 | 0 | 0 | 1 |
| 0 | 1 | 0 | 0 | 1 | 0 |
| 0 | 1 | 1 | 0 | 1 | 1 |
| 1 | 0 | 0 | 1 | 0 | 0 |
| 1 | 0 | 1 | 1 | 1 | 0 |
| 1 | 1 | 0 | 1 | 0 | 1 |
| 1 | 1 | 1 | 1 | 1 | 1 |

**Problem 9.** The *generalized Fredkin gate* $\mathrm{F}(x_1,x_2,\ldots,x_n,x_{n+1},x_{n+2})$ is the mapping of the boolean pattern

$$(x_1,x_2,\ldots,x_n,x_{n+1},x_{n+2}) \to (x_1,x_2,\ldots,x_n,x_{n+2},x_{n+1})$$

if and only if the boolean product $x_1 \cdot x_2 \cdot \ldots \cdot x_n = 1$ ($\cdot$ is the bitwise AND operation), otherwise the boolean pattern passes unchanged. Let $n = 2$ and $(x_1,x_2,x_3,x_4) = (1,1,0,1)$. Find the output.

**Solution 9.** Since $x_1 \cdot x_2 = 1$, we find

| $x_1$ | $x_2$ | $x_3$ | $x_4$ | $y_1$ | $y_2$ | $y_3$ | $y_4$ |
|---|---|---|---|---|---|---|---|
| 1 | 1 | 0 | 1 | 1 | 1 | 1 | 0 |

**Problem 10.** Is the gate ($a,b,c \in \{0,1\}$)

$$(a,b,c) \to (a, a \cdot b \oplus c, \bar{a} \cdot \bar{c} \oplus \bar{b})$$

reversible?

**Solution 10.** We set $x = a$, $y = a \cdot b \oplus c$ and $z = \bar{a} \cdot \bar{c} \oplus \bar{b}$. We find the truth table

| $a$ | $b$ | $c$ | $x$ | $y$ | $z$ |
|---|---|---|---|---|---|
| 0 | 0 | 0 | 0 | 0 | 0 |
| 0 | 0 | 1 | 0 | 1 | 1 |
| 0 | 1 | 0 | 0 | 0 | 1 |
| 0 | 1 | 1 | 0 | 1 | 0 |
| 1 | 0 | 0 | 1 | 0 | 1 |
| 1 | 0 | 1 | 1 | 1 | 1 |
| 1 | 1 | 0 | 1 | 1 | 0 |
| 1 | 1 | 1 | 1 | 0 | 0 |

Thus from the truth table we see that the gate is reversible.

**Problem 11.** Show that one Fredkin gate
$$(a, b, c) \rightarrow (a, \bar{a} \cdot b + a \cdot c, \bar{a} \cdot c + a \cdot b)$$
is sufficient to implement the XOR gate. Assume that either $\bar{b}$ or $\bar{c}$ are available.

**Solution 11.** Choosing $b = \bar{c}$ (equivalently $c = \bar{b}$) we find that the Fredkin gate yields
$$(a, b, c) \rightarrow (a, \bar{a} \cdot b + a \cdot \bar{b}, \bar{a} \cdot c + a \cdot \bar{c}) \equiv (a, a \oplus b, a \oplus c).$$

Thus we can apply the Fredkin gate to $(a, b, \bar{b})$ and use the second bit to obtain $a \oplus b$ or equivalently apply the Fredkin gate to $(a, \bar{c}, c)$ and use the third bit to obtain $a \oplus c$.

**Problem 12.** Consider the $2 \times 2$ identity matrix and the Pauli spin matrices $\sigma_1$, $\sigma_2$, $\sigma_3$ using the following notation

$$\sigma_{00} = \tau_{00} = I_2 = \begin{pmatrix} 1 & 0 \\ 0 & 1 \end{pmatrix},$$

$$\sigma_{01} = \tau_{01} = \sigma_1 = \begin{pmatrix} 0 & 1 \\ 1 & 0 \end{pmatrix},$$

$$\sigma_{10} = \tau_{10} = \sigma_3 = \begin{pmatrix} 1 & 0 \\ 0 & -1 \end{pmatrix},$$

$$\sigma_{11} = \sigma_2 = \begin{pmatrix} 0 & -i \\ i & 0 \end{pmatrix},$$

$$\tau_{11} = i\sigma_2 = \begin{pmatrix} 0 & 1 \\ -1 & 0 \end{pmatrix}.$$

Let $n$ be a positive integer. If $\mathbf{v}, \mathbf{w} \in \mathbb{Z}_2^n$ with

$$\mathbf{v} = \begin{pmatrix} v_1 \\ \vdots \\ v_n \end{pmatrix}, \qquad \mathbf{w} = \begin{pmatrix} w_1 \\ \vdots \\ w_n \end{pmatrix}$$

and

$$\mathbf{b} := \begin{pmatrix} \mathbf{v} \\ \mathbf{w} \end{pmatrix} \in \mathbb{Z}_2^{2n}$$

we define

$$\sigma_{\mathbf{b}} := \sigma_{v_1 w_1} \otimes \sigma_{v_2 w_2} \otimes \cdots \otimes \sigma_{v_n w_n}$$

and

$$\tau_{\mathbf{b}} := \tau_{v_1 w_1} \otimes \tau_{v_2 w_2} \otimes \cdots \otimes \tau_{v_n w_n}.$$

Thus we can associate a bit string $\mathbf{b}$ with each $\sigma_{\mathbf{b}}$ and vice versa.
(i) Let $n = 3$ and

$$\sigma_{\mathbf{b}_1} = \sigma_1 \otimes \sigma_3 \otimes \sigma_1, \qquad \sigma_{\mathbf{b}_2} = I_2 \otimes \sigma_3 \otimes \sigma_1.$$

Find the corresponding bit strings $\mathbf{b}_1$ and $\mathbf{b}_2$ for the given $\sigma_{\mathbf{b}_1}$ and $\sigma_{\mathbf{b}_2}$. Then XOR the two bit strings and find the corresponding $\sigma_{\mathbf{b}_3}$. Calculate the matrix product $\sigma_{\mathbf{b}_1} \sigma_{\mathbf{b}_2}$. Discuss.

**Solution 12.**　(i) We obtain the vectors

$$\mathbf{b}_1 = \begin{pmatrix} 0 \\ 1 \\ 0 \\ 1 \\ 0 \\ 1 \end{pmatrix}, \qquad \mathbf{b}_2 = \begin{pmatrix} 0 \\ 1 \\ 0 \\ 0 \\ 0 \\ 1 \end{pmatrix}.$$

The XOR operation provides the bit string

$$\mathbf{b}_3 = \begin{pmatrix} 0 \\ 0 \\ 0 \\ 1 \\ 0 \\ 0 \end{pmatrix}$$

with the corresponding $8 \times 8$ matrix

$$\sigma_{\mathbf{b}_3} = \sigma_1 \otimes I_2 \otimes I_2$$

which is $\sigma_{\mathbf{b}_3} = \sigma_{\mathbf{b}_1} \sigma_{\mathbf{b}_2}$.

**Problem 13.** Let $x \in \{0,1\}$ and $|0\rangle$, $|1\rangle$ be the standard basis in $\mathbb{C}^2$. Consider the boolean function $f(x) = \bar{x}$. Find a $4 \times 4$ permutation matrix $P$ such that

$$|x\rangle \otimes |0\rangle \xrightarrow{P} |x\rangle \otimes |f(x)\rangle \equiv |x\rangle \otimes |\bar{x}\rangle.$$

**Solution 13.** We have to satisfy

$$P(|0\rangle \otimes |0\rangle) = |0\rangle \otimes |1\rangle \rightarrow P\begin{pmatrix} 1 \\ 0 \\ 0 \\ 0 \end{pmatrix} = \begin{pmatrix} 0 \\ 1 \\ 0 \\ 0 \end{pmatrix}$$

$$P(|1\rangle \otimes |0\rangle) = |1\rangle \otimes |1\rangle \rightarrow P\begin{pmatrix} 0 \\ 0 \\ 1 \\ 0 \end{pmatrix} = \begin{pmatrix} 0 \\ 0 \\ 0 \\ 1 \end{pmatrix}.$$

A solution is

$$P = \begin{pmatrix} 0 & 0 & 0 & 1 \\ 1 & 0 & 0 & 0 \\ 0 & 1 & 0 & 0 \\ 0 & 0 & 1 & 0 \end{pmatrix}.$$

**Problem 14.** Consider the reversible gate (Feynman gate)

$$x_1 \rightarrow x_1, \qquad x_2 \rightarrow x_1 \oplus x_2.$$

The inverse function is given by $(x_1, x_2) \rightarrow (x_1, x_1 \oplus x_2)$. Let $|0\rangle$, $|1\rangle$ be the standard basis in the Hilbert space $\mathbb{C}^2$. Find the unitary matrix which implements ($x_1, x_2 \in \{0,1\}$)

$$|x_1\rangle \otimes |x_2\rangle \mapsto |x_1\rangle \otimes |x_1 \oplus x_2\rangle.$$

**Solution 14.** We have

$$|0\rangle \otimes |0\rangle \mapsto |0\rangle \otimes |0\rangle, \quad |0\rangle \otimes |1\rangle \mapsto |0\rangle \otimes |1\rangle$$

$$|1\rangle \otimes |0\rangle \mapsto |1\rangle \otimes |1\rangle, \quad |1\rangle \otimes |1\rangle \mapsto |1\rangle \otimes |0\rangle.$$

This provides the $4 \times 4$ permutation matrix

$$U = \begin{pmatrix} 1 & 0 & 0 & 0 \\ 0 & 1 & 0 & 0 \\ 0 & 0 & 0 & 1 \\ 0 & 0 & 1 & 0 \end{pmatrix} = \begin{pmatrix} 1 & 0 \\ 0 & 1 \end{pmatrix} \oplus \begin{pmatrix} 0 & 1 \\ 1 & 0 \end{pmatrix}.$$

which is the CNOT-gate and $\oplus$ denotes the direct sum.

**Problem 15.**    Let $x_1, x_2 \in \{0, 1\}$ and $\oplus$ be the XOR operation. Then

$$(x_1, x_2) \mapsto (x_1 \oplus 1, x_1 \oplus x_2)$$

is a 2-bit reversible gate since

$$(0, 0) \mapsto (1, 0), \quad (0, 1) \mapsto (1, 1), \quad (1, 0) \mapsto (0, 1), \quad (1, 1) \mapsto (0, 0).$$

Let $|0\rangle, |1\rangle$ be the standard basis in $\mathbb{C}^2$. Find the $4 \times 4$ permutation matrix $P$ such that

$$P(|x_1\rangle \otimes |x_2\rangle) = |x_1 \oplus 1\rangle \otimes |x_1 \oplus x_2\rangle.$$

**Solution 15.**    We calculate the Kronecker products of the vectors. This provides the four equations for $P$

$$P \begin{pmatrix} 1 \\ 0 \\ 0 \\ 0 \end{pmatrix} = \begin{pmatrix} 0 \\ 0 \\ 1 \\ 0 \end{pmatrix}, \quad P \begin{pmatrix} 0 \\ 1 \\ 0 \\ 0 \end{pmatrix} = \begin{pmatrix} 0 \\ 0 \\ 0 \\ 1 \end{pmatrix},$$

$$P \begin{pmatrix} 0 \\ 0 \\ 1 \\ 0 \end{pmatrix} = \begin{pmatrix} 0 \\ 1 \\ 0 \\ 0 \end{pmatrix}, \quad P \begin{pmatrix} 0 \\ 0 \\ 0 \\ 1 \end{pmatrix} = \begin{pmatrix} 1 \\ 0 \\ 0 \\ 0 \end{pmatrix}.$$

Consequently we obtain the $4 \times 4$ permutation matrix

$$P = \begin{pmatrix} 0 & 0 & 0 & 1 \\ 0 & 0 & 1 & 0 \\ 1 & 0 & 0 & 0 \\ 0 & 1 & 0 & 0 \end{pmatrix}$$

with the eigenvalues $+1, -1, +i, -i$.

**Problem 16.**    Given the $4 \times 4$ permutation matrix

$$U = \begin{pmatrix} 0 & 1 & 0 & 0 \\ 0 & 0 & 0 & 1 \\ 1 & 0 & 0 & 0 \\ 0 & 0 & 1 & 0 \end{pmatrix}$$

with the eigenvalues $+1, -1, +i, -i$. Find the corresponding boolean function.

**Solution 16.** Since the matrix has $4 = 2^2$ rows, the function $\mathbf{f} : \{0,1\}^2 \to \{0,1\}^2$ has two arguments. The first column (i.e. the column numbered 0) has a 1 in the third row (the row numbered 2) for which $b^{-1}(0) = (0,0)$ and $b^{-1}(2) = (1,0)$. Thus $(0,0) \to (1,0)$. From the second column $(0,1) \to (0,0)$. The third column provides $(1,0) \to (1,1)$ and the fourth column $(1,1) \to (0,1)$. Thus we have the map

$$(0,0) \mapsto (1,0), \qquad (0,1) \mapsto (0,0), \qquad (1,0) \mapsto (1,1), \qquad (1,1) \mapsto (0,1).$$

The right hand side provides the boolean expression

$$\mathbf{f}(x_1,x_2) = (\overline{x_1} \cdot \overline{x_2} + x_1 \cdot \overline{x_2}, x_1 \cdot \overline{x_2} + x_1 \cdot x_2) = (\overline{x_2}, x_1).$$

**Problem 17.** Given the boolean function $f(x_1,x_2) = x_1 \cdot \bar{x}_2$. Thus the map is ($x_1, x_2, y \in \{0,1\}$)

$$|x_1\rangle \otimes |x_2\rangle \otimes |y\rangle \mapsto |x_1\rangle \otimes |x_2\rangle \otimes |y \oplus (x_1 \cdot \bar{x}_2)\rangle$$

with

$$|0\rangle \otimes |0\rangle \otimes |0\rangle \mapsto |0\rangle \otimes |0\rangle \otimes |0\rangle, \quad |0\rangle \otimes |0\rangle \otimes |1\rangle \to |0\rangle \otimes |0\rangle \otimes |1\rangle$$

$$|0\rangle \otimes |1\rangle \otimes |0\rangle \mapsto |0\rangle \otimes |1\rangle \otimes |0\rangle, \quad |0\rangle \otimes |1\rangle \otimes |1\rangle \mapsto |0\rangle \otimes |1\rangle \otimes |1\rangle$$

$$|1\rangle \otimes |0\rangle \otimes |0\rangle \mapsto |1\rangle \otimes |0\rangle \otimes |1\rangle, \quad |1\rangle \otimes |0\rangle \otimes |1\rangle \mapsto |1\rangle \otimes |0\rangle \otimes |0\rangle$$

$$|1\rangle \otimes |1\rangle \otimes |0\rangle \mapsto |1\rangle \otimes |1\rangle \otimes |0\rangle, \quad |1\rangle \otimes |1\rangle \otimes |1\rangle \mapsto |1\rangle \otimes |1\rangle \otimes |1\rangle.$$

Find the $8 \times 8$ permutation matrix for this map.

**Solution 17.** The map leads to the $8 \times 8$ permutation matrix

$$U_f = I_4 \oplus \begin{pmatrix} 0 & 1 \\ 1 & 0 \end{pmatrix} \oplus I_2$$

where $\oplus$ denotes the direct sum and $I_n$ is the $n \times n$ identity matrix.

**Programming Problems**

**Problem 1.** Consider the reversible 3-input/3-output gate given by

$$x_1' = x_1 \oplus x_3$$
$$x_2' = x_1 \oplus x_2$$
$$x_3' = (x_1 \cdot x_2) \oplus (x_1 \cdot x_3) \oplus (x_2 \cdot x_3).$$

The inverse is given by

$$x_1 = x_1' \cdot x_2' \cdot \overline{x_3'} + \overline{x_1'} \cdot x_3' + \overline{x_2'} \cdot x_3'$$
$$x_2 = \overline{x_1'} \cdot x_2' \cdot \overline{x_3'} + x_1' \cdot x_3' + \overline{x_2'} \cdot x_3'$$
$$x_3 = x_1' \cdot \overline{x_2'} \cdot \overline{x_3'} + \overline{x_1'} \cdot x_3' + x_2' \cdot x_3'.$$

Give an implementation in C++ utilizing the `bitset` class.

**Solution 1.** In the bitset class & is the AND operation, | the OR operation, ^ the XOR operation and ~ the NOT operation.

```
// reversiblegate.cpp
#include <iostream>
#include <string>
#include <bitset>
using namespace std;

int main(void)
{
bitset<1> x1(string("1"));
bitset<1> x2(string("0"));
bitset<1> x3(string("1"));
cout << "x1 = " << x1 << endl;
cout << "x2 = " << x2 << endl;
cout << "x3 = " << x3 << endl;
bitset<1> x1p; bitset<1> x2p; bitset<1> x3p;
x1p = x1^x3; x2p = x1^x2;
x3p = (x1 & x2)^(x1 & x3)^(x2 & x3);
cout << "x1p = " << x1p << endl;
cout << "x2p = " << x2p << endl;
cout << "x3p = " << x3p << endl;
x1 = (x1p & x2p & (~x3p)) | ((~x1p) & x3p) | ((~x2p) & x3p);
x2 = ((~x1p) & x2p & (~x3p)) | (x1p & x3p) | ((~x2p) & x3p);
x3 = (x1p & (~x2p) & (~x3p)) | (~x1p & x3p) | (x2p & x3p);
cout << "x1 = " << x1 << endl;
cout << "x2 = " << x2 << endl;
cout << "x3 = " << x3 << endl;
return 0;
}
```

**Problem 2.** Given the boolean function $f(x_1, x_2) = x_1 \cdot \bar{x}_2$. Thus the map is ($x_1, x_2, y \in \{0,1\}$)

$$|x_1\rangle \otimes |x_2\rangle \otimes |y\rangle \mapsto |x_1\rangle \otimes |x_2\rangle \otimes |y \oplus (x_1 \cdot \bar{x}_2)\rangle$$

with

$$|0\rangle \otimes |0\rangle \otimes |0\rangle \mapsto |0\rangle \otimes |0\rangle \otimes |0\rangle, \quad |0\rangle \otimes |0\rangle \otimes |1\rangle \to |0\rangle \otimes |0\rangle \otimes |1\rangle$$

$$|0\rangle \otimes |1\rangle \otimes |0\rangle \mapsto |0\rangle \otimes |1\rangle \otimes |0\rangle, \quad |0\rangle \otimes |1\rangle \otimes |1\rangle \mapsto |0\rangle \otimes |1\rangle \otimes |1\rangle$$
$$|1\rangle \otimes |0\rangle \otimes |0\rangle \mapsto |1\rangle \otimes |0\rangle \otimes |1\rangle, \quad |1\rangle \otimes |0\rangle \otimes |1\rangle \mapsto |1\rangle \otimes |0\rangle \otimes |0\rangle$$
$$|1\rangle \otimes |1\rangle \otimes |0\rangle \mapsto |1\rangle \otimes |1\rangle \otimes |0\rangle, \quad |1\rangle \otimes |1\rangle \otimes |1\rangle \mapsto |1\rangle \otimes |1\rangle \otimes |1\rangle.$$

This leads to the $8 \times 8$ permutation matrix

$$U_f = I_4 \oplus \begin{pmatrix} 0 & 1 \\ 1 & 0 \end{pmatrix} \oplus I_2$$

where $\oplus$ denotes the direct sum and $I_n$ is the $n \times n$ identity matrix. Give a C++ implementation of this map.

**Solution 2.** The function `main` first finds the permutation matrix implementing the example: $f(x_1, x_2) = x_1 \cdot \overline{x}_2$. Then the map (truth table) is printed. Finally we recreate the map from the permutation matrix, which is the reversible map

$$g(x_1, x_2, x_3) = (x_1, x_2, (x_1 \cdot \overline{x}_2) \oplus x_3) = (x_1, x_2, x_1 \cdot \overline{x}_2 \cdot \overline{x_3} + \overline{x_1} \cdot x_3 + x_2 \cdot x_3).$$

The program counts from 0, i.e. $x_0$. The C++ program is

```cpp
// quantumgates.cpp
#include <bitset>
#include <iostream>
#include <list>
#include <map>
#include <vector>
#include "symbolicc++.h"
using namespace std;

const int n=3;

// a class to provide ordering of bitsets
// so that they can be used in maps
template <const size_t n> class cmpbst
{
 public:
  bool operator()(const bitset<n> &b1,const bitset<n> &b2)
  {
   size_t i;
   for(i=0;i<n;++i) if(b1[i] != b2[i]) return (b1[i] < b2[i]);
   return false;
  }
};

// for a given reversible boolean map, find the corresponding
// permutation matrix
```

```
template <const size_t n>
Symbolic permutation(const map<bitset<n>,bitset<n>,cmpbst<n> > &m)
{
 unsigned int N = (1 << n);
 Symbolic P = Symbolic("P",N,N)*0;
 typename map<bitset<n>, bitset<n> >::const_iterator i;
 for(i=m.begin();i!=m.end();++i)
   P(i->second.to_ulong(),i->first.to_ulong()) = 1;
 return P;
}


// simplifies a sum of products form using resolution the
// products are represented by bitsets and the sum is the
// list of bitsets
template <const size_t n>
list<pair<bitset<n>,bitset<n> > > simplify(const list<bitset<n> > &s)
{
 bool change = true;
 // a list which indicates whether bitsets were used in resolution
 // or need to be copied to the next round
 list<bool> copy;
 list<bool>::iterator ci1, ci2;
 // each bitset is stored with a mask which indicates which bits
 // may be used for resolution, once a bit is used it will be masked
 list<pair<bitset<n>,bitset<n> > > r, t1, t2, *tp1=&t1, *tp2=&t2, *tpp;
 typename list<bitset<n> >::const_iterator li;
 typename list<pair<bitset<n>, bitset<n> > >::const_iterator lpi1, lpi2;
 for(li=s.begin();li!=s.end();++li)
 {
   t1.push_back(make_pair(*li,bitset<n>()));
   // initially all bitsets propagate
   copy.push_back(true);
 }
 while(!tp1->empty())
 {
  // track whether resolution has been applied
  // if no change is recorded, we are done
  change = false;
  for(lpi1=tp1->begin(),ci1=copy.begin();lpi1!=tp1->end();++lpi1,++ci1)
  {
   // search for a second bitset which differs from this bitset
   // in exactly one place (taking into account the masks)
   for(lpi2=lpi1,ci2=ci1;lpi2!=tp1->end();++lpi2,++ci2)
   {
    // only compare if the masks are the same
    if(lpi1->second==lpi2->second)
    {
     // XOR finds the differing bits which are then masked
```

```
      bitset<n> diff = ((lpi1->first ^ lpi2->first) & ~lpi1->second);
      // only one bit differs so apply resolution
      if(diff.count()==1)
      {
      // mask the bit which has been used
      tp2->push_back(make_pair(lpi1->first,lpi1->second | diff));
      change = true;
      // these bitsets have been used in resolution, don't copy them
      *ci1 = *ci2 = false;
      }
    }
  }
  if(*ci1) r.push_back(*lpi1);
 }
 // reset the variables for the next application of resolution
 tpp = tp1; tp1 = tp2; tp2 = tpp; tp2->clear();
 copy.clear(); copy.resize(tp1->size(),true);
}
r.unique();
return r;
}

// find a symbolic expression for a given boolean map
template <const size_t n>
Symbolic expression(const map<bitset<n>,bitset<n>,cmpbst<n> > &m)
{
 size_t j, k;
 Symbolic S("S",n), NOT("NOT"), x("x",n);
 vector<list<bitset<n> > > terms(n);
 vector<list<pair<bitset<n>, bitset<n> > > > simplified(n);
 typename map<bitset<n>,bitset<n> >::const_iterator i;
 typename list<pair<bitset<n>,bitset<n> > >::iterator li;
 // for each y_j, record all values of x_1,...,x_n such that y_j=1
 for(i=m.begin();i!=m.end();++i)
  for(j=0;j<n;++j) if(i->second[j]) terms[j].push_back(i->first);
 // construct each symbolic expression for y_j
 for(j=0;j<n;++j)
 {
  S(j) = 0;
  // find a smaller set of terms
  simplified[j] = simplify(terms[j]);
  for(li=simplified[j].begin();li!=simplified[j].end();++li)
  {
   Symbolic P = 1;
   for(k=0;k<n;++k)
    if(!li->second[k])
    {
     // this is the usual construction of a product for the
```

```
     // sum of products form generated from a truth table
     if(li->first[k]) P *= x(k); else P *= NOT[x(k)];
     }
   S(j) += P;
  }
 }
 return S;
}


// determine the reversible boolean map from a permutation matrix
template <const size_t n>
map<bitset<n>,bitset<n>,cmpbst<n> > booleanmap(const Symbolic &permutation
{
 size_t i, j;
 map<bitset<n>, bitset<n>, cmpbst<n> > m;
 for(i=0;i<(1<<n);++i)
   for(j=0;j<(1<<n);++j)
     if(permutation(i,j)!=0) m[bitset<n>(j)] = bitset<n>(i);
 return m;
}


// reverse the contents of a bitset
template <const int n> bitset<n> reverse(const bitset<n> &b)
{
 size_t i;
 bitset<n> r;
 for(i=0;i<n;++i) r[n-i-1] = b[i];
 return r;
}


int main(void)
{
 int i1, i2, i3;
 bitset<3> a, b;
 map<bitset<3>, bitset<3>, cmpbst<3> > f, g;
 map<bitset<3>, bitset<3>, cmpbst<3> >::const_iterator i;
 Symbolic P;
 for(i1=0;i1<2;++i1)
   for(i2=0;i2<2;++i2)
     for(i3=0;i3<2;++i3)
     {
     a[0] = b[0] = i1; a[1] = b[1] = i2; a[2] = i3;
     b[2] = a[2]^(a[0] & (!a[1]));
     f[a] = b;
     }
 P = permutation(f); cout << P << endl;
 g = booleanmap<3>(P);
 for(i=g.begin();i!=g.end();++i)
```

```
  cout << reverse<3>(i->first) << " -> " << reverse<3>(i->second) << endl;
  cout << expression(f) << endl;
  return 0;
}
```

The output is

```
[1 0 0 0 0 0 0 0]
[0 0 0 0 0 1 0 0]
[0 0 1 0 0 0 0 0]
[0 0 0 1 0 0 0 0]
[0 0 0 0 1 0 0 0]
[0 1 0 0 0 0 0 0]
[0 0 0 0 0 0 1 0]
[0 0 0 0 0 0 0 1]
000 -> 000
001 -> 001
010 -> 010
011 -> 011
100 -> 101
101 -> 100
110 -> 110
111 -> 111
[                  x0                  ]
[                  x1                  ]
[x0*NOT[x1]*NOT[x2]+NOT[x0]*x2+x1*x2]
```

# 6.3   Supplementary Problems

**Problem 1.**  Given the reversible gate as truth table

| $x_1$ | $x_2$ | $x_3$ | $f_1$ | $f_2$ | $f_3$ |
|-------|-------|-------|-------|-------|-------|
| 0 | 0 | 0 | 0 | 0 | 1 |
| 0 | 0 | 1 | 0 | 1 | 0 |
| 0 | 1 | 0 | 0 | 1 | 1 |
| 0 | 1 | 1 | 1 | 0 | 0 |
| 1 | 0 | 0 | 1 | 0 | 1 |
| 1 | 0 | 1 | 1 | 1 | 0 |
| 1 | 1 | 0 | 1 | 1 | 1 |
| 1 | 1 | 1 | 0 | 0 | 0 |

Show that

$$f_1 = \overline{x}_1 \cdot x_2 \cdot x_3 + x_1 \cdot \overline{x}_2 \cdot \overline{x}_3 + x_1 \cdot \overline{x}_2 \cdot x_3 + x_1 \cdot x_2 \cdot \overline{x}_3$$

$$f_2 = \overline{x}_1 \cdot \overline{x}_2 \cdot x_3 + \overline{x}_1 \cdot x_2 \cdot \overline{x}_3 + x_1 \cdot \overline{x}_2 \cdot x_3 + x_1 \cdot x_2 \cdot \overline{x}_3$$
$$f_3 = \overline{x}_1 \cdot \overline{x}_2 \cdot \overline{x}_3 + \overline{x}_1 \cdot x_2 \cdot \overline{x}_3 + x_1 \cdot \overline{x}_2 \cdot \overline{x}_3 + x_1 \cdot x_2 \cdot \overline{x}_3.$$

**Problem 2.**  Let $x_1, x_2 \in \{0, 1\}$ and $\oplus$ be the XOR operation. Then

$$(x_1, x_2) \mapsto (x_1 \oplus 1, x_1 \oplus x_2)$$

is a 2-bit reversible gate since

$$(0,0) \mapsto (1,0), \quad (0,1) \mapsto (1,1), \quad (1,0) \mapsto (0,1), \quad (1,1) \mapsto (0,0).$$

Let $|0\rangle$, $|1\rangle$ be the standard basis in $\mathbb{C}^2$. Find the $4 \times 4$ permutation matrix $P$ such that
$$P(|x_1\rangle \otimes |x_2\rangle) = |x_1 \oplus 1\rangle \otimes |x_1 \oplus x_2\rangle.$$

**Problem 3.**  The *Toffoli gate* $\mathrm{T}(x_1, x_2, x_3)$ has 3 inputs $(x_1, x_2, x_3)$ and three outputs $(y_1, y_2, y_3)$ and is given by

$$(x_1, x_2, x_3) \rightarrow (x_1, x_2, x_3 \oplus (x_1 \cdot x_2))$$

where $x_1, x_2, x_3 \in \{0, 1\}$, $\oplus$ is the XOR operation and $\cdot$ the AND operation. Show that the truth table is given by

| $x_1$ | $x_2$ | $x_3$ | $y_1$ | $y_2$ | $y_3$ |
|---|---|---|---|---|---|
| 0 | 0 | 0 | 0 | 0 | 0 |
| 0 | 0 | 1 | 0 | 0 | 1 |
| 0 | 1 | 0 | 0 | 1 | 0 |
| 0 | 1 | 1 | 0 | 1 | 1 |
| 1 | 0 | 0 | 1 | 0 | 0 |
| 1 | 0 | 1 | 1 | 0 | 1 |
| 1 | 1 | 0 | 1 | 1 | 1 |
| 1 | 1 | 1 | 1 | 1 | 0 |

**Problem 4.**  Consider the 3-input/3-output gate given by

$$x_1' = x_1 \oplus x_3, \quad x_2' = x_1 \oplus x_2, \quad x_3' = (x_1 \cdot x_2) \oplus (x_1 \cdot x_3) \oplus (x_2 \cdot x_3).$$

Give the truth table. Show that the gate is invertible. Find the inverse function.

# Chapter 7

# Unitary Transforms and Quantum Gates

## 7.1 Introduction

Quantum gates are realised as *unitary operators*. Let $\mathcal{H}$ denote a Hilbert space. A linear operator $U$ in a Hilbert space is unitary if

$$U^* = U^{-1}.$$

In other words, $U$ is unitary if and only if $U$ is invertible and $U^{-1} = U^*$. An equivalent form of definition is: $U$ is unitary iff

$$\langle U\mathbf{x}, U\mathbf{y} \rangle = \langle \mathbf{x}, \mathbf{y} \rangle, \qquad \text{for all} \quad \mathbf{x}, \mathbf{y} \in \mathcal{H}.$$

Thus we have

$$U^*U = UU^* = I$$

where $*$ denotes the adjoint and $I$ is the identity operator. The composite of two unitary operators is again unitary.

Here we consider $n \times n$ unitary matrices. If $U_1$, $U_2$ are $n \times n$ unitary matrices, then $U_1 U_2$ is an $n \times n$ unitary matrix. The $n \times n$ unitary matrices form a group under matrix multiplication. All the eigenvalues of a unitary matrix have absolute value equal to 1, i.e. $|\lambda| = 1$. Thus $|\det(U)| = 1$.

175

All $n \times n$ unitary matrices form a group under matrix multiplication. An important subgroup are all $n \times n$ unitary matrices with $\det(U) = 1$. If $\det(U_1) = 1$ and $\det(U_2) = 1$, then $\det(U_1 U_2) = 1$. Another important subgroup are all $n \times n$ *permutation matrices*. A number of quantum gates are given as permutation matrices such as

$$U_{CNOT} = \begin{pmatrix} 1 & 0 & 0 & 0 \\ 0 & 1 & 0 & 0 \\ 0 & 0 & 0 & 1 \\ 0 & 0 & 1 & 0 \end{pmatrix}, \quad U_{SWAP} = \begin{pmatrix} 1 & 0 & 0 & 0 \\ 0 & 0 & 1 & 0 \\ 0 & 1 & 0 & 0 \\ 0 & 0 & 0 & 1 \end{pmatrix}.$$

If $K$ is a hermitian matrix, then $\exp(iK)$ is a unitary matrix. Thus if $\hat{H}$ is a hermitian matrix describing the Hamilton operator we find that the exponential function

$$\exp(-i\hat{H}t/\hbar)$$

is a unitary matrix. Let $A$ be an arbitrary $n \times n$ matrix over $\mathbb{C}$. Then

$$U \exp(A) U^* = \exp(U A U^*).$$

If $U$ and $V$ are unitary, then $U \otimes V$ and $U \oplus V$ are unitary, where $\otimes$ is the Kronecker product and $\oplus$ the direct sum.
The square roots of a unitary matrix are not necessarily unitary matrices again.
Let $\mathbf{v}$ be a normalized state in $\mathbb{C}^n$. Then $U\mathbf{v}$ is a normalized state.
Let $\mathbf{v}_1, \mathbf{v}_2, \ldots, \mathbf{v}_n$ be an orthonormal basis in $\mathbb{C}^n$. Then

$$V = \begin{pmatrix} \mathbf{v}_1 & \mathbf{v}_2 & \cdots & \mathbf{v}_n \end{pmatrix}$$

is a unitary matrix. Let $|\lambda_j| = 1$. Then

$$U = \sum_{j=1}^{n} \lambda_j \mathbf{v}_j \mathbf{v}_j^*$$

is a unitary matrix.
An important unitary matrix is the *Fourier matrix* given by

$$F = \frac{1}{\sqrt{n}} \begin{pmatrix} 1 & 1 & 1 & \cdots & 1 \\ 1 & \omega & \omega^2 & \cdots & \omega^{n-1} \\ 1 & \omega^2 & \omega^4 & \cdots & \omega^{2(n-1)} \\ \vdots & \vdots & \vdots & \ddots & \vdots \\ 1 & \omega^{n-1} & \omega^{2(n-1)} & \cdots & \omega^{(n-1)(n-1)} \end{pmatrix}$$

where $\omega^n = 1$.

## 7.2   Solved Problems

**Problem 1.**   Let $U$ be a $2 \times 2$ unitary matrix and $I_2$ be the $2 \times 2$ identity matrix. Is the $4 \times 4$ matrix

$$V(\alpha) = \begin{pmatrix} 0 & 0 \\ 0 & 1 \end{pmatrix} \otimes U + \begin{pmatrix} e^{i\alpha} & 0 \\ 0 & 0 \end{pmatrix} \otimes I_2, \qquad \alpha \in \mathbb{R}$$

unitary?

**Solution 1.**   Yes, $V(\alpha)$ is unitary. Since

$$\begin{pmatrix} 0 & 0 \\ 0 & 1 \end{pmatrix} \begin{pmatrix} e^{i\alpha} & 0 \\ 0 & 0 \end{pmatrix} = \begin{pmatrix} e^{i\alpha} & 0 \\ 0 & 0 \end{pmatrix} \begin{pmatrix} 0 & 0 \\ 0 & 1 \end{pmatrix} = \begin{pmatrix} 0 & 0 \\ 0 & 0 \end{pmatrix}$$

and $UU^* = I_2$ we obtain

$$V(\alpha)V^*(\alpha) = \begin{pmatrix} 0 & 0 \\ 0 & 1 \end{pmatrix} \otimes I_2 + \begin{pmatrix} 1 & 0 \\ 0 & 0 \end{pmatrix} \otimes I_2 = I_2 \otimes I_2.$$

**Problem 2.**   (i) Let

$$M := \frac{1}{\sqrt{2}} \begin{pmatrix} 1 & i & 0 & 0 \\ 0 & 0 & i & 1 \\ 0 & 0 & i & -1 \\ 1 & -i & 0 & 0 \end{pmatrix}.$$

Is the matrix $M$ unitary? In quantum computing $M$ is called the *magic gate*.
(ii) Let

$$U_H := \frac{1}{\sqrt{2}} \begin{pmatrix} 1 & 1 \\ 1 & -1 \end{pmatrix}, \qquad U_S := \begin{pmatrix} 1 & 0 \\ 0 & i \end{pmatrix}$$

and

$$U_{CNOT2} = \begin{pmatrix} 1 & 0 & 0 & 0 \\ 0 & 0 & 0 & 1 \\ 0 & 0 & 1 & 0 \\ 0 & 1 & 0 & 0 \end{pmatrix}.$$

Show that the matrix $M$ can be written as

$$M = U_{CNOT2}(I_2 \otimes U_H)(U_S \otimes U_S).$$

(iii) Let $SO(4)$ be the special orthogonal Lie group. Let $SU(2)$ be the special unitary Lie group. Show that for every real orthogonal matrix $U \in SO(4)$, the matrix $MUM^{-1}$ is the Kronecker product of two 2-dimensional special unitary matrices, i.e.

$$MUM^{-1} \in SU(2) \otimes SU(2).$$

**Solution 2.**   (i) Since $MM^* = I_4$ we find that $M$ is unitary.
(ii) We obtain

$$M = \begin{pmatrix} 1 & 0 & 0 & 0 \\ 0 & 0 & 0 & 1 \\ 0 & 0 & 1 & 0 \\ 0 & 1 & 0 & 0 \end{pmatrix} \begin{pmatrix} \frac{1}{\sqrt{2}} & \frac{1}{\sqrt{2}} & 0 & 0 \\ \frac{1}{\sqrt{2}} & -\frac{1}{\sqrt{2}} & 0 & 0 \\ 0 & 0 & \frac{1}{\sqrt{2}} & \frac{1}{\sqrt{2}} \\ 0 & 0 & \frac{1}{\sqrt{2}} & -\frac{1}{\sqrt{2}} \end{pmatrix} \begin{pmatrix} 1 & 0 & 0 & 0 \\ 0 & i & 0 & 0 \\ 0 & 0 & i & 0 \\ 0 & 0 & 0 & -1 \end{pmatrix}.$$

(iii) We show that for every $A \otimes B \in SU(2) \otimes SU(2)$, we have

$$M^{-1}(A \otimes B)M \in SO(4).$$

Now every matrix $A \in SU(2)$ can be written as $R_z(\alpha)R_y(\theta)R_z(\beta)$ for some $\alpha, \beta, \theta \in \mathbb{R}$, where

$$R_y(\theta) = \begin{pmatrix} \cos(\theta/2) & \sin(\theta/2) \\ -\sin(\theta/2) & \cos(\theta/2) \end{pmatrix}, \qquad R_z(\alpha) = \begin{pmatrix} e^{-i\alpha/2} & 0 \\ 0 & e^{i\alpha/2} \end{pmatrix}.$$

Therefore any matrix $A \otimes B \in SU(2) \otimes SU(2)$ can be written as a product of the matrices of the form $V \otimes I_2$ or $I_2 \otimes V$, where $V$ is either $R_y(\theta)$ or $R_z(\alpha)$. Next we have to show that $M^{-1}(V \otimes I_2)M$ and $M^{-1}(I_2 \otimes V)M$ are in $SO(4)$. We have

$$M^{-1}(R_y(\theta) \otimes I_2)M = \begin{pmatrix} \cos(\theta/2) & 0 & 0 & -\sin(\theta/2) \\ 0 & \cos(\theta/2) & \sin(\theta/2) & 0 \\ 0 & -\sin(\theta/2) & \cos(\theta/2) & 0 \\ \sin(\theta/2) & 0 & 0 & \cos(\theta/2) \end{pmatrix}$$

$$M^{-1}(R_z(\alpha) \otimes I_2)M = \begin{pmatrix} \cos(\alpha/2) & \sin(\alpha/2) & 0 & 0 \\ -\sin(\alpha/2) & \cos(\alpha/2) & 0 & 0 \\ 0 & 0 & \cos(\alpha/2) & -\sin(\alpha/2) \\ 0 & 0 & \sin(\alpha/2) & \cos(\alpha/2) \end{pmatrix}.$$

We have similar equations for the cases of $I_2 \otimes R_y(\theta)$ and $I_2 \otimes R_z(\alpha)$. Since the mapping

$$A \otimes B \to M^{-1}(A \otimes B)M$$

is one-to-one (invertible) and the Lie groups $SU(2) \otimes SU(2)$ and $SO(4)$ have the same topological dimension, we conclude that the mapping is an isomorphism between these two Lie groups.

**Problem 3.**   (i) Let $A := |0\rangle\langle 0| - |1\rangle\langle 1|$ in the Hilbert space $\mathbb{C}^2$. Calculate

$$U_H A U_H |0\rangle, \qquad U_H A U_H |1\rangle$$

where $U_H$ is the *Walsh-Hadamard transform*. The unitary transform $U_H$ is defined by

$$U_H |k\rangle = \frac{1}{\sqrt{2}}(|0\rangle + (-1)^k |1\rangle), \qquad k \in \{0, 1\}.$$

(ii) Calculate
$$(U_H \otimes U_H)U_{CNOT}(U_H \otimes U_H)|j,k\rangle$$
where $|j,k\rangle \equiv |j\rangle \otimes |k\rangle$ with $j,k \in \{0,1\}$, and the answer is in the form of a ket $|m,n\rangle$ with $m,n \in \{0,1\}$. The unitary transform
$$U_{CNOT} := |0\rangle\langle 0| \otimes I_2 + |1\rangle\langle 1| \otimes U_{NOT}$$
is the *controlled NOT* operation and the unitary transform
$$U_{NOT} = |0\rangle\langle 1| + |1\rangle\langle 0|$$
is the *NOT operation.*

**Solution 3.** (i) Let $j \in \{0,1\}$. Then
$$U_H A U_H|j\rangle = \frac{1}{\sqrt{2}}U_H A(|0\rangle + (-1)^j|1\rangle) = \frac{1}{\sqrt{2}}U_H(|0\rangle + (-1)^{j+1}|1\rangle)$$
$$= \frac{1}{\sqrt{2}}U_H(|0\rangle + (-1)^{\bar{j}}|1\rangle) = |\bar{j}\rangle$$
where $\bar{j} := 1 - j$. In other words $U_H A U_H = U_{NOT}$.
(ii) Straightforward calculation yields

$(U_H \otimes U_H)U_{CNOT}(U_H \otimes U_H)|j,k\rangle$
$= \frac{1}{2}(U_H \otimes U_H)U_{CNOT}\left((|0\rangle + (-1)^j|1\rangle) \otimes (|0\rangle + (-1)^k|1\rangle)\right)$
$= \frac{1}{2}(U_H \otimes U_H)(|00\rangle + (-1)^k|01\rangle + (-1)^j|11\rangle + (-1)^{j+k}|10\rangle)$
$= \frac{1}{2}(U_H \otimes U_H)(|0\rangle \otimes (|0\rangle + (-1)^k|1\rangle) + (-1)^j|1\rangle \otimes (|1\rangle + (-1)^k|0\rangle))$
$= \frac{1}{2}(U_H \otimes U_H)(|0\rangle \otimes (|0\rangle + (-1)^k|1\rangle) + (-1)^{j+k}|1\rangle \otimes (|0\rangle + (-1)^k|1\rangle))$
$= \frac{1}{2}(U_H \otimes U_H)(|0\rangle + (-1)^{j+k}|1\rangle) \otimes (|0\rangle + (-1)^k|1\rangle)$
$= |j \oplus k, k\rangle$

where $\oplus$ is the XOR operation. In other words we have the *controlled NOT* operation, where the control qubit is the second qubit and the target qubit is the first qubit.

**Problem 4.** Consider the linear operator (Hamilton operator)
$$H := i\hbar\omega(|0\rangle\langle 1| - |1\rangle\langle 0|)$$
operating in the Hilbert space $\mathbb{C}^2$, where $\{|0\rangle, |1\rangle\}$ is an orthonormal basis in $\mathbb{C}^2$ and $\omega$ is a real parameter (frequency).

(i) Is $H$ self-adjoint?
(ii) Find the eigenvalues and corresponding normalized eigenvectors of $H$.
(iii) Find the unitary matrix $U(t) := \exp(-iHt/\hbar)$. Find the values of $t$ such that $U(t)$ performs the *NOT operation*

$$U(t)|0\rangle \to |1\rangle, \qquad U(t)|1\rangle \to |0\rangle.$$

(iv) Calculate $U(t = \pi/4\omega)$ and $(U(t = \pi/4\omega))^2$.

**Solution 4.**   (i) The adjoint of an operator can be obtained by simply swapping the labels of the corresponding bra and ket vectors in the sum, and taking the complex conjugate of all complex coefficients. Thus

$$H^* = \overline{i}\hbar\omega(|1\rangle\langle 0| - |0\rangle\langle 1|) = -i\hbar\omega(|1\rangle\langle 0| - |0\rangle\langle 1|) = H$$

i.e. $H$ is self-adjoint. We can determine $H^*$ as follows. Let

$$H^* = a_{00}|0\rangle\langle 0| + a_{01}|0\rangle\langle 1| + a_{10}|1\rangle\langle 0| + a_{11}|1\rangle\langle 1|, \qquad a_{00}, a_{01}, a_{10}, a_{11} \in \mathbb{C}.$$

The bra vector corresponding to the ket $H|y\rangle$ is $\langle y|H^*$. We require that $\langle H^*y|x\rangle = \langle y|Hx\rangle$ for all $|x\rangle = x_0|0\rangle + x_1|1\rangle$ and $|y\rangle = y_0|0\rangle + y_1|1\rangle$. We find

$$H|x\rangle = i\hbar\omega(x_1|0\rangle - x_0|1\rangle)$$
$$H^*|y\rangle = (y_0 a_{00} + y_1 a_{01})|0\rangle + (y_0 a_{10} + y_1 a_{11})|1\rangle$$
$$\langle y|Hx\rangle = i\hbar\omega(x_1\overline{y_0} - x_0\overline{y_1})$$
$$\langle H^*y|x\rangle = x_0(\overline{y_0 a_{00}} + \overline{y_1 a_{01}}) + x_1(\overline{y_0 a_{10}} + \overline{y_1 a_{11}}).$$

Since $\langle H^*y|x\rangle = \langle y|Hx\rangle$ for all $|x\rangle$ and $|y\rangle$, we obtain

$$i\hbar\omega\overline{y_0} = (\overline{y_0 a_{10}} + \overline{y_1 a_{11}}), \qquad -i\hbar\omega\overline{y_1} = (\overline{y_0 a_{00}} + \overline{y_1 a_{01}}).$$

Consequently $a_{00} = 0$, $a_{01} = i\hbar\omega$, $a_{10} = -i\hbar\omega$, $a_{11} = 0$.
(ii) The eigenvalue equation for $H$ is

$$H(a|0\rangle + b|1\rangle) = \lambda(a|0\rangle + b|1\rangle).$$

Thus we have the two equations $-i\hbar\omega a = \lambda b$, $i\hbar\omega b = \lambda a$. If $\lambda = 0$ we have $a = 0$ and $b = 0$. Therefore we only have to consider $\lambda \neq 0$. Obviously we may assume $b \neq 0$ (thus $a \neq 0$). We obtain

$$\lambda = -\frac{i\hbar\omega a}{b}.$$

Hence $ib^2 = -ia^2$ so that $b = \pm ia$. Using $|a|^2 + |b|^2 = 1$ we find $|a| = \frac{1}{\sqrt{2}}$. We obtain the eigenvalues and corresponding orthonormal eigenvectors

$$\lambda_1 = -\hbar\omega, \quad \frac{1}{\sqrt{2}}(|0\rangle + i|1\rangle), \qquad \lambda_2 = \hbar\omega, \quad \frac{1}{\sqrt{2}}(|0\rangle - i|1\rangle).$$

(iii) We find $H^n$ ($n \in \mathbb{N}$) by observing that

$$H^2 = (\hbar\omega)^2(|0\rangle\langle0|+|1\rangle\langle1|) = (\hbar\omega)^2 I_2, \qquad H^3 = (\hbar\omega)^2 H, \qquad H^4 = (\hbar\omega)^4 I_2.$$

Thus

$$H^n = \begin{cases} (\hbar\omega)^{n-1}H & n \text{ odd} \\ (\hbar\omega)^n I_2 & n \text{ even} \end{cases}.$$

Since $U(t) := \exp(-iHt/\hbar)$ we have

$$U(t) = \sum_{j=0}^{\infty} \frac{(-\frac{it}{\hbar})^j H^j}{j!} = \sum_{j=0}^{\infty} \frac{(-i\omega t)^{2j}}{(2j)!} I_2 + \frac{1}{\hbar\omega} \sum_{j=0}^{\infty} \frac{(-i\omega t)^{2j+1}}{(2j+1)!} H$$

$$= \sum_{j=0}^{\infty} \frac{(-1)^j(\omega t)^{2j}}{(2j)!} I_2 - i\frac{1}{\hbar\omega} \sum_{j=0}^{\infty} \frac{(-1)^j(\omega t)^{2j+1}}{(2j+1)!} H$$

$$= \cos(\omega t)I_2 - \frac{i}{\hbar\omega}\sin(\omega t)H$$

$$= \cos(\omega t)(|0\rangle\langle0| + |1\rangle\langle1|) + \sin(\omega t)(|0\rangle\langle1| - |1\rangle\langle0|).$$

For the NOT operation we use $U(t = \pi/2\omega) = |0\rangle\langle1| - |1\rangle\langle0|$. The unitary transforms $U((2k+1)\pi/2\omega)$, $k \in \mathbb{N}_0$ implement the NOT operation.
(iv) We have

$$U(t = \pi/4\omega) = \frac{1}{\sqrt{2}}(|0\rangle - |1\rangle)\langle0| + \frac{1}{\sqrt{2}}(|0\rangle + |1\rangle)\langle1|$$

$$U(t = \pi/4\omega)^2 = U(t = \pi/2\omega) = |0\rangle\langle1| - |1\rangle\langle0|.$$

Thus we find $(U(t = \pi/4\omega))^2 = U(t = \pi/2\omega)$, i.e. $U(t = \pi/4\omega)$ acts as the *square root* of the *NOT operation*. Traditionally in quantum computation we use $U_{NOT} = |0\rangle\langle1| + |1\rangle\langle0|$. In this case for the $\sqrt{\text{NOT}}$ operation we use

$$U_{\sqrt{NOT}} = \frac{1}{2}(1+i)(|0\rangle\langle0| + |1\rangle\langle1|) + \frac{1}{2}(1-i)(|0\rangle\langle1| + |1\rangle\langle0|).$$

**Problem 5.** Consider the Hilbert space $\mathbb{C}^n$. Let $\mathbf{e}_1, \mathbf{e}_2, \ldots, \mathbf{e}_n$ be the standard basis in $\mathbb{C}^n$, $S_n$ be the *symmetric group* of order $n!$ and $U_\sigma$ be the unitary matrix on $\otimes^n \mathbb{C}^n$ such that

$$U_\sigma(\mathbf{e}_1 \otimes \cdots \otimes \mathbf{e}_n) := \mathbf{e}_{\sigma(1)} \otimes \cdots \otimes \mathbf{e}_{\sigma(n)}$$

where $\sigma \in S_n$. We define the matrix ("antisymmetrization operator") in the Hilbert space $\otimes^n \mathbb{C}^n$ by

$$\Pi_n := \frac{1}{n!} \sum_{\sigma \in S_n} \text{sgn}(\sigma)U_\sigma$$

where sgn is the signature of the permutation $\sigma \in S_n$. The matrices $\Pi_n$ are projection matrices. Find $\Pi_2$ and $\Pi_3$.

**Solution 5.**    (i) For $n = 2$ we have the unitary matrices

$$U_{11,22} = \begin{pmatrix} 1 & 0 \\ 0 & 1 \end{pmatrix} \otimes \begin{pmatrix} 1 & 0 \\ 0 & 1 \end{pmatrix}, \qquad U_{12,21} = \begin{pmatrix} 0 & 1 \\ 1 & 0 \end{pmatrix} \otimes \begin{pmatrix} 0 & 1 \\ 1 & 0 \end{pmatrix}$$

and therefore

$$\Pi_2 = \frac{1}{2}(U_{11,22} - U_{12,21}) = \frac{1}{2}\begin{pmatrix} 1 & 0 & 0 & -1 \\ 0 & 1 & -1 & 0 \\ 0 & -1 & 1 & 0 \\ -1 & 0 & 0 & 1 \end{pmatrix}.$$

**Problem 6.**    (i) Let $\sigma_1, \sigma_2, \sigma_3$ be the *Pauli spin matrices*. Find

$$R_{1x}(\alpha) := \exp(-i\alpha(\sigma_1 \otimes I_2)), \qquad R_{1y}(\alpha) := \exp(-i\alpha(\sigma_2 \otimes I_2))$$

where $\alpha \in \mathbb{R}$ and $I_2$ denotes the $2 \times 2$ unit matrix.
(ii) Consider the special case $R_{1x}(\alpha = \pi/2)$ and $R_{1y}(\alpha = \pi/4)$. Calculate the matrix $R_{1x}(\pi/2)R_{1y}(\pi/4)$. Discuss.

**Solution 6.**    (i) We have

$$\exp(-i\alpha(\sigma_1 \otimes I_2)) := \sum_{k=0}^{\infty} \frac{(-i\alpha(\sigma_1 \otimes I_2))^k}{k!}.$$

Since $\sigma_1^2 = I_2$ we have $(\sigma_1 \otimes I_2)^2 = I_2 \otimes I_2$. Thus we find

$$\exp(-i\alpha(\sigma_1 \otimes I_2)) = (I_2 \otimes I_2)\cos\alpha + e^{-i\pi/2}(\sigma_1 \otimes I_2)\sin(\alpha)$$

where we used $\exp(-i\pi/2) = -i$. Analogously, we find

$$\exp(-i\alpha(\sigma_2 \otimes I_2)) = (I_2 \otimes I_2)\cos(\alpha) + e^{-i\pi/2}(\sigma_2 \otimes I_2)\sin(\alpha)$$

since $(\sigma_2 \otimes I_2)^2 = I_2 \otimes I_2$.
(ii) Since $\sin(\pi/2) = 1$, $\cos(\pi/2) = 0$ we arrive at

$$R_{1x}(\pi/2) = e^{-i\pi/2}(\sigma_1 \otimes I_2).$$

From $\sin(\pi/4) = \sqrt{2}/2$, $\cos(\pi/4) = \sqrt{2}/2$ it follows that

$$R_{1y}(\pi/4) = \frac{1}{\sqrt{2}}(I_2 \otimes I_2) + \frac{1}{\sqrt{2}}e^{-i\pi/2}(\sigma_2 \otimes I_2).$$

Thus

$$R_{1x}(\pi/2)R_{1y}(\pi/4) = \frac{e^{-i\pi/2}}{\sqrt{2}}(\sigma_1 \otimes I_2) + \frac{e^{-i\pi/2}}{\sqrt{2}}(\sigma_3 \otimes I_2)$$

where we used that $\sigma_1\sigma_2 = i\sigma_3$. Therefore

$$R_{1x}(\pi/2)R_{1y}(\pi/4) = \frac{e^{-i\pi/2}}{\sqrt{2}}(\sigma_1 + \sigma_3) \otimes I_2$$

where

$$\frac{1}{\sqrt{2}}(\sigma_1 + \sigma_3) = \frac{1}{\sqrt{2}}\begin{pmatrix} 1 & 1 \\ 1 & -1 \end{pmatrix}$$

is the *Walsh-Hadamard gate*. All the single operations are in the Lie group $SU(2)$ whose determinant is $+1$, while the determinant of the Walsh-Hadamard gate is $-1$. Thus the overall phase is unavoidable.

**Problem 7.** Consider the Hilbert space $\mathbb{C}^{2^n}$. Let $\{ |0\rangle, |1\rangle, \ldots, |2^n-1\rangle \}$ be an orthonormal basis in this Hilbert space. We define the linear operator

$$U_{QFT} := \frac{1}{\sqrt{2^n}} \sum_{j=0}^{2^n-1} \sum_{k=0}^{2^n-1} e^{-i2\pi kj/2^n} |k\rangle\langle j|. \tag{1}$$

This transform is called the *quantum Fourier transform*. Show that $U_{QFT}$ is unitary. In other words show that $U_{QFT}U_{QFT}^* = I_{2^n}$, where we use the *completeness relation*

$$I_{2^n} = \sum_{j=0}^{2^n-1} |j\rangle\langle j|.$$

Thus $I_{2^n}$ is the $2^n \times 2^n$ unit matrix.

**Solution 7.** From the definition (1) we find

$$U_{QFT}^* = \frac{1}{\sqrt{2^n}} \sum_{j=0}^{2^n-1} \sum_{k=0}^{2^n-1} e^{i2\pi kj/2^n} |j\rangle\langle k|$$

where * denotes the adjoint. Therefore

$$U_{QFT}U_{QFT}^* = \frac{1}{2^n} \sum_{j=0}^{2^n-1} \sum_{k=0}^{2^n-1} \sum_{l=0}^{2^n-1} \sum_{m=0}^{2^n-1} e^{i2\pi(kj-lm)/2^n} |j\rangle\langle k|l\rangle\langle m|$$

$$= \frac{1}{2^n} \sum_{j=0}^{2^n-1} \sum_{k=0}^{2^n-1} \sum_{m=0}^{2^n-1} e^{i2\pi(kj-km)/2^n} |j\rangle\langle m|.$$

We have for $j = m$, $e^{i2\pi(kj-km)/2^n} = 1$. Thus for $j, m = 0, 1, \ldots, 2^n - 1$

$$\sum_{k=0}^{2^n-1} (e^{i2\pi(j-m)/2^n})^k = 2^n, \quad j = m$$

$$\sum_{k=0}^{2^n-1} (e^{i2\pi(j-m)/2^n})^k = \frac{1 - e^{i2\pi(j-m)}}{1 - e^{i2\pi(j-m)/2^n}} = 0, \quad j \neq m.$$

Thus

$$U_{QFT}U^*_{QFT} = \sum_{j=0}^{2^n-1} |j\rangle\langle j| = I_{2^n}.$$

**Problem 8.**   Apply the quantum Fourier transform to the state in the Hilbert space $\mathbb{C}^8$

$$\frac{1}{2}\sum_{j=0}^{7}\cos(2\pi j/8)|j\rangle$$

where the quantum Fourier transform is given by

$$U_{QFT} = \frac{1}{2\sqrt{2}}\sum_{j=0}^{7}\sum_{k=0}^{7}e^{-i2\pi kj/8}|k\rangle\langle j|.$$

We use $\{ |j\rangle : j = 0, 1, \ldots, 7 \}$ as an orthonormal basis in the Hilbert space $\mathbb{C}^8$, where $|7\rangle = |111\rangle \equiv |1\rangle \otimes |1\rangle \otimes |1\rangle$.

**Solution 8.**   We use *Euler's identity* $e^{i\theta} \equiv \cos(\theta) + i\sin(\theta)$ and

$$\sum_{k=0}^{N-1} e^{i2\pi k(n-m)/N} = N\delta_{nm}.$$

Thus we have

$$\hat{x}(k) = \sum_{j=0}^{7} e^{-i2\pi kj/8}\cos(2\pi j/8) = \frac{1}{2}\sum_{j=0}^{7}\left(e^{i2\pi(1-k)j/8} + e^{-i2\pi(1+k)j/8}\right)$$

$$= 4(\delta_{k1} + \delta_{k7})$$

and

$$U_{QFT}\frac{1}{2}\sum_{j=0}^{7}\cos(2\pi j/8)|j\rangle = \frac{1}{2\sqrt{8}}\sum_{k=0}^{7}\hat{x}(k)|k\rangle = \frac{1}{\sqrt{2}}(|1\rangle + |7\rangle).$$

**Problem 9.** Let

$$U_{IA} := \sum_{j=0}^{2^n-1} \sum_{k=0}^{2^n-1} \left( \frac{2}{2^n} - \delta_{jk} \right) |k\rangle\langle j|. \qquad (1)$$

$U_{IA}$ is called the *inversion about average operator*. Show that $U_{IA}$ is unitary. In other words show that $U_{IA}U_{IA}^* = I_{2^n}$.
Hint: Use the *completeness relation*

$$\sum_{j=0}^{2^n-1} |j\rangle\langle j| = I_{2^n}.$$

**Solution 9.** From (1) we find

$$U_{IA}^* = \sum_{j=0}^{2^n-1} \sum_{k=0}^{2^n-1} \left( \frac{2}{2^n} - \delta_{jk} \right) |k\rangle\langle j| = U_{IA}.$$

Thus

$$U_{IA}U_{IA}^* = U_{IA}^2 = \sum_{j,k,l,m=0}^{2^n-1} \left( \frac{2}{2^n} - \delta_{jk} \right) \left( \frac{2}{2^n} - \delta_{lm} \right) |k\rangle\langle j|m\rangle\langle l|$$

$$= \sum_{j,k,l=0}^{2^n-1} \left( \frac{2}{2^n} - \delta_{jk} \right) \left( \frac{2}{2^n} - \delta_{lj} \right) |k\rangle\langle l|$$

where we used that $\langle j|m\rangle = \delta_{jm}$. Furthermore, we find

$$\sum_{j=0}^{2^n-1} \left( \frac{2}{2^n} - \delta_{jk} \right) \left( \frac{2}{2^n} - \delta_{lj} \right) = \sum_{j=0}^{2^n-1} \left( \frac{4}{2^{2n}} - \delta_{jk}\frac{2}{2^n} - \delta_{lj}\frac{2}{2^n} + \delta_{jk}\delta_{lj} \right)$$

$$= \frac{4}{2^n} - \frac{2}{2^n} - \frac{2}{2^n} + \sum_{j=0}^{2^n-1} \delta_{jk}\delta_{lk} = \delta_{lk} \sum_{j=0}^{2^n-1} \delta_{jk}$$

$$= \delta_{lk}.$$

Therefore

$$U_{IA}U_{IA}^* = \sum_{j=0}^{2^n-1} |j\rangle\langle j| = I_{2^n}.$$

**Problem 10.** Let $\{\, |0\rangle, |1\rangle \,\}$ be an orthonormal basis in the two-dimensional Hilbert space $\mathbb{C}^2$ and

$$U_H|k\rangle := \frac{1}{\sqrt{2}} \left( |0\rangle + (-1)^k|1\rangle \right), \qquad k \in \{0,1\}$$

$$U_{PS(\theta)} := |00\rangle\langle 00| + |01\rangle\langle 01| + |10\rangle\langle 10| + e^{i\theta}|11\rangle\langle 11|$$
$$U_{CNOT} := |00\rangle\langle 00| + |01\rangle\langle 01| + |10\rangle\langle 11| + |11\rangle\langle 10|.$$

(i) From these definitions show that $U_H U_H = I_2$.
(ii) Calculate

$$(I_2 \otimes U_H)U_{PS(\pi)}(I_2 \otimes U_H)|ab\rangle$$

and $(I_2 \otimes U_H)U_{CNOT}(I_2 \otimes U_H)|ab\rangle$, where $a, b \in \{0, 1\}$. What is the use of these transforms?

**Solution 10.**    (i) An arbitrary state in the Hilbert space $\mathbb{C}^2$ can be written as

$$|\psi\rangle := a|0\rangle + b|1\rangle$$

where $a, b \in \mathbb{C}$ and $|a|^2 + |b|^2 = 1$. We find

$$U_H U_H |\psi\rangle = U_H \frac{1}{\sqrt{2}}(a|0\rangle + a|1\rangle + b|0\rangle - b|1\rangle) = \frac{1}{2}(2a|0\rangle + 2b|1\rangle)$$
$$= a|0\rangle + b|1\rangle.$$

Thus, $U_H U_H = I_2$.
(ii) We find

$$(I_2 \otimes U_H)U_{PS(\pi)}(I_2 \otimes U_H)|ab\rangle = (I_2 \otimes U_H)U_{PS(\pi)}\frac{1}{\sqrt{2}}|a\rangle \otimes (|0\rangle + (-1)^b|1\rangle)$$

$$= (I_2 \otimes U_H)\frac{1}{\sqrt{2}}|a\rangle \otimes (|0\rangle + (-1)^{a+b}|1\rangle)$$

$$= \frac{1}{2}|a, a \oplus b\rangle$$

where $a \oplus b = a + b$ (modulo 2) is the XOR operation. We obtain

$$(I_2 \otimes U_H)U_{CNOT}(I_2 \otimes U_H)|ab\rangle = (I_2 \otimes U_H)U_{CNOT}\frac{1}{\sqrt{2}}|a\rangle \otimes (|0\rangle + (-1)^b|1\rangle)$$

$$= \begin{cases} (I_2 \otimes U_H)\frac{1}{\sqrt{2}}|a\rangle \otimes (|0\rangle + (-1)^b|1\rangle) & a = 0 \\ (I_2 \otimes U_H)\frac{1}{\sqrt{2}}|a\rangle \otimes (|1\rangle + (-1)^b|0\rangle) & a = 1 \end{cases}$$

$$= (I_2 \otimes U_H)\frac{1}{\sqrt{2}}|a\rangle \otimes (-1)^{ab}(|0\rangle + (-1)^b|1\rangle)$$

$$= (-1)^{ab}|ab\rangle.$$

The first computation is $U_{CNOT}$, the second is $U_{PS(\pi)}$.

**Problem 11.**    The *XOR gate* is given by

$$U_{XOR}(|m\rangle \otimes |n\rangle) = |m\rangle \otimes |m \oplus n\rangle$$

where $m, n \in \{0, 1\}$ and $\oplus$ denotes addition modulo 2. The transformation has the following properties: (a) it is unitary and thus reversible, (b) it is hermitian, (c) $m \oplus n = 0$ if and only if $m = n$. The first index denotes the state of the control qubit and the second index denotes the state of the target qubit.

A generalized quantum XOR gate (*GXOR gate*) acts on two $d$-dimensional quantum systems $(d > 2)$. In analogy with qubits one calls these two systems *qudits*. The basis states $|m\rangle$ of each qudit are labeled by elements in the ring $\mathbb{Z}_d$ which we denote by the numbers, $m = 0, 1, \ldots, d - 1$, with the usual rules for addition and multiplication modulo $d$. We define two operators

$$U_{GXOR1}(|m\rangle \otimes |n\rangle) := |m\rangle \otimes |m \oplus n\rangle$$

and

$$U_{GXOR2}(|m\rangle \otimes |n\rangle) := |m\rangle \otimes |m \ominus n\rangle$$

where $m \ominus n := (m - n)$ modulo $d$. Discuss the properties of these two operators.

**Solution 11.**   For $U_{GXOR1}$ we find that the operator is unitary but not hermitian for $d > 2$. Therefore it is no longer its own inverse. We have to obtain the inverse of the $U_{GXOR1}$ gate by iteration, i.e.

$$U_{GXOR1}^{-1} = U_{GXOR1}^{d-1} = U_{GXOR1}^{\dagger} \neq U_{GXOR1}.$$

For the operator $U_{GXOR2}$ we find that in the special case for $d = 2$ it reduces to the XOR-gate. Furthermore, the operator is unitary, hermitian and $m \ominus n = 0$ modulo $d$ if and only if $m = n$.

**Problem 12.**   Given an orthonormal basis in $\mathbb{C}^N$ denoted by

$$|\phi_0\rangle, |\phi_1\rangle, \ldots, |\phi_{N-1}\rangle.$$

(i) Show that

$$U := \sum_{k=0}^{N-2} |\phi_k\rangle\langle\phi_{k+1}| + |\phi_{N-1}\rangle\langle\phi_0|$$

is a unitary matrix. Find $\text{tr}(U)$.

(ii) Find $U^N$.

(iii) Does $U$ depend on the chosen basis? Prove or disprove.

Hint. Consider $N = 2$, the standard basis $(1, 0)^T$, $(0, 1)^T$ and the basis $\frac{1}{\sqrt{2}}(1, 1)^T$, $\frac{1}{\sqrt{2}}(1, -1)^T$.

(iv) Show that the set

$$\{U, U^2, \ldots, U^N\}$$

forms a *commutative group* (*abelian group*) under matrix multiplication. The set is a subgroup of the group of all permutation matrices.

(v) Assume that the set given above is the standard basis. Show that the unitary matrix $U$ is given by the permutation matrix

$$
U = \begin{pmatrix} 0 & 1 & 0 & \cdots & 0 \\ 0 & 0 & 1 & \cdots & 0 \\ \vdots & \vdots & \vdots & \vdots & \vdots \\ 0 & 0 & 0 & \cdots & 1 \\ 1 & 0 & 0 & \cdots & 0 \end{pmatrix}.
$$

**Solution 12.** (i) Since $\langle \phi_j | \phi_k \rangle = \delta_{jk}$ we have

$$
UU^* = \left( \sum_{k=0}^{N-2} |\phi_k\rangle\langle\phi_{k+1}| + |\phi_{N-1}\rangle\langle\phi_0| \right) \left( \sum_{k=0}^{N-2} |\phi_{k+1}\rangle\langle\phi_k| + |\phi_0\rangle\langle\phi_{N-1}| \right)
$$

$$
= \sum_{k=0}^{N-1} |\phi_k\rangle\langle\phi_k| = I_N.
$$

Obviously we have $\operatorname{tr}(U) = 0$ since the terms $|\phi_k\rangle\langle\phi_k|$ do not appear in the sum (i.e. we calculate the trace in the basis $\{ |\phi_0\rangle, \ldots, |\phi_{N-1}\rangle \}$).

(ii) We notice that $U$ maps $|\phi_k\rangle$ to $|\phi_{k-1}\rangle$. Applying this $N$ times and using modulo $N$ arithmetic we obtain (i.e. $U^N$ maps $|\phi_k\rangle$ to $|\phi_{k-N}\rangle$)

$$
U^N = I_N.
$$

(iii) For the standard basis in $\mathbb{C}^2$ $\{ (1,0)^T, (0,1)^T \}$ we obtain

$$
U_{std} = \begin{pmatrix} 0 & 1 \\ 1 & 0 \end{pmatrix}.
$$

For the basis in $\mathbb{C}^2$ $\{ \frac{1}{\sqrt{2}}(1,1)^T, \frac{1}{\sqrt{2}}(1,-1)^T \}$ we obtain the Pauli spin matrix $\sigma_3$. Obviously the two unitary matrices are different. Of course there is a unitary matrix $V$ such that $U_{std} = V^{-1}\sigma_3 V$.

(iv) Since $U_N = I_N = U^0$ we have that

$$
U^s U^t = U^{s+t} = U^{s+t \bmod N}.
$$

Thus the set of matrices $\{U, U^2, \ldots, U^N\}$ forms an abelian group under matrix multiplication, because $\{0, 1, \ldots, N-1\}$ forms a group under addition modulo $N$. The two groups are isomorphic.

(v) Let $\mathbf{e}_j$ denote the element of the standard basis in $\mathbb{C}^n$ with a 1 in the $j$th position (numbered from 0) and 0 in all other positions. Then $U$ is given by

$$
U = \sum_{k=0}^{N-2} \mathbf{e}_k \mathbf{e}_{k+1}^T + \mathbf{e}_{N-1}\mathbf{e}_0^T.
$$

In the product $\mathbf{e}_k \mathbf{e}_{k+1}^T$, $\mathbf{e}_k$ denotes the row and $\mathbf{e}_{k+1}^T$ denotes the column in the matrix $U$. Thus we obtain the matrix described above.

**Problem 13.** (i) Let $\sigma_1$, $\sigma_2$ and $\sigma_3$ be the Pauli spin matrices and $I_2$ be the $2 \times 2$ unit matrix. Find

$$(\sigma_3 \otimes \sigma_3 \otimes \sigma_3 \otimes \sigma_3)(\sigma_1 \otimes \sigma_1 \otimes I_2 \otimes I_2)(\sigma_3 \otimes \sigma_3 \otimes \sigma_3 \otimes \sigma_3),$$

$$(\sigma_3 \otimes \sigma_3 \otimes \sigma_3 \otimes \sigma_3)(I_2 \otimes \sigma_1 \otimes \sigma_1 \otimes I_2)(\sigma_3 \otimes \sigma_3 \otimes \sigma_3 \otimes \sigma_3),$$

$$(\sigma_3 \otimes \sigma_3 \otimes \sigma_3 \otimes \sigma_3)(I_2 \otimes I_2 \otimes \sigma_1 \otimes \sigma_1)(\sigma_3 \otimes \sigma_3 \otimes \sigma_3 \otimes \sigma_3).$$

(ii) Replace $\sigma_1$ by $\sigma_2$ in the expressions given above and calculate the expressions.
(iii) Given the one-dimensional *XY-model* with open boundary conditions

$$\hat{H}_{XY} = - \sum_{j=-N/2+1}^{N/2-1} \left( \frac{1+\gamma}{2}\sigma_{1,j}\sigma_{1,j+1} + \frac{1-\gamma}{2}\sigma_{2,j}\sigma_{2,j+1} \right) - \lambda \sum_{j=-N/2+1}^{N/2} \sigma_{3,j}$$

where the parameter $\lambda$ is the intensity of the magnetic field applied in the $z$-direction and the parameter $\gamma$ determines the degree of anisotropy of the spin-spin interaction, which is restricted to the $xy$-plane in spin space. Find

$$\left( \prod_{j=-N/2+1}^{N/2} \sigma_{3,j} \right) \hat{H}_{XY} \left( \prod_{j=-N/2+1}^{N/2} \sigma_{3,j} \right).$$

**Solution 13.** (i) Since $\sigma_3^2 = I_2$ and $\sigma_3\sigma_1\sigma_3 = -\sigma_1$ we find for the first expression

$$(\sigma_3 \otimes \sigma_3 \otimes \sigma_3 \otimes \sigma_3)(\sigma_1 \otimes \sigma_1 \otimes I_2 \otimes I_2)(\sigma_3 \otimes \sigma_3 \otimes \sigma_3 \otimes \sigma_3) = \sigma_1 \otimes \sigma_1 \otimes I_2 \otimes I_2.$$

Analogously, we find

$$(\sigma_3 \otimes \sigma_3 \otimes \sigma_3 \otimes \sigma_3)(I_2 \otimes \sigma_1 \otimes \sigma_1 \otimes I_2)(\sigma_3 \otimes \sigma_3 \otimes \sigma_3 \otimes \sigma_3) = I_2 \otimes \sigma_1 \otimes \sigma_1 \otimes I_2$$

$$(\sigma_3 \otimes \sigma_3 \otimes \sigma_3 \otimes \sigma_3)(I_2 \otimes I_2 \otimes \sigma_1 \otimes \sigma_1)(\sigma_3 \otimes \sigma_3 \otimes \sigma_3 \otimes \sigma_3) = I_2 \otimes I_2 \otimes \sigma_1 \otimes \sigma_1.$$

(ii) Replacing $\sigma_1$ by $\sigma_2$ and using $\sigma_3\sigma_2\sigma_3 = -\sigma_2$ yields

$$(\sigma_3 \otimes \sigma_3 \otimes \sigma_3 \otimes \sigma_3)(\sigma_2 \otimes \sigma_2 \otimes I_2 \otimes I_2)(\sigma_3 \otimes \sigma_3 \otimes \sigma_3 \otimes \sigma_3) = \sigma_2 \otimes \sigma_2 \otimes I_2 \otimes I_2$$

$$(\sigma_3 \otimes \sigma_3 \otimes \sigma_3 \otimes \sigma_3)(I_2 \otimes \sigma_2 \otimes \sigma_2 \otimes I_2)(\sigma_3 \otimes \sigma_3 \otimes \sigma_3 \otimes \sigma_3) = I_2 \otimes \sigma_2 \otimes \sigma_2 \otimes I_2$$

$$(\sigma_3 \otimes \sigma_3 \otimes \sigma_3 \otimes \sigma_3)(I_2 \otimes I_2 \otimes \sigma_2 \otimes \sigma_2)(\sigma_3 \otimes \sigma_3 \otimes \sigma_3 \otimes \sigma_3) = I_2 \otimes I_2 \otimes \sigma_2 \otimes \sigma_2.$$

(iii) Using the results from (i) and (ii) and extending from $N = 4$ to arbitrary $N$, we find

$$\left( \prod_{j=-N/2+1}^{N/2} \sigma_{3,j} \right) \hat{H}_{XY} \left( \prod_{j=-N/2+1}^{N/2} \sigma_{3,j} \right) = \hat{H}_{XY}.$$

From (ii) and (iii) we find that the Hamilton operator $\hat{H}_{XY}$ is invariant under this transformation.

**Problem 14.**    (i) Consider the product state $|D\rangle \otimes |P\rangle$, where $|D\rangle$ is a state to describe a $m$-qubit data register and $|P\rangle$ is a state to describe an $n$-qubit program register. Let $G$ be a unitary operator acting on this product state

$$|D\rangle \otimes |P\rangle \to G(|D\rangle \otimes |P\rangle).$$

The unitary operator is implemented as follows. A unitary operator $U$ acting on the $m$-qubits of the data register, is said to be implemented by this gate array if there exists a state $|P_U\rangle$ of the program register such that

$$G(|D\rangle \otimes |P_U\rangle) = (U|D\rangle) \otimes |P'_U\rangle$$

for all states $|D\rangle$ of the data register and some state $|P'_U\rangle$ of the program register. Show that $|P'_U\rangle$ does not depend on $|D\rangle$.
(ii) Suppose distinct (up to a global phase) unitary operators $U_1, \ldots, U_N$ are implemented by some programmable quantum gate array. Show that the corresponding programs $|P_1\rangle, \ldots, |P_N\rangle$ are mutually orthogonal.

**Solution 14.**    (i) Consider

$$G(|D_1\rangle \otimes |P\rangle) = (U|D_1\rangle) \otimes |P'_1\rangle, \qquad G(|D_2\rangle \otimes |P\rangle) = (U|D_2\rangle) \otimes |P'_2\rangle.$$

Taking the scalar product of these two equations and using $G^\dagger G = I$, $U^\dagger U = I$ and $\langle P|P\rangle = 1$ we find

$$\langle D_1|D_2\rangle = \langle D_1|D_2\rangle \langle P'_1|P'_2\rangle.$$

If $\langle D_1|D_2\rangle \neq 0$ we find $\langle P'_1|P'_2\rangle = 1$. Thus $|P'_1\rangle = |P'_2\rangle$. Consequently, there is no $|D\rangle$ dependence of $|P'_U\rangle$. What happens for $\langle D_1|D_2\rangle = 0$ ?
(ii) Suppose that $|P\rangle$ and $|Q\rangle$ are programs which implement unitary operators $U_p$ and $U_q$ which are distinct up to global phase changes. Then for an arbitrary data state $|D\rangle$ we have

$$G(|D\rangle \otimes |P\rangle) = (U_p|D\rangle) \otimes |P'\rangle$$

$$G(|D\rangle \otimes |Q\rangle) = (U_q|D\rangle) \otimes |Q'\rangle$$

where $|P'\rangle$ and $|Q'\rangle$ are states of the program register. Taking the scalar product of these two equations and using $G^\dagger G = I$, $\langle D|D\rangle = 1$ we obtain

$$\langle Q|P\rangle = \langle Q'|P'\rangle \langle D|U_q^\dagger U_p|D\rangle.$$

Suppose that $\langle Q'|P'\rangle \neq 0$. Then we have

$$\frac{\langle Q|P\rangle}{\langle Q'|P'\rangle} = \langle D|U_q^\dagger U_p|D\rangle.$$

The left-hand side of this equation has no $|D\rangle$ dependence. Thus we have $U_q^\dagger U_p = cI$ for some complex number $c$. It follows that we can only have $\langle P'|Q'\rangle \neq 0$ if $U_p$ and $U_q$ are the same up to a global phase. However we assumed that this is not the case and therefore $\langle Q'|P'\rangle = 0$. Hence

$$\langle Q|P\rangle = 0.$$

This means the states $|Q\rangle$ and $|P\rangle$ are orthogonal.

**Problem 15.** Consider three two-dimensional Hilbert spaces $\mathcal{H}_1$, $\mathcal{H}_2$ and $\mathcal{H}_3$. Consider the normalized product state

$$|\psi\rangle = \sum_{j=0}^{1}\sum_{k=0}^{1}\sum_{\ell=0}^{1} c_{jk\ell}|j\rangle \otimes |k\rangle \otimes |\ell\rangle$$

in the product Hilbert space $\mathcal{H}_1 \otimes \mathcal{H}_2 \otimes \mathcal{H}_3$. Let $U_1$, $U_2$, $U_3$ be unitary operators acting in these Hilbert spaces. By the *First Fundamental Theorem* of invariant theory applied to $U_1$, $U_2$, $U_3$, any polynomial in $c_{jk\ell}$ which is invariant under the action on $|\psi\rangle$ of the local unitary transformation $U_1 \otimes U_2 \otimes U_3$ is a sum of homogeneous polynomials of even degree (say $2r$). For $r = 1$ we have

$$P_{\sigma_1\sigma_2}(\mathbf{c}) = \sum_{j_1=0}^{1}\sum_{k_1=0}^{1}\sum_{\ell_1=0}^{1} c_{j_1 k_1 \ell_1} c^*_{j_1 k_{\sigma_1(1)} \ell_{\sigma_2(1)}}$$

where $\sigma_1$ and $\sigma_2$ are permutations of 1. We denote by $e$ the identity permutation. For $r = 2$ we have

$$P_{\sigma_1\sigma_2}(\mathbf{c}) = \sum_{j_1=0}^{1}\sum_{k_1=0}^{1}\sum_{\ell_1=0}^{1}\sum_{j_2=0}^{1}\sum_{k_2=0}^{1}\sum_{\ell_2=0}^{1} c_{j_1 k_1 \ell_1} c_{j_2 k_2 \ell_2} c^*_{j_1 k_{\sigma_1(1)} \ell_{\sigma_2(1)}} c^*_{j_2 k_{\sigma_1(2)} \ell_{\sigma_2(2)}}.$$

(i) Calculate the invariants.
(ii) Describe the connection with the partial traces

$$\rho_1 := \mathrm{tr}_{23}(|\psi\rangle\langle\psi|), \quad \rho_2 := \mathrm{tr}_{31}(|\psi\rangle\langle\psi|), \quad \rho_3 := \mathrm{tr}_{12}(|\psi\rangle\langle\psi|)$$

of the density operator $\rho := |\psi\rangle\langle\psi|$.

**Solution 15.**    (i) Obviously for the case $r = 1$ (degree 2) we only have the identity permutation, i.e. $\sigma_1 = \sigma_2 = e$ with $e(1) = 1$, $e(2) = 2$. Thus we find only one invariant, namely

$$I_0 = \sum_{j=0}^{1}\sum_{k=0}^{1}\sum_{\ell=0}^{1} c_{jk\ell}c_{jk\ell}^* = \langle\psi|\psi\rangle = 1$$

which is the normalization condition. For the case $r = 2$ (degree 4) we find four linearly independent quartic invariants since $e(1) = 1$, $e(2) = 2$, $\sigma(1) = 2$, $\sigma(2) = 1$. Thus

$$I_1 = P_{ee}(\mathbf{c}) = \sum_{j_1=0}^{1}\sum_{k_1=0}^{1}\sum_{\ell_1=0}^{1}\sum_{j_2=0}^{1}\sum_{k_2=0}^{1}\sum_{\ell_2=0}^{1} c_{j_1 k_1 \ell_1}c_{j_1 k_1 \ell_1}^* c_{j_2 k_2 \ell_2}c_{j_2 k_2 \ell_2}^* = \langle\psi|\psi\rangle^2$$

$$I_2 = P_{e\sigma}(\mathbf{c}) = \sum_{j_1=0}^{1}\sum_{k_1=0}^{1}\sum_{\ell_1=0}^{1}\sum_{j_2=0}^{1}\sum_{k_2=0}^{1}\sum_{\ell_2=0}^{1} c_{j_1 k_1 \ell_1}c_{j_1 k_1 \ell_2}^* c_{j_2 k_2 \ell_2}c_{j_2 k_2 \ell_1}^*$$

$$I_3 = P_{\sigma e}(\mathbf{c}) = \sum_{j_1=0}^{1}\sum_{k_1=0}^{1}\sum_{\ell_1=0}^{1}\sum_{j_2=0}^{1}\sum_{k_2=0}^{1}\sum_{\ell_2=0}^{1} c_{j_1 k_1 \ell_1}c_{j_1 k_2 \ell_1}^* c_{j_2 k_2 \ell_2}c_{j_2 k_1 \ell_2}^*$$

$$I_4 = P_{\sigma\sigma}(\mathbf{c}) = \sum_{j_1=0}^{1}\sum_{k_1=0}^{1}\sum_{\ell_1=0}^{1}\sum_{j_2=0}^{1}\sum_{k_2=0}^{1}\sum_{\ell_2=0}^{1} c_{j_1 k_1 \ell_1}c_{j_1 k_2 \ell_2}^* c_{j_2 k_2 \ell_2}c_{j_2 k_1 \ell_1}^* .$$

(ii) We find the invariants $I_2 = \text{tr}(\rho_3^2)$, $I_3 = \text{tr}(\rho_2^2)$, $I_4 = \text{tr}(\rho_1^2)$.

**Problem 16.**    Consider two Hilbert spaces $\mathcal{H}_{reg}$ and $\mathcal{H}_{sys}$ and the product state

$$|\psi\rangle = (\alpha|0^{reg}\rangle + \beta|1^{reg}\rangle) \otimes |0^{sys}\rangle$$

in the Hilbert space $\mathcal{H}_{reg} \otimes \mathcal{H}_{sys}$, where $reg$ stands for register and $sys$ for system. Consider the swap operation (swap gate)

$$U_{swap}((\alpha|0^{reg}\rangle + \beta|1^{reg}\rangle) \otimes |0^{sys}\rangle) = |0^{reg}\rangle \otimes (\alpha|0^{sys}\rangle + \beta|1^{sys}\rangle).$$

Discuss the operation on physical grounds.

**Solution 16.**    Creating such a superposition could violate conservation laws (for example charge) and in this case is forbidden by superselection rules.

**Problem 17.**    The *Toffoli gate* is the unitary operator acting as

$$U_T|a, b, c\rangle = |a, b, a \cdot b + c\rangle \equiv |a\rangle \otimes |b\rangle \otimes |b + c\rangle$$

in the Hilbert space $\mathbb{C}^8$, where $a, b, c \in \{0, 1\}$ and $ab$ denotes the AND operation of $a$ and $b$. The addition $+$ is modulo 2.
(i) Find the truth table.
(ii) Find the matrix representation for the standard basis.
The Toffoli gate is an extension of the CNOT gate.

**Solution 17.**   (i) We have the truth table

| $a$ | $b$ | $c$ | $a$ | $b$ | $ab + c$ |
|---|---|---|---|---|---|
| 0 | 0 | 0 | 0 | 0 | 0 |
| 0 | 0 | 1 | 0 | 0 | 1 |
| 0 | 1 | 0 | 0 | 1 | 0 |
| 0 | 1 | 1 | 0 | 1 | 1 |
| 1 | 0 | 0 | 1 | 0 | 0 |
| 1 | 0 | 1 | 1 | 0 | 1 |
| 1 | 1 | 0 | 1 | 1 | 1 |
| 1 | 1 | 1 | 1 | 1 | 0 |

(ii) The matrix representation of the Toffoli gate is given by the $8 \times 8$ permutation matrix

$$I_6 \oplus \begin{pmatrix} 0 & 1 \\ 1 & 0 \end{pmatrix}$$

where $\oplus$ denotes the direct sum.

**Problem 18.**   The *Fredkin gate* is the unitary operator acting as

$$U_F |c, x, y\rangle = |c, cx + \bar{c}y, \bar{c}x + cy\rangle$$

in the Hilbert space $\mathbb{C}^8$, where $c, x, y \in \{0, 1\}$.
(i) Consider the cases $c = 0$ and $c = 1$.
(ii) Find the matrix representation for the standard basis.

**Solution 18.**   (i) For $c = 0$ we have $\bar{c} = 1$. Therefore

$$cx = 0, \quad \bar{c}x = x, \quad cy = 0, \quad \bar{c}y = y.$$

Thus

$$U_F |0, x, y\rangle = |0, y, x\rangle.$$

For $c = 1$ we have $\bar{c} = 0$. Therefore $cx = x$, $\bar{c}x = 0$, $cy = y$, $\bar{c}y = 0$. Thus

$$U_F |1, x, y\rangle = |1, x, y\rangle.$$

Consequently $c$ is a control bit. If $c = 0$ then $x$ and $y$ swap around. If $c = 1$ then $x$ and $y$ stay the same.

(ii) The matrix representation of the Fredkin gate is given by the $8 \times 8$ permutation matrix

$$(1) \oplus U_{NOT} \oplus I_5$$

where $\oplus$ denotes the direct sum.

**Problem 19.**    Consider the $8 \times 8$ matrix

$$U(\alpha) = \frac{e^{i\alpha}}{\sqrt{2}}(I_2 \otimes I_2 \otimes I_2 + i\sigma_1 \otimes \sigma_1 \otimes \sigma_1)$$

where $\alpha \in \mathbb{R}$.
(i) Show that $U(\alpha)$ is unitary.
(ii) Consider the standard basis $|0\rangle$, $|1\rangle$ and the product state

$$|\psi\rangle = |\downarrow\rangle \otimes |\downarrow\rangle \otimes |\downarrow\rangle \equiv \begin{pmatrix} 0 \\ 1 \end{pmatrix} \otimes \begin{pmatrix} 0 \\ 1 \end{pmatrix} \otimes \begin{pmatrix} 0 \\ 1 \end{pmatrix}.$$

Calculate the state $U|\psi\rangle$.
(iii) Consider $U(\alpha = 0)$ and the unitary $8 \times 8$ diagonal matrix

$$V = \text{diag}(e^{i3\phi/2},\ 1,\ 1,\ 1,\ 1,\ 1,\ 1,\ e^{-i3\phi/2}).$$

Calculate the state $VU(\alpha = 0)|\psi\rangle$.
(iv) Calculate the state $U(\alpha = 0)VU(\alpha = 0)|\psi\rangle$.
(v) Let $|\xi_1\rangle = |\downarrow\rangle \otimes |\downarrow\rangle \otimes |\downarrow\rangle$, $|\xi_2\rangle = |\uparrow\rangle \otimes |\uparrow\rangle \otimes |\uparrow\rangle$. Calculate the probabilities

$$|\langle \xi_1 | U(\alpha = 0)VU(\alpha = 0)|\psi\rangle|^2, \qquad |\langle \xi_2 | U(\alpha = 0)VU(\alpha = 0)|\psi\rangle|^2.$$

**Solution 19.**    (i) Since $\sigma_1^* = \sigma_1$ we have

$$U^*(\alpha) = \frac{e^{-i\alpha}}{\sqrt{2}}(I_2 \otimes I_2 \otimes I_2 - i\sigma_1 \otimes \sigma_1 \otimes \sigma_1).$$

Since $\sigma_1^2 = I_2$ we obtain

$$U^*(\alpha)U(\alpha) = \frac{1}{2}(I_2 \otimes I_2 \otimes I_2 + I_2 \otimes I_2 \otimes I_2) = I_2 \otimes I_2 \otimes I_2 = I_8.$$

Thus $U$ is a unitary matrix.
(ii) We find

$$U(\alpha)|\psi\rangle = \frac{e^{i\alpha}}{\sqrt{2}}(I_2 \otimes I_2 \otimes I_2 + i\sigma_1 \otimes \sigma_1 \otimes \sigma_1)(|\downarrow\rangle \otimes |\downarrow\rangle \otimes |\downarrow\rangle)$$

$$= \frac{e^{i\alpha}}{\sqrt{2}}(|\downarrow\rangle \otimes |\downarrow\rangle \otimes |\downarrow\rangle + i|\uparrow\rangle \otimes |\uparrow\rangle \otimes |\uparrow\rangle)$$

$$= \frac{e^{i\alpha}}{\sqrt{2}}(|\downarrow\rangle \otimes |\downarrow\rangle \otimes |\downarrow\rangle + e^{i\pi/2}|\uparrow\rangle \otimes |\uparrow\rangle \otimes |\uparrow\rangle).$$

This is the *GHZ state*.
(iii) We find

$$VU(\alpha = 0)|\psi\rangle = \frac{1}{\sqrt{2}}(e^{-i3\phi/2}|\downarrow\rangle \otimes |\downarrow\rangle \otimes |\downarrow\rangle + ie^{i3\phi/2}|\uparrow\rangle \otimes |\uparrow\rangle \otimes |\uparrow\rangle))$$

$$= \frac{1}{\sqrt{2}}(\cos(3\phi/2)(|\downarrow\rangle \otimes |\downarrow\rangle \otimes |\downarrow\rangle + i|\uparrow\rangle \otimes |\uparrow\rangle \otimes |\uparrow\rangle)$$
$$-i\sin(3\phi/2)(|\downarrow\rangle \otimes |\downarrow\rangle \otimes |\downarrow\rangle - i|\uparrow\rangle \otimes |\uparrow\rangle \otimes |\uparrow\rangle)).$$

(iv) Using the result from (iii) we find

$$U(\alpha = 0)VU(\alpha = 0)|\psi\rangle = i\cos(\frac{3}{2}\phi)|\uparrow\rangle \otimes |\uparrow\rangle \otimes |\uparrow\rangle - i\sin(\frac{3}{2}\phi)|\downarrow\rangle \otimes |\downarrow\rangle \otimes |\downarrow\rangle.$$

(v) We obtain for the probabilities

$$|\langle\xi_1|U(\alpha = 0)VU(\alpha = 0)|\psi\rangle|^2 = \sin^2(3\phi/2) \equiv \frac{1}{2} - \frac{1}{2}\cos(3\phi)$$

and

$$|\langle\xi_2|U(\alpha = 0)VU(\alpha = 0)|\psi\rangle|^2 = \cos^2(3\phi/2) \equiv \frac{1}{2} + \frac{1}{2}\cos(3\phi).$$

Draw $|\langle\xi_1|U(\alpha = 0)VU(\alpha = 0)|\psi\rangle|^2$ and $|\langle\xi_2|U(\alpha = 0)VU(\alpha = 0)|\psi\rangle|^2$ as functions of $\phi$.

**Problem 20.** A *quantum 2-torus* is based on a $C^*$-algebra $\mathcal{A}$ generated by the elements $U_1$ and $U_2$ with the relations

$$U_1^{-1} = U_1^*, \qquad U_2^* = U_2^{-1}, \qquad U_1U_2 = zU_2U_1$$

where $z$ is a fixed complex unit. The algebra $\mathcal{A}$ will be commutative if and only if $z = 1$, and in this case it describes the classical 2-torus. If $z$ is different from 1 then $\mathcal{A}$ describes a purely quantum object. This space is called a quantum torus. A non commutative $d$-torus $T_\theta^d$ is a $C^*$-algebra generated by $d$-unitaries $U_1, U_2, \ldots, U_d$ subject to the relations

$$U_\alpha U_\beta = e^{2\pi i\theta_{\alpha\beta}}U_\beta U_\alpha, \qquad \alpha, \beta = 1, 2, \ldots, d$$

where $\theta = (\theta_{\alpha\beta})$ is a skew-symmetric matrix with real entries. Consider the $C^*$-algebra given by the $2 \times 2$ matrices.
(i) Let

$$U_1 = \begin{pmatrix} 0 & 1 \\ 1 & 0 \end{pmatrix}, \qquad U_2 = \begin{pmatrix} 0 & 1 \\ -1 & 0 \end{pmatrix}.$$

Can we find $z \in \mathbb{C}$ such that $U_1U_2 = zU_2U_1$ ?

(ii) Let

$$U_1 = \frac{1}{\sqrt{2}} \begin{pmatrix} 1 & 1 \\ 1 & -1 \end{pmatrix}, \qquad U_2 = \begin{pmatrix} 0 & 1 \\ 1 & 0 \end{pmatrix}.$$

Can we find $z \in \mathbb{C}$ such that $U_1 U_2 = z U_2 U_1$ ?

**Solution 20.**    (i) Since

$$U_1 U_2 = \begin{pmatrix} -1 & 0 \\ 0 & 1 \end{pmatrix}, \qquad U_2 U_1 = \begin{pmatrix} 1 & 0 \\ 0 & -1 \end{pmatrix}$$

we have $U_1 U_2 = e^{i\pi} U_2 U_1$.
(ii) Since

$$U_1 U_2 = \frac{1}{\sqrt{2}} \begin{pmatrix} 1 & 1 \\ -1 & 1 \end{pmatrix}, \qquad U_2 U_1 = \frac{1}{\sqrt{2}} \begin{pmatrix} 1 & -1 \\ 1 & 1 \end{pmatrix}$$

we cannot find $z \in \mathbb{C}$ such that $U_1 U_2 = z U_2 U_1$, but we find the unitary matrix

$$Z = \begin{pmatrix} 0 & 1 \\ -1 & 0 \end{pmatrix} = i\sigma_2$$

such that $U_1 U_2 = Z U_2 U_1$.

**Problem 21.**    Let $\alpha \in \mathbb{R}$. Find the unitary matrix

$$U(\alpha) = \exp(-i\alpha(\sigma_1 \otimes \sigma_1 + \sigma_2 \otimes \sigma_2 + \sigma_3 \otimes \sigma_3)).$$

**Solution 21.**    Since

$$[\sigma_1 \otimes \sigma_1, \sigma_2 \otimes \sigma_2] = 0_4, \quad [\sigma_1 \otimes \sigma_1, \sigma_3 \otimes \sigma_3] = 0_4, \quad [\sigma_2 \otimes \sigma_2, \sigma_3 \otimes \sigma_3] = 0_4$$

we can write

$$U(\alpha) = \exp(-i\alpha(\sigma_1 \otimes \sigma_1)) \exp(-i\alpha(\sigma_2 \otimes \sigma_2)) \exp(-i\alpha(\sigma_3 \otimes \sigma_3)).$$

It follows that

$$U(\alpha) = (\cos^3(\alpha) - i\sin^3(\alpha))I_4 - i\sin(\alpha)\cos(\alpha)e^{i\alpha}(\sigma_1 \otimes \sigma_1 + \sigma_2 \otimes \sigma_2 + \sigma_3 \otimes \sigma_3).$$

**Problem 22.**    The *Hadamard gate* is defined by

$$U_H|0\rangle = \frac{1}{\sqrt{2}}(|0\rangle + |1\rangle), \qquad U_H|1\rangle = \frac{1}{\sqrt{2}}(|0\rangle - |1\rangle)$$

and the *CNOT gate* is defined as

$$U_{CNOT}(|a\rangle \otimes |b\rangle) = |a\rangle \otimes |a \oplus b\rangle$$

where $a, b \in \{0, 1\}$ and $\oplus$ is the XOR operation. Calculate the state

$$U_{CNOT}(U_H \otimes I)(|0\rangle \otimes |0\rangle), \quad U_{CNOT}(U_H \otimes I)(|0\rangle \otimes |1\rangle)$$
$$U_{CNOT}(U_H \otimes I)(|1\rangle \otimes |0\rangle), \quad U_{CNOT}(U_H \otimes I)(|1\rangle \otimes |1\rangle)$$

and discuss.

**Solution 22.** We obtain

$$U_{CNOT}(U_H \otimes I)(|0\rangle \otimes |0\rangle) = \frac{1}{\sqrt{2}}(|0\rangle \otimes |0\rangle + |1\rangle \otimes |1\rangle)$$

$$U_{CNOT}(U_H \otimes I)(|0\rangle \otimes |1\rangle) = \frac{1}{\sqrt{2}}(|0\rangle \otimes |1\rangle + |1\rangle \otimes |0\rangle)$$

$$U_{CNOT}(U_H \otimes I)(|1\rangle \otimes |0\rangle) = \frac{1}{\sqrt{2}}(|0\rangle \otimes |0\rangle - |1\rangle \otimes |1\rangle)$$

$$U_{CNOT}(U_H \otimes I)(|1\rangle \otimes |1\rangle) = \frac{1}{\sqrt{2}}(|0\rangle \otimes |1\rangle - |1\rangle \otimes |0\rangle).$$

These are the *Bell states*. Thus we generated the Bell states (which are maximally entangled) from non-entangled states.

**Problem 23.** Consider the unitary $2 \times 2$ matrix

$$U = \frac{1}{\sqrt{2}} \begin{pmatrix} 1 & -1 \\ 1 & 1 \end{pmatrix}.$$

Calculate the *logarithm* of $U$, i.e. $\log(U)$ using

$$\log(U) = \int_0^1 (U - I_2)(t(U - I_2) + I_2)^{-1} dt$$

to find $B$ given by $U = \exp(B)$. This equation can be applied if the matrix $U$ has no eigenvalues on $\mathbb{R}^-$ (the closed negative real axis). Set $B = iK$. Find $K$.

**Solution 23.** The eigenvalues of $U$ are given by

$$\lambda_1 = \frac{1}{\sqrt{2}}(1 + i), \qquad \lambda_2 = \frac{1}{\sqrt{2}}(1 - i).$$

Thus the condition to apply the equation is satisfied. We consider first the general case $U = (u_{jk})$ and then simplify to $u_{11} = u_{22} = 1/\sqrt{2}$ and $u_{21} = -u_{12} = 1/\sqrt{2}$. We obtain

$$t(U - I_2) + I_2 = \begin{pmatrix} 1 + t(u_{11} - 1) & tu_{12} \\ tu_{21} & 1 + t(u_{22} - 1) \end{pmatrix}$$

and

$$\det(t(U - I_2) + I_2) = d(t) = 1 + t(-2 + \text{tr}(U)) + t^2(1 - \text{tr}(U) + \det(U)).$$

Let $X \equiv \det(U) - \text{tr}(U) + 1$. Then

$$(U - I_2)(t(U - I_2) + I_2)^{-1} = \frac{1}{d(t)} \begin{pmatrix} tX + u_{11} - 1 & u_{12} \\ u_{21} & tX + u_{22} - 1 \end{pmatrix}.$$

With $u_{11} = u_{22} = 1/\sqrt{2}$, $u_{21} = -u_{12} = 1/\sqrt{2}$ we obtain

$$d(t) = 1 + t(-2 + \sqrt{2}) + t^2(2 - \sqrt{2})$$

and $X \equiv \det(U) - \text{tr}(U) + 1 = 2 - \sqrt{2}$. Thus the matrix takes the form

$$\frac{1}{d(t)} \begin{pmatrix} t(2 - \sqrt{2}) + 1/\sqrt{2} - 1 & -1/\sqrt{2} \\ 1/\sqrt{2} & t(2 - \sqrt{2}) + 1/\sqrt{2} - 1 \end{pmatrix}.$$

Since

$$\int_0^1 \frac{1}{d(t)} dt = \frac{2}{\sqrt{2}} \arctan\left( \frac{2(2 - \sqrt{2})t + \sqrt{2} - 2}{\sqrt{2}} \right) \Bigg|_0^1 = \sqrt{2}\frac{\pi}{4}$$

and

$$\int_0^1 \frac{t}{d(t)} dt = \frac{1}{\sqrt{2}} \frac{\pi}{4}$$

we obtain

$$B = \begin{pmatrix} 0 & -\pi/4 \\ \pi/4 & 0 \end{pmatrix}, \qquad K = -iB = \begin{pmatrix} 0 & i\pi/4 \\ -i\pi/4 & 0 \end{pmatrix}.$$

**Problem 24.**  Let $\{ |0\rangle, |1\rangle, \ldots, |n - 1\rangle \}$ be an orthonormal basis in the Hilbert space $\mathbb{C}^n$.
(i) Is the linear operator

$$Z_n := \sum_{j=0}^{n-1} \exp(2\pi i j/n) |j\rangle\langle j|$$

unitary?
(ii) Can the operator $Z_n$ be expressed as the exponent of a hermitian operator?

**Solution 24.**  (i) Since

$$Z_n^* = \sum_{j=0}^{n-1} \exp(-2\pi i j/n) |j\rangle\langle j|$$

we find $Z_n Z_n^* = I$. Therefore $Z_n$ is unitary.

(ii) Since $Z_n$ is unitary we can find a hermitian operator, say $\hat{K}$, such that $Z_n = \exp(i\hat{K})$. We obviously find

$$Z_n = \exp(2\pi i\hat{\theta}/n), \qquad \hat{\theta} := \sum_{j=0}^{n-1} j|j\rangle\langle j|.$$

The operator $\hat{\theta}$ is the $SU(2)$ *phase operator*.

**Problem 25.**   The four *Bell states* are given by

$$|\phi^+\rangle = \frac{1}{\sqrt{2}}(|0\rangle \otimes |0\rangle + |1\rangle \otimes |1\rangle), \qquad |\phi^-\rangle = \frac{1}{\sqrt{2}}(|0\rangle \otimes |0\rangle - |1\rangle \otimes |1\rangle)$$

$$|\psi^+\rangle = \frac{1}{\sqrt{2}}(|0\rangle \otimes |1\rangle + |1\rangle \otimes |0\rangle), \qquad |\psi^-\rangle = \frac{1}{\sqrt{2}}(|0\rangle \otimes |1\rangle - |1\rangle \otimes |0\rangle).$$

Show that the Bell states can be transformed to each other under local unitary transformations (i.e. the Kronecker product $U \otimes V$ of two unitary $2 \times 2$ matrices $U$ and $V$). Hint. Consider the Pauli spin matrices and the $2 \times 2$ identity matrix.

**Solution 25.**   We find

$$|\phi^-\rangle = (I_2 \otimes \sigma_3)|\phi^+\rangle = (\sigma_3 \otimes I_2)|\phi^+\rangle$$
$$|\psi^+\rangle = (I_2 \otimes \sigma_1)|\phi^+\rangle = (\sigma_1 \otimes I_2)|\phi^+\rangle$$
$$|\psi^-\rangle = (I_2 \otimes (-i\sigma_2))|\phi^+\rangle = (i\sigma_2 \otimes I_2)|\phi^+\rangle.$$

**Problem 26.**   Consider the unitary matrix

$$V = \begin{pmatrix} 0 & 0 & 1 \\ 0 & 1 & 0 \\ 1 & 0 & 0 \end{pmatrix}.$$

Find the hermitian matrix $K$ such that $V = \exp(iK)$.

**Solution 26.**   The matrix $V$ is also hermitian. We calculate the eigenvalues and normalized eigenvectors of $V$. From the normalized eigenvectors we construct a unitary matrix $W$ such that $W^*VW$ is a diagonal matrix. Then

$$W^*VW = W^*e^{iK}W = e^{iW^*KW} = e^{iL}$$

where $L = W^*KW$. Since $W^*VW$ is a diagonal matrix $L$ is also a diagonal matrix. Thus $(W^*VW)_{jj} = e^{i\ell_{jj}}$. Finally we find $K$ from $K = WLW^*$.

Now the eigenvalues of $V$ are given by $\lambda_1 = 1$, $\lambda_2 = 1$, $\lambda_3 = -1$ with the corresponding normalized eigenvectors

$$\begin{pmatrix} 0 \\ 1 \\ 0 \end{pmatrix}, \quad \frac{1}{\sqrt{2}}\begin{pmatrix} 1 \\ 0 \\ 1 \end{pmatrix}, \quad \frac{1}{\sqrt{2}}\begin{pmatrix} 1 \\ 0 \\ -1 \end{pmatrix}.$$

This leads to the unitary (orthogonal) matrix

$$W = \begin{pmatrix} 0 & 1/\sqrt{2} & 1/\sqrt{2} \\ 1 & 0 & 0 \\ 0 & 1/\sqrt{2} & -1/\sqrt{2} \end{pmatrix}.$$

Thus we obtain the diagonal matrix

$$W^*VW = \begin{pmatrix} 1 & 0 & 0 \\ 0 & 1 & 0 \\ 0 & 0 & -1 \end{pmatrix} = e^{iL}.$$

From $W^*VW = \exp(iL)$ we find the equations $1 = e^{i\ell_{11}}$, $1 = e^{i\ell_{22}}$, $-1 = e^{i\ell_{33}}$. The solution is $\ell_{11} = \ell_{22} = 0$, $\ell_{33} = \pi$. Hence we obtain the hermitian matrix

$$K = WLW^* = \frac{\pi}{2}\begin{pmatrix} 1 & 0 & -1 \\ 0 & 0 & 0 \\ -1 & 0 & 1 \end{pmatrix}.$$

**Problem 27.**    Let $U$ be an $n \times n$ unitary matrix. Show that if the bipartite states $|\psi\rangle, |\phi\rangle \in \mathbb{C}^n \otimes \mathbb{C}^m$ satisfy $|\phi\rangle = (U \otimes I_m)|\psi\rangle$, then the ranks of the corresponding reduced density matrices satisfy

$$r(\rho_1^\psi) \geq r(\rho_1^\phi), \qquad r(\rho_2^\psi) \geq r(\rho_2^\phi).$$

**Solution 27.**    We consider the *Schmidt decomposition* of the state $|\psi\rangle$

$$|\psi\rangle = \sum_{j=1}^{s} \sqrt{\lambda_j^\psi}|j\rangle \otimes |j\rangle, \qquad \lambda_j^\psi > 0, \quad s \leq \min(n, m)$$

where $s$ is the number of non vanishing terms in the Schmidt decomposition. We write the unitary operator as

$$U = \sum_{j=1}^{n} |\mu_j\rangle\langle j|$$

where $|\mu_j\rangle \in \mathbb{C}^n$. Then we find that

$$\rho_1^\psi = \sum_{j=1}^{s} |j\rangle\langle j|, \qquad \rho_1^\phi = U\rho_1^\psi U^* = \sum_{j=1}^{s} |\mu_j\rangle\langle \mu_j|.$$

Thus $r(\rho_1^\phi) \le s$. The second inequality follows from the fact that for any bipartite state $r(\rho_1) = r(\rho_2)$.

**Problem 28.** Consider the $4 \times 4$ hermitian matrix

$$R := \sigma_1 \otimes \sigma_1 + \sigma_2 \otimes \sigma_2 + \sigma_3 \otimes \sigma_3$$

and the Hamilton operator

$$\hat{H}(t) = \frac{1}{2}\hbar\omega(t)R.$$

Let

$$\phi = \frac{1}{2}\int_0^T \omega(t)dt$$

and

$$U(T) = \exp\left(-i\int_0^T \hat{H}(t)dt/\hbar\right) = \exp\left(-\frac{1}{2}\left(\int_0^T \omega(t)dt\right)R\right)$$

$$= \exp(-i\phi R).$$

Calculate $U(T)$ and express it using the *swap gate*

$$U_{sw} = \begin{pmatrix} 1 & 0 & 0 & 0 \\ 0 & 0 & 1 & 0 \\ 0 & 1 & 0 & 0 \\ 0 & 0 & 0 & 1 \end{pmatrix}.$$

**Solution 28.** Note that $U_{sw} = I_4 + R$. We have

$$U(T) = e^{-i\phi R} = \frac{1}{4}e^{-i\phi}(3I_4 + R) + \frac{1}{4}e^{-3i\phi}(I_4 - R)$$

$$= e^{i\phi}\left(\cos(2\phi)I_4 - \frac{i}{2}\sin(2\phi)(I_4 + R)\right)$$

$$= e^{i\phi}(\cos(2\phi)I_4 - i\sin(2\phi)U_{sw}).$$

**Problem 29.** Consider the *Bell basis* in the form

$$\frac{1}{\sqrt{2}}\begin{pmatrix} 1 \\ 0 \\ 0 \\ -i \end{pmatrix}, \quad \frac{1}{\sqrt{2}}\begin{pmatrix} 0 \\ 1 \\ -i \\ 0 \end{pmatrix}, \quad \frac{1}{\sqrt{2}}\begin{pmatrix} 0 \\ -i \\ 1 \\ 0 \end{pmatrix}, \quad \frac{1}{\sqrt{2}}\begin{pmatrix} -i \\ 0 \\ 0 \\ 1 \end{pmatrix}.$$

From the Bell states we form the matrix

$$B = \frac{1}{\sqrt{2}} \begin{pmatrix} 1 & 0 & 0 & -i \\ 0 & 1 & -i & 0 \\ 0 & -i & 1 & 0 \\ -i & 0 & 0 & 1 \end{pmatrix}.$$

(i) Show that $B$ is unitary. Calculate $B^2, B^3, B^4, \ldots, B^8$ and show that $B^k$ ($k = 1, 2 \ldots, 8$) form a group under matrix multiplication.
(ii) Let $\gamma_1$ and $\gamma_4$ the Dirac gamma matrices

$$\gamma_1 = \begin{pmatrix} 0 & 0 & 0 & -i \\ 0 & 0 & -i & 0 \\ 0 & i & 0 & 0 \\ i & 0 & 0 & 0 \end{pmatrix}, \quad \gamma_4 = \begin{pmatrix} 1 & 0 & 0 & 0 \\ 0 & 1 & 0 & 0 \\ 0 & 0 & -1 & 0 \\ 0 & 0 & 0 & -1 \end{pmatrix}.$$

Calculate $\exp(\pi\gamma_4\gamma_1/4)$ and show that $\exp(\pi\gamma_4\gamma_1/4) = B$.
(iii) Let $T_2 = B$ and

$$T_1 = \exp(-i\pi I_2 \otimes \sigma_3/4), \qquad T_3 = \exp(-i\pi\sigma_3 \otimes I_2/4).$$

Calculate $T_1$ and $T_3$. Show that $T_1T_2T_1 = T_2T_1T_2$, $T_3T_2T_3 = T_2T_3T_2$ and $T_1T_3 = T_3T_1$, i.e. we have a *braid like relation*.

**Solution 29.** (i) We have $B^*B = I_4$. Thus $B^* = B^{-1}$ and $B$ is unitary. We define

$$N := \begin{pmatrix} 0 & 0 & 0 & 1 \\ 0 & 0 & 1 & 0 \\ 0 & 1 & 0 & 0 \\ 1 & 0 & 0 & 0 \end{pmatrix}.$$

Thus $N^2 = I_4$ and we can write $B = \frac{1}{\sqrt{2}}(I_4 - iN)$. We find

$$B^2 = \frac{1}{2}(I_4 - iN)(I_4 - iN) = -iN$$

$$B^3 = \frac{1}{\sqrt{2}}(I_4 - iN)(-iN) = -\frac{1}{\sqrt{2}}(I_4 + iN)$$

$$B^4 = B^2B^2 = (-iN)(-iN) = -I_4$$

$$B^5 = -B$$

$$B^6 = -B^2 = iN$$

$$B^7 = -B^3 = \frac{1}{\sqrt{2}}(I_4 + iN)$$

$$B^8 = B^4B^4 = I_4.$$

(ii) We have

$$\gamma_4\gamma_1 = \begin{pmatrix} 0 & 0 & 0 & -i \\ 0 & 0 & -i & 0 \\ 0 & -i & 0 & 0 \\ -i & 0 & 0 & 0 \end{pmatrix} = -iN = B^2.$$

Thus $\gamma_4\gamma_1 = -iN$ and $(\gamma_4\gamma_1)^2 = -I_4$, $(\gamma_4\gamma_1)^3 = iN$, $(\gamma_4\gamma_1)^4 = I_4$. Using this result we find

$$\exp(\pi\gamma_4\gamma_1/4) = I_4\cos(\pi/4) - iN\sin(\pi/4) = \frac{1}{\sqrt{2}}(I_4 - iN) = B.$$

(iii) Using that $\exp(-i\pi/2) = -i$ we find

$$T_3T_2T_3 = \frac{1}{\sqrt{2}} \begin{pmatrix} -i & 0 & 0 & -i \\ 0 & -i & -i & 0 \\ 0 & -i & i & 0 \\ -i & 0 & 0 & i \end{pmatrix}.$$

Using that $e^{-i\pi/4} - e^{i\pi/4} = -i\sqrt{2}$ and $e^{-i\pi/4} + e^{i\pi/4} = \sqrt{2}$ we find

$$T_2T_3T_2 = \frac{1}{\sqrt{2}} \begin{pmatrix} -i & 0 & 0 & -i \\ 0 & -i & -i & 0 \\ 0 & -i & i & 0 \\ -i & 0 & 0 & i \end{pmatrix}.$$

Thus $T_2T_3T_2 = T_3T_2T_3$. Analogously we find $T_1T_2T_1 = T_2T_1T_2$. Obviously $T_1T_3 = T_3T_1$ since $T_1$ and $T_3$ are diagonal matrices.

**Problem 30.** Consider the *swap gate*

$$U_{swap} = \begin{pmatrix} 1 & 0 & 0 & 0 \\ 0 & 0 & 1 & 0 \\ 0 & 1 & 0 & 0 \\ 0 & 0 & 0 & 1 \end{pmatrix}.$$

Can the swap gate be written as

$$U_{swap} = R \begin{pmatrix} 1 & 0 & 0 & 0 \\ 0 & 1 & 0 & 0 \\ 0 & 0 & -1 & 0 \\ 0 & 0 & 0 & 1 \end{pmatrix} R^{-1}$$

where $R$ is an orthogonal matrix?

**Solution 30.** The matrix $U_{swap}$ is not only unitary but also hermitian with eigenvalues $+1, +1, +1, -1$. To construct the matrix $R$ we just find the normalized eigenvectors of $U$. This yields

$$R = R^{-1} = R^* = \begin{pmatrix} 1 & 0 & 0 & 0 \\ 0 & 1/\sqrt{2} & 1/\sqrt{2} & 0 \\ 0 & 1/\sqrt{2} & -1/\sqrt{2} & 0 \\ 0 & 0 & 0 & 1 \end{pmatrix}.$$

**Problem 31.**    Let $d \geq 2$ and $|0\rangle, |1\rangle, \ldots, |d-1\rangle$ be an orthonormal basis in the Hilbert space $\mathbb{C}^d$. Let $|\psi\rangle, |\phi\rangle$ be two normalized states in $\mathbb{C}^d$. Let

$$S = \sum_{j,k=0}^{d-1} ((|j\rangle\langle k|) \otimes (|k\rangle\langle j|)).$$

Show that $S(|\psi\rangle \otimes |\phi\rangle) = |\phi\rangle \otimes |\psi\rangle$. Thus $S$ is a *swap operator*.

**Solution 31.**    We have

$$S(|\psi\rangle \otimes |\phi\rangle) = \sum_{j,k=0}^{d-1} ((|j\rangle\langle k|) \otimes (|k\rangle\langle j|))(|\psi\rangle \otimes |\phi\rangle)$$

$$= \sum_{j,k=0}^{d-1} \langle k|\psi\rangle\langle j|\phi\rangle(|j\rangle \otimes |k\rangle) = \sum_{j,k=0}^{d-1} \langle j|\psi\rangle\langle k|\phi\rangle(|k\rangle \otimes |j\rangle)$$

$$= |\phi\rangle \otimes |\psi\rangle.$$

**Problem 32.**    Let $\sigma_1, \sigma_2, \sigma_3$ be the Pauli spin matrices and

$$\sigma_{j,1} = I_2 \otimes \cdots \otimes I_2 \otimes \sigma_1 \otimes I_2 \otimes \cdots \otimes I_2$$

where $\sigma_1$ is a the $j$ position (counting from left to right) and $j = 1, 2, \ldots, N$. Analogously we have $\sigma_{j,2}$ and $\sigma_{j,3}$. Let

$$\sigma_{j,\pm} := \sigma_{j,1} \pm i\sigma_{j,2}.$$

Consider the unitary matrix

$$U := \exp\left(i \sum_{j=1}^{n} \chi_j \sigma_{j,+} \sigma_{j,-}\right).$$

Find $U\sigma_{j,1}U^*$, $U\sigma_{j,-}U^*$, $U\sigma_{j,3}U^*$.

**Solution 32.**    We obtain

$$U\sigma_{j,1}U^* = e^{i\chi_j}\sigma_{j,+}, \quad U\sigma_{j,-}U^* = e^{-i\chi_j}\sigma_{j,-}, \quad U\sigma_{j,3}U^* = \sigma_{j,3}.$$

**Problem 33.** Let $\sigma_1$, $\sigma_3$ be the Pauli spin matrices and

$$\Pi = \frac{1}{2}(I_2 + \sigma_3) \otimes I_2.$$

Show that $\Pi$ is a projection matrix. Show that the $4 \times 4$ matrix

$$U = \frac{1}{2}(I_2 + \sigma_3) \otimes I_2 + \frac{1}{2}(I_2 - \sigma_3) \otimes \sigma_1$$

is unitary.

**Solution 33.** We have $\Pi = \Pi^*$ and $\Pi^2 = \Pi$ and

$$U = I_2 \oplus \begin{pmatrix} 0 & 1 \\ 1 & 0 \end{pmatrix}.$$

Thus $U = U^*$, $U^2 = I_4$ and $U$ is unitary.

**Problem 34.** The vectors

$$\mathbf{v}_1 = \frac{1}{\sqrt{2}} \begin{pmatrix} 1 \\ 0 \\ 1 \end{pmatrix}, \quad \mathbf{v}_2 = \begin{pmatrix} 0 \\ 1 \\ 0 \end{pmatrix}, \quad \mathbf{v}_3 = \frac{1}{\sqrt{2}} \begin{pmatrix} 1 \\ 0 \\ -1 \end{pmatrix}$$

form an orthonormal basis in the Hilbert space $\mathbb{C}^3$. Find the unitary matrices $U_{12}$, $U_{23}$, $U_{31}$ such that $U_{12}\mathbf{v}_1 = \mathbf{v}_2$, $U_{23}\mathbf{v}_2 = \mathbf{v}_3$, $U_{31}\mathbf{v}_3 = \mathbf{v}_1$. Then calculate $U_{31}U_{23}U_{12}$ and the matrix $V = \lambda_1\mathbf{v}_1\mathbf{v}_1^* + \lambda_2\mathbf{v}_2\mathbf{v}_2^* + \lambda_3\mathbf{v}_3\mathbf{v}_3^*$, where the complex numbers $\lambda_1$, $\lambda_2$, $\lambda_3$ satisfy $\lambda_1\bar{\lambda}_1 = 1$, $\lambda_2\bar{\lambda}_2 = 1$, $\lambda_3\bar{\lambda}_3 = 1$. Show that the matrix is unitary.

**Solution 34.** We obtain

$$U_{12} = \begin{pmatrix} 0 & 1 & 0 \\ 1/\sqrt{2} & 0 & 1/\sqrt{2} \\ 1/\sqrt{2} & 0 & -1/\sqrt{2} \end{pmatrix}, \quad U_{23} = \begin{pmatrix} 0 & 1/\sqrt{2} & 1/\sqrt{2} \\ 1 & 0 & 0 \\ 0 & -1/\sqrt{2} & 1/\sqrt{2} \end{pmatrix},$$

$$U_{31} = I_2 \oplus (-1).$$

Obviously $U_{31}U_{23}U_{12} = I_3$. Now

$$V = \frac{\lambda_1}{2} \begin{pmatrix} 1 \\ 0 \\ 1 \end{pmatrix} (1 \ 0 \ 1) + \lambda_2 \begin{pmatrix} 0 \\ 1 \\ 0 \end{pmatrix} (0 \ 1 \ 0) + \frac{\lambda_3}{2} \begin{pmatrix} 1 \\ 0 \\ -1 \end{pmatrix} (1 \ 0 \ -1).$$

It follows that

$$V = \frac{1}{2} \begin{pmatrix} \lambda_1 + \lambda_3 & 0 & \lambda_1 - \lambda_3 \\ 0 & 2\lambda_2 & 0 \\ \lambda_1 - \lambda_3 & 0 & \lambda_1 + \lambda_3 \end{pmatrix}.$$

and $VV^* = I_3$, i.e. the matrix $V$ is unitary.

**Problem 35.**    Let $\phi \in \mathbb{R}$. Consider the $4 \times 4$ matrix

$$
A(\phi) = \begin{pmatrix} 0 & 1 & 0 & 0 \\ 0 & 0 & 1 & 0 \\ 0 & 0 & 0 & 1 \\ e^{i\phi} & 0 & 0 & 0 \end{pmatrix}.
$$

(i) Is the matrix unitary?
(ii) Find the eigenvalues and normalized eigenvectors of $A(\phi)$.
(iii) Can the matrix be written as the Kronecker product of two $2 \times 2$ matrices?

**Solution 35.**    (i) Yes. We have $A(\phi)A^*(\phi) = I_4$.
(ii) The eigenvalues are $ie^{i\phi/4}$, $-e^{i\phi/4}$, $-ie^{i\phi/4}$, $e^{i\phi/4}$ with the corresponding normalized eigenvectors

$$
\frac{1}{2}\begin{pmatrix} 1 \\ ie^{i\phi/4} \\ -e^{i\phi/2} \\ -ie^{3i\phi/4} \end{pmatrix}, \quad \frac{1}{2}\begin{pmatrix} 1 \\ -e^{i\phi/4} \\ e^{i\phi/2} \\ -e^{3i\phi/4} \end{pmatrix}, \quad \frac{1}{2}\begin{pmatrix} 1 \\ -ie^{i\phi/4} \\ -e^{i\phi/2} \\ ie^{3i\phi/4} \end{pmatrix}, \quad \frac{1}{2}\begin{pmatrix} 1 \\ e^{i\phi/4} \\ e^{i\phi/2} \\ e^{3i\phi/4} \end{pmatrix}.
$$

(iii) No.

**Problem 36.**    (i) Let $A$ be an $n \times n$ matrix over $\mathbb{C}$ and $\Pi$ be an $m \times m$ projection matrix. Let $z \in \mathbb{C}$. Calculate $\exp(z(A \otimes \Pi))$.
(ii) Let $A_1$, $A_2$ be $n \times n$ matrices over $\mathbb{C}$. Let $\Pi_1$, $\Pi_2$ be $m \times m$ projection matrices with $\Pi_1\Pi_2 = 0$. Calculate $\exp(z(A_1 \otimes \Pi_1 + A_2 \otimes \Pi_2))$.
(iii) Use the result from (ii) to find the unitary matrix $U(t) = \exp(-i\hat{H}t/\hbar)$, where $\hat{H} = \hbar\omega(A_1 \otimes \Pi_1 + A_2 \otimes \Pi_2)$ and we assume that $A_1$ and $A_2$ are hermitian matrices.
(iv) Apply the result of (iii) to

$$
A_1 = \sigma_1, \quad \Pi_1 = \frac{1}{2}\begin{pmatrix} 1 & 1 \\ 1 & 1 \end{pmatrix}, \quad A_2 = \sigma_3, \quad \Pi_2 = \frac{1}{2}\begin{pmatrix} 1 & -1 \\ -1 & 1 \end{pmatrix}.
$$

**Solution 36.**    (i) We find $\exp(z(A \otimes \Pi)) = I_n \otimes I_m + (e^{zA} - I_n) \otimes \Pi$.
(ii) We obtain

$$
\exp(z(A_1 \otimes \Pi_1 + A_2 \otimes \Pi_2)) = I_n \otimes I_m + (e^{zA_1} - I_n) \otimes \Pi_1 + (e^{zA_2} - I_n) \otimes \Pi_2.
$$

(iii) Since $U(t) = \exp(-i\omega t(A_1 \otimes \Pi_1 + A_2 \otimes \Pi_2))$ with $z = -i\omega t$ we obtain the unitary matrix

$$
U(t) = I_n \otimes I_m + (e^{-i\omega t A_1} - I_n) \otimes \Pi_1 + (e^{-i\omega t A_2} - I_n) \otimes \Pi_2.
$$

(iv) With $m = n = 2$, $A_1 = \sigma_1$ and $A_2 = \sigma_3$ we find the unitary matrix

$$U(t) = I_2 \otimes I_2 + (e^{-i\omega t\sigma_1} - I_2) \otimes \Pi_1 + (e^{-i\omega t\sigma_3} - I_2) \otimes \Pi_2.$$

With $\exp(z\sigma_1) = \cosh(z)I_2 + \sinh(z)\sigma_1$, $\exp(z\sigma_3) = \cosh(z)I_2 + \sinh(z)\sigma_3$ and $z = -i\omega t$ and $\sinh(-i\omega t) \equiv -i\sin(\omega t)$, $\cosh(-i\omega t) \equiv \cos(\omega t)$ we obtain

$$U(t) = I_2 \otimes I_2 + (I_2 \cos(\omega t) - i\sigma_1 \sin(\omega t) - I_2) \otimes \Pi_1$$
$$+(I_2 \cos(\omega t) - i\sigma_3 \sin(\omega t) - I_2) \otimes \Pi_2.$$

## Programming Problems

**Problem 1.**  Given the standard basis $e_1$, $e_2$, $e_3$, $e_4$ in $\mathbb{C}^4$. Consider the unitary matrices

$$U_{CNOT} = \begin{pmatrix} 1 & 0 & 0 & 0 \\ 0 & 1 & 0 & 0 \\ 0 & 0 & 0 & 1 \\ 0 & 0 & 1 & 0 \end{pmatrix}, \quad V_H = U_H \otimes I_2$$

where $U_H$ is the Hadamard gate

$$U_H = \frac{1}{\sqrt{2}} \begin{pmatrix} 1 & 1 \\ 1 & -1 \end{pmatrix}.$$

Show that $U_{CNOT}V_H e_j$, $(j = 1, 2, 3, 4)$ will provide the Bell states.

**Solution 1.**  The Maxima program will do the job

```
/* BellUnitary.mac */
e1: matrix([1],[0],[0],[0]); e2: matrix([0],[1],[0],[0]);
e3: matrix([0],[0],[1],[0]); e4: matrix([0],[0],[0],[1]);
UCNOT: matrix([1,0,0,0],[0,1,0,0],[0,0,0,1],[0,0,1,0]);
UH: matrix([1/sqrt(2),1/sqrt(2)],[1/sqrt(2),-1/sqrt(2)]);
I2: matrix([1,0],[0,1]);
KUHI2: kronecker_product(UH,I2);
b1: UCNOT . KUHI2 . e1; b2: UCNOT . KUHI2 . e2;
b3: UCNOT . KUHI2 . e3; b4: UCNOT . KUHI2 . e4;
```

The output is

$$\frac{1}{\sqrt{2}} \begin{pmatrix} 1 \\ 0 \\ 0 \\ 1 \end{pmatrix}, \quad \frac{1}{\sqrt{2}} \begin{pmatrix} 0 \\ 1 \\ 1 \\ 0 \end{pmatrix}, \quad \frac{1}{\sqrt{2}} \begin{pmatrix} 0 \\ 1 \\ -1 \\ 0 \end{pmatrix}, \quad \frac{1}{\sqrt{2}} \begin{pmatrix} 1 \\ 0 \\ 0 \\ -1 \end{pmatrix}.$$

**Problem 2.**    Consider the normalized vector in $\mathbb{C}^4$

$$\mathbf{v} = \frac{1}{2} \begin{pmatrix} i & -i & i & -i \end{pmatrix}^T.$$

Show that $U = I_4 - 2\mathbf{v}\mathbf{v}^*$ is a unitary matrix. Find the eigenvalues of $U$.

**Solution 2.**    The following Maxima program will do the job.

```
/* unitary.mac */
v: (1/2)*matrix([%i],[-%i],[%i],[-%i]);
vT: transpose(v); vTC: conjugate(vT);
I4: matrix([1,0,0,0],[0,1,0,0],[0,0,1,0],[0,0,0,1]);
U: I4-2*(v . vTC); UT: transpose(U); UTC: conjugate(UT);
R: U . UTC;
eigenvalues(U);
```

The eigenvalues are $-1$ (1 times) and $+1$ (3 times).


# 7.3    Supplementary Problems


**Problem 1.**    (i) Let $\phi \in \mathbb{R}$. Show that the matrix

$$A(\phi) = \begin{pmatrix} 0 & e^{-i\phi} \\ e^{i\phi} & 0 \end{pmatrix}$$

hermitian and unitary. Find the eigenvalues and normalized eigenvectors of $A(\phi)$. Let $I_2$ be the $2 \times 2$ unit matrix. Find the eigenvalues of $A(\phi) \otimes I_2$.

**Problem 2.**    Let $U$ be a unitary and hermitian $n \times n$ matrix. Show that

$$\Pi_+ = \frac{1}{2}(I_n + U), \quad \Pi_- = \frac{1}{2}(I_n - U)$$

are projection matrices. Show that $\Pi_+\Pi_- = 0_n$.

**Problem 3.**    Find all the square roots of the $2 \times 2$ identity matrix. For example the Pauli spin matrices $\sigma_1$, $\sigma_2$, $\sigma_3$ are solutions.

# Chapter 8

# Entropy

## 8.1 Introduction

For any density operator (where $\langle \psi_j | \psi_k \rangle = \delta_{jk}$)

$$\rho = \sum_{j=1}^{n} \lambda_j |\psi_j\rangle\langle\psi_j|, \quad \lambda_j \geq 0, \quad \sum_{j=1}^{n} \lambda_j = 1$$

the *von Neumann entropy* is defined as

$$S(\rho) := -\text{tr}(\rho \log(\rho))$$

or equivalently

$$S(\rho) := -\sum_{j=1}^{n} \lambda_j \log(\lambda_j)$$

with

$$0\log(0) = 0, \quad 1\log(1) = 0.$$

Thus the von Neumann entropy is equal to the Shannon entropy of the eigenvalues. We have $S(\rho) \geq 0$ with equality iff $\rho$ is a *pure state*, i.e. $\rho = |\psi\rangle\langle\psi|$. Furthermore we have the inequality

$$S(\rho) \leq \log(n)$$

where $n$ is the dimension of $\rho$. We find equality iff

$$\rho = \frac{1}{n} I$$

where $I$ is the identity operator. The entropy is unchanged under unitary transformation

$$S(U \rho U^*) = S(\rho).$$

For a joint system $AB$ we have

$$S(\rho_{AB}) \leq S(\rho_A) + S(\rho_B).$$

For density operators $\rho$ and $\sigma$ we have the *quantum relative entropy*

$$S(\rho||\sigma) := \operatorname{tr}(\rho \log(\rho)) - \operatorname{tr}(\rho \log(\sigma)).$$

We have *Klein's inequality*

$$S(\rho||\sigma) \geq 0$$

with equality iff $\rho = \sigma$.

Let $A$, $B$ be $n \times n$ hermitian matrices acting in the Hilbert space $\mathbb{C}^n$. Assume that the eigenvalues of $A$ are pairwise different and analogously for $B$. Then the normalized eigenvectors $|\alpha_j\rangle$ $(j = 1, \ldots, n)$ of $A$ form an orthonormal basis in $\mathbb{C}^n$ and analogously for $B$ the normalized eigenvectors $|\beta_j\rangle$ $(j = 1, \ldots, n)$ form an orthonormal basis in $\mathbb{C}^n$. Let $|\psi\rangle$ be a normalized state in $\mathbb{C}^n$. Then there are $n$ possible outcomes for measurements of each observable and the probabilities $p_j(A, |\psi\rangle)$, $p_j(B, |\psi\rangle)$ $(j = 1, \ldots, n)$ are given by

$$p_j(A, |\psi\rangle) := |\langle \psi | \alpha_j \rangle|^2, \qquad p_j(B, |\psi\rangle) := |\langle \psi | \beta_j \rangle|^2.$$

Let $H_{|\psi\rangle}(X)$ be the *Shannon information entropy*

$$H_{|\psi\rangle}(X) := -\sum_{j=1}^{n} p_j(X, |\psi\rangle) \ln(p_j(X, |\psi\rangle))$$

corresponding to the probability distribution $\{p_j(X, |\psi\rangle)\}$ $(j = 1, \ldots, n)$. The (Maassen-Uffink) *entropic uncertainty relation* is given by

$$H_{|\psi\rangle}(A) + H_{|\psi\rangle}(B) \geq -2 \ln( \max_{1 \leq j,k \leq n} |\langle \alpha_j | \beta_k \rangle|) > 0.$$

Note that the right-hand side does not involve the state $|\psi\rangle$.

The (Landau-Pollak) uncertainty relation states that

$$\arccos(\sqrt{P_A}) + \arccos(\sqrt{P_B}) \geq \arccos( \max_{1 \leq j,k \leq n} |\langle \alpha_j | \beta_k \rangle|)$$

where

$$P_A := \max_{1 \leq j \leq n} p_j(A, |\psi\rangle), \qquad P_B := \max_{1 \leq j \leq n} p_j(B, |\psi\rangle).$$

## 8.2   Solved Problems

**Problem 1.**   (i) Consider the density matrix (pure state)

$$\rho = \frac{1}{2} \begin{pmatrix} 1 & 1 \\ 1 & 1 \end{pmatrix} \equiv \frac{1}{\sqrt{2}} \begin{pmatrix} 1 \\ 1 \end{pmatrix} \frac{1}{\sqrt{2}} (1 \quad 1).$$

Find the von Neumann entropy $S(\rho) = -\mathrm{tr}(\rho \log_2(\rho))$.
(ii) Consider the density matrix (mixed state)

$$\rho = \begin{pmatrix} 1/2 & 0 \\ 0 & 1/2 \end{pmatrix}.$$

Find the von Neumann entropy $S(\rho) = -\mathrm{tr}(\rho \log_2(\rho))$.

**Solution 1.**   (i) The eigenvalues of $\rho$ are 0 and 1. Hence $S(\rho) = 0$.
(ii) Since $\log_2(1/2) = -1$ we obtain $S(\rho) = 1$. Note that $\rho$ describes a mixed state.

**Problem 2.**   Let $W$ be a positive semidefinite $n \times n$ matrix over $\mathbb{C}$ with $\mathrm{tr}(W) = 1$. Let $\lambda_1, \lambda_2, \ldots, \lambda_n$ be the eigenvalues (which obviously are real and nonnegative) and $\mathbf{w}_1, \mathbf{w}_2, \ldots, \mathbf{w}_n$ be the corresponding normalized eigenvectors of $W$. We can assume that the normalized eigenvectors form an orthonormal basis in $\mathbb{C}^n$. If eigenvalues are degenerate and the corresponding normalized eigenvectors are not orthogonal we can apply the Gram-Schmidt algorithm. Calculate $\mathrm{tr}(W \ln(W))$.

**Solution 2.**   Let $U$ be a unitary matrix such that $U^{-1}WU$ is a diagonal matrix. Note that $U^{-1} = U^*$. Obviously the unitary matrix $U$ is constructed from the normalized eigenvectors of $W$. Then we have

$$\mathrm{tr}(W \ln W) = \mathrm{tr}(U^{-1}(W \ln(W))U) = \mathrm{tr}(U^{-1}WUU^{-1}(\ln(W))U)$$
$$= \mathrm{tr}(U^{-1}WU \ln(U^{-1}WU)).$$

The diagonal elements of $U^{-1}WU$ are the nonnegative real eigenvalues of $W$. Consequently

$$\mathrm{tr}(W \ln(W)) = \sum_{j=1}^{n} \lambda_j \ln(\lambda_j).$$

Note that $0 \cdot \ln(0) = 0$ if an eigenvalue of $W$ is 0.

**Problem 3.**   Let $W$ be a positive semidefinite $n \times n$ matrix over $\mathbb{C}$ with $\mathrm{tr}(W) = 1$. Let $\lambda_1, \lambda_2, \ldots, \lambda_n$ be the eigenvalues (which obviously are real and nonnegative) and $\mathbf{w}_1, \mathbf{w}_2, \ldots, \mathbf{w}_n$ be the corresponding normalized eigenvectors (column vectors) of $W$. We can assume that the normalized eigenvectors form an orthonormal basis in $\mathbb{C}^n$. If eigenvalues are degenerate and the corresponding normalized eigenvectors are not orthogonal we can apply the Gram-Schmidt algorithm. Let $\widetilde{W}$ be a positive definite $n \times n$ matrix over $\mathbb{C}$ with $\mathrm{tr}(\widetilde{W}) = 1$. Let $\widetilde{\lambda}_1, \widetilde{\lambda}_2, \ldots, \widetilde{\lambda}_n$ be the eigenvalues (which obviously are real and positive) and $\widetilde{\mathbf{w}}_1, \widetilde{\mathbf{w}}_2, \ldots, \widetilde{\mathbf{w}}_n$ be the corresponding normalized eigenvectors of $\widetilde{W}$. We can assume that the normalized eigenvectors form an orthonormal basis in $\mathbb{C}^n$. If eigenvalues are degenerate and the corresponding eigenvectors are not orthogonal we can apply the Gram-Schmidt algorithm. Calculate

$$\mathrm{tr}(W \ln(\widetilde{W})).$$

What happens if we allow the matrix $\widetilde{W}$ to be positive semidefinite?

**Solution 3.**    Let $\widetilde{U}$ be a unitary matrix such that $\widetilde{U}^{-1}\widetilde{W}\widetilde{U}$ is a diagonal matrix. Note that $\widetilde{U}^{-1} = \widetilde{U}^*$. Obviously the unitary matrix $\widetilde{U}$ is constructed from the normalized eigenvectors of $\widetilde{W}$. Then we have

$$\mathrm{tr}(W \ln(\widetilde{W})) = \mathrm{tr}(\widetilde{U}^{-1}(W \ln(\widetilde{W}))\widetilde{U}) = \mathrm{tr}(\widetilde{U}^{-1}W\widetilde{U}\widetilde{U}^{-1}(\ln \widetilde{W})\widetilde{U})$$
$$= \mathrm{tr}(\widetilde{U}^{-1}W\widetilde{U} \ln(\widetilde{U}^{-1}\widetilde{W}\widetilde{U})) = \mathrm{tr}(\widetilde{U}^{-1}W\widetilde{U} \ln(\widetilde{W}_D))$$

where $\widetilde{W}_D = \widetilde{U}^{-1}\widetilde{W}\widetilde{U}$ is a diagonal matrix. Using the spectral representation of $W$

$$W = \sum_{j=1}^{n} \lambda_j \mathbf{w}_j \mathbf{w}_j^*$$

we obtain

$$\mathrm{tr}(W \ln(\widetilde{W})) = \sum_{j=1}^{n} \lambda_j \mathrm{tr}(\widetilde{U}^{-1}(\mathbf{w}_j \mathbf{w}_j^*)\widetilde{U} \ln(\widetilde{W}_D)).$$

Since

$$\widetilde{U} = \sum_{\ell=1}^{n} \widetilde{\mathbf{w}}_\ell \mathbf{e}_\ell^*, \qquad \widetilde{U}^* = \sum_{k=1}^{n} \mathbf{e}_k \widetilde{\mathbf{w}}_k^*$$

where $\{\, \mathbf{e}_k \ : \ k = 1, 2, \ldots, n \,\}$ is the standard basis in $\mathbb{C}^n$ we obtain

$$\mathrm{tr}(W \ln(\widetilde{W})) = \sum_{j=1}^{n} \lambda_j \mathrm{tr}(\widetilde{U}^{-1}(\mathbf{w}_j \mathbf{w}_j^*)\widetilde{U} \ln(\widetilde{W}_D))$$

$$= \sum_{j=1}^{n} \sum_{k=1}^{n} \sum_{\ell=1}^{n} \lambda_j \mathrm{tr}(\mathbf{e}_k \widetilde{\mathbf{w}}_k^* (\mathbf{w}_j \mathbf{w}_j^*) \widetilde{\mathbf{w}}_\ell \mathbf{e}_\ell^* \ln(\widetilde{W}_D)).$$

To calculate the trace we use the standard basis $\{\,\mathbf{e}_j \,:\, j = 1, 2, \ldots, n\,\}$. Thus

$$
\begin{aligned}
\operatorname{tr}(W \ln(\widetilde{W})) &= \sum_{j=1}^{n}\sum_{k=1}^{n}\sum_{\ell=1}^{n}\sum_{r=1}^{n} \lambda_j (\mathbf{e}_r^* \mathbf{e}_k \widetilde{\mathbf{w}}_k^* (\mathbf{w}_j \mathbf{w}_j^*) \widetilde{\mathbf{w}}_\ell \mathbf{e}_\ell^* \ln(\widetilde{W}_D) \mathbf{e}_r) \\
&= \sum_{j=1}^{n}\sum_{k=1}^{n}\sum_{\ell=1}^{n}\sum_{r=1}^{n} \lambda_j (\delta_{rk} \widetilde{\mathbf{w}}_k^* (\mathbf{w}_j \mathbf{w}_j^*) \widetilde{\mathbf{w}}_\ell \mathbf{e}_\ell^* \ln(\widetilde{W}_D) \mathbf{e}_r) \\
&= \sum_{j=1}^{n}\sum_{k=1}^{n}\sum_{\ell=1}^{n} \lambda_j (\widetilde{\mathbf{w}}_k^* (\mathbf{w}_j \mathbf{w}_j^*) \widetilde{\mathbf{w}}_\ell \mathbf{e}_\ell^* \ln(\widetilde{W}_D) \mathbf{e}_k) \\
&= \sum_{j=1}^{n}\sum_{k=1}^{n}\sum_{\ell=1}^{n} \lambda_j (\widetilde{\mathbf{w}}_k^* (\mathbf{w}_j \mathbf{w}_j^*) \widetilde{\mathbf{w}}_\ell \delta_{\ell k} \ln(\widetilde{\lambda}_k)) \\
&= \sum_{j=1}^{n}\sum_{k=1}^{n} \lambda_j (\widetilde{\mathbf{w}}_k^* (\mathbf{w}_j \mathbf{w}_j^*) \widetilde{\mathbf{w}}_k \ln(\widetilde{\lambda}_k)).
\end{aligned}
$$

Consequently

$$
\operatorname{tr}(W \ln(\widetilde{W})) = \sum_{j=1}^{n}\sum_{k=1}^{n} \lambda_j (\widetilde{\mathbf{w}}_k^* \mathbf{w}_j)(\mathbf{w}_j^* \widetilde{\mathbf{w}}_k) \ln(\widetilde{\lambda}_k).
$$

Note that $\widetilde{\mathbf{w}}_k^* \mathbf{w}_j = (\mathbf{w}_j^* \widetilde{\mathbf{w}}_k)^*$. Can the condition on $\widetilde{W}$ be extended to positive semidefinite, i.e. some of the eigenvalues of $\widetilde{W}$ could be zero? Assume that one eigenvalue of $\widetilde{W}$ is 0, say $\widetilde{\lambda}_p = 0$. Thus we should have

$$
\sum_{j=1}^{n} \lambda_j \widetilde{\mathbf{w}}_p^* (\mathbf{w}_j \mathbf{w}_j^*) \widetilde{\mathbf{w}}_p = \widetilde{\mathbf{w}}_p^* W \widetilde{\mathbf{w}}_p = 0
$$

in order to apply $0 \cdot \ln 0 = 0$. This is in general not true. Consider, for example the density matrices

$$
W = \begin{pmatrix} 1/2 & 0 \\ 0 & 1/2 \end{pmatrix}, \qquad \widetilde{W} = \begin{pmatrix} 1 & 0 \\ 0 & 0 \end{pmatrix}.
$$

**Problem 4.** The *relative entropy of entanglement* for bipartite states (say, $A$ and $B$), where the quantum state is described by the density matrix $W$, is defined as

$$
E_r(W) := \min_{\widetilde{W} \in D} S(W \| \widetilde{W})
$$

where $D$ is the set of all convex combinations of separable density matrices (i.e. $\widetilde{W} = W_A \otimes W_B$) and

$$
S(W \| \widetilde{W}) := \operatorname{tr}(W(\log_2(W) - \log_2(\widetilde{W}))) \equiv \operatorname{tr}(W \log_2(W) - W \log_2(\widetilde{W}))
$$

is the *quantum relative entropy*. Let $\widetilde{W}_{min}$ denote the separable state that minimizes the relative entropy. Thus to calculate $E_r(W)$ is to find the state $\widetilde{W}_{min}$. Consider the Bell state

$$|\Phi^+\rangle = \frac{1}{\sqrt{2}}(|0\rangle \otimes |0\rangle + |1\rangle \otimes |1\rangle)$$

and thus $W = |\Phi^+\rangle\langle\Phi^+|$. Let

$$\widetilde{W} = \frac{1}{2}(|0\rangle\langle 0| + |1\rangle\langle 1|) \otimes \frac{1}{2}(|0\rangle\langle 0| + |1\rangle\langle 1|).$$

Calculate $S(W\|\widetilde{W})$.

**Solution 4.**   Using the standard basis in $\mathbb{C}^4$ we obtain the density matrices

$$W = |\Phi^+\rangle\langle\Phi^+| = \frac{1}{2}\begin{pmatrix} 1 & 0 & 0 & 1 \\ 0 & 0 & 0 & 0 \\ 0 & 0 & 0 & 0 \\ 1 & 0 & 0 & 1 \end{pmatrix}$$

and

$$\widetilde{W} = \frac{1}{4}I_4$$

where $I_4$ is the $4 \times 4$ identity matrix. To calculate $\text{tr}(W\log_2(W))$ and $\text{tr}(W\log_2(\widetilde{W}))$ we need the eigenvalues and normalized eigenvectors of $W$ and $\widetilde{W}$. For the density matrix $W$ we find the eigenvalues $\lambda_1 = 1$, $\lambda_2 = 0$, $\lambda_3 = 0$, $\lambda_4 = 0$ with the corresponding normalized eigenvectors

$$\mathbf{w}_1 = \frac{1}{\sqrt{2}}\begin{pmatrix} 1 \\ 0 \\ 0 \\ 1 \end{pmatrix}, \quad \mathbf{w}_2 = \frac{1}{\sqrt{2}}\begin{pmatrix} 1 \\ 0 \\ 0 \\ -1 \end{pmatrix}, \quad \mathbf{w}_3 = \begin{pmatrix} 0 \\ 1 \\ 0 \\ 0 \end{pmatrix}, \quad \mathbf{w}_4 = \begin{pmatrix} 0 \\ 0 \\ 1 \\ 0 \end{pmatrix}.$$

Thus we find

$$\text{tr}(W\log_2(W)) = \sum_{j=1}^{4} \lambda_j \log_2(\lambda_j) = 0$$

where we used $0\log_2(0) = 0$ and $1\log_2(1) = 0$. For $\widetilde{W}$ we obtain the eigenvalues $\widetilde{\lambda}_1 = 1/4$, $\widetilde{\lambda}_2 = 1/4$, $\widetilde{\lambda}_3 = 1/4$, $\widetilde{\lambda}_4 = 1/4$ and the corresponding normalized eigenvectors (standard basis)

$$\widetilde{\mathbf{w}}_1 = \begin{pmatrix} 1 \\ 0 \\ 0 \\ 0 \end{pmatrix}, \quad \widetilde{\mathbf{w}}_2 = \begin{pmatrix} 0 \\ 1 \\ 0 \\ 0 \end{pmatrix}, \quad \widetilde{\mathbf{w}}_3 = \begin{pmatrix} 0 \\ 0 \\ 1 \\ 0 \end{pmatrix}, \quad \widetilde{\mathbf{w}}_4 = \begin{pmatrix} 0 \\ 0 \\ 0 \\ 1 \end{pmatrix}.$$

Since

$$\mathrm{tr}(W \log_2(\widetilde{W})) = \sum_{j=1}^{4}\sum_{k=1}^{4} \lambda_j(\widetilde{\mathbf{w}}_k^*\mathbf{w}_j)(\mathbf{w}_j^*\widetilde{\mathbf{w}}_k) \log_2(\lambda_k)$$

and

$$\mathbf{w}_1^*\widetilde{\mathbf{w}}_1 = \frac{1}{\sqrt{2}}, \quad \mathbf{w}_1^*\widetilde{\mathbf{w}}_2 = 0, \quad \mathbf{w}_1^*\widetilde{\mathbf{w}}_3 = 0, \quad \mathbf{w}_1^*\widetilde{\mathbf{w}}_4 = \frac{1}{\sqrt{2}}$$

$$\mathbf{w}_2^*\widetilde{\mathbf{w}}_1 = \frac{1}{\sqrt{2}}, \quad \mathbf{w}_2^*\widetilde{\mathbf{w}}_2 = 0, \quad \mathbf{w}_2^*\widetilde{\mathbf{w}}_3 = 0, \quad \mathbf{w}_2^*\widetilde{\mathbf{w}}_4 = -\frac{1}{\sqrt{2}}$$

$$\mathbf{w}_3^*\widetilde{\mathbf{w}}_1 = 0, \quad \mathbf{w}_3^*\widetilde{\mathbf{w}}_2 = 1, \quad \mathbf{w}_3^*\widetilde{\mathbf{w}}_3 = 0, \quad \mathbf{w}_3^*\widetilde{\mathbf{w}}_4 = 0$$

$$\mathbf{w}_4^*\widetilde{\mathbf{w}}_1 = 0, \quad \mathbf{w}_4^*\widetilde{\mathbf{w}}_2 = 0, \quad \mathbf{w}_4^*\widetilde{\mathbf{w}}_3 = 1, \quad \mathbf{w}_4^*\widetilde{\mathbf{w}}_4 = 0$$

we obtain $\mathrm{tr}(W \log_2(\widetilde{W})) = -2$. Consequently, $S(W\|\widetilde{W}) = 2$.

**Problem 5.** Let $\rho_{AB}$ be a density matrix defined on a $(N \times N)$-dimensional Hilbert space $\mathcal{H} \otimes \mathcal{H}$. The *classical information capacity* is defined as

$$C(\rho) := \log_2(N) + S(\rho_B) - S(\rho_{AB})$$

where $\rho_B$ is the reduced quantum state obtained by $\rho_B = \mathrm{tr}_A(\rho_{AB})$ and $S(\rho)$ is the von Neumann entropy of a quantum state (density matrix) $S(\rho) = -\mathrm{tr}(\rho \log_2(\rho))$. Consider the Bell state

$$|\psi\rangle = \frac{1}{\sqrt{2}}(|0\rangle \otimes |0\rangle + |1\rangle \otimes |1\rangle).$$

Calculate the density matrix $\rho_{AB}$, the reduced density matrix $\rho_B$ and then the classical information capacity.

**Solution 5.** We find the density matrix (pure state)

$$\rho_{AB} = |\psi\rangle\langle\psi| = \frac{1}{2}\begin{pmatrix} 1 & 0 & 0 & 1 \\ 0 & 0 & 0 & 0 \\ 0 & 0 & 0 & 0 \\ 1 & 0 & 0 & 1 \end{pmatrix}.$$

Thus we obtain the density matrix (mixed state)

$$\rho_B = \begin{pmatrix} 1/2 & 0 \\ 0 & 1/2 \end{pmatrix}.$$

With $N = 2$ and $S(\rho_{AB}) = 0$ it follows that

$$C(\rho) = \log_2(N) + S(\rho_B) - S(\rho_{AB}) = 2.$$

**Problem 6.**   The *quantum relative entropy* between two density operators $\rho$ and $\sigma$ is defined by

$$S_b(\rho\|\sigma) := \mathrm{tr}(\rho\log_b(\rho) - \rho\log_b(\sigma)) = -S_b(\rho) - \mathrm{tr}(\rho\log_b(\sigma)).$$

Here $S_b$ is the von Neumann entropy where the log is taken with the base $b$. Show that $S_b(\rho\|\sigma) \geq 0$. This inequality is known as *Klein's inequality*.

**Solution 6.**   Consider the term $\rho\log_b(\sigma)$ where $\rho$ and $\sigma$ are density operators on a finite-dimensional Hilbert space of dimension $n$. Let $\lambda_j$ and $|\phi_j\rangle$ be the corresponding eigenvalues and (orthonormal) eigenstates of $\rho$. Similarly, let $\mu_k$ and $|\psi_k\rangle$ be the corresponding eigenvalues and (orthonormal) eigenstates of $\sigma$. Thus we have

$$\rho\log_2(\sigma) = \left(\sum_{j=1}^{n}\lambda_j|\phi_j\rangle\langle\phi_j|\right)\left(\sum_{k=1}^{n}\log_b\mu_k|\psi_k\rangle\langle\psi_k|\right)$$

$$= \sum_{j,k=1}^{n}\lambda_j\log_b(\mu_k)\langle\phi_j|\psi_k\rangle|\phi_j\rangle\langle\psi_k|.$$

Taking the trace using the basis $\{|\psi_k\rangle,\ k = 1, 2, \ldots, n\}$ yields

$$\mathrm{tr}(\rho\log_b(\sigma)) = \sum_{j,k,l=1}^{n}\lambda_j(\log_b(\mu_k))\langle\phi_j|\psi_k\rangle\langle\psi_l|\phi_j\rangle\langle\psi_k|\psi_l\rangle$$

$$= \sum_{j,k=1}^{n}\lambda_j(\log_b(\mu_k))\langle\phi_j|\psi_k\rangle\langle\psi_k|\phi_j\rangle$$

$$= \sum_{j,k=1}^{n}\lambda_j(\log_b(\mu_k))|\langle\phi_j|\psi_k\rangle|^2.$$

Thus we obtain

$$S_b(\rho\|\sigma) = \sum_{j=1}^{n}\lambda_j\log_b(\lambda_j) - \sum_{j,k=1}^{n}\lambda_j(\log_b(\mu_k))|\langle\phi_j|\psi_k\rangle|^2$$

$$= \sum_{j=1}^{n}\lambda_j\log_b(\lambda_j) - \sum_{j=1}^{n}\lambda_j\log_b\left(\prod_{k=1}^{n}\mu_k^{|\langle\phi_j|\psi_k\rangle|^2}\right)$$

$$= \sum_{j=1}^{n}\lambda_j\log_b(\lambda_j) - \sum_{j=1}^{n}\lambda_j\log_b(\nu_j)$$

where

$$\nu_j := \left(\prod_{k=1}^{n}\mu_k^{|\langle\phi_j|\psi_k\rangle|^2}\right).$$

If $\nu_j = 0$ for some $j$ where $\lambda_j \neq 0$ then $S_B(\rho||\sigma) \geq 0$ trivially. We assume $\nu_j = 0$ if and only if $\lambda_j = 0$. Since

$$\sum_{k=1}^{n} \mu_k = \sum_{k=1}^{n} |\langle \phi_j | \psi_k \rangle|^2 = \sum_{j=1}^{n} |\langle \phi_j | \psi_k \rangle|^2 = 1$$

we find that

$$0 \leq \nu_j \leq 1, \qquad \alpha := \sum_{j=1}^{n} \nu_j \leq 1.$$

We assume that the $\lambda_j$ are ordered in non-decreasing order and $m \leq n$ is chosen such that $\lambda_j = 0$ iff $j > m$. We determine the maximum value of

$$\sum_{j=1}^{m} \lambda_j \log_b(x_j), \qquad \sum_{j=1}^{m} x_j = \alpha$$

which can be formulated as a *Lagrange multiplier problem*. Thus we find the critical points of the function

$$f(x_1, \ldots, x_m) := \sum_{j=1}^{m} \lambda_j \log_b(x_j) - \theta \left( \sum_{j=1}^{m} x_j - \alpha \right)$$

where $\theta$ is the Lagrange multiplier. We obtain

$$\frac{\partial f}{\partial x_j} = \frac{\lambda_j}{x_j \ln(b)} - \theta.$$

Since $\partial f / \partial x_j = \partial f / \partial x_k = 0$ for the *critical points* we have, since $\lambda_j \neq 0$ and $\lambda_k \neq 0$, $x_k = \lambda_k x_j / \lambda_j$. Inserting $x_k \neq 0$ into the constraint yields

$$\sum_{j=1}^{n} x_j = x_k + \sum_{\substack{j=1 \\ j \neq k}}^{n} \frac{\lambda_j}{\lambda_k} x_k = \alpha.$$

Thus $x_k = \alpha \lambda_k$, since $\sum_{j=1}^{n} \lambda_j = 1$. Since $\sum_{j=1}^{m} \lambda_j \log_b(x_j)$ is unbounded from below we have a maximum. Consequently

$$S_b(\rho||\sigma) = \sum_{j=1}^{n} \lambda_j \log_b(\lambda_j) - \sum_{j=1}^{n} \lambda_j \log_b(\nu_j)$$

$$\geq \sum_{j=1}^{n} \lambda_j \log_b(\lambda_j) - \sum_{j=1}^{n} \lambda_j \log_b(\alpha \lambda_j)$$

$$= -\log_b(\alpha) \geq 0$$

since $0 \leq \alpha \leq 1$.

**Problem 7.**    Show that for the quantum relative entropy

$$S_b(\rho\|\sigma) := \text{tr}(\rho \log_b(\rho) - \rho \log_b(\sigma)) = -S_b(\rho) - \text{tr}(\rho \log_b(\sigma))$$

the equality

$$S_b(\rho\|\rho_A \otimes \rho_B) = S_b(\rho_A) + S_b(\rho_B) - S_b(\rho)$$

holds.

**Solution 7.**    From $\sigma = \rho_A \otimes \rho_B$ we find that the eigenvalues of $\sigma$ are the products of eigenvalues $\lambda_{A,j}$ of $\rho_A$ and $\lambda_{B,k}$ of $\rho_B$. The corresponding (orthonormal) eigenstates are $|\phi_{A,j}\rangle \otimes |\phi_{B,k}\rangle$ composed of the orthonormal eigenstates of $\rho_A$ and $\rho_B$ corresponding to $\lambda_{A,j}$ and $\lambda_{B,k}$. Consequently

$$
\begin{aligned}
S_b(\rho\|\rho_A \otimes \rho_B) &= -S_b(\rho) - \text{tr}(\rho \log_b(\rho_A \otimes \rho_B)) \\
&= -S_b(\rho) - \text{tr}_A\left(\text{tr}_B\left(\rho \log_b(\rho_A \otimes \rho_B)\right)\right) \\
&= -S_b(\rho) - \text{tr}_A\left(\rho_A \sum_{j=1}^{m} \log_b(\lambda_{A,j})|\phi_{A,j}\rangle\langle\phi_{A,j}|\right) \\
&\quad -\text{tr}_B\left(\rho_B \sum_{k=1}^{n} \log_b(\lambda_{B,k})|\phi_{B,k}\rangle\langle\phi_{B,k}|\right) \\
&= S_b(\rho_A) + S_b(\rho_B) - S_b(\rho).
\end{aligned}
$$

From Klein's inequality, $S(\rho\|\rho_A \otimes \rho_B) \geq 0$, we obtain the property of *subadditivity* for the von Neumann entropy

$$S_b(\rho) \leq S_b(\rho_A) + S_b(\rho_B).$$

**Problem 8.**    An $n \times n$ density matrix $\rho$ is a positive semidefinite matrix such that $\text{tr}(\rho) = 1$. The nonnegative eigenvalues of $\rho$ are the probabilities of the physical states described by the corresponding eigenvectors. The entropy of the statistical state described by the density matrix $\rho$ is defined by

$$S(\rho) := -\text{tr}(\rho \ln(\rho)).$$

For the $n \times n$ hermitian matrix $H$ (Hamilton operator) the statistical average of the energy $E$ is defined by

$$E := \text{tr}(H\rho).$$

Let

$$\psi(\rho) := \text{tr}(H\rho) - \text{tr}(\rho \ln(\rho)).$$

(i) Show that $\ln(\mathrm{tr}(e^H)) = \max\{\,\mathrm{tr}(H\rho) + S(\rho)\,\}$.
(ii) Show that $-S(\rho) = \max\{\,\mathrm{tr}(H\rho) - \ln(\mathrm{tr}(e^H))\,\}$.

**Solution 8.** (i) For every $n \times n$ hermitian matrix $Q$, and for each $\epsilon \in \mathbb{R}$ in a neighbourhood of $0$, consider the differentiable function

$$f(\epsilon) = \psi(e^{-i\epsilon Q}\rho e^{i\epsilon Q}).$$

Since $e^{i\epsilon Q}$ is a unitary matrix and the trace is invariant under unitary transformation we have

$$f(\epsilon) = \mathrm{tr}(He^{-i\epsilon Q}\rho e^{i\epsilon Q}) - \mathrm{tr}(\rho\ln(\rho)).$$

By the extremum condition it follows that

$$\frac{df(0)}{d\epsilon} = i\mathrm{tr}(Q[H, \rho]) = 0$$

where $[\,,\,]$ denotes the commutator. Since $Q$ is arbitrary, we conclude that $[H, \rho] = 0_n$ and therefore $e^H$ and $\rho$ also commute. Using $\mathrm{tr}(\rho) = 1$, we obtain

$$\begin{aligned}
\mathrm{tr}(H\rho) + S(\rho) &= \mathrm{tr}(\rho(\ln(e^H) - \ln(\rho))) \\
&= \ln(\mathrm{tr}(e^H)) - \mathrm{tr}(\rho(\ln(\mathrm{tr}(e^H)) + \ln(\rho) - \ln(e^H))).
\end{aligned}$$

Since $e^H$ and $\rho$ commute, it follows that

$$\mathrm{tr}(H\rho) + S(\rho) = \ln(\mathrm{tr}(e^H)) - \mathrm{tr}(e^H\rho e^{-H}\ln(\mathrm{tr}(e^H)\rho e^{-H})).$$

Setting $C := \mathrm{tr}(e^H)\rho e^{-H}$ we obtain

$$\ln(\mathrm{tr}(e^H)) - \mathrm{tr}(e^H\rho e^{-H}\ln(\mathrm{tr}(e^H)\rho e^{-H})) =$$

$$\ln(\mathrm{tr}(e^H)) - (\mathrm{tr}e^H)^{-1}\mathrm{tr}(e^H(C\ln(C) - C + I_n)).$$

Since $x\ln(x) - x + 1 \geq 0$ for $x \geq 0$, we conclude that the last expression is less than or equal to $\ln(\mathrm{tr}(e^H))$ and equality occurs only if $C = I_n$. This means

$$\rho = \frac{e^H}{\mathrm{tr}(e^H)}.$$

If $\rho = e^H/\mathrm{tr}(e^H)$ we find that $\psi(e^H/\mathrm{tr}(e^H)) = \ln(\mathrm{tr}(e^H))$.
(ii) Since

$$\mathrm{tr}((H + kI_n)\rho) - \ln(\mathrm{tr}(e^{H+kI_n})) = \mathrm{tr}(H\rho) - \ln(\mathrm{tr}(e^H)), \quad k \in \mathbb{R}$$

we may assume that $\mathrm{tr}(e^H) = 1$. Following an argument similar to the one in (i) we can show that the maximum of $\mathrm{tr}(H\rho) - \ln(\mathrm{tr}(e^H))$ for hermitian matrices $H$ occurs when $[H,\rho] = 0_n$. Thus $[\rho, e^H] = 0_n$. Since $\text{tr}(\rho) - \text{tr}(e^H) = 0_n$, we have

$$-\text{tr}(H\rho) + \text{tr}(\rho \ln(\rho)) = \text{tr}(e^H \rho e^{-H} \ln(\rho e^{-H})) - \text{tr}(e^H \rho e^{-H}) + \text{tr}(e^H)$$
$$= \text{tr}(e^H (Z \ln(Z) - Z + I_n))$$
$$\geq 0$$

where $Z := \rho e^{-H}$. Hence the maximum occurs when $H = \ln(\rho)$.

**Problem 9.**   Consider the normalized states $|\psi_k\rangle$, $k = 0, 1, \ldots, N-1$ in the Hilbert space $\mathbb{C}^N$. A positive operator valued measure is specified by a decomposition of the identity matrix $I_N$ into $M$ positive semidefinite matrices $P_m$, i.e.

$$I_N = \sum_{m=0}^{M-1} P_m.$$

The mutual information is defined by

$$I = \sum_{n=0}^{N-1} \sum_{m=0}^{M-1} p_{nm} \log_N \left( \frac{p_{nm}}{p_{n\cdot} p_{\cdot m}} \right)$$

where $p_{nm} := \langle \psi_n | P_m | \psi_n \rangle$ are the joint probabilities and

$$p_{n\cdot} := \sum_{m=0}^{M-1} p_{nm}, \qquad p_{\cdot m} := \sum_{n=0}^{N-1} p_{nm}$$

are their marginals. Let $M = N = 2$ and

$$P_0 = \frac{1}{2} \begin{pmatrix} 1 & 1 \\ 1 & 1 \end{pmatrix}, \qquad P_1 = \frac{1}{2} \begin{pmatrix} 1 & -1 \\ -1 & 1 \end{pmatrix}$$

$$|\psi_0\rangle = \frac{1}{\sqrt{2}} \begin{pmatrix} 1 \\ -1 \end{pmatrix}, \qquad |\psi_1\rangle = \begin{pmatrix} 0 \\ 1 \end{pmatrix}.$$

Find $p_{nm}$, $p_{n\cdot}$, $p_{\cdot m}$ and then $I$.

**Solution 9.**   Straightforward calculation yields

$$p_{00} = \langle \psi_0 | P_0 | \psi_0 \rangle = 0, \qquad p_{10} = \langle \psi_1 | P_0 | \psi_1 \rangle = \frac{1}{2}$$

$$p_{01} = \langle \psi_0 | P_1 | \psi_0 \rangle = 1, \qquad p_{11} = \langle \psi_1 | P_1 | \psi_1 \rangle = \frac{3}{2}.$$

Thus

$$p_{0\cdot} = 1, \qquad p_{1\cdot} = 1, \qquad p_{\cdot 0} = \frac{1}{2}, \qquad p_{\cdot 1} = \frac{3}{2}$$

and $I = 0$.

**Problem 10.**   The *Kullback-Leibler distance* between two probability mass functions

$$\mathbf{w} = (w_1, \ldots, w_n), \qquad \widetilde{\mathbf{w}} = (\widetilde{w}_1, \ldots, \widetilde{w}_n)$$

is defined by

$$D(\mathbf{w} \parallel \widetilde{\mathbf{w}}) := \sum_{j=1}^{n} w_j \log \left( \frac{w_j}{\widetilde{w}_j} \right).$$

Show that the Kullback-Leibler distance between two mixtures densities

$$\sum_{j=1}^{n} w_j f_j, \qquad \sum_{j=1}^{n} \widetilde{w}_j \widetilde{f}_j$$

has the upper bound

$$D \left( \sum_{j=1}^{n} w_j f_j \parallel \sum_{j=1}^{n} \widetilde{w}_j \widetilde{f}_j \right) \leq D(\mathbf{w} \parallel \widetilde{\mathbf{w}}) + \sum_{j=1}^{n} w_j D(f_j \parallel \widetilde{f}_j)$$

with equality if and only if

$$\frac{w_j f_j}{\sum_{j=1}^{n} w_j f_j} = \frac{\widetilde{w}_j \widetilde{f}_j}{\sum_{j=1}^{n} \widetilde{w}_j \widetilde{f}_j}$$

for all $j$.

**Solution 10.**   Utilizing the *log-sum inequality* we have

$$D \left( \sum_{j=1}^{n} w_j f_j \parallel \sum_{j=1}^{n} \widetilde{w}_j \widetilde{f}_j \right) = \int \left( \sum_{j=1}^{n} w_j f_j \right) \log \left( \frac{\sum_{j=1}^{n} w_j f_j}{\sum_{j=1}^{n} \widetilde{w}_j \widetilde{f}_j} \right)$$

$$\leq \int \sum_{j=1}^{n} w_j f_j \log \left( \frac{w_j f_j}{\widetilde{w}_j \widetilde{f}_j} \right)$$

$$= \sum_{j=1}^{n} w_j \log \left( \frac{w_j}{\widetilde{w}_j} \right) + \sum_{j=1}^{n} w_j \int f_j \log \left( \frac{f_j}{\widetilde{f}_j} \right)$$

$$= D(\mathbf{w} \parallel \widetilde{\mathbf{w}}) + \sum_{j=1}^{n} w_j D(f_j \parallel \widetilde{f}_j).$$

**Problem 11.**   Let $A$, $B$ be $n \times n$ hermitian matrices acting in the Hilbert space $\mathbb{C}^n$. Assume that the eigenvalues of $A$ are pairwise different and analogously for $B$. Then the normalized eigenvectors $|\alpha_j\rangle$ ($j = 1, \ldots, n$) of $A$ form an orthonormal basis in $\mathbb{C}^n$ and analogously for $B$ the normalized eigenvectors $|\beta_j\rangle$ ($j = 1, \ldots, n$) form an orthonormal basis in $\mathbb{C}^n$. Let $|\psi\rangle$ be a normalized state in $\mathbb{C}^n$. Then there are $n$ possible outcomes for measurements of each observable and the probabilities $p_j(A, |\psi\rangle)$, $p_j(B, |\psi\rangle)$ ($j = 1, \ldots, n$) are given by

$$p_j(A, |\psi\rangle) := |\langle \psi | \alpha_j \rangle|^2, \qquad p_j(B, |\psi\rangle) := |\langle \psi | \beta_j \rangle|^2.$$

Let $H_{|\psi\rangle}(X)$ be the *Shannon information entropy*

$$H_{|\psi\rangle}(X) := -\sum_{j=1}^{n} p_j(X, |\psi\rangle) \ln(p_j(X, |\psi\rangle))$$

corresponding to the probability distribution $\{p_j(X, |\psi\rangle)\}$ ($j = 1, \ldots, n$). The (Maassen-Uffink) *entropic uncertainty relation* is given by

$$H_{|\psi\rangle}(A) + H_{|\psi\rangle}(B) \geq -2\ln(\max_{1 \leq j,k \leq n} |\langle \alpha_j | \beta_k \rangle|) > 0.$$

Note that the right-hand side does not involve the state $|\psi\rangle$.
Let

$$A = \sigma_1 = \begin{pmatrix} 0 & 1 \\ 1 & 0 \end{pmatrix}, \quad B = \sigma_3 = \begin{pmatrix} 1 & 0 \\ 0 & -1 \end{pmatrix}, \quad |\psi\rangle = \begin{pmatrix} \cos(\theta) \\ \sin(\theta) \end{pmatrix}.$$

Calculate the left and right-hand side of the entropic uncertainty relation. Is the entropic uncertainty relation tight for this case?

**Solution 11.**    (i) The eigenvalues and eigenvectors of $A = \sigma_1$ are given by

$$+1 \mapsto |\alpha_1\rangle = \frac{1}{\sqrt{2}} \begin{pmatrix} 1 \\ 1 \end{pmatrix}, \qquad -1 \mapsto |\alpha_2\rangle = \frac{1}{\sqrt{2}} \begin{pmatrix} 1 \\ -1 \end{pmatrix}.$$

The eigenvalues and eigenvectors of $B = \sigma_3$ are given by

$$+1 \mapsto |\beta_1\rangle = \begin{pmatrix} 1 \\ 0 \end{pmatrix}, \qquad -1 \mapsto |\beta_2\rangle = \begin{pmatrix} 0 \\ 1 \end{pmatrix}.$$

Thus

$$|\langle \alpha_1 | \beta_1 \rangle| = \frac{1}{\sqrt{2}}, \quad |\langle \alpha_1 | \beta_2 \rangle| = \frac{1}{\sqrt{2}}, \quad |\langle \alpha_2 | \beta_1 \rangle| = \frac{1}{\sqrt{2}}, \quad |\langle \alpha_2 | \beta_2 \rangle| = \frac{1}{\sqrt{2}}$$

and the right-hand side of the inequality becomes

$$-2\ln(\max_{1 \leq j,k \leq n} |\langle \alpha_j | \beta_k \rangle|) = -2\ln\left(\frac{1}{\sqrt{2}}\right) = \ln(2).$$

Now

$$p_1(A, |\psi\rangle) = |\langle\alpha_1|\psi\rangle|^2 = \frac{1}{2}|\cos(\theta) + \sin(\theta)|^2 = \frac{1}{2}(1 + \sin(2\theta))$$

$$p_2(A, |\psi\rangle) = |\langle\alpha_2|\psi\rangle|^2 = \frac{1}{2}|\cos(\theta) - \sin(\theta)|^2 = \frac{1}{2}(1 - \sin(2\theta))$$

and

$$p_1(B, |\psi\rangle) = |\langle\beta_1|\psi\rangle|^2 = \cos^2(\theta)$$

$$p_2(B, |\psi\rangle) = |\langle\beta_2|\psi\rangle|^2 = \sin^2(\theta).$$

Thus

$$H_{|\psi\rangle}(A) = \ln(2) - \frac{1}{2}(1 + \sin(2\theta))\ln(1 + \sin(2\theta)) - \frac{1}{2}(1 - \sin(2\theta))\ln(1 - \sin(2\theta))$$

$$H_{|\psi\rangle}(B) = -\cos^2(\theta)\ln(\cos^2(\theta)) - \sin^2(\theta)\ln(\sin^2(\theta)).$$

Note that with $\theta = 0$ the left-hand side reduces to $\ln(2)$, i.e. we have an equality.

## Programming Problems

**Problem 1.** Consider the density matrix

$$\rho = \begin{pmatrix} 5/12 & 1/6 & 1/6 \\ 1/6 & 1/6 & 1/6 \\ 1/6 & 1/6 & 5/12 \end{pmatrix}.$$

Show that we have a mixed state. Find the von Neumann entropy.

**Solution 1.** We evaluate $\rho^2$ and show that $\rho^2 \neq \rho$. Then we calculate the eigenvalues and the von Neumann entropy.

```
/* mixed.mac */
rho: matrix([5/12,1/6,1/6],[1/6,1/6,1/6],[1/6,1/6,5/12]);
rho2: rho . rho;
if rho=rho2 then print("pure state")
else print("mixed state");
R: eigenvalues(rho);
R: part(R,1);
x1: part(R,1); x2: part(R,2); x3: part(R,3);
if x1>0 then t1: x1;
if x2>0 then t2: x2;
if x3>0 then t3: x3;
t: -t1*log(t1)-t2*log(t2)-t3*log(t3);
t: ratsimp(t);
```

The eigenvalues are

$$\frac{1}{4}, \quad \frac{9 + \sqrt{57}}{24}, \quad \frac{9 - \sqrt{57}}{24}.$$

**Problem 2.**   Consider the two $3 \times 3$ hermitian matrices

$$K_1 = \begin{pmatrix} 0 & 0 & -i \\ 0 & 0 & 0 \\ i & 0 & 0 \end{pmatrix}, \quad K_2 = \begin{pmatrix} 1 & 0 & 0 \\ 0 & 0 & 0 \\ 0 & 0 & -1 \end{pmatrix}$$

and the normalized state in $\mathbb{C}^3$

$$\psi = \frac{1}{\sqrt{3}} \begin{pmatrix} 1 & 1 & 1 \end{pmatrix}^T.$$

Find the left and right-hand side of the entropic inequality.

**Solution 2.**   The eigenvalues of $K_1$ are $-1$, $0$ $+1$ we the corresponding normalized eigenvectors

$$k_{11} = \frac{1}{\sqrt{2}} \begin{pmatrix} 1 \\ 0 \\ -i \end{pmatrix}, \quad k_{12} = \begin{pmatrix} 0 \\ 1 \\ 0 \end{pmatrix}, \quad k_{13} = \frac{1}{\sqrt{2}} \begin{pmatrix} 1 \\ 0 \\ i \end{pmatrix}.$$

The eigenvalues of $K_2$ are $-1$, $0$ $+1$ we the corresponding normalized eigenvectors

$$k_{21} = \begin{pmatrix} 1 \\ 0 \\ 0 \end{pmatrix}, \quad k_{22} = \begin{pmatrix} 0 \\ 1 \\ 0 \end{pmatrix}, \quad k_{23} = \begin{pmatrix} 0 \\ 0 \\ 1 \end{pmatrix}.$$

Now we apply the Maxima program

```
/* entropic.mac */
k11: matrix([1],[0],[-%i])/sqrt(2);
k11T: transpose(k11); k11TC: conjugate(k11T);
k12: matrix([0],[1],[0]);
k12T: transpose(k12); k12TC: conjugate(k12T);
k13: matrix([1],[0],[%i])/sqrt(2);
k13T: transpose(k13); k13TC: conjugate(k13T);
k21: matrix([1],[0],[0]);
k21T: transpose(k21); k21TC: conjugate(k21T);
k22: matrix([0],[1],[0]);
k22T: transpose(k22); k22TC: conjugate(k22T);
k23: matrix([0],[0],[1]);
k23T: transpose(k23); k23TC: conjugate(k23T);
sc1121: abs(k11TC . k21); sc1122: abs(k11TC . k22);
```

```
sc1123: abs(k11TC . k23); sc1221: abs(k12TC . k21);
sc1222: abs(k12TC . k22); sc1223: abs(k12TC . k23);
sc1321: abs(k13TC . k21); sc1322: abs(k13TC . k22);
sc1323: abs(k13TC . k23);
m: max(sc1121,sc1122,sc1123,sc1221,sc1222,
        sc1223,sc1321,sc1322,sc1323);
psi: matrix([1],[1],[1])/sqrt(3);
psiT: transpose(psi); psiTC: conjugate(psiT);
p11: (abs(psiTC . k11))^2;
p12: (abs(psiTC . k12))^2;
p13: (abs(psiTC . k13))^2;
p21: (abs(psiTC . k21))^2;
p22: (abs(psiTC . k22))^2;
p23: (abs(psiTC . k23))^2;
LHS: -p11*log(p11)-p12*log(p12)-p13*log(p13)-p21*log(p21)
     -p22*log(p22)-p23*log(p23);
RHS: -2*log(m);
```

The output 0 for the right-hand side which is obvious since $K_1$ and $K_2$ have a common eigenvector, namely (0 1 0). For the left-hand side we find $2\log(3)$.

## 8.3   Supplementary Problems

**Problem 1.**   Consider the Hilbert space $\mathbb{C}^n$. Let $A$, $B$ be two hermitian $n \times n$ matrices (observable). Assume that $A$ and $B$ have non-degenerate eigenvalues with the corresponding normalized eigenvectors $|a_1\rangle$, $|a_2\rangle$, ..., $|a_n\rangle$ and $|b_1\rangle$, $|b_2\rangle$, ..., $|b_n\rangle$, respectively. The entropic uncertainty relation is an inequality given by

$$S_{(A)} + S_{(B)} \geq S_{(AB)}$$

where

$$S_{(A)} = -\sum_{j=1}^{n} |\langle\psi|a_j\rangle|^2 \ln(|\langle\psi|a_j\rangle|^2), \quad S_{(B)} = -\sum_{j=1}^{n} |\langle\psi|b_j\rangle|^2 \ln(|\langle\psi|b_j\rangle|^2),$$

and $S_{(AB)}$ is a positive constant which gives the lower bound of the right-hand side of the inequality. Consider the Hilbert space $\mathbb{C}^2$. Let

$$A = \sigma_1, \quad B = \sigma_2, \quad |\psi\rangle = \begin{pmatrix} \cos(\theta) \\ \sin(\theta) \end{pmatrix}.$$

Find $S_{(A)}$, $S_{(B)}$ and $S_{(A)} + S_{(B)}$.

**Problem 2.**    Consider the Hilbert space $\mathbb{C}^n$ and $|\psi\rangle \in \mathbb{C}^n$. Let $A$ and $B$ $n \times n$ hermitian matrices (observable) with non-degenerate eigenvalues and corresponding normalized eigenvectors $|u_j\rangle$, $|v_j\rangle$ $(j = 1, \ldots, n)$. The entropic uncertainty relation is an inequality of the form

$$S_{(A)} + S_{(B)} \geq S_{AB}$$

where

$$S_{(A)} = -\sum_{j=1}^{n} |\langle\psi|u_j\rangle|^2 \ln(|\langle\psi|u_j\rangle|^2), \quad S_{(A)} = -\sum_{j=1}^{n} |\langle\psi|v_j\rangle|^2 \ln(|\langle\psi|v_j\rangle|^2)$$

and $S_{AB}$ is a positive constant providing the lower bound of the right-hand side of the inequality. Let

$$A = \sigma_1 = \begin{pmatrix} 0 & 1 \\ 1 & 0 \end{pmatrix}, \quad B = \sigma_3 = \begin{pmatrix} 1 & 0 \\ 0 & -1 \end{pmatrix}$$

and

$$|\psi\rangle = \begin{pmatrix} \cos(\theta) \\ \sin(\theta) \end{pmatrix}.$$

Calculate $S_{(A)}$ and $S_{(B)}$.

**Problem 3.**    Consider the $4 \times 4$ spin matrix

$$S_2 = \frac{1}{2} \begin{pmatrix} 0 & -i\sqrt{3} & 0 & 0 \\ i\sqrt{3} & 0 & -2i & 0 \\ 0 & 2i & 0 & -i\sqrt{3} \\ 0 & 0 & i\sqrt{3} & 0 \end{pmatrix}$$

the normalized entangled vector

$$|\psi\rangle = \frac{1}{2} \begin{pmatrix} 1 \\ 1 \\ 1 \\ -1 \end{pmatrix}.$$

Find the left and right-hand side of the entropic inequality.

# Chapter 9

# Measurement

## 9.1  Introduction

In quantum measurement models we consider what kind of measurements can be made on quantum systems as well as how to determine the probability that a measurement yields a given result. The effect that measurement has on the state of a quantum system is also important.

The pure states of a quantum system, $S$, are described by normalized vectors $|\psi\rangle$ which are elements of a Hilbert space $\mathcal{H}$ that describes $S$. The pure states of a quantum mechanical system are rays in a Hilbert space $\mathcal{H}$ (i.e., unit vectors with an arbitrary phase). The concept of a state as a ray in a Hilbert space leads to the probability interpretation in quantum mechanics. Given a physical system in the state $\psi$, the probability that it is in the state $|\chi\rangle$ is

$$|\langle\psi|\chi\rangle|^2.$$

Obviously

$$0 \leq |\langle\psi|\chi\rangle|^2 \leq 1.$$

While the phase of a vector $|\psi\rangle$ has no physical significance, the relative phase of two vectors does. Consider the Schrödinger equation with time independent $\hat{H}$

$$i\hbar\frac{d}{dt}|\psi(t)\rangle = \hat{H}|\psi(t)\rangle$$

and the initial state $|\psi(t = 0)\rangle$. Then

$$|\psi(t)\rangle = \exp(-i\hat{H}t/\hbar)|\psi(t = 0)\rangle$$

and

$$|\langle\psi(0)|e^{-i\hat{H}t/\hbar}|\psi(0)\rangle|^2$$

is the probability to find the state $|\psi(t)\rangle$ in the initial state $|\psi(0)\rangle$.

A *positive operator-valued measure (POVM)* is a collection

$$\{\, E_j \ : \ j = 1, 2, \ldots, n \,\}$$

of nonnegative (positive semi-definite) operators, satisfying

$$\sum_{j=1}^{n} E_j = I$$

where $I$ is the identity operator. In other words a partition of unity (identity operator) by nonnegative operators is called a positive operator-valued measure (POVM). When a state $|\psi\rangle$ is subjected to such a POVM, outcome $j$ occurs with probability

$$p(j) = \langle\psi|E_j|\psi\rangle.$$

For example consider a qubit system

$$E_1 = |0\rangle\langle0|, \quad E_2 = |1\rangle\langle1|, \quad |\psi\rangle = \frac{1}{\sqrt{2}}(|0\rangle + |1\rangle).$$

Since $\langle0|0\rangle = \langle1|1\rangle = 1$ and $\langle0|1\rangle = \langle1|0\rangle = 0$ we find

$$p(1) = \langle\psi|E_1|\psi\rangle = \frac{1}{2}, \qquad p(2) = \langle\psi|E_2|\psi\rangle = \frac{1}{2}.$$

Measurement can be generalized in the sense that an ancilla system (in a well defined state), identified by the Hilbert space $\mathcal{H}_A$, is introduced and allowed to interact with the quantum system identified by the Hilbert space $\mathcal{H}$. The ancilla system is subsequently measured, which may disturb the original system.

## 9.2  Solved Problems

**Problem 1.**  Consider the state

$$|\psi\rangle = \frac{1}{\sqrt{3}}|00\rangle + \sqrt{\frac{2}{3}}|11\rangle \equiv \frac{1}{\sqrt{3}}|0\rangle \otimes |0\rangle + \sqrt{\frac{2}{3}}|1\rangle \otimes |1\rangle$$

and the product state $|\phi\rangle = |11\rangle \equiv |1\rangle \otimes |1\rangle$. Find $p := |\langle\phi|\psi\rangle|^2$, i.e. the probability of finding $|\psi\rangle$ in the state $|\phi\rangle$.

**Solution 1.**  Since $\langle 11|00\rangle = 0$ and $\langle 11|11\rangle = 1$ we obtain $p = 2/3$.

**Problem 2.**  Consider the hermitian matrix (Hamilton operator)

$$\hat{H} = \hbar\omega S_1$$

where $S_1$ is the $3 \times 3$ spin-1 matrix

$$S_1 = \frac{1}{\sqrt{2}}\begin{pmatrix} 0 & 1 & 0 \\ 1 & 0 & 1 \\ 0 & 1 & 0 \end{pmatrix}.$$

(i) Calculate $\exp(-i\hat{H}t/\hbar)$.
(ii) Consider the normalized vector in $\mathbb{C}^3$

$$|\psi(0)\rangle = \frac{1}{\sqrt{3}}\begin{pmatrix} 1 \\ 1 \\ 1 \end{pmatrix}.$$

Calculate $|\psi(t)\rangle = \exp(-i\hat{H}t/\hbar)|\psi(0)\rangle$.
(iii) Find the probability of finding $\psi(t)$ in the initial state $\psi(0)$, i.e.

$$|\langle\psi(t)|\psi(0)\rangle|^2.$$

**Solution 2.**  (i) We have

$$S_1^2 = \frac{1}{2}\begin{pmatrix} 1 & 0 & 1 \\ 0 & 2 & 0 \\ 1 & 0 & 1 \end{pmatrix}, \qquad S_1^3 = \frac{1}{\sqrt{2}}\begin{pmatrix} 0 & 1 & 0 \\ 1 & 0 & 1 \\ 0 & 1 & 0 \end{pmatrix} = S_1.$$

Using this result we find

$$\exp(-i\hat{H}t/\hbar) = I_3 - \frac{i}{\sqrt{2}}\sin(\omega t)A + \frac{1}{2}(\cos(\omega t) - 1)A^2$$

$$= \begin{pmatrix} \frac{1}{2}\cos(\omega t) + \frac{1}{2} & -\frac{i}{\sqrt{2}}\sin(\omega t) & \frac{1}{2}\cos(\omega t) - \frac{1}{2} \\ -\frac{i}{\sqrt{2}}\sin(\omega t) & \cos(\omega t) & -\frac{i}{\sqrt{2}}\sin(\omega t) \\ \frac{1}{2}\cos(\omega t) - \frac{1}{2} & -\frac{i}{\sqrt{2}}\sin(\omega t) & \frac{1}{2}\cos(\omega t) + \frac{1}{2} \end{pmatrix}.$$

(ii) We obtain

$$|\psi(t)\rangle = \exp(-i\hat{H}t/\hbar)|\psi(0)\rangle = \frac{1}{\sqrt{3}}\begin{pmatrix} \cos(\omega t) - \frac{i}{\sqrt{2}}\sin(\omega t) \\ \cos(\omega t) - i\sqrt{2}\sin(\omega t) \\ \cos(\omega t) - \frac{i}{\sqrt{2}}\sin(\omega t) \end{pmatrix}.$$

(iii) We find the probability

$$|\langle\psi(t)|\psi(0)\rangle|^2 = \frac{1}{9}|3\cos(\omega t) - i2\sqrt{2}\sin(\omega t)|^2 = 1 - \frac{1}{9}\sin^2(\omega t).$$

**Problem 3.**   Consider the states (standard basis)

$$|0\rangle := \begin{pmatrix} 1 \\ 0 \end{pmatrix}, \qquad |1\rangle := \begin{pmatrix} 0 \\ 1 \end{pmatrix}$$

in the Hilbert space $\mathbb{C}^2$ and the Bell state

$$|\psi\rangle = \frac{1}{\sqrt{2}}(|0\rangle \otimes |1\rangle - |1\rangle \otimes |0\rangle)$$

in the Hilbert space $\mathbb{C}^4$. Let $(\alpha, \beta \in \mathbb{R})$

$$|\alpha\rangle := \cos(\alpha)|0\rangle + \sin(\alpha)|1\rangle, \qquad |\beta\rangle := \cos(\beta)|0\rangle + \sin(\beta)|1\rangle$$

be states in $\mathbb{C}^2$. Find the probability

$$p(\alpha, \beta) := |(\langle\alpha| \otimes \langle\beta|)|\psi\rangle|^2.$$

Discuss $p$ as a function of $\alpha$ and $\beta$.

**Solution 3.**   Since $\langle 0|0\rangle = \langle 1|1\rangle = 1$, $\langle 0|1\rangle = \langle 1|0\rangle = 0$ it follows that

$$(\langle 0| \otimes \langle 1|)(|0\rangle \otimes |1\rangle)) = 1, \quad ((\langle 1| \otimes \langle 0|)(|1\rangle \otimes |0\rangle)) = 1.$$

We find
$$p(\alpha, \beta) = \frac{1}{2}(\cos(\alpha)\sin(\beta) - \sin(\alpha)\cos(\beta))^2.$$

Using a trigonometric identity we arrive at

$$p(\alpha, \beta) = \frac{1}{2}\sin^2(\alpha - \beta).$$

Thus $p(\alpha, \beta) \leq 1/2$ for all $\alpha$, $\beta$ since $\sin^2(\phi) \leq 1$ for all $\phi \in \mathbb{R}$. For example, if $\alpha = \beta$ we have $p = 0$. If $\alpha - \beta = \pi/2$ we have $p = 1/2$.

**Problem 4.** Consider the normalized entangled state

$$|\psi\rangle = \frac{1}{\sqrt{2}}(|01\rangle - |10\rangle) \equiv \frac{1}{\sqrt{2}}(|0\rangle \otimes |1\rangle - |1\rangle \otimes |0\rangle)$$

and $\langle 0| \otimes I_2$, where $I_2$ is the $2 \times 2$ unit matrix. Find $(\langle 0| \otimes I_2)|\psi\rangle$. Discuss.

**Solution 4.** Since $\langle 0|0\rangle = 1$, $\langle 0|1\rangle = 0$ and $I_2|1\rangle = |1\rangle$, we obtain

$$(\langle 0| \otimes I_2)|\psi\rangle = \frac{1}{\sqrt{2}}|1\rangle.$$

The first system is measured with probability $1/2$ and the system collapses to the state $|1\rangle$ (*partial measurement*).

**Problem 5.** Let $\{ |0\rangle, |1\rangle \}$ denote an orthonormal basis in $\mathbb{C}^2$. In other words

$$\langle 0|1\rangle = \langle 1|0\rangle = 0, \qquad \langle 0|0\rangle = \langle 1|1\rangle = 1.$$

(i) Show that for $\alpha, \beta \in \mathbb{R}$,

$$A := \alpha|0\rangle\langle 0| + \beta|1\rangle\langle 1|$$

is an observable. Describe the measurement outcomes and associated probabilities when measuring the first qubit of the two qubit system described by

$$|\psi\rangle = \frac{1}{\sqrt{2}}(|0\rangle \otimes |0\rangle + |1\rangle \otimes |1\rangle)$$

where the measurement is described by $A$.
(ii) Let $|\phi\rangle$ be the state of the system after the measurement in (i). Describe the measurement outcomes and associated probabilities when measuring the second qubit of $A$.

**Solution 5.** (i) To show that $A$ is an observable it suffices to show that $A^* = A$. Using

$$(|0\rangle\langle 0|)^* = \langle 0|^*|0\rangle^* = |0\rangle\langle 0|, \quad (|0\rangle\langle 1|)^* = \langle 1|^*|0\rangle^* = |1\rangle\langle 0|,$$

$$(|1\rangle\langle 0|)^* = \langle 0|^*|1\rangle^* = |0\rangle\langle 1|, \quad (|1\rangle\langle 1|)^* = \langle 1|^*|1\rangle^* = |1\rangle\langle 1|$$

we find that

$$A^* = (\alpha|0\rangle\langle 0| + \beta|1\rangle\langle 1|)^* = (\alpha|0\rangle\langle 0|)^* + (\beta|1\rangle\langle 1|)^* = \overline{\alpha}|0\rangle\langle 0| + \overline{\beta}|1\rangle\langle 1|$$

$$= \alpha|0\rangle\langle 0| + \beta|1\rangle\langle 1| = A$$

since $\overline{\alpha} = \alpha$ for $\alpha \in \mathbb{R}$. Since we work in $\mathbb{C}^2$, $A$ has two eigenvalues. From

$$A|0\rangle = (\alpha|0\rangle\langle 0| + \beta|1\rangle\langle 1|)|0\rangle = \alpha|0\rangle\langle 0|0\rangle + \beta|1\rangle\langle 1|0\rangle = \alpha|0\rangle$$

and

$$A|1\rangle = (\alpha|0\rangle\langle0| + \beta|1\rangle\langle1|)|1\rangle = \alpha|0\rangle\langle0|1\rangle + \beta|1\rangle\langle1|1\rangle = \beta|1\rangle$$

the two eigenvalues (i.e. measurement outcomes) are $\alpha$ and $\beta$ with corresponding orthonormal eigenstates $|0\rangle$ and $|1\rangle$. Thus for measuring the first qubit we consider the observable $A \otimes I_2$ with eigenvalues $\alpha$ (eigenstates $|0\rangle \otimes |0\rangle$ and $|0\rangle \otimes |1\rangle$) and $\beta$ (eigenstates $|1\rangle \otimes |0\rangle$ and $|1\rangle \otimes |1\rangle$). This is not the only choice for the corresponding eigenstates, but is a convenient one. It will be useful to calculate some scalar products in advance

$$(|0\rangle \otimes |0\rangle)^*|\psi\rangle = (\langle0| \otimes \langle0|)\frac{1}{\sqrt{2}}(|0\rangle \otimes |1\rangle - |1\rangle \otimes |0\rangle)$$

$$= \frac{1}{\sqrt{2}}(\langle0|0\rangle \otimes \langle0|1\rangle - \langle0|1\rangle \otimes \langle0|0\rangle) = 0,$$

$$(|0\rangle \otimes |1\rangle)^*|\psi\rangle = \frac{1}{\sqrt{2}}, \quad (|1\rangle \otimes |0\rangle)^*|\psi\rangle = -\frac{1}{\sqrt{2}}, \quad (|1\rangle \otimes |1\rangle)^*|\psi\rangle = 0.$$

We need to consider two possibilities, namely $\alpha = \beta$ and $\alpha \neq \beta$.
For $\alpha = \beta$ there is only one measurement outcome: $\alpha$. The corresponding projection operator onto the eigenspace is determined from the eigenstates

$$\Pi_\alpha := (|0\rangle \otimes |0\rangle)(|0\rangle \otimes |0\rangle)^* + (|0\rangle \otimes |1\rangle)(|0\rangle \otimes |1\rangle)^*$$
$$+(|1\rangle \otimes |0\rangle)(|1\rangle \otimes |0\rangle)^* + (|1\rangle \otimes |1\rangle)(|1\rangle \otimes |1\rangle)^*$$
$$= |0\rangle\langle0| \otimes |0\rangle\langle0| + |0\rangle\langle0| \otimes |1\rangle\langle1| + |1\rangle\langle1| \otimes |0\rangle\langle0| + |1\rangle\langle1| \otimes |1\rangle\langle1|$$
$$= I_2 \otimes I_2 = I_4$$

i.e. the identity operator. The probability of obtaining the measurement outcome $\alpha$ is

$$p_\alpha = \|\Pi_\alpha|\psi\rangle\|^2 = \||\psi\rangle\|^2 = \langle\psi|\psi\rangle = \frac{1}{\sqrt{2}}(\langle0| \otimes \langle0| - \langle1| \otimes \langle1|)|\psi\rangle$$

$$= \frac{1}{2} - \left(-\frac{1}{2}\right) = 1.$$

The state after measurement is the projected state $\Pi_\alpha|\psi\rangle = I_4|\psi\rangle = |\psi\rangle$ which when normalized yields $|\phi\rangle = |\psi\rangle$ since $\langle\psi|\psi\rangle = 1$.
$\alpha \neq \beta$: Measurement outcome $\alpha$: The corresponding projection operator onto the eigenspace is determined from the eigenstates

$$\Pi_\alpha := (|0\rangle \otimes |0\rangle)(|0\rangle \otimes |0\rangle)^* + (|0\rangle \otimes |1\rangle)(|0\rangle \otimes |1\rangle)^*$$
$$= |0\rangle\langle0| \otimes |0\rangle\langle0| + |0\rangle\langle0| \otimes |1\rangle\langle1|$$
$$= |0\rangle\langle0| \otimes I_2.$$

Note that

$$\Pi_\alpha|\psi\rangle = \frac{1}{\sqrt{2}}(|0\rangle\langle 0| \otimes I_2)(|0\rangle \otimes |1\rangle - |1\rangle \otimes |0\rangle)$$

$$= \frac{1}{\sqrt{2}}(|0\rangle\langle 0|0\rangle) \otimes |1\rangle - (|0\rangle\langle 0|1\rangle) \otimes |0\rangle = \frac{1}{\sqrt{2}}|0\rangle \otimes |1\rangle.$$

The probability of obtaining the measurement outcome $\alpha$ is

$$p_\alpha = \|\Pi_\alpha|\psi\rangle\|^2 = (\Pi_\alpha|\psi\rangle)^*\Pi_\alpha|\psi\rangle = \frac{1}{\sqrt{2}}(\langle 0| \otimes \langle 1|)\frac{1}{\sqrt{2}}(|0\rangle \otimes |1\rangle) = \frac{1}{2}.$$

The state after measurement is the projected state $\Pi_\alpha|\psi\rangle = \frac{1}{\sqrt{2}}|0\rangle \otimes |1\rangle$ which when normalized yields

$$|\phi\rangle = \frac{\Pi_\alpha|\psi\rangle}{\|\Pi_\alpha|\psi\rangle\|} = \frac{\Pi_\alpha|\psi\rangle}{\sqrt{p_\alpha}} = |0\rangle \otimes |1\rangle.$$

Measurement outcome $\beta$: The corresponding projection onto the eigenspace is determined from the eigenstates

$$\Pi_\beta := (|1\rangle \otimes |0\rangle)(|1\rangle \otimes |0\rangle)^* + (|1\rangle \otimes |1\rangle)(|1\rangle \otimes |1\rangle)^*$$

$$= |1\rangle\langle 1| \otimes I_2.$$

Note that $\Pi_\beta|\psi\rangle = -\frac{1}{\sqrt{2}}|1\rangle \otimes |0\rangle$. The probability of obtaining the measurement outcome $\beta$ is

$$p_\beta = \|\Pi_\beta|\psi\rangle\|^2 = \frac{1}{2}.$$

The state after measurement is the projected state

$$\Pi_\beta|\psi\rangle = \frac{1}{\sqrt{2}}|1\rangle \otimes |0\rangle$$

which when normalized yields

$$|\phi\rangle = \frac{\Pi_\beta|\psi\rangle}{\|\Pi_\beta|\psi\rangle\|} = \frac{\Pi_\alpha|\psi\rangle}{\sqrt{p_\beta}} = -|1\rangle \otimes |0\rangle.$$

(ii) For measuring the second qubit we consider the observable $I_2 \otimes A$ with eigenvalues $\alpha$ (eigenstates $|0\rangle \otimes |0\rangle$ and $|1\rangle \otimes |0\rangle$) and $\beta$ (eigenstates $|0\rangle \otimes |1\rangle$ and $|1\rangle \otimes |1\rangle$). This is not the only choice for the corresponding eigenstates, but is a convenient one. The measurement of the second qubit depends on the results of the first measurement. Thus we need to consider three cases $\alpha = \beta$, and the two outcomes $\alpha$ and $\beta$ when $\alpha \neq \beta$.
$\alpha = \beta$: We have

$$|\phi\rangle = |\psi\rangle = \frac{1}{\sqrt{2}}(|0\rangle \otimes |1\rangle - |1\rangle \otimes |0\rangle).$$

There is only one measurement outcome: $\alpha$. The corresponding projection onto the eigenspace is determined from the eigenstates

$$
\begin{aligned}
\Pi_\alpha &:= (|0\rangle \otimes |0\rangle)(|0\rangle \otimes |0\rangle)^* + (|1\rangle \otimes |0\rangle)(|1\rangle \otimes |0\rangle)^* \\
&\quad + (|0\rangle \otimes |1\rangle)(|0\rangle \otimes |1\rangle)^* + (|1\rangle \otimes |1\rangle)(|1\rangle \otimes |1\rangle)^* \\
&= |0\rangle\langle 0| \otimes |0\rangle\langle 0| + |1\rangle\langle 1| \otimes |0\rangle\langle 0| + |0\rangle\langle 0| \otimes |1\rangle\langle 1| + |1\rangle\langle 1| \otimes |1\rangle\langle 1| \\
&= I_2 \otimes I_2 = I_4
\end{aligned}
$$

i.e. the identity operator. As above, the probability of the measurement outcome $\alpha$ is $p_{\alpha,\alpha} = 1$.

$\alpha \neq \beta$: First measurement outcome was $\alpha$: We have $|\phi\rangle = |0\rangle \otimes |1\rangle$.
Measurement outcome $\alpha$: The corresponding projection onto the eigenspace is determined from the eigenstates

$$
\Pi_{\alpha,\alpha} := (|0\rangle \otimes |0\rangle)(|0\rangle \otimes |0\rangle)^* + (|1\rangle \otimes |0\rangle)(|1\rangle \otimes |0\rangle)^* = I_2 \otimes |0\rangle\langle 0|.
$$

Note that $\Pi_\alpha|\phi\rangle = 0$. The probability of obtaining the measurement outcome $\alpha$ is $p_{\alpha,\alpha} = 0$.
Measurement outcome $\beta$: The corresponding projection onto the eigenspace is determined from the eigenstates

$$
\Pi_{\alpha,\beta} = I_2 \otimes |1\rangle\langle 1|.
$$

Note that $\Pi_{\alpha,\beta}|\phi\rangle = |0\rangle \otimes |1\rangle = |\phi\rangle$. The probability of obtaining the measurement outcome $\beta$ is $p_{\alpha,\beta} = 1$.
First measurement outcome was $\beta$: We have

$$
|\phi\rangle = -|1\rangle \otimes |0\rangle.
$$

Measurement outcome $\alpha$: The corresponding projection onto the eigenspace is determined from the eigenstates

$$
\Pi_{\beta,\alpha} = I_2 \otimes |0\rangle\langle 0|.
$$

Note that $\Pi_\alpha|\phi\rangle = -|1\rangle \otimes |0\rangle = |\phi\rangle$. The probability of obtaining the measurement outcome $\alpha$ is $p_{\beta,\alpha} = 1$.
Measurement outcome $\beta$: The corresponding projection operator onto the eigenspace is determined from the eigenstates

$$
\Pi_{\beta,\beta} = I_2 \otimes |1\rangle\langle 1|.
$$

Note that $\Pi_{\beta,\beta}|\phi\rangle = 0$. The probability of obtaining the measurement outcome $\beta$ is $p_{\beta,\beta} = 0$.

Tabulating the probabilities for $\alpha \neq \beta$ we find

| Outcomes | $\alpha, \alpha$ | $\alpha, \beta$ | $\beta, \alpha$ | $\beta, \beta$ |
|---|---|---|---|---|
| **Probability** | $p_\alpha p_{\alpha,\alpha} = 0$ | $p_\alpha p_{\alpha,\beta} = \frac{1}{2}$ | $p_\beta p_{\beta,\alpha} = \frac{1}{2}$ | $p_\beta p_{\beta,\beta} = 0$ |

Consequently, for $\alpha \neq \beta$, the probability that the two measurement outcomes are the same is 0 (impossible) and the probability that the two measurement outcomes are different is 1 (certain).

**Problem 6.** Assume that Alice operates a device that prepares a quantum system and Bob does subsequent measurement on the system and records the results. The preparation device indicates the state the system is prepared in. A preparation readout event $j$, where $j = 1, 2, \dots, m$ of the preparation device is associated with a linear non-negative definite operator $\Lambda_j$ acting on the state space of the system. The operators $\Lambda_j$ need not be orthogonal to each other. The measurement device has a readout event $k$, where $k = 1, 2, \dots, n$ that shows the result of the measurement. A measurement device is associated with a measurement device operator $\Gamma_k$ which is also linear and non-negative definite. For a *von Neumann measurement* this operator would be a pure state projector. Let

$$\Lambda := \sum_{j=1}^{m} \Lambda_j, \qquad \Gamma := \sum_{k=1}^{n} \Gamma_k.$$

Give an interpretation of the following probabilities

$$p(j, k) = \frac{\operatorname{tr}(\Lambda_j \Gamma_k)}{\operatorname{tr}(\Lambda \Gamma)} \tag{1}$$

$$p(j) = \frac{\operatorname{tr}(\Lambda_j \Gamma)}{\operatorname{tr}(\Lambda \Gamma)} \tag{2}$$

$$p(k) = \frac{\operatorname{tr}(\Lambda \Gamma_k)}{\operatorname{tr}(\Lambda \Gamma)} \tag{3}$$

$$p(k|j) = \frac{\operatorname{tr}(\Lambda_j \Gamma_k)}{\operatorname{tr}(\Lambda_j \Gamma)} \tag{4}$$

$$p(j|k) = \frac{\operatorname{tr}(\Lambda_j \Gamma_k)}{\operatorname{tr}(\Lambda \Gamma_k)}. \tag{5}$$

**Solution 6.** Expression (1) is the probability associated with a particular point $(j, k)$ in the sample space. Expression (2) is the probability that, if an experiment chosen at random has a recorded combined event, this event includes preparation event $j$. Expression (3) is the probability that the recorded combined event includes the measurement event $k$. Expression (4) is the probability that, if the recorded combined event includes event $j$, it also includes event $k$. Thus it is the probability that the event recorded by Bob is the detection of the state corresponding to $\Gamma_k$ if the state prepared by Alice in the experiment corresponds to $\Lambda_j$. This expression can be used for prediction. To calculate the required probability from the operator $\Lambda_j$ associated with the preparation event $j$, every possible operator $\Gamma_k$ must be known, that is, the mathematical description of the operation of the measuring device must be known. Analogously, (5) is the probability that the state prepared by Alice corresponds to $\Lambda_j$ if the event recorded by Bob is the detection of the state corresponding to $\Gamma_k$. This expression can be used for retrodiction if $\Gamma_k$ and all the $\Lambda_j$ operators of the preparation device are known.

**Problem 7.**    Let $A$ be an $n \times n$ hermitian matrix. Then the eigenvalues $\lambda_j$, $j = 1, 2, \ldots, n$ are real. Assume that all eigenvalues are distinct. The matrix $A$ can be written as (*spectral representation*)

$$A := \sum_{j=1}^{n} \lambda_j P_j, \qquad P_j := |u_j\rangle\langle u_j| \tag{1}$$

where $|u_j\rangle$ are the normalized eigenvectors of $A$ with eigenvalue $\lambda_j$. For the projectors $P_j$ we have $P_j P_k = \delta_{jk} P_j$. Every observable $A$ defines a *projective measurement*. A state $|\psi\rangle$ in $\mathbb{C}^n$ subject to projective measurement by observable (1) goes into the state

$$\frac{P_j|\psi\rangle}{\sqrt{\langle\psi|P_j|\psi\rangle}}$$

with *probability*

$$p(j) = \langle\psi|P_j|\psi\rangle \equiv \langle\psi|u_j\rangle\langle u_j|\psi\rangle = |\langle\psi|u_j\rangle|^2.$$

The eigenvalues $\lambda_j$ are registered as the measured value. If the system is subjected to the same measurement immediately after a projective measurement, the same outcome occurs with certainty. The *expectation* of the measured value is

$$\langle A\rangle = \sum_{j=1}^{n} \lambda_j p(j) = \langle\psi|A|\psi\rangle.$$

(i) Let

$$A = \sigma_2 = \begin{pmatrix} 0 & -i \\ i & 0 \end{pmatrix}.$$

Find the spectral representation of $A$.

(ii) Let

$$|\psi\rangle = \frac{1}{\sqrt{2}} \begin{pmatrix} 1 \\ 1 \end{pmatrix}.$$

Calculate the probabilities corresponding to the eigenvalues $\lambda_1$ and $\lambda_2$ of $A$

$$p(\lambda_1) = \langle \psi | P_{\lambda_1} | \psi \rangle, \qquad p(\lambda_2) = \langle \psi | P_{\lambda_2} | \psi \rangle.$$

**Solution 7.** (i) The eigenvalues of $A$ are $\lambda_1 = 1$ and $\lambda_2 = -1$. The corresponding eigenvectors are

$$|u_1\rangle = \frac{1}{\sqrt{2}} \begin{pmatrix} 1 \\ i \end{pmatrix}, \qquad |u_2\rangle = \frac{1}{\sqrt{2}} \begin{pmatrix} 1 \\ -i \end{pmatrix}.$$

Thus

$$P_{\lambda_1} = |u_1\rangle\langle u_1| = \frac{1}{2} \begin{pmatrix} 1 \\ i \end{pmatrix} (1 \;\; -i) = \frac{1}{2} \begin{pmatrix} 1 & -i \\ i & 1 \end{pmatrix}$$

$$P_{\lambda_2} = |u_2\rangle\langle u_2| = \frac{1}{2} \begin{pmatrix} 1 \\ -i \end{pmatrix} (1 \;\; i) = \frac{1}{2} \begin{pmatrix} 1 & i \\ -i & 1 \end{pmatrix}$$

with $I_2 = P_{\lambda_1} + P_{\lambda_2}$ and $A = P_{\lambda_1} - P_{\lambda_2}$.
(ii) We have

$$P_{\lambda_1}|\psi\rangle = \frac{1}{2} \begin{pmatrix} 1 & -i \\ i & 1 \end{pmatrix} \frac{1}{\sqrt{2}} \begin{pmatrix} 1 \\ 1 \end{pmatrix} = \frac{1}{2\sqrt{2}} \begin{pmatrix} 1-i \\ 1+i \end{pmatrix}$$

$$P_{\lambda_2}|\psi\rangle = \frac{1}{2} \begin{pmatrix} 1 & i \\ -i & 1 \end{pmatrix} \frac{1}{\sqrt{2}} \begin{pmatrix} 1 \\ 1 \end{pmatrix} = \frac{1}{2\sqrt{2}} \begin{pmatrix} 1+i \\ 1-i \end{pmatrix}.$$

Thus

$$p(\lambda_1) = \langle \psi | P_{\lambda_1} | \psi \rangle = \frac{1}{2}, \qquad p(\lambda_2) = \langle \psi | P_{\lambda_2} | \psi \rangle = \frac{1}{2}.$$

**Problem 8.** Let $(\theta \in \mathbb{R})$

$$P(\theta) := e^{i\theta}|0\rangle\langle 0| + e^{-i\theta}|1\rangle\langle 1| \equiv e^{i\theta}(|0\rangle\langle 0| + e^{-i2\theta}|1\rangle\langle 1|)$$

denote the *phase change transform* on a single qubit.
(i) Calculate $(\phi \in \mathbb{R})$

$$|s(\theta, \phi)\rangle := P\left(\frac{\pi}{4} - \frac{\phi}{2}\right) U_H P\left(\frac{\theta}{2}\right) U_H |0\rangle.$$

(ii) Determine the probability that the state $|s(\theta, \phi)\rangle$ is in the state

(a) $|0\rangle$,      (b) $|1\rangle$,      (c) $|s(\theta', \phi')\rangle$.

The real parameters $\theta$ and $\phi$ can be interpreted as spherical co-ordinates which define any qubit on the unit sphere called the *Bloch sphere*.

**Solution 8.**    (i) We have

$$
\begin{aligned}
|s(\theta,\phi)\rangle &= P\left(\frac{\pi}{4} - \frac{\phi}{2}\right) U_H P\left(\frac{\theta}{2}\right) U_H |0\rangle \\
&= P\left(\frac{\pi}{4} - \frac{\phi}{2}\right) U_H P\left(\frac{\theta}{2}\right) \frac{1}{\sqrt{2}}(|0\rangle + |1\rangle) \\
&= P\left(\frac{\pi}{4} - \frac{\phi}{2}\right) U_H \frac{1}{\sqrt{2}}(e^{i\theta/2}|0\rangle + e^{-i\theta/2}|1\rangle) \\
&= P\left(\frac{\pi}{4} - \frac{\phi}{2}\right) \frac{1}{2}\left(e^{i\theta/2}(|0\rangle + |1\rangle) + e^{-i\theta/2}(|0\rangle - |1\rangle)\right) \\
&= P\left(\frac{\pi}{4} - \frac{\phi}{2}\right) (\cos(\theta/2)|0\rangle + i\sin(\theta/2)|1\rangle) \\
&= e^{i\frac{\pi}{4}} \left(e^{-i\frac{\phi}{2}}\cos(\theta/2)|0\rangle + e^{i\frac{\phi}{2}}\sin(\theta/2)|1\rangle\right) \\
&= e^{i(\frac{\pi}{4} - \frac{\phi}{2})} \left(\cos(\theta/2)|0\rangle + e^{i\phi}\sin(\theta/2)|1\rangle\right).
\end{aligned}
$$

The most general state of a single qubit is described by three real parameters $\theta, \phi, \sigma \in \mathbb{R}$

$$
e^{i\sigma} \left(\cos(\theta/2)|0\rangle + e^{i\phi}\sin(\theta/2)|1\rangle\right).
$$

The parameter $\sigma$ represents the *global phase*, and can be ignored since it cannot be detected in the measurement model. The same applies to the global phase $\exp(i(\pi/4 - \phi/2))$ in the derivation. Thus $\theta$ and $\phi$ can be used to define any single qubit $|s(\theta,\phi)\rangle$.
(ii) For the probabilities (*a*) we have

$$
|\langle 0|s(\theta,\phi)\rangle|^2 = \cos^2(\theta/2).
$$

For the probability (b) we have

$$
|\langle 1|s(\theta,\phi)\rangle|^2 = \sin^2(\theta/2).
$$

For the probability $|\langle s(\theta',\phi')|s(\theta,\phi)\rangle|^2$ we find

$$
|\langle s(\theta',\phi')|s(\theta,\phi)\rangle|^2 = \left| \cos(\theta/2)\cos(\theta'/2)e^{\frac{i}{2}(\phi'-\phi)} + \sin(\theta/2)\sin(\theta'/2)e^{\frac{i}{2}(\phi-\phi')} \right|^2
$$

where we used $\langle 0|0\rangle = \langle 1|1\rangle = 1$ and $\langle 0|1\rangle = \langle 1|0\rangle = 0$. It follows that

$$
|\langle s(\theta',\phi')|s(\theta,\phi)\rangle|^2 =
$$

$$
\cos^2((\phi'-\phi)/2)\cos^2((\theta'-\theta)/2) + \sin^2((\phi'-\phi)/2)\cos^2((\theta'+\theta)/2).
$$

If $\theta' = \theta$ and $\phi' = \phi$ we find 1 for the probability.

**Problem 9.** Consider the finite-dimensional Hilbert space $\mathbb{C}^n$ with $n > 2$. Consider an orthonormal basis

$$\{\,|0\rangle,\ |1\rangle,\ \ldots\ ,|n-1\rangle\,\}.$$

Let $E$ be any projector in it, and $E_j := |j\rangle\langle j|$, where $j = 0, 1, \ldots, n-1$. Let the probability of obtaining 1 when measuring $E$ be $P(E)$. Then

$$P(I) = 1, \quad 0 \le P(E) \le 1, \quad P(0) = 0, \quad E_j E_k = \delta_{jk} E_j.$$

$$P(E_0 + E_1 + \cdots + E_{n-1}) = P(E_0) + P(E_1) + \cdots + P(E_{n-1}). \quad (1)$$

A state $s$ is determined by the function $P(E)$ which satisfies (1). *Gleason's theorem* states that for any $P(E)$ which satisfies (1) there exists a density matrix $\rho$ such that

$$P(E) = \mathrm{tr}(\rho E).$$

In other words, $s$ is described by the density matrix $\rho$. Show that Gleason's theorem does not hold in two-dimensional Hilbert spaces.

**Solution 9.** In the two-dimensional Hilbert space consider the eigenvalue equation

$$(\boldsymbol{\sigma} \cdot \mathbf{n})|\mathbf{m}\rangle = |\mathbf{m}\rangle$$

where $\boldsymbol{\sigma} \cdot \mathbf{n} := \sigma_1 n_1 + \sigma_2 n_2 + \sigma_3 n_3$, $\mathbf{n}$ is the unit vector in $\mathbb{R}^3$ ($\|\mathbf{n}\| = 1$) with the parameter representation

$$\mathbf{n} := (\sin(\theta)\cos(\phi), \sin(\theta)\sin(\phi), \cos(\theta))$$

where $0 \le \theta \le \pi$, $0 \le \phi < 2\pi$ and

$$|\mathbf{m}\rangle := \begin{pmatrix} \cos(\theta/2) \\ e^{i\phi}\sin(\theta/2) \end{pmatrix}.$$

The projector onto $|\mathbf{m}\rangle$ is given by

$$E_{\mathbf{m}} \equiv |\mathbf{m}\rangle\langle\mathbf{m}| = \begin{pmatrix} \cos^2(\theta/2) & \frac{1}{2}e^{-i\phi}\sin(\theta) \\ \frac{1}{2}e^{i\phi}\sin(\theta) & \sin^2(\theta/2) \end{pmatrix} = E(\theta, \phi)$$

since $\cos(\theta/2)\sin(\theta/2) \equiv \frac{1}{2}\sin(\theta)$. Equation (1) holds with

$$P(E_{\mathbf{m}} + E_{-\mathbf{m}}) = P(E_{\mathbf{m}}) + P(E_{-\mathbf{m}}) = P(I) = 1, \quad E_{\mathbf{m}}E_{-\mathbf{m}} = 0.$$

It is not difficult to find probability distribution functions $P_{\mathbf{m}} = P(\theta, \phi)$ such that no density matrix $\rho$ exists. An example is

$$P(\theta, \phi) = \frac{1}{2} + \frac{\cos^3(\theta)}{2}.$$

**Problem 10.**    Consider the two qubits in the Hilbert space $\mathbb{C}^2$

$$|\psi_1\rangle := \cos(\theta_1/2)|0\rangle + \sin(\theta_1/2)e^{i\phi_1}|1\rangle$$

$$|\psi_2\rangle := \cos(\theta_2/2)|0\rangle + \sin(\theta_2/2)e^{i\phi_2}|1\rangle.$$

(i) Find the product state $|\psi\rangle = |\psi_1\rangle \otimes |\psi_2\rangle$ in $\mathbb{C}^4$.
(ii) Consider the *qutrit state* in the Hilbert space $\mathbb{C}^3$

$$|\phi\rangle = \frac{1}{\sqrt{3}}(|0\rangle + |1\rangle + |2\rangle).$$

To encode the state $|\psi_1\rangle \otimes |\psi_2\rangle$ we use the state $|\phi\rangle$ and perform projective measurements on the state $|\phi\rangle \otimes (|\psi_1\rangle \otimes |\psi_2\rangle)$ given by the projection operators $P_0$, $P_1$, $P_2$, $P_3$ acting in the Hilbert space $\mathbb{C}^3 \otimes \mathbb{C}^4$

$$\begin{aligned}
P_0 := {} & |0\rangle\langle 0| \otimes (|1\rangle \otimes |0\rangle\langle 1| \otimes \langle 0|) \\
& + |1\rangle\langle 1| \otimes (|0\rangle \otimes |1\rangle\langle 0| \otimes \langle 1|) + |2\rangle\langle 2| \otimes (|1\rangle \otimes |1\rangle\langle 1| \otimes \langle 1|) \\
P_1 := {} & |0\rangle\langle 0| \otimes (|0\rangle \otimes |1\rangle\langle 0| \otimes \langle 1|) \\
& + |1\rangle\langle 1| \otimes (|1\rangle \otimes |1\rangle\langle 1| \otimes \langle 1|) + |2\rangle\langle 2| \otimes (|0\rangle \otimes |0\rangle\langle 0| \otimes \langle 0|) \\
P_2 := {} & |0\rangle\langle 0| \otimes (|1\rangle \otimes |1\rangle\langle 1| \otimes \langle 1|) \\
& + |1\rangle\langle 1| \otimes (|0\rangle \otimes |0\rangle\langle 0| \otimes \langle 0|) + |2\rangle\langle 2| \otimes (|1\rangle \otimes |0\rangle\langle 1| \otimes \langle 0|) \\
P_3 := {} & |0\rangle\langle 0| \otimes (|0\rangle \otimes |0\rangle\langle 0| \otimes \langle 0|) \\
& + |1\rangle\langle 1| \otimes (|1\rangle \otimes |0\rangle\langle 1| \otimes \langle 0|) + |2\rangle\langle 2| \otimes (|0\rangle \otimes |1\rangle\langle 0| \otimes \langle 1|).
\end{aligned}$$

Find the probability $p_0 := (\langle\phi| \otimes \langle\psi|)P_0(|\phi\rangle \otimes |\psi\rangle)$.

**Solution 10.**    (i) We have

$$\begin{aligned}
|\psi\rangle = {} & |\psi_1\rangle \otimes |\psi_2\rangle \\
= {} & \cos(\theta_1/2)\cos(\theta_2/2)|0\rangle \otimes |0\rangle + \sin(\theta_1/2)e^{i\phi_1}\cos(\theta_2/2)|1\rangle \otimes |0\rangle \\
& + \cos(\theta_1/2)\sin(\theta_2/2)e^{i\phi_2}|0\rangle \otimes |1\rangle + \sin(\theta_1/2)e^{i\phi_1}\sin(\theta_2/2)e^{i\phi_2}|1\rangle \otimes |1\rangle.
\end{aligned}$$

(ii) Using (i) we have

$$\begin{aligned}
P_0(|\phi\rangle \otimes |\psi\rangle) = {} & \frac{1}{\sqrt{3}}|0\rangle \otimes \sin(\theta_1/2)e^{i\phi_1}\cos(\theta_2/2)|1\rangle \otimes |0\rangle \\
& + \frac{1}{\sqrt{3}}|1\rangle \otimes \cos(\theta_1/2)\sin(\theta_2/2)e^{i\phi_2}|0\rangle \otimes |1\rangle \\
& + \frac{1}{\sqrt{3}}|2\rangle \otimes \sin(\theta_1/2)e^{i\phi_1}\sin(\theta_2/2)e^{i\phi_2}|1\rangle \otimes |1\rangle.
\end{aligned}$$

Then

$$(\langle\phi| \otimes \langle\psi|)P_0(|\phi\rangle \otimes |\psi\rangle) = \frac{1}{3}\sin(\theta_1/2)e^{i\phi_1}\cos(\theta_2/2)\langle\psi|(|1\rangle \otimes |0\rangle)$$

$$+ \frac{1}{3}\cos(\theta_1/2)\sin(\theta_2/2)e^{i\phi_2}\langle\psi|(|0\rangle \otimes |1\rangle)$$

$$+ \frac{1}{3}\sin(\theta_1/2)e^{i\phi_1}\sin(\theta_2/2)e^{i\phi_2}\langle\psi|(|1\rangle \otimes |1\rangle).$$

Since

$$\langle\psi|(|1\rangle \otimes |0\rangle) = e^{-i\phi_1}\sin(\theta_1/2)\cos(\theta_2/2)$$
$$\langle\psi|(|0\rangle \otimes |1\rangle) = e^{-i\phi_2}\cos(\theta_1/2)\sin(\theta_2/2)$$
$$\langle\psi|(|1\rangle \otimes |1\rangle) = e^{-i\phi_1}e^{-i\phi_2}\sin(\theta_1/2)\sin(\theta_2/2)$$

we obtain the probability

$$(\langle\phi| \otimes \langle\psi|)P_0(|\phi\rangle \otimes |\psi\rangle) = \frac{1}{3}\left(1 - \cos^2(\theta_1/2)\cos^2(\theta_2/2)\right)$$

where we used $\sin^2(\alpha) + \cos^2(\alpha) \equiv 1$.

**Problem 11.** Let $B$ be an observable with $k$ possible measurement outcomes $(b_j)$

$$B = \sum_{j=1}^{k} b_j\Pi_j, \qquad \sum_{j=1}^{k}\Pi_j = I$$

where $\Pi_j$ denotes mutually orthogonal projection operators. The measurement of a system described by the density operator $\rho$ yields the following *orthogonal measurement*:

1. The outcome (eigenvalue) $b_j$ is obtained with probability

$$p(b_j) = \mathrm{tr}(\Pi_j\rho) = \mathrm{tr}(\Pi_j\rho\Pi_j)$$

where we used the cyclic invariance of the trace and $\Pi_j^2 = \Pi_j$.

2. The expectation value (average measurement value) is given by $\mathrm{tr}(\rho B)$.

3. The state of the system after measurement is in the measured state of the system (i.e. the system is projected onto the state corresponding to the measurement outcome)

$$\rho_{b_j} = \frac{\Pi_j\rho\Pi_j}{p(b_j)}.$$

Discuss orthogonal measurement of the $W$ state

$$|W\rangle\langle W| = \frac{1}{3}\begin{pmatrix} 0 & 0 & 0 & 0 & 0 & 0 & 0 & 0 \\ 0 & 1 & 1 & 0 & 1 & 0 & 0 & 0 \\ 0 & 1 & 1 & 0 & 1 & 0 & 0 & 0 \\ 0 & 0 & 0 & 0 & 0 & 0 & 0 & 0 \\ 0 & 1 & 1 & 0 & 1 & 0 & 0 & 0 \\ 0 & 0 & 0 & 0 & 0 & 0 & 0 & 0 \\ 0 & 0 & 0 & 0 & 0 & 0 & 0 & 0 \\ 0 & 0 & 0 & 0 & 0 & 0 & 0 & 0 \end{pmatrix}$$

with respect to the observable $B$ given by the diagonal matrix

$$\text{diag}(0,1,0,0,0,0,0,0) + 2\text{diag}(0,0,1,0,0,0,0,0) + 3\text{diag}(0,0,0,0,1,0,0,0).$$

**Solution 11.**    The measurement outcomes are $b_1 = 0$, $b_2 = 1$, $b_3 = 2$, $b_4 = 3$ (eigenvalues) with corresponding projection operators

$$\Pi_1 = \text{diag}(1,0,0,1,0,1,1,1), \quad \Pi_2 = \text{diag}(0,1,0,0,0,0,0,0),$$
$$\Pi_3 = \text{diag}(0,0,1,0,0,0,0,0), \quad \Pi_4 = \text{diag}(0,0,0,0,1,0,0,0).$$

The probability of the outcome $b_2 = 1$ is given by

$$p_2 := \text{tr}(\Pi_2 W \Pi_2) = \frac{1}{3}.$$

Similarly $p_1 = 0$ and $p_3 = p_4 = \frac{1}{3}$. However, the density operators

$$\rho_1 := \frac{1}{3}\text{diag}(0,1,1,0,1,0,0,0)$$

$$\rho_2 := \frac{1}{6}\begin{pmatrix} 0 & 0 & 0 & 0 & 0 & 0 & 0 & 0 \\ 0 & 2 & 1 & 0 & 1 & 0 & 0 & 0 \\ 0 & 1 & 2 & 0 & 1 & 0 & 0 & 0 \\ 0 & 0 & 0 & 0 & 0 & 0 & 0 & 0 \\ 0 & 1 & 1 & 0 & 2 & 0 & 0 & 0 \\ 0 & 0 & 0 & 0 & 0 & 0 & 0 & 0 \\ 0 & 0 & 0 & 0 & 0 & 0 & 0 & 0 \\ 0 & 0 & 0 & 0 & 0 & 0 & 0 & 0 \end{pmatrix}$$

yield the same probabilities for the measurement outcomes with respect to the measurement $B$. For the measurement outcome $b_2$ the state of the system becomes $\text{diag}(0,1,0,0,0,0,0,0)$, for $b_3$ $\text{diag}(0,0,1,0,0,0,0,0)$ and for $b_4$ $\text{diag}(0,0,0,0,1,0,0,0)$. It is not possible to obtain the measurement outcome $b_1$.

**Problem 12.**    Measurement can be generalized in the sense that an ancilla system (in a well defined state), identified by the Hilbert space $\mathcal{H}_A$, is introduced and allowed to interact with the quantum system identified by the Hilbert space $\mathcal{H}$. The ancilla system is subsequently measured, which may disturb the original system. Let $\rho$ be a density operator on $\mathcal{H}$ and $B$ an observable on $\mathcal{H}_A$ with $k$ possible measurement outcomes

$$B = \sum_{j=1}^{k} b_j \Pi_j, \qquad \sum_{j=1}^{k} \Pi_j = I$$

where $\Pi_j$ denotes mutually orthogonal projection operators. Thus $I \otimes B$ is an observable on the product Hilbert space $\mathcal{H} \otimes \mathcal{H}_A$. Let $|b\rangle$ be a normalized eigenvector of $B$, and consequently an eigenvector of only one of the $\Pi_j$. The generalized measurement of a system described by the density operator $\rho$ yields the following

1. The system $\rho$ is extended with the ancilla in the normalized pure state $|b\rangle$ which gives the density operator $\rho \otimes |b\rangle\langle b|$.
2. The two systems interact via a unitary operator $U$, i.e. the system is transformed according to $U(\rho \otimes |b\rangle\langle b|)U^*$.
3. The outcome $b_j$ is obtained with probability

$$p(b_j) = \text{tr}((I \otimes \Pi_j)U(\rho \otimes |b\rangle\langle b|)U^*(I \otimes \Pi_j)).$$

4. The expectation value (average measurement value) is given by

$$\text{tr}(U(\rho \otimes |b\rangle\langle b|)U^*(I \otimes B)).$$

5. The state of the system after measurement is in the measured state of the system (i.e. the system is projected onto the state corresponding to the measurement outcome)

$$\sigma_{b_j} = \frac{(I \otimes \Pi_j)U(\rho \otimes |b\rangle\langle b|)U^*(I \otimes \Pi_j)}{p(b_j)}.$$

6. The state of the original system after discarding the ancilla system is given by

$$\rho_{b_j} = \text{tr}_{\mathcal{H}_A}(\sigma_{b_j}).$$

Let $\phi_B$ denote an orthonormal basis for $\mathcal{H}_A$ of normalized eigenvectors of $B$, and so $|b\rangle \in \phi_B$. Thus we write the unitary operator $U$ as

$$U = \sum_{|j\rangle,|k\rangle \in \phi_B} U_{jk} \otimes |j\rangle\langle k|.$$

Describe generalized measurement in terms of the operators $U_{jk}$ where $|j\rangle, |k\rangle \in \phi_B$.

**Solution 12.** We find that the constraint $UU^* = U^*U = I$ yields

$$U^*U = \sum_{|j\rangle,|k\rangle,|l\rangle,|m\rangle \in \phi_B} U_{jk}^* U_{lm} \otimes |k\rangle\langle j|l\rangle\langle m| = \sum_{|j\rangle,|k\rangle,|m\rangle \in \phi_B} U_{jk}^* U_{jm} \otimes |k\rangle\langle m|$$
$$= I$$

and

$$UU^* = \sum_{|j\rangle,|k\rangle,|l\rangle,|m\rangle \in \phi_B} U_{jk} U_{lm}^* \otimes |j\rangle\langle k|m\rangle\langle l| = \sum_{|j\rangle,|k\rangle,|l\rangle \in \phi_B} U_{jk} U_{lk}^* \otimes |j\rangle\langle l|$$
$$= I.$$

Equating to $I$ we find

$$\sum_{|j\rangle\in\phi_B} U_{jk}^* U_{jm} = \delta_{km} I, \qquad \sum_{|k\rangle\in\phi_B} U_{jk} U_{lk}^* = \delta_{jl} I.$$

Consequently we find

$$(I \otimes \Pi_j) U (\rho \otimes |b\rangle\langle b|) U^* (I \otimes \Pi_j)$$

$$= (I \otimes \Pi_j) \left( \sum_{|m\rangle,|n\rangle\in\phi_B} U_{mb}\rho U_{nb}^* \otimes |m\rangle\langle n| \right) (I \otimes \Pi_j)$$

$$= \sum_{|m\rangle,|n\rangle\in\phi_B} (U_{mb}\rho U_{nb}^*) \otimes (\Pi_j|m\rangle\langle n|\Pi_j).$$

Applying the definition of $\rho_{b_j}$ yields

$$\rho_{b_j} = \sum_{|m\rangle\in\phi_B, \Pi_j|m\rangle=|m\rangle} \frac{1}{p(b_j)} U_{mb}\rho U_{mb}^*$$

where

$$p(b_j) = \mathrm{tr}\left( \sum_{|m\rangle\in\phi_B, \Pi_j|m\rangle=|m\rangle} U_{mb}\rho U_{mb}^* \right).$$

When the $b_j$ are non-degenerate we find

$$p(b_j) = \mathrm{tr}\left( U_{mb}\rho U_{mb}^* \right).$$

Thus for a given $|b\rangle$ the measurement is described by the operators $U_{mb}$ for $|m\rangle \in \phi_B$.

**Problem 13.** Assuming that the $b_j$ are non-degenerate (i.e. that measurement yields maximal information) and that $|j\rangle$ is the eigenvector of $\Pi_j$ (for the eigenvalue 1) we have $k$ operators $U_m$ where $m = 1, 2, \ldots, k$ and the *generalized measurement*

1. $\sum_{m=1}^{k} U_m^* U_m = I.$
2. The outcome $b_j$ is obtained with probability

$$p(b_j) = \mathrm{tr}\left( U_j\rho U_j^* \right).$$

3. The state of the system $\rho$ after the measurement outcome $b_j$ is given by

$$\rho_{b_j} = \frac{1}{p(b_j)} U_j\rho U_j^*.$$

Consider the Hilbert space $\mathcal{H} = \mathbb{C}^2$ and the ancilla Hilbert space $\mathcal{H}_A = \mathbb{C}^2$. We construct a generalized measurement described by $\alpha, \beta \in \mathbb{C}$

$$U_0 = \alpha|0\rangle\langle 0| + \beta|1\rangle\langle 1|, \qquad U_1 = \alpha|1\rangle\langle 1| + \beta|0\rangle\langle 0|$$

where $\{|0\rangle, |1\rangle\}$ forms an orthonormal basis in $\mathbb{C}^2$ and $|\alpha|^2 + |\beta|^2 = 1$. Construct the generalized measurement for $U_0$ and $U_1$.

**Solution 13.** We have

$$\begin{aligned}
U_0^* U_0 + U_1^* U_1 &= (\overline{\alpha}|0\rangle\langle 0| + \overline{\beta}|1\rangle\langle 1|)(\alpha|0\rangle\langle 0| + \beta|1\rangle\langle 1|) \\
&\quad + (\overline{\alpha}|1\rangle\langle 1| + \overline{\beta}|0\rangle\langle 0|)(\alpha|1\rangle\langle 1| + \beta|0\rangle\langle 0|) \\
&= |0\rangle\langle 0| + |1\rangle\langle 1| = I_2.
\end{aligned}$$

For $\alpha = 0$ or $\beta = 0$, $U_0$ and $U_1$ describe an orthogonal measurement. To construct the unitary operator

$$U := U_{00} \otimes |0\rangle\langle 0| + U_{01} \otimes |0\rangle\langle 1| + U_{10} \otimes |1\rangle\langle 0| + U_{11} \otimes |1\rangle\langle 1|$$

we use the commutators given by

$$[U_0, U_0^*] = [U_1, U_1^*] = [U_0, U_1] = [U_0, U_1^*] = 0_2$$

and set $U_{00} := U_0$, $U_{01} := U_1^*$, $U_{10} := U_1$, $U_{11} := -U_0^*$ to satisfy that $U$ is unitary. One orthogonal measurement that implements the generalized measurement is described by the projection operators

$$\Pi_1 = |0\rangle\langle 0|, \qquad \Pi_2 = |1\rangle\langle 1|$$

and the corresponding observable

$$B := -1\Pi_1 + 1\Pi_2.$$

Thus the generalized measurement of $\rho$, a density operator on $\mathcal{H}$, may proceed as follows:
1. Perform $U$ on the density matrix $\rho \otimes |0\rangle\langle 0|$.
2. Measure the ancillary system with respect to the observable $B$.

**Problem 14.** (i) Consider the finite-dimensional Hilbert space $\mathbb{C}^d$. A symmetric informatically complete positive operator valued measure (SIC-POVM) consists of $d^2$ outcomes that are subnormalized projection matrices $\Pi_j$ onto pure states

$$\Pi_j = \frac{1}{d}|\psi_j\rangle\langle\psi_j|$$

for $j, k = 1, \ldots, d^2$ such that

$$|\langle\psi_k|\psi_k\rangle|^2 = \frac{1 + d\delta_{jk}}{d + 1}.$$

Consider the case $d = 2$. Show that the normalized vectors

$$|\psi_1\rangle = \begin{pmatrix} \sqrt{(3+\sqrt{3})/6} \\ e^{i\pi/4}\sqrt{(3-\sqrt{3})/6} \end{pmatrix}$$

$$|\psi_2\rangle = \begin{pmatrix} \sqrt{(3+\sqrt{3})/6} \\ -e^{i\pi/4}\sqrt{(3-\sqrt{3})/6} \end{pmatrix}$$

$$|\psi_3\rangle = \begin{pmatrix} e^{i\pi/4}\sqrt{(3-\sqrt{3})/6} \\ \sqrt{(3+\sqrt{3})/6} \end{pmatrix}$$

$$|\psi_4\rangle = \begin{pmatrix} -e^{i\pi/4}\sqrt{(3-\sqrt{3})/6} \\ \sqrt{(3+\sqrt{3})/6} \end{pmatrix}$$

satisfy this condition.

(ii) Consider the matrices $\sigma_1$, $-i\sigma_2$, $\sigma_3$. Find $\sigma_1|\psi_1\rangle$, $-i\sigma_2|\psi_1\rangle$, $\sigma_3|\psi_1\rangle$.

(iii) Let $d = 2$ and

$$S_d :=$$

$$\sum_{j=1}^{d} |j\rangle \otimes |j\rangle \otimes \langle j| \otimes \langle j| + \sum_{k>j=1} \frac{1}{\sqrt{2}}(|j\rangle \otimes |k\rangle + |k\rangle \otimes |j\rangle) \otimes \frac{1}{\sqrt{2}}(\langle j| \otimes \langle k| + \langle k| \otimes \langle j|)$$

where $|1\rangle$, $|2\rangle$ denotes the standard basis in $\mathbb{C}^2$, i.e.

$$|1\rangle = \begin{pmatrix} 1 \\ 0 \end{pmatrix}, \qquad |2\rangle = \begin{pmatrix} 0 \\ 1 \end{pmatrix}.$$

Show that

$$\sum_{j=1}^{d^2} (|\psi_j\rangle \otimes |\psi_j\rangle)(\langle\psi_j| \otimes \langle\psi_j|) = \frac{2d}{d+1}.$$

**Solution 14.**   (i) Since

$$(\sqrt{(3+\sqrt{3})/6})^2 = 1/2 + \sqrt{3}/6, \quad (\sqrt{(3-\sqrt{3})/6})^2 = 1/2 - \sqrt{3}/6,$$

$$\sqrt{(3+\sqrt{3})/6}\sqrt{(3-\sqrt{3})/6} = 1/\sqrt{6}$$

we find that the condition is satisfied.

(ii) We obtain $\sigma_1|\psi_1\rangle = |\psi_3\rangle$, $-i\sigma_2|\psi_1\rangle = |\psi_4\rangle$, $\sigma_3|\psi_1\rangle = |\psi_2\rangle$. Thus using the Pauli spin matrices we can generate $|\psi_2\rangle$, $|\psi_3\rangle$, $|\psi_4\rangle$ from $|\psi_1\rangle$.

(iii) We obtain

$$S_2 = \begin{pmatrix} 1 & 0 & 0 & 0 \\ 0 & 1/2 & 1/2 & 0 \\ 0 & 1/2 & 1/2 & 0 \\ 0 & 0 & 0 & 1 \end{pmatrix}.$$

Can one find a SIC-POVM in $\mathbb{C}^4$ using the states from (i) and the Kronecker product?

**Problem 15.** Let $\sigma_1$, $\sigma_2$, $\sigma_3$ be the Pauli spin matrices. Consider the $2 \times 2$ matrix over the complex numbers

$$\Pi(\mathbf{n}) := \frac{1}{2} \left( I_2 + \sum_{j=1}^{3} n_j \sigma_j \right)$$

where $\mathbf{n} := (n_1, n_2, n_3)$ ($n_j \in \mathbb{R}$) is a unit vector, i.e. $n_1^2 + n_2^2 + n_3^2 = 1$.
(i) Describe the property of $\Pi(\mathbf{n})$, i.e. find $\Pi^*(\mathbf{n})$, $\mathrm{tr}(\Pi(\mathbf{n}))$ and $\Pi^2(\mathbf{n})$, where tr denotes the trace. The trace is the sum of the diagonal elements of a square matrix.
(ii) Find the vector

$$\Pi(\mathbf{n}) \begin{pmatrix} e^{i\phi} \cos(\theta) \\ \sin(\theta) \end{pmatrix}.$$

Discuss.

**Solution 15.** (i) For the Pauli matrices we have $\sigma_1^* = \sigma_1$, $\sigma_2^* = \sigma_2$, $\sigma_3^* = \sigma_3$. Thus $\Pi(\mathbf{n}) = \Pi^*(\mathbf{n})$. Since $\mathrm{tr}(\sigma_1) = \mathrm{tr}(\sigma_2) = \mathrm{tr}(\sigma_3) = 0$ and the trace operation is linear, we obtain $\mathrm{tr}(\Pi(\mathbf{n})) = 1$. Since $\sigma_1^2 = \sigma_2^2 = \sigma_3^2 = I_2$ and for the anti-commutators

$$\sigma_1\sigma_2 + \sigma_2\sigma_1 = 0_2, \quad \sigma_2\sigma_3 + \sigma_3\sigma_2 = 0_2, \quad \sigma_3\sigma_1 + \sigma_1\sigma_3 = 0_2$$

the expression

$$\Pi^2(\mathbf{n}) = \frac{1}{4} \left( I_2 + \sum_{j=1}^{3} n_j \sigma_j \right)^2 = \frac{1}{4} I_2 + \frac{1}{2} \sum_{j=1}^{3} n_j \sigma_j + \frac{1}{4} \sum_{j=1}^{3} \sum_{k=1}^{3} n_j n_k \sigma_j \sigma_k$$

simplifies to

$$\Pi^2(\mathbf{n}) = \frac{1}{4} I_2 + \frac{1}{2} \sum_{j=1}^{3} n_j \sigma_j + \frac{1}{4} \sum_{j=1}^{3} n_j^2 I_2.$$

Using $n_1^2 + n_2^2 + n_3^2 = 1$ we obtain $\Pi^2(\mathbf{n}) = \Pi(\mathbf{n})$.
(ii) We find

$$\Pi(\mathbf{n}) \begin{pmatrix} e^{i\phi} \cos(\theta) \\ \sin(\theta) \end{pmatrix} = \frac{1}{2} \begin{pmatrix} (1 + n_3)e^{i\phi} \cos(\theta) + (n_1 - in_2) \sin(\theta) \\ (n_1 + in_2)e^{i\phi} \cos(\theta) + (1 - n_3) \sin(\theta) \end{pmatrix}.$$

## Programming Problems

**Problem 1.**    Let $|0\rangle$, $|1\rangle$ be the standard basis in $\mathbb{C}^2$. Consider the Bell state

$$|\psi\rangle = \frac{1}{\sqrt{2}}(|0\rangle_A \otimes |0\rangle_B + |1\rangle_A \otimes |1\rangle_B)$$

where $A$ refers to Alice and $B$ refers to Bob. Let

$$\Pi_0 = |0\rangle\langle 0| = \begin{pmatrix} 1 & 0 \\ 0 & 0 \end{pmatrix}, \quad \Pi_1 = |1\rangle\langle 1| = \begin{pmatrix} 0 & 0 \\ 0 & 1 \end{pmatrix}$$

be two projection matrices with $\Pi_0\Pi_1 = 0_2$. Measurement of the first qubit (Alice) provides

$$p_1(0) = \langle\psi|(\Pi_0 \otimes I_2)^*(\Pi_0 \otimes I_2)|\psi\rangle = \frac{1}{\sqrt{2}}(\langle 0| \otimes \langle 0|)\frac{1}{\sqrt{2}}(|0\rangle \otimes |0\rangle) = \frac{1}{2}.$$

Hence the post-measurement state $|\phi\rangle$ is given by

$$|\phi\rangle = \frac{1}{\sqrt{p_1(0)}}(\Pi_0 \otimes I_2)|\psi\rangle = |0\rangle \otimes |0\rangle = \begin{pmatrix} 1 \\ 0 \\ 0 \\ 0 \end{pmatrix}.$$

This state is not entangled. The measurement of qubit two (Bob) will then result with certainty in the same result

$$p_2(0) = \langle\phi|(I_2 \otimes \Pi_0)^*(I_2 \otimes \Pi_0)|\phi\rangle = 1.$$

Give a Maxima implementation of this calculation.

**Solution 1.**    Owing to the structure of the vectors and matrices the conjugate complex operation could be avoided.

```
/* AliceBob.mac */
e1: matrix([1,0]); e1T: transpose(e1);
e2: matrix([0,1]); e2T: transpose(e2);
psi: (kronecker_product(e1,e1) + kronecker_product(e2,e2))/sqrt(2);
psiT: transpose(psi);
I2: matrix([1,0],[0,1]);
Pi0: matrix([1,0],[0,0]); Pi0T: transpose(Pi0);
Pi1: matrix([0,0],[0,1]); Pi1T: transpose(Pi1);
p10: psiT . transpose(kronecker_product(Pi0,I2))
     . kronecker_product(Pi0,I2) . psi;
phi: (kronecker_product(Pi0,I2) . psi)/sqrt(p10);
phiT: transpose(phi);
p20: phiT . transpose(kronecker_product(I2,Pi0))
     . kronecker_product(I2,Pi0) . phi;
```

**Problem 2.** Consider the standard basis $|0\rangle$, $|1\rangle$, the Bell state

$$|\psi\rangle = \frac{1}{\sqrt{2}}(|0\rangle \otimes |0\rangle + |1\rangle \otimes |1\rangle).$$

and the projection matrices

$$\Pi_a = \frac{1}{2}(I_2 - \sigma_1), \quad \Pi_b = \frac{1}{2}(I_2 + \sigma_3).$$

Find

$$\langle\psi|(\Pi_a \otimes \Pi_b)|\psi\rangle.$$

**Solution 2.** A Maxima implementation is

```
/* POVM.mac */
e0: matrix([1],[0]);
e1: matrix([0],[1]);
I2: matrix([1,0],[0,1]);
sig1: matrix([0,1],[1,0]);
sig2: matrix([0,-%i],[%i,0]);
sig3: matrix([1,0],[0,-1]);
psi: (kronecker_product(e0,e0)+kronecker_product(e1,e1))/sqrt(2);
psiT: transpose(psi);
Pia: (I2 + sig1)/2;
Pib: (I2 - sig3)/2;
R: psiT . kronecker_product(Pia,Pib) . psi;
```

The output is 1/4.

## 9.3 Supplementary Problems

**Problem 1.** Let $\hat{H}$ be a hermitian $n \times n$ matrix describing the Hamilton operator and acting in the Hilbert space $\mathbb{C}^n$. Let $A$, $B$ be $n \times n$ hermitian matrices and $|\psi\rangle \in \mathbb{C}^n$. One defines (*quantum correlation function*)

$$Q(|\psi\rangle) := \frac{1}{2}\langle\psi|(A(t)B - AB(t) + BA(t) - B(t)A)|\psi\rangle$$

where

$$A(t) = e^{i\hat{H}t/\hbar}Ae^{-i\hat{H}t/\hbar}, \quad B(t) = e^{i\hat{H}t/\hbar}Be^{-i\hat{H}t/\hbar}.$$

(i) Let

$$\hat{H} = \hbar\omega\sigma_2, \quad A = \sigma_1, \quad B = \sigma_3, \quad |\psi\rangle = \begin{pmatrix} \cos(\theta) \\ \sin(\theta) \end{pmatrix}.$$

Find $Q(|\psi\rangle)$.

(ii) Let

$$\hat{H} = \hbar\omega\sigma_2 \otimes \sigma_2, \quad A = \sigma_1 \otimes \sigma_1, \quad B = \sigma_3 \otimes \sigma_3,$$

$$|\psi\rangle = \begin{pmatrix} \cos(\phi_1) \\ \sin(\phi_1)\cos(\phi_2) \\ \sin(\phi_1)\sin(\phi_2)\cos(\phi_3) \\ \sin(\phi_1)\sin(\phi_2)\sin(\phi_3) \end{pmatrix}.$$

Find $Q(|\psi\rangle)$.

**Problem 2.**   Consider

$$|\psi_1\rangle = \frac{1}{\sqrt{2}}\begin{pmatrix} 1 \\ 1 \end{pmatrix}, \quad |\psi_2\rangle = \frac{1}{\sqrt{2}}\begin{pmatrix} 1 \\ -1 \end{pmatrix}$$

and the Bell matrix

$$U = \frac{1}{\sqrt{2}}\begin{pmatrix} 1 & 0 & 0 & 1 \\ 0 & 1 & 1 & 0 \\ 0 & 1 & -1 & 0 \\ 1 & 0 & 0 & -1 \end{pmatrix}.$$

Find

$$U(|\psi_1\rangle \otimes |\psi_2\rangle), \quad ((\langle\psi_1| \otimes \langle\psi_2|)U(|\psi_1\rangle \otimes |\psi_2\rangle)), \quad |((\langle\psi_1| \otimes \langle\psi_2|)U(|\psi_1\rangle \otimes |\psi_2\rangle))|^2.$$

# Chapter 10

# Entanglement

## 10.1 Introduction

Entanglement is the characteristic trait of quantum mechanics which enforces its entire departure from classical lines of thought. Let $\mathcal{H}_1$ and $\mathcal{H}_2$ be two finite-dimensional Hilbert spaces and let $|\psi\rangle \in \mathcal{H}_1 \otimes \mathcal{H}_2$. Then $|\psi\rangle$ is said to be *disentangled*, *separable* or a *product state* if there exist states $|\psi_1\rangle \in \mathcal{H}_1$ and $|\psi_2\rangle \in \mathcal{H}_2$ such that

$$|\psi\rangle = |\psi_1\rangle \otimes |\psi_2\rangle$$

otherwise $|\psi\rangle$ is said to be *entangled*. For example the normalized state in $\mathbb{C}^4$

$$\frac{1}{2} \begin{pmatrix} 1 \\ 1 \\ 1 \\ 1 \end{pmatrix} = \frac{1}{\sqrt{2}} \begin{pmatrix} 1 \\ 1 \end{pmatrix} \otimes \frac{1}{\sqrt{2}} \begin{pmatrix} 1 \\ 1 \end{pmatrix}$$

is a product state. For example, a polarization entangled state is

$$\frac{1}{\sqrt{2}}(|H\rangle \otimes |V\rangle + e^{i\phi}|V\rangle \otimes |H\rangle)$$

where $H$ denotes horizontal polarization and $V$ vertical polarization. This is one of the Bell states.

Let $|0\rangle$, $|1\rangle$ be an orthonormal basis in $\mathbb{C}^2$. One defines the four *Bell states* as

$$|\Phi^+\rangle = \frac{1}{\sqrt{2}}(|0\rangle \otimes |0\rangle + |1\rangle \otimes |1\rangle), \quad |\Phi^-\rangle = \frac{1}{\sqrt{2}}(|0\rangle \otimes |0\rangle - |1\rangle \otimes |1\rangle),$$

$$|\Psi^+\rangle = \frac{1}{\sqrt{2}}(|0\rangle \otimes |1\rangle + |1\rangle \otimes |0\rangle), \quad |\Psi^-\rangle = \frac{1}{\sqrt{2}}(|0\rangle \otimes |1\rangle - |1\rangle \otimes |0\rangle).$$

If we select the standard basis for $|0\rangle$, $|1\rangle$, then we have

$$\frac{1}{\sqrt{2}}\begin{pmatrix} 1 \\ 0 \\ 0 \\ 1 \end{pmatrix}, \quad \frac{1}{\sqrt{2}}\begin{pmatrix} 1 \\ 0 \\ 0 \\ -1 \end{pmatrix}, \quad \frac{1}{\sqrt{2}}\begin{pmatrix} 0 \\ 1 \\ 1 \\ 0 \end{pmatrix}, \quad \frac{1}{\sqrt{2}}\begin{pmatrix} 0 \\ 1 \\ -1 \\ 0 \end{pmatrix}.$$

The Bell states are fully entangled. They also form an orthonormal basis in $\mathbb{C}^4$.

A density operator is said to be separable if there exists $m \in \mathbb{N}$ and density operators $\rho_{1,j}$, $\rho_{2,j}$ $(j = 1, \ldots, m)$ such that

$$\rho = \sum_{j=1}^{m} p_j \rho_{1,j} \otimes \rho_{2,j}, \quad \sum_{j=1}^{m} p_j = 1, \quad p_j \in [0, 1].$$

There are several measures of entanglement, for example the von Neumann entropy, the tangle and the Schmidt number. An entanglement measure $E$ has to satisfy several requirements. For example, if the density matrix $\rho$ is separable then

$$E(\rho) = 0.$$

The entanglement of a maximally entangled state of two $n$-dimensional systems should be given by $\log(n)$. There should be no increase in entanglement under *LOCC*, i.e. local operations to the density matrix $\rho$ and classically communicating cannot increase the entanglement of $\rho$. The entanglement measure should be a convex function.

Consider the Hilbert space $\mathcal{H} = \mathbb{C}^n$ and the product Hilbert space $\mathcal{H} \otimes \mathcal{H}$. Let $A$ be an arbitrary $n \times n$ matrix over $\mathbb{C}$ and $I_n$ the $n \times n$ identity matrix. Consider the following definition. A normalized vector $\psi \in \mathcal{H} \otimes \mathcal{H}$ is called *maximally entangled*, if its reduced density matrix is maximally mixed, i.e. a multiple of $I_n$

$$\langle \psi | (A \otimes I_n) | \psi \rangle = \dim(\mathcal{H})^{-1} \mathrm{tr}(A).$$

## 10.2 Solved Problems

**Problem 1.** Consider the Hilbert space $\mathbb{C}^2 \otimes \mathbb{C}^2 \cong \mathbb{C}^4$ and the unitary $2 \times 2$ matrix

$$U(\theta, \phi) := \begin{pmatrix} \cos(\theta/2) & e^{-i\phi}\sin(\theta/2) \\ -e^{i\phi}\sin(\theta/2) & \cos(\theta/2) \end{pmatrix}.$$

Show that the state in $\mathbb{C}^4$

$$\left( U(\theta_1, \phi_1) \otimes U(\theta_2, \phi_2) \right) \begin{pmatrix} 1 \\ 0 \\ 0 \\ 0 \end{pmatrix}$$

is not entangled.

**Solution 1.** We have

$$\left( U(\theta_1, \phi_1) \otimes U(\theta_2, \phi_2) \right) \begin{pmatrix} 1 \\ 0 \\ 0 \\ 0 \end{pmatrix} = U(\theta_1, \phi_1) \begin{pmatrix} 1 \\ 0 \end{pmatrix} \otimes U(\theta_2, \phi_2) \begin{pmatrix} 1 \\ 0 \end{pmatrix}.$$

**Problem 2.** Can the *Bell state*

$$\frac{1}{\sqrt{2}}(|01\rangle - |10\rangle) \equiv \frac{1}{\sqrt{2}}(|0\rangle \otimes |1\rangle - |1\rangle \otimes |0\rangle)$$

in the Hilbert space $\mathbb{C}^4$ be written as a product state?

**Solution 2.** This state cannot be written as product state. Assume that

$$(c_0|0\rangle + c_1|1\rangle) \otimes (d_0|0\rangle + d_1|1\rangle) = \frac{1}{\sqrt{2}}(|0\rangle \otimes |1\rangle - |1\rangle \otimes |0\rangle), \qquad c_0, c_1, d_0, d_1 \in \mathbb{C}$$

where $|c_0|^2 + |c_1|^2 = 1$ and $|d_0|^2 + |d_1|^2 = 1$. Then we obtain the system of four equations

$$c_0 d_0 = 0, \quad c_0 d_1 = \frac{1}{\sqrt{2}}, \quad c_1 d_0 = -\frac{1}{\sqrt{2}}, \quad c_1 d_1 = 0.$$

This set of equations admits no solution. Thus the Bell state cannot be written as a product state. The Bell state is *entangled*.

**Problem 3.** Let $|0\rangle$, $|1\rangle$ be an arbitrary orthonormal basis. Can the state in $\mathbb{C}^4$

$$|\psi\rangle = \frac{1}{\sqrt{2}}|0\rangle \otimes |0\rangle + \frac{1}{\sqrt{8}}|0\rangle \otimes |1\rangle + \frac{1}{\sqrt{8}}|1\rangle \otimes |0\rangle + \frac{1}{\sqrt{4}}|1\rangle \otimes |1\rangle$$

be written as a product state?

**Solution 3.**    From $(c_1|0\rangle + c_2|1\rangle) \otimes (d_1|0\rangle + d_2|1\rangle) = |\psi\rangle$ we obtain the two conditions

$$c_1 d_1 c_2 d_2 = \frac{1}{\sqrt{8}}, \qquad c_1 d_1 c_2 d_2 = \frac{1}{8}.$$

Consequently we have a contradiction and the state $|\psi\rangle$ cannot be written as a product state.

**Problem 4.**    Consider the Hilbert space $\mathbb{C}^2 \otimes \mathbb{C}^2$ and the unitary $2 \times 2$ matrix

$$U(\theta, \phi) := \begin{pmatrix} \cos(\theta/2) & e^{-i\phi} \sin(\theta/2) \\ -e^{i\phi} \sin(\theta/2) & \cos(\theta/2) \end{pmatrix}.$$

Show that the state

$$\left( U(\theta_1, \phi_1) \otimes U(\theta_2, \phi_2) \right) \frac{1}{\sqrt{2}} \begin{pmatrix} 1 \\ 0 \\ 0 \\ 1 \end{pmatrix}$$

is entangled.

**Solution 4.**    We use the fact that the vector $(x_1, x_2, x_3, x_4)^T \in \mathbb{C}^4$ is separable if and only if $x_1 x_4 = x_2 x_3$. We obtain

$$U(\theta_1, \phi_1)(1,0)^T \otimes U(\theta_2, \phi_2)(1,0)^T + U(\theta_1, \phi_1)(0,1)^T \otimes U(\theta_2, \phi_2)(0,1)^T$$

$$= \begin{pmatrix} \cos(\theta_1/2)\cos(\theta_2/2) + e^{-i(\phi_1+\phi_2)}\sin(\theta_1/2)\sin(\theta_2/2) \\ \cos(\theta_2/2)e^{-i\phi_1}\sin(\theta_1/2) - \cos(\theta_1/2)e^{i\phi_2}\sin(\theta/2/2) \\ \cos(\theta_1/2)e^{-i\phi_2}\sin(\theta_2/2) - \cos(\theta_2/2)e^{i\phi_1}\sin(\theta_1/2) \\ \cos(\theta_1/2)\cos(\theta_2/2) + e^{i(\phi_1+\phi_2)}\sin(\theta_1/2)\sin(\theta_2/2) \end{pmatrix}.$$

Hence $x_1 x_4 \neq x_2 x_3$ and the state is entangled.

**Problem 5.**    Consider the state

$$|\psi\rangle = \frac{1}{2}(|0\rangle \otimes |0\rangle + e^{i\phi_1}|0\rangle \otimes |1\rangle + e^{i\phi_2}|1\rangle \otimes |0\rangle + e^{i\phi_3}|1\rangle \otimes |1\rangle).$$

(i) Let $\phi_3 = \phi_1 + \phi_2$. Is the state $|\psi\rangle$ a product state?
(ii) Let $\phi_3 = \phi_1 + \phi_2 + \pi$. Is the state $|\psi\rangle$ a product state?

**Solution 5.**    (i) We have a product state, i.e.

$$\frac{1}{\sqrt{2}}(|0\rangle + e^{i\phi_1}|1\rangle) \otimes \frac{1}{\sqrt{2}}(|0\rangle + e^{i\phi_2}|1\rangle).$$

(ii) We do not have a product state we have a maximally entangled state.

**Problem 6.** Can we find $2 \times 2$ matrices $S_1$ and $S_2$ such that

$$(S_1 \otimes S_2) \left( \begin{pmatrix} 1 \\ 0 \end{pmatrix} \otimes \begin{pmatrix} 1 \\ 0 \end{pmatrix} \right) = \frac{1}{\sqrt{2}} \begin{pmatrix} 1 \\ 0 \\ 0 \\ 1 \end{pmatrix}. \tag{1}$$

**Solution 6.** From (1) we find

$$\left( S_1 \begin{pmatrix} 1 \\ 0 \end{pmatrix} \right) \otimes \left( S_2 \begin{pmatrix} 1 \\ 0 \end{pmatrix} \right) = \frac{1}{\sqrt{2}} \begin{pmatrix} 1 \\ 0 \\ 0 \\ 1 \end{pmatrix}.$$

Thus

$$\begin{pmatrix} s_{11}^{(1)} \\ s_{21}^{(1)} \end{pmatrix} \otimes \begin{pmatrix} s_{11}^{(2)} \\ s_{21}^{(2)} \end{pmatrix} = \frac{1}{\sqrt{2}} \begin{pmatrix} 1 \\ 0 \\ 0 \\ 1 \end{pmatrix}$$

or

$$\begin{pmatrix} s_{11}^{(1)} s_{11}^{(2)} \\ s_{11}^{(1)} s_{21}^{(2)} \\ s_{21}^{(1)} s_{11}^{(2)} \\ s_{21}^{(1)} s_{21}^{(2)} \end{pmatrix} = \frac{1}{\sqrt{2}} \begin{pmatrix} 1 \\ 0 \\ 0 \\ 1 \end{pmatrix}.$$

Thus we have the four conditions

$$s_{11}^{(1)} s_{11}^{(2)} = \frac{1}{\sqrt{2}}, \quad s_{11}^{(1)} s_{21}^{(2)} = 0, \quad s_{21}^{(1)} s_{11}^{(2)} = 0, \quad s_{21}^{(1)} s_{21}^{(2)} = \frac{1}{\sqrt{2}}$$

which are not compatible. Thus no $S_1$ and $S_2$ exist such that (1) is satisfied.

**Problem 7.** In the Hilbert space $\mathbb{C}^4$ we can test whether a state is entangled or not by calculating the von Neumann entropy.
(i) Consider the Hilbert space $\mathcal{H}_A \otimes \mathcal{H}_B$, where $\mathcal{H}_A = \mathcal{H}_B = \mathbb{C}^2$ and the state

$$|\psi\rangle := \frac{1}{2} \begin{pmatrix} 1 \\ -1 \\ -1 \\ 1 \end{pmatrix}.$$

Calculate the density matrices using the partial trace

$$\rho_A := \text{tr}_{\mathcal{H}_B}(|\psi\rangle\langle\psi|), \qquad \rho_B := \text{tr}_{\mathcal{H}_A}(|\psi\rangle\langle\psi|)$$

and

$$-\mathrm{tr}(\rho_A \log_2(\rho_A)), \qquad -\mathrm{tr}(\rho_B \log_2(\rho_B))$$

where $-\mathrm{tr}(\rho_A \log_2(\rho_A))$ denotes the *von Neumann entropy*.
(ii) Consider the *Bell state*

$$|\psi\rangle := \frac{1}{\sqrt{2}} \begin{pmatrix} 1 \\ 0 \\ 0 \\ -1 \end{pmatrix}.$$

Calculate $\rho_A := \mathrm{tr}_{\mathcal{H}_B}(|\psi\rangle\langle\psi|)$, $-\mathrm{tr}(\rho_A \log_2(\rho_A))$.
(iii) Consider the state

$$|\psi\rangle := \frac{1}{2}(U_1 \otimes U_2) \begin{pmatrix} 1 \\ -1 \\ -1 \\ 1 \end{pmatrix} \equiv \frac{1}{2}(U_1 \otimes U_2) \frac{1}{\sqrt{2}} \begin{pmatrix} 1 \\ -1 \end{pmatrix} \otimes \frac{1}{\sqrt{2}} \begin{pmatrix} 1 \\ -1 \end{pmatrix}$$

where $U_1$ and $U_2$ are unitary matrices acting on $\mathbb{C}^2$. Calculate

$$\rho_A := \mathrm{tr}_{\mathcal{H}_B}(|\psi\rangle\langle\psi|), \qquad -\mathrm{tr}(\rho_A \log_2(\rho_A)).$$

(iv) Consider the state

$$|\psi\rangle := \frac{1}{\sqrt{2}}(U_1 \otimes U_2) \begin{pmatrix} 1 \\ 0 \\ 0 \\ -1 \end{pmatrix}$$

where $U_1$ and $U_2$ are unitary matrices acting on $\mathbb{C}^2$. Calculate

$$\rho_A := \mathrm{tr}_{\mathcal{H}_B}(|\psi\rangle\langle\psi|), \qquad -\mathrm{tr}(\rho_A \log_2(\rho_A)).$$

**Solution 7.**    (i) We choose the standard basis in $\mathbb{C}^2$ to calculate the trace. For the density matrix $\rho$ we find

$$\rho \equiv |\psi\rangle\langle\psi| = \frac{1}{4} \begin{pmatrix} 1 & -1 & -1 & 1 \\ -1 & 1 & 1 & -1 \\ -1 & 1 & 1 & -1 \\ 1 & -1 & -1 & 1 \end{pmatrix}.$$

Therefore

$$\rho_A = \begin{pmatrix} 1 & 0 \\ 0 & 1 \end{pmatrix} \otimes (1 \quad 0)\, |\psi\rangle\langle\psi|\, \begin{pmatrix} 1 & 0 \\ 0 & 1 \end{pmatrix} \otimes \begin{pmatrix} 1 \\ 0 \end{pmatrix}$$

$$+ \begin{pmatrix} 1 & 0 \\ 0 & 1 \end{pmatrix} \otimes (0 \quad 1) |\psi\rangle\langle\psi| \begin{pmatrix} 1 & 0 \\ 0 & 1 \end{pmatrix} \otimes \begin{pmatrix} 0 \\ 1 \end{pmatrix}$$

$$= \begin{pmatrix} 1 & 0 & 0 & 0 \\ 0 & 0 & 1 & 0 \end{pmatrix} |\psi\rangle\langle\psi| \begin{pmatrix} 1 & 0 \\ 0 & 0 \\ 0 & 1 \\ 0 & 0 \end{pmatrix} + \begin{pmatrix} 0 & 1 & 0 & 0 \\ 0 & 0 & 0 & 1 \end{pmatrix} |\psi\rangle\langle\psi| \begin{pmatrix} 0 & 0 \\ 1 & 0 \\ 0 & 0 \\ 0 & 1 \end{pmatrix}$$

$$= \frac{1}{2} \begin{pmatrix} 1 & -1 \\ -1 & 1 \end{pmatrix}.$$

Analogously

$$\rho_B = \frac{1}{4} \begin{pmatrix} 1 & -1 \\ -1 & 1 \end{pmatrix} + \frac{1}{4} \begin{pmatrix} 1 & -1 \\ -1 & 1 \end{pmatrix} = \frac{1}{2} \begin{pmatrix} 1 & -1 \\ -1 & 1 \end{pmatrix}.$$

In this case, $\rho_A = \rho_B$. We diagonalise $\rho_A$. The eigenvalues are 0 and 1 with corresponding orthonormal eigenvectors $\frac{1}{\sqrt{2}}(1,1)^T$ and $\frac{1}{\sqrt{2}}(1,-1)^T$, respectively. Thus

$$-\text{tr}(\rho_A \log_2(\rho_A)) = -\text{tr}\left( \frac{1}{2} \begin{pmatrix} 1 & 1 \\ 1 & -1 \end{pmatrix} \begin{pmatrix} 0 & 0 \\ 0 & 1 \end{pmatrix} \begin{pmatrix} 1 & 1 \\ 1 & -1 \end{pmatrix} \right.$$

$$\left. \times \frac{1}{2} \begin{pmatrix} 1 & 1 \\ 1 & -1 \end{pmatrix} \log_2 \begin{pmatrix} 0 & 0 \\ 0 & 1 \end{pmatrix} \begin{pmatrix} 1 & -1 \\ 1 & 1 \end{pmatrix} \right)$$

$$= -\text{tr}\left( \frac{1}{2} \begin{pmatrix} 1 & 1 \\ 1 & -1 \end{pmatrix} \begin{pmatrix} 0\log_2 0 & 0 \\ 0 & 1\log_2 1 \end{pmatrix} \begin{pmatrix} 1 & -1 \\ 1 & 1 \end{pmatrix} \right)$$

$$= -\text{tr}\begin{pmatrix} 0 & 0 \\ 0 & 0 \end{pmatrix} = 0$$

where $0\log_2(0) = 0$ and $1\log_2(1) = 0$. Hence the state $|\psi\rangle$ is not entangled.
(ii) We choose the standard basis in $\mathbb{C}^2$ to calculate the trace. We have

$$\rho = |\psi\rangle\langle\psi| = \frac{1}{2} \begin{pmatrix} 1 & 0 & 0 & -1 \\ 0 & 0 & 0 & 0 \\ 0 & 0 & 0 & 0 \\ -1 & 0 & 0 & 1 \end{pmatrix}.$$

Thus

$$\rho_A = \begin{pmatrix} 1 & 0 \\ 0 & 1 \end{pmatrix} \otimes (1 \quad 0) |\psi\rangle\langle\psi| \begin{pmatrix} 1 & 0 \\ 0 & 1 \end{pmatrix} \otimes \begin{pmatrix} 1 \\ 0 \end{pmatrix}$$

$$+ \begin{pmatrix} 1 & 0 \\ 0 & 1 \end{pmatrix} \otimes (0 \quad 1) |\psi\rangle\langle\psi| \begin{pmatrix} 1 & 0 \\ 0 & 1 \end{pmatrix} \otimes \begin{pmatrix} 0 \\ 1 \end{pmatrix}$$

$$= \frac{1}{2} \begin{pmatrix} 0 & 0 \\ 0 & 1 \end{pmatrix} + \frac{1}{2} \begin{pmatrix} 1 & 0 \\ 0 & 0 \end{pmatrix} = \frac{1}{2} \begin{pmatrix} 1 & 0 \\ 0 & 1 \end{pmatrix}.$$

Therefore

$$-\text{tr}(\rho_A \log_2(\rho_A)) = -\text{tr}\left(\frac{1}{2}\begin{pmatrix} 1 & 0 \\ 0 & 1 \end{pmatrix} \log_2\left(\frac{1}{2}\begin{pmatrix} 1 & 0 \\ 0 & 1 \end{pmatrix}\right)\right) = 1$$

where $\log_2(1/2) = -1$. The Bell state $|\psi\rangle$ is entangled.
(iii) We choose the basis

$$\left\{ U_2 \begin{pmatrix} 1 \\ 0 \end{pmatrix}, U_2 \begin{pmatrix} 0 \\ 1 \end{pmatrix} \right\}$$

to calculate the partial trace. We have

$$|\psi\rangle\langle\psi| = \frac{1}{4}(U_1 \otimes U_2) \begin{pmatrix} 1 & -1 & -1 & 1 \\ -1 & 1 & 1 & -1 \\ -1 & 1 & 1 & -1 \\ 1 & -1 & -1 & 1 \end{pmatrix} (U_1^* \otimes U_2^*)$$

$$= \frac{1}{4}(U_1 \otimes U_2)\left(\begin{pmatrix} 1 & -1 \\ -1 & 1 \end{pmatrix} \otimes \begin{pmatrix} 1 & -1 \\ -1 & 1 \end{pmatrix}\right)(U_1^* \otimes U_2^*).$$

Therefore

$$\rho_A = \begin{pmatrix} 1 & 0 \\ 0 & 1 \end{pmatrix} \otimes ((1 \quad 0)\, U_2^*)\, |\psi\rangle\langle\psi| \begin{pmatrix} 1 & 0 \\ 0 & 1 \end{pmatrix} \otimes U_2 \begin{pmatrix} 1 \\ 0 \end{pmatrix}$$

$$+ \begin{pmatrix} 1 & 0 \\ 0 & 1 \end{pmatrix} \otimes ((0 \quad 1)\, U_2^*)\, |\psi\rangle\langle\psi| \begin{pmatrix} 1 & 0 \\ 0 & 1 \end{pmatrix} \otimes U_2 \begin{pmatrix} 0 \\ 1 \end{pmatrix}$$

$$= \frac{1}{4}U_1 \begin{pmatrix} 1 & -1 \\ -1 & 1 \end{pmatrix} U_1^* \otimes \left((1 \quad 0) \begin{pmatrix} 1 & -1 \\ -1 & 1 \end{pmatrix} \begin{pmatrix} 1 \\ 0 \end{pmatrix}\right)$$

$$+ \frac{1}{4}U_1 \begin{pmatrix} 1 & -1 \\ -1 & 1 \end{pmatrix} U_1^* \otimes \left((0 \quad 1) \begin{pmatrix} 1 & -1 \\ -1 & 1 \end{pmatrix} \begin{pmatrix} 0 \\ 1 \end{pmatrix}\right)$$

$$= \frac{1}{4}U_1 \begin{pmatrix} 1 & -1 \\ -1 & 1 \end{pmatrix} U_1^* + \frac{1}{4}U_1 \begin{pmatrix} 1 & -1 \\ -1 & 1 \end{pmatrix} U_1^*$$

$$= \frac{1}{2}U_1 \begin{pmatrix} 1 & -1 \\ -1 & 1 \end{pmatrix} U_1^*.$$

We diagonalise $\rho_A$. The eigenvalues are 0 and 1 with corresponding orthonormal eigenvectors $\frac{1}{\sqrt{2}}U_1 (1 \quad 1)^T$ and $\frac{1}{\sqrt{2}}U_1 (1 \quad -1)^T$, respectively. Thus

$$-\text{tr}(\rho_A \log_2(\rho_A)) = -\text{tr}\left(\frac{1}{2}U_1 \begin{pmatrix} 1 & 1 \\ 1 & -1 \end{pmatrix} U_1^* \begin{pmatrix} 0 & 0 \\ 0 & 1 \end{pmatrix} U_1 \begin{pmatrix} 1 & 1 \\ 1 & -1 \end{pmatrix} U_1^* \right.$$

$$\times \frac{1}{2}U_1 \begin{pmatrix} 1 & 1 \\ 1 & -1 \end{pmatrix} U_1^* \log_2 \begin{pmatrix} 0 & 0 \\ 0 & 1 \end{pmatrix} U_1^* \begin{pmatrix} 1 & -1 \\ 1 & 1 \end{pmatrix} U_1^* \right)$$

$$= -\text{tr}\begin{pmatrix} 0 & 0 \\ 0 & 0 \end{pmatrix} = 0.$$

(iv) We choose the basis

$$\left\{ U_2 \begin{pmatrix} 1 \\ 0 \end{pmatrix}, U_2 \begin{pmatrix} 0 \\ 1 \end{pmatrix} \right\}$$

to calculate the partial trace. We have

$$|\psi\rangle\langle\psi| = \frac{1}{4}(U_1 \otimes U_2) \begin{pmatrix} 1 & 0 & 0 & -1 \\ 0 & 0 & 0 & 0 \\ 0 & 0 & 0 & 0 \\ -1 & 0 & 0 & 1 \end{pmatrix} (U_1^* \otimes U_2^*)$$

and therefore

$$U_1^* \rho_A U_1 = \begin{pmatrix} 1 & 0 \\ 0 & 1 \end{pmatrix} \otimes (1 \ 0) \begin{pmatrix} 1 & 0 & 0 & -1 \\ 0 & 0 & 0 & 0 \\ 0 & 0 & 0 & 0 \\ -1 & 0 & 0 & -1 \end{pmatrix} \begin{pmatrix} 1 & 0 \\ 0 & 1 \end{pmatrix} \otimes \begin{pmatrix} 1 \\ 0 \end{pmatrix}$$

$$+ \begin{pmatrix} 1 & 0 \\ 0 & 1 \end{pmatrix} \otimes (0 \ 1) \begin{pmatrix} 1 & 0 & 0 & -1 \\ 0 & 0 & 0 & 0 \\ 0 & 0 & 0 & 0 \\ -1 & 0 & 0 & -1 \end{pmatrix} \begin{pmatrix} 1 & 0 \\ 0 & 1 \end{pmatrix} \otimes \begin{pmatrix} 0 \\ 1 \end{pmatrix}$$

$$= \frac{1}{2} \begin{pmatrix} 0 & 0 \\ 0 & 1 \end{pmatrix} + \frac{1}{2} \begin{pmatrix} 1 & 0 \\ 0 & 0 \end{pmatrix} = \frac{1}{2} \begin{pmatrix} 1 & 0 \\ 0 & 1 \end{pmatrix}.$$

Thus

$$\rho_A = U_1(U_1^* \rho_A U_1)U_1^* = \frac{1}{2} U_1 \begin{pmatrix} 1 & 0 \\ 0 & 1 \end{pmatrix} U_1^*.$$

We choose the basis

$$\left\{ U_1 \begin{pmatrix} 1 \\ 0 \end{pmatrix}, U_1 \begin{pmatrix} 0 \\ 1 \end{pmatrix} \right\}$$

to calculate the trace. Thus

$$-\text{tr}(\rho_A \log_2(\rho_A)) = -\text{tr}\left( \frac{1}{2} U_1 \begin{pmatrix} 1 & 0 \\ 0 & 1 \end{pmatrix} U_1^* \log_2(1/2) U_1 \begin{pmatrix} 1 & 0 \\ 0 & 1 \end{pmatrix} U_1^* \right)$$

$$= -\text{tr}\left( \frac{1}{2} U_1 \begin{pmatrix} \log_2(1/2) & 0 \\ 0 & \log_2(1/2) \end{pmatrix} U_1^* \right) = 1$$

where we used the cyclic invariance of the trace, $\log_2(1/2) = -1$ and that $U_1$ is a unitary matrix, i.e. $U_1 U_1^* = I_2$.

**Problem 8.** Let $\mathcal{H}_A$ and $\mathcal{H}_B$ be two finite-dimensional Hilbert spaces over $\mathbb{C}$. Let $|\psi\rangle$ denote a pure state in the Hilbert space $\mathcal{H}_A \otimes \mathcal{H}_B$. Let $\{|0\rangle, |1\rangle\}$ denote an orthonormal basis in $\mathbb{C}^2$. The *Schmidt number* (also called the *Schmidt rank*) of $|\psi\rangle \in \mathcal{H}_A \otimes \mathcal{H}_B$ over $\mathcal{H}_A \otimes \mathcal{H}_B$ is the smallest non-negative integer $\mathrm{Sch}(|\psi\rangle, \mathcal{H}_A, \mathcal{H}_B)$ such that $|\psi\rangle$ can be written as

$$|\psi\rangle = \sum_{j=1}^{\mathrm{Sch}(|\psi\rangle, \mathcal{H}_A, \mathcal{H}_B)} |\psi_j\rangle_A \otimes |\psi_j\rangle_B$$

where $|\psi_j\rangle_A \in \mathcal{H}_A$ and $|\psi_j\rangle_B \in \mathcal{H}_B$. Let

$$|\psi\rangle = \sum_{j=1}^{\min(d_1, d_2)} \lambda_j |j\rangle_A \otimes |j\rangle_B$$

be the *Schmidt decomposition* of $|\psi\rangle$ over $\mathcal{H}_A \otimes \mathcal{H}_B$, where $d_1$ and $d_2$ are the dimensions of the subsystems. Then the Schmidt number is the number of non-zero $\lambda_j$. The $\lambda_j^2$ are the eigenvalues of the matrix $\mathrm{tr}_B(|\psi\rangle\langle\psi|)$. A separable state has Schmidt number 1 and an entangled state has Schmidt number greater than 1.

Let $f : \{0,1\}^2 \to \{0,1\}$ be a boolean function. We define the state

$$|\psi_f\rangle := \frac{1}{2} \sum_{a,b \in \{0,1\}} (-1)^{f(a,b)} |a\rangle \otimes |b\rangle. \tag{1}$$

For $f$ we select the AND, OR and XOR operations. The AND, OR and XOR operations are given by

| $a$ | $b$ | AND$(a,b)$ | OR$(a,b)$ | XOR$(a,b)$ |
|-----|-----|-----------|-----------|------------|
| 0 | 0 | 0 | 0 | 0 |
| 0 | 1 | 0 | 1 | 1 |
| 1 | 0 | 0 | 1 | 1 |
| 1 | 1 | 1 | 1 | 0 |

Find the Schmidt numbers of $|\psi_{AND}\rangle$, $|\psi_{OR}\rangle$ and $|\psi_{XOR}\rangle$ over $\mathbb{C}^2 \otimes \mathbb{C}^2$.

**Solution 8.**  From (1) we obtain

$$|\psi_{AND}\rangle = \frac{1}{2}((-1)^{0 \cdot 0}|00\rangle + (-1)^{0 \cdot 1}|01\rangle + (-1)^{1 \cdot 0}|10\rangle + (-1)^{1 \cdot 1}|11\rangle)$$

$$= \frac{1}{2}(|00\rangle + |01\rangle + |10\rangle - |11\rangle)$$

where $\cdot$ denotes the AND operation. Analogously we find for the OR and XOR operations

$$|\psi_{OR}\rangle = \frac{1}{2}(|00\rangle - |01\rangle - |10\rangle - |11\rangle)$$

$$|\psi_{XOR}\rangle = \frac{1}{2}(|00\rangle - |01\rangle - |10\rangle + |11\rangle).$$

Next we take the partial trace of $|\psi_{AND}\rangle\langle\psi_{AND}|$. We obtain

$$
\begin{aligned}
\text{tr}_B(|\psi_{AND}\rangle\langle\psi_{AND}|) &= (I_2 \otimes \langle 0|)|\psi_{AND}\rangle\langle\psi_{AND}|(I_2 \otimes |0\rangle) \\
&\quad +(I_2 \otimes \langle 1|)|\psi_{AND}\rangle\langle\psi_{AND}|(I_2 \otimes |1\rangle) \\
&= \frac{1}{4}(2|0\rangle\langle 0| + 2|1\rangle\langle 1|) \\
&= \frac{1}{2}I_2.
\end{aligned}
$$

In the above calculation we used the fact that

$$(I_2 \otimes \langle 0|)|ab\rangle\langle cd|(I_2 \otimes |0\rangle) + (I_2 \otimes \langle 1|)|ab\rangle\langle cd|(I_2 \otimes |1\rangle) = \delta_{bd}|a\rangle\langle c|$$

where $\delta_{bd}$ denotes the Kronecker delta and $|ab\rangle \equiv |a\rangle \otimes |b\rangle$. Similarly we find

$$\text{tr}_B(|\psi_{OR}\rangle\langle\psi_{OR}|) = \frac{1}{2}I_2$$

$$
\begin{aligned}
\text{tr}_B(|\psi_{XOR}\rangle\langle\psi_{XOR}|) &= \frac{1}{2}(|0\rangle\langle 0| - |0\rangle\langle 1| - |1\rangle\langle 0| + |1\rangle\langle 1|) \\
&= \frac{1}{2}(|0\rangle - |1\rangle)(\langle 0| - \langle 1|).
\end{aligned}
$$

Clearly the eigenvalues of $\text{tr}_B(|\psi_{AND}\rangle\langle\psi_{AND}|)$ and $\text{tr}_B(|\psi_{OR}\rangle\langle\psi_{OR}|)$ are $\frac{1}{2}$. Thus

$$\text{Sch}(|\psi_{AND}\rangle, \mathbb{C}^2, \mathbb{C}^2) = 2, \qquad \text{Sch}(|\psi_{OR}\rangle, \mathbb{C}^2, \mathbb{C}^2) = 2.$$

The eigenvalues of $\text{tr}_B(|\psi_{XOR}\rangle\langle\psi_{XOR}|)$ are 0 and 1. Thus

$$\text{Sch}(|\psi_{XOR}\rangle, \mathbb{C}^2, \mathbb{C}^2) = 1.$$

We note that $|\psi_{XOR}\rangle = \frac{1}{2}(|0\rangle - |1\rangle) \otimes (|0\rangle - |1\rangle)$.

**Problem 9.** One particularly interesting state in quantum computing is the *Greenberger-Horne-Zeilinger state* (GHZ state). This state of three qubits acts in the Hilbert space $\mathbb{C}^8$ and is given by

$$|\psi\rangle = \frac{1}{\sqrt{2}}\left[\begin{pmatrix} 1 \\ 0 \end{pmatrix} \otimes \begin{pmatrix} 1 \\ 0 \end{pmatrix} \otimes \begin{pmatrix} 1 \\ 0 \end{pmatrix} + \begin{pmatrix} 0 \\ 1 \end{pmatrix} \otimes \begin{pmatrix} 0 \\ 1 \end{pmatrix} \otimes \begin{pmatrix} 0 \\ 1 \end{pmatrix}\right].$$

(i) Find the density matrix $\rho = |\psi\rangle\langle\psi|$.

(ii) Let $\sigma_0 \equiv I_2$, $\sigma_1$, $\sigma_2$ and $\sigma_3$ be the Pauli spin matrices, where $I_2$ is the $2 \times 2$ unit matrix. Show that $\rho$ can be written as a linear combination in terms of Kronecker products of Pauli matrices (including $\sigma_0$), i.e.

$$\rho = \frac{1}{2^3}\sum_{j_1=0}^{3}\sum_{j_2=0}^{3}\sum_{j_3=0}^{3} c_{j_1,j_2,j_3}\sigma_{j_1} \otimes \sigma_{j_2} \otimes \sigma_{j_3}.$$

**Solution 9.**    (i) We find the dual state

$$\langle\psi| = \frac{1}{\sqrt{2}}(1\ 0\ 0\ 0\ 0\ 0\ 0\ 1).$$

Thus

$$\rho = \frac{1}{2}\begin{pmatrix} 1 & 0 & 0 & 0 & 0 & 0 & 0 & 1 \\ 0 & 0 & 0 & 0 & 0 & 0 & 0 & 0 \\ 0 & 0 & 0 & 0 & 0 & 0 & 0 & 0 \\ 0 & 0 & 0 & 0 & 0 & 0 & 0 & 0 \\ 0 & 0 & 0 & 0 & 0 & 0 & 0 & 0 \\ 0 & 0 & 0 & 0 & 0 & 0 & 0 & 0 \\ 0 & 0 & 0 & 0 & 0 & 0 & 0 & 0 \\ 1 & 0 & 0 & 0 & 0 & 0 & 0 & 1 \end{pmatrix}.$$

(ii) We find

$$\rho = \frac{1}{8}(I_2 \otimes I_2 \otimes I_2 + I_2 \otimes \sigma_3 \otimes \sigma_3 + \sigma_3 \otimes I_2 \otimes \sigma_3 + \sigma_3 \otimes \sigma_3 \otimes I_2$$
$$+ \sigma_1 \otimes \sigma_1 \otimes \sigma_1 - \sigma_1 \otimes \sigma_2 \otimes \sigma_2 - \sigma_2 \otimes \sigma_1 \otimes \sigma_2 - \sigma_2 \otimes \sigma_2 \otimes \sigma_1)$$

with $I_8 = I_2 \otimes I_2 \otimes I_2$.

**Problem 10.**    Consider a symmetric matrix $A$ over $\mathbb{R}$

$$A = \begin{pmatrix} a_{11} & a_{12} & a_{13} & a_{14} \\ a_{12} & a_{22} & a_{23} & a_{24} \\ a_{13} & a_{23} & a_{33} & a_{34} \\ a_{14} & a_{24} & a_{34} & a_{44} \end{pmatrix}$$

and the *Bell basis*

$$|\Phi^+\rangle = \frac{1}{\sqrt{2}}\begin{pmatrix} 1 \\ 0 \\ 0 \\ 1 \end{pmatrix}, \qquad |\Phi^-\rangle = \frac{1}{\sqrt{2}}\begin{pmatrix} 1 \\ 0 \\ 0 \\ -1 \end{pmatrix},$$

$$|\Psi^+\rangle = \frac{1}{\sqrt{2}}\begin{pmatrix} 0 \\ 1 \\ 1 \\ 0 \end{pmatrix}, \qquad |\Psi^-\rangle = \frac{1}{\sqrt{2}}\begin{pmatrix} 0 \\ 1 \\ -1 \\ 0 \end{pmatrix}.$$

The Bell basis forms an orthonormal basis in $\mathbb{R}^4$. Let $\tilde{A}$ denote the matrix $A$ in the Bell basis. What is the condition on the entries $a_{ij}$ such that the matrix $A$ is diagonal in the Bell basis?

**Solution 10.** Obviously we have $\tilde{a}_{ij} = \tilde{a}_{ji}$ i.e. the matrix $\tilde{A}$ is also symmetric. Straightforward calculation yields

$$\tilde{a}_{11} = (\Phi^+)^T A \Phi^+ = \frac{1}{2}(a_{11} + 2a_{14} + a_{44})$$

$$\tilde{a}_{12} = (\Phi^+)^T A \Phi^- = \frac{1}{2}(a_{11} - a_{44})$$

$$\tilde{a}_{13} = (\Phi^+)^T A \Psi^+ = \frac{1}{2}(a_{12} + a_{13} + a_{24} + a_{34})$$

$$\tilde{a}_{14} = (\Phi^+)^T A \Psi^- = \frac{1}{2}(a_{12} - a_{13} + a_{24} - a_{34})$$

$$\tilde{a}_{22} = (\Phi^-)^T A \Phi^- = \frac{1}{2}(a_{11} - 2a_{14} + a_{44})$$

$$\tilde{a}_{23} = (\Phi^-)^T A \Psi^+ = \frac{1}{2}(a_{12} + a_{13} - a_{24} - a_{23})$$

$$\tilde{a}_{24} = (\Phi^-)^T A \Psi^- = \frac{1}{2}(a_{12} - a_{13} - a_{24} + a_{34})$$

$$\tilde{a}_{33} = (\Psi^+)^T A \Psi^+ = \frac{1}{2}(a_{22} + 2a_{23} + a_{33})$$

$$\tilde{a}_{34} = (\Psi^+)^T A \Psi^- = \frac{1}{2}(a_{22} - a_{33})$$

$$\tilde{a}_{44} = (\Psi^-)^T A \Psi^- = \frac{1}{2}(a_{22} - 2a_{23} + a_{33}).$$

The condition that the matrix $\tilde{A}$ should be diagonal leads to

$$a_{11} - a_{44} = 0, \qquad a_{22} - a_{33} = 0$$

and $a_{12} = a_{13} = a_{24} = a_{34} = 0$ with the entries $a_{14}$ and $a_{23}$ arbitrary. Thus the matrix $A$ has the form

$$A = \begin{pmatrix} a_{11} & 0 & 0 & a_{14} \\ 0 & a_{22} & a_{23} & 0 \\ 0 & a_{23} & a_{22} & 0 \\ a_{14} & 0 & 0 & a_{11} \end{pmatrix}.$$

**Problem 11.** Consider a bipartite qutrit system $\mathcal{H}_A = \mathcal{H}_B = \mathbb{C}^3$ with an arbitrary orthonormal basis $\{\,|0\rangle, |1\rangle, |2\rangle\,\}$ in $\mathcal{H}_A$ and $\mathcal{H}_B$, respectively.
(i) Find the *antisymmetric subspace* $\mathcal{H}_-$ on $\mathcal{H} \otimes \mathcal{H}_B$.
(ii) Find an arbitrary antisymmetric state on $\mathcal{H}^{\otimes n}$.

**Solution 11.** (i) The antisymmetric subspace $\mathcal{H}_-$ on $\mathcal{H}_A \otimes \mathcal{H}_B$ is defined as

$$\mathcal{H}_- := \text{span}_{\mathbb{C}}\{\,|01\rangle - |10\rangle,\ |12\rangle - |21\rangle,\ |20\rangle - |02\rangle\,\} \subset \mathcal{H}_A \otimes \mathcal{H}_B.$$

Problems and Solutions

264    Problems and Solutions

(ii) An antisymmetric state on $\mathcal{H}^{\otimes n}$ is given by

$$|\psi\rangle = \sum_{j_1,j_2,\ldots,j_n=0}^{2} \sum_{k_1,k_2,\ldots,k_n=0}^{2} a_{j_1,j_2,\ldots,j_n;k_1,k_2,\ldots,k_n}|j_1,\ldots,j_n;k_1,\ldots,k_n\rangle$$

where

$$a_{j_1,j_2,\ldots,j_n;k_1,k_2,\ldots,k_n} := \left(\frac{1}{\sqrt{2}}\right)^n \sum_{i_1,i_2,\ldots,i_n=0}^{2} b_{i_1,i_2,\ldots,i_n} \prod_{m=1}^{n} \epsilon_{i_m j_m k_m}$$

and $\epsilon$ is the *Levi-Civita symbol*, i.e. $\epsilon_{ijk} = 1$ for $(i,j,k) = (1,2,3)$ and its even permutations, and $-1$ for odd permutations and $0$ otherwise.

**Problem 12.**  Let $\sigma_2$ be the second Pauli spin matrix. Then

$$\sigma_2 \otimes \sigma_2 = \begin{pmatrix} 0 & -i \\ i & 0 \end{pmatrix} \otimes \begin{pmatrix} 0 & -i \\ i & 0 \end{pmatrix} = \begin{pmatrix} 0 & 0 & 0 & -1 \\ 0 & 0 & 1 & 0 \\ 0 & 1 & 0 & 0 \\ -1 & 0 & 0 & 0 \end{pmatrix}.$$

Find the normalized state ($\gamma \in \mathbb{R}$)

$$e^{i\gamma\sigma_2\otimes\sigma_2} \begin{pmatrix} 1 \\ 0 \\ 0 \\ 0 \end{pmatrix} \equiv e^{i\gamma\sigma_2\otimes\sigma_2} \left( \begin{pmatrix} 1 \\ 0 \end{pmatrix} \otimes \begin{pmatrix} 1 \\ 0 \end{pmatrix} \right).$$

Is the state entangled? Discuss.

**Solution 12.**  Straightforward calculations provide

$$e^{i\gamma\sigma_2\otimes\sigma_2} = I_4(1 - \frac{1}{2!}\gamma^2 + \frac{1}{4!}\gamma^4 - \cdots) + i\sigma_2 \otimes \sigma_2(\gamma - \frac{1}{3!}\gamma^3 + \cdots)$$
$$= I_4 \cos(\gamma) + \sigma_2 \otimes \sigma_2(i\sin(\gamma))$$

Consequently

$$e^{i\gamma\sigma_2\otimes\sigma_2} \begin{pmatrix} 1 \\ 0 \\ 0 \\ 0 \end{pmatrix} = \begin{pmatrix} \cos(\gamma) \\ 0 \\ 0 \\ -i\sin(\gamma) \end{pmatrix}.$$

If $\gamma = \pi/4$, then the state is entangled and if $\gamma = 0$ the state is not entangled (a product state).

**Problem 13.**  Consider the density matrix (*Werner state*) in $\mathbb{C}^4$

$$\rho_w := r|\phi^+\rangle\langle\phi^+| + \frac{1-r}{4}I_4$$

where $|\phi^+\rangle = \frac{1}{\sqrt{2}}(1,0,0,1)^T$ is the Bell state, and $0 \leq r \leq 1$.
(i) Find $\text{tr}(\rho_w)$ and the eigenvalues of $\rho_w$.
(ii) Determine the *concurrence*

$$C(\rho_w) := \max\{\lambda_1 - \lambda_2 - \lambda_3 - \lambda_4, 0\}$$

where $\lambda_1 \geq \lambda_2 \geq \lambda_3 \geq \lambda_4$ are the eigenvalues of $\rho_w$.

**Solution 13.**  (i) We have

$$\rho_w = \begin{pmatrix} (1+r)/4 & 0 & 0 & r/2 \\ 0 & (1-r)/4 & 0 & 0 \\ 0 & 0 & (1-r)/4 & 0 \\ r/2 & 0 & 0 & (1+r)/4 \end{pmatrix}.$$

Thus $\text{tr}(\rho_w) = 1$. The eigenvalues of $\rho_w$ are $(1+r)/4 + r/2 = (1+3r)/4$ and $(1-r)/4$ with multiplicity 3.
(ii) From (i) it follows that

$$\lambda_1 - \lambda_2 - \lambda_3 - \lambda_4 = (1+3r)/4 - 3(1-r)/4 = (3r-1)/2.$$

The concurrence is

$$C(\rho_w) = \max\{(3r-1)/2, 0\}.$$

If $r = 0$ we have $C(\rho_w) = 0$ and if $r = 1$ then $C(\rho_w) = 1$. For $r = \frac{1}{2}$ we find $C(\rho_w) = \frac{1}{4}$.

**Problem 14.**  Let $\rho$ be a density matrix over $\mathbb{C}^2 \otimes \mathbb{C}^2 = \mathbb{C}^4$. We define the *entanglement of formation* as

$$E_f(\rho) := \min_{\{p_k,|\psi_k\rangle\}} \sum_{j=0}^{|\{p_k,|\psi_k\rangle\}|} p_j S(\text{tr}_{\mathbb{C}^2}(|\psi_j\rangle\langle\psi_j|))$$

where $\{p_k, |\psi_k\rangle\}$ indicates that the minimum should be taken over all mixtures which realize $\rho$. $|\{p_k, |\psi_k\rangle\}|$ is the number of pure states comprising the mixture and

$$S(\sigma) := -\text{tr}(\sigma \log_2(\sigma))$$

is the *von Neumann entropy*. The minimum is taken over all mixtures

$$\{(p_0, |\psi_0\rangle), (p_1, |\psi_1\rangle), \ldots\}$$

which realize $\rho$ where the cardinality of the set is obviously determined by the mixture and is finite. We can calculate $E_f(\rho)$ from

$$E_f(\rho) = h\left(\frac{1+\sqrt{1-C(\rho)^2}}{2}\right)$$

where

$$C(\rho) := \max\left\{\sqrt{\lambda_1} - \sqrt{\lambda_2} - \sqrt{\lambda_3} - \sqrt{\lambda_4}, 0\right\}$$

is the *concurrence*, $\lambda_1 \geq \lambda_2 \geq \lambda_3 \geq \lambda_4$ are the eigenvalues of

$$\rho(\sigma_2 \otimes \sigma_2)\rho^*(\sigma_2 \otimes \sigma_2)$$

and

$$h(p) := -p\log_2(p) - (1-p)\log_2(1-p)$$

is the *Shannon entropy*. Find $E_f(\rho)$ for the *Werner state*

$$\rho_w := \frac{5}{8}|\phi^+\rangle\langle\phi^+| + \frac{1}{8}\left(|\phi^-\rangle\langle\phi^-| + |\psi^+\rangle\langle\psi^+| + |\psi^-\rangle\langle\psi^-|\right) = \frac{1}{2}|\phi^+\rangle\langle\phi^+| + \frac{1}{8}I_4$$

where $|\phi^+\rangle = \frac{1}{\sqrt{2}}(1,0,0,1)^T$ is a Bell state.

**Solution 14.**   We have

$$\rho_w = \frac{1}{8}\begin{pmatrix} 3 & 0 & 0 & 2 \\ 0 & 1 & 0 & 0 \\ 0 & 0 & 1 & 0 \\ 2 & 0 & 0 & 3 \end{pmatrix}.$$

Hence

$$\tilde{\rho}_w := (\sigma_2 \otimes \sigma_2)\rho_w^*(\sigma_2 \otimes \sigma_2) = \rho_w$$

where

$$\sigma_2 \otimes \sigma_2 = \begin{pmatrix} 0 & 0 & 0 & -1 \\ 0 & 0 & 1 & 0 \\ 0 & 1 & 0 & 0 \\ -1 & 0 & 0 & 0 \end{pmatrix}.$$

Thus

$$\rho_w(\sigma_2 \otimes \sigma_2)\rho_w^*(\sigma_2 \otimes \sigma_2) = \rho_w^2 = \frac{1}{64}\begin{pmatrix} 13 & 0 & 0 & 12 \\ 0 & 1 & 0 & 0 \\ 0 & 0 & 1 & 0 \\ 12 & 0 & 0 & 13 \end{pmatrix}.$$

The eigenvalues are $\frac{25}{64}, \frac{1}{64}, \frac{1}{64}$ and $\frac{1}{64}$. The concurrence is

$$C(\rho) = \max\left\{\frac{5}{8} - \frac{1}{8} - \frac{1}{8} - \frac{1}{8}, 0\right\} = \frac{1}{4}.$$

This result is consistent with solution 11 when $r = \frac{1}{2}$. Thus $E_f(\rho) = 0.1176$.

**Problem 15.** Let $\mathcal{H}_A$ and $\mathcal{H}_B$ denote two finite-dimensional Hilbert spaces. Consider the Hamilton operator

$$\hat{H} = X_A \otimes X_B$$

where the linear operator $X_A = X_A^{-1}$ acts on $\mathcal{H}_A$ and the linear operator $X_B = X_B^{-1}$ acts on $\mathcal{H}_B$. Consequently $\hat{H} = \hat{H}^{-1}$. Let $|\psi\rangle \in \mathcal{H}_A \otimes \mathcal{H}_B$. The *von Neumann entropy* is given by

$$E(|\psi\rangle) := -\text{tr}_A(\rho_A \log_2(\rho_A))$$

where $\rho_A = \text{tr}_B(|\psi\rangle\langle\psi|)$. The *entanglement capability* of $\hat{H}$ is defined as

$$E(\hat{H}) := \max_{|\psi\rangle \in \mathcal{H}_A \otimes \mathcal{H}_B} \Gamma(t)|_{t\to 0}$$

where

$$\Gamma(t) := \frac{dE(\exp(-i\hat{H}t)|\psi(0)\rangle)}{dt}$$

is the *state entanglement rate*.
(i) Show that

$$\Gamma(t) = i\text{tr}_A\left(\text{tr}_B([\hat{H}, |\psi\rangle\langle\psi|] \log_2(\rho_A))\right)$$

where $[\,,\,]$ denotes the commutator.
(ii) Show that an upper bound on $\Gamma(t)|_{t\to 0}$ is given by $\Gamma(t)|_{t\to 0} \leq 1.9123$.

**Solution 15.** Let $\rho_{AB}(t) := |\psi(t)\rangle\langle\psi(t)|$ and $\rho_A(t) := \text{tr}_B(\rho_{AB}(t))$. We have

$$\rho_{AB}(t) = \exp(-i\hat{H}t)\rho_{AB}(0)\exp(i\hat{H}t)$$

and the time evolution of $\rho_{AB}(t)$ (*von Neumann equation*) is given by

$$i\frac{d\rho_{AB}(t)}{dt} = [\hat{H}, \rho_{AB}(t)].$$

Thus

$$i\frac{d\rho_A(t)}{dt} = \text{tr}_B[\hat{H}, \rho_{AB}(t)].$$

It follows that

$$\Gamma(t) = -\frac{d}{dt}\text{tr}_A(\rho_A \log_2(\rho_A)) = -\text{tr}_A(\frac{d}{dt}\rho_A \log_2(\rho_A))$$

$$= -\text{tr}_A\left(\frac{d\rho_A}{dt}\log_2(\rho_A) + \rho_A\frac{d}{dt}\log_2(\rho_A)\right) = -\text{tr}_A\left(\frac{d\rho_A}{dt}\log_2(\rho_A)\right)$$

$$= i\text{tr}_A(\text{tr}_B[\hat{H}, \rho_{AB}] \log_2(\rho_A))$$

since

$$\text{tr}_A \left( \rho_A \frac{d}{dt} \log_2(\rho_A) \right) = 0.$$

Let

$$|\psi(0)\rangle = \sum_{j=1}^{\text{Sch}(|\psi(0)\rangle)} \sqrt{\lambda_j} |\phi_j\rangle \otimes |\eta_j\rangle$$

be a *Schmidt decomposition* of $|\psi(0)\rangle$ over $\mathcal{H}_A \otimes \mathcal{H}_B$, where $\lambda_j > 0$ with

$$\sum_{j=1}^{\text{Sch}(|\psi(0)\rangle)} \lambda_j = 1$$

and $\{ |\phi_1\rangle, \ldots, |\phi_{\text{Sch}(|\psi(0)\rangle)}\rangle \}$, $\{ |\eta_1\rangle, \ldots, |\eta_{\text{Sch}(|\psi(0)\rangle)}\rangle \}$ are orthonormal sets of states. $\text{Sch}(|\psi(0)\rangle)$ denotes the Schmidt rank of $|\psi(0)\rangle$ over $\mathcal{H}_A \otimes \mathcal{H}_B$. Thus

$$\text{tr}_B \left( [\hat{H}, \rho_{AB}(0)] \right) = \sum_{j=1}^{\text{Sch}(|\psi(0)\rangle)} I \otimes \langle \eta_j | \left[ \hat{H}, \rho_{AB}(0) \right] I \otimes |\eta_j\rangle$$

$$= \sum_{j=1}^{\text{Sch}(|\psi(0)\rangle)} I \otimes \langle \eta_j | [X_A \otimes X_B, \rho_{AB}(0)] I \otimes |\eta_j\rangle$$

$$= \sum_{m=1}^{\text{Sch}(|\psi(0)\rangle)} \sum_{n=1}^{\text{Sch}(|\psi(0)\rangle)} \sqrt{\lambda_m \lambda_n} \langle \eta_n | X_B | \eta_m \rangle X_A |\phi_m\rangle\langle\phi_n|$$

$$- \sum_{m=1}^{\text{Sch}(|\psi(0)\rangle)} \sum_{n=1}^{\text{Sch}(|\psi(0)\rangle)} \sqrt{\lambda_m \lambda_n} |\phi_m\rangle\langle\phi_n| X_A \langle \eta_n | X_B | \eta_m \rangle$$

$$= \sum_{m=1}^{\text{Sch}(|\psi(0)\rangle)} \sum_{n=1}^{\text{Sch}(|\psi(0)\rangle)} \sqrt{\lambda_m \lambda_n} \langle \eta_n | X_B | \eta_m \rangle [X_A, |\phi_m\rangle\langle\phi_n|]$$

where we used the result

$$\rho_{AB}(0) = \sum_{m=1}^{\text{Sch}(|\psi(0)\rangle)} \sum_{n=1}^{\text{Sch}(|\psi(0)\rangle)} \sqrt{\lambda_m \lambda_n} (|\phi_m\rangle \otimes |\eta_m\rangle)(\langle\phi_n| \otimes \langle\eta_n|).$$

Since

$$\rho_A(0) = \sum_{j=1}^{\text{Sch}(|\psi(0)\rangle)} \lambda_j |\phi_j\rangle\langle\phi_j|, \quad \log_2(\rho_A(0)) = \sum_{j=1}^{\text{Sch}(|\psi(0)\rangle)} \log_2(\lambda_j |\phi_j\rangle\langle\phi_j|)$$

we find

$$\Gamma(t)|_{t\to 0} = i\text{tr}_A(\text{tr}_B[\hat{H}, \rho_{AB}(0)] \log_2 \rho_A(0))$$

$$= i \sum_{j=1}^{\text{Sch}(|\psi(0)\rangle)} \langle \phi_j | \text{tr}_B[\hat{H}, \rho_{AB}(0)] \log_2(\rho_A(0)) | \phi_j \rangle$$

$$= i \sum_{m=1}^{\text{Sch}(|\psi(0)\rangle)} \sum_{n=1}^{\text{Sch}(|\psi(0)\rangle)} \sqrt{\lambda_m \lambda_n} \log_2 \lambda_n \langle \eta_n | X_B | \eta_m \rangle \langle \phi_n | X_A | \phi_m \rangle$$

$$- i \sum_{m=1}^{\text{Sch}(|\psi(0)\rangle)} \sum_{n=1}^{\text{Sch}(|\psi(0)\rangle)} \sqrt{\lambda_m \lambda_n} \log_2 \lambda_m \langle \eta_n | X_B | \eta_m \rangle \langle \phi_n | X_A | \phi_m \rangle$$

$$= i \sum_{m=1}^{\text{Sch}(|\psi(0)\rangle)} \sum_{n=1}^{\text{Sch}(|\psi(0)\rangle)} \sqrt{\lambda_m \lambda_n} \log_2 \frac{\lambda_n}{\lambda_m} \langle \eta_n | X_B | \eta_m \rangle \langle \phi_n | X_A | \phi_m \rangle$$

$$\leq \sum_{m=1}^{\text{Sch}(|\psi(0)\rangle)} \sum_{n=1}^{\text{Sch}(|\psi(0)\rangle)} \sqrt{\lambda_m \lambda_n} \left| \log_2 \frac{\lambda_n}{\lambda_m} \right| |\langle \eta_n | X_B | \eta_m \rangle| |\langle \phi_n | X_A | \phi_m \rangle|$$

$$= \sum_{m=1}^{\text{Sch}(|\psi(0)\rangle)} \sum_{n=1}^{\text{Sch}(|\psi(0)\rangle)} (\lambda_m + \lambda_n) |\langle \eta_n | X_B | \eta_m \rangle| |\langle \phi_n | X_A | \phi_m \rangle|$$

$$\times \sqrt{\frac{\lambda_n}{\lambda_m + \lambda_n} \frac{\lambda_m}{\lambda_m + \lambda_n}} \left| \log_2 \frac{\lambda_n}{\lambda_m} \right|$$

$$\leq \sum_{m=1}^{\text{Sch}(|\psi(0)\rangle)} \sum_{n=1}^{\text{Sch}(|\psi(0)\rangle)} (\lambda_m + \lambda_n) |\langle \eta_n | X_B | \eta_m \rangle| |\langle \phi_n | X_A | \phi_m \rangle|$$

$$\times \max_{x \in (0,1)} \sqrt{x(1-x)} \log_2 \left( \frac{x}{1-x} \right)$$

$$\leq 2 \max_{x \in (0,1)} \sqrt{x(1-x)} \log_2 \left( \frac{x}{1-x} \right)$$

$$\approx 1.9123$$

where we used

$$\sum_{m=1}^{\text{Sch}(|\psi(0)\rangle)} \sum_{n=1}^{\text{Sch}(|\psi(0)\rangle)} \lambda_m |\langle \eta_n | X_B | \eta_m \rangle| |\langle \phi_n | X_A | \phi_m \rangle|$$

$$\leq \sum_{m=1}^{\text{Sch}(|\psi(0)\rangle)} \sum_{n=1}^{\text{Sch}(|\psi(0)\rangle)} |\langle \eta_n | X_B | \eta_m \rangle| |\langle \phi_n | X_A | \phi_m \rangle| \leq 1$$

since $X_A^2 = I$ and $X_B^2 = I$.

**Problem 16.**   Consider the Hamilton operator

$$\hat{H} = \mu_1 \sigma_1 \otimes \sigma_1 + \mu_2 \sigma_2 \otimes \sigma_2, \qquad \mu_1, \mu_2 \in \mathbb{R}$$

where $\sigma_1$ and $\sigma_2$ are Pauli spin matrices.

(i) Calculate the eigenvalues and eigenvectors of $\hat{H}$. Are the eigenvectors entangled?

(ii) Let $|\psi\rangle \in \mathbb{C}^4$. The *von Neumann entropy* is given by

$$E(|\psi\rangle) := -\text{tr}(\rho_A \log_2(\rho_A))$$

where $\rho_A := \text{tr}_{\mathbb{C}^2}(|\psi\rangle\langle\psi|)$. The *entanglement capability* of $\hat{H}$ is defined as

$$E(\hat{H}) := \max_{|\psi\rangle \in \mathbb{C}^4} \Gamma(t)|_{t \to 0}$$

where

$$\Gamma(t) := \frac{dE(\exp(-i\hat{H}t)|\psi(0)\rangle)}{dt}$$

is the *state entanglement rate*. Show that $E(\hat{H}) = \alpha(\mu_1 + \mu_2)$, where

$$\alpha = 2 \max_{x \in (0,1)} \sqrt{x(1-x)} \log_2\left(\frac{x}{1-x}\right).$$

**Solution 16.**    (i) The matrix representation of the Hamilton operator $\hat{H}$ is given by

$$\hat{H} = \begin{pmatrix} 0 & 0 & 0 & \mu_1 - \mu_2 \\ 0 & 0 & \mu_1 + \mu_2 & 0 \\ 0 & \mu_1 + \mu_2 & 0 & 0 \\ \mu_1 - \mu_2 & 0 & 0 & 0 \end{pmatrix}.$$

The eigenvalues are $\mu_1 - \mu_2$ with corresponding eigenvector

$$|\phi^+\rangle = \frac{1}{\sqrt{2}}(1, 0, 0, 1)^T$$

$\mu_2 - \mu_1$ with corresponding eigenvector

$$|\phi^-\rangle = \frac{1}{\sqrt{2}}(1, 0, 0, -1)^T$$

$\mu_1 + \mu_2$ with corresponding eigenvector

$$|\psi^+\rangle = \frac{1}{\sqrt{2}}(0, 1, 1, 0)^T$$

and $-\mu_x - \mu_y$ with corresponding eigenvector

$$|\psi^-\rangle = \frac{1}{\sqrt{2}}(0, 1, -1, 0)^T.$$

Clearly all four eigenvectors are entangled (Bell basis).
(ii) Consider

$$|\psi_{max}\rangle := \begin{pmatrix} 0 \\ \sqrt{x_0} \\ -i\sqrt{1-x_0} \\ 0 \end{pmatrix}$$

$$= (\sqrt{x_0} - i\sqrt{1-x_0})\frac{1}{2}\begin{pmatrix} 0 \\ 1 \\ 1 \\ 0 \end{pmatrix} + (\sqrt{x_0} + i\sqrt{1-x_0})\frac{1}{2}\begin{pmatrix} 0 \\ 1 \\ -1 \\ 0 \end{pmatrix}$$

where $x_0 \in (0,1)$ satisfies

$$\alpha = 2\sqrt{x_0(1-x_0)}\log_2\left(\frac{x_0}{1-x_0}\right).$$

Now we have

$$\exp(-i\hat{H}t)|\psi_{max}\rangle = \begin{pmatrix} 0 \\ \cos(t(\mu_1+\mu_2)) - \sqrt{1-x_0}\sin(t(\mu_1+\mu_2)) \\ -i\left(\sqrt{x_0}\sin(t(\mu_1+\mu_2)) + \sqrt{1-x_0}\cos(t(\mu_1+\mu_2))\right) \\ 0 \end{pmatrix}.$$

Defining

$$a_1 := \sqrt{x_0}\cos(t(\mu_1+\mu_2)) - \sqrt{1-x_0}\sin(t(\mu_1+\mu_2))$$
$$a_2 := \sqrt{x_0}\sin(t(\mu_1+\mu_2)) + \sqrt{1-x_0}\cos(t(\mu_1+\mu_2))$$

and

$$\rho_{max}(t) := \exp(-i\hat{H}t)|\psi_{max}\rangle\langle\psi_{max}|\exp(i\hat{H}t)$$

we find

$$\rho_{max}(t) = \begin{pmatrix} 0 & 0 & 0 & 0 \\ 0 & a_1^2 & ia_1\bar{a}_2 & 0 \\ 0 & -ia_2\bar{a}_1 & a_2^2 & 0 \\ 0 & 0 & 0 & 0 \end{pmatrix}, \qquad \mathrm{tr}_{\mathbb{C}^2}(\rho_{max}(t)) = \begin{pmatrix} a_1^2 & 0 \\ 0 & a_2^2 \end{pmatrix}.$$

Thus

$$\Gamma(t) = -\frac{d}{dt}\left(a_1^2\log_2(a_1^2) + a_2^2\log_2(a_2^2)\right)$$

$$= -\left(4a_1\frac{da_1}{dt}\log_2(a_1) + \frac{2}{\ln(2)}a_1\frac{da_1}{dt} + 4a_2\frac{da_2}{dt}\log_2(a_2) + \frac{2}{\ln(2)}a_2\frac{da_2}{dt}\right)$$

$$= 4\mu a_1 a_2(\log_2(a_1/a_2))$$

$$= 4\mu\left(\sqrt{x_0(1-x_0)}\cos(2t\mu) + \frac{1}{2}(2x_0-1)\sin(2t\mu)\right)$$

$$\times \log_2\left(\frac{\sqrt{x_0}\cos(2t\mu) - \sqrt{1-x_0}\sin(2t\mu)}{\sqrt{x_0}\sin(2t\mu) + \sqrt{1-x_0}\cos(2t\mu)}\right)$$

where $\mu \equiv \mu_1 + \mu_2$ and we used that $a_1$ and $a_2$ satisfy the system of linear differential equations with constant coefficients

$$\frac{da_1}{dt} = -a_2, \qquad \frac{da_2}{dt} = a_1.$$

Since $\hat{H}$ is asymptotically equivalent to $(\mu_1 + \mu_2)\sigma_1 \otimes \sigma_1$ and

$$E((\mu_1 + \mu_2)\sigma_1 \otimes \sigma_1) \le (\mu_1 + \mu_2)\alpha$$

and using

$$\Gamma(0) = (\mu_1 + \mu_2)2\sqrt{x_0(1 - x_0)}\log_2\left(\frac{x_0}{1 - x_0}\right) = \alpha(\mu_1 + \mu_2)$$

we find $E(\hat{H}) = \alpha(\mu_1 + \mu_2)$.

**Problem 17.**   Consider the orthonormal basis $\{\,|0\rangle, |1\rangle, \ldots, |n-1\rangle\,\}$ in the Hilbert space $\mathbb{C}^n$. We assume in the following that this is the standard basis. Consider the states (*coherent states*)

$$|\beta\rangle = \left(1 - \sum_{k=1}^{n-1} x_k\right)^{1/2} |0\rangle + \sum_{k=1}^{n-1} \sqrt{x_k}\,e^{i\phi_k}|k\rangle$$

where $\phi_k \in [0, 2\pi)$, $0 \le x_k \le 1$ and with the constraints

$$0 \le x_j \le 1 - \sum_{k=j+1}^{n-1} x_k, \qquad j = 1, 2, \ldots, n-2.$$

The Lebesgue measure is given by

$$d\mu(\beta) = \frac{n!}{(2\pi)^{n-1}} \prod_{j=1}^{n-1} dx_j d\phi_j.$$

(i) Let $n = 4$. Then the state $|\beta\rangle$ is given by

$$|\beta\rangle = \begin{pmatrix} (1 - x_1 - x_2 - x_3)^{1/2} \\ \sqrt{x_1}\,e^{i\phi_1} \\ \sqrt{x_2}\,e^{i\phi_2} \\ \sqrt{x_3}\,e^{i\phi_3} \end{pmatrix}.$$

Show that this state is normalized.
(ii) Calculate the density matrix $\rho = |\beta\rangle\langle\beta|$.
(iii) Show that the coherent states $|\beta\rangle$ satisfy

$$\int_\Omega d\mu(\beta)|\beta\rangle\langle\beta| = I_4$$

where $d\mu(\beta)$ is the uniform measure given above and $\Omega$ the domain for $\phi_j$ $(j = 1, 2, 3)$ and $x_k$ $(k = 1, 2, 3)$ described above. $I_4$ is the $4 \times 4$ unit matrix. This equation is called the *resolution of identity* and a coherent state must satisfy this condition.

(iv) Find the reduced density matrix from $|\beta\rangle$ and a condition for entanglement.

**Solution 17.** (i) Taking the scalar product we have

$$\langle \beta | \beta \rangle = (1 - x_1 - x_2 - x_3) + x_1 + x_2 + x_3 = 1.$$

Thus the state is normalized.
(ii) We find the $4 \times 4$ matrix

$$\rho = |\beta\rangle\langle\beta| =$$

$$\begin{pmatrix} d^2 & d\sqrt{x_1}e^{-i\phi_1} & d\sqrt{x_2}e^{-i\phi_2} & d\sqrt{x_3}e^{-i\phi_3} \\ d\sqrt{x_1}e^{i\phi_1} & x_1 & \sqrt{x_1}\sqrt{x_2}e^{i(\phi_1-\phi_2)} & \sqrt{x_1}\sqrt{x_3}e^{i(\phi_1-\phi_3)} \\ d\sqrt{x_2}e^{i\phi_2} & \sqrt{x_1}\sqrt{x_2}e^{i(\phi_2-\phi_1)} & x_2 & \sqrt{x_2}\sqrt{x_3}e^{i(\phi_2-\phi_3)} \\ d\sqrt{x_3}e^{i\phi_3} & \sqrt{x_3}\sqrt{x_1}e^{i(\phi_3-\phi_1)} & \sqrt{x_3}\sqrt{x_2}e^{i(\phi_3-\phi_2)} & x_3 \end{pmatrix}$$

where $d := (1 - x_1 - x_2 - x_3)^{1/2}$.
(iii) Since

$$\int_{\phi=0}^{2\pi} e^{i\phi} = 0, \qquad \int_{\phi_3=0}^{2\pi}\int_{\phi_2=0}^{2\pi}\int_{\phi_1=0}^{2\pi} d\phi_3 d\phi_2 d\phi_1 = (2\pi)^3$$

and

$$\int_{x_3=0}^{1}\int_{x_2=0}^{1-x_3}\int_{x_1=0}^{1-x_2-x_3} dx_3 dx_2 dx_1 = \frac{1}{6}$$

$$\int_{x_3=0}^{1}\int_{x_2=0}^{1-x_3}\int_{x_1=0}^{1-x_2-x_3} dx_3 dx_2 dx_1 x_1 = \frac{1}{24}$$

$$\int_{x_3=0}^{1}\int_{x_2=0}^{1-x_3}\int_{x_1=0}^{1-x_2-x_3} dx_3 dx_2 dx_1 x_2 = \frac{1}{24}$$

$$\int_{x_3=0}^{1}\int_{x_2=0}^{1-x_3}\int_{x_1=0}^{1-x_2-x_3} dx_3 dx_2 dx_1 x_3 = \frac{1}{24}$$

we find (1).
(iv) Let $|0\rangle_4$, $|1\rangle_4$, $|2\rangle_4$, $|3\rangle_4$ be the standard basis in $\mathbb{C}^4$ and $|0\rangle_2$, $|1\rangle_2$ be the standard basis in $\mathbb{C}^2$. Then we can write $|0\rangle_4 = |0\rangle_2 \otimes |0\rangle_2$, $|1\rangle_4 = |0\rangle_2 \otimes |1\rangle_2$, $|2\rangle_4 = |1\rangle_2 \otimes |0\rangle_2$ and $|3\rangle_4 = |1\rangle_2 \otimes |1\rangle_2$ with the coefficients

$$c_{00} = (1 - x_1 - x_2 - x_3)^{1/2}, \qquad c_{01} = \sqrt{x_1}e^{i\phi_1},$$

$$c_{10} = \sqrt{x_2}e^{i\phi_2}, \qquad c_{11} = \sqrt{x_3}e^{i\phi_3}$$

which leads to the $2 \times 2$ matrix

$$C = \begin{pmatrix} (1 - x_1 - x_2 - x_3)^{1/2} & \sqrt{x_1}e^{i\phi_1} \\ \sqrt{x_2}e^{i\phi_2} & \sqrt{x_3}e^{i\phi_3} \end{pmatrix}.$$

The reduced density matrix is

$$CC^\dagger = \begin{pmatrix} 1 - x_2 - x_3 & d\sqrt{x_2}e^{-i\phi_2} + \sqrt{x_1}\sqrt{x_3}e^{i\phi_1}e^{-i\phi_3} \\ d\sqrt{x_2}e^{i\phi_2} + \sqrt{x_1}\sqrt{x_3}e^{-i\phi_1}e^{i\phi_3} & x_2 + x_3 \end{pmatrix}$$

where $d := (1 - x_1 - x_2 - x_3)^{1/2}$. We obtain

$$\det(CC^\dagger) = x_3 d^2 + x_1 x_2 - 2\sqrt{x_1 x_2 x_3} d \cos(\phi_1 + \phi_2 - \phi_3).$$

The state $|\beta\rangle$ is not entangled if $\det(C^\dagger C) = 0$.

**Problem 18.**  Consider the pure state

$$|\psi\rangle := \alpha|00\rangle + \beta|11\rangle$$

in the Hilbert space $\mathbb{C}^2 \otimes \mathbb{C}^2$, where $\alpha, \beta \in \mathbb{C}$ and $|\alpha|^2 + |\beta|^2 = 1$. Let $\rho := |\psi\rangle\langle\psi|$ be the corresponding density matrix.
(i) Find $-\mathrm{tr}(\rho_1 \log_2(\rho_1))$, where $\rho_1 := \mathrm{tr}_{\mathbb{C}^2}(\rho)$.
(ii) Let $\tilde\rho$ be a density matrix for a disentangled state on $\mathbb{C}^2 \otimes \mathbb{C}^2$. Find the *fidelity* (also called *Uhlmann's transition probability*)

$$\mathcal{F}(\rho, \tilde\rho) := \left[ \mathrm{tr}\sqrt{\sqrt{\rho}\tilde\rho\sqrt{\rho}} \right]^2.$$

(iii) Show that the minimum over $\tilde\rho$ of the *modified Bures metric*

$$D_B(\rho, \tilde\rho) := 2 - 2\mathcal{F}(\rho, \tilde\rho)$$

is given by $4|\alpha|^2(1 - |\alpha|^2)$ at $\sigma := |\alpha|^2|00\rangle\langle00| + |\beta|^2|11\rangle\langle11|$. The *Bures metric* is defined as

$$D_{Bures}(\rho, \tilde\rho) := 2 - 2\sqrt{\mathcal{F}(\rho, \tilde\rho)}.$$

(iv) Compare the result in (iii) with the result from (i).

**Solution 18.**  (i) We find that

$$\rho = |\alpha|^2|00\rangle\langle00| + |\beta|^2|11\rangle\langle11| + \alpha\bar\beta|00\rangle\langle11| + \beta\bar\alpha|11\rangle\langle00|.$$

Taking the partial trace over the first qubit in $\mathbb{C}^2$ yields

$$\rho_1 = \mathrm{tr}_{\mathbb{C}^2}(\rho) = |\alpha|^2|0\rangle\langle0| + |\beta|^2|1\rangle\langle1|.$$

Thus

$$-\text{tr}(\rho_1 \log_2(\rho_1)) = -|\alpha|^2 \log_2(|\alpha|^2) - (1 - |\alpha|^2) \log_2(1 - |\alpha|^2).$$

(ii) Since $\rho$ is a pure state we have $\sqrt{\rho} = \rho$ and

$$\mathcal{F}(\rho, \widetilde{\rho}) = \left[\text{tr}\sqrt{\sqrt{\rho}\widetilde{\rho}\sqrt{\rho}}\right]^2 = \left[\text{tr}\sqrt{\rho\widetilde{\rho}\rho}\right]^2$$

$$= \left[\text{tr}\sqrt{|\psi\rangle\langle\psi|\widetilde{\rho}|\psi\rangle\langle\psi|}\right]^2 = \left[\text{tr}\sqrt{\langle\psi|\widetilde{\rho}|\psi\rangle|\psi\rangle\langle\psi|}\right]^2$$

$$= |\langle\psi|\widetilde{\rho}|\psi\rangle|\,(\text{tr}(\rho))^2 = |\langle\psi|\widetilde{\rho}|\psi\rangle|.$$

(iii) From (ii) we have

$$D_B(\rho, \sigma) = 2 - 2\mathcal{F}(\rho, \sigma) = 2 - 2|\langle\psi|\sigma|\psi\rangle|.$$

For $\sigma = |\alpha|^2|00\rangle\langle00| + |\beta|^2|11\rangle\langle11|$ we find

$$D_B(\rho, \sigma) = 2 - 2(|\alpha^4| + |\beta|^4) = 2 - 2(|\alpha^4| + (1 - |\alpha|^2)^2) = 4|\alpha|^2(1 - |\alpha|^2).$$

Obviously $\sigma$ is not entangled. For $|\alpha|^2 = 0$ or $|\alpha|^2 = 1$ it is immediately clear that we have a minimum. Thus consider $0 < |\alpha|^2 < 1$. Now let $\nu$ be any fixed density matrix in $\mathbb{C}^4$ and $\lambda \in [0, 1]$. Thus the convex function

$$\sigma(\lambda) := \lambda\sigma + (1 - \lambda)\nu$$

is also a density matrix. It follows that

$$\frac{d}{d\lambda}D_B(\rho, \sigma(\lambda))\bigg|_{\lambda=1} = -2\frac{d}{d\lambda}|\lambda\langle\psi|\sigma|\psi\rangle + (1 - \lambda)\langle\psi|\nu|\psi\rangle|\bigg|_{\lambda=1}$$

$$= -2\frac{d}{d\lambda}|\lambda(|\alpha|^4 + |\beta|^4) + (1 - \lambda)\langle\psi|\nu|\psi\rangle|\bigg|_{\lambda=1}$$

$$= \begin{cases} -2(|\alpha|^4 + |\beta|^4 - \langle\psi|\nu|\psi\rangle) & |\alpha|^4 + |\beta|^4 \geq 0 \\ +2(|\alpha|^4 + |\beta|^4 - \langle\psi|\nu|\psi\rangle) & |\alpha|^4 + |\beta|^4 < 0 \end{cases}$$

$$= -2(|\alpha|^4 + |\beta|^4 - \langle\psi|\nu|\psi\rangle)$$

$$= -2((|\alpha|^2 + |\beta|^2)^2 - 2|\alpha|^2|\beta|^2 - \langle\psi|\nu|\psi\rangle)$$

$$= -2(-2|\alpha|^2|\beta|^2 + 1 - \langle\psi|\nu|\psi\rangle)$$

where we used that $\langle\psi|\nu|\psi\rangle$ is real. If $\nu$ is sufficiently close to $\rho = |\psi\rangle\langle\psi|$ then

$$1 - \langle\psi|\nu|\psi\rangle < 2|\alpha|^2|\beta|^2$$

and $D_B(\rho, \sigma(\lambda))$ is increasing around $\sigma$. Thus we have found the minimum $4|\alpha|^2(1 - |\alpha|^2)$.

(iv) For $|\alpha| \in [0, 1]$ we find

$$4|\alpha|^2(1 - |\alpha|^2) \leq -|\alpha|^2 \log_2 |\alpha|^2 - (1 - |\alpha|^2) \log_2(1 - |\alpha|^2).$$

**Problem 19.**    The two-point *Hubbard model* with cyclic boundary conditions is given by

$$\hat{H} = t(c_{1\uparrow}^\dagger c_{2\uparrow} + c_{1\downarrow}^\dagger c_{2\downarrow} + c_{2\uparrow}^\dagger c_{1\uparrow} + c_{2\downarrow}^\dagger c_{1\downarrow}) + U(n_{1\uparrow}n_{1\downarrow} + n_{2\uparrow}n_{2\downarrow})$$

where

$$n_{j\uparrow} := c_{j\uparrow}^\dagger c_{j\uparrow}, \quad n_{j\downarrow} := c_{j\downarrow}^\dagger c_{j\downarrow}, \quad j = 1, 2.$$

The *Fermi operators* $c_{j\uparrow}^\dagger, c_{j\downarrow}^\dagger, c_{j\uparrow}, c_{j\downarrow}$ obey the *anti-commutation relations*

$$[c_{j,\sigma}^\dagger, c_{k,\sigma'}]_+ = \delta_{\sigma\sigma'}\delta_{jk}I, \quad [c_{j,\sigma}^\dagger, c_{k,\sigma'}^\dagger]_+ = [c_{j,\sigma}, c_{k,\sigma'}]_+ = 0.$$

$\hat{H}$ commutes with the total number operator $\hat{N}$, and the total spin operator $\hat{S}_z$ in the $z$ direction

$$\hat{N} := \sum_{j=1}^{2}(c_{j\uparrow}^\dagger c_{j\uparrow} + c_{j\downarrow}^\dagger c_{j\downarrow}), \quad \hat{S}_z := \frac{1}{2}\sum_{j=1}^{2}(c_{j\uparrow}^\dagger c_{j\uparrow} - c_{j\downarrow}^\dagger c_{j\downarrow}).$$

We consider the subspace with two electrons, $N = 2$ and $S_z = 0$. A basis for 2 particles with total spin 0 is

$$|s_1\rangle := c_{1\uparrow}^\dagger c_{1\downarrow}^\dagger|0\rangle, \quad |s_2\rangle := c_{1\uparrow}^\dagger c_{2\downarrow}^\dagger|0\rangle, \quad |s_3\rangle := c_{2\uparrow}^\dagger c_{1\downarrow}^\dagger|0\rangle, \quad |s_4\rangle := c_{2\uparrow}^\dagger c_{2\downarrow}^\dagger|0\rangle$$

where $\langle 0|0\rangle = 1$.
(i) Find the matrix representation of $\hat{H}$ in this basis.
(ii) Can the matrix representation of $\hat{H}$ be written in the form

$$\hat{H} = A_1 \otimes I_2 + I_2 \otimes A_2$$

where $A_1$ and $A_2$ are $2 \times 2$ matrices and $I_2$ is the $2 \times 2$ identity matrix?

**Solution 19.**    (i) Applying $\hat{H}$ to the basis gives

$$\hat{H}|s_1\rangle = t|s_2\rangle + t|s_3\rangle + U|s_1\rangle$$
$$\hat{H}|s_2\rangle = t|s_1\rangle + t|s_4\rangle$$
$$\hat{H}|s_3\rangle = t|s_1\rangle + t|s_4\rangle$$
$$\hat{H}|s_4\rangle = t|s_2\rangle + t|s_3\rangle + U|s_4\rangle.$$

Identifying $|s_i\rangle$ with elements $\mathbf{e}_i$ of the standard basis in $\mathbb{C}^4$ yields the matrix representation of $\hat{H}$

$$\hat{H} = \begin{pmatrix} U & t & t & 0 \\ t & 0 & 0 & t \\ t & 0 & 0 & t \\ 0 & t & t & U \end{pmatrix}.$$

(ii) Suppose a Hamilton operator $\hat{K}$ can be written as $\hat{K} = A_1 \otimes I_2 + I_2 \otimes A_2$ where $A_1, A_2 \in M^2$ and $I_2$ is the $2 \times 2$ identity matrix. Then we have

$$\exp(-i\hat{K}\tau/\hbar) = \exp(-i\tau A_1/\hbar \otimes I_2 - i\tau I_2/\hbar \otimes A_2)$$
$$= \exp(-i\tau A_1/\hbar) \otimes \exp(-i\tau A_2/\hbar).$$

In this case separable states remain separable under time evolution in the model, and entangled states remain entangled under time evolution in the model. For the matrix representation of $\hat{H}$, however we have

$$\hat{H} = tV_{NOT} \otimes I_2 + tI_2 \otimes V_{NOT} + \text{diag}(U, 0, 0, U), \qquad V_{NOT} := \begin{pmatrix} 0 & 1 \\ 1 & 0 \end{pmatrix}.$$

The diagonal matrix $\text{diag}(U, 0, 0, U)$ cannot be written in the form $A_1 \otimes I_2 + I_2 \otimes A_2$. Thus we conclude that almost all initial separable states evolve into entangled states under the time evolution of the model.

**Problem 20.** Find the matrix representation of the two-point Hubbard model in the basis

$$\left\{ \frac{1}{\sqrt{2}}(c_{1\downarrow}^\dagger c_{1\uparrow}^\dagger|0\rangle + c_{2\downarrow}^\dagger c_{2\uparrow}^\dagger|0\rangle), \frac{1}{\sqrt{2}}(c_{1\downarrow}^\dagger c_{2\uparrow}^\dagger|0\rangle + c_{2\downarrow}^\dagger c_{1\uparrow}^\dagger|0\rangle), \right.$$

$$\left. \frac{1}{\sqrt{2}}(c_{1\downarrow}^\dagger c_{1\uparrow}^\dagger|0\rangle - c_{2\downarrow}^\dagger c_{2\uparrow}^\dagger|0\rangle), \frac{1}{\sqrt{2}}(c_{1\downarrow}^\dagger c_{2\uparrow}^\dagger|0\rangle - c_{2\downarrow}^\dagger c_{1\uparrow}^\dagger|0\rangle) \right\}.$$

**Solution 20.** The two-point Hubbard model admits a discrete symmetry under the change $1 \to 2$, $2 \to 1$. Thus we have a finite group with two elements. We obtain two irreducible representations. The group-theoretical reduction leads to the two invariant subspaces

$$\left\{ \frac{1}{\sqrt{2}}(c_{1\downarrow}^\dagger c_{1\uparrow}^\dagger|0\rangle + c_{2\downarrow}^\dagger c_{2\uparrow}^\dagger|0\rangle), \frac{1}{\sqrt{2}}(c_{1\downarrow}^\dagger c_{2\uparrow}^\dagger|0\rangle + c_{2\downarrow}^\dagger c_{1\uparrow}^\dagger|0\rangle) \right\},$$

$$\left\{ \frac{1}{\sqrt{2}}(c_{1\downarrow}^\dagger c_{1\uparrow}^\dagger|0\rangle - c_{2\downarrow}^\dagger c_{2\uparrow}^\dagger|0\rangle), \frac{1}{\sqrt{2}}(c_{1\downarrow}^\dagger c_{2\uparrow}^\dagger|0\rangle - c_{2\downarrow}^\dagger c_{1\uparrow}^\dagger|0\rangle) \right\}.$$

These four states can be considered as the Bell states. In the Bell basis the matrix representation of the Hubbard model is given by

$$\begin{pmatrix} U & 2t & 0 & 0 \\ 2t & 0 & 0 & 0 \\ 0 & 0 & 0 & 0 \\ 0 & 0 & 0 & U \end{pmatrix} = \begin{pmatrix} U & 2t \\ 2t & 0 \end{pmatrix} \oplus \begin{pmatrix} 0 & 0 \\ 0 & U \end{pmatrix}$$

where $\oplus$ denotes the direct sum.

**Problem 21.** The two-point Hubbard model with cyclic boundary conditions is given by

$$\hat{H} = t(c_{1\uparrow}^\dagger c_{2\uparrow} + c_{1\downarrow}^\dagger c_{2\downarrow} + c_{2\uparrow}^\dagger c_{1\uparrow} + c_{2\downarrow}^\dagger c_{1\downarrow}) + U(n_{1\uparrow}n_{1\downarrow} + n_{2\uparrow}n_{2\downarrow}).$$

Find the time evolution of the initial state

$$|\psi(0)\rangle = \frac{1}{\sqrt{2}}(c_{1\uparrow}^\dagger c_{1\downarrow}^\dagger - c_{2\uparrow}^\dagger c_{2\downarrow}^\dagger)|0\rangle$$

under the two-point Hubbard model. When is the state $|\psi(\tau)\rangle$ entangled?

**Solution 21.** Solving the Schrödinger equation

$$i\hbar \frac{d}{d\tau}|\psi(\tau)\rangle = \hat{H}|\psi(\tau)\rangle$$

we find $|\psi(\tau)\rangle = e^{-iU\tau/\hbar}|\psi(0)\rangle$. Consequently, the condition for separability is given by

$$\exp\left(-2i\frac{\tau}{\hbar}U\right) = 0.$$

This equation cannot be satisfied. Thus $|\psi(\tau)\rangle$ is entangled for all $\tau$.

**Problem 22.** An arbitrary pure state in the Hilbert space $\mathbb{C}^4$ can be written as

$$|\psi\rangle = \begin{pmatrix} \cos(\theta_3) \\ \sin(\theta_3)\cos(\theta_2)e^{i\phi_3} \\ \sin(\theta_3)\sin(\theta_2)\cos(\theta_1)e^{i\phi_2} \\ \sin(\theta_3)\sin(\theta_2)\sin(\theta_1)e^{i\phi_1} \end{pmatrix}$$

where $\theta_k \in [0, \pi/2]$, and $\phi_k \in [0, 2\pi)$ for $k = 1, 2, 3$.
(i) Find values for $\theta_k$ and $\phi_k$ ($k = 1, 2, 3$) so that we obtain the unentangled state (product state)

$$|\alpha\rangle = \begin{pmatrix} 1 \\ 0 \\ 0 \\ 0 \end{pmatrix} \equiv \begin{pmatrix} 1 \\ 0 \end{pmatrix} \otimes \begin{pmatrix} 1 \\ 0 \end{pmatrix}.$$

(ii) Find values for $\theta_k$ and $\phi_k$ ($k = 1, 2, 3$) so that we obtain the entangled (Bell) state

$$|\beta\rangle = \frac{1}{\sqrt{2}} \begin{pmatrix} 1 \\ 0 \\ 0 \\ 1 \end{pmatrix}.$$

**Solution 22.** (i) We obtain the unentangled state by setting $\theta_3 = 0$, $\theta_2 = 0$ and $\theta_1 = 0$.

(ii) We obtain the Bell state by setting $\theta_3 = \pi/4$, $\theta_2 = \pi/2$, $\theta_1 = \pi/2$ and $\phi_3 = \phi_2 = \phi_1 = 0$.

**Problem 23.** A completely entangled state $|\Psi^{AB}\rangle$ of an $(N \times N)$-dimensional Hilbert space $\mathcal{H} \otimes \mathcal{H}$ can be written as

$$|\Psi^{AB}\rangle = \frac{1}{\sqrt{N}} \sum_{k=0}^{N-1} |\psi_k^A\rangle \otimes |\psi_k^B\rangle$$

where $\{ |\psi_k\rangle : k = 0, 1, \ldots, N-1 \}$ is a orthonormal basis of the Hilbert space $\mathcal{H}$. We define the linear operators (*unitary depolarizers*)

$$U_{jk} := \sum_{\ell=0}^{N-1} e^{(2\pi i/N)j\ell} |\psi_{\ell \bmod N}\rangle\langle\psi_{\ell+k \bmod N}|$$

where $j, k = 0, 1, \ldots, N-1$.
(i) Calculate

$$\frac{1}{N} \sum_{j=0}^{N-1} \sum_{k=0}^{N-1} U_{jk} X U_{jk}^\dagger, \qquad \frac{1}{N} \mathrm{tr}(U_{jk} U_{\ell m}^\dagger)$$

where $X$ is an arbitrary linear operator defined on the Hilbert space $\mathcal{H}$.
(ii) Discuss the set

$$\{ |\Psi_{jk}^{AB}\rangle = (U_{jk}^A \otimes I_B)|\Psi^{AB}\rangle \ : \ j, k = 0, 1, \ldots, N-1 \}.$$

**Solution 23.** (i) We obtain

$$\frac{1}{N} \sum_{j=0}^{N-1} \sum_{k=0}^{N-1} U_{jk} X U_{jk}^\dagger = (\mathrm{tr}(X)) I$$

where $I$ is the identity operator in $\mathcal{H}$. We obtain

$$\frac{1}{N} \mathrm{tr}(U_{jk} U_{\ell m}^\dagger) = \delta_{j\ell} \delta_{km}.$$

(ii) The set is an orthonormal basis of the Hilbert space $\mathcal{H} \otimes \mathcal{H}$.

**Problem 24.** Consider the *GHZ state* (Greenberger-Horne-Zeilinger state)

$$|GHZ\rangle = \frac{1}{\sqrt{2}}(|0\rangle \otimes |0\rangle \otimes |0\rangle + |1\rangle \otimes |1\rangle \otimes |1\rangle)$$

and the *W state*

$$|W\rangle = \frac{1}{\sqrt{3}}(|0\rangle \otimes |0\rangle \otimes |1\rangle + |0\rangle \otimes |1\rangle \otimes |0\rangle + |1\rangle \otimes |0\rangle \otimes |0\rangle).$$

(i) Calculate the states

$$(\sqrt{2}\langle 0| \otimes I_2 \otimes I_2)|GHZ\rangle, \quad (\sqrt{2}I_2 \otimes \langle 0| \otimes I_2)|GHZ\rangle$$

$$(\sqrt{2}I_2 \otimes I_2 \otimes \langle 0|)|GHZ\rangle, \quad (\sqrt{2}\langle 1| \otimes I_2 \otimes I_2)|GHZ\rangle$$

$$(\sqrt{2}I_2 \otimes \langle 1| \otimes I_2)|GHZ\rangle, \quad (\sqrt{2}I_2 \otimes I_2 \otimes \langle 1|)|GHZ\rangle$$

and discuss.
(iii) Calculate

$$\left(\frac{\sqrt{3}}{\sqrt{2}}\langle 0| \otimes I_2 \otimes I_2\right)|W\rangle, \quad \left(\frac{\sqrt{3}}{\sqrt{2}}I_2 \otimes \langle 0| \otimes I_2\right)|W\rangle,$$

$$\left(\frac{\sqrt{3}}{\sqrt{2}}I_2 \otimes I_2 \otimes \langle 0|\right)|W\rangle, \quad (\sqrt{3}\langle 1| \otimes I_2 \otimes I_2)|W\rangle,$$

$$(\sqrt{3}I_2 \otimes \langle 1| \otimes I_2)|W\rangle, \quad (\sqrt{3}I_2 \otimes I_2 \otimes \langle 1|)|W\rangle$$

and discuss.

**Solution 24.**   (i) We find

$$(\sqrt{2}\langle 0| \otimes I_2 \otimes I_2)|GHZ\rangle = |0\rangle \otimes |0\rangle, \quad (\sqrt{2}I_2 \otimes \langle 0| \otimes I_2)|GHZ\rangle = |0\rangle \otimes |0\rangle$$

$$(\sqrt{2}I_2 \otimes I_2 \otimes \langle 0||GHZ\rangle = |0\rangle \otimes |0\rangle, \quad (\sqrt{2}\langle 1| \otimes I_2 \otimes I_2)|GHZ\rangle = |1\rangle \otimes |1\rangle$$

$$(\sqrt{2}I_2 \otimes \langle 1| \otimes I_2)|GHZ\rangle = |1\rangle \otimes |1\rangle, \quad (\sqrt{2}I_2 \otimes I_2 \otimes \langle 1|)|GHZ\rangle = |1\rangle \otimes |1\rangle.$$

Thus all the two-particle states are not entangled after measurement of the third state, although the $|GHZ\rangle$ state is entangled. The $|GHZ\rangle$ state is usually referred to as maximally entangled in several senses, e.g. it violates Bell inequalities maximally. However, from the result above we find that the state is maximally fragile, i.e. if one particle is lost or projected onto the computational basis $\{|0\rangle, |1\rangle\}$, then all entanglement is destroyed.
(ii) We find

$$\left(\frac{\sqrt{3}}{\sqrt{2}}\langle 0| \otimes I_2 \otimes I_2\right)|W\rangle = \frac{1}{\sqrt{2}}(|0\rangle \otimes |1\rangle + |1\rangle \otimes |0\rangle)$$

$$\left(\frac{\sqrt{3}}{\sqrt{2}}I_2 \otimes \langle 0| \otimes I_2\right)|W\rangle = \frac{1}{\sqrt{2}}(|0\rangle \otimes |1\rangle + |1\rangle \otimes |0\rangle)$$

$$\left(\frac{\sqrt{3}}{\sqrt{2}}I_2 \otimes I_2 \otimes \langle 0|\right)|W\rangle = \frac{1}{\sqrt{2}}(|0\rangle \otimes |1\rangle + |1\rangle \otimes |0\rangle).$$

Thus in this case the projected states are (maximally) entangled (Bell states). However, for the other cases we find

$$(\sqrt{3}\langle 1| \otimes I_2 \otimes I_2)|W\rangle = |0\rangle \otimes |0\rangle$$
$$(\sqrt{3}I_2 \otimes \langle 1| \otimes I_2)|W\rangle = |0\rangle \otimes |0\rangle$$
$$(\sqrt{3}I_2 \otimes I_2 \otimes \langle 1|)|W\rangle = |0\rangle \otimes |0\rangle.$$

Thus these states are not entangled.

**Problem 25.** Let $\sigma_1$, $\sigma_2$ and $\sigma_3$ be the Pauli spin matrices. Consider the Hamilton operator

$$\hat{H} = \frac{1}{2}\epsilon(\sigma_3 \otimes I_2 + I_2 \otimes \sigma_3) - \Delta(\sigma_1 \otimes \sigma_1)$$

where $\epsilon > 0$ and $\Delta > 0$. Find the eigenvalues and normalized eigenvectors of $\hat{H}$. Are the eigenvectors entangled?

**Solution 25.** Let

$$|\uparrow\rangle = \begin{pmatrix} 1 \\ 0 \end{pmatrix}, \qquad |\downarrow\rangle = \begin{pmatrix} 0 \\ 1 \end{pmatrix}$$

and $|\uparrow\uparrow\rangle = |\uparrow\rangle \otimes |\uparrow\rangle$ etc.. Then the eigenvalues and normalized eigenvectors are given by

$$E_0 = -\sqrt{\epsilon^2 + \Delta^2}, \qquad |0\rangle = \frac{1}{\sqrt{1+a^2}}(|\downarrow\downarrow\rangle + a|\uparrow\uparrow\rangle)$$

$$E_1 = -\Delta, \qquad |1\rangle = \frac{1}{\sqrt{2}}(|\uparrow\downarrow\rangle + |\downarrow\uparrow\rangle)$$

$$E_2 = +\Delta, \qquad |2\rangle = \frac{1}{\sqrt{2}}(-|\uparrow\downarrow\rangle + |\downarrow\uparrow\rangle)$$

$$E_3 = +\sqrt{\epsilon^2 + \Delta^2}, \qquad |3\rangle = \frac{1}{\sqrt{1+a^2}}(-a|\downarrow\downarrow\rangle + |\uparrow\uparrow\rangle)$$

where

$$a := \frac{\sqrt{\epsilon^2 + \Delta^2} - \epsilon}{\Delta}.$$

The eigenstates are entangled.

**Problem 26.** Consider the Hilbert space $\mathcal{H} = \mathbb{C}^n$ and the product space $\mathcal{H} \otimes \mathcal{H}$. Let $A$ be an arbitrary $n \times n$ matrix over $\mathbb{C}$ and $I_n$ the $n \times n$ identity matrix. Consider the following definition. A normalized vector $\psi \in \mathcal{H} \otimes \mathcal{H}$ is called *maximally entangled*, if its reduced density matrix is maximally mixed, i.e. a multiple of $I_n$

$$\langle\psi|(A \otimes I_n)|\psi\rangle = \dim(\mathcal{H})^{-1}\text{tr}(A). \tag{1}$$

(i) Show that the *Bell states* in $\mathbb{C}^4$

$$\Phi^+ = \frac{1}{\sqrt{2}}\begin{pmatrix}1\\0\\0\\1\end{pmatrix}, \quad \Phi^- = \frac{1}{\sqrt{2}}\begin{pmatrix}1\\0\\0\\-1\end{pmatrix}, \quad \Psi^+ = \frac{1}{\sqrt{2}}\begin{pmatrix}0\\1\\1\\0\end{pmatrix}, \quad \Psi^- = \frac{1}{\sqrt{2}}\begin{pmatrix}0\\1\\-1\\0\end{pmatrix}$$

satisfy equation (1), where $\mathcal{H} = \mathbb{C}^2$.
(ii) Calculate the left and right-hand side of equation (1) for the vector
$(1\ 0\ 0\ 0)$. Discuss.

**Solution 26.**    (i) We have $\dim(\mathcal{H}) = n = 2$ and $\mathrm{tr}(A) = a_{11} + a_{22}$. For
the left-hand side we have

$$(A \otimes I_n) = \begin{pmatrix} a_{11} & 0 & a_{12} & 0 \\ 0 & a_{11} & 0 & a_{12} \\ a_{21} & 0 & a_{22} & 0 \\ 0 & a_{21} & 0 & a_{22} \end{pmatrix}.$$

Thus for the Bell state $\Psi^+$ we have

$$\frac{1}{2}(1\ 0\ 0\ 1)\begin{pmatrix} a_{11} & 0 & a_{12} & 0 \\ 0 & a_{11} & 0 & a_{12} \\ a_{21} & 0 & a_{22} & 0 \\ 0 & a_{21} & 0 & a_{22} \end{pmatrix}\begin{pmatrix}1\\0\\0\\1\end{pmatrix} = \frac{1}{2}(a_{11} + a_{22}).$$

Analogously we prove for the other Bell states that equation (1) is satisfied.
(ii) For the vector $(1\ 0\ 0\ 0)$ we obtain for the left hand side $a_{11}$. Thus
equation (1) is not satisfied. This state is not entangled.

**Problem 27.**    Let $|H\rangle$ ($|V\rangle$) indicate the state of a horizontal (vertical)
polarized photon. Suppose we have the product state

$$|\psi\rangle_{1234} = \frac{1}{\sqrt{2}}(|H\rangle_1 \otimes |V\rangle_2 - |V\rangle_1 \otimes |H\rangle_2) \otimes \frac{1}{\sqrt{2}}(|H\rangle_3 \otimes |V\rangle_4 - |V\rangle_3 \otimes |H\rangle_4).$$

Thus we have a product state of two polarization entangled pairs. One
photon out of each pair (2 and 3) is directed to the two inputs of a *polarizing
beam splitter*. What is the output?

**Solution 27.**    Since the polarizing beam splitter transmits horizontally
polarized photons and reflects vertically polarized photons, coincidence detection between the two polarizing beam splitter outputs implies that either
both photons 2 and 3 are both horizontally polarized or both vertically polarized. Thus the state (1) is projected onto a two-dimensional subspace
spanned by

$$|V\rangle_1 \otimes |H\rangle_2 \otimes |H\rangle_3 \otimes |V\rangle_4, \qquad |H\rangle_1 \otimes |V\rangle_2 \otimes |V\rangle_3 \otimes |H\rangle_4.$$

After the polarizing beam splitter, the renormalized state corresponding to a fourfold coincidence is

$$|\phi\rangle_{12'3'4} = \frac{1}{\sqrt{2}}(|H\rangle_1 \otimes |V\rangle_{2'} \otimes |V\rangle_{3'} \otimes |H\rangle_4 + |V\rangle_1 \otimes |H\rangle_{2'} \otimes |H\rangle_{3'} \otimes |V\rangle_4).$$

This is a *GHZ state* of four particles, which can exhibit nonlocal behaviour according to the GHZ theorem.

**Problem 28.** An *entanglement witness* $W$ on the product Hilbert space $\mathcal{H}_A \otimes \mathcal{H}_B$ is a linear operator on $\mathcal{H}_A \otimes \mathcal{H}_B$ such that $W$ is not positive semi-definite and

$$\langle \psi | W | \psi \rangle \geq 0$$

for all separable $|\psi\rangle = |\psi_A\rangle \otimes |\psi_B\rangle$, where $|\psi_A\rangle \in \mathcal{H}_A$ and $|\psi_B\rangle \in \mathcal{H}_B$. Show that

$$W = I_4 - 2|\Phi^+\rangle\langle\Phi^+| = \begin{pmatrix} 0 & 0 & 0 & -1 \\ 0 & 1 & 0 & 0 \\ 0 & 0 & 1 & 0 \\ -1 & 0 & 0 & 0 \end{pmatrix}$$

is an entanglement witness.

**Solution 28.** The eigenvalues of $W$ are $-1$ and $1$ with multiplicity 3, with corresponding eigenvectors given by the Bell basis. Clearly $W$ is not positive semi-definite. Using $|\psi\rangle = |\psi_A\rangle \otimes |\psi_B\rangle$, with $|\psi_A\rangle, |\psi_B\rangle \in \mathbb{C}^2$, we find

$$\begin{aligned}
\langle\psi|W|\psi\rangle &= \langle\psi_A|\psi_A\rangle\langle\psi_B|\psi_B\rangle - |\langle\psi_A|0\rangle\langle\psi_B|0\rangle + \langle\psi_A|1\rangle\langle\psi_B|1\rangle|^2 \\
&= \left(|\langle\psi_A|0\rangle|^2 + |\langle\psi_A|1\rangle|^2\right)\left(|\langle\psi_B|0\rangle|^2 + |\langle\psi_B|1\rangle|^2\right) \\
&\quad - |\langle\psi_A|0\rangle\langle\psi_B|0\rangle + \langle\psi_A|1\rangle\langle\psi_B|1\rangle|^2 \\
&\geq \left(|\langle\psi_A|0\rangle|^2 + |\langle\psi_A|1\rangle|^2\right)\left(|\langle\psi_B|0\rangle|^2 + |\langle\psi_B|1\rangle|^2\right) \\
&\quad - \left(|\langle\psi_A|0\rangle\langle\psi_B|0\rangle| + |\langle\psi_A|1\rangle\langle\psi_B|1\rangle|\right)^2 \\
&= \left(|\langle\psi_A|0\rangle\langle\psi_B|1\rangle| - |\langle\psi_A|1\rangle\langle\psi_B|0\rangle|\right)^2 \geq 0.
\end{aligned}$$

Thus the operator $W$ is an entanglement witness. Furthermore,

$$\text{tr}(W|\Phi^+\rangle\langle\Phi^+|) = \langle\Phi^+|W|\Phi^+\rangle = -1.$$

Consequently $W$ is a witness for the entangled Bell state $|\Phi^+\rangle$.

**Problem 29.** Consider the distillation under 1-local operations of a pure state in the Hilbert space $\mathbb{C}^4$

$$|\psi\rangle := \alpha|00\rangle + \beta|11\rangle$$

with $|\alpha|^2 + |\beta|^2 = 1$. If this state can be distilled, it will be transformed into the Bell state

$$|\Phi^+\rangle = \frac{1}{\sqrt{2}}|00\rangle + \frac{1}{\sqrt{2}}|11\rangle.$$

Find a generalized measurement on the first qubit which can be used to distill $|\Phi^+\rangle$ from $|\psi\rangle$.

**Solution 29.** We could attempt to construct a separable unitary transformation to perform this transform, however one component of a generalized measurement on the first qubit will achieve this for $k \in \mathbb{C}$, $k \neq 0$

$$U_0 := \frac{k}{\alpha}|0\rangle\langle0| + \frac{k}{\beta}|1\rangle\langle1|$$

which transforms the density operator $\rho := |\psi\rangle\langle\psi|$ to

$$\frac{(U_0 \otimes I)\rho(U_0 \otimes I)^*}{\text{tr}(U_0 \otimes I)\rho(U_0 \otimes I)^*} = |\Phi^+\rangle\langle\Phi^+|$$

with probability

$$p_0 := \text{tr}((U_0 \otimes I)\rho(U_0 \otimes I)^*) = 2|k|^2.$$

Of course this is only possible for $|\alpha^2|$, $|\beta|^2 \neq 0$, i.e. for $\alpha = 0$ or $\beta = 0$ there is no distillable entanglement. To complete the generalized measurement we must find $U_1$ such that $U_0^*U_0 + U_1^*U_1 = I$. Since $U_0^*U_0$ is clearly positive semidefinite we can use the polar decomposition $U_1 = U_1 H_1$, where $U_1$ is unitary and $H_1$ is positive semidefinite. We obtain

$$U_1 = U_1\sqrt{I - U_0^*U_0} = U_1\left(\sqrt{1 - \left|\frac{k}{\alpha}\right|^2}|0\rangle\langle0| + \sqrt{1 - \left|\frac{k}{\beta}\right|^2}|1\rangle\langle1|\right).$$

Consequently it is necessary that $k \leq \min\{|\alpha|^2, |\beta|^2\}$.

**Problem 30.** Consider the GHZ state (Greenberger-Horne-Zeilinger)

$$|\psi\rangle := \frac{1}{\sqrt{2}}(|001\rangle + |110\rangle).$$

We consider the basis $\alpha = \{|L\rangle, |R\rangle\}$ and the basis $\beta = \{|H\rangle, |V\rangle\}$ described by

$$\alpha: \quad |0\rangle = \frac{1}{\sqrt{2}}(|L\rangle + |R\rangle), \quad |1\rangle = \frac{1}{\sqrt{2}}(|L\rangle - |R\rangle),$$

$$\beta: \quad |0\rangle = \frac{1}{\sqrt{2}}(|H\rangle + |V\rangle), \quad |1\rangle = \frac{i}{\sqrt{2}}(|H\rangle - |V\rangle).$$

(i) Express $|\psi\rangle$ in terms of the bases $\alpha$ and $\beta$, where one qubit is in the basis $\alpha$ and the remaining two in the basis $\beta$.

(ii) Use (i) to predict the measurement outcomes where all qubits are measured in the basis $\alpha$. Compare with the actual outcomes.

**Solution 30.**  (i) Expressing only one qubit of $|\psi\rangle$ in the basis $\alpha$ and the rest in the basis $\beta$ basis yields

$$
\begin{aligned}
|\psi\rangle &= \tfrac{i}{2}(|LHH\rangle + |RVH\rangle - |LVV\rangle - |RHV\rangle) & \alpha\beta\beta \\
&= \tfrac{i}{2}(|HLH\rangle + |VRH\rangle - |VLV\rangle - |HRV\rangle) & \beta\alpha\beta \\
&= \tfrac{1}{2}(|HVL\rangle + |VHL\rangle - |HHR\rangle - |VVR\rangle) & \beta\beta\alpha
\end{aligned}
$$

Measuring two qubits in the $\beta$ basis as given in the previous equations allows us to deduce the result of measuring the other qubit in the $\alpha$ basis. Let $\beta_j$ denote the result after measuring qubit $j$ in the $\beta$ basis. As an example, from the first equality, if $\beta_2 = \beta_3$ then the first qubit is $|L\rangle$, and $|R\rangle$ when $\beta_2 \neq \beta_3$. Thus we construct the following table.

| Outcomes in $\beta$ basis | Outcomes in $\alpha$ basis |
|---|---|
| $\beta_1 = \beta_2 = \beta_3$ | LLR |
| $\beta_1 = \beta_2 \neq \beta_3$ | RRR |
| $\beta_1 = \beta_3 \neq \beta_2$ | RLL |
| $\beta_2 = \beta_3 \neq \beta_1$ | LRL |

(ii) We find in the $\alpha$ basis

$$|\psi\rangle = \tfrac{1}{2}(|LLL\rangle + |RRL\rangle - |LRR\rangle - |RLR\rangle).$$

None of the results obtained are consistent with the outcomes in the table deduced in (i).

**Problem 31.**  Consider the state in $\mathbb{C}^8$

$$|\psi\rangle = \sum_{j_1,j_2,j_3=0}^{1} c_{j_1 j_2 j_3} |j_1\rangle \otimes |j_2\rangle \otimes |j_3\rangle$$

where

$$|0\rangle = \begin{pmatrix} 1 \\ 0 \end{pmatrix}, \qquad |1\rangle = \begin{pmatrix} 0 \\ 1 \end{pmatrix}.$$

Let $R, S, T$ be $2 \times 2$ matrices over $\mathbb{C}$

$$
R = \begin{pmatrix} r_{11} & r_{12} \\ r_{21} & r_{22} \end{pmatrix}, \qquad
S = \begin{pmatrix} s_{11} & s_{12} \\ s_{21} & s_{22} \end{pmatrix}, \qquad
T = \begin{pmatrix} t_{11} & t_{12} \\ t_{21} & t_{22} \end{pmatrix}
$$

where $\det(R) = 1$, $\det(S) = 1$, $\det(T) = 1$. This means that $R$, $S$ and $T$ are elements of the Lie group $SL(2, \mathbb{C})$. Let

$$\sum_{j_1,j_2,j_3=0}^{1} d_{j_1 j_2 j_3} |j_1\rangle \otimes |j_2\rangle \otimes |j_3\rangle = (R \otimes S \otimes T) \sum_{j_1,j_2,j_3=0}^{1} c_{j_1 j_2 j_3} |j_1\rangle \otimes |j_2\rangle \otimes |j_3\rangle.$$
(1)

We define the *hyperdeterminant* of $C = (c_{j_1 j_2 j_3})$ with $j_1, j_2, j_3 \in \{0,1\}$ as

$$\mathrm{Det}(C) := (c_{000}^2 c_{111}^2 + c_{001}^2 c_{110}^2 + c_{010}^2 c_{101}^2 + c_{011}^2 c_{100}^2)$$
$$-2(c_{000}c_{001}c_{110}c_{111} + c_{000}c_{010}c_{101}c_{111} + c_{000}c_{011}c_{100}c_{111}$$
$$+c_{001}c_{010}c_{101}c_{110} + c_{001}c_{011}c_{110}c_{100} + c_{010}c_{011}c_{101}c_{100})$$
$$+4(c_{000}c_{011}c_{101}c_{110} + c_{001}c_{010}c_{100}c_{111}).$$

Show that

$$\mathrm{Det}(C) = \mathrm{Det}(D).$$
(2)

Owing to (2) the quantity $\mathrm{Det}(C)$ is called an *invariant*.

**Solution 31.**    Note that $\det(R \otimes S \otimes T) = 1$. We find eight equations for $d_{000}$, $d_{001}$, $\dots$, $d_{111}$. From $\det(R) = \det(S) = \det(T) = 1$ it follows that $r_{11}r_{22} - r_{12}r_{21} = 1$, $s_{11}s_{22} - s_{12}s_{21} = 1$, $t_{11}t_{22} - t_{12}t_{21} = 1$. Inserting $d_{j_1 j_2 j_3}$ into the right-hand side of (2) and inserting these conditions it follows that $\mathrm{Det}(D) = \mathrm{Det}(C)$.

**Problem 32.**    The *hyperdeterminant* of a $2 \times 2 \times 2$ hypermatrix $C = (c_{ijk})$ ($i, j, k \in \{0,1\}$) is defined by

$$\mathrm{Det}(C) := -\frac{1}{2} \sum_{i,j,k,m,n,p=0}^{1} \sum_{i',j',k',m',n',p'=0}^{1} \epsilon_{ii'} \epsilon_{jj'} \epsilon_{kk'} \epsilon_{mm'} \epsilon_{nn'} \epsilon_{pp'} c_{ijk} c_{i'j'm} c_{npk} c_{n'p'm'}$$

where $\epsilon_{00} = \epsilon_{11} = 0$, $\epsilon_{01} = 1$, $\epsilon_{10} = -1$.
(i) Calculate $\mathrm{Det}(C)$.
(ii) Consider the three qubit state

$$|\psi\rangle = \sum_{i,j,k=0}^{1} c_{ijk} |i\rangle \otimes |j\rangle \otimes |k\rangle.$$

The *three tangle* $\tau_3$ is a measure of entanglement and is defined for the three qubit state $|\psi\rangle$ as

$$\tau_{123} := 4|\mathrm{Det}(C)|$$

where $C = (c_{ijk})$. Find the three tangle for the GHZ state

$$|GHZ\rangle = \frac{1}{\sqrt{2}}(|0\rangle \otimes |0\rangle \otimes |0\rangle + |1\rangle \otimes |1\rangle \otimes |1\rangle)$$

and the W state

$$|W\rangle = \frac{1}{\sqrt{3}}(|0\rangle \otimes |0\rangle \otimes |1\rangle + |0\rangle \otimes |1\rangle \otimes |0\rangle + |1\rangle \otimes |0\rangle \otimes |0\rangle).$$

**Solution 32.** (i) We obtain

$$\mathrm{Det}(C) := (c_{000}^2 c_{111}^2 + c_{001}^2 c_{110}^2 + c_{010}^2 c_{101}^2 + c_{011}^2 c_{100}^2)$$
$$-2(c_{000}c_{001}c_{110}c_{111} + c_{000}c_{010}c_{101}c_{111} + c_{000}c_{011}c_{100}c_{111}$$
$$+ c_{001}c_{010}c_{101}c_{110} + c_{001}c_{011}c_{110}c_{100} + c_{010}c_{011}c_{101}c_{100})$$
$$+4(c_{000}c_{011}c_{101}c_{110} + c_{001}c_{010}c_{100}c_{111}).$$

(ii) For the GHZ state we obtain $c_{000} = 1/\sqrt{2}$, $c_{111} = 1/\sqrt{2}$. All other coefficients are zero. Thus we find for the three tangle $\tau_3 = 1$. For the W state we have $c_{001} = c_{010} = c_{100} = \frac{1}{\sqrt{3}}$. Thus $\tau_3 = 0$. Using this measure of entanglement the W state is not entangled. Note that the $W$-state cannot be written as a product state.

**Problem 33.** Let $|0\rangle$, $|1\rangle$ be an orthonormal basis in $\mathbb{C}^2$. Consider the normalized state

$$|\psi\rangle = \sum_{j,k=0}^{1} c_{jk}|j\rangle \otimes |k\rangle$$

in the Hilbert space $\mathbb{C}^4$ and the $2 \times 2$ matrix $C = (c_{jk})$. Using the 4 coefficients $c_{jk}$ ($j,k \in \{0,1\}$) we form a multilinear polynomial $p$ in two variables $x_1, x_2$

$$p(x_1, x_2) = c_{00} + c_{01}x_1 + c_{10}x_2 + c_{11}x_1 x_2. \tag{1}$$

Show that determinant $\det(C) = c_{00}c_{11} - c_{01}c_{10}$ is the unique irreducible polynomial (up to sign) of content one in the 4 unknowns $c_{jk}$ that vanishes whenever the system of equations

$$p = \frac{\partial p}{\partial x_1} = \frac{\partial p}{\partial x_2} = 0 \tag{2}$$

has a solution $(x_1^*, x_2^*)$ in $\mathbb{C}^2$.

**Solution 33.** Inserting (2) into (1) provides the three equations

$$c_{00} + c_{01}x_1 + c_{10}x_2 + c_{11}x_1 x_2 = 0, \quad c_{01} + c_{11}x_2 = 0, \quad c_{10} + c_{11}x_1 = 0.$$

Multiplication of the first equation with $c_{11}$ and inserting the second and third equation yields

$$c_{00}c_{11} + c_{01}c_{11}x_1 + c_{10}c_{11}x_2 + c_{11}x_1 c_{11}x_2 = c_{00}c_{11} - c_{01}c_{10} - c_{10}c_{01} + c_{10}c_{01}$$
$$= c_{00}c_{11} - c_{01}c_{10}$$
$$= 0.$$

**Problem 34.**  Consider the state

$$|\psi\rangle = \cos(\alpha)|00\rangle + \sin(\alpha)|11\rangle, \qquad 0 < \alpha < \pi/4$$

where $\alpha$ is called the *Schmidt angle*.
(i) Find the eigenvalues of the density matrix $|\psi\rangle\langle\psi|$.
(ii) Find the partially traced density matrix (we find when we trace over one of the subsystems).
(iii) Show that the partially traced density matrix has two unequal and non-zero eigenvalues $\lambda_1 = \cos^2(\alpha)$ and $\lambda_2 = \sin^2(\alpha)$.
(iv) Calculate the von Neumann entropy for the corresponding density matrix. Show that the entropy grows monotonically with the Schmidt angle.

**Solution 34.**  (i) We obtain for the density matrix

$$\rho(\alpha) = \begin{pmatrix} \cos^2(\alpha) & 0 & 0 & \sin(2\alpha)/2 \\ 0 & 0 & 0 & 0 \\ 0 & 0 & 0 & 0 \\ \sin(2\alpha)/2 & 0 & 0 & \sin^2(\alpha) \end{pmatrix}$$

with the eigenvalues 0 (triple) and 1. Thus the eigenvalues are independent of $\alpha$.
(ii) We obtain the diagonal matrix

$$\rho_2 = \begin{pmatrix} \cos^2(\alpha) & 0 \\ 0 & \sin^2(\alpha) \end{pmatrix}.$$

(iii) From (ii) we see that the eigenvalues are $\cos^2(\alpha)$ and $\sin^2(\alpha)$.
(iv) Thus for the entropy we find

$$S(\rho_2) = -\cos^2(\alpha)\log_2(\cos^2(\alpha)) - \sin^2(\alpha)\log_2(\sin^2(\alpha))$$

which is monotonically increasing for $0 < \cos^2(\alpha) < 1/2$, i.e. $0 < \alpha < \pi/4$.

**Problem 35.**  Let $\sigma_1$, $\sigma_2$, $\sigma_3$ be the Pauli spin matrices. We form the nine $4 \times 4$ matrices

$$\Sigma_{jk} := \sigma_j \otimes \sigma_k, \qquad j,k = 1,2,3.$$

Note that $[\Sigma_{jk}, \Sigma_{mn}] = 0_4$. The *variance* of an hermitian operator $\hat{O}$ and a wave vector $|\phi\rangle$ is defined by

$$V_{\hat{O}}(|\phi\rangle) := \langle\phi|(\hat{O})^2|\phi\rangle - (\langle\phi|\hat{O}|\phi\rangle)^2.$$

The *remoteness* for a given normalized state $|\psi\rangle$ in $\mathbb{C}^4$ is defined by

$$R(|\psi\rangle) = \sum_{j=1}^{3}\sum_{k=1}^{3}\left(\langle\psi|(\Sigma_{jk})^2|\psi\rangle - (\langle\psi|\Sigma_{jk}|\psi\rangle)^2\right).$$

Find the remoteness for the Bell states

$$|\phi^+\rangle = \frac{1}{\sqrt{2}}(|0\rangle \otimes |0\rangle + |1\rangle \otimes |1\rangle), \quad |\phi^-\rangle = \frac{1}{\sqrt{2}}(|0\rangle \otimes |0\rangle - |1\rangle \otimes |1\rangle)$$

$$|\psi^+\rangle = \frac{1}{\sqrt{2}}(|0\rangle \otimes |1\rangle + |1\rangle \otimes |0\rangle), \quad |\psi^-\rangle = \frac{1}{\sqrt{2}}(|0\rangle \otimes |1\rangle - |1\rangle \otimes |0\rangle).$$

**Solution 35.** We have for all $j, k = 1, 2, 3$

$$\Sigma_{jk}^2 = I_2 \otimes I_2 = I_4.$$

Thus $\langle \psi | (\Sigma_{jk})^2 | \psi \rangle = 1$, where $|\psi\rangle$ is one of the Bell states. Now

$$\langle \psi | (\sigma_1 \otimes \sigma_1) | \psi \rangle = \langle \psi | (\sigma_2 \otimes \sigma_2) | \psi \rangle = \langle \psi | (\sigma_3 \otimes \sigma_3) | \psi \rangle = 1.$$

All the other matrices of $\Sigma_{jk}$ yield 0 for all Bell states. Thus we find for the remoteness of all Bell states $R = 6$.

**Problem 36.** A general pure state $|\Psi\rangle$ of two qubits can be written as

$$|\Psi\rangle = e^{i\phi_0} \cos\theta_0 |00\rangle + e^{i\phi_1} \sin(\theta_0)\cos(\theta_1)|01\rangle$$
$$+ e^{i\phi_2} \sin(\theta_0)\sin(\theta_1)\cos(\theta_2)|10\rangle + e^{i\phi_3}\sin(\theta_0)\sin(\theta_1)\sin(\theta_2)|11\rangle \quad (1)$$

where $\phi_j$ and $\theta_k$ are chosen uniformly according to the *Haar measure*

$$d\mu = \frac{1}{(2\pi)^4} d(\sin(\theta_0))^6 d(\sin(\theta_1))^4 d(\sin(\theta_2))^2 d\phi_0 d\phi_1 d\phi_2 d\phi_3 \quad (2)$$

with

$$0 \le \phi_j < 2\pi, \qquad 0 \le \theta_k < \frac{\pi}{2} \quad (3)$$

where $j = 0, 1, 2, 3$ and $k = 0, 1, 2$. An extra overall random phase $e^{i\phi_0}$ is included to maintain consistency with $SU(n)$, where $n = 4$. For a pure state of two qubits the *tangle* $\tau$, is defined as

$$\tau := 4\det(\rho_A) \quad (4)$$

where $\rho_A$ is the *reduced density matrix* obtained when qubit $B$ has been traced over (or vice versa, permuting $A$ and $B$). The tangle $\tau$ is an entanglement measure.

(i) Find $\tau$ for $|\psi\rangle$. Then find $\tau$ for the four Bell states and the unentangled state $|00\rangle$.

(ii) Using the Haar measure find $\tau$ for a randomly selected pure state.

**Solution 36.**    (i) From the state $|\Psi\rangle$ we obtain the $4 \times 4$ density matrix

$$
\rho = |\Psi\rangle\langle\Psi| = \begin{pmatrix} \psi_0\psi_0^* & \psi_0\psi_1^* & \psi_0\psi_2^* & \psi_0\psi_3^* \\ \psi_1\psi_0^* & \psi_1\psi_1^* & \psi_1\psi_2^* & \psi_1\psi_3^* \\ \psi_2\psi_0^* & \psi_2\psi_1^* & \psi_2\psi_2^* & \psi_2\psi_3^* \\ \psi_3\psi_0^* & \psi_3\psi_1^* & \psi_3\psi_2^* & \psi_3\psi_3^* \end{pmatrix}
\tag{5}
$$

where

$$
\psi_0 = e^{i\phi_0}\cos(\theta_0), \quad \psi_1 = e^{i\phi_1}\sin(\theta_0)\cos(\theta_1)
$$

$$
\psi_2 = e^{i\phi_2}\sin(\theta_0)\sin(\theta_1)\cos(\theta_2), \quad \psi_3 = e^{i\phi_3}\sin(\theta_0)\sin(\theta_1)\sin(\theta_2). \tag{6}
$$

Using the basis

$$
\begin{pmatrix} 1 \\ 0 \end{pmatrix} \otimes I_2, \qquad \begin{pmatrix} 0 \\ 1 \end{pmatrix} \otimes I_2 \tag{7}
$$

where $I_2$ is the $2 \times 2$ unit matrix we find the $2 \times 2$ matrix

$$
\rho_A = \begin{pmatrix} \psi_0\psi_0^* + \psi_2\psi_2^* & \psi_0\psi_1^* + \psi_2\psi_3^* \\ \psi_1\psi_0^* + \psi_3\psi_2^* & \psi_1\psi_1^* + \psi_3\psi_3^* \end{pmatrix}. \tag{8}
$$

It follows that

$$
\det(\rho_A) = (\psi_0\psi_0^* + \psi_2\psi_2^*)(\psi_1\psi_1^* + \psi_3\psi_3^*) - (\psi_1\psi_0^* + \psi_3\psi_2^*)(\psi_0\psi_1^* + \psi_2\psi_3^*). \tag{9}
$$

Therefore

$$
\det(\rho_A) = \psi_0\psi_0^*\psi_3\psi_3^* + \psi_1\psi_1^*\psi_2\psi_2^* - \psi_0\psi_1^*\psi_2^*\psi_3 - \psi_0^*\psi_1\psi_2\psi_3^*. \tag{10}
$$

Inserting (6) into (10) we get

$$
\det(\rho_A) = \cos^2(\theta_0)\sin^2(\theta_0)\sin^2(\theta_1)\sin^2(\theta_2) + \sin^4(\theta_0)\cos^2(\theta_1)\sin^2(\theta_1)\cos^2(\theta_2)
$$
$$
- (e^{i(\phi_0 - \phi_1 - \phi_2 - \phi_3)} + e^{i(-\phi_0 + \phi_1 + \phi_2 - \phi_3)})
$$
$$
\times \sin^3(\theta_0)\cos(\theta_0)\sin^2(\theta_1)\cos(\theta_1)\sin(\theta_2)\cos(\theta_2).
$$

It follows that the four Bell states have the maximum possible entanglement, i.e. $\tau = 1$. The product state $|00\rangle$ has $\tau = 0$.
(ii) From (2) we find

$$
\frac{48}{(2\pi)^4}\cos(\theta_0)(\sin(\theta_0))^5\cos(\theta_1)(\sin\theta_1)^3\cos(\theta_2)\sin(\theta_2)d\theta_0 d\theta_1 d\theta_2 d\phi_0 d\phi_1 d\phi_2 d\phi_3
$$

and

$$
\int_{SU(4)} d\mu = 1
$$

i.e. the Haar measure is normalized. Here we made use of

$$
\int_0^{2\pi} d\phi = 2\pi, \quad \text{and} \quad \int_0^{\pi/2} \sin^k(x)\cos(x)dx = \frac{1}{k+1}
$$

where $k = 1, 2, \ldots$. Integrating $\det(\rho_A)$ (or $\det(\rho_B)$) over the Haar measure gives $\langle \tau \rangle = \frac{2}{5}$, where we used

$$\int_0^{\pi/2} \sin^m(x) \cos^n(x) dx = \frac{m-1}{m+n} \int_0^{\pi/2} \sin^{m-2}(x) \cos^n(x) dx$$

$$\int_0^{\pi/2} \sin^m(x) \cos^n(x) dx = \frac{n-1}{m+n} \int_0^{\pi/2} \sin^m(x) \cos^{n-2}(x) dx$$

and

$$\int_0^{\pi/2} \sin(x) \cos(x) dx = \frac{1}{2}.$$

A randomly selected pure state of two qubits might thus be expected to have 0.4 tangle units of entanglement.

## Programming Problems

**Problem 1.** Let $|\psi\rangle$ be a given state in the Hilbert space $\mathbb{C}^n$. Let $X$ and $Y$ be two $n \times n$ hermitian matrices. We define the *correlation* for a given state $|\psi\rangle$ as

$$C_{XY}(|\psi\rangle) := \langle\psi|XY|\psi\rangle - \langle\psi|X|\psi\rangle\langle\psi|Y|\psi\rangle.$$

Let $n = 4$,

$$X = \begin{pmatrix} 0 & 0 & 0 & 1 \\ 0 & 0 & 1 & 0 \\ 0 & 1 & 0 & 0 \\ 1 & 0 & 0 & 0 \end{pmatrix}, \quad Y = \begin{pmatrix} 1 & 0 & 0 & 0 \\ 0 & 0 & 0 & 1 \\ 0 & 0 & 1 & 0 \\ 0 & 1 & 0 & 0 \end{pmatrix}$$

and consider the Bell state

$$|\psi\rangle = \frac{1}{\sqrt{2}} \begin{pmatrix} 1 \\ 0 \\ 0 \\ 1 \end{pmatrix}.$$

Find the correlation.

**Solution 1.** Since $X|\psi\rangle = |\psi\rangle$ we have $\langle\psi|X|\psi\rangle = 1$ and $\langle\psi|X = \langle\psi|$. Thus

$$\langle\psi|XY|\psi\rangle - \langle\psi|X|\psi\rangle\langle\psi|Y|\psi\rangle = \langle\psi|Y|\psi\rangle - \langle\psi|Y|\psi\rangle = 0.$$

A Maxima implementation is

```
/* CorrelationBell.mac */
X: matrix([0,0,0,1],[0,0,1,0],[0,1,0,0],[1,0,0,0]);
Y: matrix([1,0,0,0],[0,0,0,1],[0,0,1,0],[0,1,0,0]);
b: matrix([1/sqrt(2)],[0],[0],[1/sqrt(2)]);
bT: transpose(b);
CXYb: bT . X . Y . b - (bT . X . b) . (bT . Y . b);
```

**Problem 2.** Consider the normalized vector in $\mathbb{C}^6$

$$\mathbf{v} = \frac{1}{2}\begin{pmatrix} 1 \\ -1 \\ 0 \\ 0 \\ 1 \\ -1 \end{pmatrix}.$$

Let $\mathbf{u}$ be a vector in $\mathbb{C}^3$ and $\mathbf{w}$ be a vector in $\mathbb{C}^2$. Can $\mathbf{v}$ be written as $\mathbf{v} = \mathbf{u} \otimes \mathbf{w}$? Can $\mathbf{v}$ be written as $\mathbf{v} = \mathbf{w} \otimes \mathbf{u}$?

**Solution 2.** In the first case we have to solve the system of equations

$$\begin{pmatrix} 1/2 \\ -1/2 \\ 0 \\ 0 \\ 1/2 \\ -1/2 \end{pmatrix} = \begin{pmatrix} u_1 w_1 \\ u_1 w_2 \\ u_2 w_1 \\ u_2 w_2 \\ u_3 w_1 \\ u_3 w_2 \end{pmatrix}.$$

In the second case we have to solve the system of equations

$$\begin{pmatrix} 1/2 \\ -1/2 \\ 0 \\ 0 \\ 1/2 \\ -1/2 \end{pmatrix} = \begin{pmatrix} w_1 u_1 \\ w_1 u_2 \\ w_1 u_3 \\ w_2 u_1 \\ w_2 u_2 \\ w_2 u_3 \end{pmatrix}.$$

For the first system the Maxima program

```
/* C6.mac */
solve([1/2-u1*w1=0,-1/2-u1*w2=0,u2*w1=0,u2*w2=0,1/2-u3*w1=0,-1/2-u3*w2],
      [u1,u2,u3,w1,w2]);
solve([1/2-w1*u1=0,-1/2-w1*u2=0,w1*u3=0,w2*u1=0,1/2-w2*u2=0,-1/2-w2*u3],
      [u1,u2,u3,w1,w2]);
```

provides the solution ($r$ arbitrary and $r \neq 0$)

$$u_1 = \frac{1}{2r}, \quad u_2 = 0, \quad u_3 = \frac{1}{2r}, \quad w_1 = r, \quad w_2 = -r.$$

Hence we can write

$$\frac{1}{2}\begin{pmatrix} 1 \\ -1 \\ 0 \\ 0 \\ 1 \\ -1 \end{pmatrix} = \frac{1}{\sqrt{2}}\begin{pmatrix} 1 \\ 0 \\ 1 \end{pmatrix} \otimes \frac{1}{\sqrt{2}}\begin{pmatrix} 1 \\ -1 \end{pmatrix}.$$

The solution set for the second system is empty, i.e. $\mathbf{v}$ cannot be written as $\mathbf{v} = \mathbf{w} \otimes \mathbf{u}$.

## 10.3  Supplementary Problems

**Problem 1.**   Consider the two Hilbert spaces $\mathcal{H}_1 = \mathcal{H}_2 = \mathbb{C}^d$ and the product Hilbert space $\mathcal{H} = \mathcal{H}_1 \otimes \mathcal{H}_2$. A state $|\psi\rangle \in \mathcal{H}$ is called maximally entangled if

$$\mathrm{tr}_{\mathcal{H}_1}(|\psi\rangle\langle\psi|) = \mathrm{tr}_{\mathcal{H}_2}(|\psi\rangle\langle\psi|) = \frac{1}{d}.$$

Apply this definition to the Bell states in $\mathcal{H} = \mathbb{C}^4$, i.e. $d = 2$

$$|\psi_1\rangle = \frac{1}{\sqrt{2}}\begin{pmatrix} 1 \\ 0 \\ 0 \\ 1 \end{pmatrix}, \quad |\psi_2\rangle = \frac{1}{\sqrt{2}}\begin{pmatrix} 1 \\ 0 \\ 0 \\ -1 \end{pmatrix},$$

$$|\psi_3\rangle = \frac{1}{\sqrt{2}}\begin{pmatrix} 0 \\ 1 \\ 1 \\ 0 \end{pmatrix}, \quad |\psi_4\rangle = \frac{1}{\sqrt{2}}\begin{pmatrix} 0 \\ 1 \\ -1 \\ 0 \end{pmatrix}.$$

**Problem 2.**   (i) The normalized states

$$\frac{1}{2}\begin{pmatrix} -1 \\ 1 \\ 1 \\ 1 \end{pmatrix}, \quad \frac{1}{2}\begin{pmatrix} 1 \\ -1 \\ 1 \\ 1 \end{pmatrix}, \quad \frac{1}{2}\begin{pmatrix} 1 \\ 1 \\ -1 \\ 1 \end{pmatrix}, \quad \frac{1}{2}\begin{pmatrix} 1 \\ 1 \\ 1 \\ -1 \end{pmatrix}$$

form an orthonormal basis in $\mathbb{C}^4$. Are the states entangled?
(ii) Do the normalized states

$$\frac{1}{\sqrt{3}}\begin{pmatrix} -1 \\ 1 \\ 1 \\ 0 \end{pmatrix}, \quad \frac{1}{\sqrt{3}}\begin{pmatrix} 0 \\ -1 \\ 1 \\ 1 \end{pmatrix}, \quad \frac{1}{\sqrt{3}}\begin{pmatrix} 1 \\ 0 \\ -1 \\ 1 \end{pmatrix}, \quad \frac{1}{\sqrt{3}}\begin{pmatrix} 1 \\ 1 \\ 0 \\ -1 \end{pmatrix}$$

form an orthonormal basis in $\mathbb{C}^4$? Are the states entangled?

**Problem 3.** Show that the GHZ state

$$|\psi\rangle = \frac{1}{\sqrt{2}}(1\ \ 0\ \ 0\ \ 0\ \ 0\ \ 0\ \ 0\ \ 1)^T$$

in $\mathbb{C}^8$ is entangled.

**Problem 4.** We consider the finite-dimensional Hilbert space $\mathcal{H} = \mathbb{C}^{2^n}$ and the normalized state

$$|\psi\rangle = \sum_{j_1,j_2,\ldots,j_n=0}^{1} c_{j_1,j_2,\ldots,j_n}|j_1\rangle \otimes |j_2\rangle \otimes \cdots \otimes |j_n\rangle$$

in this Hilbert space. Here $|0\rangle$, $|1\rangle$ denotes the standard basis. Let $\epsilon_{jk}$ $(j,k=0,1)$ be defined by $\epsilon_{00} = \epsilon_{11} = 0$, $\epsilon_{01} = 1$, $\epsilon_{10} = -1$. Let $n$ be even or $n = 3$. Then an $n$-tangle can be introduced by

$$\tau_{1\ldots n} = 2\left| \sum_{\substack{\alpha_1,\ldots,\alpha_n=0 \\ \delta_1,\ldots,\delta_n=0}}^{1} c_{\alpha_1\ldots\alpha_n} c_{\beta_1\ldots\beta_n} c_{\gamma_1\ldots\gamma_n} c_{\delta_1\ldots\delta_n} \right.$$

$$\left. \times \epsilon_{\alpha_1\beta_1}\epsilon_{\alpha_2\beta_2}\cdots\epsilon_{\alpha_{n-1}\beta_{n-1}}\epsilon_{\gamma_1\delta_1}\epsilon_{\gamma_2\delta_2}\cdots\epsilon_{\gamma_{n-1}\delta_{n-1}}\epsilon_{\alpha_n\gamma_n}\epsilon_{\beta_n\delta_n} \right|.$$

Consider the case $n = 4$ and a state $|\psi\rangle$ with $c_{0000} = 1/\sqrt{2}$, $c_{1111} = 1/\sqrt{2}$ and all other coefficients are 0. Show that $\tau_{1234} = 1$.

# Chapter 11

# Bell Inequality

## 11.1 Introduction

Bell's theorem states that, according to quantum mechanics, the value of a certain combination of correlations for experiments of two distant systems can be higher than the highest value allowed by any local-realistic theory of the type proposed by Einstein, Podolsky and Rosen, in which local properties of a system determine the result of any experiment on that system. The most discussed Bell inequality, the Clauser-Horne-Shimony-Holt (CHSH) inequality states that in any local-realistic theory the absolute value of a combination of four correlations is bounded by 2.

Consider a system with two distant particles $i$ and $j$. Let $A$ and $a$ ($B$ and $b$) be physical observables taking values $-1$ and $1$ referring to local experiments on particle $i$ ($j$). The correlation $C(A, B)$ of $A$ and $B$ is defined as

$$C(A, B) := P_{AB}(1, 1) - P_{AB}(1, -1) - P_{AB}(-1, 1) + P_{AB}(-1, -1)$$

where $P_{AB}(1, -1)$ denotes the joint probability of obtaining $A = 1$ and $B = -1$ when $A$ and $B$ are measured. In any theory in which local variables of particle $i$ ($j$) determine the result of local experiments on particle $i$ ($j$), the absolute value of a particular combination of correlations is bounded by 2

$$|C(A, B) - mC(A, b) - nC(a, B) - mnC(a, b)| \leq 2$$

where $m$ and $n$ can either be $-1$ or $1$. The CHSH inequality holds for any local-realistic theory, where $m, n \in \{-1, 1\}$. For a two particle system in a quantum pure state $|\psi\rangle$, the quantum correlation of $A$ and $B$ is defined

$$C_Q(A, B) := \langle \psi | \hat{A}\hat{B} | \psi \rangle$$

where $\hat{A}$ and $\hat{B}$ are the self-adjoint operators which represent the observables $A$ and $B$. For certain choices of $\hat{A}$, $\hat{a}$, $\hat{B}$, $\hat{b}$, and $|\psi\rangle$, the quantum correlation violates the CHSH inequality.

Let $\mathcal{H}_A$ and $\mathcal{H}_B$ be finite-dimensional Hilbert spaces. Let $\mathcal{H}$ be the Hilbert space $\mathcal{H} = \mathcal{H}_A \otimes \mathcal{H}_B$, i.e. $\mathcal{H}$ is the tensor product of the two Hilbert spaces $\mathcal{H}_A$ and $\mathcal{H}_B$. Let $\hat{A}_1$, $\hat{A}_2$ be hermitian operators (matrices) in $\mathcal{H}_A$ with

$$\hat{A}_1^2 = I_A, \qquad \hat{A}_2^2 = I_A$$

and let $\hat{B}_1$, $\hat{B}_2$ be hermitian operators (matrices) in $\mathcal{H}_B$ with

$$\hat{B}_1^2 = I_B, \qquad \hat{B}_2^2 = I_B.$$

Let $|\psi\rangle$ be a normalized state in the product Hilbert space $\mathcal{H}_A \otimes \mathcal{H}_B$. The generalized *Bell inequality* is given by

$$|\langle \psi | \hat{A}_1 \otimes \hat{B}_1 | \psi \rangle + \langle \psi | \hat{A}_1 \otimes \hat{B}_2 | \psi \rangle + \langle \psi | \hat{A}_2 \otimes \hat{B}_1 | \psi \rangle - \langle \psi | \hat{A}_2 \otimes \hat{B}_2 | \psi \rangle| \leq 2.$$

If $|\psi\rangle$ can be written as a product state $|\psi\rangle = |\phi_A\rangle \otimes |\phi_B\rangle$ then the Bell inequality is not violated.

The Pauli spin matrices $\sigma_1$, $\sigma_2$, $\sigma_3$ admit the eigenvalues $+1$ and $-1$ and the $2 \times 2$ identity matrix admits the eigenvalue $+1$. Hence all Kronecker products of these matrices admits these eigenvalues.
Let

$$A_1 = \sigma_3 \otimes I_2, \quad A_2 = \sigma_1 \otimes I_2, \quad A_3 = \sigma_2 \otimes I_2$$

$$B_1 = I_2 \otimes \frac{1}{\sqrt{3}}(\sigma_3 + \sigma_1 - \sigma_2), \quad B_2 = I_2 \otimes \frac{1}{\sqrt{3}}(\sigma_3 - \sigma_1 + \sigma_2)$$

$$B_3 = I_2 \otimes \frac{1}{\sqrt{3}}(-\sigma_3 + \sigma_1 + \sigma_2), \quad B_4 = I_2 \otimes \frac{1}{\sqrt{3}}(-\sigma_3 - \sigma_1 - \sigma_2)$$

$$\Sigma := A_1(B_1 + B_2 - B_3 - B_4) + A_2(B_1 - B_2 + B_3 - B_4) + A_3(B_1 - B_2 - B_3 + B_4)$$

and $|\psi\rangle = \frac{1}{\sqrt{2}}(|0\rangle_A \otimes |0\rangle_B + |1\rangle_A \otimes |1\rangle_B)$. Then

$$\langle \psi | \Sigma | \psi \rangle = 4\sqrt{3}.$$

# 11.2 Solved Problems

**Problem 1.** Let $\mathcal{H}_A$ and $\mathcal{H}_B$ be finite-dimensional Hilbert spaces. Let $\mathcal{H}$ be the Hilbert space $\mathcal{H} = \mathcal{H}_A \otimes \mathcal{H}_B$, i.e. $\mathcal{H}$ is the tensor product of the two Hilbert spaces $\mathcal{H}_A$ and $\mathcal{H}_B$. Let $|\psi\rangle$ be a normalized vector (pure state) in $\mathcal{H}$. Let $X$ be an observable (described as a hermitian matrix $\hat{X}$) in $\mathcal{H}$. Then $\langle\psi|\hat{X}|\psi\rangle$ defines the expectation values. The following three conditions are equivalent when applied to pure states.

1. *Factorisability:* $|\psi\rangle = |\alpha\rangle \otimes |\beta\rangle$, where $|\alpha\rangle \in \mathcal{H}_A$ and $|\beta\rangle \in \mathcal{H}_B$ with $|\alpha\rangle$ and $|\beta\rangle$ normalized.

2. The generalized *Bell inequality:* Let $\hat{A}_1$, $\hat{A}_2$ be hermitian operators (matrices) in $\mathcal{H}_A$ with

$$\hat{A}_1^2 = I_A, \qquad \hat{A}_2^2 = I_A$$

where $I_A$ is the identity operator in $\mathcal{H}_A$. Let $\hat{B}_1$, $\hat{B}_2$ be hermitian operators (matrices) in $\mathcal{H}_B$ with

$$\hat{B}_1^2 = I_B, \qquad \hat{B}_2^2 = I_B$$

where $I_B$ is the identity operator in $\mathcal{H}_B$. Thus the eigenvalues of $\hat{A}_1$, $\hat{A}_2$, $\hat{B}_1$ and $\hat{B}_2$ can only be $\pm 1$. The generalized Bell inequality is

$$|\langle\psi|\hat{A}_1 \otimes \hat{B}_1|\psi\rangle + \langle\psi|\hat{A}_1 \otimes \hat{B}_2|\psi\rangle + \langle\psi|\hat{A}_2 \otimes \hat{B}_1|\psi\rangle - \langle\psi|\hat{A}_2 \otimes \hat{B}_2|\psi\rangle| \le 2.$$

3. *Statistical independence:* For all hermitian operators $\hat{A}$ on $\mathcal{H}_A$ and $\hat{B}$ on $\mathcal{H}_B$ with the conditions given above

$$\langle\psi|\hat{A} \otimes \hat{B}|\psi\rangle = \langle\psi|\hat{A} \otimes I_B|\psi\rangle\langle\psi|I_A \otimes \hat{B}|\psi\rangle.$$

(i) Show that condition 3 follows from condition 1.
(ii) Show that condition 2 follows from condition 3.

**Solution 1.** (i) Consider the product state $|\psi\rangle = |\alpha\rangle \otimes |\beta\rangle$. Then

$$\langle\psi|(\hat{A} \otimes \hat{B})|\psi\rangle = ((\langle\beta| \otimes \langle\alpha|)(\hat{A} \otimes \hat{B})(|\alpha\rangle \otimes |\beta\rangle)) = \langle\alpha|\hat{A}|\alpha\rangle\langle\beta|\hat{B}|\beta\rangle$$
$$= \langle\psi|\hat{A} \otimes I_B|\psi\rangle\langle\psi|I_A \otimes \hat{B}|\psi\rangle.$$

(ii) We use the shortcut notation $\langle\hat{A}_1 \otimes \hat{B}_1\rangle \equiv \langle\psi|\hat{A}_1 \otimes \hat{B}_1|\psi\rangle$ etc. Using statistical independence we have

$$|\langle\hat{A}_1 \otimes \hat{B}_1\rangle + \langle\hat{A}_1 \otimes \hat{B}_2\rangle + \langle\hat{A}_2 \otimes \hat{B}_1\rangle - \langle\hat{A}_2 \otimes \hat{B}_2\rangle| =$$

$$|\langle\hat{A}_1 \otimes I_B\rangle(\langle\hat{I}_A \otimes \hat{B}_1\rangle + \langle\hat{I}_A \otimes \hat{B}_2\rangle) + \langle\hat{A}_2 \otimes I_B\rangle(\langle I_A \otimes \hat{B}_1\rangle - \langle I_A \otimes \hat{B}_2\rangle)|.$$

Using the fact that $|\langle \hat{A}_1 \otimes I_B \rangle| \leq 1$ and $|\langle \hat{A}_2 \otimes I_B \rangle| \leq 1$ we have

$$|\langle \hat{A}_1 \otimes \hat{B}_1 \rangle + \langle \hat{A}_1 \otimes \hat{B}_2 \rangle + \langle \hat{A}_2 \otimes \hat{B}_1 \rangle - \langle \hat{A}_2 \otimes \hat{B}_2 \rangle|$$
$$\leq |\langle I_A \otimes \hat{B}_1 \rangle + \langle I_A \otimes \hat{B}_2 \rangle| + |\langle I_A \otimes \hat{B}_1 \rangle - \langle I_A \otimes \hat{B}_2 \rangle|$$
$$\leq \max(((\langle I_A \otimes \hat{B}_1 \rangle + \langle I_A \otimes \hat{B}_2 \rangle) + (\langle I_A \otimes \hat{B}_1 \rangle - \langle I_A \otimes \hat{B}_2 \rangle),$$
$$(\langle I_A \otimes \hat{B}_1 \rangle + \langle I_A \otimes \hat{B}_2 \rangle) - (\langle I_A \otimes \hat{B}_1 \rangle - \langle I_A \otimes \hat{B}_2 \rangle),$$
$$-(\langle \hat{I}_A \otimes \hat{B}_1 \rangle + \langle I_A \otimes \hat{B}_1 \rangle) + (\langle I_A \otimes \hat{B}_1 \rangle - \langle I_A \otimes \hat{B}_2 \rangle),$$
$$-(\langle I_A \otimes \hat{B}_1 \rangle + \langle I_A \otimes \hat{B}_2 \rangle) - (\langle I_A \otimes \hat{B}_1 \rangle - \langle I_A \otimes \hat{B}_1 \rangle))$$
$$= \max(2\langle I_A \otimes \hat{B}_1 \rangle, 2\langle I_A \otimes \hat{B}_2 \rangle, -2\langle I_A \otimes \hat{B}_2 \rangle, -2\langle I_A \otimes \hat{B}_1 \rangle) \leq 2$$

where we also used $|\langle I_A \otimes \hat{B}_1 \rangle| \leq 1$, $|\langle I_A \otimes \hat{B}_2 \rangle| \leq 1$.

**Problem 2.**    Let $\mathcal{H}_A = \mathcal{H}_B = \mathbb{C}^2$. Let $\{\, |0\rangle, |1\rangle \,\}$ be the standard basis in $\mathbb{C}^2$. Consider the entangled state in $\mathcal{H} = \mathbb{C}^4$ (*EPR state*)

$$|\psi\rangle = \frac{1}{\sqrt{2}}(|0\rangle \otimes |1\rangle - |1\rangle \otimes |0\rangle)$$

which is one of the Bell states. Show that this state and the operators

$$\hat{A}_1 := \sigma_1, \qquad \hat{A}_2 := \sigma_2$$

$$\hat{B}_1 := \frac{1}{\sqrt{2}}(\sigma_1 + \sigma_2), \qquad \hat{B}_2 := \frac{1}{\sqrt{2}}(\sigma_1 - \sigma_2)$$

violate the Bell inequality.

**Solution 2.**    We have

$$\hat{A}_1|0\rangle = |1\rangle, \qquad \hat{A}_1|1\rangle = |0\rangle$$

$$\hat{A}_2|0\rangle = i|1\rangle, \qquad \hat{A}_2|1\rangle = -i|0\rangle$$

$$\hat{B}_1|0\rangle = \frac{1}{\sqrt{2}}(|1\rangle + i|1\rangle), \qquad \hat{B}_1|1\rangle = \frac{1}{\sqrt{2}}(|0\rangle - i|0\rangle)$$

$$\hat{B}_2|0\rangle = \frac{1}{\sqrt{2}}(|1\rangle - i|1\rangle), \qquad \hat{B}_2|1\rangle = \frac{1}{\sqrt{2}}(|0\rangle - i|0\rangle).$$

Using $\langle 0|0\rangle = \langle 1|1\rangle = 1$ and $\langle 0|1\rangle = \langle 1|0\rangle = 0$, we find

$$|\langle\psi|\hat{A}_1 \otimes \hat{B}_1|\psi\rangle + \langle\psi|\hat{A}_1 \otimes \hat{B}_2|\psi\rangle + \langle\psi|\hat{A}_2 \otimes \hat{B}_1|\psi\rangle - \langle\psi|\hat{A}_2 \otimes \hat{B}_2|\psi\rangle| = 2\sqrt{2}.$$

Thus the Bell inequality is violated since $2\sqrt{2} > 2$.

**Problem 3.** Let $\sigma_1$, $\sigma_2$ and $\sigma_3$ be the Pauli spin matrices. Consider the *Bell operator* defined by

$$B := \frac{1}{2} \sum_{j,k=1}^{3} (a_j(c_k + d_k) + b_j(c_k - d_k))\sigma_j \otimes \sigma_k$$

where $\mathbf{a}, \mathbf{b}, \mathbf{c}, \mathbf{d}$ are real unit vectors in $\mathbb{R}^3$.
(i) Calculate the matrix $B^2$.
(ii) Consider the Bell state

$$|\Phi^+\rangle = \frac{1}{\sqrt{2}} \begin{pmatrix} 1 \\ 0 \\ 0 \\ 1 \end{pmatrix}.$$

Calculate the density operator $\rho = |\Phi^+\rangle\langle\Phi^+|$ and then $\text{tr}(\rho B)$. Discuss.
(iii) Let

$$|e_1\rangle = \begin{pmatrix} 1 \\ 0 \\ 0 \\ 0 \end{pmatrix}.$$

Calculate the density operator $\rho = |e_1\rangle\langle e_1|$ and then $\text{tr}(\rho B)$. Compare to (ii).

**Solution 3.** (i) Since

$$(\sigma_j \otimes \sigma_k)(\sigma_m \otimes \sigma_n) \equiv (\sigma_j\sigma_m) \otimes (\sigma_k\sigma_n)$$

we have

$$B^2 = \frac{1}{4} \sum_{j,k=1}^{3} \sum_{m,n=1}^{3} (a_j(c_k + d_k) + b_j(c_k - d_k))(a_m(c_n + d_n)$$
$$+ b_m(c_n - d_n))(\sigma_j\sigma_m) \otimes (\sigma_k\sigma_n).$$

Since $\sigma_j^2 = I_2$,

$$\sigma_1\sigma_2 = -\sigma_2\sigma_1 = i\sigma_3, \quad \sigma_2\sigma_3 = -\sigma_3\sigma_2 = i\sigma_1, \quad \sigma_3\sigma_1 = -\sigma_1\sigma_3 = i\sigma_2$$

and $B^2$ is hermitian (since $B$ is hermitian) we obtain

$$B^2 = I_2 \otimes I_2 + (a_2b_3 - a_3b_2)(c_2d_3 - c_3d_2)\sigma_1 \otimes \sigma_1$$
$$+ (a_1b_3 - b_1a_3)(c_1d_3 - d_1c_3)\sigma_2 \otimes \sigma_2$$
$$+ (a_1b_2 - a_2b_1)(c_1d_2 - c_2d_1)\sigma_3 \otimes \sigma_3$$
$$+ (a_2b_3 - b_2a_3)(c_3d_1 - c_1d_3)\sigma_1 \otimes \sigma_2$$

$$+(a_1b_3 - b_1a_3)(c_3d_2 - c_2d_3)\sigma_2 \otimes \sigma_1$$
$$+(a_2b_3 - a_3b_2)(c_1d_2 - c_2d_1)\sigma_1 \otimes \sigma_3$$
$$+(a_1b_2 - a_2b_1)(c_2d_3 - c_3d_2)\sigma_3 \otimes \sigma_1$$
$$+(a_1b_3 - a_3b_1)(c_2d_1 - c_1d_2)\sigma_2 \otimes \sigma_3$$
$$+(a_1b_2 - a_2b_1)(c_3d_1 - c_1d_3)\sigma_3 \otimes \sigma_2.$$

(ii) We obtain

$$\rho = |\Phi^+\rangle\langle\Phi^+| = \frac{1}{2}\begin{pmatrix} 1 & 0 & 0 & 1 \\ 0 & 0 & 0 & 0 \\ 0 & 0 & 0 & 0 \\ 1 & 0 & 0 & 1 \end{pmatrix}.$$

Thus $\rho$ can be written as a sum of Kronecker products

$$\rho = \frac{1}{2}\left(\begin{pmatrix} 1 & 0 \\ 0 & 0 \end{pmatrix} \otimes \begin{pmatrix} 1 & 0 \\ 0 & 0 \end{pmatrix} + \begin{pmatrix} 0 & 0 \\ 0 & 1 \end{pmatrix} \otimes \begin{pmatrix} 0 & 0 \\ 0 & 1 \end{pmatrix}\right.$$
$$\left. + \begin{pmatrix} 0 & 1 \\ 0 & 0 \end{pmatrix} \otimes \begin{pmatrix} 0 & 1 \\ 0 & 0 \end{pmatrix} + \begin{pmatrix} 0 & 0 \\ 1 & 0 \end{pmatrix} \otimes \begin{pmatrix} 0 & 0 \\ 1 & 0 \end{pmatrix}\right).$$

Since

$$\mathrm{tr}\left(\begin{pmatrix} 1 & 0 \\ 0 & 0 \end{pmatrix}\sigma_j\right) = 0, \qquad \mathrm{tr}\left(\begin{pmatrix} 0 & 0 \\ 0 & 1 \end{pmatrix}\sigma_j\right) = 0$$

for $j = 1, 2$ and

$$\mathrm{tr}\left(\begin{pmatrix} 1 & 0 \\ 0 & 0 \end{pmatrix}\sigma_3\right) = 1, \qquad \mathrm{tr}\left(\begin{pmatrix} 0 & 0 \\ 0 & 1 \end{pmatrix}\sigma_3\right) = -1,$$

$$\mathrm{tr}\left(\begin{pmatrix} 0 & 1 \\ 0 & 0 \end{pmatrix}\sigma_1\right) = 1, \qquad \mathrm{tr}\left(\begin{pmatrix} 0 & 0 \\ 1 & 0 \end{pmatrix}\sigma_1\right) = 1,$$

$$\mathrm{tr}\left(\begin{pmatrix} 0 & 1 \\ 0 & 0 \end{pmatrix}\sigma_2\right) = i, \qquad \mathrm{tr}\left(\begin{pmatrix} 0 & 0 \\ 1 & 0 \end{pmatrix}\sigma_2\right) = -i$$

$$\mathrm{tr}\left(\begin{pmatrix} 0 & 1 \\ 0 & 0 \end{pmatrix}\sigma_3\right) = 0, \qquad \mathrm{tr}\left(\begin{pmatrix} 0 & 0 \\ 1 & 0 \end{pmatrix}\sigma_3\right) = 0$$

we find

$$\mathrm{tr}(\rho B) = \frac{1}{2}\left(a_1(c_1 + d_1) + b_1(c_1 - d_1) - a_2(c_2 + d_2) - b_2(c_2 - d_2)\right.$$
$$\left. + a_3(c_3 + d_3) + b_3(c_3 - d_3)\right).$$

(iii) We have

$$\rho = |e_1\rangle\langle e_1| = \begin{pmatrix} 1 & 0 & 0 & 0 \\ 0 & 0 & 0 & 0 \\ 0 & 0 & 0 & 0 \\ 0 & 0 & 0 & 0 \end{pmatrix} = \begin{pmatrix} 1 & 0 \\ 0 & 0 \end{pmatrix} \otimes \begin{pmatrix} 1 & 0 \\ 0 & 0 \end{pmatrix}.$$

Thus

$$\mathrm{tr}(\rho B) = \frac{1}{2}(a_3(c_3 + d_3) + b_3(c_3 - d_3)).$$

**Problem 4.**    Consider the Pauli spin matrices $\boldsymbol{\sigma} = (\sigma_1, \sigma_2, \sigma_3)$. Let $\mathbf{q}$, $\mathbf{r}$, $\mathbf{s}$, $\mathbf{t}$ be unit vectors in $\mathbb{R}^3$. We define

$$Q := \mathbf{q} \cdot \boldsymbol{\sigma}, \quad R := \mathbf{r} \cdot \boldsymbol{\sigma}, \quad S := \mathbf{s} \cdot \boldsymbol{\sigma}, \quad T := \mathbf{t} \cdot \boldsymbol{\sigma}.$$

Calculate the matrix $(Q \otimes S + R \otimes S + R \otimes T - Q \otimes T)^2$ and express the result using commutators.

**Solution 4.**    Using that for $j, k = 1, 2, 3$ we have

$$\sigma_j \sigma_k = \delta_{jk} I_2 + i \sum_{\ell=1}^{3} \epsilon_{jk\ell} \sigma_\ell$$

with $\epsilon_{123} = \epsilon_{231} = \epsilon_{312} = 1$, $\epsilon_{321} = \epsilon_{213} = \epsilon_{132} = -1$ and 0 otherwise we obtain

$$(Q \otimes S + R \otimes S + R \otimes T - Q \otimes T)^2 \equiv 4I_2 \otimes I_2 + [Q, R] \otimes [S, T].$$

**Problem 5.**    Let $X_1$ and $X_2$ be $m \times m$ hermitian matrices with

$$X_1^2 = X_2^2 = I_m.$$

Let $Y_1$ and $Y_2$ be $n \times n$ hermitian matrices with

$$Y_1^2 = Y_2^2 = I_n.$$

(i) What can be said about the eigenvalues of $X_1$, $X_2$, $Y_1$ and $Y_2$?
(ii) Consider the so-called *Bell operator*

$$B := X_1 \otimes (Y_1 + Y_2) + X_2 \otimes (Y_1 - Y_2).$$

Calculate $B^2$. Express the result using commutators.
(iii) What can be said about the eigenvalues of $B^2$ and $B$?

**Solution 5.**    (i) From the eigenvalue equation $X_1 \mathbf{x}_1 = \lambda \mathbf{x}_1$ we obtain

$$X_1(X_1 \mathbf{x}_1) = X_1(\lambda \mathbf{x}_1) = \lambda X_1 \mathbf{x}_1 = \lambda^2 \mathbf{x}_1.$$

Consequently $I_m \mathbf{x}_1 = \lambda^2 \mathbf{x}_1$ and therefore $1 = \lambda^2$. Thus the eigenvalues can only be $\pm 1$. Analogously, we find that the eigenvalues for the matrices $X_2$, $Y_1$ and $Y_2$ can only be $\pm 1$.

(ii) Since $X_1^2 = X_2^2 = I_m$ and $Y_1^2 = Y_2^2 = I_n$ we have

$$
\begin{aligned}
B^2 &= I_m \otimes (Y_1 + Y_2)(Y_1 + Y_2) + (X_1 X_2) \otimes (Y_1 + Y_2)(Y_1 - Y_2) \\
&\quad + (X_2 X_1) \otimes (Y_1 - Y_2)(Y_1 + Y_2) + I_m \otimes (Y_1 - Y_2)(Y_1 - Y_2) \\
&= I_m \otimes (2I_n + Y_1 Y_2 + Y_2 Y_1) + X_1 X_2 \otimes (Y_2 Y_1 - Y_1 Y_2) \\
&\quad + X_2 X_1 \otimes (Y_1 Y_2 - Y_2 Y_1) + I_m \otimes (2I_n - Y_1 Y_2 - Y_2 Y_1) \\
&= 4 I_m \otimes I_n - (X_1 X_2) \otimes [Y_1, Y_2] + (X_2 X_1) \otimes [Y_1, Y_2] \\
&= 4 I_m \otimes I_n - [X_1, X_2] \otimes [Y_1, Y_2] \\
&= 4 I_m \otimes I_n + (i[X_1, X_2]) \otimes (i[Y_1, Y_2]).
\end{aligned}
$$

(iii) Note that the commutator $[A, B]$ of two hermitian $d \times d$ matrices $A$, $B$ is in general not hermitian. However $i[A, B]$ is hermitian. Thus we can find an $m \times m$ unitary matrix $U$ and an $n \times n$ unitary matrix $V$ such that

$$
U(i[X_1, X_2])U^* = \text{diag}(\alpha_1, \alpha_2, \ldots, \alpha_m)
$$

$$
V(i[Y_1, Y_2])V^* = \text{diag}(\beta_1, \beta_2, \ldots, \beta_n)
$$

with $\alpha_j, \beta_j \in \mathbb{R}$. It follows that

$$
\begin{aligned}
(U \otimes V)B^2(U^* \otimes V^*) &= 4 I_m \otimes I_n + (U(i[X_1, X_2])U^*) \otimes (V(i[Y_1, Y_2])V^*) \\
&= 4 I_m \otimes I_n + \text{diag}(\alpha_1, \ldots, \alpha_m) \otimes \text{diag}(\beta_1, \ldots, \beta_n).
\end{aligned}
$$

The real eigenvalues $\alpha_j$ of the hermitian matrix $i[X_1, X_2]$ are restricted by $-2 \le \alpha_j \le 2$ for $j = 1, 2, \ldots, m$. Analogously, the real eigenvalues $\beta_k$ of the hermitian matrix $i[Y_1, Y_2]$ are restricted by $-2 \le \beta_k \le 2$ for $k = 1, 2, \ldots, n$. Thus we have $-4 \le \alpha_j \beta_k \le 4$. The eigenvalues of $B^2$ are therefore given by

$$
4 + \alpha_j \beta_k, \qquad j = 1, 2, \ldots, m, \quad k = 1, 2, \ldots, n.
$$

It follows that the eigenvalues of $B$ are

$$
|\lambda_{jk}| = \sqrt{4 + \alpha_j \beta_k}
$$

with $j = 1, 2, \ldots, m$ and $k = 1, 2, \ldots, n$.

**Problem 6.**  Let $\mathbf{n}$, $\mathbf{m}$ be unit vectors in $\mathbb{R}^3$. Consider the spin singlet state (entangled state)

$$
|\psi\rangle = \frac{1}{\sqrt{2}} \left( \begin{pmatrix} 1 \\ 0 \end{pmatrix} \otimes \begin{pmatrix} 0 \\ 1 \end{pmatrix} - \begin{pmatrix} 0 \\ 1 \end{pmatrix} \otimes \begin{pmatrix} 1 \\ 0 \end{pmatrix} \right).
$$

(i) Show that the quantum mechanical expectation values $E(\mathbf{n}, \mathbf{m})$

$$
E(\mathbf{n}, \mathbf{m}) = \langle \psi | (\boldsymbol{\sigma} \cdot \mathbf{n}) \otimes (\boldsymbol{\sigma} \cdot \mathbf{m}) | \psi \rangle
$$

is given by

$$E(\mathbf{n}, \mathbf{m}) = -\mathbf{m} \cdot \mathbf{n} = -\|\mathbf{n}\| \cdot \|\mathbf{m}\| \cos(\phi) = -\cos(\phi_{\mathbf{n},\mathbf{m}}) \qquad (1)$$

where $\phi_{\mathbf{n},\mathbf{m}}$ are the angles between the two quantization directions $\mathbf{n}$ and $\mathbf{m}$. We write $\cos(\phi_{\mathbf{n},\mathbf{m}})$ to indicate that $\phi_{\mathbf{m},\mathbf{n}}$ is the angle between $\mathbf{m}$ and $\mathbf{n}$.

(ii) The *CHSH inequality* is given by

$$|E(\mathbf{n}, \mathbf{m}) - E(\mathbf{n}, \mathbf{m}')| + |E(\mathbf{n}', \mathbf{m}') + E(\mathbf{n}', \mathbf{m})| \le 2. \qquad (2)$$

Insert (1) into (2) and then find the angles, where the inequality is maximally violated.

**Solution 6.** (i) Note that $\boldsymbol{\sigma} \cdot \mathbf{n} = n_1 \sigma_1 + n_2 \sigma_2 + n_3 \sigma_3$. Straightforward calculation yields the result.

(ii) We obtain

$$|\cos(\phi_{\mathbf{n},\mathbf{m}}) - \cos(\phi_{\mathbf{n},\mathbf{m}'})| + |\cos(\phi_{\mathbf{n}',\mathbf{m}'}) + \cos(\phi_{\mathbf{n}',\mathbf{m}})| \le 2.$$

The maximal violation is $2\sqrt{2}$, achieved by the angles

$$\phi_{\mathbf{n},\mathbf{m}'} = 3\pi/4, \qquad \phi_{\mathbf{n},\mathbf{m}} = \phi_{\mathbf{n}',\mathbf{m}'} = \phi_{\mathbf{n}',\mathbf{m}} = \pi/4$$

where $\cos(3\pi/4) = -1/\sqrt{2}$ and $\cos(\pi/4) = 1/\sqrt{2}$. Angles which violate the inequality (2) are called *Bell angles*.

**Problem 7.** Consider four observers: Alice (A), Bob (B), Charlie (C) and Dora (D) each having one qubit. Every observer is allowed to choose between two dichotomic observables. Denote the outcome of observer $X$'s measurement by $X_i$ ($X = A, B, C, D$) with $i = 1, 2$. Under the assumption of local realism, each outcome can either take the value $+1$ or $-1$. The correlations between the measurement outcomes of all four observers can be represented by the product $A_i B_j C_k D_l$, where $i, j, k, l = 1, 2$. In a local realistic theory, the correlation function of the measurement performed by all four observers is the average of $A_i B_j C_k D_l$ over many runs of the experiment

$$Q(A_i B_j C_k D_l) := \langle \psi | A_i B_j C_k D_l | \psi \rangle.$$

The Mermin-Ardehali-Belinskii-Klyshko inequality is given by

$$Q(A_1 B_1 C_1 D_1) - Q(A_1 B_1 C_1 D_2) - Q(A_1 B_1 C_2 D_1) - Q(A_1 B_2 C_1 D_1)$$
$$-Q(A_2 B_1 C_1 D_1) - Q(A_1 B_1 C_2 D_2) - Q(A_1 B_2 C_1 D_2) - Q(A_2 B_1 C_1 D_2)$$
$$-Q(A_1 B_2 C_2 D_1) - Q(A_2 B_1 C_2 D_1) - Q(A_2 B_2 C_1 D_1) + Q(A_2 B_2 C_2 D_2)$$
$$+Q(A_2 B_2 C_2 D_1) + Q(A_2 B_2 C_1 D_2) + Q(A_2 B_1 C_2 D_2) + Q(A_1 B_2 C_2 D_2) \le 4.$$

Each observer $X$ measures the spin of each qubit by projecting it either along $\mathbf{n}_1^X$ or $\mathbf{n}_2^X$. Every observer can independently choose between two arbitrary directions. For a four qubit state $|\psi\rangle$, the correlation functions are thus given by

$$Q(A_i B_j C_k D_l) = \langle\psi|(\mathbf{n}_i^A \cdot \boldsymbol{\sigma}) \otimes (\mathbf{n}_j^B \cdot \boldsymbol{\sigma}) \otimes (\mathbf{n}_k^C \cdot \boldsymbol{\sigma}) \otimes (\mathbf{n}_l^D \cdot \boldsymbol{\sigma})|\psi\rangle$$

where $\cdot$ denotes the scalar product, i.e. $\mathbf{n}_j^X \cdot \boldsymbol{\sigma} := n_{j1}^X \sigma_1 + n_{j2}^X \sigma_2 + n_{j3}^X \sigma_3$. Let

$$\mathbf{n}_1^A = \begin{pmatrix} 1 \\ 0 \\ 0 \end{pmatrix}, \quad \mathbf{n}_2^A = \begin{pmatrix} 0 \\ 0 \\ 1 \end{pmatrix}, \quad \mathbf{n}_1^B = \begin{pmatrix} 0 \\ 1 \\ 0 \end{pmatrix}, \quad \mathbf{n}_2^B = \begin{pmatrix} 0 \\ 0 \\ 1 \end{pmatrix},$$

$$\mathbf{n}_1^C = \begin{pmatrix} 0 \\ 1 \\ 0 \end{pmatrix}, \quad \mathbf{n}_2^C = \begin{pmatrix} 0 \\ 0 \\ 1 \end{pmatrix}, \quad \mathbf{n}_1^D = \frac{1}{\sqrt{2}}\begin{pmatrix} -1 \\ 0 \\ 1 \end{pmatrix}, \quad \mathbf{n}_2^D = \frac{1}{\sqrt{2}}\begin{pmatrix} 1 \\ 0 \\ 1 \end{pmatrix}.$$

Show that the Mermin-Ardehali-Belinskii-Klyshko inequality is violated for the state

$$|\psi\rangle = \frac{1}{2\sqrt{2}}(|0000\rangle - |0011\rangle - |0101\rangle + |0110\rangle + |1001\rangle + |1010\rangle + |1100\rangle + |1111\rangle)$$

where

$$|0\rangle = \begin{pmatrix} 1 \\ 0 \end{pmatrix}, \quad |1\rangle = \begin{pmatrix} 0 \\ 1 \end{pmatrix}.$$

and $|0000\rangle \equiv |0\rangle \otimes |0\rangle \otimes |0\rangle \otimes |0\rangle$ etc..

**Solution 7.**   For the first term we have

$$Q(A_1 B_1 C_1 D_1) = \frac{1}{\sqrt{2}}\langle\psi|\sigma_1 \otimes \sigma_2 \otimes \sigma_2 \otimes (-\sigma_1 + \sigma_3)|\psi\rangle$$

$$= -\frac{1}{\sqrt{2}}\langle\psi|(\sigma_1 \otimes \sigma_2 \otimes \sigma_2 \otimes \sigma_1)|\psi\rangle$$

$$+ \frac{1}{\sqrt{2}}\langle\psi|(\sigma_1 \otimes \sigma_2 \otimes \sigma_2 \otimes \sigma_3)|\psi\rangle.$$

Using $\langle\psi|(\sigma_1 \otimes \sigma_2 \otimes \sigma_2 \otimes \sigma_1)|\psi\rangle = -1$, $\langle\psi|(\sigma_1 \otimes \sigma_2 \otimes \sigma_2 \otimes \sigma_3)|\psi\rangle = 0$ we obtain for the first term

$$Q(A_1 B_1 C_1 D_1) = \frac{1}{\sqrt{2}}.$$

Analogously we calculate the other terms. Summing up the terms we find the value $4\sqrt{2}$. Since $4\sqrt{2} > 4$ the Mermin-Ardehali-Belinskii-Klyshko inequality is violated by the state $|\psi\rangle$.

**Programming Problem**

**Problem 1.** Let $\hat{A} = \sigma_1$, $\hat{A}_2 = \sigma_2$

$$\hat{B}_1 = \frac{1}{\sqrt{2}}(\sigma_1 + \sigma_2), \quad \hat{B}_2 = \frac{1}{\sqrt{2}}(\sigma_1 - \sigma_2).$$

(i) Consider the entangled state

$$|\psi\rangle = (1 \ \ 1 \ \ 1 \ \ -1)^T.$$

Is the Bell inequality violated?
(ii) Consider the product state

$$|\phi\rangle = \frac{1}{2}\begin{pmatrix} 1 \\ 1 \\ 1 \\ 1 \end{pmatrix} \equiv \frac{1}{\sqrt{2}}\begin{pmatrix} 1 \\ 1 \end{pmatrix} \otimes \frac{1}{\sqrt{2}}\begin{pmatrix} 1 \\ 1 \end{pmatrix}.$$

Is the Bell inequality violated?

**Solution 1.** We apply the Maxima program

```
/* Bellinequality.mac */
A1: matrix([0,1],[1,0]); A2: matrix([0,-%i],[%i,0]);
B1: (A1+A2)/sqrt(2);    B2: (A1-A2)/sqrt(2);
K11: kronecker_product(A1,B1);
K12: kronecker_product(A1,B2);
K21: kronecker_product(A2,B1);
K22: kronecker_product(A2,B2);
psi: matrix([1],[1],[1],[-1])/2;
psiT: transpose(psi);
S: psiT . K11 . psi + psiT . K12 . psi
   + psiT . K21 . psi - psiT . K22 . psi;
S: ratsimp(S);
Sabs: abs(S);
phi: matrix([1],[1],[1],[1])/2;
phiT: transpose(phi);
S: phiT . K11 . phi + phiT . K12 . phi
   + phiT . K21 . phi - phiT . K22 . phi;
S: ratsimp(S);
Sabs: abs(S);
```

In both cases the Bell inequality is not violated which is obvious for the second case since it is a product state.

## 11.3   Supplementary Problems

**Problem 1.**   Let $\sigma_1$, $\sigma_2$ and $\sigma_3$ be the Pauli spin matrices. Consider the *Bell operator* defined by

$$B := \frac{1}{2} \sum_{j,k=1}^{3} (a_j(c_k + d_k) + b_j(c_k - d_k))\sigma_j \otimes \sigma_k$$

where $\mathbf{a}, \mathbf{b}, \mathbf{c}, \mathbf{d}$ are real unit vectors in $\mathbb{R}^3$.
(i) Consider the orthonormal basis in $\mathbb{C}^4$

$$\mathbf{v}_1 = \frac{1}{2}\begin{pmatrix} -1 \\ 1 \\ 1 \\ 1 \end{pmatrix}, \quad \mathbf{v}_2 = \frac{1}{2}\begin{pmatrix} 1 \\ -1 \\ 1 \\ 1 \end{pmatrix}, \quad \mathbf{v}_3 = \frac{1}{2}\begin{pmatrix} 1 \\ 1 \\ -1 \\ 1 \end{pmatrix}, \quad \mathbf{v}_4 = \frac{1}{2}\begin{pmatrix} 1 \\ 1 \\ 1 \\ -1 \end{pmatrix}.$$

Calculate the density operators $\rho_j = \mathbf{v}_j \mathbf{v}_j^*$ and then $\text{tr}(\rho_j B)$. Discuss.
(ii) Consider the basis in $\mathbb{C}^4$

$$\mathbf{u}_1 = \frac{1}{\sqrt{3}}\begin{pmatrix} -1 \\ 1 \\ 1 \\ 0 \end{pmatrix}, \quad \mathbf{u}_2 = \frac{1}{\sqrt{3}}\begin{pmatrix} 0 \\ -1 \\ 1 \\ 1 \end{pmatrix}, \quad \mathbf{u}_3 = \frac{1}{\sqrt{3}}\begin{pmatrix} 1 \\ 0 \\ -1 \\ 1 \end{pmatrix}, \quad \mathbf{u}_4 = \frac{1}{\sqrt{3}}\begin{pmatrix} 1 \\ 1 \\ 0 \\ -1 \end{pmatrix}.$$

Calculate the density operators $\rho_j = \mathbf{u}_j \mathbf{u}_j^*$ and $\text{tr}(\rho_j B)$. Compare to (i).

**Problem 2.**   Consider the Hilbert space $\mathcal{H}_A = \mathbb{C}^2$ and the Pauli spin matrices

$$\hat{A}_1 = \sigma_1 = \begin{pmatrix} 0 & 1 \\ 1 & 0 \end{pmatrix}, \quad \hat{A}_2 = \sigma_2 = \begin{pmatrix} 0 & -i \\ i & 0 \end{pmatrix}$$

the Hilbert space $\mathcal{H}_B = \mathbb{C}^3$ with the hermitian matrices

$$\hat{B}_1 = \begin{pmatrix} -1 & 0 & 0 \\ 0 & 1 & 0 \\ 0 & 0 & -1 \end{pmatrix}, \quad \hat{B}_2 = \begin{pmatrix} 0 & 0 & 1 \\ 0 & -1 & 0 \\ 1 & 0 & 0 \end{pmatrix}$$

with $\hat{B}_1^2 = I_3$, $\hat{B}_2^2 = I_3$ and the normalized state in $\mathbb{C}^6$

$$|\psi\rangle = \frac{1}{\sqrt{2}}(1 \ \ 0 \ \ 0 \ \ 0 \ \ 0 \ \ -1)^T.$$

Is the Bell inequality violated?

# Chapter 12

# Teleportation

## 12.1 Introduction

Teleportation is the transmission of quantum information using a classical channel and entanglement. It demonstrates the use of entanglement as a communication resource. The simplest case is to consider the teleportation of a single qubit using two bits of classical communication and one entangled pair (EPR-pair). Quantum teleportation is the disembodied transport of an unknown quantum state from one place to another. The key idea is that two distant operators, Alice at a sending station and Bob at a receiving terminal, share an entangled quantum bipartite state and exploit its nonlocal character as a quantum resource. The resource state can be the *singlet state* of a pair of spin-$\frac{1}{2}$ particles (Bell state)

$$|\psi\rangle = \frac{1}{\sqrt{2}}(|0\rangle \otimes |1\rangle - |1\rangle \otimes |0\rangle).$$

Particle 1 is given to Alice and particle 2 is given to Bob. Alice intends to transport an unknown state of a third spin-$\frac{1}{2}$ particle to Bob. She performs a complete projective measurement on the joint system consisting of particle 1 and 3 and then conveys its outcome to Bob via a classical communication channel. As a consequence of Alice's measurement, the total-spin state of the three-particle system collapses. Owing to the entanglement, this involves a breakdown of the spin-$\frac{1}{2}$ state of Bob's particle 2. Nevertheless, Bob makes use of the information transmitted classically by Alice

307

to transform his reduced state into an output that is an accurate replica of the original unknown input.

## 12.2   Solved Problems

**Problem 1.**   Consider the following states $(a, b \in \mathbb{C})$

$$|\psi\rangle := a|0\rangle + b|1\rangle, \quad |a|^2 + |b|^2 = 1 \tag{1}$$

$$|\phi\rangle := |\psi\rangle \otimes \frac{1}{\sqrt{2}}(|00\rangle + |11\rangle). \tag{2}$$

(i) Show that the state $|\phi\rangle$ can be written as

$$|\phi\rangle = \frac{1}{2\sqrt{2}}(|00\rangle + |11\rangle) \otimes (a|0\rangle + b|1\rangle) + \frac{1}{2\sqrt{2}}(|00\rangle - |11\rangle) \otimes (a|0\rangle - b|1\rangle)$$

$$+ \frac{1}{2\sqrt{2}}(|01\rangle + |10\rangle) \otimes (a|1\rangle + b|0\rangle) + \frac{1}{2\sqrt{2}}(|01\rangle - |10\rangle) \otimes (a|1\rangle - b|0\rangle).$$

(ii) Describe how measurement of the first two qubits of $|\phi\rangle$ can be used to obtain $|\psi\rangle$ as the last qubit. Alice has the first qubit of $|\phi\rangle$ and Alice and Bob share the second and third qubits of $|\phi\rangle$ (an EPR-pair).

**Solution 1.**   (i) Inserting (1) into (2) we obtain

$$|\phi\rangle = \frac{1}{\sqrt{2}}(a|000\rangle + a|011\rangle + b|100\rangle + b|111\rangle).$$

On the other hand we have

$$\frac{1}{2\sqrt{2}}(|00\rangle + |11\rangle) \otimes (a|0\rangle + b|1\rangle) + \frac{1}{2\sqrt{2}}(|00\rangle - |11\rangle) \otimes (a|0\rangle - b|1\rangle)$$

$$+ \frac{1}{2\sqrt{2}}(|01\rangle + |10\rangle) \otimes (a|1\rangle + b|0\rangle) + \frac{1}{2\sqrt{2}}(|01\rangle - |10\rangle) \otimes (a|1\rangle - b|0\rangle)$$

$$= \frac{1}{2\sqrt{2}}(a|000\rangle + a|110\rangle + b|001\rangle + b|111\rangle)$$

$$+ \frac{1}{2\sqrt{2}}(a|000\rangle - a|110\rangle - b|001\rangle + b|111\rangle)$$

$$+ \frac{1}{2\sqrt{2}}(a|011\rangle + a|101\rangle + b|010\rangle + b|100\rangle)$$

$$+ \frac{1}{2\sqrt{2}}(a|011\rangle - a|101\rangle - b|010\rangle + b|100\rangle)$$

$$= \frac{1}{\sqrt{2}}(a|000\rangle + a|011\rangle + b|100\rangle + b|111\rangle)$$

$$= |\phi\rangle.$$

(ii) We measure in the *Bell basis*

$$\left\{\frac{1}{\sqrt{2}}(|00\rangle + |11\rangle), \ \frac{1}{\sqrt{2}}(|00\rangle - |11\rangle), \ \frac{1}{\sqrt{2}}(|01\rangle + |10\rangle), \ \frac{1}{\sqrt{2}}(|01\rangle - |10\rangle)\right\}.$$

From the state $|\phi\rangle$ we can see that the first two qubits are in each of the Bell states with equal probability. Thus if we measure the first two qubits in the Bell basis we obtain a result corresponding to each of the Bell states and can perform a transform to obtain $|\psi\rangle$ in the last qubit as follows

| Bell State | Transform |
|---|---|
| $\frac{1}{\sqrt{2}}(|00\rangle + |11\rangle)$ | $I_2$ |
| $\frac{1}{\sqrt{2}}(|00\rangle - |11\rangle)$ | $|0\rangle\langle 0| - |1\rangle\langle 1|$ |
| $\frac{1}{\sqrt{2}}(|01\rangle + |10\rangle)$ | $U_{NOT}$ |
| $\frac{1}{\sqrt{2}}(|01\rangle - |10\rangle)$ | $|0\rangle\langle 1| - |1\rangle\langle 0|$ |

After measurement and applying the corresponding transform we obtain $|\psi\rangle$ as the last qubit. So if Alice and Bob initially share the entangled pair

$$\frac{1}{\sqrt{2}}(|00\rangle + |11\rangle) \equiv \frac{1}{\sqrt{2}}(|0\rangle_A \otimes |0\rangle_B + |1\rangle_A \otimes |1\rangle_B).$$

Alice can perform a measurement in the Bell basis on her qubit and her part of the entangled pair and sends the result (two bits) to Bob who applies the corresponding transform to his part of the entangled pair. The state $|\psi\rangle$ is thus *teleported* from Alice's qubit to Bob's qubit. Note that the Bell basis is obtained by applying $U_{CNOT}(U_H \otimes I_2)$ to the computational basis $\{|00\rangle, |01\rangle, |10\rangle, |11\rangle\}$. The transforms are unitary and therefore invertible. Thus we can also measure the first two qubits in the computational basis after applying the unitary matrix

$$(U_H \otimes I_2)U_{CNOT}.$$

**Problem 2.** Consider the state in the Hilbert space $\mathcal{H} = \mathbb{C}^{16}$

$$|\psi_0\rangle = |0101\rangle \equiv |0\rangle \otimes |1\rangle \otimes |0\rangle \otimes |1\rangle$$

where $\{|0\rangle, |1\rangle\}$ is the standard basis in $\mathbb{C}^2$. Let

$$|\psi_1\rangle = B|\psi_0\rangle = \frac{1}{\sqrt{2}}(|0101\rangle + |0110\rangle)$$

$$|\psi_2\rangle = U|\psi_1\rangle = \frac{1}{\sqrt{2}}(|0101\rangle + |1010\rangle)$$

$$|\psi_3\rangle = S|\psi_2\rangle = \frac{1}{\sqrt{2}}(|0101\rangle - |1010\rangle)$$

$$|\psi_4\rangle = U^*|\psi_3\rangle = \frac{1}{\sqrt{2}}(|0101\rangle - |0110\rangle)$$

$$|\psi_5\rangle = B^*|\psi_4\rangle = -|0110\rangle.$$

Find the $16 \times 16$ unitary matrices $B$, $U$, $S$ which perform these transformations.

**Solution 2.**   From the above equations we find the following

$$B|0101\rangle = \frac{1}{\sqrt{2}}(|0101\rangle + |0110\rangle)$$

$$U\frac{1}{\sqrt{2}}(|0101\rangle + |0110\rangle) = \frac{1}{\sqrt{2}}(|0101\rangle + |1010\rangle)$$

$$S\frac{1}{\sqrt{2}}(|0101\rangle + |1010\rangle) = \frac{1}{\sqrt{2}}(|0101\rangle - |1010\rangle)$$

$$U^*\frac{1}{\sqrt{2}}(|0101\rangle - |1010\rangle) = \frac{1}{\sqrt{2}}(|0101\rangle - |0110\rangle)$$

$$B^*\frac{1}{\sqrt{2}}(|0101\rangle - |0110\rangle) = -|0110\rangle.$$

A unitary transform maps an orthonormal basis to an orthonormal basis. The above equations do not determine $B$, $U$ and $S$ uniquely. For simplicity let $B$, $U$ and $S$ act as the identity on subspaces for which the unitary transformations are not constrained by the above equations. For $B$ we have

$$B|0101\rangle = |01\rangle \otimes \frac{1}{\sqrt{2}}(|01\rangle + |10\rangle), \qquad B|0110\rangle = |01\rangle \otimes \frac{1}{\sqrt{2}}(|10\rangle - |01\rangle).$$

One solution is

$$B = \frac{1}{\sqrt{2}}I_4 \otimes (|\gamma\rangle\langle 01| + |\delta\rangle\langle 10| + |\alpha\rangle\langle 00| + |\beta\rangle\langle 11|)$$

where

$$|\alpha\rangle = |00\rangle + |11\rangle, \quad |\beta\rangle = |00\rangle - |11\rangle, \quad |\gamma\rangle = |01\rangle + |10\rangle, \quad |\delta\rangle = |10\rangle - |01\rangle.$$

This means that $B$ maps from the computational basis to the Bell basis in the second two qubits. For $U$ we have

$$U\left(|01\rangle \otimes \frac{1}{\sqrt{2}}(|01\rangle + |10\rangle)\right) = \frac{1}{\sqrt{2}}(|0101\rangle + |1010\rangle)$$

and

$$U\left(|01\rangle \otimes \frac{1}{\sqrt{2}}(|01\rangle - |10\rangle)\right) = \frac{1}{\sqrt{2}}(|0101\rangle - |1010\rangle).$$

We rewrite these equations in the simpler form

$$U|0101\rangle = |0101\rangle, \qquad U|0110\rangle = |1010\rangle.$$

A solution for $U$ is then

$$U = I_{16} + (|1010\rangle - |0110\rangle)(\langle 0110| - \langle 1010|)$$

i.e. $U$ is the identity except on the subspace spanned by $|0110\rangle$ and $|1010\rangle$, where $U$ swaps $|0110\rangle$ and $|1010\rangle$. For $S$ we have

$$S\frac{1}{\sqrt{2}}(|0101\rangle + |1010\rangle) = \frac{1}{\sqrt{2}}(|0101\rangle - |1010\rangle).$$

A solution for $S$ is $S = I_{16} - 2|1010\rangle\langle 1010|$ i.e. $S$ is the identity except for changing the sign of $|1010\rangle$.

**Problem 3.** Let $|\psi\rangle := a|0\rangle + b|1\rangle$ be an arbitrary qubit state. Let $|\phi\rangle$ be another arbitrary qubit state. Let $U$ be a unitary operator which acts on two qubits.
(i) Determine the implications of measuring the first two qubits of

$$|\theta\rangle := |\psi\rangle \otimes \frac{1}{\sqrt{2}}(I_2 \otimes U)\left((|00\rangle + |11\rangle) \otimes |\phi\rangle\right)$$

with respect to the Bell basis. How can we obtain $U(|\psi\rangle \otimes |\phi\rangle)$ as the last two qubits?
(ii) Alice has $|\psi\rangle$ and Bob has $|\phi\rangle$. Describe how $U$ can be applied to $|\psi\rangle \otimes |\phi\rangle$ using only classical communication and prior shared entanglement. After the computation, Alice must still have the first qubit of $U(|\psi\rangle \otimes |\phi\rangle)$ and Bob must still have the second qubit of $U(|\psi\rangle \otimes |\phi\rangle)$.

**Solution 3.** (i) We have

$$|\theta\rangle = a|0\rangle \otimes \frac{1}{\sqrt{2}}(I_2 \otimes U)\left((|00\rangle + |11\rangle) \otimes |\phi\rangle\right)$$

$$+ b|1\rangle \otimes \frac{1}{\sqrt{2}}(I_2 \otimes U)\left((|00\rangle + |11\rangle) \otimes |\phi\rangle\right)$$

$$= a|0\rangle \otimes \frac{1}{\sqrt{2}}\left(|0\rangle \otimes U(|0\rangle \otimes |\phi\rangle) + |1\rangle \otimes U(|1\rangle \otimes |\phi\rangle)\right)$$

$$+ b|1\rangle \otimes \frac{1}{\sqrt{2}}\left(|0\rangle \otimes U(|0\rangle \otimes |\phi\rangle) + |1\rangle \otimes U(|1\rangle \otimes |\phi\rangle)\right)$$

$$= \frac{1}{\sqrt{2}} (|00\rangle \otimes U(a|0\rangle \otimes |\phi\rangle) + |01\rangle \otimes U(a|1\rangle \otimes |\phi\rangle)))$$

$$+ \frac{1}{\sqrt{2}} (|10\rangle \otimes U(b|0\rangle \otimes |\phi\rangle) + |11\rangle \otimes U(b|1\rangle \otimes |\phi\rangle))).$$

Expanding $|00\rangle$, $|01\rangle$, $|10\rangle$ and $|11\rangle$ in the Bell basis for the first two qubits yields the state

$$|\theta\rangle = \frac{1}{2\sqrt{2}} (|00\rangle + |11\rangle) \otimes U((a|0\rangle + b|1\rangle) \otimes |\phi\rangle)$$

$$+ \frac{1}{2\sqrt{2}} (|00\rangle - |11\rangle) \otimes U((a|0\rangle - b|1\rangle) \otimes |\phi\rangle)$$

$$+ \frac{1}{2\sqrt{2}} (|01\rangle + |10\rangle) \otimes U((a|1\rangle + b|0\rangle) \otimes |\phi\rangle)$$

$$+ \frac{1}{2\sqrt{2}} (|01\rangle - |10\rangle) \otimes U((a|1\rangle - b|0\rangle) \otimes |\phi\rangle) .$$

We measure in the *Bell basis*

$$\left\{ \frac{1}{\sqrt{2}} (|00\rangle + |11\rangle), \frac{1}{\sqrt{2}} (|00\rangle - |11\rangle), \frac{1}{\sqrt{2}} (|01\rangle + |10\rangle), \frac{1}{\sqrt{2}} (|01\rangle - |10\rangle) \right\}.$$

From $|\theta\rangle$ we can see that the first two qubits are in each of the Bell states with equal probability. If we make a measurement we obtain a result corresponding to each of the Bell states and can perform a transform to obtain $U(|\psi\rangle \otimes |\phi\rangle)$ in the last two qubits as follows

| Bell State | Transform |
|---|---|
| $\frac{1}{\sqrt{2}}(|00\rangle + |11\rangle)$ | $I_2$ |
| $\frac{1}{\sqrt{2}}(|00\rangle - |11\rangle)$ | $U((|0\rangle\langle 0| - |1\rangle\langle 1|) \otimes I_2) U^*$ |
| $\frac{1}{\sqrt{2}}(|01\rangle + |10\rangle)$ | $U(U_{NOT} \otimes I_2) U^*$ |
| $\frac{1}{\sqrt{2}}(|01\rangle - |10\rangle)$ | $U((|0\rangle\langle 1| - |1\rangle\langle 0|) \otimes I_2) U^*$ |

Thus after measurement and applying the corresponding transform we obtain $U(|\psi\rangle \otimes |\phi\rangle)$ as the last two qubits. Thus if Alice and Bob initially share the entangled state

$$\frac{1}{\sqrt{2}} (|00\rangle + |11\rangle)$$

Bob applies $U$ to his two qubits. Then Alice can perform a measurement in the Bell basis on her qubit and her part of the entangled pair and sends the result (two bits) to Bob who applies the corresponding transform from the table to his part of the entangled pair. Thus with probability $\frac{1}{4}$ Bob can begin the computation $U(|\psi\rangle \otimes |\phi\rangle)$ without knowing the state $|\psi\rangle$ and still obtain the correct result after Alice measures her two qubits. With probability $\frac{3}{4}$ he still has to apply a transform which is independent of $|\psi\rangle$. (ii) Alice teleports $|\psi\rangle$ to Bob with one entangled pair, Bob performs the computation $U(|\psi\rangle \otimes |\phi\rangle)$ on his two qubits and then teleports the first qubit back to Alice with a second entangled pair. Thus 4 bits of communication are used in this scheme (Alice sends two to Bob, and then Bob sends two to Alice). Alice and Bob can perform $U_{CNOT}$ even though their qubits are spatially separated if they have prior entanglement.

**Problem 4.** In quantum teleportation we start with the following state in the Hilbert space $\mathbb{C}^8$

$$|\psi\rangle \otimes |0\rangle \otimes |0\rangle \equiv (a|0\rangle + b|1\rangle) \otimes |0\rangle \otimes |0\rangle \equiv |\psi 00\rangle$$

where $|a|^2 + |b|^2 = 1$. The quantum circuit for teleportation is given by

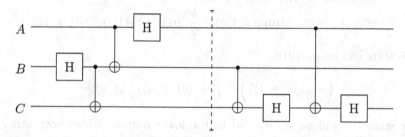

where $A$ is the input $|\psi\rangle$, $B$ the input $|0\rangle$ and $C$ the input $|0\rangle$. Study what happens when we feed the product state $|\psi 00\rangle$ into the quantum circuit. From the circuit we have the following eight $8 \times 8$ unitary matrices (left to right)

$$U_1 = I_2 \otimes U_H \otimes I_2, \quad U_2 = I_2 \otimes U_{XOR},$$
$$U_3 = U_{XOR} \otimes I_2, \quad U_4 = U_H \otimes I_2 \otimes I_2,$$
$$U_5 = I_2 \otimes U_{XOR}, \quad U_6 = I_2 \otimes I_2 \otimes U_H,$$
$$U_7 = I_4 \oplus U_{NOT} \oplus U_{NOT}, \quad U_8 = I_2 \otimes I_2 \otimes U_H$$

where $\oplus$ denotes the direct sum of matrices, $U_H$ denotes the Hadamard gate, $U_{XOR}$ denotes the $XOR$-gate and

$$U_{NOT} := \begin{pmatrix} 0 & 1 \\ 1 & 0 \end{pmatrix}.$$

(i) Find $U_8 U_7 U_6 U_5 U_4 U_3 U_2 U_1 |\psi 00\rangle$.
(ii) Write a program which implements and verifies the teleportation algorithm.

**Solution 4.**    (i) Applying the first four unitary matrices to the input state we obtain

$$U_4 U_3 U_2 U_1 |\psi 00\rangle$$

$$= \frac{a}{2}(|000\rangle + |100\rangle + |011\rangle + |111\rangle) + \frac{b}{2}(|010\rangle - |110\rangle + |001\rangle - |101\rangle).$$

This state can be rewritten as

$$U_4 U_3 U_2 U_1 |\psi 00\rangle = \frac{1}{\sqrt{2}}(|0\rangle + |1\rangle) \otimes (\frac{a}{\sqrt{2}}(|0\rangle \otimes |0\rangle + |1\rangle \otimes |1\rangle))$$

$$+ \frac{1}{\sqrt{2}}(|0\rangle - |1\rangle) \otimes (\frac{b}{\sqrt{2}}(|0\rangle \otimes |1\rangle + |1\rangle \otimes |0\rangle)).$$

Applying all eight unitary matrices to the input state we obtain

$$U_8 U_7 U_6 U_5 U_4 U_3 U_2 U_1 |\psi 00\rangle$$

$$= \frac{a}{2}(|000\rangle + |100\rangle + |010\rangle + |110\rangle) + \frac{b}{2}(|011\rangle + |111\rangle + |001\rangle + |101\rangle).$$

This state can be rewritten as

$$\left( \frac{1}{\sqrt{2}}(|0\rangle + |1\rangle) \right) \otimes \left( \frac{1}{\sqrt{2}}(|0\rangle + |1\rangle) \right) \otimes |\psi\rangle.$$

The state $|\psi\rangle$ will be transferred to the lower output, where both other outputs will come out in the state $(|0\rangle + |1\rangle)/\sqrt{2}$. If the two upper outputs are measured in the standard basis ($|0\rangle$ versus $|1\rangle$), two random classical bits will be obtained in addition to the quantum state $|\psi\rangle$ on the lower output.

(ii) The implementation in SymbolicC++ is as follows. The **Matrix** class of SymbolicC++ includes the method **kron** for the Kronecker product of two matrices and the method **dsum** for the direct sum of two matrices. The overloaded operators * and + are used for matrix multiplication and addition. The identity matrix is also implemented. Thus the code for the quantum circuit is as follows.

```
// teleport.cpp
#include <iostream>
#include "symbolicc++.h"
using namespace std;

Symbolic Hadamard(const Symbolic &v)
{
 Symbolic H("",2,2);
 Symbolic sqrt12 = sqrt(1/Symbolic(2));
 H(0,0) = sqrt12; H(0,1) =  sqrt12;
 H(1,0) = sqrt12; H(1,1) = -sqrt12;
```

```
 return (H*v);
}

Symbolic XOR(const Symbolic &v)
{
 Symbolic X("",4,4);
 X(0,0) = 1; X(0,1) = 0; X(0,2) = 0; X(0,3) = 0;
 X(1,0) = 0; X(1,1) = 1; X(1,2) = 0; X(1,3) = 0;
 X(2,0) = 0; X(2,1) = 0; X(2,2) = 0; X(2,3) = 1;
 X(3,0) = 0; X(3,1) = 0; X(3,2) = 1; X(3,3) = 0;
 return (X*v);
}

Symbolic Bell(const Symbolic &v)
{
 Symbolic I("",2,2), H("",2,2), X("",4,4);
 Symbolic sqrt12 = sqrt(1/Symbolic(2));
 I = I.identity();
 H(0,0) = sqrt12; H(0,1) =  sqrt12;
 H(1,0) = sqrt12; H(1,1) = -sqrt12;
 Symbolic UH = kron(H,I);
 X(0,0) = 1; X(0,1) = 0; X(0,2) = 0; X(0,3) = 0;
 X(1,0) = 0; X(1,1) = 1; X(1,2) = 0; X(1,3) = 0;
 X(2,0) = 0; X(2,1) = 0; X(2,2) = 0; X(2,3) = 1;
 X(3,0) = 0; X(3,1) = 0; X(3,2) = 1; X(3,3) = 0;
 return (X*(UH*v));
}

Symbolic Swap(const Symbolic &v)
{
 Symbolic S("",4,4);
 S(0,0) = 1; S(0,1) = 0; S(0,2) = 0; S(0,3) = 0;
 S(1,0) = 0; S(1,1) = 0; S(1,2) = 0; S(1,3) = 1;
 S(2,0) = 0; S(2,1) = 0; S(2,2) = 1; S(2,3) = 0;
 S(3,0) = 0; S(3,1) = 1; S(3,2) = 0; S(3,3) = 0;
 return XOR(S*XOR(v));
}

Symbolic Teleport(const Symbolic &v)
{
 Symbolic result;
 Symbolic NOT("",2,2),H("",2,2),I("",2,2),X("",4,4);
 Symbolic sqrt12 = sqrt(1/Symbolic(2));
 NOT(0,0) = 0; NOT(0,1) = 1; NOT(1,0) = 1; NOT(1,1) = 0;
 H(0,0) = sqrt12; H(0,1) =  sqrt12;
 H(1,0) = sqrt12; H(1,1) = -sqrt12;
 I = I.identity();
 X(0,0) = 1; X(0,1) = 0; X(0,2) = 0; X(0,3) = 0;
```

```
X(1,0) = 0; X(1,1) = 1; X(1,2) = 0; X(1,3) = 0;
X(2,0) = 0; X(2,1) = 0; X(2,2) = 0; X(2,3) = 1;
X(3,0) = 0; X(3,1) = 0; X(3,2) = 1; X(3,3) = 0;
Symbolic U1 = kron(I,kron(H,I)); Symbolic U2 = kron(I,X);
Symbolic U3 = kron(X,I); Symbolic U4 = kron(H,kron(I,I));
Symbolic U5 = kron(I,X); Symbolic U6 = kron(I,kron(I,H));
Symbolic U7 = dsum(I,dsum(I,dsum(NOT,NOT)));
Symbolic U8 = kron(I,kron(I,H));
result = U8*(U7*(U6*(U5*(U4*(U3*(U2*(U1*v))))))));
return result;
}


// The outcome after measuring value for qubit.
// Since the probabilities may be symbolic this function
// cannot simulate a measurement where random outcomes
// have the correct distribution
Symbolic Measure(const Symbolic &v,unsigned int qubit,
                 unsigned int value)
{
 int i,len,skip = 1-value;
 Symbolic result(v);
 Symbolic D;
 len = v.rows()/int(pow(2.0,qubit+1.0));
 for(i=0;i<v.rows();i++)
 {
  if(!(i%len)) skip = 1-skip;
  if(skip) result(i) = 0; else D += result(i)*result(i);
 }
 return result/sqrt(D);
}


// for output clarity
ostream &print(ostream &o,const Symbolic &v)
{
 char *b2[2]={"|0>","|1>"};
 char *b4[4]={"|00>","|01>","|10>","|11>"};
 char *b8[8]={"|000>","|001>","|010>","|011>",
              "|100>","|101>","|110>","|111>"};
 char **b;
 if(v.rows()==2) b=b2;
 if(v.rows()==4) b=b4;
 if(v.rows()==8) b=b8;
 for(int i=0;i<v.rows();i++)
  if(v(i)!=0) o << "+(" << v(i) << ")" << b[i];
 return o;
}


int main(void)
```

```
{
  Symbolic zero("",2),one("",2);
  Symbolic zz("",4),zo("",4),oz("",4),oo("",4),qreg;
  Symbolic tp00,tp01,tp10,tp11,psiGHZ;
  Symbolic a("a"), b("b");
  Symbolic sqrt12 = sqrt(1/Symbolic(2));
  zero(0) = 1; zero(1) = 0; one(0)  = 0; one(1)  = 1;
  zz = kron(zero,zero); zo = kron(zero,one);
  oz = kron(one,zero);  oo = kron(one,one);
  cout << "UH|0> = "; print(cout,Hadamard(zero))<< endl;
  cout << "UH|1> = "; print(cout,Hadamard(one)) << endl;
  cout << endl;
  cout << "UXOR|00> = "; print(cout,XOR(zz)) << endl;
  cout << "UXOR|01> = "; print(cout,XOR(zo)) << endl;
  cout << "UXOR|10> = "; print(cout,XOR(oz)) << endl;
  cout << "UXOR|11> = "; print(cout,XOR(oo)) << endl;
  cout << endl;
  cout << "UBELL|00> = "; print(cout,Bell(zz)) << endl;
  cout << "UBELL|01> = "; print(cout,Bell(zo)) << endl;
  cout << "UBELL|10> = "; print(cout,Bell(oz)) << endl;
  cout << "UBELL|11> = "; print(cout,Bell(oo)) << endl;
  cout << endl;
  cout << "USWAP|00> = "; print(cout,Swap(zz)) << endl;
  cout << "USWAP|01> = "; print(cout,Swap(zo)) << endl;
  cout << "USWAP|10> = "; print(cout,Swap(oz)) << endl;
  cout << "USWAP|11> = "; print(cout,Swap(oo)) << endl;
  cout << endl;
  qreg=kron(a*zero+b*one,kron(zero,zero));
  cout << "UTELEPORT("; print(cout,qreg) << ") = ";
   print(cout,qreg=Teleport(qreg)) << endl;
  cout << "Results after measurement of first 2 qubits:" << endl;
  tp00 = Measure(Measure(qreg,0,0),1,0);
  tp01 = Measure(Measure(qreg,0,0),1,1);
  tp10 = Measure(Measure(qreg,0,1),1,0);
  tp11 = Measure(Measure(qreg,0,1),1,1);
  Equations simplify = (a*a==1-b*b,1/sqrt(1/Symbolic(4))==2);
  tp00 = tp00.subst_all(simplify);
  tp01 = tp01.subst_all(simplify);
  tp10 = tp10.subst_all(simplify);
  tp11 = tp11.subst_all(simplify);
  cout << " |00> : " ; print(cout,tp00) << endl;
  cout << " |01> : " ; print(cout,tp01) << endl;
  cout << " |10> : " ; print(cout,tp10) << endl;
  cout << " |11> : " ; print(cout,tp11) << endl;
  cout << endl;
  psiGHZ=kron(zz,zero)*sqrt12+kron(oo,one)*sqrt12;
  cout << "Greenberger-Horne-Zeilinger state : ";
  print(cout,psiGHZ) << endl;
```

```
cout << "Measuring qubit 0 as 1 yields : ";
print(cout,Measure(psiGHZ,0,1)) << endl;
cout << "Measuring qubit 1 as 1 yields : ";
print(cout,Measure(psiGHZ,1,1)) << endl;
cout << "Measuring qubit 2 as 0 yields : ";
print(cout,Measure(psiGHZ,2,0)) << endl;
return 0;
}
```

The program generates the following output

```
UTELEPORT(+(a)|000>+(b)|100>) =
    +(1/2*a)|000>+(1/2*b)|001>+(1/2*a)|010>
    +(1/2*b)|011>+(1/2*a)|100>+(1/2*b)|101>
    +(1/2*a)|110>+(1/2*b)|111>
Results after measurement of first 2 qubits:
|00> : +(a)|000>+(b)|001>
|01> : +(a)|010>+(b)|011>
|10> : +(a)|100>+(b)|101>
|11> : +(a)|110>+(b)|111>
```

# 12.3   Supplementary Problems

**Problem 1.**  Let $v_1$, $v_2$, $v_3$ be normalized states in $\mathbb{C}^2$. Find the unitary $8 \times 8$ matrices $U_1$, $U_2$, $U_3$ such

$$U_1(v_1 \otimes v_2 \otimes v_3) = v_2 \otimes v_3 \otimes v_1$$

$$U_2(v_2 \otimes v_3 \otimes v_1) = v_3 \otimes v_1 \otimes v_2$$

$$U_3(v_3 \otimes v_1 \otimes v_2) = v_1 \otimes v_2 \otimes v_3.$$

Discuss

**Problem 2.**  Let $|0\rangle$, $|1\rangle$ be the standard basis in $\mathbb{C}^2$. Construct a unitary $8 \times 8$ matrix $U$ applying the generalized Gram-Schmidt technique such that

$$U\left( \begin{pmatrix} \cos(\theta) \\ \sin(\theta) \end{pmatrix} \otimes |0\rangle \otimes |0\rangle \right) =$$

$$\frac{1}{\sqrt{2}}(|0\rangle + |1\rangle) \otimes \begin{pmatrix} \cos(\theta) \\ \sin(\theta) \end{pmatrix} \otimes \frac{1}{\sqrt{2}}(|0\rangle + |1\rangle).$$

# Chapter 13

# Cloning

## 13.1  Introduction

*Cloning* is the duplication of information. Cloning is necessarily a physical process. Obviously we have to find unitary transformations for the cloning process. The *no cloning theorem* is a result of quantum mechanics which forbids the creation of identical copies of an arbitrary unknown quantum state. However, approximate copies still have many uses in quantum computing. Owing to the superposition principle of quantum mechanics it is not possible in general to clone an arbitrary quantum state. As a consequence any successful attempt to clone a state will destroy the original state in the process. Quantum cloning machines are devices for approximately cloning arbitrary quantum states. Suppose we want to clone the state $|\psi\rangle \in \mathcal{H}$ to obtain

$$|\psi\rangle \otimes |\psi\rangle.$$

The approximate cloning operation yields some mixed state $\rho$ in the product Hilbert space $\mathcal{H} \otimes \mathcal{H}$. The fidelity of the cloning process for $|\psi\rangle$ is given by

$$(\langle\psi| \otimes \langle\psi|)\rho(|\psi\rangle \otimes |\psi\rangle).$$

Optimal cloning attempts to maximize the average fidelity (closeness to $|\psi\rangle \otimes |\psi\rangle$). In the case of symmetric cloning we assume

$$\langle\psi|\mathrm{tr}_{\mathcal{H}}(\rho)|\psi\rangle$$

is independent of the system on which the partial trace is performed. Thus for symmetric cloning we need only use $\langle\psi|\mathrm{tr}_{\mathcal{H}}(\rho)|\psi\rangle$ as the measure of fidelity. We provide exercises describing what types of information can be cloned accurately and techniques for cloning certain types of information.

## 13.2    Solved Problems

**Problem 1.**    Let $a \in \{0,1\}$ and $|0\rangle$, $|1\rangle$ be the standard basis. Find a unitary $4 \times 4$ matrix that maps

$$U : |a\rangle \otimes |a\rangle \mapsto |a\rangle \otimes |\bar{a}\rangle$$

where $\bar{a}$ denotes the NOT operation applied to $a$.

**Solution 1.**    Obviously the permutation matrix

$$U = \begin{pmatrix} 0 & 1 & 0 & 0 \\ 1 & 0 & 0 & 0 \\ 0 & 0 & 0 & 1 \\ 0 & 0 & 1 & 0 \end{pmatrix} = \begin{pmatrix} 0 & 1 \\ 1 & 0 \end{pmatrix} \oplus \begin{pmatrix} 0 & 1 \\ 1 & 0 \end{pmatrix}$$

provides such a map, where $\oplus$ denotes the direct sum.

**Problem 2.**    The *CNOT gate* maps ($a, b \in \{0,1\}$)

$$|a\rangle \otimes |b\rangle \rightarrow |a\rangle \otimes |a \oplus b\rangle$$

where $\oplus$ is the XOR operation. Show that the CNOT gate can be used to clone a bit.

**Solution 2.**    Setting $b = 0$ we obtain from the CNOT gate

$$|a\rangle \otimes |0\rangle \rightarrow |a\rangle \otimes |a\rangle$$

since $a \oplus 0 = a$ for all $a$. Thus we have cloned a bit.

**Problem 3.**    Let

$$\mathbf{x} = \begin{pmatrix} x_1 \\ x_2 \end{pmatrix}, \qquad x_1 x_1^* + x_2 x_2^* = 1$$

be an arbitrary normalized vector in $\mathbb{C}^2$. Can we construct a $4 \times 4$ unitary matrix $U$ such that

$$U\left(\begin{pmatrix} x_1 \\ x_2 \end{pmatrix} \otimes \begin{pmatrix} 1 \\ 0 \end{pmatrix}\right) = \begin{pmatrix} x_1 \\ x_2 \end{pmatrix} \otimes \begin{pmatrix} x_1 \\ x_2 \end{pmatrix} ? \tag{1}$$

Prove or disprove this equation.

**Solution 3.** Such a matrix does not exist. This can be seen as follows. From the right-hand side of (1) we have

$$\begin{pmatrix} x_1 \\ x_2 \end{pmatrix} \otimes \begin{pmatrix} x_1 \\ x_2 \end{pmatrix} = \left( \begin{pmatrix} x_1 \\ 0 \end{pmatrix} + \begin{pmatrix} 0 \\ x_2 \end{pmatrix} \right) \otimes \left( \begin{pmatrix} x_1 \\ 0 \end{pmatrix} + \begin{pmatrix} 0 \\ x_2 \end{pmatrix} \right) =$$

$$\begin{pmatrix} x_1 \\ 0 \end{pmatrix} \otimes \begin{pmatrix} x_1 \\ 0 \end{pmatrix} + \begin{pmatrix} x_1 \\ 0 \end{pmatrix} \otimes \begin{pmatrix} 0 \\ x_2 \end{pmatrix} + \begin{pmatrix} 0 \\ x_2 \end{pmatrix} \otimes \begin{pmatrix} x_1 \\ 0 \end{pmatrix} + \begin{pmatrix} 0 \\ x_2 \end{pmatrix} \otimes \begin{pmatrix} 0 \\ x_2 \end{pmatrix}.$$

On the other hand, from the left-hand side of (1) we find

$$U \left( \begin{pmatrix} x_1 \\ x_2 \end{pmatrix} \otimes \begin{pmatrix} 1 \\ 0 \end{pmatrix} \right) = U \left( \begin{pmatrix} x_1 \\ 0 \end{pmatrix} \otimes \begin{pmatrix} 1 \\ 0 \end{pmatrix} + \begin{pmatrix} 0 \\ x_2 \end{pmatrix} \otimes \begin{pmatrix} 1 \\ 0 \end{pmatrix} \right)$$

$$= \begin{pmatrix} x_1 \\ 0 \end{pmatrix} \otimes \begin{pmatrix} x_1 \\ 0 \end{pmatrix} + \begin{pmatrix} 0 \\ x_2 \end{pmatrix} \otimes \begin{pmatrix} 0 \\ x_2 \end{pmatrix}$$

where we used the linearity of the unitary matrix $U$. Comparing these two equations we find a contradiction. This is the *no cloning theorem*.

However equation (1) does hold when

$$\begin{pmatrix} x_1 \\ 0 \end{pmatrix} \otimes \begin{pmatrix} 0 \\ x_2 \end{pmatrix} + \begin{pmatrix} 0 \\ x_2 \end{pmatrix} \otimes \begin{pmatrix} x_1 \\ 0 \end{pmatrix} = \begin{pmatrix} 0 \\ 0 \end{pmatrix} \otimes \begin{pmatrix} 0 \\ 0 \end{pmatrix}.$$

Therefore $x_1 x_2 = 0$. Thus at least one of $x_1$ and $x_2$ must be zero. It is still possible to clone elements of a known orthonormal basis.

**Problem 4.** Let $|\psi\rangle$, $|s\rangle$, $|\phi\rangle$ be normalized states in a Hilbert space $\mathcal{H}$. Let $U$ be a unitary operator, i.e. $U^{-1} = U^*$ in the product Hilbert space $\mathcal{H} \otimes \mathcal{H}$ such that

$$U(|\psi\rangle \otimes |s\rangle) = |\psi\rangle \otimes |\psi\rangle$$
$$U(|\phi\rangle \otimes |s\rangle) = |\phi\rangle \otimes |\phi\rangle.$$

Show that $\langle \phi|\psi \rangle = \langle \phi|\psi \rangle^2$. Find solutions to this equation.

**Solution 4.** Taking the scalar product of these two equations with $U^* = U^{-1}$ and $\langle s|s \rangle = 1$ we obtain

$$(\langle \psi| \otimes \langle s|)U^*U(|\phi\rangle \otimes |s\rangle) = (\langle \psi| \otimes \langle \psi|)(|\phi\rangle \otimes |\phi\rangle)$$
$$(\langle \psi| \otimes \langle s|)(|\phi\rangle \otimes |s\rangle) = \langle \psi|\phi \rangle \langle \psi|\phi \rangle$$
$$\langle \psi|\phi \rangle = \langle \psi|\phi \rangle^2.$$

The equation can be satisfied if $\langle\psi|\phi\rangle = 0$ ($|\psi\rangle$ and $|\phi\rangle$ are orthonormal to each other) or $\langle\psi|\phi\rangle = 1$, i.e. $|\psi\rangle = |\phi\rangle$.

**Problem 5.**    Let $\{\,|0\rangle\,,|1\rangle\,\}$ be a basis in $\mathbb{C}^2$. Let $|\psi\rangle$ be an arbitrary qubit. Is there a unitary transformation such that

$$|\psi\rangle \otimes |\psi\rangle \to |\psi\rangle \otimes |0\rangle\,?$$

**Solution 5.**    Such a unitary transformation does not exist. For an arbitrary qubit $|\psi\rangle$ the product states $|\psi\rangle \otimes |\psi\rangle$ span a three-dimensional subspace of the four-dimensional Hilbert space $\mathbb{C}^4$ of two qubits. However, the product states $|\psi\rangle \otimes |0\rangle$ span only a two-dimensional subspace, as $|0\rangle$ is a fixed state. Thus the unitary transform would take a system with *von Neumann entropy* $\log_2(3)$ to one with von Neumann entropy $\log_2(2)$. Since the system is closed (we have a unitary transformation), this decrease of entropy is therefore a violation of the *second law of thermodynamics*. Thus the second law of thermodynamics implies that such a unitary transformation does not exist.

**Problem 6.**    Consider the approximate cloning of a qubit $|\psi\rangle \in \mathbb{C}^2$ by simply measuring the qubit with respect to a randomly chosen orthonormal basis $\{\,|0\rangle,|1\rangle\,\}$, and then using the state $|0\rangle \otimes |0\rangle$ or $|1\rangle \otimes |1\rangle$ corresponding to the measurement outcome. Determine the average fidelity of this cloning process.

**Solution 6.**    The state $|\psi\rangle$ is represented by a density operator

$$|\psi\rangle\langle\psi| = \frac{1}{2}(I_2 + \mathbf{n} \cdot \boldsymbol{\sigma})$$

where $\mathbf{n}$ is a unit vector in $\mathbb{R}^3$ and $\mathbf{n}\cdot\boldsymbol{\sigma} := n_1\sigma_1 + n_2\sigma_2 + n_3\sigma_3$. The randomly chosen orthonormal basis can be represented by the density operators

$$|0\rangle\langle0| = \frac{1}{2}(I_2 + \mathbf{m} \cdot \boldsymbol{\sigma})$$

$$|1\rangle\langle1| = I_2 - |0\rangle\langle0| = \frac{1}{2}(I_2 - \mathbf{m} \cdot \boldsymbol{\sigma})$$

where $\mathbf{m}$ is a unit vector in $\mathbb{R}^3$. Measuring $|\psi\rangle$ in the basis $\{\,|0\rangle,|1\rangle\,\}$ yields $|0\rangle$ with probability

$$|\langle0|\psi\rangle|^2 = \frac{1}{2}(1 + \mathbf{n} \cdot \mathbf{m})$$

and $|1\rangle$ with probability

$$|\langle1|\psi\rangle|^2 = \frac{1}{2}(1 - \mathbf{n} \cdot \mathbf{m})$$

where $|\langle 0|\psi\rangle|^2 = \langle 0|\psi\rangle\langle\psi|0\rangle$. Thus we construct a two qubit system described by the density operator

$$\frac{1}{2}(1 + \mathbf{n}\cdot\mathbf{m})|0\rangle\langle 0| \otimes |0\rangle\langle 0| + \frac{1}{2}(1 - \mathbf{n}\cdot\mathbf{m})|1\rangle\langle 1| \otimes |1\rangle\langle 1|.$$

If we trace out either qubit we are left with the same density operator, i.e. this is symmetric cloning. The fidelity for $|\psi\rangle$ (symmetric cloning) is given by

$$\left[\frac{1}{2}(1 + \mathbf{n}\cdot\mathbf{m})\right]^2 + \left[\frac{1}{2}(1 - \mathbf{n}\cdot\mathbf{m})\right]^2 = \frac{1}{2}(1 + (\mathbf{n}\cdot\mathbf{m})^2).$$

To determine the average fidelity we integrate over all $\mathbf{m} \in \mathbb{R}^3$ with $\|\mathbf{m}\| = 1$. Thus we use *spherical coordinates*

$$\mathbf{m} = (\sin(\theta)\cos(\phi), \sin(\theta)\sin(\phi), \cos(\theta)), \qquad \phi \in [0, 2\pi], \quad \theta \in [0, \pi].$$

Then

$$\mathbf{n}\cdot\mathbf{m} = n_1\sin(\theta)\cos(\phi) + n_2\sin(\theta)\sin(\phi) + n_3\cos(\theta).$$

Consequently the average fidelity is given by

$$\frac{1}{4\pi}\int_{\|\mathbf{m}\|=1} \frac{1}{2}(1 + (\mathbf{n}\cdot\mathbf{m})^2)\, d\mathbf{m}$$

$$= \frac{1}{4\pi}\int_0^\pi \int_0^{2\pi} \frac{1}{2}(1 + \mathbf{n}\cdot\mathbf{m})^2)\sin(\theta)d\phi\, d\theta$$

$$= \frac{1}{2} + \frac{1}{8}\int_0^\pi (n_1^2\sin^3(\theta) + n_2^2\sin^3(\theta) + 2n_3^2\cos^2(\theta)\sin(\theta))d\theta$$

$$= \frac{1}{2} + \frac{1}{8}\int_0^\pi (2\cos^2(\theta)\sin(\theta) + (n_1^2 + n_2^2)(\sin(\theta) - 3\cos^2(\theta)\sin(\theta)))\, d\theta$$

$$= \frac{2}{3}$$

where we used $n_3^2 = 1 - n_1^2 - n_2^2$. The average fidelity is independent of $|\psi\rangle$, i.e. the cloning process in universal. However the cloning process is not optimal.

## Programming Problem

**Problem 1.** Show that there is a unitary $4 \times 4$ matrix such that

$$U\left(\frac{1}{\sqrt{2}}\begin{pmatrix}1\\1\end{pmatrix} \otimes \frac{1}{\sqrt{2}}\begin{pmatrix}1\\-1\end{pmatrix}\right) = \frac{1}{\sqrt{2}}\begin{pmatrix}1\\1\end{pmatrix} \otimes \frac{1}{\sqrt{2}}\begin{pmatrix}1\\1\end{pmatrix}.$$

**Solution 1.**  The unitary matrix

$$U = \begin{pmatrix} 1 & 0 \\ 0 & 1 \end{pmatrix} \otimes \begin{pmatrix} 0 & -1 \\ 1 & 0 \end{pmatrix}$$

will do the job.

```
/* cloning.mac */
v1: matrix([1],[1])/sqrt(2); v2: matrix([1],[-1])/sqrt(2);
I2: matrix([1,0],[0,1]);
V: matrix([0,-1],[1,0]);
U: kronecker_product(I2,V);
v1v2k: kronecker_product(v1,v2);
v1v2kU: U . v1v2k;
v1v1k: kronecker_product(v1,v1);
if(v1v1k=v1v2kU) then print("U does the cloning")
else print("U does not do the cloning");
```

# 13.3   Supplementary Problems

**Problem 1.**   Let $a, b \in \{0, 1\}$.
(i) Can one find a unitary $4 \times 4$ matrix such that

$$U(|a\rangle \otimes |b\rangle) = |a\rangle \otimes |a \cdot b\rangle$$

for all $a, b \in \{0, 1\}$? Here $\cdot$ denotes the AND operation.
(ii) Can one find a unitary $4 \times 4$ matrix such that

$$U(|a\rangle \otimes |b\rangle) = |a\rangle \otimes |a + b\rangle$$

for all $a, b \in \{0, 1\}$? Here $+$ denotes the OR operation.

**Problem 2.**   Let

$$\mathbf{x} = \begin{pmatrix} x_1 \\ x_2 \\ x_3 \end{pmatrix}, \qquad x_1 x_1^* + x_2 x_2^* + x_3 x_3^* = 1$$

be an arbitrary normalized vector in $\mathbb{C}^2$. Can we construct a $9 \times 9$ unitary matrix $U$ such that

$$U\left( \begin{pmatrix} x_1 \\ x_2 \\ x_3 \end{pmatrix} \otimes \begin{pmatrix} 1 \\ 0 \\ 0 \end{pmatrix} \right) = \begin{pmatrix} x_1 \\ x_2 \\ x_3 \end{pmatrix} \otimes \begin{pmatrix} x_1 \\ x_2 \\ x_3 \end{pmatrix} ? \tag{1}$$

# Chapter 14

# Quantum Algorithms

## 14.1 Introduction

An *algorithm* is a precise description of how to realize a given objective, for example solving a computational problem. We distinguish between *classical* and *quantum* algorithms where quantum physical resources are used. Quantum algorithms run on quantum computers and thus utilizing unitary transformations such as the quantum Fourier transform and superposition. Some quantum algorithms also utilize entanglement. Quantum algorithms include: quantum counting, quantum phase estimation algorithm, Deutsch's algorithm, Deutsch-Josza algorithm, Simon's algorithm, Shor's algorithm, Gover's algorithm and hidden subgroup problem. Quantum counting algorithm is a quantum algorithm for counting the number of solutions for a given search problem. The quantum phase estimation algorithm is used to find the eigenphase of a normalized eigenvector of a unitary gate (unitary matrix) given a quantum state proportional to the normalized eigenvector and access to the unitary gate. Deutsch's problem leads to the simplest quantum algorithm. Consider the Boolean functions $f$ that map $\{\,0,\,1\,\}$ to $\{\,0,\,1\,\}$. There are exactly four such functions: two constant functions ($f(0) = f(1) = 0$ and $f(0) = f(1) = 1$) and two balanced functions ($f(0) = 0, f(1) = 1$ and $f(0) = 1, f(1) = 0$). In Deutsch's problem one is allowed to evaluate the function only once and we are required to deduce from the result whether $f$ is constant or balanced. Thus we are asked for the global property of $f$. A generalization is the Deutsch-Josza problem. One considers the boolean functions

$$f : \{0,1\}^n \rightarrow \{0,1\}$$

in the following way. Assume that, for one of these functions, it is promised that it either constant or balanced (i.e. has an equal number of 0's outputs as 1's) and consider the goal of determining which of the two properties the function actually has. Given a boolean function $f : \{0,1\}^n \rightarrow \{0,1\}$. Assume that $f$ is known to be invariant under some $n$-bit XOR mask $b$. Simon's problem is to determine $b$. Shor's algorithm solves the discrete logarithm problem and the integer factorization problem. Grover's algorithm searches an unsorted database with $N$ entries for a marked entry. The algorithms of Deutsch, Simon, Shor and others can be formulated group theoretically as a hidden subgroup problem. Let $f$ be a function from a finitely generated group $G$ to a finite set such that $f$ is constant on the cosets of a subgroup $K$ and distinct on each on each coset. The cosets of $K$ are the sets

$$g \cdot K := \{ g \cdot k \ : \ k \in K \}, \quad g \in G.$$

The cosets partition the group $G$, i.e. the union of all cosets is the set of the group $G$ and every two cosets are equal or their intersection is empty. Thus one writes

$$K = \{ k \in G \ : \ f(k \cdot g) = f(g), \ \forall g \in G \}.$$

The problem is, for a given $f$ and $G$ determine the hidden subgroup $K$.

The *quantum Fourier transform* plays an important role in a number of quantum algorithms. The quantum Fourier transform on the additive group of integers modulo $2^m$ is the mapping

$$|a\rangle \overset{F_{2^m}}{\rightarrow} \sum_{y=0}^{2^m-1} \exp(2\pi i a y/2^m)|y\rangle$$

where $a \in \{0,1,\ldots,2^m-1\}$. The state on the right-hand side is unentangled. The quantum Fourier transform is used in Shor's algorithm. Given positive integers $a$ and $N$ which are relatively prime and such that $a < N$. The goal is to find the minimum positive integer $r$ such that $a^r \bmod N = 1$.

## 14.2 Solved Problems

**Problem 1.** In classical communication complexity Alice is provided with a binary string

$$\mathbf{x} = x_0 x_1 \cdots x_{n-1}$$

of length $n$ and Bob is provided with a binary string

$$\mathbf{y} = y_0 y_1 \cdots y_{n-1}$$

of length $n$. Alice has to determine a boolean function

$$f : \{0, 1\}^n \times \{0, 1\}^n \to \{0, 1\}$$

with the least communication between herself and Bob.
(i) Consider the *parity function*

$$f(\mathbf{x}, \mathbf{y}) = x_0 \oplus x_1 \oplus \cdots \oplus x_{n-1} \oplus y_0 \oplus y_1 \oplus \cdots \oplus y_{n-1}$$

where $\oplus$ is the *XOR operation*, i.e.

$$0 \oplus 0 = 0, \quad 0 \oplus 1 = 1, \quad 1 \oplus 0 = 1, \quad 1 \oplus 1 = 0.$$

How many bits has Bob to send to Alice so that she can determine $f$?
(ii) Consider the inner product modulo-2 function

$$f(\mathbf{x}, \mathbf{y}) = (x_0 \cdot y_0) \oplus (x_1 \cdot y_1) \oplus \cdots \oplus (x_{n-1} \cdot y_{n-1})$$

where $\cdot$ denotes the *AND operation*, i.e.

$$0 \cdot 0 = 0, \quad 0 \cdot 1 = 0, \quad 1 \cdot 0 = 0, \quad 1 \cdot 1 = 1.$$

What is the minimum number of bits Bob has to send to Alice so that she can compute this function?

**Solution 1.** (i) Obviously Bob has to send only one bit, the one he finds by computing $y_0 \oplus y_1 \oplus \cdots \oplus y_{n-1}$.
(ii) Bob must send all $n$ bits in order for Alice to compute $f$.

**Problem 2.** Find all $x_A, x_B, x_C \in \{0, 1\}$ such that $x_A + x_B + x_C = 1 \mod 2$. We use the mapping $f_1 : \{0, 1\} \to U(2)$

$$f_1(0) := U_H, \qquad f_1(1) := I_2$$

where $U_H$ is the Walsh-Hadamard transform and $U(2)$ denotes the unitary group over $\mathbb{C}^2$. Thus we can map from the triple $(x_A, x_B, x_C)$ to linear operators acting on three qubits

$$f_3(x_A, x_B, x_C) := f_1(x_A) \otimes f_1(x_B) \otimes f_1(x_C).$$

Let

$$|\psi\rangle := \frac{1}{2}(|001\rangle + |010\rangle + |100\rangle - |111\rangle).$$

For each triple $(x_A, x_B, x_C)$ found in the first part of the problem, calculate

$$|\phi\rangle := f_3(x_A, x_B, x_C)|\psi\rangle.$$

Let $s_A, s_B, s_C$ denote the result (0 or 1) of measuring the first, second and third qubit, respectively of $|\phi\rangle$ in the computational basis. In each case determine

$$s_A + s_B + s_C \bmod 2, \qquad x_A \cdot x_B \cdot x_C.$$

**Solution 2.**   We have

$$(x_A, x_B, x_C) \in \{\, (0,0,1), (0,1,0), (1,0,0), (1,1,1) \,\}.$$

We note the symmetry of the state $|\psi\rangle$ with respect to qubit ordering. Thus we need only to calculate the transform for $(0,0,1)$ and $(1,1,1)$. For $(1,1,1)$ we have $f_3(1,1,1)|\psi\rangle = (I_2 \otimes I_2 \otimes I_2)|\psi\rangle = |\psi\rangle$. Measuring the qubits yields

$$(s_A, s_B, s_C) \in \{\, (0,0,1), (0,1,0), (1,0,0), (1,1,1) \,\}$$

with equal probability. In each case we find $s_A + s_B + s_C = 1 \bmod 2$. For $(0,0,1)$ we have $f_3(0,0,1) = U_H \otimes U_H \otimes I_2$. Since

$$|\psi\rangle = \frac{1}{2}(|01\rangle + |10\rangle) \otimes |0\rangle + \frac{1}{2}(|00\rangle - |11\rangle) \otimes |1\rangle$$

we obtain

$$f_3(0,0,1)|\psi\rangle = \frac{1}{2}(|00\rangle - |11\rangle) \otimes |0\rangle + \frac{1}{2}(|01\rangle + |10\rangle) \otimes |1\rangle.$$

We find that measuring the qubits yields

$$(s_A, s_B, s_C) \in \{\, (0,0,0), (0,1,1), (1,0,1), (1,1,0) \,\}$$

with equal probability. In each case we find $s_A + s_B + s_C = 0 \bmod 2$.

| $(x_A, x_B, x_C)$ | $x_A \cdot x_B \cdot x_C$ | $s_A + s_B + s_C \bmod 2$ |
|:---:|:---:|:---:|
| (0,0,1) | 0 | 0 |
| (0,1,0) | 0 | 0 |
| (1,0,0) | 0 | 0 |
| (1,1,1) | 1 | 1 |

We find that $s_A + s_B + s_C = x_A \cdot x_B \cdot x_C \bmod 2$. Suppose Alice, Bob and Carol each have a bit string $(x_{A,1}, \ldots, x_{A,n})$, $(x_{B,1}, \ldots, x_{B,n})$ and $(x_{C,1}, \ldots, x_{C,n})$, respectively. They want to calculate

$$f(\mathbf{x}_A, \mathbf{x}_B, \mathbf{x}_C) = \sum_{j=1}^{n} (x_{A,j} \cdot x_{B,j} \cdot x_{C,j}) \bmod 2$$

sharing (communicating) as little information as possible. If Alice, Bob and Carol share $n$ triplets of qubits in the state $|\psi\rangle$ they can calculate $s_{A,1}, \ldots, s_{A,n}, s_{B,1}, \ldots, s_{B,n}$ and $s_{C,1}, \ldots, s_{C,n}$ respectively as above. Thus

$$f(\mathbf{x}_A, \mathbf{x}_B, \mathbf{x}_C) = \sum_{j=1}^{n} (s_{A,j} + s_{B,j} + s_{C,j}) \bmod 2.$$

If Alice, Bob and Carol calculate

$$S_{A|B|C} = \sum_{j=1}^{n} S_{A|B|C,j} \bmod 2.$$

Bob and Carol need only to send one bit each ($S_B$ and $S_C$) to Alice for Alice to compute $f(\mathbf{x}_A, \mathbf{x}_B, \mathbf{x}_C) = S_A + S_B + S_C$, for any $n$. In other words the *communication complexity* is 2. Classically, for $n \geq 3$, three bits of communication are required.

**Problem 3.** (i) Find all $x, y, z \in \{0, 1, 2, 3\}$ such that

$$x + y + z = 0 \bmod 2. \tag{1}$$

What are the possible values of the function

$$f(x, y, z) := \frac{(x + y + z) \bmod 4}{2}$$

when the condition (1) holds?
(ii) Now use the binary representation for $x = x_1 x_0$, $y = y_1 y_0$ and $z = z_1 z_0$ where $x_0, x_1, y_0, y_1, z_0, z_1 \in \{0, 1\}$. Describe the condition $x + y + z = 0 \bmod 2$ in terms of $x_0, x_1, y_0, y_1, z_0$ and $z_1$.
(iii) We use the map

$$f_1(0) = I_2, \qquad f_1(1) = U_H.$$

Thus we can map from the triple $(x_0, y_0, z_0)$ to linear operators acting on three qubits

$$f_3(x_0, y_0, z_0) = f_1(x_0) \otimes f_1(y_0) \otimes f_1(z_0).$$

Let

$$|\psi\rangle := \frac{1}{2}(|000\rangle - |011\rangle - |101\rangle - |110\rangle).$$

For each triple $(x_0, y_0, z_0)$ found in part (i) calculate

$$|\phi\rangle := f_3(x_0, y_0, z_0)|\psi\rangle.$$

Let $s_x, s_y, s_z$ denote the result (0 or 1) of measuring the first, second and third qubit, respectively of $|\phi\rangle$ in the computational basis. In each case determine

$$s_x + s_y + s_z \bmod 2, \qquad x_0 + y_0 + z_0.$$

**Solution 3.**    (i) Obviously $x + y + z$ must be even. Thus the sum includes only an even number (0 or 2) of odd numbers. Thus we have the nine combinations

$$(0,0,0), (0,1,1), (0,0,2), (1,1,2), (0,2,2), (0,1,3), (2,2,2), (1,2,3), (0,3,3).$$

(ii) Let $(x, y, z)$ be an element of the set of all permutations of elements of the above set. When $x + y + z$ is even, $(x + y + z) \bmod 4 \in \{0, 2\}$. Now when $x + y + z = 0 \bmod 2$ then $f(x, y, z) \in \{0, 1\}$. Since $x + y + z = 0 \bmod 2$ the least significant bit of the sum must be zero. The least significant bit is given by $x_0 \oplus y_0 \oplus z_0 = 0$. We find that

$$f(x, y, z) = x_1 \oplus y_1 \oplus z_1 \oplus (x_0 + y_0 + z_0).$$

XOR is denoted by "$\oplus$" and OR is denoted by "+". Thus we have

$$(x_0, y_0, z_0) \in \{ (0,0,0), (0,1,1), (1,0,1), (1,1,0) \}.$$

(iii) We note the symmetry of the state $|\psi\rangle$ with respect to the qubit ordering. Thus we need only calculate the transform for $(0,0,0)$ and $(0,1,1)$. For $(0,0,0)$ we have

$$f_3(0,0,0)|\psi\rangle = I_2 \otimes I_2 \otimes I_2|\psi\rangle = |\psi\rangle.$$

Measuring the qubits yields

$$(s_x, s_y, s_z) \in \{ (0,0,0), (0,1,1), (1,0,1), (1,1,0) \}$$

with equal probability. In each case we find $s_x + s_y + s_z = 0 \bmod 2$. For $(0,1,1)$ we have $f_3(0,1,1) = I_2 \otimes U_H \otimes U_H$. Note that $|\psi\rangle$ can be written as

$$|\psi\rangle = \frac{1}{2}|0\rangle \otimes (|00\rangle - |11\rangle) - \frac{1}{2}|1\rangle \otimes (|01\rangle + |10\rangle).$$

Therefore

$$f_3(0,1,1)|\psi\rangle = \frac{1}{2}|0\rangle \otimes (|01\rangle + |10\rangle) - \frac{1}{2}|1\rangle \otimes (|00\rangle - |11\rangle).$$

We find that measuring the qubits yields

$$(s_x, s_y, s_z) \in \{(0,1,0), (1,0,0), (0,0,1), (1,1,1)\}$$

with equal probability. In each case $s_x + s_y + s_z = 1 \bmod 2$.

| $(x_0, y_0, z_0)$ | $x_0 + y_0 + z_0$ | $s_x + s_y + s_z \bmod 2$ |
|---|---|---|
| (0,0,0) | 0 | 0 |
| (0,1,1) | 1 | 1 |
| (1,0,1) | 1 | 1 |
| (1,1,0) | 1 | 1 |

We find that $(s_x + s_y + s_z \bmod 2) = x_0 + y_0 + z_0$. Thus for three parties to calculate $f(x, y, z)$, where each party has one of the $x$, $y$ and $z$, it is sufficient for each party to send one bit ($x_1 \oplus s_x$ or $y_1 \oplus s_y$ or $z_1 \oplus s_z$) to the other parties to calculate $f(x, y, z)$. In other words each party can calculate

$$x_1 \oplus s_x \oplus y_1 \oplus s_y \oplus z_1 \oplus s_z = x_1 \oplus y_1 \oplus z_1 \oplus (x_0 + y_0 + z_0) = f(x, y, z)$$

after communication. In other words three bits broadcast to all parties are sufficient to calculate $f(x, y, z)$, the *communication complexity* is 3 bits. Classically it is necessary that 4 bits be broadcast.

**Problem 4.** (i) Determine the eigenvalues and eigenvectors of

$$A(x) := (1 - x)I_2 + xU_{NOT}, \qquad x \in \{0, 1\}.$$

(ii) Show that the unitary transform

$$U_f = |0f(0)\rangle\langle 00| + |0\overline{f(0)}\rangle\langle 01| + |1f(1)\rangle\langle 10| + |1\overline{f(1)}\rangle\langle 11|$$

where $f : \{0,1\} \to \{0,1\}$ is a boolean function and $\overline{x}$ denotes the boolean negation of $x$, can be written as

$$U_f = |0\rangle\langle 0| \otimes A(f(0)) + |1\rangle\langle 1| \otimes A(f(1)).$$

(iii) Calculate

$$U_f \left( I_2 \otimes \frac{1}{\sqrt{2}} (|0\rangle - |1\rangle) \right).$$

Consider the cases $f(0) = f(1)$ and $f(0) \neq f(1)$.

**Solution 4.** (i) We have $A(0) = I_2$ and $A(1) = U_{NOT}$. Thus $A_0$ has eigenvalues 1 (twice), and $A(1)$ has eigenvalues 1 and $-1$. We tabulate the eigenvalues and corresponding eigenvectors of $A(x)$

| eigenvalue | eigenvector |
|------------|-------------|
| 1 | $\frac{1}{\sqrt{2}}(|0\rangle + |1\rangle)$ |
| $(-1)^x$ | $\frac{1}{\sqrt{2}}(|0\rangle - |1\rangle)$ |

(ii) We have

$$
\begin{aligned}
U_f ={} & (1 - f(0))|00\rangle\langle00| + f(0)|01\rangle\langle00| \\
& + f(0)|00\rangle\langle01| + (1 - f(0))|01\rangle\langle01| \\
& + (1 - f(1))|10\rangle\langle10| + f(1)|11\rangle\langle10| \\
& + f(1)|10\rangle\langle11| + (1 - f(1))|11\rangle\langle11| \\
={} & |0\rangle\langle0| \otimes ((1 - f(0))|0\rangle\langle0| + f(0)|1\rangle\langle0|) \\
& + |0\rangle\langle0| \otimes (f(0)|0\rangle\langle1| + (1 - f(0))|1\rangle\langle1|) \\
& + |1\rangle\langle1| \otimes ((1 - f(1))|0\rangle\langle0| + f(1)|1\rangle\langle0|) \\
& + |1\rangle\langle1| \otimes (f(1)|0\rangle\langle1| + (1 - f(1))|1\rangle\langle1|) \\
={} & |0\rangle\langle0| \otimes ((1 - f(0))(|0\rangle\langle0| + |1\rangle\langle1|) + f(0)(|0\rangle\langle1| + |1\rangle\langle0|)) \\
& + |1\rangle\langle1| \otimes ((1 - f(1))(|0\rangle\langle0| + |1\rangle\langle1|) + f(1)(|0\rangle\langle1| + |1\rangle\langle0|)) \\
={} & |0\rangle\langle0| \otimes A(f(0)) + |1\rangle\langle1| \otimes A(f(1)).
\end{aligned}
$$

(iii) We find

$$
\begin{aligned}
U_f\left(I_2 \otimes \frac{1}{\sqrt{2}}(|0\rangle - |1\rangle)\right) ={} & |0\rangle\langle0| \otimes A(f(0))\frac{1}{\sqrt{2}}(|0\rangle - |1\rangle) \\
& + |1\rangle\langle1| \otimes A(f(1))\frac{1}{\sqrt{2}}(|0\rangle - |1\rangle) \\
={} & |0\rangle\langle0| \otimes (-1)^{f(0)}\frac{1}{\sqrt{2}}(|0\rangle - |1\rangle) \\
& + |1\rangle\langle1| \otimes (-1)^{f(1)}\frac{1}{\sqrt{2}}(|0\rangle - |1\rangle) \\
={} & (-1)^{f(0)}|0\rangle\langle0| \otimes \frac{1}{\sqrt{2}}(|0\rangle - |1\rangle) \\
& + (-1)^{f(1)}|1\rangle\langle1| \otimes \frac{1}{\sqrt{2}}(|0\rangle - |1\rangle) \\
={} & \left((-1)^{f(0)}\left(|0\rangle\langle0| + (-1)^{f(0)+f(1)}|1\rangle\langle1|\right) \otimes I_2\right) \\
& \times \left(I_2 \otimes \frac{1}{\sqrt{2}}(|0\rangle - |1\rangle)\right).
\end{aligned}
$$

Thus when $f(0) = f(1)$ we apply the identity operator to the first qubit and when $f(0) \neq f(1)$ we apply a phase change to the first qubit. The eigenvalues $(-1)^{f(0)}$ and $(-1)^{f(1)}$ are said to *kick back* to the first qubit. A phase change combined with two Walsh-Hadamard transforms in the appropriate order implements a NOT gate.

**Problem 5.** (i) Alice and Bob share $n$ entangled pairs of the form $\frac{1}{\sqrt{2}}(|00\rangle + |11\rangle)$. We can write their shared state of $2n$ qubits in the form of the generalized Bell state

$$|\psi\rangle = \frac{1}{\sqrt{2^n}} \sum_{j=0}^{2^n-1} |j\rangle \otimes |j\rangle \tag{1}$$

where the first $n$ qubits belong to Alice and the second $n$ qubits belong to Bob. Furthermore Alice has $2^n$ bits $a_0, \ldots, a_{2^n-1}$ and Bob has $2^n$ bits $b_0, \ldots, b_{2^n-1}$. Let the unitary operators $U_{PA}$ and $U_{PB}$ act on the computational basis as follows

$$U_{PA}|j\rangle = (-1)^{a_j}|j\rangle, \qquad j = 0, 1, \ldots, 2^n - 1$$

$$U_{PB}|j\rangle = (-1)^{b_j}|j\rangle, \qquad j = 0, 1, \ldots, 2^n - 1.$$

Let

$$|\phi\rangle := (U_{PA} \otimes U_{PB})|\psi\rangle. \tag{2}$$

Calculate the state

$$\left(\bigotimes_n U_H\right) \otimes \left(\bigotimes_n U_H\right) |\phi\rangle. \tag{3}$$

(ii) For each of the cases

$$(a) \quad a_0 = b_0, a_1 = b_1, \ldots, a_{2^n-1} = b_{2^n-1}$$

$$(b) \quad \sum_{k=0}^{2^n-1} |a_k - b_k| = 2^{n-1}$$

determine when measurement of the first $n$ qubits in the computational basis yields the same result as measurement of the second $n$ qubits in the computational basis.

**Solution 5.** (i) From (1) and (2) we obtain

$$|\phi\rangle = \frac{1}{\sqrt{2^n}} \sum_{j=0}^{2^n-1} (-1)^{a_j+b_j} |j\rangle \otimes |j\rangle.$$

Thus we find for (3)

$$\left(\bigotimes_n U_H\right) \otimes \left(\bigotimes_n U_H\right) |\phi\rangle = \frac{1}{(2\sqrt{2})^n} \sum_{j=0}^{2^n-1} \sum_{k=0}^{2^n-1} \sum_{l=0}^{2^n-1} (-1)^{a_j+b_j+j*k+j*l} |k\rangle \otimes |l\rangle$$

since

$$\left(\bigotimes_n U_H\right)|j\rangle = \bigotimes_{s=0}^{n-1} U_H|j_s\rangle$$

$$= \bigotimes_{s=0}^{n-1} U_H \frac{1}{\sqrt{2}}(|0\rangle + (-1)^{j_s}|1\rangle)$$

$$= \sum_{k=0}^{2^n-1} (-1)^{j_0 k_0 + j_1 k_1 + \cdots + j_{n-1} k_{n-1}}|k\rangle$$

where we decompose $j$ and $k$ as follows

$$j = j_0 + j_1 2 + j_2 4 + \cdots + j_{n-1} 2^{n-1}, \qquad k = k_0 + k_1 2 + k_2 4 + \cdots + k_{n-1} 2^{n-1}$$

and

$$j * k := (j_0 \cdot k_0) \oplus (j_1 \cdot k_1) \oplus \cdots \oplus (j_{n-1} \cdot k_{n-1})$$
$$= j_0 k_0 + j_1 k_1 + \cdots + j_{n-1} k_{n-1} \bmod 2.$$

(ii) For the case $(a)$ we have for $k = l$

$$\frac{1}{(2\sqrt{2})^n} \sum_{j=0}^{2^n-1} (-1)^{a_j + b_j + j*k + j*l} = \frac{1}{(2\sqrt{2})^n} \sum_{j=0}^{2^n-1} (-1)^{j*k + j*l}$$

$$= \frac{1}{(2\sqrt{2})^n} \sum_{j=0}^{2^n-1} (-1)^{j*(k+l)}$$

$$= \frac{1}{(2\sqrt{2})^n} 2^n$$

$$= 2^{-n/2}.$$

In other words the probability of measuring $|k\rangle \otimes |k\rangle$ for a given $k$ is $2^{-n}$. Furthermore

$$\sum_{k=0}^{2^n-1} 2^{-n} = 2^{-n} \sum_{k=0}^{2^n-1} 1 = 1.$$

For the case $(b)$ we find when $k = l$

$$\sum_{j=0}^{2^n-1} (-1)^{a_j + b_j + j*k + j*l} = \sum_{j=0}^{2^n-1} (-1)^{a_j + b_j} = 0.$$

Thus if condition $(a)$ holds measuring the $2n$ qubits in the computational basis always yields $|j\rangle$ and $|j\rangle$, i.e. the first $n$ qubits always yield exactly the same result as the second $n$ qubits.

If condition $(b)$ holds then measuring the $2n$ qubits in the computational basis yields $|j\rangle$ and $|k\rangle$ where $j \neq k$, i.e. the first $n$ qubits never yield the same result as the second $n$ qubits.

**Problem 6.** (i) Show that the vectors

$$|0_H\rangle := \frac{1}{\sqrt{2}}(|0\rangle + |1\rangle), \qquad |1_H\rangle := \frac{1}{\sqrt{2}}(|0\rangle - |1\rangle)$$

form an orthonormal basis for $\mathbb{C}^2$.

(ii) Determine the probabilities associated with finding $|0\rangle$ in the states $|0_H\rangle$ and $|1_H\rangle$.

(iii) Determine how to obtain $|0_H\rangle$ and $|1_H\rangle$ using only measurement and the phase change operation

$$U_{PS} := |0\rangle\langle 0| - |1\rangle\langle 1|.$$

(iv) Let $f : \{0,1\} \to \{0,1\}$ be a boolean function and

$$U_f := |0f(0)\rangle\langle 00| + |0\overline{f(0)}\rangle\langle 01| + |1f(1)\rangle\langle 10| + |1\overline{f(1)}\rangle\langle 11|.$$

Determine in terms of $|0_H\rangle$ and $|1_H\rangle$

$$(a)\ \ U_f(|0_H\rangle \otimes |0_H\rangle), \qquad (b)\ \ U_f(|0_H\rangle \otimes |1_H\rangle).$$

These techniques are used to solve *Deutsch's problem*.

**Solution 6.** (i) First we demonstrate the linear independence of the vectors

$$a|0\rangle + b|1\rangle = \frac{1}{\sqrt{2}}(a+b)|0_H\rangle + \frac{1}{\sqrt{2}}(a-b)|1_H\rangle$$

$$a|0_H\rangle + b|1_H\rangle = \frac{1}{\sqrt{2}}(a+b)|0\rangle + \frac{1}{\sqrt{2}}(a-b)|1\rangle.$$

Thus for $a|0_H\rangle + b|1_H\rangle = 0$ it follows that $a = b = 0$.

(ii) We find

$$\langle 0_H|0_H\rangle = \frac{1}{2}(\langle 0|0\rangle + \langle 0|1\rangle + \langle 1|0\rangle + \langle 1|1\rangle) = 1$$

$$\langle 1_H|1_H\rangle = \frac{1}{2}(\langle 0|0\rangle - \langle 0|1\rangle - \langle 1|0\rangle + \langle 1|1\rangle) = 1$$

$$\langle 0_H|1_H\rangle = \frac{1}{2}(\langle 0|0\rangle - \langle 0|1\rangle + \langle 1|0\rangle - \langle 1|1\rangle) = 0$$

$$|\langle 0|0_H\rangle|^2 = \frac{1}{2}|(\langle 0|0\rangle + \langle 0|1\rangle|^2 = \frac{1}{2}$$

$$|\langle 0|1_H\rangle|^2 = \frac{1}{2}|(\langle 0|0\rangle - \langle 0|1\rangle|^2 = \frac{1}{2}.$$

Thus measurement projects the state $|0\rangle$ onto $|0_H\rangle$ and $|1_H\rangle$ with equal probability.

(iii) Starting with $|0\rangle$, we can obtain $|0_H\rangle$ and $|1_H\rangle$ by measurement in the $|0_H\rangle$ and $|1_H\rangle$ basis and applying $U_{PS}$ as follows

| Desired state | Measure | Transform |
|:---:|:---:|:---:|
| $|0_H\rangle$ | $|0_H\rangle$ | $I_2$ |
| $|0_H\rangle$ | $|1_H\rangle$ | $U_{PS}$ |
| $|1_H\rangle$ | $|0_H\rangle$ | $U_{PS}$ |
| $|1_H\rangle$ | $|1_H\rangle$ | $I_2$ |

(iv) For (a) we have

$$|0_H\rangle \otimes |0_H\rangle = \frac{1}{2}(|00\rangle + |01\rangle + |10\rangle + |11\rangle).$$

Thus

$$U_f|0_H\rangle \otimes |0_H\rangle = \frac{1}{2}(|0f(0)\rangle + |0\overline{f(0)}\rangle + |1f(1)\rangle + |1\overline{f(1)}\rangle)$$

$$= \frac{1}{2}\Big((1 - f(0))|00\rangle + f(0)|01\rangle + f(0)|00\rangle + (1 - f(0))|01\rangle$$

$$+ (1 - f(1))|10\rangle + f(1)|11\rangle + f(1)|10\rangle + (1 - f(1))|11\rangle\Big)$$

$$= |0_H\rangle \otimes |0_H\rangle.$$

For (b) we have

$$|0_H\rangle \otimes |1_H\rangle = \frac{1}{2}(|00\rangle - |01\rangle + |10\rangle - |11\rangle).$$

Thus

$$U_f|0_H\rangle \otimes |1_H\rangle = \frac{1}{2}(|0f(0)\rangle - |0\overline{f(0)}\rangle + |1f(1)\rangle - |1\overline{f(1)}\rangle)$$

$$= \frac{1}{2}((1 - f(0))|00\rangle + f(0)|01\rangle - f(0)|00\rangle - (1 - f(0))|01\rangle$$

$$+ (1 - f(1))|10\rangle + f(1)|11\rangle - f(1)|10\rangle - (1 - f(1))|11\rangle)$$

$$= \frac{1}{2}((1 - 2f(0))|00\rangle - (1 - 2f(0))|01\rangle$$

$$+ (1 - 2f(1))|10\rangle - (1 - 2f(1))|11\rangle)$$

$$= \frac{1}{2}((-1)^{f(0)}|00\rangle - (-1)^{f(0)}|01\rangle$$

$$-(-1)^{f(1)}|10\rangle - (-1)^{f(1)}|11\rangle)$$
$$= \frac{1}{2}\left((-1)^{f(0)}|0\rangle \otimes (|0\rangle - |1\rangle) + (-1)^{f(1)}|1\rangle(|0\rangle - |1\rangle)\right)$$
$$= \frac{1}{\sqrt{2}}(-1)^{f(0)}((|0\rangle - (-1)^{f(0)+f(1)}|1\rangle) \otimes |1_H\rangle)$$
$$= \frac{1}{\sqrt{2}}(-1)^{f(0)}((|0\rangle - (-1)^{f(0)\oplus f(1)}|1\rangle) \otimes |1_H\rangle)$$
$$= (-1)^{f(0)}|f(0) \oplus f(1)_H\rangle \otimes |1_H\rangle.$$

Note that $f(0) \oplus f(1)$ is 0 when $f$ is constant, and 1 when $f$ is balanced. Thus by determining $f(0) \oplus f(1)$ we have solved *Deutsch's problem* .

**Problem 7.** Consider the following quantum game $G_n$ with $n \geq 3$ players. Each player $P_j$ $(j = 0, 1, \dots, n-1)$ receives a single input bit $x_j$ and has to produce a single output bit $y_j$. It is known that there is an even number of 1s among the inputs. The players are not allowed to communicate after receiving their inputs. Then they are challenged to produce a collective output that contains an even number of 1s if and only if the number of 1s in the input is divisible by 4. Therefore, we require that

$$\sum_{j=0}^{n-1} y_j \equiv \frac{1}{2} \sum_{j=0}^{n-1} x_j \pmod{2}$$

provided that

$$\sum_{j=0}^{n-1} x_j \equiv 0 \pmod{2}.$$

We call $\mathbf{x} = x_0 x_1 \cdots x_{n-1}$ the question and $\mathbf{y} = y_0 y_1 \cdots y_{n-1}$ the answer. Show that if the $n$-players are allowed to share prior entanglement, then they can always win the game $G_n$.

**Solution 7.** We define the following $n$-qubit entangled state in the Hilbert space $\mathbb{C}^{2^n}$

$$|\psi_+\rangle := \frac{1}{\sqrt{2}}(|00\cdots 0\rangle + |11\cdots 1\rangle)$$

$$|\psi_-\rangle := \frac{1}{\sqrt{2}}(|00\cdots 0\rangle - |11\cdots 1\rangle).$$

The *Walsh-Hadamard transform* is given by

$$U_H|0\rangle \to \frac{1}{\sqrt{2}}|0\rangle + \frac{1}{\sqrt{2}}|1\rangle, \qquad U_H|1\rangle \to \frac{1}{\sqrt{2}}|0\rangle - \frac{1}{\sqrt{2}}|1\rangle.$$

Furthermore consider the unitary transformation

$$U_S|0\rangle \to |0\rangle, \qquad U_S|1\rangle \to e^{i\pi/2}|1\rangle$$

where $e^{i\pi/2} = i$. If the unitary transformation $U_S$ is applied to any two qubits of $|\psi_+\rangle$, while other qubits are left undisturbed, then

$$U_S|\psi_+\rangle = |\psi_-\rangle$$

and if $U_S$ is applied to any two qubits of $|\psi_-\rangle$, then

$$U_S|\psi_-\rangle = |\psi_+\rangle.$$

Therefore, if the qubits of $|\psi_+\rangle$ are distributed among $n$ players, and if exactly $m$ of them apply $S$ to their qubit, the resulting state will be $|\psi_+\rangle$ if $m \equiv 0 \pmod 4$ and $|\psi_-\rangle$ if $m \equiv 2 \pmod 4$. The effect of applying the Walsh-Hadamard transform to each qubit in $|\psi_+\rangle$ is to produce an equal superposition of all classical $n$-bit strings that contain an even number of 1s, whereas the effect of applying the Walsh-Hadamard transform to each qubit in $|\psi_-\rangle$ is to produce an equal superposition of all classical $n$-bits that contain an odd number of 1s. Thus

$$(U_H \otimes U_H \otimes \cdots \otimes U_H)|\psi_+\rangle = \frac{1}{\sqrt{2^{n-1}}} \sum_{\Delta(\mathbf{y})=0\,(mod\,2)} |y_0 y_1 \cdots y_{n-1}\rangle$$

$$(U_H \otimes U_H \otimes \cdots \otimes U_H)|\psi_-\rangle = \frac{1}{\sqrt{2^{n-1}}} \sum_{\Delta(\mathbf{y})=1\,(mod\,2)} |y_0 y_1 \cdots y_{n-1}\rangle$$

where

$$\Delta(\mathbf{y}) := \sum_{j=0}^{n-1} y_j$$

denotes the *Hamming weight* of $\mathbf{y}$. Consequently the strategy is as follows: At the beginning the state $|\psi_+\rangle$ is produced and its $n$-qubits are distributed among the $n$ players. After the separation each player $A_j$ receives input bit $x_j$ and does the following

1. If $x_j = 1$, $A_j$ applies the unitary transformation $U_S$ to his qubit; otherwise he/she does nothing.

2. He/she applies $U_H$ to this qubit.

3. He/she measures his/her qubit in order to obtain $y_j$.

4. He/she produces $y_j$ as his/her output.

An even number of players will apply $U_S$ to their qubit. If that number is divisible by 4, which means that $\frac{1}{2}\sum_{j=0}^{n-1} x_j$ is even, then the states reverts to $|\psi_+\rangle$ after step 1 and therefore to a superposition of all $|y_0 y_1 \cdots y_{n-1}\rangle$

such that $\Delta(\mathbf{y}) \equiv 0 \pmod 2$ after step 2. It follows that $\sum_{j=0}^{n-1} y_i$, the number of players who measure and output 1, is even. If the number of players who apply $S$ to their qubit is congruent to 2 modulo 4, which means that $\frac{1}{2} \sum_{j=0}^{n-1} x_j$ is odd, then the state evolves to $|\psi_-\rangle$ after step 1 and therefore to a superposition of all $|\mathbf{y}\rangle \equiv |y_0 y_1 \cdots y_{n-1}\rangle$ such that $\Delta(\mathbf{y}) \equiv 1 \pmod 2$ after step 2. In this case $\sum_{j=0}^{n-1} y_j$ is odd. In either case, (1) is satisfied at the end of the protocol.

**Problem 8.** Let $x_0, x_1, y_0, y_1 \in \{0,1\}$ where Alice has $x_0$ and $x_1$ and Bob has $y_0$ and $y_1$. Alice and Bob want to calculate the boolean function

$$g(x_0, x_1, y_0, y_1) := x_1 \oplus y_1 \oplus (x_0 \cdot y_0)$$

where $\oplus$ denotes the XOR operation and $\cdot$ denotes the AND operation. Furthermore Alice and Bob share an EPR pair (Bell state)

$$\frac{1}{\sqrt{2}}(|00\rangle - |11\rangle).$$

Alice applies the unitary matrix

$$U_R\left(-\frac{\pi}{16} + x_0 \frac{\pi}{4}\right) \otimes I_2$$

to her qubit of the EPR pair and Bob applies the unitary matrix

$$I_2 \otimes U_R\left(-\frac{\pi}{16} + y_0 \frac{\pi}{4}\right)$$

to his qubit of the EPR pair, where

$$U_R(\theta) := \begin{pmatrix} \cos(\theta) & -\sin(\theta) \\ \sin(\theta) & \cos(\theta) \end{pmatrix}.$$

Let $a$ denote the result of Alice measuring her qubit of the EPR pair and let $b$ denote the result of Bob measuring his qubit of the EPR pair. Find the probability that $a \oplus b = x_0 \cdot y_0$, where $\oplus$ denotes the boolean XOR operation and $\cdot$ denotes the boolean AND operation.

**Solution 8.** We define $|\psi\rangle$ to be the state of the EPR-pair after Alice and Bob apply their transforms. Consequently

$$|\psi\rangle := U_R\left(-\frac{\pi}{16} + x_0 \frac{\pi}{4}\right) \otimes U_R\left(-\frac{\pi}{16} + y_0 \frac{\pi}{4}\right) \frac{1}{\sqrt{2}}(|00\rangle - |11\rangle).$$

Thus

$$|\psi\rangle = \frac{1}{\sqrt{2}}\left(\left(\cos\left(-\frac{\pi}{16} + x_0 \frac{\pi}{4}\right)|0\rangle + \sin\left(-\frac{\pi}{16} + x_0 \frac{\pi}{4}\right)|1\rangle\right)\right.$$

$$\otimes \left( \cos \left( -\frac{\pi}{16} + y_0 \frac{\pi}{4} \right) |0\rangle + \sin \left( -\frac{\pi}{16} + y_0 \frac{\pi}{4} \right) |1\rangle \right)$$

$$- \left( -\sin \left( -\frac{\pi}{16} + x_0 \frac{\pi}{4} \right) |0\rangle + \cos \left( -\frac{\pi}{16} + x_0 \frac{\pi}{4} \right) |1\rangle \right)$$

$$\otimes \left( -\sin \left( -\frac{\pi}{16} + y_0 \frac{\pi}{4} \right) |0\rangle + \cos \left( -\frac{\pi}{16} + y_0 \frac{\pi}{4} \right) |1\rangle \right) \right)$$

$$= \frac{1}{\sqrt{2}} \left( \cos \left( -\frac{\pi}{8} + (x_0 + y_0) \frac{\pi}{4} \right) |00\rangle + \sin \left( -\frac{\pi}{8} + (x_0 + y_0) \frac{\pi}{4} \right) |01\rangle \right.$$

$$\left. + \sin \left( -\frac{\pi}{8} + (x_0 + y_0) \frac{\pi}{4} \right) |10\rangle - \cos \left( -\frac{\pi}{8} + (x_0 + y_0) \frac{\pi}{4} \right) |11\rangle \right).$$

Thus we find for the probabilities of obtaining $a$ and $b$

| $a$ | $b$ | $a \oplus b$ | $P(a,b)$ |
|---|---|---|---|
| 0 | 0 | 0 | $\frac{1}{2} \cos^2 \left( -\pi/8 + (x_0 + y_0)\pi/4 \right)$ |
| 0 | 1 | 1 | $\frac{1}{2} \sin^2 \left( -\pi/8 + (x_0 + y_0)\pi/4 \right)$ |
| 1 | 0 | 1 | $\frac{1}{2} \sin^2 \left( -\pi/8 + (x_0 + y_0)\pi/4 \right)$ |
| 1 | 1 | 0 | $\frac{1}{2} \cos^2 \left( -\pi/8 + (x_0 + y_0)\pi/4 \right)$ |

Next we find the probability that

$$a \oplus b = x_0 \cdot y_0$$

for given $x_0$ and $y_0$

| $x_0$ | $y_0$ | $x_0 \cdot y_0$ | $P(a \oplus b = x_0 \cdot y_0)$ |
|---|---|---|---|
| 0 | 0 | 0 | $P(a=0,b=0) + P(a=1,b=1) = \cos^2(\pi/8)$ |
| 0 | 1 | 0 | $P(a=0,b=0) + P(a=1,b=1) = \cos^2(\pi/8)$ |
| 1 | 0 | 0 | $P(a=0,b=0) + P(a=1,b=1) = \cos^2(\pi/8)$ |
| 1 | 1 | 1 | $P(a=1,b=0) + P(a=0,b=1) = \cos^2(\pi/8)$ |

We find the probability

$$P(a \oplus b = x_0 \cdot y_0) = \cos^2(\pi/8).$$

**Problem 9.**   Let $G$ be a finite group and let $\rho : G \to GL(n, \mathbb{C})$ be a representation of $G$ where $n \in \mathbb{N}$. Show that

$$C := \sum_{g \in G} \rho(g)^* \rho(g)$$

is positive definite. Thus the positive definite matrices $C^{-1}$, $\sqrt{C}$ and $\sqrt{C}^{-1}$ exist. Is $\rho_C : G \to GL(n, \mathbb{C})$ defined by

$$\rho_C(g) := \sqrt{C} \rho(g) \sqrt{C}^{-1}$$

a unitary representation of $G$?

**Solution 9.** Clearly $C$ is hermitian. Since $\rho(g)$ is invertible, we have $\rho(g)\mathbf{x} \neq 0$ for all $\mathbf{x} \neq \mathbf{0}$, $\mathbf{x} \in \mathbb{C}^n$. Thus

$$\mathbf{x}^* C \mathbf{x} = \sum_{g \in G} \mathbf{x}^* \rho(g)^* \rho(g) \mathbf{x} = \sum_{g \in G} \|\rho(g)\mathbf{x}\|^2 > 0.$$

It follows that $C$ is positive definite and the positive definite matrices $C^{-1}$, $\sqrt{C}$ and $\sqrt{C}^{-1}$ exist. Obviously $\rho_C(g)$ is invertible, then the inverse is given by $\rho_C(g)^{-1} = \sqrt{C}\rho(g)^{-1}\sqrt{C}^{-1}$. Let $g_1, g_2 \in G$ then

$$\rho_C(g_1)\rho_C(g_2) = \sqrt{C}\rho(g_1)\sqrt{C}^{-1}\sqrt{C}\rho(g_2)\sqrt{C}^{-1} = \sqrt{C}\rho(g_1)\rho(g_2)\sqrt{C}^{-1}$$
$$= \sqrt{C}\rho(g_1 \cdot g_2)\sqrt{C}^{-1}$$

where $\cdot$ is the group operation. Thus $\rho_C$ provides a representation for $G$. We also have for all $g' \in G$

$$\rho(g')^* C \rho(g') = \sum_{g \in G} \rho(g')^* \rho(g)^* \rho(g) \rho(g') = \sum_{g \in G} (\rho(g)\rho(g'))^* \rho(g)\rho(g')$$
$$= \sum_{g \in G} (\rho(g \cdot g'))^* \rho(g \cdot g') = \sum_{g \in G} \rho(g)^* \rho(g) = C$$

where we used that $\{g' \cdot g : g \in G\} = G$. Now we find

$$\rho_C(g)^* \rho_C(g) = \sqrt{C}^{-1*} \rho(g)^* \sqrt{C}^* \sqrt{C}\rho(g)\sqrt{C}^{-1} = \sqrt{C}^{-1}\rho(g)^* C\rho(g)\sqrt{C}^{-1}$$
$$= \sqrt{C}^{-1}C\sqrt{C}^{-1} = I_n$$

where $I_n$ is the identity operator on $\mathbb{C}^n$. Thus $\rho_C$ is a unitary representation for $G$.

**Problem 10.** Let $G$ be a finite abelian group with identity $0$ and let $f : G \to S$ for some finite set $S$. We use the orthonormal bases

$$\{ |g\rangle_G : g \in G \} \subset \mathbb{C}^{|G|} \quad \text{and} \quad \{ |s\rangle_S : s \in S \} \subset \mathbb{C}^{|S|}.$$

Find a unitary operator $U_f$ such that

$$U_f(|g\rangle_G \otimes |s_0\rangle_S) = |g\rangle_G \otimes |f(g)\rangle_S$$

where $s_0 \in S$.

**Solution 10.** Let $h_g : S \setminus \{s_0\} \to S \setminus \{f(g)\}$ be arbitrary one to one functions for each $g \in G$. One solution is

$$U_f = \sum_{g \in G} |g\rangle_G\langle g|_G \otimes \left( |f(g)\rangle_S\langle s_0|_S + \sum_{s \in S \setminus \{s_0\}} |h_g(s)\rangle_S\langle s|_S \right).$$

**Problem 11.**    Let $G$ be a finite abelian group with identity 0 and let $f : G \to S$ for some finite set $S$. We use the orthonormal bases

$$\{ |g\rangle \; : \; g \in G \} \subset \mathbb{C}^{|G|}.$$

Discuss the quantum Fourier transform over the group $G$.

**Solution 11.**    Since $G$ is a finite abelian group the irreducible representations of $G$ are of the form $\chi : G \to \mathbb{C}$. There are $|G|$ such representations labeled $\chi_g$ for $g \in G$. Since $\chi$ is a group homomorphism we have $\chi(g_1 \cdot g_2) = \chi(g_1)\chi(g_2)$ for all $g_1, g_2 \in G$. Irreducible representations $\chi_1 \neq \chi_2$ have the property that

$$\sum_{g \in G} \chi_1(g)\overline{\chi_2(g)} = 0.$$

The trivial irreducible representation $\chi_1(g) = 1$ for all $g \in G$ provides

$$\sum_{g \in G} \chi(g)\overline{\chi_1(g)} = \sum_{g \in G} \chi(g) = 0$$

for nontrivial $\chi$. The transform over the group structure is

$$U_F(G) := \frac{1}{\sqrt{|G|}} \sum_{u \in G} \sum_{v \in G} \overline{\chi_u(v)} |u\rangle\langle v|$$

where the $\chi_l : G \to \mathbb{C}$ are the $|G|$ irreducible representations of $G$, $\chi_l(g)$ is a $|G|$-th root of unity. For $G = \mathbb{Z}_N$, the group of integers $\{0, 1, \ldots, N-1\}$ with addition modulo $N$, we obtain the quantum Fourier transform.

**Problem 12.**    Let $G$ be a finite abelian group with identity 0, and $K$ be a subgroup of $G$. The cosets of $K$, i.e. $g \cdot K := \{ g \cdot k \; : \; k \in K \}$ for $g \in G$, partition $G$. Now suppose $f : G \to X$, for some finite set $X$, with the property $f(g_1) = f(g_2)$ if and only if $g_1 \cdot K = g_2 \cdot K$. From $f$ and $G$ we wish to find the hidden subgroup $K$. As the computational basis we use the orthonormal basis $B_G = \{ |g\rangle_G \; : \; g \in G \} \subset \mathbb{C}^{|G|}$ and the orthonormal basis $B_X = \{ |x\rangle_X \; : \; x \in X \} \subset \mathbb{C}^{|X|}$. We begin with the state

$$\frac{1}{\sqrt{|G|}} \sum_{g \in G} |g\rangle_G \otimes |f(g)\rangle_X.$$

Show how measurement of the second system in the basis $B_X$ and then performing the quantum Fourier transform over the group structure for the first system can be used to find $K$.

**Solution 12.** We can factor out the second state according to the cosets of $K$

$$\frac{1}{\sqrt{|G|}} \sum_{g \cdot K} \left( \sum_{k \in K} |g \cdot k\rangle_G \right) \otimes |f(g)\rangle_X.$$

Measuring the second state in the basis $B_X$ projects the first state, for some $g$, onto

$$\frac{1}{\sqrt{|K|}} \sum_{k \in K} |g \cdot k\rangle_G.$$

Applying $U_F(G)$ yields

$$U_F(G) \frac{1}{\sqrt{|K|}} \sum_{k \in K} |g \cdot k\rangle_G = \frac{1}{\sqrt{|G| \cdot |K|}} \sum_{u \in G} \left( \sum_{k \in K} \overline{\chi_u(g \cdot k)} \right) |u\rangle_G$$

$$= \frac{1}{\sqrt{|G| \cdot |K|}} \sum_{u \in G} \overline{\chi_u(g)} \left( \sum_{k \in K} \overline{\chi_u(k)} \right) |u\rangle_G$$

since the irreducible representations are group homomorphisms. Due to the fact that the irreducible representation $\chi_u$ of $G$ can also be considered as an irreducible representation of the subgroup $K$ we have

$$\sum_{k \in K} \chi(k) = 0$$

for nontrivial $\chi$ so that

$$U_F(G) \frac{1}{\sqrt{|K|}} \sum_{k \in K} |g \cdot k\rangle_G = \frac{1}{\sqrt{|G| \cdot |K|}} \sum_{u \in G} \overline{\chi_u(g)} \left( \sum_{k \in K} \overline{\chi_u(k)} \right) |u\rangle_G$$

$$= \sqrt{\frac{|K|}{|G|}} \sum_{\substack{u \in G \\ \chi_u(k)=1 \, \forall k \in K}} \overline{\chi_u(g)} |u\rangle_G.$$

In the last result we find non-zero probability amplitudes for $|u\rangle_G$ when $\chi_u(k) = 1$ for all $k \in K$. Measurement of the state yields $u \in G$ with $\chi_u(k) = 1 \, \forall k \in K$. We can consequently test the different $K$ against $\chi_u$. If $K$ has a generator $k_0$, then $\chi_u(k_0) = 1$ so that measurement provides the possibility of recovering the generator of $K$. In general, the process must be repeated since we may have found a generator for a subgroup.

**Problem 13.** (i) The CHSH game is a game between a referee from one side and two player (named Alice and Bob) from the other side. Alice and Bob are separated and not allowed to communicate till the game is over. Let $a, b, x, y \in \{0, 1\}$. The game starts with a referee selecting two bits $x$

and $y$ uniformly at random. The referee then sends $x$ to Alice and $y$ to Bob. Alice sends back to the referee a bit $a$ and Bob sends back a bits $b$. So Bob's response bit $b$ cannot depend on Alice's input and vice versa Alice's response bit $a$ cannot depend on Bob's input. After the bits $a$ and $b$ has been transferred to the referee. Then he tests whether the boolean equation is satisfied

$$x \cdot y = a \oplus b.$$

If it is satisfied Alice and Bob win and the referee loses. If not the referee wins and Alice and Bob lose. What is the best strategy for Alice and Bob to win?

(ii) Let $|0\rangle$, $|1\rangle$ be the standard basis in $\mathbb{C}^2$, $I_2$ be the $2 \times 2$ identity matrix,

$$R_A = \begin{pmatrix} \cos(\pi/8) & -\sin(\pi/8) \\ \sin(\pi/8) & \cos(\pi/8) \end{pmatrix}, \quad R_B = \begin{pmatrix} \cos(\pi/8) & \sin(\pi/8) \\ -\sin(\pi/8) & \cos(\pi/8) \end{pmatrix}$$

and

$$U_A = R_A \otimes I_2, \quad U_B = I_2 \otimes R_B, \quad U_{AB} = R_A \otimes R_B$$

where $A$ refers to Alice and $B$ refers to Bob. Note that $R_A R_B = I_2$. Consider the Bell state

$$|\psi\rangle = \frac{1}{\sqrt{2}}(|0\rangle_A \otimes |0\rangle_B + |1\rangle_A \otimes |1\rangle_B).$$

If Alice receives $x = 0$ and Bob receives $y = 0$, then calculate

$$|\widetilde{\psi}\rangle = (I_2 \otimes I_2)|\psi\rangle, \quad p_{00} = |\langle\psi||\widetilde{\psi}\rangle|^2.$$

If Alice receives $x = 0$ and Bob receives $y = 1$, then calculate

$$|\widetilde{\psi}\rangle = (I_2 \otimes R_B)|\psi\rangle, \quad p_{01} = |\langle\psi||\widetilde{\psi}\rangle|^2.$$

If Alice receives $x = 1$ and Bob receives $y = 0$, then calculate

$$|\widetilde{\psi}\rangle = (R_A \otimes I_2)|\psi\rangle, \quad p_{10} = |\langle\psi||\widetilde{\psi}\rangle|^2.$$

If Alice receives $x = 1$ and Bob receives $y = 1$, then calculate

$$|\widetilde{\psi}\rangle = (R_A \otimes R_B)|\psi\rangle, \quad p_{11} = |\langle\psi||\widetilde{\psi}\rangle|^2.$$

Calculate the probability

$$p = \frac{1}{4}p_{00} + \frac{1}{4}p_{01} + \frac{1}{4}p_{10} + \frac{1}{4}p_{11}$$

and show that $p > 3/4$. Note that

$$\cos(\pi/8) = \frac{1}{2}\sqrt{2 + \sqrt{2}}, \quad \sin(\pi/8) = \frac{1}{2}\sqrt{2 - \sqrt{2}},$$

$$\cos^2(\pi/8) - \sin^2(\pi/8) = \frac{1}{2}\sqrt{2}.$$

**Solution 13.** (i) The truth table is

| $x$ | $y$ | $a$ | $b$ | $x \cdot y$ | $a \oplus b$ | $w$ |
|---|---|---|---|---|---|---|
| 0 | 0 | 0 | 0 | 0 | 0 | 1 |
| 0 | 0 | 0 | 1 | 0 | 1 | 0 |
| 0 | 0 | 1 | 0 | 0 | 1 | 0 |
| 0 | 0 | 1 | 1 | 0 | 0 | 1 |
| 0 | 1 | 0 | 0 | 0 | 0 | 1 |
| 0 | 1 | 0 | 1 | 0 | 1 | 0 |
| 0 | 1 | 1 | 0 | 0 | 1 | 0 |
| 0 | 1 | 1 | 1 | 0 | 0 | 1 |
| 1 | 0 | 0 | 0 | 0 | 0 | 1 |
| 1 | 0 | 0 | 1 | 0 | 1 | 0 |
| 1 | 0 | 1 | 0 | 0 | 1 | 0 |
| 1 | 0 | 1 | 1 | 0 | 0 | 1 |
| 1 | 1 | 0 | 0 | 1 | 0 | 0 |
| 1 | 1 | 0 | 1 | 1 | 1 | 1 |
| 1 | 1 | 1 | 0 | 1 | 1 | 1 |
| 1 | 1 | 1 | 1 | 1 | 0 | 0 |

From the truth table we find the following. If Alice always submits $a = 0$ and Bob always submits $b = 0$ the winning probability for Alice and Bob is $3/4$. Analogously, if Alice always submits $a = 1$ and Bob always submits $b = 1$ the winning probability for Alice and Bob is $3/4$.

(ii) Obviously for $(x = 0, y = 0)$ we obtain $p_{00} = 1$. For $(x = 0, y = 1)$ we obtain $p_{01} = \cos^2(\pi/8)$. For $(x = 1, y = 0)$ we obtain $p_{10} = \cos^2(\pi/8)$ For $(x = 1, y = 1)$ we have

$$(R_A \otimes R_B)|\psi\rangle = (R_A \otimes R_B)\frac{1}{\sqrt{2}}(|0\rangle \otimes |0\rangle + |1\rangle \otimes |1\rangle)$$

$$= \frac{1}{\sqrt{2}}((R_A|0\rangle) \otimes (R_B|0\rangle) + (R_A|1\rangle) \otimes (R_B|1\rangle))$$

$$= \frac{1}{\sqrt{2}}(\cos(\pi/4)|0\rangle \otimes |0\rangle - \sin(\pi/4)|0\rangle \otimes |1\rangle$$

$$+ \sin(\pi/4)|1\rangle \otimes |0\rangle + \cos(\pi/4)|1\rangle \otimes |1\rangle)$$

where we utilized that

$$\cos^2(\pi/8) - \sin^2(\pi/8) = \cos(\pi/4) = \sin(\pi/4), \quad 2\sin(\pi/8)\cos(\pi/8) = \sin(\pi/4).$$

It follows that

$$|\langle\psi|(R_A \otimes R_B)|\psi\rangle|^2 = \frac{1}{2}.$$

Hence

$$p = \frac{1}{4}(p_{00} + p_{01} + p_{10} + p_{11}) = \frac{3}{8} + \frac{1}{2}\cos^2(\pi/8) = \frac{5}{8} + \frac{\sqrt{2}}{8} \approx 0.8018$$

## Programming Problems

**Problem 1.**   Consider two bitstrings $b1$, $b2$ of the same length $n$. We define the scalar product of the two bitstrings as

$$b1 \star b2 = (b1[0] \cdot b2[0]) \oplus (b1[1] \cdots b2[0]) \oplus \cdot \oplus (b1[n-1] \cdot b2[n-1])$$

where $\oplus$ is the XOR operation and $\cdot$ is the AND operation. Give a C++ implementation utilizing the `bitset` class.

**Solution 1.**   The C++ program is

```
// scalarproduct.cpp
// c++ -std=c++11 -o scalarproduct scalarproduct.cpp
#include <iostream>
#include <string>
#include <bitset>
using namespace std;

int main(void)
{
// least significant bit on the right-hand side
const int n=8;
bitset<n> b1(string("10001010"));
bitset<n> b2(string("00110111"));
cout << b2[0] << endl << endl;
int temp[n];
int j;
for(j=0;j<n;j++)
{
temp[j] = b1[j] & b2[j];
cout << temp[j] << endl;
}
cout << endl;
int scalar = temp[0] ^ temp[1];
for(int j=2;j<n;j++)
{
scalar = scalar ^ temp[j];
}
cout << scalar << endl;
return 0;
}
```

**Problem 2.**   Let $a, b, x, y \in \{0, 1\}$. Find all solutions of the boolean equation

$$a \cdot b = x \oplus y.$$

Utilize the `bitset` class of C++.

**Solution 2.** The C++ program provides the solutions

```
// bitequation.cpp
#include <iostream>
#include <bitset>
using namespace std;

int main(void)
{
bitset<1> a;
bitset<1> b;
bitset<1> x;
bitset<1> y;
int j1,j2,j3,j4;
cout << "The solutions are: " << endl;
for(j1=0;j1<=1;j1++)
 for(j2=0;j2<=1;j2++)
  for(j3=0;j3<=1;j3++)
   for(j4=0;j4<=1;j4++)
   {
   a = j1; b = j2; x = j3; y = j4;
   if((a & b)==(x ^ y))
   { cout << "a = " << j1 << " " << "b = " << j2 << " "
          << "x = " << j3 << " " << "y = " << j4 << endl;
   }
   }
return 0;
}
```

provides the solutions

```
The solutions are:
a = 0 b = 0 x = 0 y = 0
a = 0 b = 0 x = 1 y = 1
a = 0 b = 1 x = 0 y = 0
a = 0 b = 1 x = 1 y = 1
a = 1 b = 0 x = 0 y = 0
a = 1 b = 0 x = 1 y = 1
a = 1 b = 1 x = 0 y = 1
a = 1 b = 1 x = 1 y = 0
```

**Problem 3.** Let $\omega := e^{2\pi i/4}$. The $4 \times 4$ quantum Fourier transform is given by

$$U = \frac{1}{2} \begin{pmatrix} 1 & 1 & 1 & 1 \\ 1 & \omega & \omega^2 & \omega^3 \\ 1 & \omega^2 & \omega^4 & \omega^6 \\ 1 & \omega^3 & \omega^6 & \omega^9 \end{pmatrix}.$$

Since $\omega^4 = 1$ we can write

$$U = \frac{1}{2} \begin{pmatrix} 1 & 1 & 1 & 1 \\ 1 & \omega & \omega^2 & \omega^3 \\ 1 & \omega^2 & 1 & \omega^2 \\ 1 & \omega^3 & \omega^2 & \omega \end{pmatrix}.$$

Show that the matrix is unitary. Then find the four eigenvalues. Apply matrix to the normalized vector

$$\phi = \frac{1}{2} \begin{pmatrix} 1 \\ 1 \\ 1 \\ 1 \end{pmatrix}.$$

Discuss.

**Solution 3.** The Maxima program

```
/* Fourier.mac */
om: exp(2*%pi*%i/4);
F: matrix([1,1,1,1],[1,om,om^2,om^3],[1,om^2,1,om^2],[1,om^3,om^2,om])/2;
F: trigexpand(F);
FT: transpose(F);
FTC: conjugate(FT);
R: F . FTC;
phi: matrix([1],[1],[1],[1])/2;
psi: F . phi;
eigenvalues(F);
```

provides us with the information that $F$ is unitary since $FF^* = I_4$. The eigenvalues are $+1$ (twice), $-1$ and $i$ and

$$F\phi = \begin{pmatrix} 1 \\ 0 \\ 0 \\ 0 \end{pmatrix}.$$

## 14.3 Supplementary Problems

**Problem 1.** Consider the standard basis

$$|1\rangle = \begin{pmatrix} 1 \\ 0 \\ 0 \end{pmatrix}, \quad |2\rangle = \begin{pmatrix} 0 \\ 1 \\ 0 \end{pmatrix}, \quad |3\rangle = \begin{pmatrix} 0 \\ 0 \\ 1 \end{pmatrix}$$

in $\mathbb{C}^3$, the unitary matrix (quantum Fourier transform)

$$U = \frac{1}{\sqrt{3}} \begin{pmatrix} 1 & 1 & 1 \\ 1 & \exp(i2\pi/3) & \exp(-i2\pi/3) \\ 1 & \exp(-i2\pi/3) & \exp(i2\pi/3) \end{pmatrix}$$

and the six permutation matrices

$$P_1 = \begin{pmatrix} 1 & 0 & 0 \\ 0 & 1 & 0 \\ 0 & 0 & 1 \end{pmatrix}, \quad P_2 = \begin{pmatrix} 0 & 1 & 0 \\ 0 & 0 & 1 \\ 1 & 0 & 0 \end{pmatrix}, \quad P_3 = \begin{pmatrix} 0 & 0 & 1 \\ 1 & 0 & 0 \\ 0 & 1 & 0 \end{pmatrix},$$

$$P_4 = \begin{pmatrix} 0 & 0 & 1 \\ 0 & 1 & 0 \\ 1 & 0 & 0 \end{pmatrix}, \quad P_5 = \begin{pmatrix} 0 & 1 & 0 \\ 1 & 0 & 0 \\ 0 & 0 & 1 \end{pmatrix}, \quad P_6 = \begin{pmatrix} 1 & 0 & 0 \\ 0 & 0 & 1 \\ 0 & 1 & 0 \end{pmatrix}.$$

The first three permutation matrices $P_1$, $P_2$, $P_3$ are the even permutation matrices with

$$\det(P_1) = \det(P_2) = \det(P_3) = 1$$

and the last three permutation matrices $P_4$, $P_5$, $P_6$ are the odd permutation matrices with

$$\det(P_4) = \det(P_5) = \det(P_6) = -1.$$

Do the following calculation. First find the normalized state

$$|\psi_1\rangle = U|2\rangle.$$

Then calculate the six normalized states

$$|\psi_k\rangle = P_k|\psi_1\rangle, \quad k = 1, 2, 3, 4, 5, 6$$

Next find $U^*$ and check it is really the inverse of $U$. Then calculate the six normalized states

$$U^*|\psi_k\rangle, \quad k = 1, 2, 3, 4, 5, 6.$$

Finally find the twelve probabilities

$$|\langle 2|U^*|\psi_k\rangle|^2, \quad |\langle 3|U^*|\psi_k\rangle|^2.$$

Discuss. This exercise plays a role for the quantum permutation algorithm.

**Problem 2.**   Suppose the input consists of $n$ qubits and the boolean function to be calculated is

$$f := \{0, 1, \ldots, 2^n - 1\} \to \{0, 1\}.$$

Show that

$$U_f := \sum_{j=0}^{2^n-1} \sum_{k=0}^{1} |j\rangle\langle j| \otimes |k \oplus f(j)\rangle\langle k|$$

is a unitary matrix (a permutation matrix).

# Chapter 15

# Quantum Error Correction

---

## 15.1 Introduction

In classical communication theory where bits are communicated, the only possible type of error that can occur is a bit flip. In the quantum case any rotation or phase change in the Hilbert space of the quantum state is an error. Thus there are an infinite number of different errors that could occur just for a single qubit. Fortunately the measurement process involves the projection of the quantum state into a compatible subspace. Thus measurement to determine the occurrence of an error reduces the error to one compatible with the measurement. Suppose the data is contained in the state $|\psi\rangle$, and the environment is described by the state $|E\rangle$. The initial state of the entire system is described by the tensor product of the states $|\psi\rangle \otimes |E\rangle$, which evolves according to some unitary operation $U$. The state $|\psi\rangle$ evolves according to the unitary operation $U_\psi$ which describes a quantum algorithm. In classical error correction codes, all that needs to be corrected are bit flips. In the quantum case errors such as bit flips, phase changes and rotations complicate the error correction techniques. Since arbitrary errors in an encoding of information cannot be corrected, only certain types of errors are assumed to occur. The types of errors depend on the implementation. For example, suppose the types of errors (which we assume are distinguishable due to an encoding) are described by the unitary basis $E_1, \ldots, E_n$ so that all errors are described by a linear combination

$$E = c_1 E_1 + \cdots + c_n E_n, \qquad E_j^\dagger E_j = I$$

where $I$ is the identity operator and $j = 1, \ldots, n$. We use the product state $|\psi\rangle \otimes |0\rangle$, where $|\psi\rangle$ is an encoded quantum state with the necessary property that it can be used to determine if any error of $E_1, \ldots, E_n$ has occurred, and the second quantum register will hold the number of the type of error which occurred. Let $S$ denote the operator for the *error syndrome*

$$S(E_j \otimes I)|\psi\rangle \otimes |0\rangle := |\psi\rangle \otimes |j\rangle.$$

Now the encoded state with errors is given by

$$(E \otimes I)(|\psi\rangle \otimes |0\rangle) = \sum_{j=1}^{n} c_j E_j |\psi\rangle \otimes |0\rangle.$$

Applying the operator for the error syndrome gives

$$S(E \otimes I)(|\psi\rangle \otimes |0\rangle) = \sum_{j=1}^{n} c_j E_j |\psi\rangle \otimes |j\rangle.$$

Measuring the second register identifies the error. Suppose the measurement corresponds to $|k\rangle$, then the error can be repaired since

$$(E_k^{-1} \otimes I)((E_k \otimes I)|\psi\rangle \otimes |k\rangle) = |\psi\rangle \otimes |k\rangle.$$

Given a normalized state in $\mathbb{C}^2$

$$|\psi\rangle = \alpha|0\rangle + \beta|1\rangle, \qquad |\alpha|^2 + |\beta|^2 = 1.$$

A *bit-flip* changes the normalized state into

$$\alpha|1\rangle + \beta|0\rangle$$

i.e. we apply the operator $|0\rangle\langle 1|$. The *phase-flip error* changes the normalized state into

$$\alpha|0\rangle - \beta|1\rangle$$

i.e. we apply the operator $|0\rangle\langle 0| + e^{i\pi}|1\rangle\langle 1|$.

Let $\sigma_1$, $\sigma_2$, $\sigma_3$ be the Pauli spin matrices and $\sigma_0 = I_2$. A general single bit error is thus a map

$$e_1 I_2 + e_2 \sigma_1 + e_3 \sigma_2 + e_4 \sigma_3.$$

Interaction with the environment maps single qubits as

$$|\psi\rangle \mapsto (e_1 \sigma_0 + e_2 \sigma_1 + e_3 \sigma_2 + e_4 \sigma_3)|\psi\rangle.$$

## 15.2 Solved Problems

**Problem 1.** Calculate the following in terms of $I_2, X, Y, Z$

(i) $XZ, \quad ZX$

(ii) $U_{CNOT}(X \otimes I_2)U_{CNOT}$

(iii) $U_{CNOT}(I_2 \otimes X)U_{CNOT}$

(iv) $U_{CNOT}(Z \otimes I_2)U_{CNOT}$

(v) $U_{CNOT}(I_2 \otimes Z)U_{CNOT}$

(vi) $U_{CNOT}(X \otimes X)U_{CNOT}$

(vii) $U_{CNOT}(Z \otimes Z)U_{CNOT}$

(viii) $U_{CNOT}U_{CNOT}$

where

$$I_2 := |0\rangle\langle 0| + |1\rangle\langle 1|, \quad X := |0\rangle\langle 1| + |1\rangle\langle 0|$$
$$Y := |0\rangle\langle 1| - |1\rangle\langle 0|, \quad Z := |0\rangle\langle 0| - |1\rangle\langle 1|$$
$$U_{CNOT} := |0\rangle\langle 0| \otimes I_2 + |1\rangle\langle 1| \otimes X.$$

**Solution 1.** Straightforward calculation yields

(i) $XZ = -Y, \quad ZX = Y$

(ii) $U_{CNOT}(X \otimes I_2)U_{CNOT} = X \otimes X$

(iii) $U_{CNOT}(I_2 \otimes X)U_{CNOT} = I_2 \otimes X$

(iv) $U_{CNOT}(Z \otimes I_2)U_{CNOT} = Z \otimes I_2$

(v) $U_{CNOT}(I_2 \otimes Z)U_{CNOT} = Z \otimes Z$

(vi) $U_{CNOT}(X \otimes X)U_{CNOT} = X \otimes I_2$

(vii) $U_{CNOT}(Z \otimes Z)U_{CNOT} = I_2 \otimes Z$

(viii) $U_{CNOT}U_{CNOT} = I_2 \otimes I_2$.

**Problem 2.** Suppose that the only errors which can occur to three qubits are described by the set of $8 \times 8$ unitary matrices

$$\{I_2 \otimes I_2 \otimes I_2, I_2 \otimes U_{NOT} \otimes U_{NOT}, I_2 \otimes U_P \otimes U_P, I_2 \otimes (U_P U_{NOT}) \otimes (U_P U_{NOT})\}$$

where $U_P := |0\rangle\langle 0| - |1\rangle\langle 1|$, $U_{NOT} := |0\rangle\langle 1| + |1\rangle\langle 0|$. A linear combination of these unitary matrices is given by

$$E := \alpha I_2 \otimes I_2 \otimes I_2 + \beta I_2 \otimes U_{NOT} \otimes U_{NOT} + \delta I_2 \otimes U_P \otimes U_P$$
$$+ \gamma I_2 \otimes (U_P U_{NOT}) \otimes (U_P U_{NOT})$$

where $\alpha, \beta, \delta, \gamma \in \mathbb{C}$. Describe how an arbitrary error $E$ on the three-qubit state

$$\frac{1}{\sqrt{2}}(|00\rangle + |11\rangle) \otimes |\psi\rangle$$

can be corrected to obtain the correct $|\psi\rangle$ as the last qubit, where

$$|\psi\rangle := a|0\rangle + b|1\rangle, \quad |a|^2 + |b|^2 = 1, \quad a, b \in \mathbb{C}.$$

**Solution 2.**   Applying the matrix ($I_8 \equiv I_2 \otimes I_2 \otimes I_2$)

$$\alpha I_8 + \beta I_2 \otimes U_{NOT} \otimes U_{NOT} + \delta I_2 \otimes U_P \otimes U_P + \gamma I_2 \otimes (U_P U_{NOT}) \otimes (U_P U_{NOT})$$

to the state

$$\frac{1}{\sqrt{2}}(|00\rangle + |11\rangle) \otimes |\psi\rangle$$

yields the state

$$\alpha \frac{1}{\sqrt{2}}(|00\rangle + |11\rangle) \otimes |\psi\rangle + \beta \frac{1}{\sqrt{2}}(|01\rangle + |10\rangle) \otimes (a|1\rangle + b|0\rangle)$$

$$+\delta \frac{1}{\sqrt{2}}(|00\rangle - |11\rangle) \otimes (a|0\rangle - b|1\rangle) + \gamma \frac{1}{\sqrt{2}}(|01\rangle - |10\rangle) \otimes (a|1\rangle - b|0\rangle).$$

Thus we measure the first two qubits in the Bell basis and apply the corresponding transform to the last qubit to obtain $|\psi\rangle$.

| measure | transform |
|---|---|
| $\frac{1}{\sqrt{2}}(|00\rangle + |11\rangle)$ | $I_2$ |
| $\frac{1}{\sqrt{2}}(|01\rangle + |10\rangle)$ | $U_{NOT}$ |
| $\frac{1}{\sqrt{2}}(|00\rangle - |11\rangle)$ | $U_P$ |
| $\frac{1}{\sqrt{2}}(|01\rangle - |10\rangle)$ | $U_{NOT}U_P$ |

**Problem 3.**   Assume that the only errors that occur in a system of qubits are isolated to individual qubits, i.e. the error in one qubit state is independent of the error in another qubit state. Hence the error for each qubit can be expressed as a linear operator $E$ on the Hilbert space $\mathbb{C}^2$. Furthermore $E$ can be expressed as a linear combination of the $2 \times 2$ identity matrix and the Pauli spin matrices $\sigma_1$, $\sigma_2$ and $\sigma_3$. Now consider a non-degenerate $n$-qubit code representing a single qubit state which can correct errors in up to $k$ qubits.
(i) Find a lower bound describing $n$.
(ii) Find the lower bound for $k = 1$.

**Hint.** The $n$-qubit states representing qubits with errors should be distinct (orthogonal) for distinct errors and distinct from the case where there are no errors.

**Solution 3.** (i) We have 3 distinct errors on a single qubit described by the Pauli matrices. Thus there are

$$3^l \binom{n}{l} \equiv 3^l \frac{n!}{l!(n-l)!}$$

distinct errors in $l$ qubits of $n$ qubits. The total number of ways to have at most $k$ errors in $n$ qubits is then given by

$$\sum_{l=0}^{k} 3^l \binom{n}{l}.$$

There are $2^n$ orthogonal states in a Hilbert space describing $n$ qubits. Since the states representing qubits ($|0\rangle$ or $|1\rangle$) with distinct errors should be orthogonal, we find

$$2 \sum_{l=0}^{k} 3^l \binom{n}{l} \leq 2^n.$$

(ii) For $k = 1$ we have the bound $2(1 + 3n) \leq 2^n$. In other words, for $k = 1$ we find $n \geq 5$.

**Problem 4.** Consider the Pauli matrices $\sigma_1$, $\sigma_2$, $\sigma_3$ and the $2 \times 2$ unit matrix. Do these matrices form a group under matrix multiplication? If so, provide a proof. If not, what set with minimal cardinality includes the Pauli matrices and forms a group under matrix multiplication?

**Solution 4.** Obviously these matrices do not form a group since $\sigma_1 \sigma_2 = i\sigma_3$ which is not a Pauli matrix. Since the factor $i$ is introduced, all factors which are powers of $i$ must be included i.e. $\pm 1$ and $\pm i$:

$$\{\pm \sigma, \pm i\sigma : \sigma \in \{I_2, \sigma_1, \sigma_2, \sigma_3\}\}.$$

Thus we have a group of cardinality 16.

**Problem 5.** Let $S$ be a set of operators closed under the hermitian conjugate (adjoint) such that the quantum code $C_S$ of the set of states

$$C_S := \{ |\psi\rangle : U|\psi\rangle = |\psi\rangle, \quad \forall U \in S \}$$

is non trivial (does not consist only of the zero state). The set $S$ is called the *stabilizer* of the code $C_S$.

(i) Show that

$$[M, N]|\psi\rangle = 0, \qquad \forall M, N \in S, \quad |\psi\rangle \in C_S.$$

(ii) Let $[E, M] = 0$ for some $M \in S$. What can be said about $E|\psi\rangle$ when $|\psi\rangle \in C_S$ ?

(iii) Let $[E, M]_+ = 0$ for some $M \in S$. What can be said about $E|\psi\rangle$ when $|\psi\rangle \in C_S$ ?

**Solution 5.**   (i) Since $MN|\psi\rangle = M|\psi\rangle = |\psi\rangle$, $NM|\psi\rangle = N|\psi\rangle = |\psi\rangle$ it follows that $[M, N]|\psi\rangle = 0$.

(ii) For $|\psi\rangle \in C_S$ we have

$$ME|\psi\rangle = EM|\psi\rangle = E|\psi\rangle.$$

Thus $E|\psi\rangle$ is an eigenstate of $M$ corresponding to the eigenvalue 1.

(iii) For $|\phi\rangle, |\psi\rangle \in C_S$ we have

$$\langle\phi|E|\psi\rangle = \langle\phi|EM|\psi\rangle = -\langle\phi|ME|\psi\rangle = -\langle\phi|M^*E|\psi\rangle = -\langle\phi|E|\psi\rangle.$$

Thus $\langle\phi|E|\psi\rangle = 0$. In other words, $E|\psi\rangle \notin C_S$. Furthermore

$$ME|\psi\rangle = -EM|\psi\rangle = -E|\psi\rangle$$

thus $E|\psi\rangle$ is an eigenstate of $M$ corresponding to the eigenvalue $-1$.

**Problem 6.**   Consider the application of the controlled NOT gate to the state $|10\rangle \equiv |1\rangle \otimes |0\rangle$. Suppose that a single qubit error $\sigma_1 \otimes I_2$ occurs before the controlled NOT operation is performed. How many errors does the resulting state have after performing the control NOT operation? Also discuss the case when the error is given by $\frac{1}{\sqrt{2}}(I_2 + \sigma_1) \otimes I_2$.

**Solution 6.**   The intended operation is $U_{CNOT}|10\rangle = |11\rangle$. The operation with error is

$$U_{CNOT}(\sigma_1 \otimes I_2)|10\rangle = U_{CNOT}|00\rangle = |00\rangle.$$

Thus we have two single qubit errors $\sigma_1 \otimes I_2$ and $I_2 \otimes \sigma_1$. For the second type of single qubit error we have

$$U_{CNOT}\frac{1}{\sqrt{2}}(I_2 + \sigma_1) \otimes I_2|10\rangle = U_{CNOT}\frac{1}{\sqrt{2}}(|10\rangle + |00\rangle) = \frac{1}{\sqrt{2}}(|00\rangle + |11\rangle).$$

Here we find that the error cannot be expressed in terms of single qubit errors.

**Problem 7.** Consider the application of the controlled NOT gate to corresponding pairs of qubits of the state $|\tilde{1}\tilde{0}\rangle$ where $|\tilde{0}\rangle$ is a code word for $|0\rangle$ and $|\tilde{1}\rangle$ is a code word for $|1\rangle$ in a quantum error correction code that can correct single qubit errors. In other words the controlled NOT gate is applied to the $i$-th qubit of $|\tilde{1}\rangle$ as control and the $i$-th qubit of $|\tilde{0}\rangle$ as target. Suppose that a single qubit error

$$\frac{1}{\sqrt{2}}(I_2 + \sigma_1)$$

on one of the qubits of $|\tilde{1}\rangle$ occurs before the controlled NOT operation is performed. Discuss the error correction of this computation.

**Solution 7.** Suppose the single qubit error occurs on the $j$-th qubit, and that the encoding $|\tilde{1}\rangle$ and $|\tilde{0}\rangle$ is $n$ qubits long. Thus the error is

$$E := \frac{1}{\sqrt{2}} \left( \bigotimes_{k=1}^{j-1} I_2 \right) \otimes (I_2 + \sigma_1) \otimes \left( \bigotimes_{k=j+1}^{2n} I_2 \right).$$

Let $\tilde{U}_{CNOT}$ denote the pairwise application of the controlled NOT operation. Then

$$\tilde{U}_{CNOT} E |\tilde{1}\tilde{0}\rangle = \tilde{U}_{CNOT} E \tilde{U}_{CNOT}^* \tilde{U}_{CNOT} |\tilde{1}\tilde{0}\rangle.$$

The controlled NOT operation is given by

$$U_{CNOT} = \frac{1}{2}(I_2 + \sigma_3) \otimes I_2 + \frac{1}{2}(I_2 - \sigma_3) \otimes \sigma_1.$$

We find the matrix

$$\frac{1}{\sqrt{2}} U_{CNOT} (I_2 + \sigma_1) U_{CNOT}^* = \frac{1}{\sqrt{2}} (I_2 \otimes I_2 + \sigma_1 \otimes \sigma_1).$$

Consequently we obtain the normalized state

$$\tilde{U}_{CNOT} E |\tilde{1}\tilde{0}\rangle = \left( \frac{1}{\sqrt{2}} \left( \bigotimes_{k=1}^{2n} I_2 \right) \right.$$

$$+ \frac{1}{\sqrt{2}} \left( \bigotimes_{k=1}^{j-1} I_2 \right) \otimes \sigma_1 \otimes \left( \bigotimes_{k=j+1}^{n+j-1} I_2 \right) \otimes \sigma_1 \otimes \left( \bigotimes_{k=n+j+1}^{2n} I_2 \right) \right)$$

$$\times \tilde{U}_{CNOT} |\tilde{1}\tilde{0}\rangle.$$

Thus there is a linear combination of no error, or a single qubit error in the first block of $n$ qubits and a single qubit error in the second block of $n$ qubits. Both errors can be independently corrected.

### Programming Problem

**Problem 1.** Consider the single qubit state $a|0\rangle + b|1\rangle$ with $a\bar{a} + b\bar{b} = 1$. Encode the single qubit state in three qubits as

$$a(|0\rangle \otimes |0\rangle \otimes |0\rangle) + b(|1\rangle \otimes |1\rangle \otimes |1\rangle)$$

i.e. $|0\rangle \mapsto |0\rangle \otimes |\rangle \otimes |0\rangle$, $|1\rangle \mapsto |1\rangle \otimes |1\rangle \otimes |1\rangle$. Give a Maxima implementation.

**Solution 1.** Utilizing the $U_{CNOT}$ gate

$$U_{CNOT} = I_2 \oplus \begin{pmatrix} 0 & 1 \\ 1 & 0 \end{pmatrix}$$

and $I_2 \otimes U_{CNOT}$, $U_{CNOT} \otimes I_2$ we can implement the encoding.

```
/* encoding.mac */
v0: matrix([1],[0]); v1: matrix([0],[1]);
psi: a*v0 + b*v1;
alpha: kronecker_product(psi,kronecker_product(v0,v0));
I2: matrix([1,0],[0,1]);
UCNOT: matrix([1,0,0,0],[0,1,0,0],[0,0,0,1],[0,0,1,0]);
U1: kronecker_product(I2,UCNOT);
U2: kronecker_product(UCNOT,I2);
beta: U1 . U2 . alpha;
```

## 15.3 Supplementary Problem

**Problem 1.** Let $|0\rangle$, $|1\rangle$ be the standard basis in $\mathbb{C}^2$. Consider the quantum bit

$$|\psi\rangle = \frac{1}{\sqrt{2}}(|0\rangle - |1\rangle)$$

which is encoded as

$$C|\psi\rangle = |\phi\rangle = \frac{1}{\sqrt{2}}(|0\rangle \otimes |0\rangle \otimes |0\rangle - |1\rangle \otimes |1\rangle \otimes |1\rangle).$$

Consider the error $E = \frac{4}{5}\sigma_1 \otimes I_2 \otimes I_2 + \frac{3}{5}I_2 \otimes \sigma_1 \otimes I_2$. Show that

$$E|\phi\rangle = \frac{4}{5\sqrt{2}}(|1\rangle \otimes |0\rangle \otimes |0\rangle - |0\rangle \otimes |1\rangle \otimes |1\rangle) + \frac{3}{5\sqrt{2}}(|0\rangle \otimes |1\rangle \otimes |0\rangle - |1\rangle \otimes |0\rangle \otimes |1\rangle).$$

Then apply the syndrome extraction to $(E|\psi\rangle) \otimes |0\rangle \otimes |0\rangle \otimes |0\rangle$.

# Chapter 16

# Quantum Cryptography

## 16.1 Introduction

Cryptography usually involves a key or keys to be used in encryption and decryption algorithms. Classical cryptography generally relies on maps that are perceived to be very difficult to invert with incomplete information. One popular algorithm due to Rivest, Shamir and Adelman is the RSA algorithm

$$n = pq, \quad M \in \{0, 1, \ldots, n-1\}, \quad ed = 1 \bmod (p-1)(q-1)$$

$$C := (M^e \bmod n), \quad M \equiv (C^d \bmod n)$$

where $p$ and $q$ are large prime numbers, $M$ is the message and $C$ is the encrypted message. If $p$, $q$ and $d$ are unknown then, in general, $C$ cannot easily be obtained from $M$. However, Shor found that $p$ and $q$ can be obtained with relative ease (and consequently also $d$) using the quantum Fourier transform.

Quantum cryptography is concerned with the secure distribution of keys using quantum communication channels. Another application is hiding classical data in quantum states. Quantum cryptographic techniques rely on physics to supply secure communication in the sense that it is possible to determine if someone has intercepted the message. This is due to the fact that measurement in quantum mechanics is associated with a disturbance (projection) of the quantum state. Entanglement can also be used to detect whether a message has been intercepted. For example, when one qubit of an EPR pair is measured the correlation is destroyed which can be tested with Bell's inequality.

## 16.2   Solved Problems

**Problem 1.**   Let $p, q$ be large prime numbers, $(e, n)$ a public key, and $(d, n)$ a private key for the RSA cryptosystem where $n = pq$. Let $M, C \in \{0, 1, \ldots, n-1\}$ with

$$C \equiv M^e \bmod n, \qquad M \equiv C^d \bmod n.$$

(i) Let $gcd(a, n) = 1$. Show that $p$ and $q$ can be determined from even $r \in \mathbb{N}$ if $a^r \equiv 1 \bmod n$, $a^{r/2} \not\equiv 1 \bmod n$, $a^{r/2} \not\equiv 1 \bmod n$.
(ii) Illustrate (i) with $p = 5$, $q = 11$ and $a = 6$.
(iii) Show that $M$ can be determined from $C$ and $t \in \mathbb{N}$, where ($t$ minimal)

$$C^{et} \equiv C \bmod n.$$

(iv) Illustrate (iii) with $p = 5$, $q = 11$, $e = 9$ and $C = 48$.

**Solution 1.**   (i) We have

$$a^r - 1 \equiv 0 \bmod n, \qquad (a^{\frac{r}{2}} - 1)(a^{\frac{r}{2}} + 1) \equiv 0 \bmod n$$

Consequently

$$[(a^{\frac{r}{2}} - 1) \bmod n][(a^{\frac{r}{2}} + 1) \bmod n] = kn = kpq$$

for some $k \in \mathbb{Z}$. Consequently one of $p$ and $q$ is given by

$$gcd(a^{\frac{r}{2}} - 1 \bmod n, n).$$

The second is found by division of $n$. The $gcd$ can be determined efficiently using the Euclidean algorithm.
(ii) The powers (from 0) of $a$ in modulo $5 \cdot 11 = 55$ arithmetic are

$$1, 6, 36, 216 \equiv 51, \equiv 306 \equiv 31, \equiv 186 \equiv 21, \equiv 126 \equiv 16, \equiv 96 \equiv 41,$$

$$\equiv 246 \equiv 26, \equiv 156 \equiv 46, \equiv 276 \equiv 1$$

where the last power is 10. Thus we use $r = 10$. Thus the periodicity of this sequence is also 10. Consequently

$$6^5 - 1 = 7775 \equiv 20 \bmod n.$$

Using the Euclidean algorithm

$$55 = 2 \cdot 20 + 15, \quad 20 = 1 \cdot 15 + 5, \quad 15 = 3 \cdot 5 + 0$$

Thus we find the $gcd(20, 55) = 5$. Consequently $p = 5$ and $q = n/5 = 11$.
(iii) We find

$$C^t \equiv M^{et} \equiv C^{det} \equiv (C^{et})^d \equiv C^d \equiv M \bmod n.$$

(iv) We find

$$C^e = 48^9 = 4^9 4^9 3^9 \equiv 14 \cdot 14 \cdot 48 \equiv 3.$$

The powers (from 0) of $C^e$ in modulo $5 \cdot 11 = 55$ arithmetic are

$$1, 3, 9, 27, 26, 23, 14, 42, 16, 48$$

i.e. we use $t = 9$. Now $C^t \equiv 3 \equiv M \bmod n$.

**Problem 2.** Apply the quantum Fourier transform to the first register of

$$|\psi\rangle := \frac{1}{\sqrt{6}}(|0\rangle \otimes |0\rangle + |1\rangle \otimes |1\rangle + |2\rangle \otimes |2\rangle + |3\rangle \otimes |0\rangle + |4\rangle \otimes |1\rangle + |5\rangle \otimes |2\rangle)$$

to find the underlying periodicity in the second register. In this case the
quantum Fourier transform is given by

$$U_{QFT} = \frac{1}{\sqrt{6}} \sum_{j,k=0}^{5} e^{-i2\pi jk/6}|j\rangle\langle k|.$$

**Solution 2.** Applying the quantum Fourier transform yields

$$(U_{QFT} \otimes I)|\psi\rangle = \frac{1}{6} \sum_{j=0}^{5} |j\rangle \otimes \left(|0\rangle + e^{-i2\pi j/6}|1\rangle + e^{-i2\pi j2/6}|2\rangle \right.$$

$$+ e^{-i2\pi j3/6}|0\rangle + e^{-i2\pi j4/6}|1\rangle + \left. e^{-i2\pi j5/6}|2\rangle\right)$$

$$= \frac{1}{6} \sum_{j=0}^{5} [1 + e^{-i\pi j}] \, |j\rangle \otimes \left(|0\rangle + e^{-i\pi j/3}|1\rangle + e^{-i2\pi j/3}|2\rangle\right)$$

$$= \frac{1}{3} \sum_{j=0}^{2} |2j\rangle \otimes \left(|0\rangle + e^{-i\pi 2j/3}|1\rangle + e^{-i4\pi j/3}|2\rangle\right).$$

Thus measuring the first register yields 0, 2 or 4. The minimum positive
value is 2. The period is $6/2 = 3$.

**Problem 3.**    Let $B_1 = \{|0\rangle, |1\rangle\}$ and $B_2 = \{|0_H\rangle, |1_H\rangle\}$ denote two orthonormal bases in $\mathbb{C}^2$ where

$$|0_H\rangle := \frac{1}{\sqrt{2}}(|0\rangle + |1\rangle), \qquad |1_H\rangle := \frac{1}{\sqrt{2}}(|0\rangle - |1\rangle).$$

Show that

$$|\psi\rangle := \frac{1}{\sqrt{2}}(|0\rangle \otimes |1\rangle - |1\rangle \otimes |0\rangle) = -\frac{1}{\sqrt{2}}(|0_H\rangle \otimes |1_H\rangle - |1_H\rangle \otimes |0_H\rangle).$$

**Solution 3.**    A simple calculation yields

$$|\psi\rangle := \frac{1}{\sqrt{2}}(|0\rangle \otimes |1\rangle - |1\rangle \otimes |0\rangle)$$

$$= \frac{1}{2\sqrt{2}}([|0_H\rangle + |1_H\rangle] \otimes [|0_H\rangle - |1_H\rangle] - [|0_H\rangle - |1_H\rangle] \otimes [|0_H\rangle + |1_H\rangle])$$

$$= -\frac{1}{\sqrt{2}}(|0_H\rangle \otimes |1_H\rangle - |1_H\rangle \otimes |0_H\rangle).$$

**Problem 4.**    Let

$$B_1 := \{\ |\psi_0\rangle := |H\rangle, \ |\psi_1\rangle := |V\rangle \ \}$$

denote an orthonormal basis in the Hilbert space $\mathbb{C}^2$. The states $|H\rangle$ and $|V\rangle$ can be identified with the horizontal and vertical polarization of a photon. Let

$$B_2 := \left\{\ |\phi_0\rangle := \frac{1}{\sqrt{2}}(|H\rangle + |V\rangle), \ |\phi_1\rangle := \frac{1}{\sqrt{2}}(|H\rangle - |V\rangle)\ \right\}$$

denote a second orthonormal basis in $\mathbb{C}^2$. These states are identified with the $45^o$ and $-45^o$ polarization of a photon. Alice sends photons randomly prepared in one of the four states $|H\rangle$, $|V\rangle$, $|\phi_0\rangle$ and $|\phi_1\rangle$ to Bob. Bob then randomly chooses a basis $B_1$ or $B_2$ to measure the polarization of the photon. All random decisions follow the uniform distribution. Alice and Bob interpret $|\psi_0\rangle$ as binary 0 and $|\psi_1\rangle$ as binary 1 in the basis $B_1$. They interpret $|\phi_0\rangle$ as binary 0 and $|\phi_1\rangle$ as binary 1 in the basis $B_2$.
(i) What is the probability that Bob measures the photon in the state prepared by Alice, i.e. what is the probability that the binary interpretation is identical for Alice and Bob?
(ii) An eavesdropper (named Eve) intercepts the photons sent to Bob and then resends a photon to Bob. Eve also detects the photon polarization in one of the bases $B_1$ or $B_2$ before resending. What is the probability that the binary interpretation is identical for Alice and Bob?

**Solution 4.** (i) The probability that Alice chooses to prepare a state from the basis $B_1$ is $\frac{1}{2}$ and from $B_2$ is $\frac{1}{2}$. Similarly the probabilities that Bob chooses to measure in the basis $B_1$ and $B_2$ are also $\frac{1}{2}$. Thus the probability that Alice and Bob measure in the same basis is $\frac{1}{4} + \frac{1}{4} = \frac{1}{2}$. To determine the correlations in the binary interpretation we consider the two cases $(a)$ Alice and Bob use the same basis and $(b)$ Alice and Bob use a different basis. The cases $(a)$ and $(b)$ have equal probability of $\frac{1}{2}$. For the case $(a)$ Alice and Bob have the same binary interpretation. For the case $(b)$ we note that

$$|\langle\psi_0|\phi_0\rangle|^2 = |\langle\psi_0|\phi_1\rangle|^2 = |\langle\psi_1|\phi_0\rangle|^2 = |\langle\psi_1|\phi_1\rangle|^2 = \frac{1}{2}.$$

In other words, if Bob uses the wrong basis he obtains the correct binary interpretation with probability $\frac{1}{2}$. Therefore the total probability that Alice and Bob have the same binary interpretation is

$$\frac{1}{2} \cdot 1 + \frac{1}{2} \cdot \frac{1}{2} = \frac{3}{4}.$$

Thus 75% of the photons sent by Alice have an identical binary interpretation shared by Alice and Bob.

(ii) From (i) the probability that Alice and Eve, Eve and Bob, as well as Alice and Bob measure in the same basis are all $\frac{1}{4} + \frac{1}{4} = \frac{1}{2}$. Also from (i) we find that if Alice and Eve work in the same basis Bob has a 75% chance of obtaining the correct result since Eve does not perturb the state of the photon. Similarly if Bob and Eve work in the same basis Bob has a 75% chance of obtaining the correct result since Bob does not perturb the state of the photon after Eve resends it. Now we consider the case when Eve uses a different basis from that of Alice and Bob. Suppose Alice sends $|\psi_0\rangle$ from $B_1$, and Eve measures in $B_2$. Thus Eve will obtain $|\phi_0\rangle$ or $|\phi_1\rangle$ with equal probability $\frac{1}{2}$. Now Bob measures in the basis $B_1$ and obtains $|\psi_0\rangle$ with probability $\frac{1}{2}$ or $|\psi_1\rangle$ with probability $\frac{1}{2}$. Thus we can construct the following table where $P_1$ is the probability that Eve obtains Alice's binary interpretation of the state correctly and $P_2$ is the probability that Bob obtains Alice's binary interpretation of the state correctly.

| Alice's basis | Eve's basis | Bob's basis | $P_1$ | $P_2$ |
|:---:|:---:|:---:|:---:|:---:|
| $B_1$ | $B_1$ | $B_1$ | 1 | 1 |
| $B_1$ | $B_1$ | $B_2$ | 1 | 1/2 |
| $B_1$ | $B_2$ | $B_1$ | 1/2 | 1/2 |
| $B_1$ | $B_2$ | $B_2$ | 1/2 | 1/2 |
| $B_2$ | $B_1$ | $B_1$ | 1/2 | 1/2 |
| $B_2$ | $B_1$ | $B_2$ | 1/2 | 1/2 |
| $B_2$ | $B_2$ | $B_1$ | 1 | 1/2 |
| $B_2$ | $B_2$ | $B_2$ | 1 | 1 |

The total probability that Bob's binary interpretation corresponds to Alice's binary interpretation is

$$\frac{1}{8}\left(1 + \frac{1}{2} + \frac{1}{2} + \frac{1}{2} + \frac{1}{2} + \frac{1}{2} + \frac{1}{2} + 1\right) = \frac{5}{8}$$

i.e. 62.5%.

**Problem 5.** (i) Consider the two-qubit singlet state in the Hilbert space $\mathbb{C}^4$

$$|\psi\rangle = \frac{1}{\sqrt{2}}(|01\rangle - |10\rangle) \equiv \frac{1}{\sqrt{2}}(|0\rangle \otimes |1\rangle - |1\rangle \otimes |0\rangle).$$

Let $U$ be a $2 \times 2$ unitary matrix with $\det(U) = 1$. Find the state $(U \otimes U)|\psi\rangle$.
(ii) Consider the state

$$|\psi\rangle = \frac{1}{2\sqrt{3}}(2|0011\rangle - |0101\rangle - |0110\rangle - |1001\rangle - |1010\rangle + 2|1100\rangle)$$

in the Hilbert space $\mathbb{C}^{16}$. This state is an extension of the two-qubit singlet state given in (i). Calculate the state $(U \otimes U \otimes U \otimes U)|\psi\rangle$.
(iii) The state given in (i) and (ii) can be extended to arbitrary $N$ ($N =$ even) as follows

$$|\psi\rangle = \frac{1}{(N/2)!\sqrt{N/2 + 1}} \sum_{\substack{permutations \\ 0...01...1}} p!\left(\frac{N}{2} - p\right)!(-1)^{N/2-p}|j_1 j_2 \ldots j_N\rangle$$

where the sum is extended over all the states obtained by permuting the state

$$|0\ldots 01\ldots 1\rangle \equiv |0\rangle \otimes \cdots \otimes |0\rangle \otimes |1\rangle \otimes \cdots \otimes |1\rangle$$

which contains the same number of 0s and 1s and $p$ is the number of 0s in the first $N/2$ positions. Thus the state is a singlet state. Let

$$U^{\otimes N} \equiv U \otimes \cdots \otimes U \qquad N - \text{times}$$

Find the state $U^{\otimes N}|\psi\rangle$.

**Solution 5.** (i) A unitary transformation for $2 \times 2$ matrices is given by

$$|0\rangle \to a|0\rangle + b|1\rangle, \qquad |1\rangle \to c|0\rangle + d|1\rangle$$

where $ad - bc = e^{i\phi}$ ($\phi \in \mathbb{R}$). We obtain

$$(U \otimes U)|\psi\rangle = \frac{e^{i\phi}}{\sqrt{2}}(|0\rangle \otimes |1\rangle - |1\rangle \otimes |0\rangle).$$

For $\phi = 0$ ($\det(U) = 1$), we obtain the eigenvalue equation $(U \otimes U)|\psi\rangle = |\psi\rangle$.
(ii) Using the results from (i) and $\det(U) = 1$, we find

$$(U \otimes U \otimes U \otimes U)|\psi\rangle = |\psi\rangle.$$

(iii) Using the result from (i), we also find $U^{\otimes N}|\psi\rangle = |\psi\rangle$. The state $|\psi\rangle$ given in (iii) can be used to distribute cryptographic keys, encode quantum information in decoherence-free subspaces, perform secret sharing, teleclone quantum states, and also for solving the liar detection and Byzantine generals problems.

**Problem 6.** Let $\rho$ denote an arbitrary $4 \times 4$ density matrix. Consider the unitary operators (bilateral rotations)

$$B_x := \frac{1}{2}\begin{pmatrix} 1 & i \\ i & 1 \end{pmatrix} \otimes \begin{pmatrix} 1 & i \\ i & 1 \end{pmatrix}, \quad B_y := \frac{1}{2}\begin{pmatrix} 1 & 1 \\ -1 & 1 \end{pmatrix} \otimes \begin{pmatrix} 1 & 1 \\ -1 & 1 \end{pmatrix},$$

$$B_z := i\begin{pmatrix} 1 & 0 \\ 0 & -i \end{pmatrix} \otimes \begin{pmatrix} 1 & 0 \\ 0 & -i \end{pmatrix}$$

and the unitary operators

$$U_1 := I_4, \ U_2 := B_x^2, \ U_3 := B_y^2, \ U_4 := B_z^2, \ U_5 := B_xB_y, \ U_6 := B_yB_z,$$

$$U_7 := B_zB_x, \ U_8 := B_yB_x, \ U_9 := U_5^2, \ U_{10} := U_6^2, \ U_{11} := U_7^2, \ U_{12} := U_8^2.$$

The mixed state $\rho_W$ is prepared by transforming 12 systems each described by the mixed state $\rho$ according to each of the operators $U_1, \ldots, U_{12}$. Calculate the density matrix

$$\rho_W := \frac{1}{12}\sum_{j=1}^{12} U_j\rho U_j^*.$$

Express $\rho_W$ in terms of the Bell basis.

**Solution 6.** An arbitrary $4 \times 4$ density matrix $\rho$ can be written in the form

$$\begin{pmatrix} a_{11} & a_{12} & a_{13} & a_{14} \\ \overline{a_{12}} & a_{22} & a_{23} & a_{24} \\ \overline{a_{13}} & \overline{a_{23}} & a_{33} & a_{34} \\ \overline{a_{14}} & \overline{a_{24}} & \overline{a_{34}} & 1 - a_{11} - a_{22} - a_{33} \end{pmatrix}$$

where $a_{11}, a_{12}, a_{13} \in \mathbb{R}$. We find

$$\rho_W = \frac{1}{6}\begin{pmatrix} 2 - 2F & 0 & 0 & 0 \\ 0 & 1 + 2F & 1 - 4F & 0 \\ 0 & 1 - 4F & 1 + 2F & 0 \\ 0 & 0 & 0 & 2 - 2F \end{pmatrix}$$

where $F := (a_{22} + a_{33} - 2\Re a_{23})/2$. In terms of the Bell basis we have

$$\rho_W = F|\psi^-\rangle\langle\psi^-| + \frac{1-F}{3}(|\psi^+\rangle\langle\psi^+| + |\phi^+\rangle\langle\phi^+| + |\phi^-\rangle\langle\phi^-|).$$

**Problem 7.** Let

$$\rho_W := F|\phi^+\rangle\langle\phi^+| + \frac{1-F}{3}(|\psi^+\rangle\langle\psi^+| + |\psi^-\rangle\langle\psi^-| + |\phi^-\rangle\langle\phi^-|)$$

where $F \in [0,1]$ is the fidelity. Consider the unitary matrices

$$U_{BXOR1} := \begin{pmatrix} 1 & 0 \\ 0 & 0 \end{pmatrix} \otimes I_2 \otimes I_2 \otimes I_2 + \begin{pmatrix} 0 & 0 \\ 0 & 1 \end{pmatrix} \otimes I_2 \otimes \begin{pmatrix} 0 & 1 \\ 1 & 0 \end{pmatrix} \otimes I_2,$$

$$U_{BXOR2} := I_2 \otimes \begin{pmatrix} 1 & 0 \\ 0 & 0 \end{pmatrix} \otimes I_2 \otimes I_2 + I_2 \otimes \begin{pmatrix} 0 & 0 \\ 0 & 1 \end{pmatrix} \otimes I_2 \otimes \begin{pmatrix} 0 & 1 \\ 1 & 0 \end{pmatrix}$$

and $U_{BXOR} = U_{BXOR1}U_{BXOR2}$ (*bilateral exclusive or*). Let

$$\rho := U_{BXOR}(\rho_W \otimes \rho_W)U_{BXOR}^*.$$

Calculate the probability $p_c$ that the last two qubits of a system described by $\rho$ are found in the same state when measured with respect to the standard basis, i.e. the last two qubits are in one of the states

$$\begin{pmatrix} 1 \\ 0 \end{pmatrix} \otimes \begin{pmatrix} 1 \\ 0 \end{pmatrix}, \qquad \begin{pmatrix} 0 \\ 1 \end{pmatrix} \otimes \begin{pmatrix} 0 \\ 1 \end{pmatrix}.$$

Determine the fidelity

$$F' := \frac{\text{tr}\,(\rho\,(|\phi^+\rangle\langle\phi^+| \otimes \Pi))}{p_c}, \qquad \Pi = \begin{pmatrix} 1 & 0 & 0 & 0 \\ 0 & 0 & 0 & 0 \\ 0 & 0 & 0 & 0 \\ 0 & 0 & 0 & 1 \end{pmatrix}.$$

It is the projection onto the space of states compatible with the measurement outcomes above. Discuss the case $1/2 < F < 1$.

**Solution 7.** The probability $p_c$ is given by

$$p_c = \text{tr}\,(\rho\,(I_4 \otimes \Pi)) = \frac{8F^2 - 4F + 5}{9}.$$

Thus the new fidelity $F'$ is given by

$$F' = \frac{10F^2 - 2F + 1}{8F^2 - 4F + 5}.$$

For $F = 1/2$ we find $F' = 1/2$ and for $F = 1$ we find $F' = 1$. We have

$$F' - F = \frac{(2F-1)(4F-1)(1-F)}{4F(2F-1)+5}.$$

Obviously for $1/2 < F < 1$ we have $F > 0$, $2F - 1 > 0$, $4F - 1 > 0$ and $1 - F > 0$. Thus $F' > F$ when $1/2 < F < 1$.

## Programming Problem

**Problem 1.**  Alice and Bob share the entangled state (one of the four Bell states)

$$|\psi\rangle = \frac{1}{\sqrt{2}}(|0\rangle_A \otimes |0\rangle_B + |1\rangle_A \otimes |1\rangle_B).$$

Alice applies $I_2 \otimes I_2$ to $|\psi\rangle$ when she wants to send the bit string 00 and sends $(I_2 \otimes I_2)|\psi\rangle$. Alice applies $\sigma_3 \otimes I_2$ to $|\psi\rangle$ when she wants to send the bit string 01 and sends $(\sigma_3 \otimes I_2)|\psi\rangle$. Alice applies $\sigma_1 \otimes I_2$ to $|\psi\rangle$ when she wants to send the bit string 10 and sends $(\sigma_1 \otimes I_2)|\psi\rangle$. Alice applies $i\sigma_2 \otimes I_2$ to $|\psi\rangle$ when she wants to send the bit string 11 and sends $(i\sigma_2 \otimes I_2)|\psi\rangle$. What states are send to Bob?

**Solution 1.**  We have

$$(I_2 \otimes I_2)|\psi\rangle = \psi$$

$$(\sigma_3 \otimes I_2)|\psi\rangle = \frac{1}{\sqrt{2}}(|0\rangle \otimes |0\rangle - |1\rangle \otimes |1\rangle)$$

$$(\sigma_1 \otimes I_2)|\psi\rangle = \frac{1}{\sqrt{2}}(|1\rangle \otimes |0\rangle + |0\rangle \otimes |1\rangle)$$

$$(i\sigma_2 \otimes I_2)|\psi\rangle = \frac{1}{\sqrt{2}}(|0\rangle \otimes |1\rangle - |1\rangle \otimes |0\rangle).$$

So we have the four Bell states which form an orthonormal basis in $\mathbb{C}^4$. The corresponding Maxima program is

```
/* AliceBob.mac */
I2: matrix([1,0],[0,1]);
sig1: matrix([0,1],[1,0]);
sig2: matrix([0,-%i],[%i,0]);
sig3: matrix([1,0],[0,-1]);
e0: matrix([1],[0]); e1: matrix([0],[1]);
psi: (kronecker_product(e0,e0)+kronecker_product(e1,e1))/sqrt(2);
t1: kronecker_product(I2,I2) . psi;
t2: kronecker_product(sig3,I2) . psi;
t3: kronecker_product(sig1,I2) . psi;
t4: kronecker_product(%i*sig2,I2) . psi;
```

## 16.3   Supplementary Problem

**Problem 1.**   Given an orthonormal bases in $\mathbb{C}^2$

$$B_{xy} = \{\, |x\rangle,\ |y\rangle\, \}, \quad B_{uv} = \{\, |u\rangle,\ |v\rangle\, \}$$

with

$$|x\rangle = \frac{1}{\sqrt{2}}(|u\rangle + |v\rangle), \quad |y\rangle = \frac{1}{\sqrt{2}}(|u\rangle - |v\rangle).$$

Thus we have mutually unbiased bases. Alice encodes her key-bits, for example as a polarized photon, and sends it to Bob.
(i) Assume that Alice has chosen the state $|x\rangle$ with the density matrix $|x\rangle\langle x|$. The state of Eve in $\mathbb{C}^2$ is $|\psi_0\rangle$ with the density matrix $|\psi_0\rangle\langle\psi_0|$. Eve applies a unitary $4 \times 4$ matrix to the product state $|x\rangle \otimes |\psi_0\rangle$

$$U(|x\rangle \otimes |\psi_0\rangle) = |B\rangle$$

so that $|B\rangle$ is an entangled state. Show that the Schmidt decomposition of the state $|B\rangle$ is of the form

$$|B\rangle = \sqrt{\alpha}|x\rangle \otimes |\xi_x\rangle + \sqrt{1-\alpha}|y\rangle \otimes |\zeta_x\rangle$$

where $|\xi_x\rangle \perp |\zeta_x\rangle$. Show that the density matrix for the post-interaction state $|X\rangle$ is of the form $\rho_x^{AE} = |X\rangle\langle X| = U(\rho_x^A \otimes \rho_0^E)U^*$.
(ii) Show that when Alice sends the state $|y\rangle$ the entangled state

$$U(|y\rangle \otimes |\psi_0\rangle) = |Y\rangle$$

is of the form

$$|Y\rangle = \sqrt{\beta}|y\rangle \otimes |\xi_y\rangle + \sqrt{1-\beta}|x\rangle \otimes |\zeta_y\rangle$$

where $|\xi_y\rangle \perp |\zeta_y\rangle$.
(iii) Study the special case where $U$ is the Bell matrix

$$U = \frac{1}{\sqrt{2}} \begin{pmatrix} 1 & 0 & 0 & 1 \\ 0 & 1 & 1 & 0 \\ 0 & 1 & -1 & 0 \\ 1 & 0 & 0 & -1 \end{pmatrix}$$

and

$$|u\rangle = \begin{pmatrix} 1 \\ 0 \end{pmatrix}, \quad |v\rangle = \begin{pmatrix} 0 \\ 1 \end{pmatrix}.$$

# Chapter 17

# Quantum Channels

## 17.1 Introduction

We consider the Hilbert space $\mathcal{H}$ of $n \times n$ matrices over $\mathbb{C}$ with the scalar product (Frobenius inner product)

$$\langle A, B \rangle := \text{tr}(AB^*)$$

with $A, B \in \mathcal{H}$. We also consider the Hilbert space $\mathbb{C}^n$ and the vec operator. Given a $n \times m$ matrix the *vec operator* stacks the column on top of each other.

A state is described using $n \times n$ density matrices $\rho$, i.e. $\text{tr}(\rho) = 1$ and $\rho \geq 0$ (positive semidefinite). For a pure state we have $\rho^2 = \rho$ and for a mixed state we have $\rho^2 \neq \rho$.

The space of trace-class operators acting in this Hilbert space is denoted by $S(\mathcal{H})$. A quantum channel from a Hilbert space $\mathcal{H}_A$ to a Hilbert space $\mathcal{H}_B$ is represented by a completely positive trace-preserving map

$$\Phi : S(\mathcal{H}_A) \to S(\mathcal{H}_B).$$

Such a positive trace-preserving map can be represented in Stinespring representation, Kraus operator representation and Choi-Jamiolkowski representation.

Let $H_n$ denote the vector space of $n \times n$ Hermitian matrices over the real numbers. We say that $\rho \in H_n$ is positive semi-definite (or $\rho \geq 0$) if $\mathbf{x}^* \rho \mathbf{x} \geq 0$ for all $\mathbf{x} \in \mathbb{C}^n$, or equivalently: all of the eigenvalues of $\rho$ are non-negative. A linear map $\psi : H_n \to H_p$ is TPCP (trace-preserving completely positive) if

1. TP (trace-preserving): $\forall \rho \in H_n$, $\mathrm{tr}(\rho) = \mathrm{tr}(\psi(\rho))$

2. CP (completely positive): $\forall m \in \mathbb{N}$, $\rho \in H_{mn}$,

$$\rho \geq 0 \quad \Rightarrow \quad (\psi \otimes I_{m \times m})(\rho) \geq 0$$

where $I_{m \times m}$ is the identity operator on $m \times m$ matrices.

Let $\mathbb{H}_n$ be the vector space of the $n \times n$ hermitian matrices and $H \in \mathbb{H}_n$. Consider a family of $n \times n$ matrices $V_1, \ldots, V_m$ over $\mathbb{C}$. Consider the completely positive map $\Psi : \mathbb{H}_n \to \mathbb{H}_n$ defined by

$$\Psi(H) = \sum_{j=1}^{m} V_j H V_j^*$$

This map is said to be a Kraus map if

$$\sum_{j=1}^{m} V_j V_j^* = I_n.$$

Then the matrices $V_1, V_2, \ldots, V_m$ are called Kraus operators.

A completely positive trace-preserving map $\Phi : S(A) \to S(B)$ can be represented in three different ways, the Stinespring representation, Kraus operator representation and Choi-Jamiolkowski representation. Stinespring's representation tells us that every quantum channel $\Phi : S(A) \to S(B)$ can be written in terms of an isometry $V$ from $A$ to the joint system $B \otimes E$ ($E$ environment) followed by a partial trace such that $\Phi(\rho) = \mathrm{tr}_E(V \rho V^\dagger)$ for all $\rho \in S(A)$. Tracing out system $B$ instead of $E$ defines a complementary channel $\phi^c(\rho) = \mathrm{tr}_B(V \rho V^\dagger)$ for all $\rho \in S(\mathcal{H}_A)$. The Choi-Jamiolkowski representation of the channel $\Phi : S(A) \to S(B)$ is the operator $J(\Phi) \in S(B \otimes A)$ that is defined as

$$J(\Phi) = |A|(\Phi \times I_A)(|\Omega\rangle\langle\Omega|) = \sum_{1 \leq j,k \leq |A|} \Phi(E_{jk} \otimes E_{jk})$$

where

$$|\Omega\rangle = \frac{1}{\sqrt{d}} \sum_{j=1}^{d} |j\rangle \otimes |j\rangle$$

and $E_{jk}$ is the elementary matrix with 1 at entry $(jk)$ and 0 otherwise. $|A| := \dim(A)$, $|B| := \dim(B)$ denote the input and output dimension of the quantum channel, respectively.

## 17.2   Solved Problems

**Problem 1.**   Let $\mathbb{H}_n$ be the vector space of $n \times n$ hermitian matrices. The adjoint (conjugate transpose) of a matrix $A \in \mathbb{C}^{n \times n}$ is denoted by $A^*$, Consider a family $V_1, V_2, \ldots, V_m$ of $n \times n$ matrices over $\mathbb{C}$. We associate with this family the completely positive map $\psi : \mathbb{H}_n \to \mathbb{H}_n$ defined by

$$\psi(X) = \sum_{j=1}^{m} V_j X V_j^*.$$

The map $\psi$ is said to be a *Kraus map* if $\psi(I_n) = I_n$, i.e.

$$\sum_{j=1}^{m} V_j V_j^* = I_n$$

and the matrices $V_1, V_2, \ldots, V_m$ are called *Kraus operators*.

Let $m = n = 2$ and

$$V_1 = \begin{pmatrix} 0 & 1 \\ 0 & 0 \end{pmatrix}, \quad V_2 = \begin{pmatrix} 0 & 0 \\ 1 & 0 \end{pmatrix}.$$

Show that $V_1$ and $V_2$ are Kraus operators and find the associated Kraus map.

**Solution 1.**   Since

$$V_1 V_1^* + V_2 V_2^* = \begin{pmatrix} 0 & 1 \\ 0 & 0 \end{pmatrix} \begin{pmatrix} 0 & 0 \\ 1 & 0 \end{pmatrix} + \begin{pmatrix} 0 & 0 \\ 1 & 0 \end{pmatrix} \begin{pmatrix} 0 & 1 \\ 0 & 0 \end{pmatrix} = \begin{pmatrix} 1 & 0 \\ 0 & 1 \end{pmatrix}$$

the matrices $V_1$ and $V_2$ are Kraus operators. The associated Kraus map is

$$\psi \begin{pmatrix} a & b \\ c & d \end{pmatrix} = \begin{pmatrix} 0 & 1 \\ 0 & 0 \end{pmatrix} \begin{pmatrix} a & b \\ c & d \end{pmatrix} \begin{pmatrix} 0 & 0 \\ 1 & 0 \end{pmatrix} + \begin{pmatrix} 0 & 0 \\ 1 & 0 \end{pmatrix} \begin{pmatrix} a & b \\ c & d \end{pmatrix} \begin{pmatrix} 0 & 1 \\ 0 & 0 \end{pmatrix}$$

$$= \begin{pmatrix} d & 0 \\ 0 & a \end{pmatrix}.$$

**Problem 2.**   Let $\psi : \mathbb{H}_n \to \mathbb{H}_n$ be a Kraus map. Thus $\psi$ is linear. Show that there exists $\Psi \in \mathbb{C}^{n \times n}$ such that for all $X \in \mathbb{H}_n$

$$\mathrm{vec}(\psi(X)) = \Psi \, \mathrm{vec}(X)$$

where 1 is an eigenvalue of $\Psi$. What is a corresponding eigenvector?

**Solution 2.** Let $V_1$, $V_2$, ..., $V_m$ be the Kraus operators associated to $\psi$. Since $\text{vec}(ABC) = (C^T \otimes A)\text{vec}(B)$ we find

$$\text{vec}(\psi(X)) = \text{vec}\left(\sum_{j=1}^{m} V_j X V_j^*\right) = \left(\sum_{j=1}^{m} \overline{V}_j \otimes V_j\right)\text{vec}(X).$$

Thus we find

$$\Psi := \sum_{j=1}^{m} \overline{V}_j \otimes V_j.$$

We also have $\psi(I_n) = I_n$ so that

$$\text{vec}(\psi(I_n)) = \Psi\text{vec}(I_n) = \text{vec}(I_n)$$

so that 1 is an eigenvalue of $\Psi$ and a corresponding eigenvector is $\text{vec}(I_n)$.

**Problem 3.** Let $L$ denote the space of linear operators on a Hilbert space $\mathcal{H}$. A linear map $\epsilon : L \to L$ is called a *positive map* if for all positive semidefinite $\rho \in L$ the operator $\epsilon(\rho)$ is also positive semidefinite. As an example we consider the transpose operation. Let $B = \{|1\rangle, |2\rangle, \ldots, |\dim(\mathcal{H})\rangle\}$ denote an orthonormal basis for the Hilbert space $\mathcal{H}$. The *transpose* of a linear operator

$$A := \sum_{j=1}^{\dim(\mathcal{H})} \sum_{k=1}^{\dim(\mathcal{H})} a_{jk}|j\rangle\langle k|$$

is given by

$$\epsilon_{T(B)}(A) := \sum_{j=1}^{\dim(\mathcal{H})} \sum_{k=1}^{\dim(\mathcal{H})} a_{jk}|k\rangle\langle j|.$$

The transpose operation $\epsilon_{T(B)}$ is a positive map.

Let $L_A$ denote the space of linear operators on a Hilbert space $\mathcal{H}_A$, $L_B$ the space of linear operators on the Hilbert space $\mathcal{H}_B$ and $L_{AB}$ the space of linear operators on the Hilbert space $\mathcal{H}_A \otimes \mathcal{H}_B$. The extension $\epsilon \otimes I$ of the linear map $\epsilon : L_A \to L_A$ is defined by

$$(\epsilon \otimes I)\left(\sum_{k=1}^{\dim(L_A)} A_k \otimes B_k\right) = \sum_{k=1}^{\dim(L_A)} \epsilon(A_k) \otimes B_k$$

where $A_k \in L_A$ and $B_k \in L_B$ for $k = 1, 2, \ldots, \dim(L_A)$. Similarly the extension $I \otimes \mu$ of $\mu : L_B \to L_B$ can be defined. A positive map $\epsilon : L \to L$ is a *completely positive map* if all possible extensions ($\epsilon \otimes I$ or $I \otimes \epsilon$) of the map to arbitrary Hilbert spaces are positive. Every completely positive map can be written in the form

$$\epsilon(\rho) = \sum_k A_k \rho A_k^*$$

where $A_k$ is a linear operator on the Hilbert space. Furthermore, if

$$\sum_k A_k A_k^* = I$$

then $\epsilon$ is trace preserving, i.e. $\text{tr}(\epsilon(\rho)) = \text{tr}(\rho)$. Is the transpose operation completely positive?

**Solution 3.** *Partial transposition* $\epsilon_{T(B)} \otimes I$ in the basis $B$ of the Hilbert space $\mathcal{H}_A$ of a state $\rho$ in the Hilbert space $\mathcal{H}_A \otimes \mathcal{H}_B$ is not completely positive. Consider the Bell state $|\Phi^+\rangle$ in the Hilbert state $\mathbb{C}^2 \otimes \mathbb{C}^2$. Then

$$|\Phi^+\rangle\langle\Phi^+| = \frac{1}{2}\begin{pmatrix} 1 & 0 & 0 & 1 \\ 0 & 0 & 0 & 0 \\ 0 & 0 & 0 & 0 \\ 1 & 0 & 0 & 1 \end{pmatrix}.$$

Thus we have

$$(\epsilon_{T(\{|0\rangle,|1\rangle\})} \otimes I)(|\Phi^+\rangle\langle\Phi^+|) = \frac{1}{2}\begin{pmatrix} 1 & 0 & 0 & 0 \\ 0 & 0 & 1 & 0 \\ 0 & 1 & 0 & 0 \\ 0 & 0 & 0 & 1 \end{pmatrix}.$$

This last matrix has as eigenvalues 1 and $-1$. Consequently it is not positive semidefinite.

**Problem 4.** Let $|0\rangle$, $|1\rangle$ be an orthonormal basis in $\mathbb{C}^2$. The *Kraus operators* are defined by

$$K_0 := (\alpha|0\rangle\langle 0| + \beta|1\rangle\langle 1|) \otimes I_2$$
$$K_1 := (\beta|1\rangle\langle 0| + \alpha|0\rangle\langle 1|) \otimes (|1\rangle\langle 0| + |0\rangle\langle 1|)$$

where $\alpha\alpha^* + \beta\beta^* = 1$.
(i) Show that $K_0^* K_0 + K_1^* K_1 = I_2 \otimes I_2$.
(ii) Let

$$|\phi\rangle = \alpha|0\rangle \otimes |0\rangle + \beta|1\rangle \otimes |1\rangle, \qquad |\psi\rangle = \frac{1}{\sqrt{2}}(|0\rangle \otimes |0\rangle + |1\rangle \otimes |1\rangle).$$

Show that $K_0|\psi\rangle\langle\psi|K_0^* + K_1|\psi\rangle\langle\psi|K_1^* = |\phi\rangle\langle\phi|$.

**Solution 4.**    (i) From $K_0^* = (\alpha^*|0\rangle\langle 0| + \beta^*|1\rangle\langle 1|) \otimes I_2$ we obtain

$$K_0^* K_0 = (\alpha\alpha^*|0\rangle\langle 0| + \beta\beta^*|1\rangle\langle 1|) \otimes I_2.$$

Since $K_1^* = (\beta^*|0\rangle\langle 1| + \alpha^*|1\rangle\langle 0|) \otimes (|0\rangle\langle 1| + |1\rangle\langle 0|)$ we have

$$K_1^* K_1 = (\beta\beta^*|0\rangle\langle 0| + \alpha\alpha^*|1\rangle\langle 1)) \otimes (|0\rangle\langle 0| + |1\rangle\langle 1|).$$

Since $|0\rangle\langle 0| + |1\rangle\langle 1| = I_2$ and $\alpha\alpha^* + \beta\beta^* = 1$ we obtain the result.
(ii) Since

$$K_0|\psi\rangle = \frac{1}{\sqrt{2}}(\alpha|0\rangle \otimes |0\rangle + \beta|1\rangle \otimes |1\rangle)$$

we have

$$A_0|\psi\rangle\langle\psi|A_0^* =$$

$$\frac{1}{2}(\alpha\alpha^*|0\rangle\langle 0|\otimes|0\rangle\langle 0|+\beta\beta^*|1\rangle\langle 1|\otimes|1\rangle\langle 1|+\alpha\beta^*|0\rangle\langle 1|\otimes|0\rangle\langle 1|+\alpha^*\beta|1\rangle\langle 0|\otimes|1\rangle\langle 1|).$$

Since

$$A_1|\psi\rangle = \frac{1}{\sqrt{2}}(\alpha|0\rangle \otimes |0\rangle + \beta|1\rangle \otimes |1\rangle) = A_0|\psi\rangle$$

we obtain $A_1|\psi\rangle\langle\psi|A_1^* = A_0|\psi\rangle\langle\psi|A_0$. Thus

$$A_0|\psi\rangle\langle\psi|A_0^* + A_1|\psi\rangle\langle\psi|A_1^* =$$

$$\alpha\alpha^*|0\rangle\langle 0| \otimes |0\rangle\langle 0| + \beta\beta^*|1\rangle\langle 1| \otimes |1\rangle\langle 1| + \alpha\beta^*|0\rangle\langle 1| \otimes |0\rangle\langle 1|+\alpha^*\beta|1\rangle\langle 0| \otimes |1\rangle\langle 0|$$

which is the density matrix $|\phi\rangle\langle\phi|$.

**Problem 5.**    Find all Kraus maps $\psi : \mathbb{H}_2 \to \mathbb{H}_2$, associated with families of 2 Kraus operators ($V_1$ and $V_2$), which provide the transformation

$$\psi \begin{pmatrix} 1 & 0 \\ 0 & 0 \end{pmatrix} = \begin{pmatrix} 0 & 0 \\ 0 & 1 \end{pmatrix}.$$

Calculate

$$\psi \begin{pmatrix} 0 & 0 \\ 0 & 1 \end{pmatrix}.$$

Is there a Kraus map associated with a single Kraus operator which also provides this transformation?

**Solution 5.**    By linearity of the Kraus map $\psi$ we have

$$\begin{pmatrix} 1 & 0 \\ 0 & 1 \end{pmatrix} = \psi \begin{pmatrix} 1 & 0 \\ 0 & 1 \end{pmatrix} = \psi \begin{pmatrix} 1 & 0 \\ 0 & 0 \end{pmatrix} + \psi \begin{pmatrix} 0 & 0 \\ 0 & 1 \end{pmatrix} = \begin{pmatrix} 0 & 0 \\ 0 & 1 \end{pmatrix} + \psi \begin{pmatrix} 0 & 0 \\ 0 & 1 \end{pmatrix}.$$

Thus

$$\psi \begin{pmatrix} 0 & 0 \\ 0 & 1 \end{pmatrix} = \begin{pmatrix} 1 & 0 \\ 0 & 0 \end{pmatrix}.$$

We construct the matrix $\Psi$ which implements $\psi$ under the vec operator. We write $\Psi = \Pi_1 + A(I_4 - \Pi_1)$, where $A$ is a $4 \times 4$ matrix such that $\psi$ is a Kraus map providing the given transformation and

$$\Pi_1 := \frac{1}{2}\text{vec}(I_2)(\text{vec}(I_2))^* = \frac{1}{2}\begin{pmatrix} 1 & 0 & 0 & 1 \\ 0 & 0 & 0 & 0 \\ 0 & 0 & 0 & 0 \\ 1 & 0 & 0 & 1 \end{pmatrix}$$

is a projection matrix. We have

$$\Psi\text{vec}\begin{pmatrix} 1 & 0 \\ 0 & 0 \end{pmatrix} = \Psi\begin{pmatrix} 1 \\ 0 \\ 0 \\ 0 \end{pmatrix} = \frac{1}{2}\begin{pmatrix} 1 \\ 0 \\ 0 \\ 1 \end{pmatrix} + \frac{1}{2}A\begin{pmatrix} 1 \\ 0 \\ 0 \\ -1 \end{pmatrix} = \begin{pmatrix} 0 \\ 0 \\ 0 \\ 1 \end{pmatrix} = \text{vec}\begin{pmatrix} 0 & 0 \\ 0 & 1 \end{pmatrix}$$

so that

$$A\begin{pmatrix} 1 \\ 0 \\ 0 \\ -1 \end{pmatrix} = \begin{pmatrix} -1 \\ 0 \\ 0 \\ 1 \end{pmatrix}.$$

We write $A = -\Pi_{-1} + B(I_4 - \Pi_{-1})$, where $B$ is a $4 \times 4$ matrix and

$$\Pi_{-1} := \frac{1}{2}\begin{pmatrix} 1 \\ 0 \\ 0 \\ -1 \end{pmatrix}\begin{pmatrix} 1 \\ 0 \\ 0 \\ -1 \end{pmatrix}^* = \frac{1}{2}\begin{pmatrix} 1 & 0 & 0 & -1 \\ 0 & 0 & 0 & 0 \\ 0 & 0 & 0 & 0 \\ -1 & 0 & 0 & 1 \end{pmatrix}$$

is a projection operator with $\Pi_{-1}\Pi_1 = \Pi_1\Pi_{-1} = 0_4$. Consequently,

$$\Psi = \begin{pmatrix} 0 & 0 & 0 & 1 \\ 0 & 0 & 0 & 0 \\ 0 & 0 & 0 & 0 \\ 1 & 0 & 0 & 0 \end{pmatrix} + B\begin{pmatrix} 0 & 0 & 0 & 0 \\ 0 & 1 & 0 & 0 \\ 0 & 0 & 1 & 0 \\ 0 & 0 & 0 & 0 \end{pmatrix} = \begin{pmatrix} 0 & b_{12} & b_{13} & 1 \\ 0 & b_{22} & b_{23} & 0 \\ 0 & b_{32} & b_{33} & 0 \\ 1 & b_{42} & b_{43} & 0 \end{pmatrix}.$$

Clearly $\Psi$ cannot be written as a Kronecker product of $2 \times 2$ matrices, so no Kraus map associated with a single Kraus operator can provide the given map. Assume $\Psi = \overline{C} \otimes C + \overline{D} \otimes D$ for some $2 \times 2$ matrices $C$ and $D$. We find the equations

$$\begin{pmatrix} 0 & b_{12} \\ 0 & b_{22} \end{pmatrix} = \overline{c_{11}}C + \overline{d_{11}}D, \quad \begin{pmatrix} b_{13} & 1 \\ b_{23} & 0 \end{pmatrix} = \overline{c_{12}}C + \overline{d_{12}}D,$$

$$\begin{pmatrix} 0 & b_{32} \\ 1 & b_{42} \end{pmatrix} = \overline{c_{21}}C + \overline{d_{21}}D, \quad \begin{pmatrix} b_{33} & 0 \\ b_{43} & 0 \end{pmatrix} = \overline{c_{22}}C + \overline{d_{22}}D.$$

Clearly the first columns (second columns) of $C$ and $D$ are linearly dependent. Thus we find

$$V_1 = C = \begin{pmatrix} 0 & c_2 \\ c_1 & 0 \end{pmatrix}, \quad V_2 = D = \begin{pmatrix} 0 & d_2 \\ d_1 & 0 \end{pmatrix}$$

where $c_1, c_2, d_1, d_2 \in \mathbb{C}$ satisfy $|c_1|^2 + |d_1|^2 = 1$, $|c_2|^2 + |d_2|^2 = 1$.

**Problem 6.**   Let $\rho_1, \rho_2 \in H_n$ be positive semi-definite matrices.
(i) Is $\rho_3 = \rho_1 + \rho_2$ positive semi-definite?
(ii) Is $\rho_4 = k\rho_1$ ($k \in \mathbb{C}$) positive semi-definite?
(iii) Let $\rho_5 \in H_n$ such that $\rho_1 + \rho_5$ is positive semi-definite. Is $\rho_5$ positive semi-definite?

**Solution 6.**   (i) Let $\mathbf{x} \in \mathbb{C}^n$, then $\mathbf{x}^* \rho_3 \mathbf{x} = \mathbf{x}^* \rho_1 \mathbf{x} + \mathbf{x}^* \rho_1 \mathbf{x} \geq 0$ since $\mathbf{x}^* \rho_1 \mathbf{x} \geq 0$ and $\mathbf{x}^* \rho_1 \mathbf{x} \geq 0$. Thus $\rho_3$ is positive semi-definite.
(ii) Let $\mathbf{x} \in \mathbb{C}^n$, then $\mathbf{x}^* \rho_4 \mathbf{x} = k\mathbf{x}^* \rho_1 \mathbf{x}$ is non-negative if and only if $k \in \mathbb{R}$ with $k \geq 0$. Thus, in general, $\rho_4 = k\rho_1$ is positive semi-definite if and only if $k \geq 0$ (the only exception is when $\rho_1$ is the zero matrix, in which case the statement is true for all $k \in \mathbb{C}$).
(iii) In general, no. Consider $\begin{pmatrix} 1 & 1 \\ 1 & 1 \end{pmatrix} = \begin{pmatrix} 1 & 0 \\ 0 & 1 \end{pmatrix} + \begin{pmatrix} 0 & 1 \\ 1 & 0 \end{pmatrix}$. All of the matrices appearing in this equation are positive semi-definite, except for $\begin{pmatrix} 0 & 1 \\ 1 & 0 \end{pmatrix}$.

**Problem 7.**   Show that a linear map $\psi : H_n \to H_p$ is a TP map if and only if $\psi^*(I_n) = I_n$, where $^*$ denotes the adjoint with respect to the Frobenius inner product and $I_n$ is the $n \times n$ identity matrix.

**Solution 7.**   The Frobenius inner product $\langle A, B \rangle := \mathrm{tr}(B^*A)$ provides

$$\mathrm{tr}(\rho) = \mathrm{tr}(I_n^* \rho) = \langle \rho, I_n \rangle$$
$$\mathrm{tr}(\psi(\rho)) = \mathrm{tr}(I^* \psi(\rho)) = \langle \psi(\rho), I_n \rangle = \langle \rho, \psi^*(I_n) \rangle$$

so that

$$\mathrm{tr}(\rho) = \mathrm{tr}(\psi(\rho)) \quad \Leftrightarrow \quad \psi^*(I_n) = I_n.$$

**Problem 8.**   Show that

$$\rho_0 := \sum_{i,j=1}^{n} (E_{ij} \otimes E_{ij}) \in H_{n^2}$$

is positive semi-definite, where $E_{ij}$ is the elementary $n \times n$ matrix with a 1 in row $i$ and column $j$ and 0 elsewhere.

**Solution 8.**   Any vector $\mathbf{x}$ in $\mathbb{C}^{n^2}$ can be written in the form

$$\mathbf{x} = \sum_{s,t=1}^{n} x_{s,t} \mathbf{e}_s \otimes \mathbf{e}_t$$

where $\{e_1, \ldots, e_n\}$ is the standard basis in $\mathbb{C}^n$ and $x_{s,t} \in \mathbb{C}$ for $s, t \in \{1, \ldots, n\}$. Since $e_s^* E_{ij} = \delta_{is} e_j^*$ we find

$$\mathbf{x}^* \rho_0 \mathbf{x} = \sum_{i,j=1}^n \overline{x_{ii}} x_{jj} = \left| \sum_{i=1}^n x_{ii} \right|^2 \geq 0.$$

**Problem 9.** An orthonormal basis, with respect to the Frobenius inner product, for $H_n$ ($n \geq 2$) is given by $B = B_1 \cup B_2$ where

$$B_1 = \left\{ \frac{1}{\sqrt{2}} (E_{jk} + E_{kj}) : j, k = 1, \ldots, n, \, j \leq k \right\}$$

$$B_2 = \left\{ \frac{i}{\sqrt{2}} (E_{jk} - E_{kj}) : j, k = 1, \ldots, n, \, j < k \right\}.$$

Express

$$\rho_0 := \sum_{i,j=1}^n (E_{ij} \otimes E_{ij}) \in H_{n^2}$$

in terms of this basis.

**Solution 9.** We find

$$\rho_0 = \sum_{F \in B_1} F \otimes F - \sum_{G \in B_2} G \otimes G.$$

**Problem 10.** Show that a linear map $\psi : H_n \to H_p$ is a CP map if and only if $(\psi \otimes I_{n \times n})(\rho_0)$ is positive semi-definite where

$$\rho_0 := \sum_{i,j=1}^n (E_{ij} \otimes E_{ij}) \in H_{n^2}.$$

**Solution 10.** If $\psi$ is completely positive, it follows immediately that $(\psi \otimes I_{n \times n})(\rho_0)$ is positive semi-definite (since $\rho_0$ is positive semi-definite). Consider the linear extension $\tilde{\psi} : M_{n \times n} \to M_{p \times p}$ of $\psi$, i.e. $\tilde{\psi}(\rho) = \psi(\rho)$ for all $\rho \in H_n$. Assume that

$$(\psi \otimes I_{n \times n})(\rho_0) = \sum_{i,j=1}^n \tilde{\psi}(E_{ij}) \otimes E_{ij}$$

is positive semi-definite. This is the *Choi-Jamiołkowski representation*. We have a spectral decomposition

$$(\psi \otimes I_{n\times n})(\rho_0) = \sum_{k=1}^{n^2} \lambda_k u_j u_k^* = \sum_{k=1}^{n^2} v_k v_k^*$$

where $\{u_1,\ldots,u_n\}$ is an orthonormal set and $v_k := \sqrt{\lambda_k}u_k$ (note that $\lambda_k \geq 0$). Define

$$P_j := I_n \otimes \mathbf{e}_j^*$$

and consider $P_i(\psi \otimes I_{n\times n})(\rho_0)P_j^*$:

$$(\psi \otimes I_{n\times n})(\rho_0) = \sum_{i,j=1}^{n} \widetilde{\psi}(E_{ij}) \otimes E_{ij} = \sum_{k=1}^{n^2} v_k v_k^*$$

$$P_i(\psi \otimes I_{n\times n})(\rho_0)P_j^* = \widetilde{\psi}(E_{ij}) = \sum_{k=1}^{n^2} P_i v_k (P_j v_k)^*$$

and define $V_k e_j := P_j v_k$ so that

$$P_i(\psi \otimes I_{n\times n})(\rho_0)P_j^* = \widetilde{\psi}(E_{ij}) = \sum_{k=1}^{n^2} V_k E_{ij} V_k^*.$$

By linear extension, the map $\psi$ is given by

$$\psi(\rho) := \sum_{k=1}^{n^2} V_k \rho V_k^*$$

which is completely positive since for $\rho \in H_{mn}$, $\rho \geq 0$

$$(\psi \otimes I_{m\times m})\rho = \sum_{k=1}^{n^2}(V_k \otimes I_m)\rho(V_k \otimes I_m)^*$$

is positive semi-definite. Consequently $\psi$ is completely positive if and only if there exists $V_1,\ldots V_{n^2}$ (Kraus operators) such that

$$\psi(\rho) := \sum_{k=1}^{n^2} V_k \rho V_k^*.$$

This is a *Kraus representation*.

**Problem 11.**   Is the map $\psi : H_n \to H_p$ given by $\psi(\rho) = \rho^T$ completely positive?

**Solution 11.**   Consider

$$\rho_0 := \sum_{i,j=1}^{n} (E_{ij} \otimes E_{ij}) \in H_{n^2}.$$

We have

$$(\psi \otimes I_{n \times n})(\rho_0) = \sum_{i,j=1}^{n} \widetilde{\psi}(E_{ij}) \otimes E_{ij} = \sum_{i,j=1}^{n} E_{ji} \otimes E_{ij}.$$

Any vector $\mathbf{x}$ in $C^{n^2}$ can be written in the form

$$\mathbf{x} = \sum_{s,t=1}^{n} x_{s,t} \mathbf{e}_s \otimes \mathbf{e}_t$$

where $\{\mathbf{e}_1, \dots, \mathbf{e}_n\}$ is the standard basis in $\mathbb{C}^n$ and $x_{s,t} \in \mathbb{C}$ for $s,t \in \{1, \dots, n\}$. Since $\mathbf{e}_s^* E_{ij} = \delta_{is} \mathbf{e}_j^*$ we find

$$\mathbf{x}^*(\psi \otimes I_{n \times n})(\rho_0)\mathbf{x} = \sum_{i,j=1}^{n} \overline{x_{ji}} x_{ij} \not\geq 0$$

in general (consider $x_{12} = -x_{21} = 1$ and all other coefficients are zero).

**Problem 12.**   Let $\psi : H_n \to H_p$ given by

$$\psi(\rho) := \sum_{k=1}^{n^2} V_k \rho V_k^*$$

be a CP map. Find the condition on $V_1, \dots, V_{n^2}$ such that $\psi$ is TP (and hence TPCP).

**Solution 12.**   The adjoint of

$$\psi(\rho) := \sum_{k=1}^{n^2} V_k \rho V_k^*$$

with respect to the Frobenius inner product is

$$\psi^*(\rho) := \sum_{k=1}^{n^2} V_k^* \rho V_k$$

and since $\psi$ is TP if and only if $\psi^*(I_n) = I_n$ we find

$$\text{tr}(\rho) = \text{tr}(\psi(\rho)) \quad \Leftrightarrow \quad \sum_{k=1}^{n^2} V_k^* V_k = I_n.$$

**Problem 13.**   Let $\psi : H_n \to H_p$ given by

$$\psi(\rho) := \sum_{k=1}^{n^2} V_k \rho V_k^*$$

be a CP map. Show that there exists a matrix $V$ such that

$$\psi(\rho) = \mathrm{tr}_m(V(\rho \otimes I_{n^2})V^*).$$

**Solution 13.**   Let

$$V = \sum_{k=1}^{n^2} (V_k \otimes E_{kk}).$$

Then

$$V(\rho \otimes I_m)V^* = \sum_{k,l=1}^{n^2} \delta_{kl}(V_k \rho V_l^*) \otimes E_{kk} = \sum_{k=1}^{n^2} (V_k \rho V_k^*) \otimes E_{kk}$$

and

$$\mathrm{tr}_{n^2}(V(\rho \otimes I_{n^2})V^*) = \sum_{k=1}^{n^2} (V_k \rho V_k^*)\mathrm{tr}(E_{kk}) = \psi(\rho).$$

This is a *Stinespring representation*.

**Problem 14.**   A minimal *Stinespring representation* of a CP map $\psi :$ $H_n \to H_p$ is a representation

$$\psi(\rho) = \mathrm{tr}_m(V(\rho \otimes I_m)V^*)$$

where $m$ is minimal. This corresponds to minimizing the number of non-zero Kraus operators $V_k$ in a Kraus representation

$$\psi(\rho) := \sum_{k=1}^{m} V_k \rho V_k^*.$$

Given a Kraus representation

$$\psi(\rho) := \sum_{k=1}^{n^2} \widetilde{V}_k \rho \widetilde{V}_k^*.$$

Consider

$$A = \sum_{k=1}^{n^2} (\mathrm{vec}(\widetilde{V}_k))(\mathrm{vec}(\widetilde{V}_k))^*.$$

The matrix $A$ is positive definite and thus has a spectral decomposition

$$A = \sum_{k=1}^{m} \lambda_k \mathbf{v}_k \mathbf{v}_k^{*}$$

where $m$ is the rank of $A$ and $\lambda_1, \ldots, \lambda_m > 0$. We find the Kraus operators $V_k$ for the minimal representation from $\mathrm{vec} V_k = \sqrt{\lambda_k} \mathbf{v}_k$. Find a minimal representation for the completely map on $H_2$ given by

$$\tilde{V}_1 = \begin{pmatrix} 0 & 1 \\ 1 & 0 \end{pmatrix}, \quad \tilde{V}_2 = \begin{pmatrix} 0 & 1 \\ 0 & 0 \end{pmatrix}, \quad \tilde{V}_3 = \begin{pmatrix} 0 & 0 \\ 1 & 0 \end{pmatrix}, \quad \tilde{V}_4 = \begin{pmatrix} 0 & 0 \\ 0 & 0 \end{pmatrix}.$$

**Solution 14.** The matrix $A$ is given by

$$A = \begin{pmatrix} 0 & 0 & 0 & 0 \\ 0 & 2 & 1 & 0 \\ 0 & 1 & 2 & 0 \\ 0 & 0 & 0 & 0 \end{pmatrix}$$

$$= 3 \cdot \frac{1}{\sqrt{2}} \begin{pmatrix} 0 \\ 1 \\ 1 \\ 0 \end{pmatrix} \frac{1}{\sqrt{2}} (0 \quad 1 \quad 1 \quad 0) + 1 \cdot \frac{1}{\sqrt{2}} \begin{pmatrix} 0 \\ 1 \\ -1 \\ 0 \end{pmatrix} \frac{1}{\sqrt{2}} (0 \quad 1 \quad -1 \quad 0)$$

with non-zero eigenvalues 1 and 3. An optimal Kraus representation is given by

$$V_1 = \sqrt{\frac{3}{2}} \begin{pmatrix} 0 & 1 \\ 1 & 0 \end{pmatrix}, \quad V_2 = \frac{1}{\sqrt{2}} \begin{pmatrix} 0 & -1 \\ 1 & 0 \end{pmatrix}.$$

We verify that the map is correct

$$\psi \begin{pmatrix} a & b \\ c & d \end{pmatrix} = \begin{pmatrix} 2d & c \\ b & 2a \end{pmatrix}$$

$$= \frac{3}{2} \begin{pmatrix} 0 & 1 \\ 1 & 0 \end{pmatrix} \begin{pmatrix} a & b \\ c & d \end{pmatrix} \begin{pmatrix} 0 & 1 \\ 1 & 0 \end{pmatrix}$$

$$+ \frac{1}{2} \begin{pmatrix} 0 & -1 \\ 1 & 0 \end{pmatrix} \begin{pmatrix} a & b \\ c & d \end{pmatrix} \begin{pmatrix} 0 & 1 \\ -1 & 0 \end{pmatrix}.$$

The corresponding minimal Stinespring representation is given by

$$V = \sum_{k=1}^{n} V_k \otimes E_{kk} = \sqrt{\frac{3}{2}} \begin{pmatrix} 0 & 1 \\ 1 & 0 \end{pmatrix} \otimes \begin{pmatrix} 1 & 0 \\ 0 & 0 \end{pmatrix} + \frac{1}{\sqrt{2}} \begin{pmatrix} 0 & -1 \\ 1 & 0 \end{pmatrix} \otimes \begin{pmatrix} 0 & 0 \\ 0 & 1 \end{pmatrix}$$

$$= \frac{1}{\sqrt{2}} \begin{pmatrix} 0 & 0 & \sqrt{3} & 0 \\ 0 & 0 & 0 & -1 \\ \sqrt{3} & 0 & 0 & 0 \\ 0 & 1 & 0 & 0 \end{pmatrix}$$

i.e.

$$\psi \begin{pmatrix} a & b \\ c & d \end{pmatrix} = \operatorname{tr}_2 V \left( \begin{pmatrix} a & b \\ c & d \end{pmatrix} \otimes \begin{pmatrix} 1 & 0 \\ 0 & 1 \end{pmatrix} \right) V^*$$

$$= \operatorname{tr}_2 \frac{1}{2} \begin{pmatrix} 3d & 0 & 3c & 0 \\ 0 & d & 0 & -c \\ 3b & 0 & 3a & 0 \\ 0 & -b & 0 & a \end{pmatrix} = \begin{pmatrix} 2d & c \\ b & 2a \end{pmatrix}.$$

**Problem 15.** Consider the Kraus operators $K_1$ and $K_2$

$$K_1 = \begin{pmatrix} 0 & 1 \\ 0 & 0 \end{pmatrix} \Rightarrow K_1^* = \begin{pmatrix} 0 & 0 \\ 1 & 0 \end{pmatrix}, \quad K_2 = \begin{pmatrix} 0 & 0 \\ 1 & 0 \end{pmatrix} \Rightarrow K_2^* = \begin{pmatrix} 0 & 1 \\ 0 & 0 \end{pmatrix}$$

and an arbitrary $2 \times 2$ matrix $A = (a_{jk})$. Then

$$K_1 A K_1^* + K_2 A K_2^* = \begin{pmatrix} a_{22} & 0 \\ 0 & a_{11} \end{pmatrix}.$$

So the trace is preserved under this transformation. Let

$$c_1^\dagger, \quad c_2^\dagger, \quad c_1, \quad c_2$$

be Fermi creation and annihilation operators, respectively. Consider the operators

$$\hat{K}_1 = \begin{pmatrix} c_1^\dagger & c_2^\dagger \end{pmatrix} \begin{pmatrix} 0 & 1 \\ 0 & 0 \end{pmatrix} = c_1^\dagger c_2, \quad \hat{K}_1^\dagger = c_2^\dagger c_1$$

$$\hat{K}_2 = \begin{pmatrix} c_1^\dagger & c_2^\dagger \end{pmatrix} \begin{pmatrix} 0 & 0 \\ 1 & 0 \end{pmatrix} = c_2^\dagger c_1, \quad \hat{K}_2^\dagger = c_1^\dagger c_2$$

and

$$\hat{A} = \begin{pmatrix} c_1^\dagger & c_2^\dagger \end{pmatrix} A \begin{pmatrix} c_1 \\ c_2 \end{pmatrix} = a_{11} c_1^\dagger + a_{12} c_1^\dagger c_2 + a_{21} c_2^\dagger c_1 + a_{22} c_2^\dagger c_2.$$

Find the operator $\hat{K}_1 \hat{A} \hat{K}_1^\dagger + \hat{K}_2 \hat{A} \hat{K}_2^\dagger$.

**Solution 15.** With $c_j^\dagger c_k + c_k c_j^\dagger = \delta_{jk} I$ we obtain

$$\hat{K}_1 \hat{A} \hat{K}_1^\dagger + \hat{K}_2 \hat{A} \hat{K}_2^\dagger = a_{11} c_2^\dagger c_2 + a_{22} c_1^\dagger c_1 - (a_{11} + a_{22}) c_1^\dagger c_1 c_2^\dagger c_2)$$

$$= \begin{pmatrix} c_1^\dagger & c_2^\dagger \end{pmatrix} \begin{pmatrix} a_{22} & 0 \\ 0 & a_{11} \end{pmatrix} \begin{pmatrix} c_1 \\ c_2 \end{pmatrix} - (a_{11} + a_{22}) c_1^\dagger c_1 c_2^\dagger c_2.$$

**Programming Problem**

**Problem 1.** Let $\mu \in [0,1]$. Apply computer algebra to show that the $2 \times 2$ matrices

$$K_0 = \frac{\sqrt{1+3\mu}}{2}I_2, \quad K_1 = \frac{\sqrt{1-\mu}}{2}\sigma_1, \quad K_2 = \frac{\sqrt{1-\mu}}{2}\sigma_2, \quad K_3 = \frac{\sqrt{1-\mu}}{2}\sigma_3$$

are Kraus operators. Show that the sixteen $4 \times 4$ matrices $K_j \otimes K_\ell$, $(j, \ell = 0, 1, 2, 3)$ are Kraus operators.

**Solution 1.** Note that the operators $K_j$ and $K_j \otimes K_\ell$ are hermitian. The Maxima program is

```
/* Kraus.mac */
I2: matrix([1,0],[0,1]);
sig1: matrix([0,1],[1,0]);
sig2: matrix([0,-%i],[%i,0]);
sig3: matrix([1,0],[0,-1]);
K0: sqrt(1+3*mu)*I2/2; K1: sqrt(1-mu)*sig1/2;
K2: sqrt(1-mu)*sig2/2; K3: sqrt(1-mu)*sig3/2;
K0TC: K0; K1TC: K1; K2TC: K2; K3TC: K3;
S: K0 . K0TC + K1 . K1TC + K2 . K2TC + K3 . K3TC;
S: ratsimp(S);
K00: kronecker_product(K0,K0); K01: kronecker_product(K0,K1);
K02: kronecker_product(K0,K2); K03: kronecker_product(K0,K3);
K10: kronecker_product(K1,K0); K11: kronecker_product(K1,K1);
K12: kronecker_product(K1,K2); K13: kronecker_product(K1,K3);
K20: kronecker_product(K2,K0); K21: kronecker_product(K2,K1);
K22: kronecker_product(K2,K2); K23: kronecker_product(K2,K3);
K30: kronecker_product(K3,K0); K31: kronecker_product(K3,K1);
K32: kronecker_product(K3,K2); K33: kronecker_product(K3,K3);
K00TC: K00; K01TC: K01; K02TC: K02; K03TC: K03;
K10TC: K10; K11TC: K11; K12TC: K12; K13TC: K13;
K20TC: K20; K21TC: K21; K22TC: K22; K23TC: K23;
K30TC: K30; K31TC: K31; K32TC: K32; K33TC: K33;
S: K00 . K00TC + K01 . K01TC + K02 . K02TC + K03 . K03TC
+ K10 . K10TC + K11 . K11TC + K12 . K12TC + K13 . K13TC
+ K20 . K20TC + K21 . K21TC + K22 . K22TC + K23 . K23TC
+ K30 . K30TC + K31 . K31TC + K32 . K32TC + K33 . K33TC;
S: ratsimp(S);
```

# 17.3   Supplementary Problems

**Problem 1.**   Let $|0\rangle$, $|1\rangle$, $|2\rangle$ be the standard basis in $\mathbb{C}^3$ and $\alpha = (2 - \sqrt{3})/4$, $\beta = 2 + \sqrt{3}$. Show that

$$K_1 = \sqrt{\alpha}(\sqrt{\beta}|0\rangle\langle 0| + |1\rangle\langle 1|) \otimes (\sqrt{\beta}|0\rangle\langle 1| + |1\rangle\langle 0|)$$
$$K_2 = \sqrt{\alpha}(|1\rangle\langle 0| + \sqrt{\beta}|0\rangle\langle 1|) \otimes (|1\rangle\langle 1| + \sqrt{\beta}|0\rangle\langle 0|)$$
$$K_3 = \sqrt{\alpha}(\sqrt{\beta}|0\rangle\langle 0| + |1\rangle\langle 2|) \otimes (\sqrt{\beta}|0\rangle\langle 2| + |1\rangle\langle 0|)$$
$$K_4 = \sqrt{\alpha}(|1\rangle\langle 0| + \sqrt{\beta}|0\rangle\langle 2|) \otimes (|1\rangle\langle 2| + \sqrt{\beta}|0\rangle\langle 0|)$$
$$K_5 = |1\rangle\langle 1| \otimes (2\sqrt{\alpha\beta}|1\rangle\langle 1| + |2\rangle\langle 2|)$$
$$K_6 = |2\rangle\langle 2| \otimes (|1\rangle\langle 1| + \sqrt{2\alpha\beta}|2\rangle\langle 2|)$$

are Kraus operators in $\mathbb{C}^9$.

**Problem 2.**   Let $\mu \in [0,1]$ and $\sigma_1$, $\sigma_2$, $\sigma_3$, $\sigma_0 = I_2$ be the Pauli spin matrices. Then the four $2 \times 2$ matrices

$$K_0 = \frac{\sqrt{1+3\mu}}{2}\sigma_0, \quad K_1 = \frac{\sqrt{1-\mu}}{2}\sigma_1, \quad K_2 = \frac{\sqrt{1-\mu}}{2}\sigma_2, \quad K_3 = \frac{\sqrt{1-\mu}}{2}\sigma_3$$

are Kraus operators. Show that the sixteen $4 \times 4$ matrices $K_j \star K_\ell$, $(j, \ell = 0, 1, 2, 3)$ are Kraus operators, where $\star$ denotes the star product.

**Problem 3.**   Let $K_j$ $(j = 1, \ldots, m)$ be $n \times n$ matrices over $\mathbb{C}$ with

$$\sum_{j=1}^{m} K_j K_j^* = I_n.$$

Show that

$$\sum_{j=1}^{m}\sum_{\ell=1}^{m}(K_j \otimes K_\ell)(K_j^* \otimes K_\ell^*) = I_n \otimes I_n \equiv I_{n^2}.$$

**Problem 4.**   Consider the $4 \times 4$ matrices with trace 1

$$A = \frac{1}{2}\begin{pmatrix} 1 & 0 & 0 & 1 \\ 0 & 0 & 0 & 0 \\ 0 & 0 & 0 & 0 \\ 1 & 0 & 0 & 1 \end{pmatrix}, \quad B = \frac{1}{2}\begin{pmatrix} 0 & 0 & 0 & 0 \\ 0 & 1 & 1 & 0 \\ 0 & 1 & 1 & 0 \\ 0 & 0 & 0 & 0 \end{pmatrix}.$$

(i) Can one find a unitary matrix $U$ such that $UAU^{-1} = B$?
(ii) Apply the vec operator to $A$ and $B$. Then find a $16 \times 16$ unitary matrix $W$ such that $W\text{vec}(A) = \text{vec}(B)$.

# Part II

# Infinite-Dimensional Hilbert Spaces

# Chapter 18

# Bose Operators and Number States

## 18.1   Introduction

Besides qubit-based quantum computing and quantum algorithms, quantum information over continuous variables is also applied and used in fields such as quantum teleportation and quantum cryptography. For continuous systems Bose operators play the central role. Consider a family of linear operators $b_j$, $b_j^\dagger$, $j = 1, 2, \ldots, m$ on an inner product space $V$, satisfying the commutation relations (*Heisenberg algebra*)

$$[b_j, b_k] = [b_j^\dagger, b_k^\dagger] = 0, \qquad [b_j, b_k^\dagger] = \delta_{jk} I \tag{1}$$

where $I$ is the identity operator. The operator $b_j^\dagger$ is called a *Bose creation operator* and the operator $b_j$ is called an *Bose annihilation operator*. The inner product space must be infinite dimensional for (1) to hold. For, if $A$ and $B$ are $n \times n$ matrices such that $[A, B] = \lambda I$, then

$$\mathrm{tr}([A, B]) = 0$$

implies $\lambda = 0$.

Let $|\mathbf{0}\rangle \equiv |00...0\rangle$ be the *vacuum state*, i.e.

$$b_j|\mathbf{0}\rangle = 0|\mathbf{0}\rangle$$

with $\langle 0|0 \rangle = 1$, $j = 1, 2, \ldots, m$. A normalized state is given by

$$\frac{1}{\sqrt{n_1! n_2! \cdots n_k!}} (b_{j_1}^\dagger)^{n_1} (b_{j_2}^\dagger)^{n_2} \cdots (b_{j_k}^\dagger)^{n_k} |0\rangle \equiv |n_1, n_2, \ldots, n_k\rangle$$

where $j_1, j_2, \ldots, j_k \in \{1, 2, \ldots, m\}$ and $n_1, n_2, \ldots, n_k \in \{0, 1, 2, \ldots\}$. The states are called *number states* (also called *Fock states*). For $b_j$ we also use the notation $I \otimes I \otimes \cdots \otimes b \otimes I \otimes I \otimes \cdots \otimes I$, where $b$ is in the $j$th position.

Let $m = 1$. Consider the number states $|n\rangle$ $(n = 0, 1, \ldots)$. Then

$$b|n\rangle = \sqrt{n}|n-1\rangle, \quad b^\dagger|n\rangle = \sqrt{n+1}|n+1\rangle, \quad b^\dagger b|n\rangle = n|n\rangle$$

and

$$\sum_{n=0}^\infty |n\rangle\langle n| = I$$

where $I$ is the identity operator and $\hat{n} = b^\dagger b$ is the *number operator*. We have

$$|n\rangle = \frac{1}{\sqrt{n!}}(b^\dagger)^n |0\rangle, \quad n = 0, 1, 2, \ldots .$$

The number states $|n\rangle$ form an orthonormal basis in the Hilbert space $\ell_2(\mathbb{N}_0)$. The Hamilton operator $\hat{H}$ of the one-dimensional harmonic oscillator can be written as

$$\hat{H} = \hbar\omega \left( b^\dagger b + \frac{1}{2} I \right)$$

with the eigenvalue equation $\hat{H}|n\rangle = (\hbar\omega + 1/2)|n\rangle$

Consider the commutation relation

$$[\hat{q}, \hat{p}] = i\hbar I$$

with $\hat{p} = -i\hbar \partial/\partial q$. The electric field $\mathbf{E}$ for a single mode in a box of length $L$ can be written as

$$\mathbf{E}(\mathbf{x}, t) = i e_\lambda \sqrt{\frac{\hbar\omega}{2\epsilon_0 L^3}} \left( b(0) e^{-i(\omega t - \mathbf{k}\cdot\mathbf{x})} - b^\dagger(0) e^{i(\omega t - \mathbf{k}\cdot\mathbf{x})} \right)$$

where $\mathbf{k} \cdot \mathbf{x} = k_1 x_1 + k_2 x_2 + k_3 x_3$.

Define the unitary operators

$$\hat{U}_\alpha = \exp(i\alpha\hat{p}/\hbar), \quad \hat{V}_\alpha = \exp(i\beta\hat{q}/\hbar).$$

Then

$$\hat{U}_\alpha \hat{V}_\beta = \exp(i\alpha\beta/\hbar) \hat{V}_\beta \hat{U}_\alpha$$

with $\alpha$ dimension meter and $\beta$ dimension kg . meter . sec$^{-1}$.

## 18.2   Solved Problems

**Problem 1.**   Consider the Hamilton operator for the one dimensional *harmonic oscillator*

$$\hat{H} = \frac{1}{2m}\hat{p}^2 + \frac{1}{2}m\omega^2\hat{q}^2$$

where $\omega$ is the frequency and $m$ the mass. We introduce the *characteristic length*

$$\ell_0 := \sqrt{\frac{\hbar}{m\omega}}$$

and define the dimensionless linear unbounded operators (Bose operators)

$$b := \frac{1}{\sqrt{2}}\left(\frac{\hat{q}}{\ell_0} + i\frac{\hat{p}}{\hbar/\ell_0}\right), \qquad b^\dagger := \frac{1}{\sqrt{2}}\left(\frac{\hat{q}}{\ell_0} - i\frac{\hat{p}}{\hbar/\ell_0}\right).$$

(i) Find the commutator $[b, b^\dagger]$.
(ii) Express $\hat{q}$ and $\hat{p}$ in terms of $b$ and $b^\dagger$, i.e. find the inverse transform.
(iii) Express $\hat{H}$ in terms of $b$ and $b^\dagger$. Find $\hat{H}|n\rangle$. Discuss

**Solution 1.**   (i) Since

$$\hat{p} := -i\hbar\frac{\partial}{\partial q}$$

we obtain $[b, b^\dagger] = I$, where $I$ is the identity operator.
(ii) We find

$$\hat{q} = \frac{1}{\sqrt{2}}\ell_0(b + b^\dagger), \qquad \hat{p} = \frac{\hbar/\ell_0}{\sqrt{2}i}(b - b^\dagger).$$

(iii) We obtain

$$\hat{H} = \hbar\omega\left(b^\dagger b + \frac{1}{2}I\right)$$

and the eigenvalue equation $\hat{H}|n\rangle = \hbar\omega(n + 1/2)|n\rangle$ with $n = 0, 1, \ldots$.

**Problem 2.**   Let $b$, $b^\dagger$ be Bose annihilation and creation operators. Find the commutator of the operators

$$\tilde{b} = \epsilon_1 I + b, \qquad \tilde{b}^\dagger = \epsilon_2 I + b^\dagger$$

where $I$ is the identity operator and $\epsilon_1$, $\epsilon_2$ are constants.

**Solution 2.**   We have

$$
\begin{aligned}
[\tilde{b}, \tilde{b}^\dagger] &= [\epsilon_1 I + b, \epsilon_2 I + b^\dagger] = [\epsilon_1 I, \epsilon_2 I] + [\epsilon_1 I, b^\dagger] + [b, \epsilon_2 I] + [b, b^\dagger] \\
&= [b, b^\dagger] \\
&= I.
\end{aligned}
$$

**Problem 3.**    (i) Let $\beta \in \mathbb{C}$. Calculate the commutators

$$[b, \beta b^\dagger - \beta^* b], \qquad [b^\dagger, \beta b^\dagger - \beta^* b].$$

(ii) Let $n \in \mathbb{N}$ and $\beta \in \mathbb{C}$. Calculate the commutator $[b^\dagger, (\beta b^\dagger - \beta^* b)^n]$.

**Solution 3.**    (i) We find

$$[b, \beta b^\dagger - \beta^* b] = \beta I, \qquad [b^\dagger, \beta b^\dagger - \beta^* b] = \beta^* I.$$

(ii) If $f$ is an analytic function we have

$$[b^\dagger, f(b)] = -\frac{df(b)}{db}.$$

Thus

$$[b^\dagger, (\beta b^\dagger - \beta^* b)^n] = n\beta^* (\beta b^\dagger - \beta^* b)^{n-1}.$$

For $n = 1$ we have $[b^\dagger, (\beta b^\dagger - \beta^* n)] = \beta^* I$.

**Problem 4.**    Let $\hat{n} := b^\dagger b$ and $\alpha \in \mathbb{R}$. Calculate the state $\exp(\alpha \hat{n})|0\rangle$.

**Solution 4.**    Since $\hat{n}|0\rangle = 0|0\rangle$ we obtain $\exp(\alpha \hat{n})|0\rangle = |0\rangle$.

**Problem 5.**    (i) Calculate the commutators $[b^2, b^\dagger b]$, $[b^2, b^{\dagger 2}]$.
(ii) Using the commutation relations and $b|0\rangle = 0|0\rangle$ calculate the state

$$bbb^\dagger b^\dagger |0\rangle.$$

Is the state normalized?
(iii) Let

$$|n\rangle := \frac{1}{\sqrt{n!}}(b^\dagger)^n |0\rangle, \quad n = 0, 1, 2, \ldots$$

be the number states. Find the operator

$$\sum_{n=0}^{\infty} |n\rangle\langle n|.$$

**Solution 5.**    (i) Using the commutation relations we obtain

$$[b^2, b^\dagger b] = 2b^2, \qquad [b^2, b^{\dagger 2}] = 2I + 4b^\dagger b.$$

(ii) Using the commutation relations and $b|0\rangle = 0|0\rangle$ we find

$$bbb^\dagger b^\dagger |0\rangle = 2|0\rangle.$$

(iii) Since $|n\rangle$ $(n = 0, 1, 2, \ldots)$ is an orthonormal basis we find

$$\sum_{n=0}^{\infty} |n\rangle\langle n| = I$$

where $I$ is the identity operator, i.e. we have the infinite-dimensional unit matrix. This is the *completeness relation*.

**Problem 6.** Using the number states $|n\rangle$ find the matrix representation of the unbounded operators $b^\dagger b$.

**Solution 6.** Since $b^\dagger b|n\rangle = n|n\rangle$ we obtain the infinite dimensional unbounded diagonal matrix diag$(0, 1, 2, \ldots)$.

**Problem 7.** Let $\hat{n} := b^\dagger b$ be the *number operator*. Calculate the commutators $[\hat{n}, b]$, $[\hat{n}, [\hat{n}, b]]$, $[\hat{n}, [\hat{n}, [\hat{n}, b]]]$. Discuss the general case for $m$ commutators.

**Solution 7.** We have $[\hat{n}, b] = -b$, $[\hat{n}, [\hat{n}, b]] = (-1)^2 b$, $[\hat{n}, [\hat{n}, [\hat{n}, b]]] = (-1)^3 b$. Obviously, for the general case with $m$ commutators we find $(-1)^m b$.

**Problem 8.** Using the number states $|n\rangle$ find the matrix representation of the unbounded operators $b^\dagger + b$.

**Solution 8.** Since $b|n\rangle = \sqrt{n}|n-1\rangle$, $b^\dagger|n\rangle = \sqrt{n+1}|n+1\rangle$ we obtain the infinite dimensional unbounded symmetric matrix

$$\begin{pmatrix} 0 & 1 & 0 & 0 & \cdots \\ 1 & 0 & \sqrt{2} & 0 & \cdots \\ 0 & \sqrt{2} & 0 & \sqrt{3} & \cdots \\ 0 & 0 & \sqrt{3} & 0 & \cdots \\ \vdots & \vdots & \vdots & & \ddots \end{pmatrix}.$$

**Problem 9.** Let $b^\dagger$, $b$ be Bose creation and annihilation operators, respectively.
(i) Calculate the commutator $[b^\dagger + b, b^\dagger b]$.
(ii) Consider the symmetric $4 \times 4$ matrices

$$A = \begin{pmatrix} 0 & 0 & 0 & 0 \\ 0 & 1 & 0 & 0 \\ 0 & 0 & 2 & 0 \\ 0 & 0 & 0 & 3 \end{pmatrix}, \quad B = \begin{pmatrix} 0 & 1 & 0 & 0 \\ 1 & 0 & \sqrt{2} & 0 \\ 0 & \sqrt{2} & 0 & \sqrt{3} \\ 0 & 0 & \sqrt{3} & 0 \end{pmatrix}.$$

Find the commutator $[A, B]$. These matrices appear when we truncate the infinite dimensional unbounded matrices $b^\dagger b$ and $b^\dagger + b$.

**Solution 9.**    (i) Using $bb^\dagger = I + b^\dagger b$ we obtain $[b^\dagger + b, b^\dagger b] = -b^\dagger + b$.
(ii) We obtain the skew-symmetric matrix

$$[A, B] = \begin{pmatrix} 0 & -1 & 0 & 0 \\ 1 & 0 & -\sqrt{2} & 0 \\ 0 & \sqrt{2} & 0 & -\sqrt{3} \\ 0 & 0 & \sqrt{3} & 0 \end{pmatrix}$$

which is a truncation of the infinite dimensional unbounded matrix $-b^\dagger + b$.

**Problem 10.**    (i) Let $\epsilon \in \mathbb{R}$. Find

$$f_c(\epsilon) = e^{-i\epsilon b^\dagger b} b^\dagger e^{i\epsilon b^\dagger b}, \quad f_a(\epsilon) = e^{-i\epsilon b^\dagger b} b e^{i\epsilon b^\dagger b}$$

i.e. $f_c(0) = b^\dagger$, $f_a(0) = b$.
(ii) Then find the $2 \times 2$ matrix $A(\epsilon)$ such that

$$\begin{pmatrix} e^{-i\epsilon b^\dagger b}(b^\dagger + b)e^{i\epsilon b^\dagger b} \\ e^{-i\epsilon b^\dagger b}(ib^\dagger - ib)e^{i\epsilon b^\dagger b} \end{pmatrix} = A(\epsilon) \begin{pmatrix} b^\dagger + b \\ ib^\dagger - ib \end{pmatrix}.$$

**Solution 10.**    Applying *parameter differentiation* and $bb^\dagger = I + b^\dagger b$ we obtain the differential equation

$$\frac{df_c}{d\epsilon} = -if_c$$

with the initial condition $f_c(0) = b^\dagger$. The solution is $f_c(\epsilon) = e^{-i\epsilon} b^\dagger$. Analogously for $f_a$ we find $df_a/d\epsilon = if_a$ with the initial condition $f_a(0) = b$. Hence $f_a(\epsilon) = e^{i\epsilon} b$.
(ii) Since

$$e^{-i\epsilon b^\dagger b}(b^\dagger + b)e^{i\epsilon b^\dagger b} = e^{-i\epsilon} b^\dagger + e^{i\epsilon} b$$
$$e^{-i\epsilon b^\dagger b}(ib^\dagger - ib)e^{i\epsilon b^\dagger b} = ie^{-i\epsilon} b^\dagger - ie^{i\epsilon} b$$

we obtain

$$\begin{pmatrix} e^{-i\epsilon} b^\dagger + e^{i\epsilon} b \\ ie^{-i\epsilon} b^\dagger - ie^{i\epsilon} b \end{pmatrix} = \begin{pmatrix} (\cos(\epsilon) - i\sin(\epsilon))b^\dagger + (\cos(\epsilon) + i\sin(\epsilon))b \\ i(\cos(\epsilon) - i\sin(\epsilon))b^\dagger - i(\cos(\epsilon) + i\sin(\epsilon))b \end{pmatrix}$$
$$= \begin{pmatrix} \cos(\epsilon) & -\sin(\epsilon) \\ \sin(\epsilon) & \cos(\epsilon) \end{pmatrix} \begin{pmatrix} b^\dagger + b \\ ib^\dagger - ib \end{pmatrix}.$$

**Problem 11.**   Consider the operators

$$\hat{A} = b_k e^{-i\phi_k} + b_k^\dagger e^{i\phi_k}, \qquad \hat{B} = b_\ell e^{-i\phi_\ell} + b_\ell^\dagger e^{i\phi_\ell}.$$

Find the commutator $[\hat{A}, \hat{B}]$. What is the condition on the phases $\phi_k$ and $\phi_\ell$ such that $[\hat{A}, \hat{B}] = 0$ for $k = \ell$?

**Solution 11.**   We have

$$
\begin{aligned}
[\hat{A}, \hat{B}] &= [b_k e^{-i\phi_k} + b_k^\dagger e^{i\phi_k}, b_\ell e^{-i\phi_\ell} + b_\ell^\dagger e^{i\phi_\ell}] \\
&= [b_k e^{-i\phi_k}, b_\ell^\dagger e^{i\phi_\ell}] + [b_k^\dagger e^{i\phi_k}, b_\ell e^{-i\phi_\ell}] \\
&= (e^{i(\phi_\ell - \phi_k)} - e^{i(\phi_k - \phi_\ell)})\delta_{\ell k} \\
&= 2i \sin(\phi_\ell - \phi_k)\delta_{\ell k}.
\end{aligned}
$$

Thus the commutator is zero for $k = \ell$ if $\sin(\phi_\ell - \phi_k) = 0$ i.e. $\phi_\ell - \phi_k = n\pi$ and $n \in \mathbb{Z}$.

**Problem 12.**   Let $f$ be an analytic function in $z$ and $z \in \mathbb{C}$. Let $\hat{n} := b^\dagger b$ be the number operator. Calculate

$$e^{z\hat{n}} f(b), \quad e^{z\hat{n}} f(b^\dagger), \quad e^{zb^\dagger} f(b), \quad e^{zb} f(b^\dagger).$$

**Solution 12.**   Since $\hat{n}b = b(\hat{n} - I)$ and $\hat{n}b^\dagger = b^\dagger(I + \hat{n})$ we find

$$
\begin{aligned}
e^{z\hat{n}} f(b) &= f(e^{-z}b)e^{z\hat{n}} \\
e^{z\hat{n}} f(b^\dagger) &= f(e^{z}b^\dagger)e^{z\hat{n}} \\
e^{zb^\dagger} f(b) &= f(b - zI)e^{zb^\dagger} \\
e^{zb} f(b^\dagger) &= f(b^\dagger + zI)e^{zb}.
\end{aligned}
$$

**Problem 13.**   Let $b^\dagger$, $b$ be Bose creation and annihilation operator with the commutator relation $[b, b^\dagger] = I$, where $I$ is the identity operator. Show that one has a representation

$$b^\dagger \mapsto z, \qquad b \mapsto \frac{d}{dz}.$$

**Solution 13.**   We calculate the commutator

$$[d/dz, z]f(z)$$

where $f$ is an analytic functions. Using the product rule we obtain

$$[d/dz, z]f(z) = \frac{d}{dz}(zf(z)) - z\frac{df(z)}{dz} = f(z) + z\frac{df(z)}{dz} - z\frac{df(z)}{dz} = f(z).$$

**Problem 14.**　Consider the differential operators

$$b = \frac{1}{\sqrt{2}}\left(x + \frac{d}{dx}\right), \qquad b^\dagger = \frac{1}{\sqrt{2}}\left(x - \frac{d}{dx}\right)$$

acting in the vector space $S(\mathbb{R})$. Find the commutator $[b, b^\dagger]$. Find the operator $\hat{N} = b^\dagger b$.

**Solution 14.**　Let $f \in S(\mathbb{R})$. We have

$$
\begin{aligned}
[b, b^\dagger]f &= bb^\dagger f - b^\dagger bf \\
&= \frac{1}{2}\left(x^2 f - \frac{d}{dx}(xf) - x\frac{df}{dx} - \frac{d^2 f}{dx^2}\right) \\
&\quad - \frac{1}{2}\left(x^2 f - \frac{d}{dx}(xf) + x\frac{df}{dx} - \frac{d^2 f}{dx^2}\right) \\
&= \frac{d}{dx}(xf) - x\frac{df}{dx} \\
&= f.
\end{aligned}
$$

We obtain the differential operator

$$\hat{N} = \frac{1}{2}\left(x^2 - 1 - \frac{d^2}{dx^2}\right).$$

**Problem 15.**　Let $\epsilon \in \mathbb{R}$ and $\epsilon > 0$, $\hat{n} = b^\dagger b$ and $\{|n\rangle : n = 0, 1, 2, \dots\}$ denote the number states.
(i) Using the number states calculate the *trace*

$$\mathrm{tr}(e^{-\epsilon\hat{n}}) \equiv \sum_{n=0}^{\infty}\langle n|e^{-\epsilon\hat{n}}|n\rangle.$$

(ii) Using the number states calculate the trace

$$\mathrm{tr}(b^\dagger b e^{-\epsilon b^\dagger b}) = \sum_{n=0}^{\infty}\langle n|b^\dagger b e^{-\epsilon b^\dagger b}|n\rangle.$$

**Solution 15.** (i) Since $\hat{n}|n\rangle = n|n\rangle$ we have $e^{-\epsilon b^\dagger b}|n\rangle = e^{-\epsilon n}|n\rangle$. Thus

$$\text{tr}(e^{-\epsilon\hat{n}}) = \sum_{n=0}^{\infty} e^{-\epsilon n} = \frac{e^\epsilon}{e^\epsilon - 1}.$$

(ii) Using $\langle n|b^\dagger b = \langle n|n$ we obtain

$$\text{tr}(b^\dagger b e^{-\epsilon b^\dagger b}) = \sum_{n=0}^{\infty} n e^{-\epsilon n} = \frac{e^\epsilon}{(e^\epsilon - 1)^2}.$$

**Problem 16.** Consider the Hamilton operator $\hat{H} = \hbar\omega b^\dagger b$.
(i) Calculate the trace $\text{tr}(e^{-\hat{H}/k_BT})$.
(ii) Consider the *density operator*

$$\rho = \frac{\exp(-\hat{H}/k_BT)}{\text{tr}(\exp(-\hat{H}/k_BT))}.$$

Let $\hat{n} := b^\dagger b$. Calculate $\bar{n} \equiv \langle\hat{n}\rangle \equiv \langle b^\dagger b\rangle \equiv \text{tr}(b^\dagger b\rho)$ and $\langle\hat{H}\rangle$.
(iii) Calculate $\text{tr}(\rho b)$, $\text{tr}(\rho b^\dagger)$.

**Solution 16.** (i) We set $\lambda := \hbar\omega/(k_BT)$. Then using the completeness relation

$$I = \sum_{n=0}^{\infty} |n\rangle\langle n|,$$

$\hat{n}|n\rangle = n|n\rangle$ and $e^{-\lambda\hat{n}}|n\rangle = e^{-\lambda n}|n\rangle$ we find

$$e^{-\hat{H}/k_BT} = e^{-\lambda b^\dagger b} = \sum_{n=0}^{\infty} e^{-\lambda b^\dagger b}|n\rangle\langle n| = \sum_{n=0}^{\infty} e^{-\lambda n}|n\rangle\langle n|.$$

Thus

$$\text{tr}(e^{-\hat{H}/k_BT}) = \sum_{n=0}^{\infty}(e^{-\lambda})^n = \frac{1}{1-e^{-\lambda}}.$$

(ii) Using the result from (i) we obtain

$$\bar{n} = \langle b^\dagger b\rangle = \frac{1}{e^\lambda - 1}$$

and therefore

$$\langle\hat{H}\rangle = \frac{\hbar\omega}{e^{\hbar\omega/k_BT} - 1}.$$

(iii) Since $\langle n|b|n\rangle = 0$, $\langle n|b^\dagger|n\rangle = 0$ we obtain

$$\langle b\rangle = \text{tr}(\rho b) = 0, \qquad \langle b^\dagger\rangle = \text{tr}(\rho b^\dagger) = 0.$$

**Problem 17.**    Consider the *Bose-Einstein density operator*

$$\rho = \frac{1}{\bar{n}+1} \sum_{n=0}^{\infty} \left( \frac{\bar{n}}{\bar{n}+1} \right)^n |n\rangle\langle n|$$

where $|n\rangle$, $n = 0, 1, 2, \ldots$ are the number states and

$$\bar{n} := \left( \exp\left( \frac{\hbar\omega}{k_B T} \right) - 1 \right)^{-1}.$$

Show that $\mathrm{tr}(\rho) = 1$.

**Solution 17.**    We set $\alpha = \hbar\omega/(k_B T)$ and use the number states as an orthonormal basis. Thus

$$\mathrm{tr}(\rho) = \sum_{m=0}^{\infty} \langle m|\rho|m\rangle = \frac{1}{\bar{n}+1} \sum_{n=0}^{\infty} \left( \frac{\bar{n}}{\bar{n}+1} \right)^n$$

where we used $\langle m|n\rangle = \langle n|m\rangle = \delta_{mn}$. Now we have

$$\frac{1}{\bar{n}+1} = 1 - e^{-\alpha}, \qquad \frac{\bar{n}}{\bar{n}+1} = e^{-\alpha}.$$

Since the sum is a *geometric series* we have

$$\sum_{n=0}^{\infty} (e^{-\alpha})^n = \frac{1}{1 - e^{-\alpha}}.$$

Thus $\mathrm{tr}(\rho) = 1$.

**Problem 18.**    Let $b$ and $b^\dagger$ be Bose annihilation and creation operators, respectively. Consider the general one-mode canonical *Bogolubov transform*

$$\tilde{b} := e^{i\phi} \cosh(s)b + e^{i\psi} \sinh(s)b^\dagger$$
$$\tilde{b}^\dagger := e^{-i\phi} \cosh(s)b^\dagger + e^{-i\psi} \sinh(s)b$$

where $s$ is a real parameter (*squeezing parameter*).
(i) Show that the operators $\tilde{b}$ and $\tilde{b}^\dagger$ satisfy the Bose commutation relations.
(ii) Find the inverse Bogolubov transform.

**Solution 18.**    (i) Since $\cosh^2(s) - \sinh^2(s) = 1$ we find

$$[\tilde{b}, \tilde{b}^\dagger] = bb^\dagger - b^\dagger b = I.$$

(ii) The transformation can be written in the matrix form

$$\begin{pmatrix} \widetilde{b} \\ \widetilde{b}^\dagger \end{pmatrix} = \begin{pmatrix} e^{i\phi}\cosh(s) & e^{i\psi}\sinh(s) \\ e^{-i\psi}\sinh(s) & e^{-i\phi}\cosh(s) \end{pmatrix} \begin{pmatrix} b \\ b^\dagger \end{pmatrix}.$$

The determinant of this matrix is $+1$. Thus the inverse transformation is given by

$$\begin{pmatrix} b \\ b^\dagger \end{pmatrix} = \begin{pmatrix} e^{-i\phi}\cosh(s) & -e^{i\psi}\sinh(s) \\ -e^{-i\psi}\sinh(s) & e^{i\phi}\cosh(s) \end{pmatrix} \begin{pmatrix} \widetilde{b} \\ \widetilde{b}^\dagger \end{pmatrix}.$$

**Problem 19.** Let $\epsilon \in \mathbb{R}$ and $f : \mathbb{C}^2 \to \mathbb{C}$ be an *entire analytic function*. If a function $f$ is analytic on the whole complex plane then $f$ is said to be entire. Show that

$$e^{\epsilon b} f(b, b^\dagger) e^{-\epsilon b} = f(b, b^\dagger + \epsilon I) \tag{1a}$$

$$e^{-\epsilon b^\dagger} f(b, b^\dagger) e^{\epsilon b^\dagger} = f(b + \epsilon I, b^\dagger). \tag{1b}$$

**Solution 19.** We have

$$e^{\epsilon b} f(b, b^\dagger) e^{-\epsilon b} = f(e^{\epsilon b} b e^{-\epsilon b}, e^{\epsilon b} b^\dagger e^{-\epsilon b}) = f(b, e^{\epsilon b} b^\dagger e^{-\epsilon b}).$$

Since $e^{\epsilon b} b^\dagger e^{-\epsilon b} = b^\dagger + \epsilon I$ we find (1a), where we used $[b, b^\dagger] = I$. A similar proof holds for (1b).

**Problem 20.** Let $f : \mathbb{C} \to \mathbb{C}$ be an entire analytic function. Show that

$$f(b^\dagger b) = \sum_{n=0}^{\infty} f(n)|n\rangle\langle n|$$

where $|n\rangle$ is the number state.

**Solution 20.** The completeness relation is given by

$$\sum_{n=0}^{\infty} |n\rangle\langle n| = I.$$

Since $b^\dagger b|n\rangle = n|n\rangle$, we have $f(b^\dagger b)|n\rangle = f(n)|n\rangle$. It follows that

$$f(b^\dagger b) = f(b^\dagger b)I = f(b^\dagger b)\sum_{n=0}^{\infty}|n\rangle\langle n| = \sum_{n=0}^{\infty} f(b^\dagger b)|n\rangle\langle n| = \sum_{n=0}^{\infty} f(n)|n\rangle\langle n|.$$

**Problem 21.**    Let $f$ be an analytic function in $x$ and $y$. Let $b^\dagger$ and $b$ be Bose creation and annihilation operators, respectively. We can define $f(b, b^\dagger)$ by its power series expansion

$$f(b, b^\dagger) := \sum_{j_1=0}^{\infty} \sum_{j_2=0}^{\infty} \cdots \sum_{j_n=0}^{\infty} f(j_1, j_2, \ldots, j_n)(b^\dagger)^{j_1} b^{j_2} \ldots b^{j_n}.$$

We can use the commutation relation for Bose operators repeatedly to rearrange the operators $b$, $b^\dagger$ so that

$$f(b, b^\dagger) = \sum_{m=0}^{\infty} \sum_{n=0}^{\infty} f_{mn}^{(n)}(b^\dagger)^m b^n.$$

We say that the function $f(b, b^\dagger)$ is in *normal order form*.
(i) Consider the functions

$$f(b, b^\dagger) = b^\dagger bb^\dagger b, \qquad g(b, b^\dagger) = b^\dagger bb^\dagger bb^\dagger b.$$

Find the normal order form for these functions.
(ii) Consider the operator $\exp(-\epsilon b^\dagger b)$, where $\epsilon$ is a real positive parameter. Find the *normal order form*.

**Solution 21.**    (i) From the commutation relations for Bose operators we find $bb^\dagger = I + b^\dagger b$. Thus

$$f(b, b^\dagger) = b^\dagger b + b^\dagger b^\dagger bb, \quad g(b, b^\dagger) = b^\dagger b^\dagger b^\dagger bbb + 3b^\dagger b^\dagger bb + b^\dagger b.$$

(ii) Using the results from (i) and $b^\dagger b(b^\dagger b)^j \equiv b^\dagger (I + b^\dagger b)^j b$ we find

$$e^{-\epsilon b^\dagger b} = \sum_{j=0}^{\infty} \frac{1}{j!}(e^{-\epsilon} - 1)^j (b^\dagger)^j b^j.$$

**Problem 22.**    Let $\{\, |n\rangle \ : \ n = 0, 1, 2, \ldots \,\}$ be the number states. We define the linear operators

$$\hat{E} := \sum_{n=0}^{\infty} |n\rangle\langle n+1|, \qquad \hat{E}^\dagger := \sum_{n=0}^{\infty} |n+1\rangle\langle n|.$$

Obviously, $\hat{E}^\dagger$ follows from $\hat{E}$.
(i) Find $\hat{E}\hat{E}^\dagger$ and $\hat{E}^\dagger\hat{E}$.
(ii) Let $f$ be an analytic function. Calculate $\hat{E}f(\hat{n})\hat{E}^\dagger$ and $\hat{E}^\dagger f(\hat{n} + I)\hat{E}$, where $\hat{n}$ is the photon number operator and $I$ is the identity operator.

**Solution 22.** (i) Using $\langle m|n \rangle = \delta_{mn}$ and the completeness relation we find

$$\hat{E}\hat{E}^\dagger = \sum_{m=0}^{\infty} |m\rangle\langle m+1| \sum_{n=0}^{\infty} |n+1\rangle\langle n| = \sum_{m=0}^{\infty} \sum_{n=0}^{\infty} |m\rangle \delta_{m+1,n+1} \langle n|$$

$$= \sum_{m=0}^{\infty} |m\rangle\langle m| = I.$$

Analogously, we find $\hat{E}^\dagger\hat{E} = I - |0\rangle\langle 0|$.

(ii) Using the Taylor expansion around 0 of an analytic function we have

$$\hat{E}f(\hat{n}) = \sum_{m=0}^{\infty} |m\rangle\langle m+1| \sum_{j=0}^{\infty} \frac{f^{(j)}(0)}{j!} \hat{n}^j.$$

Applying $\hat{n}|m\rangle = m|m\rangle$, $\hat{n}^j|m\rangle = m^j|m\rangle$ and

$$(\hat{n}+I)|m\rangle = (m+1)|m\rangle, \quad (\hat{n}+I)^j|m\rangle = (m+1)^j|m\rangle$$

we obtain $\hat{E}f(\hat{n}) = f(\hat{n}+I)\hat{E}$. Thus $\hat{E}f(\hat{n})\hat{E}^\dagger = f(\hat{n}+I)$. Analogously

$$\hat{E}^\dagger f(\hat{n}+I)\hat{E} = f(\hat{n}).$$

**Problem 23.** Consider the *Susskind-Glogower canonical phase states*

$$|\phi\rangle := \sum_{n=0}^{\infty} e^{i\phi n}|n\rangle$$

where $|n\rangle$ are the number states. Let

$$\hat{L} := \sum_{n=1}^{\infty} |n-1\rangle\langle n|$$

be the non unitary number-lowering operator. Find $\hat{L}|\phi\rangle$.

**Solution 23.** Since $\langle m|n \rangle = \delta_{mn}$ we have $\hat{L}|\phi\rangle = e^{i\phi}|\phi\rangle$. This means that $|\phi\rangle$ is an eigenstate of the operator $\hat{L}$.

**Problem 24.** Let $b^\dagger$ and $b$ be Bose creation and annihilation operators, respectively. Consider the operator

$$e^{\alpha_1 b^2 + \alpha_2 (b^\dagger)^2 + \alpha_3 (bb^\dagger + b^\dagger b)} \tag{1}$$

where $\alpha_1, \alpha_2, \alpha_3 \in \mathbb{R}$. Let $\epsilon \in \mathbb{R}$ be an arbitrary real parameter. Find the smooth functions $f_0$, $f_1$, $f_2$ and $f_3$, depending on $\epsilon$, such that

$$e^{\epsilon(\alpha_1 b^2 + \alpha_2(b^\dagger)^2 + \alpha_3(bb^\dagger + b^\dagger b))} = e^{f_0(\epsilon)I + f_1(\epsilon)(b^\dagger)^2} e^{f_2(\epsilon)b^\dagger b} e^{f_3(\epsilon)b^2} \tag{2}$$

where $I$ denotes the identity operator. Then set $\epsilon = 1$. Solve the problem using *parameter differentiation* with respect to $\epsilon$. We find a system of ordinary differential equations for the functions $f_0$, $f_1$, $f_2$ and $f_3$.

**Solution 24.**    Differentiating the left-hand side of (2) with respect to $\epsilon$ yields

$$(\alpha_1 b^2 + \alpha_2(b^\dagger)^2 + \alpha_3(bb^\dagger + b^\dagger b))e^{\epsilon(\alpha_1 b^2 + \alpha_2(b^\dagger)^2 + \alpha_3(bb^\dagger + b^\dagger b))} =$$

$$e^{f_0(\epsilon)I + f_1(\epsilon)(b^\dagger)^2} \left( \frac{df_0}{d\epsilon} I + \frac{df_1}{d\epsilon}(b^\dagger)^2 \right) e^{f_2(\epsilon)b^\dagger b} e^{f_3(\epsilon)b^2}$$

$$+ e^{f_0(\epsilon)I + f_1(\epsilon)(b^\dagger)^2} e^{f_2(\epsilon)b^\dagger b} \left( \frac{df_2}{d\epsilon} b^\dagger b \right) e^{f_3(\epsilon)b^2}$$

$$+ e^{f_0(\epsilon)I + f_1(\epsilon)(b^\dagger)^2} e^{f_2(\epsilon)b^\dagger b} e^{f_3(\epsilon)b^2} \left( \frac{df_3}{d\epsilon} b^2 \right).$$

Owing to the identity $e^{\epsilon \hat{X}} e^{-\epsilon \hat{X}} = I$, we have

$$e^{-\epsilon(\alpha_1 b^2 + \alpha_2(b^\dagger)^2 + \alpha_3(bb^\dagger + b^\dagger b))} = e^{-f_3(\epsilon)b^2} e^{-f_2(\epsilon)b^\dagger b} e^{-f_0(\epsilon)I - f_1(\epsilon)(b^\dagger)^2}.$$

From the last two equation we obtain

$$\alpha_1 b^2 + \alpha_2(b^\dagger)^2 + \alpha_3(bb^\dagger + b^\dagger b) = \frac{df_0}{d\epsilon} I + \frac{df_1}{d\epsilon} e^{-f_3 b^2} e^{-f_2 b^\dagger b} (b^\dagger)^2 e^{f_2 b^\dagger b} e^{f_3 b^2}$$

$$+ \frac{df_2}{d\epsilon} e^{-f_3 b^2} b^\dagger b e^{f_3 b^2} + \frac{df_3}{d\epsilon} b^2.$$

Since

$$e^{-f_2(\epsilon)b^\dagger b}(b^\dagger)^2 e^{f_2(\epsilon)b^\dagger b} = (b^\dagger)^2 e^{-2f_2(\epsilon)}$$

$$e^{-f_3(\epsilon)b^2}(b^\dagger)^2 e^{f_3(\epsilon)b^2} = (b^\dagger)^2 + 4f_3^2(\epsilon)b^2 - 2f_3(\epsilon)(I + 2b^\dagger b)$$

$$e^{-f_3(\epsilon)b^2} b^\dagger b e^{f_3(\epsilon)b^2} = b^\dagger b - 2f_3(\epsilon)b^2$$

we find

$$\alpha_1 b^2 + \alpha_2(b^\dagger)^2 + 2\alpha_3 b^\dagger b + \alpha_3 I$$

$$= \frac{df_0}{d\epsilon} I + \frac{df_1}{d\epsilon} e^{-2f_2}((b^\dagger)^2 + 4f_3^2 b^2 - f_3(2I + 4b^\dagger b)) + \frac{df_2}{d\epsilon}(b^\dagger b - 2f_3 b^2) + \frac{df_3}{d\epsilon} b^2$$

where we used

$$e^{-\mu \hat{A}} \hat{B} e^{\mu \hat{A}} = \hat{B} - \mu[\hat{A}, \hat{B}] + \frac{1}{2!} \mu^2 [\hat{A}, [\hat{A}, \hat{B}]] + \cdots.$$

and $bb^\dagger = I + b^\dagger b$. Separating out terms with $I$, $b^2$, $(b^\dagger)^2$, and $b^\dagger b$ we find

$$\frac{df_0}{d\epsilon} - \alpha_3 - 2f_3 \frac{df_1}{d\epsilon} e^{-2f_2} = 0$$

$$4f_3^2 \frac{df_1}{d\epsilon} e^{-2f_2} + \frac{df_3}{d\epsilon} - 2\frac{df_2}{d\epsilon} f_3 - \alpha_1 = 0$$

$$\frac{df_1}{d\epsilon} - \alpha_2 e^{2f_2} = 0$$

$$4\frac{df_1}{d\epsilon} f_3 e^{-2f_2} - \frac{df_2}{d\epsilon} + 2\alpha_3 = 0.$$

Using this system of equations we can cast the system of nonlinear differential equations in the form

$$\frac{df_0}{d\epsilon} = \alpha_3 + 2\alpha_2 f_3, \qquad \frac{df_1}{d\epsilon} = \alpha_2 e^{2f_2}$$

$$\frac{df_2}{d\epsilon} = 2\alpha_3 + 4\alpha_2 f_3, \qquad \frac{df_3}{d\epsilon} = \alpha_1 + 4\alpha_3 f_3 + 4\alpha_2 f_3^2$$

with the initial conditions $f_j(0) = 0$ for $j = 0, 1, 2, 3$. We first solve the fourth equation which is a *Riccati equation* and then insert it into the third and second equation to find $f_2$ and $f_0$. Finally we solve for $f_1$. The integration yields

$$f_0(\epsilon) = \frac{1}{2} \ln(\cosh(2\lambda\epsilon) - (\alpha_3/\lambda)\sinh(2\lambda\epsilon))$$

$$f_1(\epsilon) = \frac{(\alpha_2/2\lambda)\sinh(2\lambda\epsilon)}{\cosh(2\lambda\epsilon) - (\alpha_3/\lambda)\sinh(2\lambda\epsilon)}$$

$$f_2(\epsilon) = -\ln(\cosh(2\lambda\epsilon) - (\alpha_3/\lambda)\sinh(2\lambda\epsilon))$$

$$f_3(\epsilon) = \frac{(\alpha_1/2\lambda)\sinh(2\lambda\epsilon)}{\cosh(2\lambda\epsilon) - (\alpha_3/\lambda)\sinh(2\lambda\epsilon)}$$

where $\lambda := \sqrt{\alpha_3 - \alpha_1\alpha_2}$. Setting $\epsilon = 1$ we have

$$e^{\alpha_1 b^2 + \alpha_2 (b^\dagger)^2 + \alpha_3 (bb^\dagger + b^\dagger b)} = \frac{1}{\sqrt{\cosh(2\lambda) - (\alpha_3/\lambda)\sinh(2\lambda)}}$$

$$\times \exp\left(\frac{(\alpha_2/\lambda)\sinh(2\lambda)}{\cosh(2\lambda) - (\alpha_3/\lambda)\sinh(2\lambda)}(b^\dagger)^2\right)$$

$$\times \exp(\ln(\cosh(2\lambda) - (\alpha_3/\lambda)\sinh(2\lambda))^{-1}b^\dagger b)$$

$$\times \exp\left(\frac{(\alpha_1/2\lambda)\sinh(2\lambda)}{\cosh(2\lambda) - (\alpha_3/\lambda)\sinh(2\lambda)}b^2\right).$$

**Problem 25.** The homogeneous *Bogolubov transform* of the Bose creation operator $b^\dagger$ and Bose annihilation operator $b$

$$\tilde{b} = \mu b + \nu b^\dagger, \qquad \mu, \nu \in \mathbb{C}$$

for a pair of complex parameters

$$\mu = |\mu| \exp(i\phi), \qquad \nu = |\nu| \exp(i\theta)$$

obeying additionally $|\mu|^2 - |\nu|^2 = 1$ is *canonical* since it leaves the commutator invariant, i.e.

$$[b, b^\dagger] = [\tilde{b}, \tilde{b}^\dagger] = I.$$

Every canonical transform can be represented as a unitary transformation

$$\tilde{b} = B(\mu, \nu) b B^\dagger(\mu, \nu).$$

The *Bogolubov unitary operator* $B(\mu, \nu)$ is defined by this relation up to an arbitrary phase factor. One choice is the normal form

$$B(\mu, \nu) = \mu^{-1/2} \exp\left(-\frac{\nu}{2\mu} b^{\dagger 2}\right) \exp\left(-\ln(\mu) b^\dagger b\right) \exp\left(\frac{\nu^*}{2\mu} b^2\right).$$

Show that the Bogolubov transform forms a continuous non-commutative group.

**Solution 25.**    Let $|\mu'|^2 - |\nu'|^2 = 1$, $|\mu''|^2 - |\nu''|^2 = 1$. Then we have

$$B(\mu', \nu') B(\mu'', \nu'') = B(\mu, \nu)$$

and $\mu = \mu'\mu'' + \nu'^*\nu''$, $\nu = \mu'^*\nu'' + \nu'\mu''$ with $|\mu|^2 - |\nu|^2 = 1$. The identity element of the group is given by $B(1, 0)$, where we used that $\ln(1) = 0$. The inverse element of $B(\mu, \nu)$ is given by

$$B^{-1}(\mu, \nu) = B^\dagger(\mu, \nu) = B(\mu^*, -\nu).$$

Obviously, the associative law also holds.

**Problem 26.**    Consider the operators

$$K_+ := b_1^\dagger b_2^\dagger, \qquad K_- := b_2 b_1, \qquad K_3 := \frac{1}{2}(b_1^\dagger b_1 + b_2^\dagger b_2 + I)$$

where $I$ is the identity operator. Find the commutators

$$[K_+, K_-], \quad [K_3, K_+], \quad [K_3, K_-].$$

The operators $K_+$, $K_-$, $K_3$ form a representation of the Lie algebra $su(1, 1)$.

**Solution 26.**    Using the commutation relations given above we find

$$[K_+, K_-] = b_1^\dagger b_2^\dagger b_2 b_1 - b_2 b_1 b_1^\dagger b_2^\dagger = b_1^\dagger b_2^\dagger b_2 b_1 - b_2 b_2^\dagger - b_1^\dagger b_1 b_2 b_2^\dagger$$
$$= -b_2 b_2^\dagger - b_1^\dagger b_1 = -I - b_2^\dagger b_2 - b_1^\dagger b_1$$
$$= -2K_3.$$

Analogously $[K_3, K_-] = -K_-$, $[K_3, K_+] = K_+$.

**Problem 27.** Consider the linear operators

$$J_+ := b_1^\dagger b_2, \quad J_- := b_2^\dagger b_1, \quad J_3 := \frac{1}{2}(b_1^\dagger b_1 - b_2^\dagger b_2)$$

where $b_1^\dagger, b_2^\dagger$ are Bose creation operators and $b_1, b_2$ are Bose annihilation operators and $I$ is the identity operator. Find the commutators

$$[J_+, J_-], \qquad [J_3, J_+], \qquad [J_3, J_-].$$

The operators $J_+$, $J_-$, $J_3$ form a representation of the semisimple Lie algebra $su(2)$.

**Solution 27.** Using the commutation relation given above we find

$$[J_+, J_-] = b_1^\dagger b_2 b_2^\dagger b_1 - b_2^\dagger b_1 b_1^\dagger b_2 = b_1^\dagger b_2 b_2^\dagger b_1 - b_2^\dagger b_2 - b_2^\dagger b_1^\dagger b_1 b_2$$
$$= -b_2^\dagger b_2 + b_1^\dagger b_1 = 2J_3.$$

Analogously $[J_3, J_-] = -J_-$, $[J_3, J_+] = J_+$.

**Problem 28.** Consider the Bose creation operators $b_1^\dagger, b_2^\dagger$ and Bose annihilation operators $b_1, b_2$ with $b_1 = b \otimes I$, $b_2 = I \otimes b$ and $b_1|00\rangle = 0|00\rangle$, $b_2|00\rangle = 0|00\rangle$, where $|00\rangle \equiv |0\rangle \otimes |0\rangle$ is the vacuum state. Consider the linear transformation

$$\widetilde{b}_1 = u_{11}b_1 + u_{12}b_2 + v_{11}b_1^\dagger + v_{12}b_2^\dagger$$
$$\widetilde{b}_2 = u_{21}b_1 + u_{22}b_2 + v_{21}b_1^\dagger + v_{22}b_2^\dagger$$
$$\widetilde{b}_1^\dagger = v_{11}^*b_1 + v_{12}^*b_2 + u_{11}^*b_1^\dagger + u_{12}^*b_2^\dagger$$
$$\widetilde{b}_2^\dagger = v_{21}^*b_1 + v_{22}^*b_2 + u_{21}^*b_1^\dagger + u_{22}^*b_2^\dagger$$

where $u_{jk}, v_{jk} \in \mathbb{C}$.
(i) Find the condition that the operators $\widetilde{b}_1$, $\widetilde{b}_2$, $\widetilde{b}_1^\dagger$, $\widetilde{b}_2^\dagger$ also satisfy the commutation relation for Bose operators.
(ii) For the vacuum state of the Bose fields $\widetilde{b}_1$, $\widetilde{b}_2$ we can write

$$|\widetilde{0}\rangle \equiv |\widetilde{00}\rangle \equiv |\widetilde{0}\rangle \otimes |\widetilde{0}\rangle = \sum_{m=0}^\infty \sum_{n=0}^\infty c_{mn}|m\rangle \otimes |n\rangle.$$

Find the recurrence relation for $c_{mn}$ from the condition

$$\widetilde{b}_1|\widetilde{0}\rangle \otimes |\widetilde{0}\rangle = 0, \qquad \widetilde{b}_2|\widetilde{0}\rangle \otimes |\widetilde{0}\rangle = 0. \tag{1}$$

**Solution 28.**    (i) From the conditions

$$[\tilde{b}_1, \tilde{b}_2] = 0, \qquad [\tilde{b}_1, \tilde{b}_2^\dagger] = 0, \qquad [\tilde{b}_1, \tilde{b}_1^\dagger] = I, \qquad [\tilde{b}_2, \tilde{b}_2^\dagger] = I$$

we find

$$u_{11}v_{21} + u_{12}v_{22} - u_{21}v_{11} - u_{22}v_{12} = 0 \tag{2a}$$

$$u_{11}u_{21}^* + u_{12}u_{22}^* - v_{11}v_{21}^* - v_{12}v_{22}^* = 0 \tag{2b}$$

$$u_{11}u_{11}^* + u_{12}u_{12}^* - v_{11}v_{11}^* - v_{12}v_{12}^* = 1 \tag{2c}$$

$$u_{21}u_{21}^* + u_{22}u_{22}^* - v_{21}v_{21}^* - v_{22}v_{22}^* = 1. \tag{2d}$$

(ii) From the conditions (1) we find

$$c_{(m+1)n}u_{11}\sqrt{m+1} + c_{m(n+1)}u_{12}\sqrt{n+1}$$

$$+c_{(m-1)n}v_{11}\sqrt{m} + c_{m(n-1)}v_{12}\sqrt{n} = 0 \tag{3a}$$

and

$$c_{(m+1)n}u_{21}\sqrt{m+1} + c_{m(n+1)}u_{22}\sqrt{n+1}$$

$$+c_{(m-1)n}v_{21}\sqrt{m} + c_{m(n-1)}v_{22}\sqrt{n} = 0. \tag{3b}$$

Let

$$\Delta_1 := u_{11}u_{22} - u_{12}u_{21}, \quad \Delta_2 := u_{11}v_{21} - u_{21}v_{11}, \quad \Delta_3 := u_{11}v_{22} - u_{21}v_{12}$$

$$\Delta_4 := u_{22}v_{11} - u_{12}v_{21}, \qquad \Delta_5 := u_{22}v_{12} - u_{12}v_{22}.$$

Multiplication of (3a) with $u_{21}$ and (3b) with $u_{11}$ and subtracting yields

$$c_{m(n+1)}\Delta_1\sqrt{n+1} = -c_{(m-1)n}\Delta_2\sqrt{m} - c_{m(n-1)}\Delta_3\sqrt{n}.$$

Multiplication of (3a) with $u_{22}$ and (3b) with $u_{12}$ and subtracting yields

$$c_{(m+1)n}\Delta_1\sqrt{m+1} = -c_{(m-1)n}\Delta_4\sqrt{m} - c_{m(n-1)}\Delta_5\sqrt{n}.$$

We assumed that $\Delta_1 \neq 0$. From (2a) we see that $\Delta_2 = \Delta_5$. Thus we have

$$c_{(2k)(2n+1)} = c_{(2k+1)(2n)} = 0$$

$$c_{(2k)(2n)} = (-1)^{n+k}\sqrt{(2n)!}\sqrt{(2k)!}$$

$$\times \sum_{\substack{0 \le s \le n \\ s \le k}} \left(\frac{\Delta_2}{\Delta_1}\right)^{2s} \left(\frac{\Delta_3}{2\Delta_1}\right)^{n-s} \left(\frac{\Delta_4}{2\Delta_1}\right)^{k-s} \frac{1}{(n-s)!(k-s)!(2s)!}c_0$$

and

$$c_{(2k+1)(2n+1)} = (-1)^{n+k+1}\sqrt{(2n+1)!}\sqrt{(2k+1)!}$$

$$\times \sum_{\substack{0 \le s \le n \\ s \le k}} \left(\frac{\Delta_2}{\Delta_1}\right)^{2s+1} \left(\frac{\Delta_3}{2\Delta_1}\right)^{n-s} \left(\frac{\Delta_4}{2\Delta_1}\right)^{k-s} \frac{1}{(n-s)!(k-s)!(2s+1)!} c_0.$$

Consequently, for the vacuum state of Bose operators $\widetilde{b}_1$ and $\widetilde{b}_2$ we find

$$|\widetilde{00}\rangle = \sum_{k=0,n=0}^{\infty} \left( c_{(2k)(2n)}|2k\rangle \otimes |2n\rangle + c_{(2k+1)(2n+1)}|2k+1\rangle \otimes |2n+1\rangle \right).$$

In operator form this can be written as

$$|\widetilde{00}\rangle = c_0 \exp\left( -\frac{\Delta_4}{2\Delta_1}(b_1^\dagger)^2 - \frac{\Delta_2}{\Delta_1}b_1^\dagger b_2^\dagger - \frac{\Delta_3}{2\Delta_1}(b_2^\dagger)^2 \right) |0\rangle \otimes |0\rangle.$$

Thus the unitary operator

$$U = \exp\left( -\frac{\Delta_4}{2\Delta_1}(b_1^\dagger)^2 - \frac{\Delta_2}{\Delta_1}b_1^\dagger b_2^\dagger - \frac{\Delta_3}{2\Delta_1}(b_2^\dagger)^2 \right)$$

is the operator of transformation of the vacuum states for the most general two-dimensional Bogolubov transformation. Thus we also have

$$|\widetilde{m}\rangle \otimes |\widetilde{n}\rangle = U(|m\rangle \otimes |n\rangle).$$

**Problem 29.** Quantum mechanically, a *phase shift* $\delta$ induced by a linear optical element on a single-mode optical field is described by the unitary operator

$$U := \exp(i\delta\hat{n})$$

where $\hat{n} := b^\dagger b$ is the *number operator* and $b$ the annihilation operator for the optical mode. Assume the optical field is in the state $|\psi\rangle$.
(i) Express $|\psi\rangle$ in the basis of photon number state representation.
(ii) Find the state $|\psi'\rangle := U|\psi\rangle$.
(iii) Find $|\Delta\psi\rangle := |\psi'\rangle - |\psi\rangle$ and the norm $\| |\Delta\psi\rangle \|$.

**Solution 29.** (i) We can write

$$|\psi\rangle = \sum_{m=0}^{\infty} c_m|m\rangle$$

in the basis of photon number state representation, where $c_m$ are the expansion coefficients.
(ii) The phase-shifted state $|\psi'\rangle$ can be written as

$$|\psi'\rangle = \exp(i\delta\hat{n}) \sum_{m=0}^{\infty} c_m|m\rangle = \sum_{m=0}^{\infty} c_m e^{i\delta m}|m\rangle.$$

406 Problems and Solutions

(iii) Thus for the difference we find

$$|\Delta\psi\rangle = |\psi'\rangle - |\psi\rangle = \sum_{m=0}^{\infty} c_m(e^{i\delta m} - 1)|m\rangle$$

and therefore

$$\| |\Delta\psi\rangle \|^2 = \langle\Delta\psi|\Delta\psi\rangle = 4\sum_{m=0}^{\infty} |c_m|^2 \sin^2(\delta m/2) = 4\sum_{m=0}^{\infty} P_m \sin^2(\delta m/2)$$

where $P_m = |c_m|^2$ is the photon number distribution for the input field.

**Problem 30.** The generator of displacements for numbers is formally defined by

$$D(k) := \int_{-\pi}^{\pi} d\phi\, e^{ik\phi}|\phi\rangle\langle\phi|$$

where

$$|\phi\rangle = \sum_{n=0}^{\infty} e^{in\phi}|n\rangle, \qquad \phi \in \mathbb{R}.$$

Show that these basis states are not normalized.

**Solution 30.** Since

$$\langle\phi| = \sum_{m=0}^{\infty} \langle m|e^{-im\phi}$$

and $\langle m|n\rangle = \delta_{mn}$ we find

$$\langle\phi|\phi\rangle = \sum_{m=0}^{\infty} 1.$$

**Problem 31.** Let $b_1$, $b_2$ be Bose annihilation operators. Show that

$$e^{\mu b_1 b_2} e^{\nu b_1^\dagger b_2^\dagger}|00\rangle = \frac{1}{1-\mu\nu} e^{\nu b_1^\dagger b_2^\dagger/(1-\mu\nu)}|00\rangle, \qquad \mu, \nu \in \mathbb{R} \tag{1}$$

where $|00\rangle \equiv |0\rangle \otimes |0\rangle$.

**Solution 31.** We solve the problem by considering the expression

$$e^{\mu b_1 b_2} e^{\nu b_1^\dagger b_2^\dagger}|00\rangle = e^{f(\mu, b_1^\dagger, b_2^\dagger)}|00\rangle$$

where $f$ is an analytic function. Differentiating both sides with respect to $\mu$ yields

$$b_1 b_2 e^{\mu b_1 b_2} e^{\nu b_1^\dagger b_2^\dagger}|00\rangle = e^f \frac{\partial f}{\partial \mu}|00\rangle.$$

Thus
$$b_1 b_2 e^f |00\rangle = e^f \frac{\partial f}{\partial \mu} |00\rangle.$$

Note that $\partial f/\partial \mu$ commutes with $\exp(f)$ since $f$ is a function of $b_1^\dagger$ and $b_2^\dagger$ only. If we multiply from the left by $\exp(-f)$ we find
$$e^{-f} b_1 b_2 e^f |00\rangle = \frac{\partial f}{\partial \mu} |00\rangle.$$

If follows that
$$e^{-f} b_1 e^f e^{-f} b_2 e^f |00\rangle = \frac{\partial f}{\partial \mu} |00\rangle.$$

Using
$$[b, g(b, b^\dagger)] = \frac{\partial g}{\partial b^\dagger} \tag{2}$$

with $g = e^f$, we obtain
$$e^{-f} b_1 e^f = e^{-f} \left( e^f b_1 + \frac{\partial e^f}{\partial b_1^\dagger} \right) = b_1 + \frac{\partial f}{\partial b_1^\dagger}$$

since $e^f$ commutes with $\partial f/\partial b_1^\dagger$. Similarly
$$e^{-f} b_2 e^f = b_2 + \frac{\partial f}{\partial b_2^\dagger}.$$

Thus we have
$$\left( b_1 + \frac{\partial f}{\partial b_1^\dagger} \right) \left( b_2 + \frac{\partial f}{\partial b_2^\dagger} \right) |00\rangle = \frac{\partial f}{\partial \mu} |00\rangle.$$

Since $b_2 |00\rangle = 0|00\rangle$ we arrive at
$$\left( b_1 \frac{\partial f}{\partial b_2^\dagger} + \frac{\partial f}{\partial b_1^\dagger} \frac{\partial f}{\partial b_2^\dagger} \right) |00\rangle = \frac{\partial f}{\partial \mu} |00\rangle.$$

Using (2) again with $g = \partial f/\partial b_2^\dagger$ we obtain
$$b_1 \frac{\partial f}{\partial b_2^\dagger} = \frac{\partial f}{\partial b_2^\dagger} b_1 + \frac{\partial^2 f}{\partial b_1^\dagger \partial b_2^\dagger}.$$

Since $b_1 |00\rangle = 0|00\rangle$ we obtain
$$\left( \frac{\partial^2 f}{\partial b_1^\dagger \partial b_2^\dagger} + \frac{\partial f}{\partial b_1^\dagger} \frac{\partial f}{\partial b_2^\dagger} \right) |00\rangle = \frac{\partial f}{\partial \mu} |00\rangle.$$

Since $f$ contains only $b_1^\dagger$ and $b_2^\dagger$ which commute, the solution of this partial differential equation must be of the form

$$f(\mu, b_1^\dagger, b_2^\dagger) = h_1(\mu)I + h_2(\mu)b_1^\dagger b_2^\dagger.$$

Thus $f(0, b_1^\dagger, b_2^\dagger) = \nu b_1^\dagger b_2^\dagger$ or $h_1(0) = 0$, $h_2(0) = \nu$ owing to (1). Inserting this ansatz into the partial differential equation and equating equal powers of $b_1^\dagger b_2^\dagger$, we find that $h_1$ and $h_2$ satisfy the nonlinear system of ordinary differential equations

$$\frac{dh_2}{d\mu} = h_2^2, \qquad \frac{dh_1}{d\mu} = h_2$$

with the solution of the initial value problem

$$h_2(\mu) = \frac{\nu}{1 - \mu\nu}, \qquad h_1(\mu) = -\ln(1 - \mu\nu)$$

and thus we find (1).

**Problem 32.**   The standard *Pauli group* for continuous variable quantum computing of $n$ coupled oscillator systems is the *Heisenberg-Weyl group* which consists of phase-space displacement operators for $n$ harmonic oscillators. This group is a continuous Lie group and can therefore only be generated by a set of continuously parameterized operators. The *Lie algebra* that generates this group is spanned by the $2n$ canonical operators $\hat{p}_j$, $\hat{q}_j$, $j = 1, 2, \ldots, n$ along with the commutation relation

$$[\hat{q}_j, \hat{p}_k] = i\hbar\delta_{jk}I.$$

For a single oscillator ($n = 1$) the algebra is spanned by the canonical operators $\{\hat{q}, \hat{p}, I\}$. We define the operators

$$X(q) := e^{-(i/\hbar)q\hat{p}}, \qquad Z(p) := e^{(i/\hbar)p\hat{q}}$$

where $q, p \in \mathbb{R}$. Let $\{|s\rangle : s \in \mathbb{R}\}$ be *position eigenstates* (in the sense of generalized functions).
(i) Calculate the states $X(q)|s\rangle$, $Z(p)|s\rangle$.
(ii) Find the commutator $[X(q), Z(p)]$.

**Solution 32.**   (i) We find in the sense of generalized functions that

$$X(q)|s\rangle = |s + q\rangle, \qquad Z(p)|s\rangle = \exp((i/\hbar)ps)|s\rangle.$$

Thus the operator $X(q)$ is a position translation operator. The operator $Z(p)$ is a *momentum boost operator*.

(ii) We obtain $X(q)Z(p) = e^{-(i/\hbar)qp}Z(p)X(q)$ and

$$[X(q), Z(p)] = (I - e^{(i/\hbar)qp})X(q)Z(p).$$

**Problem 33.** Let $r \in \mathbb{R}$. Find $\epsilon_1$, $\epsilon_2$ and $\epsilon_3$ such that

$$e^{r(b_1^\dagger b_2^\dagger - b_1 b_2)} \equiv e^{\epsilon_1 b_1^\dagger b_2^\dagger} e^{\epsilon_2 (b_1^\dagger b_1 + b_2^\dagger b_2 + I)} e^{\epsilon_3 b_1 b_2}.$$

**Solution 33.** Using the fact that the operators

$$K_+ := -b_1 b_2, \qquad K_- := b_1^\dagger b_2^\dagger, \qquad K_3 := -\frac{1}{2}(b_1^\dagger b_1 + b_2^\dagger b_2 + I)$$

form the semi-simple Lie algebra

$$[K_3, K_+] = K_+, \qquad [K_3, K_-] = -K_-, \qquad [K_+, K_-] = 2K_3$$

and

$$e^{r(K_+ + K_-)} \equiv e^{K_- \tanh(r)} e^{2\ln(\cosh(r))K_3} e^{K_+ \tanh(r)}$$

we find $\epsilon_1 = \tanh(r)$, $\epsilon_2 = -\ln(\cosh(r))$, $\epsilon_3 = -\tanh(r)$.

**Problem 34.** To build a simple quantum computer one could use the following optical gates

$$U_S := \exp(i\pi b^\dagger b) \qquad\qquad \textit{phase modulator}$$
$$U_B := \exp\left(\frac{\pi}{4}(b_1^\dagger b_2 - b_1 b_2^\dagger)\right) \qquad \textit{quantum beam splitter}$$
$$U_F := \exp\left(\frac{\chi}{2}b_3^\dagger b_3(b_1^\dagger b_2 - b_1 b_2^\dagger)\right) \qquad \textit{Fredkin gate}$$

(i) Calculate the state $U_S|n\rangle$.
(ii) Calculate the state $U_B|01\rangle$.
(iii) Calculate the states $U_F|011\rangle$, $U_F|101\rangle$, $U_F|xy0\rangle$ with $\chi = \pi$ and $x, y \in \{0, 1\}$.

**Solution 34.** (i) Since $b^\dagger b|n\rangle = n|n\rangle$ we obtain

$$e^{i\pi b^\dagger b}|n\rangle = e^{i\pi n}|n\rangle = (-1)^n|n\rangle.$$

(ii) Since

$$(b_1^\dagger b_2 - b_1 b_2^\dagger)|01\rangle = |10\rangle, \qquad (b_1^\dagger b_2 - b_1 b_2^\dagger)|10\rangle = -|01\rangle$$

we find

$$U_B|01\rangle = \frac{1}{\sqrt{2}}(|01\rangle + |10\rangle).$$

where we used $\sin(\pi/4) = 1/\sqrt{2}$ and $\cos(\pi/4) = 1/\sqrt{2}$.

(iii) Since

$$b_3^\dagger b_3(b_1^\dagger b_2 - b_1 b_2^\dagger)|011\rangle = |101\rangle, \qquad b_3^\dagger b_3(b_1^\dagger b_2 - b_1 b_2^\dagger)|101\rangle = -|011\rangle$$

and $b_3^\dagger b_3(b_1^\dagger b_2 - b_1 b_2^\dagger)|xy0\rangle = 0|xy0\rangle$ we find the states

$$U_F|101\rangle = -|011\rangle, \qquad U_F|011\rangle = |101\rangle, \qquad U_F|xy0\rangle = |xy0\rangle$$

where we used that $b|0\rangle = 0|0\rangle$ and $b|1\rangle = |0\rangle$. Thus $b_3^\dagger b_3$ plays the role of a control operator.

**Problem 35.**  Let $\epsilon$ be a real parameter, $\sigma_3$ the Pauli spin matrix and $I_2$ the $2 \times 2$ identity matrix. Calculate

$$f(\epsilon) = e^{\epsilon \sigma_3 \otimes (b - b^\dagger)}(I_2 \otimes b^\dagger b)e^{-\epsilon \sigma_3 \otimes (b - b^\dagger)} \tag{1}$$

using *parameter differentiation* and then solving the differential equation with the corresponding initial values (operators), i.e. $f(0) = I_2 \otimes b^\dagger b$.

**Solution 35.**  From (1) we obtain $f(0) = I_2 \otimes b^\dagger b$. Now

$$\frac{df}{d\epsilon} = e^{\epsilon \sigma_3 \otimes (b-b^\dagger)}(\sigma_3 \otimes ((b - b^\dagger)b^\dagger b - b^\dagger b(b - b^\dagger)))e^{-\epsilon \sigma_3 \otimes (b-b^\dagger)}$$

$$= e^{\epsilon \sigma_3 \otimes (b-b^\dagger)}(\sigma_3 \otimes (b + b^\dagger))e^{-\epsilon \sigma_3 \otimes (b-b^\dagger)}.$$

Thus

$$\frac{df(0)}{d\epsilon} = \sigma_3 \otimes (b + b^\dagger).$$

Since $(b - b^\dagger)(b + b^\dagger) - (b + b^\dagger)(b - b^\dagger) = 2I$ and $\sigma_3^2 = I_2$ we obtain for the second order derivative

$$\frac{d^2 f}{d\epsilon^2} = 2I_2 \otimes I.$$

Thus the solution of this second order linear differential equation is

$$f(\epsilon) = \epsilon^2 I_2 \otimes I + C_1 \epsilon + C_2.$$

Inserting the two initial values (operators) yields

$$C_1 = \sigma_3 \otimes (b + b^\dagger), \qquad C_2 = I_2 \otimes b^\dagger b.$$

It follows that

$$f(\epsilon) = \epsilon^2 I_2 \otimes I + \epsilon \sigma_3 \otimes (b + b^\dagger) + I_2 \otimes b^\dagger b.$$

**Problem 36.**   Let $b$, $b^\dagger$ be Bose operators. Find the eigenvalues of the operator $(-1)^{b^\dagger b}$. This unitary operator is defined by $e^{i\pi b^\dagger b}$. Why is this operator called the *parity operator*?

**Solution 36.**   Let $|n\rangle$ be a number state. Then $b^\dagger b|n\rangle = n|n\rangle$, $b^\dagger b b^\dagger b|n\rangle = n^2|n\rangle$ etc. Thus

$$e^{i\pi b^\dagger b}|n\rangle = e^{i\pi n}|n\rangle$$

where $n = 0, 1, 2, \ldots, \infty$. It follows that

$$\langle m|e^{i\pi n}|n\rangle = e^{i\pi n}\langle m|n\rangle = e^{i\pi n}\delta_{mn}.$$

Since $e^{i\pi n} = +1$ if $n$ is even and $e^{i\pi n} = -1$ if $n$ is odd we find that the eigenvalues of $(-1)^{b^\dagger b}$ are $+1$ and $-1$ both infinitely degenerate.

**Problem 37.**   The hermitian cosine $\hat{C}_{SG}$ and sine $\hat{S}_{SG}$ operators introduced by Susskind and Glogower are given in the number state basis by

$$\hat{C}_{SG} := \frac{1}{2}\sum_{n=0}^{\infty}(|n\rangle\langle n+1| + |n+1\rangle\langle n|)$$

$$\hat{S}_{SG} := \frac{1}{2i}\sum_{n=0}^{\infty}(|n\rangle\langle n+1| - |n+1\rangle\langle n|).$$

Solve the eigenvalue equations $\hat{C}_{SG}|c\rangle_{SG} = c|c\rangle_{SG}$, $\hat{S}_{SG}|s\rangle_{SG} = s|s\rangle_{SG}$.

**Solution 37.**   We find

$$|c\rangle_{SG} = \sqrt{\frac{2}{\pi}}\sqrt[4]{1-c^2}\sum_{n=0}^{\infty}U_n(c)|n\rangle, \quad |s\rangle_{SG} = \sqrt{\frac{2}{\pi}}\sqrt[4]{1-s^2}\sum_{n=0}^{\infty}i^n U_n(s)|n\rangle$$

where $U_n(x)$ for $x = c$ or $x = s$ are the *Chebyshev polynomials* of second kind and $x \in [-1, 1]$. They obey the recursion formula

$$U_{n+1}(x) - 2xU_n(x) + U_{n-1}(x) = 0$$

where $U_0(x) = 1$ and $U_1(x) = 2x$.

**Problem 38.**   Let $|0\rangle$ be the vacuum state, i.e. $b|0\rangle = 0|0\rangle$, and $\epsilon \in \mathbb{R}$. Calculate the state

$$\exp(\epsilon b^\dagger \otimes b^\dagger)(|0\rangle \otimes |0\rangle).$$

Sometimes one also writes $\exp(\epsilon b_1^\dagger b_2^\dagger)|0\rangle|0\rangle$.

**Solution 38.**   Since $b^\dagger|n\rangle = \sqrt{n+1}|n+1\rangle$ for $n = 0,1,2,\ldots$ we obtain

$$(b^\dagger \otimes b^\dagger)(|n\rangle \otimes |n\rangle) = (b^\dagger|n\rangle) \otimes (b^\dagger|n\rangle) = (n+1)|n+1\rangle \otimes |n+1\rangle.$$

Now

$$\exp(\epsilon b^\dagger \otimes b^\dagger)|0\rangle \otimes |0\rangle = (I \otimes I + \frac{\epsilon}{1!}b^\dagger \otimes b^\dagger + \frac{\epsilon^2}{2!}(b^\dagger)^2 \otimes (b^\dagger)^2 + \cdots)|0\rangle \otimes |0\rangle$$

$$= |0\rangle \otimes |0\rangle + \epsilon|1\rangle \otimes |1\rangle + \epsilon^2|2\rangle \otimes |2\rangle + \cdots$$

$$= \sum_{j=0}^{\infty} \epsilon^j |j\rangle \otimes |j\rangle.$$

What is the condition on $\epsilon$ so that the series converges? This means: what is the condition that the state can be normalized?

**Problem 39.**   Let $b, b^\dagger$ be Bose annihilation and creation operators, respectively. Let $\epsilon \in \mathbb{R}$ and $|0\rangle$ be the vacuum state, i.e. $b|0\rangle = 0|0\rangle$.
(i) Calculate the state $b \exp(\epsilon b^\dagger)|0\rangle$.
(ii) Calculate the state $b^n \exp(\epsilon b^\dagger)|0\rangle$, $(n = 2,3,\ldots)$.

**Solution 39.**   (i) Since

$$bb^\dagger|0\rangle = |0\rangle, \qquad b(b^\dagger)^2|0\rangle = 2b^\dagger|0\rangle, \qquad b(b^\dagger)^3|0\rangle = 3(b^\dagger)^2|0\rangle$$

and in general $b(b^\dagger)^n|0\rangle = n(b^\dagger)^{n-1}|0\rangle$ we obtain

$$b\exp(\epsilon b^\dagger)|0\rangle = \epsilon|0\rangle + \frac{\epsilon^2}{2!}2b^\dagger|0\rangle + \frac{\epsilon^3}{3!}3(b^\dagger)^2|0\rangle + \cdots$$

$$= \epsilon(I + \epsilon b^\dagger + \frac{\epsilon^2}{2!}(b^\dagger)^2 + \cdots)|0\rangle$$

$$= \epsilon \exp(\epsilon b^\dagger)|0\rangle.$$

(ii) Using the result from (i) we have

$$b^n \exp(\epsilon b^\dagger)|0\rangle = \epsilon^n \exp(\epsilon b^\dagger)|0\rangle.$$

**Problem 40.**   Let $|0\rangle$ be the vacuum state. Calculate the state

$$\frac{1}{2}(b^\dagger \otimes I - I \otimes b^\dagger)(b^\dagger \otimes I + I \otimes b^\dagger)(|0\rangle \otimes |0\rangle)$$

where $I$ is the identity operator. Discuss.

**Solution 40.**   Since $(b^\dagger \otimes I)(I \otimes b^\dagger) = (I \otimes b^\dagger)(b^\dagger \otimes I)$ we obtain the state

$$\frac{1}{\sqrt{2}}(|2\rangle \otimes |0\rangle - |0\rangle \otimes |2\rangle).$$

The photons interfere constructively or destructively. Complete destructive interference implies that the affected outgoing mode is in the vacuum state.

**Problem 41.**   Consider the operators

$$J_x = \frac{1}{2}(b_1^\dagger b_2 + b_2^\dagger b_1), \qquad J_y = \frac{1}{2}i(b_1^\dagger b_2 - b_2^\dagger b_1), \qquad J_z = \frac{1}{2}(b_1^\dagger b_1 - b_2^\dagger b_2)$$

and $J_+ := J_x + iJ_y = b_1^\dagger b_2$, $J_- := J_x - iJ_y = b_2^\dagger b_1$. Let $\phi \in \mathbb{R}$. Find

$$e^{-iJ_z\phi}b_1^\dagger e^{iJ_z\phi}, \quad e^{-iJ_z\phi}b_2^\dagger e^{iJ_z\phi}, \quad e^{-iJ_y\phi}b_1^\dagger e^{iJ_y\phi}, \quad e^{-iJ_y\phi}b_2^\dagger e^{iJ_y\phi}.$$

**Solution 41.**   We obtain

$$\exp(-iJ_z\phi)b_1^\dagger \exp(iJ_z\phi) = b_1^\dagger e^{-i\phi/2}$$
$$\exp(-iJ_z\phi)b_2^\dagger \exp(iJ_z\phi) = b_2^\dagger e^{-i\phi/2}$$
$$\exp(-iJ_y\phi)b_1^\dagger \exp(iJ_y\phi) = b_1^\dagger \cos(\phi/2) + b_2^\dagger \sin(\phi/2)$$
$$\exp(-iJ_y\phi)b_2^\dagger \exp(iJ_y\phi) = b_2^\dagger \cos(\phi/2) - b_1^\dagger \sin(\phi/2).$$

**Problem 42.**   Let $\{\,|n\rangle : n = 0, 1, 2, \dots\,\}$ be number states. Consider the linear operator

$$T_{13} := \sum_{n=0}^{\infty}(|n\rangle \otimes I \otimes I)(I \otimes I \otimes \langle n|)$$

in the product (infinite-dimensional) Hilbert space $\mathcal{H} = \mathcal{H}_1 \otimes \mathcal{H}_2 \otimes \mathcal{H}_3$ with $\mathcal{H}_1 = \mathcal{H}_2 = \mathcal{H}_3$. Here $I$ denotes the identity operator. Apply the operator $T_{13}$ to the state $I \otimes I \otimes |\psi\rangle$.

**Solution 42.**   We find

$$T_{13}(I \otimes I \otimes |\psi\rangle) = \sum_{n=0}^{\infty}(|n\rangle \otimes I \otimes I)(I \otimes I \otimes \langle n|\psi\rangle I)$$
$$= \sum_{n=0}^{\infty}((\langle n|\psi\rangle|n\rangle)) \otimes I \otimes I$$
$$= |\psi\rangle \otimes I \otimes I$$

where we used

$$|\psi\rangle = \sum_{n=0}^{\infty}\langle n|\psi\rangle|n\rangle.$$

The operator $T_{13}$ can be considered as a *transfer operator*.

**Problem 43.** A *beam splitter* can be realized by means of a linear medium where the *polarization vector* is proportional to the incoming *electric field*

$$\hat{P} = \chi \hat{E}$$

with $\chi \equiv \chi^{(1)}$ denoting the first order (linear) susceptibility. We consider the incoming field excited only in the relevant spatial modes $b_1$ and $b_2$ (at the same frequency $\omega$)

$$\hat{E}(r,t) = i\sqrt{\frac{\hbar\omega}{2\epsilon_0 V}} \left( (b_1 + b_2)e^{i(\mathbf{k}\cdot\mathbf{r} - \omega t)} + h.c. \right)$$

where *h.c.* denotes the hermitian conjugate. The interaction Hamilton operator contains only the resonant terms

$$\hat{H}_I = -\hat{P}\cdot\hat{E} = -\chi\hat{E}^2 = \frac{\chi\hbar\omega}{2\epsilon_0 V}(b_1^\dagger b_2 + b_1 b_2^\dagger)$$

where $\cdot$ denotes the scalar product. The evolution operator (in the interaction picture) of the whole device is expressed as

$$U := \exp\left( i \arctan\left(\sqrt{\frac{1-\tau}{\tau}}\right) \left(b_1^\dagger b_2 + b_1 b_2^\dagger\right)\right)$$

where $\tau$, given by

$$\tau = \left(1 + \tan^2\left(\frac{\chi\hbar\omega}{2\epsilon_0 V}\right)\right)^{-1}$$

represents the *transmissivity* of the beam splitter.
(i) Calculate $\tilde{b}_1 = U^\dagger b_1 U$, $\tilde{b}_2 = U^\dagger b_2 U$.
(ii) Find a rotation of the phase frame by $3\pi/2$.

**Solution 43.** (i) Straightforward calculation yields

$$\tilde{b}_1 = U^\dagger b_1 U = -i\tau^{1/2}b_1 + (1-\tau)^{1/2}b_2$$

$$\tilde{b}_2 = U^\dagger b_2 U = i(1-\tau)^{1/2}b_1 + \tau^{1/2}b_2.$$

(ii) A rotation of the phase frame can be obtained by the substitution $b_1 \to -ib_1$. Then we obtain

$$\tilde{b}_1 = \tau^{1/2}b_1 + (1-\tau)^{1/2}b_2, \qquad \tilde{b}_2 = \tau^{1/2}b_2 - (1-\tau)^{1/2}b_1.$$

**Problem 44.** Owing to their helical wave fronts the electromagnetic field of photons having an orbital angular momentum has a phase singularity.

There the intensity has to vanish resulting in a doughnut-like intensity distribution. These light fields can be described using Laguerre Gaussian ($LG_{pl}$) modes with two indices $p$ and $l$. The $p$-index ($p = 0, 1, 2, \ldots$) identifies the non-axial radial nodes observed in the transversal plane and the $l$-index ($l = 0, \pm1, \pm2, \ldots$) the number of the $2\pi$-phase shifts along a closed path around the beam center. The index $l$ is also called the topological winding number. It describes the helical structure of the wave front around a wave front singularity or dislocation. The index $l$ also determines the amount of orbital angular momentum in units of $\hbar$ carried by one photon. When the pump beam is a $LG_{l_0 p_0}$ mode, under conditions of collinear phase-matching, the two-photon state at the output of the nonlinear crystal can be written as a coherent superposition of eigenstates of the orbital angular-momentum operator that are correlated in orbital angular momentum, i.e., $l_1 + l_2 = l_0$, where $l_1$ and $l_2$ refer to the orbital angular momentum eigenvalues for the signal and idler photons. A photon state described by a $LG$ mode can be written as

$$|lp\rangle := \int dq \, LG_{lp}(q) b^\dagger(q) |0\rangle$$

where the mode function in the spatial frequency domain is given by

$$LG_{lp}(\rho, \phi) := \left( \frac{\omega_0 p!}{2\pi(|l| + p)!} \right)^{1/2} \left( \frac{\omega_0 \rho_k}{\sqrt{2}} \right)^{|l|} L_p^{|l|} \left( \frac{\rho_k^2 \omega_0^2}{2} \right) \exp\left( -\frac{\rho_k^2 \omega_0^2}{4} \right)$$
$$\times \exp\left( il\phi_k + i \left( p - \frac{|l|}{2} \right) \pi \right)$$

with $\rho_k$ and $\phi_k$ being the modulus and phase, respectively, of the transverse coordinate $q$. The functions $L_p^{|l|}$ are the *associated Laguerre polynomials* and $\omega_0$ is the beam width. Find the state $|lp\rangle$ for $l = p = 0$.

**Solution 44.** Since the associated Laguerre polynomial $L_0^0$ is given by $L_0^0(x) = 1$ we obtain

$$LG_{00} = \left( \frac{\omega_0}{2\pi} \right)^{1/2} \exp\left( -\frac{\rho_k^2 \omega_0^2}{4} \right).$$

Thus for $LG_{00}$ we find a Gaussian.

**Programming Problem**

**Problem 1.** Find the normal ordering of $bb^\dagger bb^\dagger$ and apply it then to the vacuum state. Find the normal ordering of $bb^\dagger b^\dagger$ and apply it then to the vacuum state. Find $(b + b^\dagger)^4$. Find the operator $b + bb^\dagger + bb^\dagger b^\dagger + bb^\dagger b^\dagger b^\dagger$ and normal ordering. Apply computer algebra.

**Solution 1.**   In the SymbolicC++ program the operator $b$ is denoted by b, the operator $b^\dagger$ by bd and the vacuum state by vs. The rules b*bd==bd*b+1 and b*vs=0 are implemented.

```
// bose2.cpp
#include <iostream>
#include "symbolicc++.h"
using namespace std;

int main(void)
{
Symbolic b("b"), bd("bd"), vs("vs");
b = ~b; bd = ~bd; vs = ~vs;  // noncommutative
Equations rules = (b*bd==bd*b+1,b*vs==0);
// example 1
Symbolic res1 = b*bd*b*bd;
cout << "res1 = " << res1.subst_all(rules) << endl;
cout << "res1*vs = " << (res1*vs).subst_all(rules) << endl;
// example 2
Symbolic res2 = b*bd*bd;
cout << "res2 = " << res2.subst_all(rules) << endl;
cout << "res2*vs = " << (res2*vs).subst_all(rules) << endl;
// example 3
Symbolic res3 = (b+bd)^4;
cout << "res3 = " << res3.subst_all(rules) << endl;
cout << "res3*vs = " << (res3*vs).subst_all(rules) << endl;
// example 4
Symbolic res4 = b + b*bd + b*bd*bd + b*bd*bd*bd;
cout << "res4 = " << res4.subst_all(rules) << endl;
cout << "res4*vs = " << (res4*vs).subst_all(rules) << endl;
return 0;
}
```

The output is

```
res1 = bd^(2)*b^(2)+3*bd*b+1
res1*vs = vs
res2 = bd^(2)*b+2*bd
res2*vs = 2*bd*vs
res3 = b^(4)+4*bd*b^(3)+6*b^(2)+6*bd^(2)*b^(2)
          +12*bd*b+4*bd^(3)*b+6*bd^(2)+bd^(4)+3
res3*vs = 3*vs+6*bd^(2)*vs+bd^(4)*vs
res4 = b+bd*b+bd^(2)*b+2*bd+bd^(3)*b+3*bd^(2)+1
res4*vs = vs+2*bd*vs+3*bd^(2)*vs
```

# 18.3 Supplementary Problems

**Problem 1.** (i) Let $b$, $b^\dagger$ be Bose creation and annihilation operators and

$$\hat{H} = \frac{1}{2}(b^\dagger b + bb^\dagger).$$

Show that $[b, \hat{H}] = b$.

(ii) Let $n$ be a positive integer. Show that

$$[b^\dagger b, (b^\dagger)^n] = n(b^\dagger)^n, \quad [b^\dagger b, b^n] = -nb^n.$$

(iii) Let $b^\dagger$, $b$ be Bose creation and annihilation operators and $z \in \mathbb{C}$. Show that

$$[b, e^{-zb^\dagger b}] = (e^z - 1)e^{-zb^\dagger b}b, \quad [b^\dagger, e^{-zb^\dagger b}] = (e^z - 1)e^{-zb^\dagger b}b^\dagger.$$

**Problem 2.** Let $f : \mathbb{C} \to \mathbb{C}$ be an analytic function.
(i) Show that $e^{zb}f(b^\dagger)|0\rangle = f(b^\dagger + zI)|0\rangle$.
(ii) Show that $e^{-zb^\dagger b}f(b^\dagger)|0\rangle = f(b^\dagger e^{-z})|0\rangle$.

**Problem 3.** Let $f, g : \mathbb{C} \to \mathbb{C}$ be analytic functions. Then

$$[f(b), f(b^\dagger)] = \sum_{j=1}^{\infty} \frac{1}{j!} \frac{d^j}{d(b^\dagger)^j} \frac{d^j}{db^j} f(b).$$

Let $\tau \in \mathbb{R}$. Use this relation to calculate the commutators

$$[b, \exp(\tau(b^\dagger)^2/2)], \quad [b^2, \exp(\tau(b^\dagger)^2/2)], \quad [b^3, \exp(\tau(b^\dagger)^2/2].$$

Then show that $b \exp(\tau(b^\dagger)^2/2)|0\rangle = \tau b^\dagger \exp(\tau(b^\dagger)^2/2)|0\rangle$.

**Problem 4.** Let $b^\dagger$, $b$ be Bose creation and annihilation operators. Show that

$$\exp(\theta b^\dagger b) = \sum_{j=0}^{\infty} \frac{(e^\theta - 1)^j}{j!} (b^\dagger)^j b^j.$$

**Problem 5.** Let $b_1^\dagger$, $b_2^\dagger$, $b_1$, $b_2$ be Bose creation and annihilation operators and let $N \geq 1$ be a positive integer. Consider the operators

$$\hat{T}_1 = \frac{\sqrt{2}}{N^{3/2}}(b_1^\dagger b_1^\dagger b_2 + b_2^\dagger b_1 b_1), \quad \hat{T}_2 = \frac{\sqrt{2}i}{N^{3/2}}(b_1^\dagger b_1^\dagger b_2 - b_2^\dagger b_1 b_1),$$

$$\hat{T}_3 = \frac{2}{N}(2b_2^\dagger b_2 - b_1^\dagger b_1).$$

Show that the commutators are given by

$$[\hat{T}_1, \hat{T}_2] = \frac{i}{N}(I - \hat{T}_3)(I + 3\hat{T}_3) + \frac{4i}{N^2}I,$$

$$[\hat{T}_3, \hat{T}_1] = \frac{4i}{N}\hat{T}_2, \quad [\hat{T}_3, \hat{T}_2] = -\frac{4i}{N}\hat{T}_1$$

where $I$ denotes the identity operator.

**Problem 6.** Let $b^\dagger$, $b$ be Bose creation and annihilation operators. Show that

$$\rho(b, b^\dagger) = (1 - e^{-\lambda})e^{-\lambda b^\dagger b}$$

is a density operator, where $\lambda \equiv \beta\hbar\omega$, $\beta = 1/(k_B T)$.

**Problem 7.** Let $|n\rangle$ be a number state and $|\beta\rangle$ be coherent state. Show that

$$|\langle n|\beta\rangle|^2 = \exp(-|\beta|^2)\frac{(|\beta|^2)^n}{n!}$$

which is a *Poisson distribution*.

**Problem 8.** Let $b^\dagger$, $b$ be Bose creation and annihilation operators. Consider the operators

$$\hat{T}_1 = (b^\dagger)^2, \quad \hat{T}_2 = -b^2, \quad \hat{T}_3 = 4\left(b^\dagger b + \frac{1}{2}I\right).$$

Show that they satisfy the commutation relations

$$[\hat{T}_1, \hat{T}_2] = T_3, \quad [\hat{T}_1, \hat{T}_3] = -8T_1, \quad [\hat{T}_2, \hat{T}_3] = 8T_2.$$

Show that

$$\exp(\epsilon((b^\dagger)^2 - b^2)) =$$

$$\exp(-\frac{1}{2}b^2 \tanh(2\epsilon)) \exp((b^\dagger b + \frac{1}{2}I) \ln(\cosh(2\epsilon)) \exp(\frac{1}{2}(b^\dagger)^2 \tanh(2\epsilon)).$$

**Problem 9.** The semi-simple Lie algebra $su(1,1)$ is given by the commutation relations $[k_1, k_2] = -ik_3$, $[k_3, k_1] = ik_2$, $[k_2, k_3] = -ik_1$, where $k_1$, $k_2$ and $k_3$ are the basis elements of the Lie algebra. Show that an infinite-dimensional matrix representation is given by

$$k_1 = \frac{1}{2}\begin{pmatrix} 0 & 1 & 0 & 0 & \cdots \\ 1 & 0 & 2 & 0 & \cdots \\ 0 & 2 & 0 & 3 & \cdots \\ 0 & 0 & 3 & 0 & \cdots \\ \vdots & \vdots & \vdots & \vdots & \ddots \end{pmatrix}, \quad k_2 = \frac{1}{2}\begin{pmatrix} 0 & i & 0 & 0 & \cdots \\ -i & 0 & 2i & 0 & \cdots \\ 0 & -2i & 0 & 3i & \cdots \\ 0 & 0 & -3i & 0 & \cdots \\ \vdots & \vdots & \vdots & \vdots & \ddots \end{pmatrix}$$

$$k_3 = \frac{1}{2}\text{diag}(1,3,5,7,\ldots).$$

**Problem 10.** Let $b^\dagger$, $b$ be the Bose creation and annihilation operators and $S_3$, $S_+$, $S_-$ be the spin-$\frac{1}{2}$ matrices. Consider the operators

$$\hat{N} = b^\dagger b \otimes I_2 + I_B \otimes S_3, \quad \hat{K} = (b^\dagger + b) \otimes (S_+ + S_-).$$

(i) Show that $[\hat{N}, b \otimes S_+ + b^\dagger \otimes S_-] = 0_B \otimes 0_2$.
(ii) Show that $[\hat{N}, \hat{K}] = 2(b^\dagger \otimes S_+ - b \otimes S_-)$.

**Problem 11.** Consider the operators

$$Q_+ = \frac{1}{2}(b_1^\dagger \otimes c + b_2 \otimes c^\dagger), \quad Q_- = \frac{1}{2}(b_1 \otimes c^\dagger - b_2^\dagger \otimes c).$$

Find $Q_+Q_-$, $Q_-Q_+$ and the anticommutator $[Q_+, Q_-]_+$.

**Problem 12.** Let

$$J_+ = b_1^\dagger b_2 \otimes I_F + I_B \otimes c^\dagger$$
$$J_- = b_2^\dagger b_1 \otimes I_F + I_B \otimes c$$
$$J_3 = (b_1^\dagger b_1 - b_2^\dagger b_2) \otimes I_F + I_B \otimes (2c^\dagger c - I_F).$$

Show that $[J_+, J_-] = J_3$, $[J_+, J_3] = -2J_+$, $[J_-, J_3] = 2J_-$.

**Problem 13.** Find the matrix representation of $b^\dagger \otimes c + b \otimes c^\dagger$ applying the basis $|n\rangle \otimes |0\rangle$, $|n\rangle \otimes c^\dagger|0\rangle$. Find the matrix representation using the basis

$$|\beta\rangle \otimes |0\rangle, \quad |\beta\rangle \otimes c^\dagger|0\rangle$$

where $|\beta\rangle$ are coherent states.

**Problem 14.** Let $c^\dagger$, $c$ be Fermi creation and annihilation operators.
(i) Show that $e^{i\pi c^\dagger c} = I_F - 2c^\dagger c$.
(ii) Show that $e^{i\pi b^\dagger b} = \text{diag}(1,-1,1,-1,\ldots)$.

**Problem 15.** Find the states

$$e^{i\pi(b^\dagger b \otimes c^\dagger c)}(|n\rangle \otimes |0\rangle), \quad e^{i\pi(b^\dagger b \otimes c^\dagger c)}(|n\rangle \otimes |0\rangle).$$

**Problem 16.** Let $b$ be a Bose operator and $I$ be the identity operator. Show that we can write

$$\delta(\beta I - b)\delta(\bar{\beta}I - b^\dagger) = \frac{1}{\pi^2}\int_\mathbb{R}\int_\mathbb{R} e^{-i\mu(\beta I - b)}e^{-i\bar{\mu}(\bar{\beta}-b^\dagger)}d(\Re(\mu))d(\Im(\mu)).$$

**Problem 17.**   Let $\sigma_1, \sigma_2, \sigma_3$ be the Pauli spin matrices and $b^\dagger$, $b$ be Bose creation and annihilation operators. Study the spectrum of the Hamilton operator

$$\hat{H} = \hbar\omega_1 I_B \otimes \sigma_3 + \hbar\omega_2 b^\dagger b \otimes I_2 + \hbar\omega_3 (b^\dagger + b) \otimes \sigma_1.$$

**Problem 18.**   Let $\hat{f}(k)$ be the *Fourier transform* of $f(x)$, where $f \in L_2(\mathbb{R}) \cap L_1(\mathbb{R})$. Show that if

$$E_x := \frac{\int_{\mathbb{R}} x^2 |f(x)|^2 dx}{\int_{\mathbb{R}} |f(x)|^2 dx} \quad \text{and} \quad E_k := \frac{\int_{\mathbb{R}} k^2 |\hat{f}(k)|^2 dk}{\int_{\mathbb{R}} |\hat{f}(k)|^2 dk}$$

exist, then $E_x E_k \geq 1/4$.

**Problem 19.**   The Fourier transform is given by

$$\hat{f}(k) = \int_{\mathbb{R}} f(x) e^{ikx} dx, \quad f(x) = \frac{1}{2\pi} \hat{f}(k) e^{-ikx} dk.$$

(i) Show that for the *Gaussian distribution* we have

$$\frac{1}{\sqrt{2\pi\sigma^2}} \exp\left(-\frac{(x-E)^2}{2\sigma^2}\right) \Leftrightarrow \exp\left(ikE - \frac{1}{2}\sigma^2 k^2\right).$$

(ii) Show that for *Poisson distribution* we have

$$\sum_{n=0}^{\infty} \frac{1}{n!} \lambda^n \exp(-\lambda)\delta(x-n) \Leftrightarrow \exp(\lambda(e^{ik} - 1))$$

(iii) Show that for *Lorentzian distribution* we have

$$\frac{\Gamma}{\pi} \frac{1}{(x-m)^2 + \Gamma^2} \Leftrightarrow \exp(imk - |k\Lambda|).$$

(iv) Show for the product of two Gaussian distributions we have

$$\exp\left(ik(E_1 + E_2) - \frac{1}{2}(\sigma_1^2 + \sigma_2^2)k^2\right)$$

# Chapter 19

# Coherent States

## 19.1 Introduction

Quantum coherent states are the closest quantum-mechanical analogue to a classical particle oscillating in a harmonic potential. Coherent states are minimum uncertainty states. Quantum computation circuits with coherent states as the logical qubits can be constructed using simple linear networks, conditional measurements and coherent superposition resource states. Coherent states are very sensitive to their environment. The output of a single mode stabilised laser can be described by a coherent state $|\beta\rangle$, where $\beta$ is a complex number which determines the average field amplitude. Harmonic oscillator coherent states can be defined in three different equivalent ways. Firstly, the coherent states are the eigenstates of the Bose annihilation operator

$$b|\beta\rangle = \beta|\beta\rangle, \qquad \beta \in \mathbb{C}.$$

Thus the spectrum of the operator $b$ fills the entire complex plane. Secondly, they are displaced vacuum states

$$|\beta\rangle = \exp(-|\beta|^2/2)\exp(\beta b^\dagger)\exp(-\beta^* b)|0\rangle$$

where $|0\rangle$ is the vacuum state with $\langle 0|0\rangle = 1$. Since

$$b|0\rangle = 0|0\rangle$$

we have

$$|\beta\rangle = \exp(-|\beta|^2/2)\exp(\beta b^\dagger)|0\rangle.$$

422  *Problems and Solutions*

Thirdly, coherent states are states of minimum uncertainty

$$\Delta p \Delta x = \frac{\hbar}{2}$$

and are thus most classical within the quantum framework.

Expressed in number states $|n\rangle$ the coherent states are given by

$$|\beta\rangle = e^{-|\beta|^2/2} \sum_{n=0}^{\infty} \frac{\beta^n}{\sqrt{n!}} |n\rangle.$$

The *displacement operator* is defined as

$$D(\beta) := \exp(\beta b^\dagger - \beta^* b).$$

Then the coherent states can be defined as

$$|\beta\rangle = D(\beta)|0\rangle.$$

We have

$$D^\dagger(\beta) = D^{-1}(\beta) = D(-\beta).$$

The completeness relation is given by

$$\frac{1}{\pi} \int_C d^2\beta |\beta\rangle\langle\beta| = I$$

where $I$ is the identity operator. Let $|\beta\rangle$, $|\gamma\rangle$ be coherent states. Then

$$\langle\gamma|\beta\rangle = \exp\left(-\frac{1}{2}(|\beta|^2 + |\gamma|^2) + \beta\gamma^*\right).$$

It follows that

$$\langle\beta| - \beta\rangle = e^{-2\beta\beta^*}.$$

Let $|n\rangle$ be number states. Then

$$\langle n|\beta\rangle = \exp\left(-\frac{1}{2}|\beta|^2\right) \frac{\beta^n}{\sqrt{n!}}.$$

Let $\hat{n} = b^\dagger b$ be the number operator. Then

$$\langle\beta|\hat{n}|\beta\rangle = \beta\beta^*.$$

## 19.2 Solved Problems

**Problem 1.** Bose creation $b^\dagger$ and annihilation $b$ operators obey the *Heisenberg algebra* $[b, b^\dagger] = I$, $[b, b] = [b^\dagger, b^\dagger] = 0$ with $b|0\rangle = 0$, where $|0\rangle$ is the vacuum state. The *coherent states* $|\beta\rangle$ can be obtained by applying the unitary *displacement operator*

$$D(\beta) := \exp(\beta b^\dagger - \beta^* b)$$

on the vacuum state $|0\rangle$, i.e.

$$|\beta\rangle := D(\beta)|0\rangle = \exp(\beta b^\dagger - \beta^* b)|0\rangle = \exp\left(-\frac{1}{2}|\beta|^2\right) \sum_{n=0}^{\infty} \frac{\beta^n}{\sqrt{n!}} |n\rangle, \quad \beta \in \mathbb{C}.$$

Show that from this definition the coherent states can also be obtained as the eigenstates of the annihilation (destruction) operator $b$, i.e. $b|\beta\rangle = \beta|\beta\rangle$.

**Solution 1.** We find for the commutator

$$[b, (\beta b^\dagger - \beta^* b)^n] = \beta n(\beta b^\dagger - \beta^* b)^{n-1}, \qquad n = 1, 2, \ldots$$

and therefore we have the commutation relation $[b, D(\beta)] = \beta D(\beta)$. Since $b|0\rangle = 0|0\rangle$ we have

$$0|0\rangle = D(\beta)b|0\rangle = (b - \beta I)D(\beta)|0\rangle = (b - \beta I)|\beta\rangle$$

where we used the commutation relation given above.

**Problem 2.** Let $D(\beta)$ be the displacement operator. Find the operators $D^\dagger(\beta)$ and $D^{-1}(\beta)$.

**Solution 2.** We find $D^\dagger(\beta) = D^{-1}(\beta) = D(-\beta)$.

**Problem 3.** Let $|\beta\rangle$ and $|\gamma\rangle$ be coherent states.
(i) Calculate $\langle\gamma|\beta\rangle$. Calculate $\langle 0|D(\beta)|0\rangle \equiv \langle 0|\beta\rangle$.
(ii) Find the probability $|\langle\gamma|\beta\rangle|^2$.

**Solution 3.** (i) Since

$$|\beta\rangle = \exp\left(-\frac{1}{2}|\beta|^2\right) \sum_{n=0}^{\infty} \frac{\beta^n}{\sqrt{n!}} |n\rangle, \qquad |\gamma\rangle = \exp\left(-\frac{1}{2}|\gamma|^2\right) \sum_{m=0}^{\infty} \frac{\gamma^m}{\sqrt{m!}} |m\rangle$$

we find

$$\langle\gamma|\beta\rangle = \exp\left(-\frac{1}{2}(|\beta|^2 + |\gamma|^2)\right) \sum_{m=0}^{\infty} \sum_{n=0}^{\infty} \frac{\gamma^{*m}}{\sqrt{m!}} \frac{\beta^n}{\sqrt{n!}} \langle m|n\rangle$$

424    *Problems and Solutions*

$$= \exp\left(-\frac{1}{2}(|\beta|^2 + |\gamma|^2)\right) \sum_{n=0}^{\infty} \frac{(\beta\gamma^*)^n}{n!}$$

$$= \exp\left(-\frac{1}{2}(|\beta|^2 + |\gamma|^2) + \beta\gamma^*\right)$$

where we used $\langle m|n \rangle = \delta_{mn}$. Using this result we find

$$\langle 0|\beta \rangle = \exp\left(-\frac{1}{2}|\beta|^2\right).$$

(ii) From (i) we obtain $|\langle \gamma|\beta \rangle|^2 = \exp(-|\beta - \gamma|^2)$. If $\gamma = \beta$ we have $|\langle \beta|\beta \rangle|^2 = 1$. If $\gamma = -\beta$ we obtain $|\langle \beta| - \beta \rangle|^2 = e^{-4|\beta|^2}$.

**Problem 4.** Let $|\beta \rangle$ be a coherent state. Let $\hat{n} := b^\dagger b$. Calculate

$$\langle \beta|\hat{n}|\beta \rangle.$$

**Solution 4.** Since $b|\beta \rangle = \beta|\beta \rangle$, $\langle \beta|b^\dagger = \langle \beta|\beta^*$ we obtain

$$\langle \beta|\hat{n}|\beta \rangle = \beta\beta^* = |\beta|^2.$$

**Problem 5.** Let $b$, $b^\dagger$ be Bose annihilation and creation operators and $\hat{x}$, $\hat{p}$ be canonical position and momentum operators which are related by the equations

$$b = \frac{1}{\sqrt{2}}(\hat{x} + i\hat{p}), \qquad b^\dagger = \frac{1}{\sqrt{2}}(\hat{x} - i\hat{p}).$$

Express the displacement operator with $\hat{x}$ and $\hat{p}$.

**Solution 5.** We obtain

$$D(\hat{x}, \hat{p}) = \exp(i(p\hat{x} - x\hat{p}))$$

where the real and complex parameters are also related by

$$\beta = \frac{1}{\sqrt{2}}(x + ip), \qquad \bar{\beta} = \frac{1}{\sqrt{2}}(x - ip).$$

**Problem 6.** Consider the Hamilton operator $\hat{H} = \hbar\omega b^\dagger b$. Let

$$U(t) := \exp(-it\hat{H}/\hbar).$$

Find the state $U(t)|\beta \rangle$, where $|\beta \rangle$ is a coherent state.

**Solution 6.** Since

$$|\beta\rangle = e^{-|\beta|^2/2} \sum_{n=0}^{\infty} \frac{\beta^n}{\sqrt{n!}} |n\rangle$$

and $b^\dagger b |n\rangle = n|n\rangle$ we find

$$U(t)|\beta\rangle = |\beta e^{-i\omega t}\rangle.$$

Thus the linear evolution of $|\beta\rangle$ is a rotation in phase space. The initial state will be revived at $\omega t = 2\pi, 4\pi, \ldots$ as expected.

**Problem 7.** Let $D(\beta) := \exp(\beta b^\dagger - \beta^* b)$ be the displacement operator.
(i) Find the operators $D(\beta)bD(-\beta)$, $D(\beta)b^\dagger D(-\beta)$.
(ii) Find the operators $D^\dagger(\beta)bD(\beta)$, $D^\dagger(\beta)b^\dagger D(\beta)$.

**Solution 7.** (i) Since $[b^\dagger, b] = -I$, $[b^\dagger, [b^\dagger, b]] = 0$ we obtain

$$D(\beta)bD(-\beta) = b - \beta I, \qquad D(\beta)b^\dagger D(-\beta) = b^\dagger - \beta^* I.$$

Note that $D(-\beta) = D^\dagger(\beta) = D^{-1}(\beta)$.
(ii) Using parameter differentiation yields

$$D^\dagger(\beta)bD(\beta) = b + \beta I, \qquad D^\dagger(\beta)b^\dagger D(\beta) = b^\dagger + \beta^* I.$$

**Problem 8.** Coherent states are defined as

$$|\beta\rangle := e^{-\beta\beta^*/2} \sum_{m=0}^{\infty} \frac{\beta^m}{\sqrt{m!}} |m\rangle$$

where $\beta$ is a complex number. We have $b|\beta\rangle = \beta|\beta\rangle$ (eigenvalue equation),

$$\sum_{m=0}^{\infty} |m\rangle\langle m| = I$$

(completeness relation) and $b^\dagger b |n\rangle = n|n\rangle$.
(i) Calculate $\langle n|\beta\rangle$ and then the probability $P_n(\beta) := |\langle n|\beta\rangle|^2$.
(ii) Let $\hat{n} := b^\dagger b$. Calculate $\langle\hat{n}\rangle := \langle\beta|\hat{n}|\beta\rangle$, $\langle\hat{n}^2\rangle := \langle\beta|\hat{n}^2|\beta\rangle$.
(iii) Calculate the *variance* $\langle(\Delta\hat{n})^2\rangle := \langle(\hat{n} - \langle\hat{n}\rangle I)^2\rangle$.

**Solution 8.** (i) Using $\langle n|m\rangle = \delta_{nm}$ we have

$$\langle n|\beta\rangle = \exp\left(-\frac{1}{2}|\beta|^2\right) \frac{\beta^n}{\sqrt{n!}}.$$

Thus we obtain

$$P_n(\beta) = \frac{(\beta\beta^*)^n \exp(-\beta\beta^*)}{n!}.$$

This is a *Poisson distribution*.

(ii) Using $b|\beta\rangle = \beta|\beta\rangle$ and therefore $\langle\beta|b^\dagger = \langle\beta|\beta^*$ we find

$$\langle\hat{n}\rangle = \langle\beta|b^\dagger b|\beta\rangle = \beta\beta^*, \qquad \langle\hat{n}^2\rangle = (\beta\beta^*)^2 + \beta\beta^*$$

where we used $bb^\dagger = b^\dagger b + I$.

(iii) Applying the results from (ii) we obtain

$$\langle(\Delta\hat{n})^2\rangle := \langle(\hat{n} - \langle\hat{n}\rangle I)^2\rangle = \beta\beta^*.$$

**Problem 9.**    Let $|0\rangle$, $|1\rangle$, ... be the number states. An arbitrary normalized state $|g\rangle$ can be expanded as

$$|g\rangle = \sum_{j=0}^{\infty} c_j|j\rangle, \qquad \sum_{j=0}^{\infty} c_j^* c_j = 1.$$

Express the state using coherent states $|\beta\rangle$. Consider the special case that $c_j = 0$ for all $j$ except for $c_n = 1$.

**Solution 9.**    We find

$$|g\rangle = \frac{1}{\pi}\int_{\mathbb{C}} e^{-|\beta|^2/2} g(\beta^*)|\beta\rangle d^2\beta$$

where $d^2\beta = d(\Re(\beta))d(\Im(\beta))$ and

$$g(\beta^*) = \sum_{j=0}^{\infty} \frac{c_j(\beta^*)^j}{\sqrt{j!}}.$$

If all $c_j$ are equal to 0 except for $c_n = 1$ we obtain

$$|g\rangle = |n\rangle = \frac{1}{\pi}\int_{\mathbb{C}} e^{-|\beta|^2/2} \frac{(\beta^*)^n}{\sqrt{n!}}|\beta\rangle d^2\beta.$$

**Problem 10.**    Let $|\beta\rangle$ be a coherent state and $|\psi\rangle$ be an arbitrary state in the Hilbert space containing $|\beta\rangle$. Show that

$$|\psi(\beta)| \leq \exp\left(\frac{1}{2}|\beta|^2\right).$$

**Solution 10.** We have the following identity (*completeness relation*)

$$\frac{1}{\pi} \int_C d^2\beta |\beta\rangle\langle\beta| = \sum_{m=0}^{\infty} |m\rangle\langle m| = I$$

from which it follows that the system of coherent states is complete. Using this equation we can expand an arbitrary state $|\psi\rangle$ with respect to the state $|\beta\rangle$

$$|\psi\rangle = \frac{1}{\pi} \int_C d^2\beta \langle\beta|\psi\rangle |\beta\rangle.$$

If the coherent state $|\beta\rangle$ is taken as $|\psi\rangle$, then this equation defines a linear dependence between the different coherent states. It follows that the system of coherent states is *supercomplete*, i.e. it contains subsystems which are complete. Using the definition for the coherent state given above we obtain

$$\langle\beta|\psi\rangle = \exp\left(-\frac{1}{2}|\beta|^2\right)\psi(\beta^*), \qquad \psi(\beta) = \sum_{n=0}^{\infty} \frac{\beta^n}{\sqrt{n!}} \langle n|\psi\rangle.$$

The inequality $|\langle n|\psi\rangle| \leq 1$ means that the function $\psi(\beta)$ for the normalization state $|\psi\rangle$ is an *entire analytic function* of the complex variables $\beta$. We also have $|\langle\beta|\psi\rangle| \leq 1$. Therefore we find a bound on the growth of $\psi(\beta)$

$$|\psi(\beta)| \leq \exp\left(\frac{1}{2}|\beta|^2\right).$$

The normalization condition can now be written as

$$\frac{1}{\pi} \int_C d^2\beta \exp(-|\beta|^2)|\psi(\beta)|^2 = \langle\psi|\psi\rangle.$$

The expansion of an arbitrary state $|\psi\rangle$ with respect to coherent states now takes the form

$$|\psi\rangle = \frac{1}{\pi} \int_C d^2\beta \exp\left(-\frac{1}{2}|\beta|^2\right)\psi(\beta^*)|\beta\rangle.$$

**Problem 11.** Coherent states $|\beta\rangle$ can be written as $|\beta\rangle = D(\beta)|0\rangle$, where $D(\beta)$ is the displacement operator and $|0\rangle$ denotes the vacuum state. Show that

$$D(\beta)D(\gamma) = \exp(i\Im(\beta\gamma^*))D(\beta+\gamma). \tag{1}$$

**Solution 11.** Since $[b, b^\dagger] = I$ we have

$$[\beta b^\dagger - \beta^* b, \gamma b^\dagger - \gamma^* b] = -[\beta b^\dagger, \gamma^* b] - [\beta^* b, \gamma b^\dagger] = (\beta\gamma^* - \beta^*\gamma)I = 2i\Im(\beta\gamma^*)I.$$

Using the *Baker-Campbell-Hausdorff formula*

$$e^A e^B = e^{A+B} e^{[A,B]/2}$$

for $[[A,B],A] = 0$ and $[[A,B],B] = 0$, we find (1). As a consequence we have

$$D(\beta)D(\gamma)D(-\beta) = e^{2i\Im(\beta\gamma^*)}D(\gamma).$$

It follows that the operators $\exp(2\pi i t)D(\beta)$ form a *group*. An element $g$ of this group is defined by the real number $t$ and a complex number $\beta$: $g(t,\beta)$. The product of two group elements $g = g_1 g_2$ is given by $g(t,\beta)$ with

$$t = t_1 + t_2 + \frac{1}{2\pi}\Im(\beta_2\beta_1^*)$$

and $\beta = \beta_1 + \beta_2$.

**Problem 12.**   Consider the displacement operator $D(\beta)$. Show that

$$D(\beta)D(\gamma) = e^{\beta\gamma^* - \beta^*\gamma}D(\gamma)D(\beta). \tag{1}$$

**Solution 12.**   The Baker-Campbell-Hausdorff formula

$$e^{A+B} = e^A e^B e^{-[A,B]/2} = e^B e^A e^{[A,B]/2}$$

for $[A,[A,B]] = [B,[A,B]] = 0$ can be applied since

$$[\beta b^\dagger - \beta^* b, \gamma b^\dagger - \gamma^* b] = (\beta\gamma^* - \beta^*\gamma)I$$

where $I$ is the identity operator. Thus (1) follows.

**Problem 13.**   Let $|\beta\rangle$ be a coherent state. Show that

$$\frac{1}{\pi}\int_C |\beta\rangle\langle\beta| d^2\beta = I$$

where $I$ is the identity operator and the integration is over the entire complex plane. Set $\beta = r\exp(i\phi)$ with $0 \leq r < \infty$ and $0 \leq \phi < 2\pi$.

**Solution 13.**   We have

$$\frac{1}{\pi}\int_C |\beta\rangle\langle\beta| d^2\beta = \frac{1}{\pi}\sum_{n=0}^{\infty}\sum_{m=0}^{\infty}\frac{|n\rangle\langle m|}{\sqrt{n!\,m!}}\int_C e^{-|\beta|^2}\beta^{*m}\beta^n d^2\beta.$$

Using $\beta = r\exp(i\phi)$ we arrive at

$$\frac{1}{\pi}\int_C |\beta\rangle\langle\beta| d^2\beta = \frac{1}{\pi}\sum_{n=0}^{\infty}\sum_{m=0}^{\infty}\frac{|n\rangle\langle m|}{\sqrt{n!\,m!}}\int_0^\infty re^{-r^2}r^{n+m}dr\int_0^{2\pi} e^{i(n-m)\phi}d\phi.$$

Since
$$\int_0^{2\pi} e^{i(n-m)\phi}d\phi = 2\pi\delta_{nm}$$
we have
$$\frac{1}{\pi}\int_C |\beta\rangle\langle\beta|d^2\beta = \sum_{n=0}^{\infty}\frac{|n\rangle\langle n|}{n!}\int_0^{\infty}e^{-s}s^n ds$$
where we set $s = r^2$ and therefore $ds = 2rdr$. Thus
$$\frac{1}{\pi}\int_C |\beta\rangle\langle\beta|d^2\beta = \sum_{n=0}^{\infty}|n\rangle\langle n| = I$$
where we used the completeness relation for the number states.

**Problem 14.** (i) The *Husimi distribution* of a coherent state $\gamma$ is given by
$$\rho_\gamma^H(\beta) := |\langle\beta|\gamma\rangle|^2.$$
Calculate $\rho_\gamma^H(\beta)$.
(ii) The Husimi distribution of the number state $|n\rangle$ is given by
$$\rho_{|n\rangle}^H(\beta) := |\langle\beta|n\rangle|^2.$$
Calculate $\rho_{|n\rangle}^H(\beta)$.
(iii) Consider the state $|n_1\rangle \otimes |n_2\rangle$. Find
$$\rho_{|n_1\rangle\otimes|n_2\rangle}^H(\beta) = |(\langle\beta_1| \otimes \langle\beta_2|)(|n_1\rangle \otimes |n_2\rangle)|^2.$$

**Solution 14.** (i) Since
$$|\beta\rangle = e^{-|\beta|^2/2}\sum_{n=0}^{\infty}\frac{\beta^n}{\sqrt{n!}}|n\rangle, \qquad |\gamma\rangle = e^{-|\gamma|^2/2}\sum_{n=0}^{\infty}\frac{\gamma^n}{\sqrt{n!}}|n\rangle$$
we find
$$\langle\beta|\gamma\rangle = e^{-|\beta|^2/2}e^{-|\gamma|^2/2}e^{\beta^*\gamma}$$
where we used $\langle m|n\rangle = \delta_{mn}$. Thus $|\langle\beta|\gamma\rangle|^2 = \exp(-|\beta - \gamma|^2)$ and the Husimi distribution of a coherent state is *Gaussian*.
(ii) Since $\langle m|n\rangle = \delta_{mn}$ we find
$$\langle\beta|n\rangle = e^{-|\beta|^2/2}\frac{\beta^{*n}}{\sqrt{n!}}$$
and hence
$$|\langle\beta|n\rangle|^2 = \frac{e^{-|\beta|^2}(|\beta|^2)^n}{n!}.$$

The Husimi distribution represents a *Poisson distribution* over the photon number states.

(iii) Since

$$((\langle \beta_1| \otimes \langle \beta_2|)(|n_1\rangle \otimes |n_2\rangle)) = \langle \beta_1|n_1\rangle \langle \beta_2|n_2\rangle$$

we obtain

$$\rho^H_{|n_1\rangle \otimes |n_2\rangle}(\beta) = \frac{e^{-|\beta_1|^2}(|\beta_1|^2)^{n_1}}{n_1!} \frac{e^{-|\beta_2|^2}(|\beta_2|^2)^{n_2}}{n_2!}.$$

**Problem 15.** In a *Kerr medium* the state evolution is governed by the interaction Hamilton operator

$$\hat{H} = \kappa(b^\dagger b)^2$$

where $\kappa$ is a coupling constant proportional to the nonlinear susceptibility of the medium. A coherent input signal state $|\beta\rangle$ evolves according to the solution of the Schrödinger equation

$$|\psi_c(t)\rangle = \exp(-i\hat{H}t)|\beta\rangle.$$

Calculate $|\psi_c(t)\rangle$ for $t = \pi/(2\kappa)$. Discuss.

**Solution 15.** Straightforward calculation yields

$$|\psi_c(t = \pi/(2\kappa))\rangle = \frac{1}{\sqrt{2}}(e^{-i\pi/4}|\beta\rangle + e^{i\pi/4}| - \beta\rangle)$$

$$= \frac{1}{\sqrt{2}}(e^{-i\pi/4}D(\beta) + e^{i\pi/4}D(-\beta))|0\rangle$$

where $D(\beta)$ is the displacement operator. The state describes a superposition of two coherent states with opposite phases. When $|\beta|$ becomes large the two components become mesoscopically distinguishable states of the radiation field. Realistic values of the Kerr nonlinear susceptibilities are quite small, thus requiring a long interaction time, or equivalently a large interaction length. Thus losses become significant and the resulting decoherence may destroy the quantum superposition.

**Problem 16.** Let $|\beta\rangle$ be a coherent state. Express the density operator

$$\rho = \frac{1}{2\pi} \int_0^{2\pi} d\phi ||\beta|e^{-i\phi}\rangle\langle|\beta|e^{-i\phi}|$$

with number states $|n\rangle$.

**Solution 16.** We obtain

$$\rho = e^{-|\beta|^2} \sum_{n=0}^{\infty} \frac{|\beta|^{2n}}{n!} |n\rangle\langle n|.$$

This is a collection of number states.

**Problem 17.** Let $b^\dagger$, $b$ be Bose creation and annihilation operators and $\hat{n} = b^\dagger b$. We can introduce *nonlinear coherent states* $B$ and $B^\dagger$ with an intensity dependent function $f(\hat{n})$ (which is an operator valued analytic function)

$$B := bf(\hat{n}) = f(\hat{n} + I)b, \qquad B^\dagger := f^\dagger(\hat{n})b^\dagger = b^\dagger f^\dagger(\hat{n} + I).$$

We consider $f$ to be real and non-negative, i.e. $f^\dagger(\hat{n}) = f(\hat{n})$.
(i) Calculate the commutator $[B, B^\dagger]$. Then consider the special case $f(\hat{n}) = I$ and $f(\hat{n}) = \hat{n}$.
(ii) Express the harmonic oscillator for $B^\dagger$ and $B$

$$\hat{H} = \frac{1}{2}(BB^\dagger + B^\dagger B)$$

using $\hat{n}$ and $f(\hat{n})$.
(iii) Find the eigenstate of the operator $B$, i.e., find the nonlinear coherent state

$$B|z\rangle_{NL} = \lambda|z\rangle_{NL}.$$

What is the condition that the state belongs to the Fock space?

**Solution 17.** (i) Since $f$ is analytic we have a Taylor expansion. We obtain

$$[B, B^\dagger] = (\hat{n} + I)f(\hat{n} + I)f(\hat{n} + I) - \hat{n}f(\hat{n})f(\hat{n}).$$

With $f(\hat{n}) = I$ we obtain the result for the Bose operators $b$ and $b^\dagger$. For the case $f(\hat{n}) = \hat{n}$ we obtain

$$[B, B^\dagger] = 3\hat{n}^2 + 3\hat{n} + I.$$

(ii) We obtain

$$\hat{H} = \frac{1}{2}((\hat{n} + I)f(\hat{n} + I)f(\hat{n} + I) + \hat{n}f(\hat{n})f(\hat{n})).$$

(iii) Using the expansion with respect to the Fock basis $\{|0\rangle, |1\rangle, \ldots\}$ and the fact that $f$ is an analytic function we obtain

$$|z\rangle_{NL} = \mathcal{N}_f(|z|^2)^{-1/2} \sum_{n=0}^{\infty} C_n z^n |n\rangle$$

where the coefficients $C_n$ are given by

$$C_n = \frac{1}{\sqrt{[nf(n)f(n)]!}}, \qquad C_0 = 1, \qquad [f(n)]! := f(n)f(n-1)\cdots f(1)$$

and the normalization constant is given by

$$\mathcal{N}_f(|z|^2) = \sum_{n=0}^{\infty} |C_n|^2 |z|^{2n}.$$

In order to have states belonging to the Fock space, it is required that $0 < \mathcal{N}_f(|z|^2) < \infty$. This implies that

$$|z| \leq \lim_{n\to\infty} n[f(n)]^2.$$

The function $f(n)$ corresponding to any nonlinear coherent state is found to be

$$f(n) = \frac{C_{n-1}}{\sqrt{n}C_n}.$$

**Problem 18.**    Let $D(\beta)$ be the displacement operator. Is

$$D(\beta)D(\beta') = D(\beta + \beta') ?$$

Prove or disprove.

**Solution 18.**    The answer is no. We find

$$D(\beta)D(\beta') = \exp\left(\frac{1}{2}(\beta\beta'^* - \beta^*\beta')\right) D(\beta + \beta').$$

Note that $\beta \in \mathbb{C}$. If $\beta, \beta' \in \mathbb{R}$ then we have $D(\beta)D(\beta') = D(\beta + \beta')$.

**Problem 19.**    Consider the coherent state $|\beta\rangle$. We define the *Schrödinger cat states*

$$|\beta_+\rangle := N_+(|\beta\rangle + |-\beta\rangle), \qquad |\beta_-\rangle := N_-(|\beta\rangle - |-\beta\rangle).$$

(i) Normalize the two states.
(ii) Calculate the probabilities $|\langle\beta_+|n\rangle|^2$, $\langle\beta_-|n\rangle|^2$ and discuss.

**Solution 19.**    (i) From $\langle\beta_+|\beta_+\rangle = 1$, $\langle\beta_-|\beta_-\rangle = 1$ and

$$\langle\gamma|\beta\rangle = \exp(-|\beta - \gamma|^2/2)$$

we obtain

$$N_+ = \frac{\exp(|\beta|^2/2)}{2\sqrt{\cosh(|\beta|^2)}}, \qquad N_- = \frac{\exp(|\beta|^2/2)}{2\sqrt{\sinh(|\beta|^2)}}.$$

(ii) Since

$$\langle n|\beta\rangle = e^{-|\beta|^2/2}\frac{\beta^n}{\sqrt{n!}}$$

we obtain the probabilities

$$|\langle\beta_+|n\rangle|^2 = \frac{1}{\cosh(|\beta|^2)}\frac{|\beta|^{2n}}{n!}(1+(-1)^n)$$
$$|\langle\beta_-|n\rangle|^2 = \frac{1}{\sinh(|\beta|^2)}\frac{|\beta|^{2n}}{n!}(1-(-1)^n).$$

**Problem 20.** Consider the linear operator

$$Z := b_1 + b_2^\dagger \equiv b \otimes I + I \otimes b \tag{1}$$

where $b_1 = b \otimes I$ and $b_2^\dagger = I \otimes b^\dagger$. Let $D_b(z) := e^{zb^\dagger - z^*b}$ ($z \in \mathbb{C}$) be the displacement operator and

$$|0\rangle\rangle := \frac{1}{\sqrt{\pi}}\sum_{n=0}^\infty (-1)^n|n\rangle \otimes |n\rangle.$$

The states $|z\rangle\rangle$ are defined by

$$|z\rangle\rangle := D_{b_1}(z)|0\rangle\rangle = D_{b_2}(z^*)|0\rangle\rangle.$$

Find the state $Z|z\rangle\rangle$. Discuss.

**Solution 20.** We have $[Z, Z^\dagger] = 0$ and

$$Z|z\rangle\rangle = z|z\rangle\rangle, \qquad z \in \mathbb{C}.$$

Thus $|z\rangle\rangle$ is an eigenstate of $Z$. For $z = 0$ the state $|0\rangle\rangle$ can be approximated by a physical (normalizable) state called the *twin beam state* - corresponding to the output of a non-degenerate optical parametric amplifier in the limit of infinite gain.

**Problem 21.** Consider the *beam splitter interaction* given by the unitary transformation

$$U_{BS} = \exp(i\theta(b_1 b_2^\dagger + b_1^\dagger b_2))$$

where $b_1$ and $b_2$ are the Bose annihilation operators. Let $|\beta\rangle$, $|\gamma\rangle$ be coherent states. Calculate

$$U_{BS}(|\gamma\rangle \otimes |\beta\rangle).$$

**Solution 21.**    We obtain

$$U_{BS}(|\gamma\rangle \otimes |\beta\rangle) = |\cos(\theta)\gamma + i\sin(\theta)\beta\rangle \otimes |\cos(\theta)\beta + i\sin(\theta)\gamma\rangle$$

where $\cos^2(\theta)$ $(\sin^2(\theta))$ is the reflectivity (transmissivity) of the beam splitter.

**Problem 22.**    The normalized *Schrödinger cat state* of a single-mode radiation field is given by

$$|\beta, \theta\rangle = \frac{1}{\sqrt{2 + 2\cos(\theta)e^{-2|\beta|^2}}}(|\beta\rangle + e^{i\theta}|-\beta\rangle).$$

Discuss the special cases $\theta = 0$, $\theta = \pi$, $\theta = \pi/2$.

**Solution 22.**    For $\theta = 0$ we obtain the even coherent states

$$|\beta, 0\rangle = \frac{1}{\sqrt{2 + 2e^{-2|\beta|^2}}}(|\beta\rangle + |-\beta\rangle).$$

For $\theta = \pi$ we obtain the odd states

$$|\beta, \pi\rangle = \frac{1}{\sqrt{2 - 2e^{-2|\beta|^2}}}(|\beta\rangle - |-\beta\rangle).$$

For $\theta = \pi/2$ we obtain the *Yurke-Stoler states*

$$|\beta, \pi/2\rangle = \frac{1}{\sqrt{2}}(|\beta\rangle + i|-\beta\rangle).$$

**Problem 23.**    We know that

$$|\beta\rangle = D(\beta)|0\rangle = e^{\beta b^\dagger - \bar{\beta}b}|0\rangle$$
$$= e^{\beta b^\dagger}e^{-\bar{\beta}b}e^{-\beta\bar{\beta}I/2}|0\rangle = e^{-\beta\bar{\beta}/2}e^{\beta b^\dagger}|0\rangle$$
$$= e^{-\beta\bar{\beta}/2}\sum_{n=0}^{\infty}\frac{\beta^n}{\sqrt{n!}}|n\rangle$$

where

$$|n\rangle = \frac{(b^\dagger)^n}{\sqrt{n!}}|0\rangle.$$

Let $\sigma_1$ be the Pauli spin matrix. Extend the calculation to the state

$$\tilde{D}(\beta)(|0\rangle \otimes \begin{pmatrix} 1 \\ 0 \end{pmatrix})$$

where

$$\tilde{D}(\beta) := \exp(\beta b^\dagger \otimes \sigma_1 - \bar{\beta} b \otimes \sigma_1).$$

**Solution 23.** First we note that since $\sigma_1^2 = I_2$ and $[b, b^\dagger] = I$

$$-[\beta b^\dagger \otimes \sigma_1, -\bar{\beta} b \otimes \sigma_1] = -\beta \bar{\beta} I_B \otimes I_2.$$

Thus we obtain

$$\tilde{D}(\beta)|0\rangle \otimes \begin{pmatrix} 1 \\ 0 \end{pmatrix} = e^{\beta b^\dagger \otimes \sigma_1} e^{-\bar{\beta} b \otimes \sigma_1} e^{-\beta \bar{\beta} I_B \otimes I_2/2}|0\rangle \otimes \begin{pmatrix} 1 \\ 0 \end{pmatrix}$$

$$= e^{-\beta \bar{\beta}/2} e^{\beta b^\dagger \otimes \sigma_x} e^{-\bar{\beta} b \otimes \sigma_1}|0\rangle \otimes \begin{pmatrix} 1 \\ 0 \end{pmatrix}$$

$$= e^{-\beta \bar{\beta}/2} e^{\beta b^\dagger \otimes \sigma_1}|0\rangle \otimes \begin{pmatrix} 1 \\ 0 \end{pmatrix}$$

$$= e^{-\beta \bar{\beta}/2} \left( \left( \sum_{n=0}^{\infty} \frac{\beta^{2n}}{\sqrt{(2n)!}}|2n\rangle \right) \otimes \begin{pmatrix} 1 \\ 0 \end{pmatrix} \right.$$

$$\left. + \left( \sum_{n=0}^{\infty} \frac{\beta^{2n+1}}{\sqrt{(2n+1)!}}|2n+1\rangle \right) \otimes \begin{pmatrix} 0 \\ 1 \end{pmatrix} \right).$$

**Problem 24.** Bose creation ($b^\dagger$) and annihilation ($b$) operators, where

$$b^\dagger = (b_1^\dagger, b_2^\dagger, \ldots, b_N^\dagger), \qquad b = (b_1, b_2, \ldots, b_N)$$

obey the *Heisenberg algebra*

$$[b_j, b_k^\dagger] = \delta_{jk} I, \quad [b_j, b_k] = [b_j^\dagger, b_k^\dagger] = 0, \quad j, k = 1, \ldots, N.$$

Coherent states, where $\mathbf{z} \in \mathbb{C}^n$, are defined as eigenvectors of the annihilation operators, that is $\mathbf{b}|\mathbf{z}\rangle = \mathbf{z}|\mathbf{z}\rangle$.
(i) Show that the normalized coherent states are given by

$$|\mathbf{z}\rangle = \exp\left(-\frac{1}{2}|\mathbf{z}|^2\right)\exp(\mathbf{z} \cdot \mathbf{b}^\dagger)|0\rangle \qquad (1)$$

where

$$|\mathbf{z}|^2 = \sum_{j=1}^{N} |z_j|^2, \qquad \mathbf{z} \cdot \mathbf{b}^\dagger = \sum_{j=1}^{N} z_j b_j^\dagger$$

and $|\mathbf{0}\rangle = |0\,0\ldots 0\rangle$ is the vacuum vector satisfying $\mathbf{b}|\mathbf{0}\rangle = 0|\mathbf{0}\rangle$.
(ii) Let $|\mathbf{w}\rangle$ be a coherent state. Find $\langle\mathbf{z}|\mathbf{w}\rangle$, $\langle\mathbf{z}|\mathbf{w}\rangle|^2$.
(iii) Calculate

$$\int_{\mathbb{R}^{2N}} d\mu(\mathbf{z})|\mathbf{z}\rangle\langle\mathbf{z}|.$$

**Solution 24.**   (i) Consider the number representation

$$|\mathbf{n}\rangle \equiv |n_1, n_2, \ldots, n_N\rangle = \frac{(b_1^\dagger)^{n_1}}{\sqrt{n_1!}}\frac{(b_2^\dagger)^{n_2}}{\sqrt{n_2!}}\cdots\frac{(b_N^\dagger)^{n_N}}{\sqrt{n_N!}}|\mathbf{0}\rangle.$$

We expand $|\mathbf{z}\rangle$ with respect to $|\mathbf{n}\rangle$ and apply

$$b_j|\mathbf{n}\rangle = \sqrt{n_j}|n_1, \cdots, n_j - 1, \cdots, n_N\rangle$$

$$b_j^\dagger|\mathbf{n}\rangle = \sqrt{n_j + 1}|n_1, \cdots, n_j + 1, \cdots, n_N\rangle$$

we find, after normalization, that $|\mathbf{z}\rangle$ is given by (1).
(ii) We find

$$\langle\mathbf{z}|\mathbf{w}\rangle = \exp\left(-\frac{1}{2}(|\mathbf{z}|^2 + |\mathbf{w}|^2 - 2\mathbf{z}^* \cdot \mathbf{w})\right)$$

and

$$|\langle\mathbf{z}|\mathbf{w}\rangle|^2 = \exp(-|\mathbf{z} - \mathbf{w}|^2).$$

(iii) Since

$$d\mu(\mathbf{z}) = \frac{1}{\pi}\prod_{j=1}^{N}d(\Re z_j)d(\Im z_j)$$

we find

$$\int_{\mathbb{R}^{2N}} d\mu(\mathbf{z})|\mathbf{z}\rangle\langle\mathbf{z}| = I$$

where $I$ is the identity operator.

## Programming Problem

**Problem 1.**   Give a computer algebra implementation that implements

$$b|\beta\rangle = \beta|\beta\rangle, \qquad \langle\beta|b^\dagger = \langle\beta|\beta^*.$$

Then find $b^2|\beta\rangle$, $\langle\beta|$, $\langle\beta|\beta\rangle$, $b|\beta\rangle$, $\langle\beta|b|\beta\rangle$.

**Solution 1.**   In the SymbolicC++ program bd denotes $b^\dagger$, b denotes $b$, cs denotes the coherent state $|\beta\rangle$ and ds denotes $\langle\beta|$ and conj denotes the complex conjugate and z denotes $\beta$.

```
// coherent.cpp
#include <iostream>
#include "symbolicc++.h"
using namespace std;

int main(void)
{
Symbolic b("b"), bd("bd"), cs("cs"), ds("ds");
b = ~b; bd = ~bd; cs = ~cs; ds = ~ds; // noncommutative
Symbolic z("z"), w("w"), conj("conj");
Equations rules =
    (b*cs[z]==z*cs[z],b*cs[w]==w*cs[w],
     ds[z]*bd==ds[z]*conj[z],ds[w]*bd==ds[w]*conj[w],
     ds[z]*cs[z]==1,ds[w]*cs[w]==1,
     ds[w]*cs[z]==exp(-(z*conj[z]+w*conj[w]-2*conj[w]*z)/2),
     ds[z]*cs[w]==exp(-(z*conj[z]+w*conj[w]-2*conj[z]*w)/2));
// example 1
Symbolic r1 = b*(b*cs[z]);
r1 = r1.subst_all(rules);
cout << r1 << endl;
r1 = r1[z==1];
cout << r1 << endl;
// example 2
cout << (ds[z]*cs[z]).subst_all(rules) << endl;
// example 3
Symbolic r2 = b*cs[z];
Symbolic r3 = ds[w]*r2;
cout << r2.subst_all(rules) << endl;
cout << r3.subst_all(rules) << endl;
return 0;
}
```

The output is

```
z^(2)*cs[z]
cs[1]
1
z*cs[z]
e^(-1/2*z*conj[z]-1/2*w*conj[w]+conj[w]*z)*z
```

## 19.3 Supplementary Problems

**Problem 1.** Let $|\beta\rangle$ ($\beta \in \mathbb{C}$) be a coherent state. Show that the projection operator $|\beta\rangle\langle\beta|$ can be expressed as

$$|\beta\rangle\langle\beta| = \exp(-\overline{\beta}\beta)\exp(\beta b^{\dagger})|0\rangle\langle 0|\exp(\overline{\beta}b).$$

Find $\mathrm{tr}(|\beta\rangle\langle\beta|$.

**Problem 2.** Let $D(\beta)$ be the displacement operator.
(i) Show that (normally ordered form of $D(\beta)$)

$$D(\beta) = \exp\left(-\frac{1}{2}|\beta|^2\right) \exp(\beta b^\dagger) \exp(-\beta^* b).$$

(ii) Show that (antinormally ordered form of $D(\beta)$)

$$D(\beta) = \exp\left(\frac{1}{2}|\beta|^2\right) \exp(-\beta^* b) \exp(\beta b^\dagger).$$

**Problem 3.** Let $\hat{n} = b^\dagger b$ and $D(\beta)$ be the displacement operator. Find the commutator

$$[\hat{n}, D(\beta)].$$

Find the commutator

$$[|n\rangle\langle n|, D(\beta)].$$

Find the commutator

$$[|n\rangle\langle n|, |\beta\rangle\langle\beta|].$$

**Problem 4.** Show that the solutions of the eigenvalue problem for the one-dimensional harmonic oscillator

$$\left(-\frac{\hbar^2}{2m}\frac{d^2}{dx^2} + \frac{m\omega^2}{2}x^2\right) u(x) = E u(x)$$

are given by

$$u_n(x) = \frac{1}{\sqrt{2^n(n!)}} \left(\frac{m\omega}{\pi\hbar}\right)^{1/4} H_n(x/x_0) \exp(-x^2/(2x_0^2))$$

with $E_n = \hbar\omega(n + 1/2)$, $x_0 := \sqrt{\hbar/(m\omega)}$ and $H_n(x/x_0)$ are the Hermite polynomials.

# Chapter 20

# Squeezed States

---

## 20.1 Introduction

Coherent states are not the most general kind of Gaussian wave packet. They are also not the most general kind of minimum-uncertainty wave packets, since the minimum uncertainty wave packet satisfies

$$\Delta q \Delta p = \hbar/2$$

which only constrains the product of the dispersions $\Delta q$ and $\Delta p$, whereas for coherent states we have that

$$(\Delta q)^2 = \hbar/(2\omega) \text{ and } (\Delta p)^2 = \hbar\omega/2.$$

For squeezed states one does not have this restriction. Unlike a coherent state, an initial squeezed state does not remain a minimum-uncertainty state in the course of time under the harmonic oscillator evolution. The product $\Delta q \Delta p$ oscillates at twice the harmonic oscillator frequency between a maximum value and a minimum value. Squeezed states possess the property that one quadrature phase has reduced fluctuations compared to the ordinary vacuum. Squeezed states of the electromagnetic field are generated by degenerate parametric down conversion in an optical cavity. The ideal squeezed state is defined as

$$|\zeta\rangle := S(\zeta)|0\rangle$$

439

where

$$S(\zeta) := \exp\left(\frac{1}{2}\zeta^* b^2 - \frac{1}{2}\zeta b^{\dagger 2}\right)$$

is the one-mode *squeezing operator* with $\zeta \in \mathbb{C}$. If we set

$$\zeta = se^{i\theta}$$

$(s \geq 0)$ then $s$ is called the *squeezing parameter*.

The operator $S(\zeta)$ is unitary, since

$$S^\dagger(\zeta) = \exp\left(\frac{\zeta}{2}(b^\dagger)^2 - \frac{\zeta^*}{2}b^2\right)$$

and therefore $S^\dagger(\zeta)S(\zeta) = I$. In disentangled form $S(\zeta)$ is given by

$$S(\zeta) = \exp\left(-\frac{1}{2}(b^\dagger)^2 e^{i\theta} \tanh(s)\right) \exp\left(-\frac{1}{2}(b^\dagger b + bb^\dagger)\ln(\cosh(s))\right)$$
$$\times \exp\left(\frac{1}{2}b^2 e^{-i\theta}\tanh(s)\right).$$

Using this result the squeezed states can be expressed with number states

$$|\zeta\rangle = \sqrt{(\operatorname{sech}(s))} \sum_{n=0}^{\infty} \frac{\sqrt{((2n)!)}}{n!}\left(-\frac{1}{2}\exp(i\theta)\tanh(s)\right)^n |2n\rangle$$

where $n = 0, 1, 2, \ldots$ and we used that $(b^\dagger b + bb^\dagger)|0\rangle = |0\rangle$. The expansion over the number state basis only contains even components. The state $|\zeta\rangle$ is normalized. We also have the eigenvalue equation

$$(b\cosh(s) + b^\dagger e^{i\phi}\sinh(s))|\zeta\rangle = 0|\zeta\rangle$$

and similarly $\langle\zeta|$ is the zero-eigenvalue left eigenstate of

$$b^\dagger \cosh(s) + be^{-i\theta}\sinh(s).$$

The overlap of two squeezed states is given by

$$\langle\xi|\zeta\rangle = \frac{1}{\sqrt{\cosh(s_1)\cosh(s_2)(1 - e^{-i(\theta_2-\theta_1)}\tanh(s_1)\tanh(s_2))}}$$

where $\xi = s_1 e^{i\theta_1}$, $\zeta = s_2 e^{i\theta_2}$.

## 20.2    Solved Problems

**Problem 1.**    Let $\zeta \in \mathbb{C}$. The squeezing operator is defined by

$$S(\zeta) := \exp\left(-\frac{\zeta}{2}(b^\dagger)^2 + \frac{\zeta^*}{2}b^2\right)$$

where $\zeta \in \mathbb{C}$. We set $\zeta = se^{i\theta}$, where $s \geq 0$ and $\theta \in \mathbb{R}$. The single mode squeezed states $|\zeta\rangle$ are defined by

$$|\zeta\rangle := S(\zeta)|0\rangle.$$

(i) Show that $S(\zeta)$ is unitary.
(ii) Find $S(\zeta)$ in the disentangled form.
(iii) Calculate the state $|\zeta\rangle = S(\zeta)|0\rangle$ using the number states.

**Solution 1.**    (i) We obtain

$$S^\dagger(\zeta) = \exp\left(\frac{\zeta}{2}(b^\dagger)^2 - \frac{\zeta^*}{2}b^2\right) = S(-\zeta).$$

Thus we have $S^\dagger(\zeta)S(\zeta) = I$.
(ii) We obtain

$$S(\zeta) = \exp\left(-\frac{1}{2}(b^\dagger)^2 e^{i\theta} \tanh(s)\right) \exp\left(-\frac{1}{2}(b^\dagger b + bb^\dagger) \ln(\cosh(s))\right)$$
$$\times \exp\left(\frac{1}{2}b^2 e^{-i\theta} \tanh(s)\right).$$

(iii) Using the result from (ii) we obtain

$$|\zeta\rangle = \sqrt{(\operatorname{sech}(s))} \sum_{n=0}^{\infty} \frac{\sqrt{((2n)!)}}{n!} \left(-\frac{1}{2}\exp(i\theta)\tanh(r)\right)^n |2n\rangle$$

where we used that $(b^\dagger b + bb^\dagger)|0\rangle = |0\rangle$.

**Problem 2.**    Consider the one-mode squeezing operator

$$S(\zeta) = e^{1/2(\zeta^* b^2 - \zeta(b^\dagger)^2)}$$

where $\zeta \in \mathbb{C}$. Calculate

$$\widetilde{b} = S^\dagger(\zeta)bS(\zeta), \quad \widetilde{b}^\dagger = S^\dagger(\zeta)b^\dagger S(\zeta).$$

**Solution 2.**    We obtain

$$\widetilde{b} = \cosh(|\zeta|)b - \frac{\zeta}{|\zeta|}\sinh(|\zeta|)b^\dagger, \quad \widetilde{b}^\dagger = \cosh(|\zeta|)b^\dagger - \frac{\zeta^*}{|\zeta|}\sinh(|\zeta|)b.$$

**Problem 3.**    Consider the single-mode squeezing operator

$$S(\zeta) = \exp\left(\frac{1}{2}\zeta(b^\dagger)^2 - \frac{1}{2}\zeta^* b^2\right).$$

Let $|0\rangle$ be the vacuum state. Find the state $S(\zeta)|0\rangle \equiv |\zeta\rangle$ using number states $|n\rangle$. Use this result to find the expectation value $\langle\zeta|b^\dagger b|\zeta\rangle$.

**Solution 3.**    Expansion over the number state basis contains only even components, i.e.

$$S(\zeta)|0\rangle \equiv |\zeta\rangle = \frac{1}{\sqrt{\mu}} \sum_{k=0}^{\infty} \left(\frac{\nu}{2\mu}\right)^k \frac{\sqrt{(2k)!}}{k!}|2k\rangle$$

where $\zeta = se^{i\theta}$, $\mu = \cosh(s)$, $\nu = e^{i\theta}\sinh(s)$. Using this result and

$$b^\dagger b|2k\rangle = 2k|2k\rangle$$

we obtain

$$\langle\zeta|b^\dagger b|\zeta\rangle = |\nu|^2.$$

**Problem 4.**    Let $S(\zeta)$ be the one-mode *squeezing operator* $(\zeta \in \mathbb{C})$

$$S(\zeta) := \exp\left(\frac{1}{2}\zeta^* b^2 - \frac{1}{2}\zeta(b^\dagger)^2\right).$$

Calculate

$$S(\chi)\begin{pmatrix} b \\ b^\dagger \end{pmatrix} S^\dagger(\zeta) \equiv \begin{pmatrix} S(\zeta)bS^\dagger(\zeta) \\ S(\zeta)b^\dagger S^\dagger(\zeta) \end{pmatrix}$$

and show that we can write

$$\begin{pmatrix} S(\zeta)bS^\dagger(\zeta) \\ S(\zeta)b^\dagger S^\dagger(\zeta) \end{pmatrix} = T(\zeta, \zeta^*)\begin{pmatrix} b \\ b^\dagger \end{pmatrix}$$

where the $2 \times 2$ matrix $T(\zeta, \zeta^*)$ depends only on $\zeta$ and $\zeta^*$.

**Solution 4.**    We have $S^\dagger(\zeta) = S(-\zeta)$. Let $\epsilon \in \mathbb{R}$. We set

$$f_1(\epsilon) = e^{\frac{1}{2}\epsilon(\zeta^* b^2 - \zeta(b^\dagger)^2)} b e^{-\frac{1}{2}\epsilon(\zeta^* b^2 - \zeta(b^\dagger)^2)}$$

and

$$f_2(\epsilon) = e^{\frac{1}{2}\epsilon(\zeta^* b^2 - \zeta(b^\dagger)^2)} b^\dagger e^{-\frac{1}{2}\epsilon(\zeta^* b^2 - \zeta(b^\dagger)^2)}$$

with the initial conditions $f_1(\epsilon = 0) = b$ and $f_2(\epsilon = 0) = b^\dagger$. We find the system of differential equations for $f_1$ and $f_2$. Using $bb^\dagger = I + b^\dagger b$ we have

$$\frac{df_1}{d\epsilon} = -\frac{1}{2}\zeta e^{\frac{1}{2}\epsilon(\zeta^* b^2 - \zeta(b^\dagger)^2)}((b^\dagger)^2 b - b(b^\dagger)^2)e^{-\frac{1}{2}\epsilon(\zeta^* b^2 - \zeta(b^\dagger)^2)}$$

$$= \zeta f_2(\epsilon).$$

Analogously, we obtain

$$\frac{df_2}{d\epsilon} = \zeta^* f_1(\epsilon).$$

In matrix notation we have the system

$$\begin{pmatrix} df_1/d\epsilon \\ df_2/d\epsilon \end{pmatrix} = \begin{pmatrix} 0 & \zeta \\ \zeta^* & 0 \end{pmatrix} \begin{pmatrix} f_1 \\ f_2 \end{pmatrix}.$$

Let

$$A = \begin{pmatrix} 0 & \zeta \\ \zeta^* & 0 \end{pmatrix}.$$

We find

$$\exp(\epsilon A) = \begin{pmatrix} \cosh(\epsilon s) & \zeta \sinh(\epsilon s)/s \\ \zeta^* \sinh(\epsilon s)/s & \cosh(\epsilon s) \end{pmatrix} = \begin{pmatrix} \cosh(\epsilon s) & e^{i\theta} \sinh(\epsilon s) \\ e^{-i\theta} \sinh(\epsilon s) & \cosh(\epsilon s) \end{pmatrix}$$

where we used $\zeta = s \exp(i\theta)$. Taking into account the initial conditions we obtain

$$\begin{pmatrix} S(\zeta)bS^\dagger(\zeta) \\ S(\zeta)b^\dagger S^\dagger(\zeta) \end{pmatrix} = \begin{pmatrix} \cosh(s) & e^{i\theta} \sinh(s) \\ e^{-i\theta} \sinh(s) & \cosh(s) \end{pmatrix} \begin{pmatrix} b \\ b^\dagger \end{pmatrix}.$$

The matrix on the right-hand side can be decomposed as

$$\begin{pmatrix} \cosh(s) & e^{i\theta} \sinh(s) \\ e^{-i\theta} \sinh(s) & \cosh(s) \end{pmatrix} \equiv$$

$$\begin{pmatrix} 0 & e^{i\theta/2} \\ e^{-i\theta/2} & 0 \end{pmatrix} \begin{pmatrix} \cosh(s) & \sinh(s) \\ \sinh(s) & \cosh(s) \end{pmatrix} \begin{pmatrix} 0 & e^{i\theta/2} \\ e^{-i\theta/2} & 0 \end{pmatrix}.$$

**Problem 5.** For generating a squeezed state of one mode we start from the Hamilton operator

$$\hat{H} = \hbar\omega b^\dagger b + i\hbar\Lambda(b^2 e^{2i\omega t} - (b^\dagger)^2 e^{-2i\omega t}).$$

A photon of the driven mode, with frequency $2\omega$, splits into two photons of the mode of interest, each with frequency $\omega$. Solve the Heisenberg equation of motion for $b$ and $b^\dagger$.

**Solution 5.** The Heisenberg equation of motion yields

$$\frac{db}{dt} = \frac{1}{i\hbar}[b, \hat{H}](t) = -i\omega b(t) - 2\Lambda b^\dagger(t)e^{-2i\omega t}$$

$$\frac{db^\dagger}{dt} = \frac{1}{i\hbar}[b^\dagger, \hat{H}](t) = i\omega b^\dagger - 2\Lambda b e^{2i\omega t}.$$

Using $b = \tilde{b}\exp(-i\omega t)$ we obtain

$$\frac{d\tilde{b}}{dt} = -2\Lambda\tilde{b}^\dagger(t), \qquad \frac{d\tilde{b}^\dagger}{dt} = -2\Lambda\tilde{b}(t).$$

Introducing the operators

$$\tilde{b}_P := \frac{1}{2}(\tilde{b} + \tilde{b}^\dagger), \qquad \tilde{b}_Q := \frac{1}{2i}(\tilde{b} - \tilde{b}^\dagger)$$

i.e.

$$\tilde{b} = \tilde{b}_P + i\tilde{b}_Q, \qquad \tilde{b}^\dagger = \tilde{b}_P - i\tilde{b}_Q$$

we finally arrive at the system of differential equations

$$\frac{db_P}{dt} = -2\Lambda b_P(t), \qquad \frac{db_Q}{dt} = 2\Lambda b_Q(t)$$

with the solution of the initial value problem

$$b_P(t) = b_P(0)e^{-2\Lambda t}, \qquad b_Q(t) = b_Q(0)e^{2\Lambda t}.$$

If the state is the vacuum state $|0\rangle$ we obtain a squeezed state. For the *electric field* $E(\mathbf{r}, t)$ we have

$$\begin{aligned} E(\mathbf{r}, t) &= i\mathcal{E}(b(t)e^{i\mathbf{k}\cdot\mathbf{r}} - b^\dagger(t)e^{-i\mathbf{k}\cdot\mathbf{r}}) \\ &= i\mathcal{E}(\tilde{b}(t)e^{i\mathbf{k}\cdot\mathbf{r}-i\omega t} - \tilde{b}^\dagger(t)e^{-i\mathbf{k}\cdot\mathbf{r}-i\omega t}) \\ &= -2\mathcal{E}(b_P(t)\sin(\mathbf{k}\cdot\mathbf{r} - \omega t) + b_Q(t)\cos(\mathbf{k}\cdot\mathbf{r} - \omega t)). \end{aligned}$$

Thus $b_P$ and $b_Q$ are the amplitudes of two quadrature components of the electric field. They are measurable by phase sensitive detection.

**Problem 6.**    Consider the *squeezing operator*

$$S(s) := \exp\left(\frac{1}{2}s(b^2 - b^{\dagger 2})\right)$$

where $s \in \mathbb{R}$. Find the operators

$$S(s)\hat{q}S^\dagger(s), \qquad S(s)\hat{p}S^\dagger(s)$$

where

$$\hat{q} := \frac{1}{\sqrt{2}}(b + b^\dagger), \qquad \hat{p} := -\frac{i}{\sqrt{2}}(b - b^\dagger).$$

**Solution 6.**    We obtain

$$S(s)\hat{q}S^\dagger(s) = e^{-s}\hat{q}, \qquad S(s)\hat{p}S^\dagger(s) = e^{s}\hat{p}.$$

**Problem 7.**   Let $\zeta \in \mathbb{C}$ and

$$\hat{G} := \frac{1}{2}(\zeta b^\dagger b^\dagger - \zeta^* bb).$$

(i) Calculate the commutators $[\hat{G}, b]$, $[\hat{G}, [\hat{G}, b]]$.
(ii) Let

$$S(\zeta) := \exp\left(\frac{1}{2}(\zeta b^\dagger b^\dagger - \zeta^* bb)\right) \equiv \exp \hat{G}.$$

Find $S(\zeta)bS(-\zeta)$, $S(\zeta)b^\dagger S(-\zeta)$.

**Solution 7.**   (i) We have

$$[\hat{G}, b] = \frac{1}{2}\zeta[b^\dagger b^\dagger, b] = -\zeta b^\dagger$$

and

$$[\hat{G}, [\hat{G}, b]] = [\hat{G}, -\zeta b^\dagger] = -\zeta[\hat{G}, b^\dagger] = |\zeta|^2 b.$$

(ii) Using the results from (i), we find

$$S(\zeta)bS(-\zeta) = (\cosh(s))b - e^{i\theta}(\sinh(s))b^\dagger$$

where $\zeta = re^{i\phi}$. Likewise, we find

$$S(\zeta)b^\dagger S(-\zeta) = (\cosh(s))b^\dagger - e^{-i\theta}(\sinh(s))b.$$

**Problem 8.**   Consider the unbounded linear operators

$$K_+ := \frac{1}{2}b^{\dagger 2}, \quad K_- := \frac{1}{2}b^2, \quad K_0 := \frac{1}{2}(b^\dagger b + \frac{1}{2}I)$$

$$A^\dagger := b^\dagger, \quad A := b$$

where $I$ is the identity operator.
(i) Show that these operators form a Lie algebra.
(ii) Consider the operator

$$P := \zeta K_+ - \zeta^* K_- + \alpha A^\dagger - \alpha^* A$$

where $\zeta$ and $\alpha$ are complex numbers. Let

$$V = e^{\beta K_+}e^{\epsilon A^\dagger}e^{\gamma K_0}e^{\nu I}e^{\delta K_-}e^{\eta A}$$

where $\beta$, $\epsilon$, $\gamma$, $\nu$, $\delta$ and $\eta$ are complex numbers. Let $e^P = V$. Find $\gamma$, $\beta$, $\delta$, $\epsilon$, $\eta$, $\nu$ as functions of $\zeta$ and $\alpha$.

**Solution 8.**    (i) For the commutators we obtain

$$[K_0, K_\pm] = \pm K_\pm, \quad [K_+, K_-] = -2K_0, \quad [K_+, A] = -A^\dagger, \quad [K_-, A^\dagger] = A$$

$$[K_0, A^\dagger] = \frac{1}{2}A^\dagger, \quad [K_0, A] = -\frac{1}{2}A, \quad [A, A^\dagger] = I.$$

This Lie algebra refers to squeezed coherent states.

(ii) We write the complex numbers $\zeta$ and $\alpha$ in terms of real numbers $\lambda$, $\mu$, $\theta$ and $\phi$ as $\zeta = re^{i\theta}$, $\alpha = se^{i\phi}$. We use the formula

$$e^A B e^{-A} = \sum_{n=0}^{\infty} \frac{[A, B]_n}{n!}$$

where the *repeated commutator* $[A, B]_n$ is defined by

$$[A, B]_n := [A, [A, B]_{n-1}]$$

with $[A, B]_0 := B$. We find

$$e^P A e^{-P} = \cosh(r)A - e^{i\theta}\sinh(r)A^\dagger + \frac{s}{r}((\cosh(r) - 1)e^{i(\theta-\phi)} - \sinh(r)e^{i\phi})$$

$$e^P A^\dagger e^{-P} = \cosh(r)A^\dagger - e^{-i\theta}\sinh(r)A + \frac{s}{r}((\cosh(r)-1)e^{i(\phi-\theta)} - \sinh(r)e^{-i\phi}).$$

The corresponding similarity transformations, induced by the operator $V$, are

$$VAV^{-1} = e^{-\gamma/2}(A - \beta A^\dagger - \epsilon I)$$
$$VA^\dagger V^{-1} = (e^{\gamma/2} - \beta\delta e^{-\gamma/2})A^\dagger + \delta e^{-\gamma/2}A + \eta I - \epsilon\delta e^{-\gamma/2}I.$$

From

$$e^P A e^{-P} = VAV^{-1}, \qquad e^P A^\dagger e^{-P} = VA^\dagger V^{-1}$$

we find, by separating out terms with $A^\dagger$, $A$ and $I$, that

$$\gamma = -2\ln(\cosh(r))$$
$$\beta = e^{i\theta}\tanh(r)$$
$$\delta = -e^{-i\theta}\tanh(r)$$
$$\epsilon = -\frac{s}{r\cosh(r)}((\cosh(r) - 1)e^{i(\theta-\phi)} - \sinh(r)e^{i\phi})$$
$$\eta = -\frac{s}{r\cosh(r)}((\cosh(r) - 1)e^{-i(\theta-\phi)} - \sinh(r)e^{-i\phi}).$$

The coefficient $\nu$ cannot be found by this method. How can we determine $\nu$? One finds

$$\nu = -\frac{s^2}{r^2\cosh(r)}((\cosh(r) - 1) + i\sin(\theta - 2\phi)(\sinh(r) - r\cosh(r))).$$

**Problem 9.** (i) Show that the operators

$$K_+ = \frac{1}{2}b^\dagger b^\dagger, \qquad K_- = \frac{1}{2}bb, \qquad K_0 = \frac{1}{2}\left(b^\dagger b + \frac{1}{2}I\right)$$

are generators of the Lie algebra $su(1,1)$.
(ii) Let $D(\beta)$ be the displacement operator and

$$S(\zeta) = \exp(\zeta K_+ - \zeta^* K_-), \qquad \zeta \in \mathbb{C}.$$

We define the state

$$|(\beta, \zeta)\rangle := D(\beta)S(\zeta)|0\rangle.$$

Show that the ordering $D(\beta)S(\zeta)$ versus $S(\zeta)D(\beta)$ is unitarily equivalent.

**Solution 9.** (i) We have $[K_0, K_+] = K_+$, $[K_0, K_-] = -K_-$, $[K_+, K_-] = -2K_0$.
(ii) We have $D(\beta)S(\zeta) = S(\zeta)D(\gamma)$, $\gamma = \beta\cosh(s) - \beta^* e^{i\theta}\sinh(s)$ and $z = se^{i\theta}$.

**Problem 10.** Consider the squeeze operator

$$S(\zeta) = \exp\left(\frac{1}{2}(\zeta^* b^2 - \zeta b^{\dagger^2})\right)$$

with $\zeta = |\zeta|\exp(i(\phi + \theta))$. Is $S(\zeta')S(\zeta'') = S(\zeta' + \zeta'')$?

**Solution 10.** The answer is no. We find

$$S(\zeta')S(\zeta'') = \exp\left(-i\phi\left(b^\dagger b + \frac{1}{2}I\right)\right)S(\zeta) \tag{1}$$

with

$$\zeta' = |\zeta'|\exp(i\theta'), \quad \mu' = \cosh|\zeta'|, \quad \nu' = \exp(i\theta')\sinh|\zeta'|$$

and $|\mu'|^2 - |\nu'|^2 = 1$ with analogous relations for $\zeta''$, $\mu''$ and $\nu''$. In (1) $\zeta$ is obtained from

$$\zeta = |\zeta|\exp(i(\phi + \theta)), \qquad |\zeta| = \tanh^{-1}\left(\frac{|\nu|}{|\mu|}\right)$$

with $\mu = \mu'\mu'' + \nu'^* \nu''$, $\nu = \mu'^* \nu'' + \nu'\mu''$ and $|\mu|^2 - |\nu|^2 = 1$.

**Problem 11.** (i) A single mode squeeze operator is defined by

$$S(z) = \exp\left(\frac{1}{2}(z(b^\dagger)^2 - z^* b^2)\right)$$

where $z \in \mathbb{C}$ with $z = re^{i\theta}$. Write $S(z)$ in the form $S(z) = \exp(iH(z))$. Find $H(z)$.

(ii) Calculate

$$R^\dagger(\phi)H(z)R(\phi)$$

where $R(\phi) = \exp(i\phi b^\dagger b)$ is a single-mode rotation. Set $\phi = \theta/2$ and thus find $H(r)$.

(iii) Use the transformation

$$b = \frac{x + ip}{\sqrt{2}}, \qquad b^\dagger = \frac{x - ip}{\sqrt{2}}$$

and express $H(r)$ in terms of $x$ and $p$.

(iv) Discuss the $H(r)$ as a Hamilton function.

(v) Write down the eigenvalue equation $H\psi(x) = E\psi$ with $p = -id/dx$. Discuss the spectrum.

**Solution 11.**    (i) Obviously we have

$$H(z) = \frac{1}{2i}(z(b^\dagger)^2 - z^* b^2).$$

(ii) We find

$$R^\dagger(\phi)H(z)R(\phi) = H(ze^{-2i\phi}).$$

Thus for $\phi = \theta/2$ we obtain $H(r)$.

(iii) Inserting the transformation yields

$$H(r) = -\frac{r}{2}(xp + px).$$

(iv) The Hamilton function $H(r) = -rxp$ leads to the Hamilton equations of motion

$$\frac{dx}{dt} = -rx, \qquad \frac{dp}{dt} = rp.$$

This describes damping of $x$ and the pumping of $p$. This is a classical picture of a squeezing process.

(v) Since

$$H(r)\psi(x) = -\frac{r}{2}(xp + px)\psi(x) = irx\frac{d\psi}{dx} + i\frac{r}{2}\psi(x)$$

it follows that the eigenvalue equation is

$$x\frac{d}{dx}\psi(x) = -\left(i\frac{E}{r} + \frac{1}{2}\right)\psi(x).$$

$H(r)$ has a purely continuous spectrum covering the whole real axis. $H(r)$ is also parity invariant ($x \to -x$). Therefore each generalized eigenvalue

$E \in \mathbb{R}$ is doubly degenerated. The eigenvectors in the sense of generalized functions are

$$\psi_{\pm}^{E}(x) = \frac{1}{\sqrt{2\pi r}} x_{\pm}^{-(iE/r+1/2)}$$

where

$$x_{+}^{\lambda} := \begin{cases} x^{\lambda} & \text{for } x \geq 0 \\ 0 & \text{for } x < 0 \end{cases}, \qquad x_{-}^{\lambda} := \begin{cases} 0 & \text{for } x \geq 0 \\ |x|^{\lambda} & \text{for } x < 0 \end{cases}$$

and $\lambda \in \mathbb{C}$.

**Problem 12.** Consider the linear operator

$$\hat{D} := \frac{1}{2}(\hat{q}\hat{p} + \hat{p}\hat{q}).$$

We set $\hbar = 1$.
(i) Find the commutators $[\hat{D}, \hat{q}]$ and $[\hat{D}, \hat{p}]$.
(ii) We consider the linear operator $S_{\epsilon} := \exp(-i\epsilon\hat{D})$, $\epsilon \in \mathbb{R}$. We define

$$S_{\epsilon}^{\dagger}\hat{q}S_{\epsilon} := \exp(i\epsilon \text{ad}\hat{D})\hat{q} \equiv \sum_{n=0}^{\infty} \frac{(i\epsilon)^{n}}{n!}(\text{ad}\hat{D})^{n}\hat{q} \qquad (1)$$

where

$$(\text{ad}\hat{D})\hat{q} := [\hat{D}, \hat{q}].$$

Calculate the operators $S_{\epsilon}^{\dagger}\hat{q}S_{\epsilon}$ and $S_{\epsilon}^{\dagger}\hat{p}S_{\epsilon}$.
(iii) Let

$$b = \frac{\sqrt{m\omega}}{\sqrt{2}}\left(\hat{q} + i\frac{\hat{p}}{m\omega}\right), \qquad b^{\dagger} = \frac{\sqrt{m\omega}}{\sqrt{2}}\left(\hat{q} - i\frac{\hat{p}}{m\omega}\right).$$

Express $\hat{D}$ in terms of $b$ and $b^{\dagger}$.
(iv) Consider the normalized state $|\epsilon\rangle := S_{\epsilon}|0\rangle$. Calculate the expectation values

$$\langle\epsilon|\hat{q}|\epsilon\rangle, \quad \langle\epsilon|\hat{p}|\epsilon\rangle, \quad \langle\epsilon|\hat{q}^{2}|\epsilon\rangle, \quad \langle\epsilon|\hat{p}^{2}|\epsilon\rangle.$$

**Solution 12.** (i) Since $[\hat{q}, \hat{p}] = iI$ we find

$$[\hat{D}, \hat{q}] = -i\hat{q}, \qquad [\hat{D}, \hat{p}] = i\hat{p}.$$

(ii) Using the result from (i) and the definition (1) we find the operator

$$S_{\epsilon}^{\dagger}\hat{q}S_{\epsilon} = e^{\epsilon}\hat{q}.$$

Using the result from (i) and the definition (1) with $\hat{q}$ replaced by $\hat{p}$, we find

$$S_{\epsilon}^{\dagger}\hat{p}S_{\epsilon} = e^{-\epsilon}\hat{p}.$$

(iii) First we express $\hat{q}$ and $\hat{p}$ in terms of $b$ and $b^\dagger$. Using the commutation relation $[\hat{q}, \hat{p}] = iI$, we obtain

$$\hat{D} = 2i(b^\dagger b^\dagger - bb).$$

(iv) Using the results from (i) through (iii) we find the expectation values

$$\langle\epsilon|\hat{q}|\epsilon\rangle = \langle 0|S_\epsilon^\dagger \hat{q} S_\epsilon|0\rangle = e^\epsilon\langle 0|\hat{q}|0\rangle = 0$$

$$\langle\epsilon|\hat{p}|\epsilon\rangle = \langle 0|S_\epsilon^\dagger \hat{p} S_\epsilon|0\rangle = e^{-\epsilon}\langle 0|\hat{p}|0\rangle = 0$$

$$\langle\epsilon|\hat{q}^2|\epsilon\rangle = e^{2\epsilon}\langle 0|\hat{q}^2|0\rangle = e^{2\epsilon}\frac{1}{2m\omega} = (\Delta q)^2$$

$$\langle\epsilon|\hat{p}^2|\epsilon\rangle = e^{-2\epsilon}\langle 0|\hat{p}^2|0\rangle = e^{-2\epsilon}\frac{m\omega}{2} = (\Delta p)^2.$$

**Problem 13.**    Consider the *uncertainty relation*

$$(\langle\psi|\hat{A}^2|\psi\rangle - \langle\psi|\hat{A}|\psi\rangle^2)(\langle\psi|\hat{B}^2|\psi\rangle - \langle\psi|\hat{B}|\psi\rangle^2) \geq \frac{1}{4}|\langle\psi|[\hat{A},\hat{B}]|\psi\rangle|^2$$

where $\hat{A}$ and $\hat{B}$ are observable, $[\hat{A},\hat{B}]$ denotes the commutator and $|\psi\rangle$ is a normalized state. Let $b^\dagger$, $b$ be Bose creation and annihilation operators and

$$\hat{A} = \frac{1}{\sqrt{2}}(ib - ib^\dagger), \qquad \hat{B} = \frac{1}{\sqrt{2}}(b + b^\dagger).$$

(i) Let $|\psi\rangle = |\beta\rangle$ be a coherent state ($\beta \in \mathbb{C}$). Find the left-hand and right-hand side of the uncertainty relation.
(ii) Let $|\psi\rangle = |n\rangle$ be a number state ($n = 0, 1, 2, \ldots$). Find the left-hand and right-hand side of the uncertainty relation.
(iii) Let $|\psi\rangle = |\zeta\rangle$ be a squeezed state ($\zeta \in \mathbb{C}$). Find the left-hand and right-hand side of the uncertainty relation.

**Solution 13.**    (i) For the commutator we find

$$[\hat{A}, \hat{B}] = iI$$

where $I$ is the identity operators. Thus with $\langle\beta|\beta\rangle = 1$ we obtain for the right-hand side

$$\frac{1}{4}|\langle\beta|iI|\beta\rangle|^2 = \frac{1}{4}.$$

Now $b|\beta\rangle = \beta|\beta\rangle$, $\langle\beta|b^\dagger = \langle\beta|\overline{\beta}$. It follows that

$$\langle\psi|\hat{A}|\psi\rangle^2 = \frac{1}{2}(-\beta^2 - \overline{\beta}^2 + 2\beta\overline{\beta})$$

$$\langle\psi|\hat{B}|\psi\rangle^2 = \frac{1}{2}(\beta^2 + \overline{\beta}^2 + 2\beta\overline{\beta})$$

$$\langle\psi|\hat{A}^2|\psi\rangle = \frac{1}{2}(1 - \beta^2 - \overline{\beta}^2 + 2\beta\overline{\beta})$$

$$\langle\psi|\hat{B}^2|\psi\rangle = \frac{1}{2}(1 + \beta^2 + \overline{\beta}^2 + 2\beta\overline{\beta})$$

and

$$\langle\beta|\hat{A}^2|\beta\rangle - \langle\beta|\hat{A}|\beta\rangle^2 = \frac{1}{2}, \quad \langle\beta|\hat{B}^2|\beta\rangle - \langle\beta|\hat{B}|\beta\rangle^2 = \frac{1}{2}.$$

It follows that the uncertainty relation is an equality for the present case.
(ii) With $\langle n|n\rangle = 1$ for the right-hand side we have again $\frac{1}{4}$. From

$$\langle n|\hat{A}|n\rangle = 0, \qquad \langle n|\hat{B}|n\rangle = 0$$

and

$$\langle n|\hat{A}^2|n\rangle = \frac{1}{2}(1 + 2n), \qquad \langle n|\hat{B}^2|n\rangle = \frac{1}{2}(1 + 2n).$$

Thus we have the inequality

$$\frac{1}{4}(1 + 2n)^2 \geq \frac{1}{4}.$$

So if $n = 0$ we have an equality.
(iii) We set $\zeta = se^{i\theta}$. From $|\zeta\rangle = S(\zeta)|0\rangle$, where $S(\zeta)$ is the squeezing operator we obtain

$$\langle\zeta|b|\zeta\rangle = \langle0|S^{-1}(\zeta)bS(\zeta)|0\rangle = \langle0|(\cosh(s)b - e^{i\theta}\sinh(s)b^\dagger)|0\rangle = 0$$

$$\langle\zeta|b^\dagger|\zeta\rangle = \langle0|S^{-1}(\zeta)b^\dagger S(\zeta)|0\rangle = \langle0|(\cosh(s)b^\dagger - e^{-i\theta}\sinh(s)b)|0\rangle = 0.$$

Now

$$\langle\zeta|b^2|\zeta\rangle = \langle0|S^{-1}(\zeta)bS(\zeta)S^{-1}(\zeta)bS(\zeta)|0\rangle = -e^{i\theta}\cosh(s)\sinh(s)$$

$$\langle\zeta|(b^\dagger)^2|\zeta\rangle = \langle0|S^{-1}(\zeta)b^\dagger S(\zeta)S^{-1}(\zeta)b^\dagger S(\zeta)|0\rangle = -e^{-i\theta}\cosh(s)\sinh(s).$$

Hence

$$\langle\zeta|\hat{A}^2|\zeta\rangle = \frac{1}{2}\cosh(2s) + \frac{1}{2}\sinh(2s)\cos(\theta)$$

and

$$\langle\zeta|\hat{B}^2|\zeta\rangle = \frac{1}{2}\cosh(2s) - \frac{1}{2}\sinh(2s)\cos(\theta)$$

and the inequality follows

$$\frac{1}{4}(\cosh^2(2s) - \sinh^2(2s)\cos^2(\theta)) \geq \frac{1}{4}.$$

With $\cosh^2(2s) - \sinh^2(2s) = 1$ we have an equality for $\theta = 0$ and $\theta = \pi$.

**Problem 14.**    Consider the operator

$$U(z) := e^{zb_1^\dagger b_2 - \bar{z}b_2^\dagger b_1}$$

where $b_1^\dagger, b_2^\dagger$ are Bose creation operators and $b_1, b_2$ are Bose annihilation operators and $z \in \mathbb{C}$. Find the operators

$$U(z)b_1U(z)^{-1}, \qquad U(z)b_2U(z)^{-1}.$$

**Solution 14.**    We have

$$U^{-1}(z) = e^{-zb_1^\dagger b_2 + \bar{z}b_2^\dagger b_1}.$$

Thus

$$U(z)b_1U(z)^{-1} = \cos(|z|)b_1 - \frac{z\sin(|z|)}{|z|}b_2$$

$$U(z)b_2U(z)^{-1} = \cos(|z|)b_2 + \frac{\bar{z}\sin(|z|)}{|z|}b_1.$$

We can write

$$(U(z)b_1U(z)^{-1}, U(z)b_2U(z)^{-1}) = (b_1, b_2) \begin{pmatrix} \cos(|z|) & \frac{\bar{z}\sin(|z|)}{|z|} \\ -\frac{z\sin(|z|)}{|z|} & \cos(|z|) \end{pmatrix}$$

where the matrix on the right-hand side is an element of the Lie group $SU(2)$.

**Problem 15.**    Consider the operator

$$U(z) := e^{zb_1^\dagger b_2^\dagger - \bar{z}b_2 b_1}$$

where $b_1^\dagger, b_2^\dagger$ are Bose creation operators and $b_1, b_2$ are Bose annihilation operators and $z \in \mathbb{C}$. Find the operators

$$U(z)b_1U(z)^{-1}, \qquad U(z)b_2^\dagger U(z)^{-1}.$$

**Solution 15.**    We have

$$U^{-1}(z) = e^{-zb_1^\dagger b_2^\dagger + \bar{z}b_2 b_1}.$$

Thus

$$U(z)b_1U(z)^{-1} = \cosh(|z|)b_1 - \frac{z\sinh(|z|)}{|z|}b_2^\dagger$$

$$U(z)b_2^\dagger U(z)^{-1} = \cosh(|z|)b_2^\dagger - \frac{\bar{z}\sinh(|z|)}{|z|}b_1.$$

We can write in matrix form

$$\left(U(z)b_1U(z)^{-1}, U(z)b_2^\dagger U(z)^{-1}\right) = (b_1, b_2^\dagger)\begin{pmatrix} \cosh(|z|) & -\frac{\bar{z}\sinh(|z|)}{|z|} \\ -\frac{z\sinh(|z|)}{|z|} & \cosh(|z|) \end{pmatrix}$$

where the matrix on the right-hand side is an element of the Lie group $SU(1,1)$.

**Problem 16.** Describe how two-mode squeezed state can be generated.

**Solution 16.** Two mode squeezed states can be generated either by entangling two independent single-mode squeezed states via a 50:50 beam splitter or by employing the non-degenerate operation of a nonlinear medium in the presence of two incoming modes. The unitary operator describing two-mode squeezing is

$$U_{12}(\zeta) = \exp(-i(\zeta b_1 b_2 + \zeta^* b_1^\dagger b_2^\dagger)/2)$$

where $\zeta \in \mathbb{C}$ is the squeezing parameter.

**Problem 17.** Consider the two-mode squeezing operator

$$S_2(\zeta) = \exp(\zeta b_1^\dagger b_2^\dagger - \zeta^* b_1 b_2).$$

Let $|0\rangle \otimes |0\rangle$ be the two-mode vacuum state. Find the normalized state

$$S_2(\zeta)(|0\rangle \otimes |0\rangle)$$

expressed in number states.

**Solution 17.** We obtain

$$S_2(\zeta)(|0\rangle \otimes |0\rangle) = \frac{1}{\sqrt{\mu}}\sum_{k=0}^{\infty}\left(\frac{\nu}{\mu}\right)^k |k\rangle \otimes |k\rangle$$

where

$$\zeta = se^{i\theta}, \quad \mu = \cosh(s), \quad \nu = e^{i\theta}\sinh(s).$$

This state is known as *two mode squeezed vacuum* or *twin beam state*.

**Problem 18.** Let $|0\rangle$ be the vacuum state. Then we define the coherent squeezed state as

$$|\beta, \zeta\rangle := D(\beta)S(\zeta)|0\rangle$$

where $D(\beta)$ is the displacement operator and $S(\zeta)$ is the squeezing operator with $\beta, \zeta \in \mathbb{C}$.

(i) Show that the state $|\beta, \zeta\rangle$ is normalized.

(ii) Show that $S(\zeta)D(\beta)|0\rangle \neq |\beta, \zeta\rangle$.

**Solution 18.**  (i) We have

$$\langle \beta, \zeta | \beta, \zeta \rangle = \langle 0 | S(-\zeta)D(-\beta)D(\beta)S(\zeta)|0\rangle = 1.$$

(ii) We have

$$
\begin{aligned}
S(\zeta)D(\alpha)|0\rangle &= S(\zeta)D(\beta)S(-\zeta)S(\zeta)|0\rangle \\
&= \exp(\beta(b^\dagger \cosh(s) + be^{-i\phi}\sinh(s)) \\
&\quad - \beta^*(b\cosh(s) + b^\dagger e^{i\theta}\sinh(s)))S(\zeta)|0\rangle \\
&= D(\beta\cosh(s) - \beta^* e^{i\theta}\sinh(s))S(\zeta)|0\rangle \\
&= |\beta\cosh(s) - \beta^* e^{i\theta}\sinh(s), \zeta\rangle
\end{aligned}
$$

where we used that

$$S(\zeta)D(\beta)S(-\zeta) =$$

$$\exp(\beta(b^\dagger \cosh(s) + be^{-i\theta}\sinh(s)) - \beta^*(b\cosh(s) + b^\dagger e^{i\theta}\sinh(s))).$$

Thus $|\beta, \zeta\rangle$ can be written as

$$|\beta, \zeta\rangle = S(\zeta)D(\beta\cosh(s) + \beta^*\sinh(s))|0\rangle.$$

**Problem 19.**  Let $\hat{n} = b^\dagger b$. Find $\langle \beta, \zeta | \hat{n} | \beta, \zeta \rangle$.

**Solution 19.**  We have

$$
\begin{aligned}
\langle \beta, \zeta | \hat{n} | \beta, \zeta \rangle &= \langle 0 | (b^\dagger \cosh(s) - be^{-\theta}\sinh(s) + \beta^* I) \\
&\quad \times (b\cosh(s) - b^\dagger e^{i\theta}\sinh(s) + \beta I)|0\rangle \\
&= \sinh^2(s) + |\beta|^2.
\end{aligned}
$$

**Problem 20.**  Let $|\alpha\rangle$ be a coherent state. The *Husimi distribution* of a quantum wave function $|\psi\rangle$ is given by

$$\rho^H_{|\psi\rangle}(\alpha) = |\langle \psi | \alpha \rangle|^2.$$

(i) A *coherent squeezed state* $|\gamma, \beta\rangle$ is defined as

$$|\zeta, \beta\rangle := D(\beta)S(\zeta)|0\rangle$$

where $D(\beta)$ is the displacement operator and $S(\zeta) := \exp((\zeta^* b^2 - \zeta(b^\dagger)^2)/2)$. The coherent squeezed states $|\zeta, \beta\rangle$ also minimize the uncertainty relation, however, the variance of both canonically coupled variables are not equal. The modulus $g$ of the complex number $\zeta = g e^{2i\theta}$ determines the strength of squeezing, $s = e^g - 1$, while the angle $\theta$ orients the squeezing axis. Find the Husimi distribution

$$\rho^H_{|\zeta,\beta\rangle}(\alpha) = |\langle \alpha | \zeta, \beta \rangle|^2$$

of a coherent squeezed state $|\zeta, \beta\rangle$.

**Solution 20.** Let $\alpha = \alpha_1 + i\alpha_2$, where $\alpha_1, \alpha_2 \in \mathbb{R}$. We obtain

$$\rho^H_{|\zeta,\beta\rangle}(\alpha) = |\langle \alpha | \zeta, \beta \rangle|^2 = \exp(-(\Re(\beta) - \alpha_1)^2/(s+1)^2 - (\Im(\beta) - \alpha_2)^2(s+1)^2).$$

**Problem 21.** Let $(r \in \mathbb{R})$

$$S(r) = \exp\left(\frac{1}{2} r(b^2 - b^{\dagger 2})\right), \qquad D(\beta) = \exp(\beta b^\dagger - \beta^* b).$$

Find the operator $S^\dagger(r) D(\beta) S(r)$.

**Solution 21.** We find

$$S^\dagger(r) D(\beta) S(r) = D(\Re(\beta) e^r + i\Im(\beta) e^{-r}).$$

**Problem 22.** Let $D(\beta)$ be the displacement operator

$$D(\beta) = \exp(\beta b^\dagger - \beta^* b)$$

and

$$S(r, \phi) = \exp(r e^{i\phi} (b^\dagger)^2/2 - r e^{-i\phi} b^2/2)$$

be the squeeze operator with the squeeze factor $r \geq 0$ and squeeze angle $\phi \in (-\pi, \pi]$. Let

$$\rho_T := \frac{1}{\bar{n} + 1} \sum_{n=0}^{\infty} \left(\frac{\bar{n}}{\bar{n} + 1}\right)^n |n\rangle \langle n|$$

be the *Bose-Einstein density operator* with the mean occupancy

$$\bar{n} = \left(\exp\left(\frac{\hbar \omega}{k_B T}\right) - 1\right)^{-1}.$$

Consider the displaced squeezed thermal state

$$\rho = D(\beta)S(r,\phi)\rho_T S^\dagger(r,\phi)D^\dagger(\beta). \tag{1}$$

The *Weyl expansion* of the density operator

$$\rho = \frac{1}{\pi}\int_{\mathbb{C}} d^2\lambda\, \chi(\lambda)D(-\lambda)$$

with $d^2\lambda = d\Re(\lambda)d\Im(\lambda)$ provides the one-to-one correspondence between the density operator $\rho$ and its characteristic function

$$\chi(\lambda) := \mathrm{tr}(\rho D(\lambda)) \equiv \sum_{n=0}^{\infty}\langle n|\rho D(\lambda)|n\rangle. \tag{2}$$

A Gaussian state has a characteristic function of the form

$$\chi(\lambda) = \exp\left(-\left(A+\frac{1}{2}\right)|\lambda|^2 - \frac{1}{2}B^*\lambda^2 - \frac{1}{2}B(\lambda^*)^2 + C^*\lambda - C\lambda^*\right)$$

with $A > 0$. Find the coefficients $A$, $B$, $C$ for the given $\rho$.

**Solution 22.**    Calculating (2) where $\rho$ is given by (1) and comparing coefficients yields

$$A = \left(\overline{n}+\frac{1}{2}\right)\cosh(2r) - \frac{1}{2}, \quad B = -\left(\overline{n}+\frac{1}{2}\right)e^{i\phi}\sinh(2r), \quad C = \beta.$$

**Problem 23.**    Let $S(\zeta)$ be the one-mode squeeze operator with $\zeta = |\zeta|e^{i\theta}$. Let $D(\beta)$ be the displacement operator. Find $S(\zeta)D(\beta)S^{-1}(\zeta)$.

**Solution 23.**    We obtain

$$S(\zeta)D(\beta)S^{-1}(\zeta) = D(\widetilde{\beta}), \quad \widetilde{\beta} = \cosh(|\zeta|)\beta + e^{i\theta}\sinh(|\zeta|)\beta^*.$$

**Programming Problem**

**Problem 1.**    Give a computer algebra implementation of squeezed states.

**Solution 1.**    The Bose creation and annihilation operators are denoted by b and bd and b*bd==1+bd*b is implemented.

```
// squeezed.cpp
#include <iostream>
#include "symbolicc++.h"
using namespace std;

int main(void)
{
Symbolic eps("eps"), zeta("zeta"), zetab("zetab");
Symbolic b("b"), bd("bd");
b = ~b; bd = ~bd;  // b and bd are noncommutative
Symbolic S, Sd;
Symbolic arg("arg");
arg = -zeta*bd*bd/2 + zetab*b*b/2;
S = exp(eps*arg);
Sd = exp(-eps*arg);
Symbolic result1("result1");
result1 = S*arg*b*Sd - S*b*arg*Sd;
result1 = result1.subst_all(b*bd==1+bd*b);
cout << result1 << endl << endl;
Symbolic result2("result2");
result2 = S*arg*bd*Sd - S*bd*arg*Sd;
result2 = result2.subst_all(b*bd==1+bd*b);
cout << result2 << endl;
return 0;
}
```

The output is

```
e^(-1/2*eps*zeta*bd^(2)+1/2*eps*zetab*b^(2))*bd*zeta*
e^(1/2*eps*zeta*bd^(2)-1/2*eps*zetab*b^(2))

e^(-1/2*eps*zeta*bd^(2)+1/2*eps*zetab*b^(2))*b*zetab*
e^(1/2*eps*zeta*bd^(2)-1/2*eps*zetab*b^(2))
```

## 20.3 Supplementary Problems

**Problem 1.** Let $|\beta\rangle$ be a coherent state and $|\zeta\rangle$ be a squeezed state. Calculate

$$\langle\beta|\zeta\rangle, \quad |\langle\beta|\zeta\rangle|^2.$$

**Problem 2.** Let $\hat{n} = b^\dagger b$ be the number operator and $S(\zeta)$ be the squeezed operator. Find the commutator

$$[\hat{n}, S(\zeta)].$$

**Problem 3.**    Consider the displacement operator, squeezing operator and rotation operator

$$D(\beta) = e^{\beta b^\dagger - \bar{\beta} b}, \quad S(\zeta) = e^{(\zeta (b^\dagger)^2 - \bar{\zeta} b^2)/2}, \quad R(\phi) = e^{-i\phi b^\dagger b}$$

where $\beta, \zeta \in \mathbb{R}$. Find the commutators

$$[D(\beta), S(\zeta)], \quad [D(\beta), R(\phi)], \quad [R(\phi), S(\zeta)].$$

**Problem 4.**    Let $b^\dagger$, $b$ be Bose creation and annihilation operators and $\gamma \in \mathbb{R}$. Calculate

$$e^{\gamma((b^\dagger)^2 - b^2)} b e^{-\gamma((b^\dagger)^2 - b^2)}, \quad e^{\gamma((b^\dagger)^2 - b^2)} b^\dagger e^{-\gamma((b^\dagger)^2 - b^2)}$$

utilizing

$$e^A B e^{-A} = B + [A, B] + \frac{1}{2!}[A, [A, B]] + \frac{1}{3!}[A, [A, [A, B]]] + \cdots$$

**Problem 5.**    Let $\zeta \in \mathbb{C}$. The squeezing operator is defined as

$$S(\zeta) = \exp\left(\frac{1}{2}\zeta(b^\dagger)^2 - \frac{1}{2}\bar{\zeta}b^2\right).$$

We set $\zeta = se^{i\theta}$. Show that

$$S^\dagger(\zeta) b S(\zeta) = \cosh(s)b + e^{i\theta}\sinh(s)b^\dagger.$$

$$S^\dagger(\zeta) b^\dagger S(\zeta) = \cosh(s)b^\dagger + e^{-i\theta}\sinh(s)b$$

$$\langle \zeta | b^\dagger b | \zeta \rangle = \langle 0 | S^\dagger(\zeta) b^\dagger S(\zeta) S^\dagger(\zeta) b S(\zeta) | 0 \rangle = \sinh^2(s).$$

**Problem 6.**    Let $D(\beta)$ be the displacement operator and $S(\zeta)$ be the squeezing operator with

$$\beta = re^{i\phi}, \quad \zeta = se^{i\theta}.$$

Show that

$$S(\zeta)D(\beta)|0\rangle = D(\beta \cosh(s) - \bar{\beta}e^{i\theta}\sinh(s))S(\zeta)|0\rangle.$$

**Problem 7.**    Let $S(\zeta)$, $D(\beta)$, $R(\alpha)$ be the squeezing operator, displacement operator and phase shift operator, where

$$R(\alpha) = \exp(i\alpha b^\dagger b).$$

Let $\rho$ be a density operator. Then

$$\tilde{\rho} = D(\beta)R(\alpha)S(\zeta)\rho S^\dagger(\zeta)R^\dagger(\alpha)D^\dagger(\beta)$$

is a density operator. Calculate $\tilde{\rho}$ for

$$\rho = |n\rangle\langle n|, \quad \rho = |\beta\rangle\langle\beta|, \quad \rho = |\zeta\rangle\langle\zeta|.$$

**Problem 8.** Consider the two-mode squeezing operator

$$S(\zeta) = \exp(\zeta b_1^\dagger b_2^\dagger - \bar{\zeta} b_1 b_2), \quad \zeta = se^{i(\theta+\pi/2)}$$

where

$$b_1^\dagger = b^\dagger \otimes I, \quad b_2^\dagger = I \otimes b^\dagger.$$

Show that

$$S(\zeta)(|0\rangle \otimes |0\rangle) = \frac{1}{\cosh(s)} \sum_{n=0}^{\infty} e^{in(\theta+\pi/2)}(\tanh(s))^n |n\rangle \otimes |n\rangle.$$

The state is mode-entangled.

**Problem 9.** Let $b_1^\dagger, b_2^\dagger, b_3^\dagger, b_1, b_2, b_3$ be Bose creation and annihilation operators. Note that

$$[b_1^\dagger b_2 - b_1 b_2^\dagger, b_2^\dagger b_3 - b_2 b_3^\dagger] = b_1^\dagger b_3 - b_1 b_3^\dagger$$
$$[b_1^\dagger b_3 - b_1 b_3^\dagger, b_1^\dagger b_2 - b_1 b_2^\dagger] = b_2^\dagger b_3 - b_3^\dagger b_2.$$

Let $\theta_1, \theta_2, \theta_3 \in \mathbb{R}$ and consider the unitary operator

$$U(\theta_1, \theta_2, \theta_3) = e^{\theta_3(b_2 b_3^\dagger - b_2^\dagger b_3)} e^{\theta_2(b_2 b_1^\dagger - b_2^\dagger b_1)} e^{\theta_1(b_2 b_3^\dagger - b_2^\dagger b_3)}.$$

Find the $3 \times 3$ matrix $M$ such that

$$U(\theta_1,\theta_2,\theta_3)\begin{pmatrix} b_1^\dagger \\ b_2^\dagger \\ b_3^\dagger \end{pmatrix} = \begin{pmatrix} Ub_1^\dagger U^\dagger \\ Ub_2^\dagger U^\dagger \\ Ub_3^\dagger U^\dagger \end{pmatrix} = M(\theta_1,\theta_2,\theta_3)\begin{pmatrix} b_1^\dagger \\ b_2^\dagger \\ b_3^\dagger \end{pmatrix}.$$

**Problem 10.** Consider the unitary operator for two-mode squeezing

$$U(\zeta) = \exp(\zeta b_1^\dagger b_2^\dagger - \bar{\zeta} b_1 b_2).$$

The Hamilton operator with $\alpha \in \mathbb{R}$ is given by

$$\hat{H} = i\alpha(b_1^\dagger b_2^\dagger - b_1 b_2).$$

The Hamilton operator indicates that two photons are simultaneously created or annihilated. Find the $4 \times 4$ matrix $M(\zeta, \bar{\zeta})$ such that

$$
U(\zeta) \begin{pmatrix} b_1 \\ b_1^\dagger \\ b_2 \\ b_2^\dagger \end{pmatrix} U^\dagger(\zeta) \equiv \begin{pmatrix} U(\zeta) b_1 U^\dagger(\zeta) \\ U(\zeta) b_1^\dagger U^\dagger(\zeta) \\ U(\zeta) b_2 U^\dagger(\zeta) \\ U(\zeta) b_2^\dagger U^\dagger(\zeta) \end{pmatrix} = M(\zeta, \bar{\zeta}) \begin{pmatrix} b_1 \\ b_1^\dagger \\ b_2 \\ b_2^\dagger \end{pmatrix} .
$$

Show that

$$
M(\zeta, \bar{\zeta}) = \begin{pmatrix} \cosh(s) & 0 & 0 & -e^{i\theta}\sinh(s) \\ 0 & \cosh(s) & -e^{-i\theta}\sinh(s) & 0 \\ 0 & -e^{i\theta}\sinh(s) & \cosh(s) & 0 \\ -e^{-i\theta}\sinh(s) & 0 & 0 & \cosh(s) \end{pmatrix}
$$

where $\zeta = se^{i\theta}$. Show that the determinant of the matrix is $+1$.

# Chapter 21

# Trace and Partial Trace

---

## 21.1 Introduction

Let $|n\rangle$ be the number states $(n = 0, 1, \ldots)$ and $|\beta\rangle$ be a coherent state $(\beta \in \mathbb{C})$. The trace of a bounded linear operator $\hat{A}$ could be calculated using the number states $|n\rangle$

$$\text{tr}(\hat{A}) = \sum_{n=0}^{\infty} \langle n|\hat{A}|n\rangle$$

or using coherent states $|\beta\rangle$

$$\text{tr}(\hat{A}) = \frac{1}{\pi} \int_{\mathbb{C}} d^2\beta \langle \beta|\hat{A}|\beta\rangle.$$

The integration is over the complex plane $\mathbb{C}$. One usually introduces polar coordinates $\beta = re^{i\phi}$ with $r \geq 0$ and $\phi \in [0, 2\pi)$.

Let $\mathcal{H}_1$, $\mathcal{H}_2$ be complex Hilbert spaces. Let $\mathcal{H}$ be their Hilbert tensor product, i.e. $\mathcal{H} = \mathcal{H}_1 \otimes \mathcal{H}_2$. Each Hilbert space admits at least one orthonormal basis. Then $\text{tr}_2$ would be the operator of taking the partial trace with respect to the Hilbert space $\mathcal{H}_2$. The linear operation $\text{tr}_2$ maps states in the Hilbert space $\mathcal{H}$ into states in the Hilbert space $\mathcal{H}_1$. Analogously we consider $\text{tr}_1$. For number states we could utilize $|n_1\rangle \otimes |n_2\rangle$, where $n_1 = 0, 1, \ldots$ and $n_2 = 0, 1, \ldots$.

## 21.2   Solved Problems

**Problem 1.**   Show that

$$\sum_{n=0}^{\infty} \langle n|\hat{A}|n\rangle = \frac{1}{\pi} \int_{\mathbb{C}} \langle \beta|\hat{A}|\beta\rangle.$$

**Solution 1.**   We have

$$|n\rangle = \frac{1}{\pi} \int_{\mathbb{C}} \langle \beta|n\rangle |\beta\rangle d^2\beta$$

and therefore

$$\langle n| = \frac{1}{\pi} \int_{\mathbb{C}} \overline{\langle \beta|n\rangle} \langle \beta| d^2\beta = \frac{1}{\pi} \int_{\mathbb{C}} \langle n|\beta\rangle \langle \beta| d^2\beta .$$

We also apply the *completeness relations*

$$I = \frac{1}{\pi} \int_{\mathbb{C}} |\beta\rangle\langle\beta| d^2\beta, \qquad I = \sum_{n=0}^{\infty} |n\rangle\langle n|.$$

Thus we have

$$\mathrm{tr}(\hat{A}) = \sum_{n=0}^{\infty} \langle n|\hat{A}|n\rangle = \sum_{n=0}^{\infty} \langle n| \left( \frac{1}{\pi} \int_{\mathbb{C}} |\beta\rangle\langle\beta| d^2\beta \right) \hat{A}|n\rangle$$

$$= \frac{1}{\pi} \int_{\mathbb{C}} \sum_{n=0}^{\infty} \langle n|\beta\rangle\langle\beta|\hat{A}|n\rangle d^2\beta$$

$$= \frac{1}{\pi^3} \int_{\mathbb{C}} \sum_{n=0}^{\infty} \left( \langle n|\gamma\rangle\langle\gamma|\beta\rangle d^2\gamma \right) \langle\beta|\hat{A}| \left( \int_{\mathbb{C}} |\delta\rangle\langle\delta|n\rangle \right) d^2\delta d^2\beta$$

$$= \frac{1}{\pi^3} \int_{\mathbb{C}} \int_{\mathbb{C}} \int_{\mathbb{C}} d^2\beta d^2\gamma d^2\delta \sum_{n=0}^{\infty} \langle n|\gamma\rangle\langle\gamma|\beta\rangle\langle\delta|n\rangle\langle\beta|\hat{A}|\delta\rangle$$

$$= \frac{1}{\pi^3} \int_{\mathbb{C}} \int_{\mathbb{C}} \int_{\mathbb{C}} d^2\beta d^2\gamma d^2\delta \left( \langle\delta| \left( \sum_{n=0}^{\infty} |n\rangle\langle n| \right) |\gamma\rangle \right) \langle\gamma|\beta\rangle$$

$$= \frac{1}{\pi^3} \int_{\mathbb{C}} \int_{\mathbb{C}} \int_{\mathbb{C}} d^2\beta d^2\gamma d^2\delta \langle\delta|\gamma\rangle\langle\delta|\gamma\rangle\langle\gamma|\beta\rangle\langle\beta|\hat{A}|\delta\rangle$$

$$= \frac{1}{\pi^2} \int_{\mathbb{C}} \int_{\mathbb{C}} d^2\beta d^2\gamma \langle\delta|\beta\rangle\langle\beta|\hat{A}|\delta\rangle$$

$$= \frac{1}{\pi} \int_{\mathbb{C}} d^2\delta \langle\delta|\hat{A}|\delta\rangle.$$

**Problem 2.** The *trace* of an analytic function $f(b, b^\dagger)$ can be calculated as

$$\text{tr}(f(b, b^\dagger)) = \sum_{n=0}^{\infty} \langle n | f(b, b^\dagger) | n \rangle$$

where $\{ |n\rangle : n = 0, 1, 2, \ldots \}$ are the number states. A second method consists of obtaining the normal order function of $f$ and integrating over the complex plane

$$\text{tr}(f(b, b^\dagger)) = \frac{1}{\pi} \int_{\mathbb{C}} \overline{f}^{(n)}(\beta, \beta^*) d^2\beta.$$

(i) Find the trace of $e^{-\epsilon b^\dagger b}$ using this second method, where $\epsilon > 0$.
(ii) Compare with the first method.

**Solution 2.** (i) The normal order form of $e^{-\epsilon b^\dagger b}$ is given by

$$e^{-\epsilon b^\dagger b} = \sum_{k=0}^{\infty} \frac{1}{k!}(e^{-\epsilon} - 1)^k (b^\dagger)^k b^k.$$

Thus we have to calculate the integral

$$\frac{1}{\pi} \int_{\mathbb{C}} \sum_{k=0}^{\infty} \frac{1}{k!}(e^{-\epsilon} - 1)^k (\beta^*)^k \beta^k d^2\beta.$$

We set $\beta = re^{i\phi}$. Thus $\beta\beta^* = r^2$. Since $d^2\beta \to d\phi r dr$ with $\phi \in [0, 2\pi)$, $r \in [0, \infty)$ and

$$\int_0^{2\pi} d\phi = 2\pi, \qquad \int_0^{\infty} re^{-ar^2} dr = \frac{1}{2a}$$

we obtain

$$\text{tr}(e^{-\epsilon b^\dagger b}) = \frac{1}{1 - e^{-\epsilon}}.$$

(ii) Using the first method we find

$$\text{tr}(e^{-\epsilon b^\dagger b}) = \sum_{n=0}^{\infty} \langle n | e^{-\epsilon b^\dagger b} | n \rangle = \sum_{n=0}^{\infty} \langle n | e^{-\epsilon n} | n \rangle$$

$$= \sum_{n=0}^{\infty} e^{-\epsilon n} \langle n | n \rangle = \sum_{n=0}^{\infty} e^{-\epsilon n}$$

$$= \frac{1}{1 - e^{-\epsilon}}.$$

Thus the first method is simpler to apply for this case.

**Problem 3.** The single-mode squeezed state $|\zeta\rangle$ is given by $|\zeta\rangle = S(\zeta)|0\rangle$, where

$$S(\zeta) = \exp\left(-\frac{\zeta}{2}(b^\dagger)^2 + \frac{\zeta^*}{2}b^2\right)$$

with $\zeta \in \mathbb{C}$. One sets $\zeta = se^{i\theta}$ with $s \geq 0$. Why does the squeezed state $|\zeta\rangle$ cannot be used to calculate the trace?

**Solution 3.** The single-mode squeezed operator $S(\zeta)$ can be written as

$$S(\zeta) = \exp\left(-\frac{1}{2}(b^\dagger)^2 e^{i\theta}\tanh(s)\right)\exp\left(-\frac{1}{2}(b^\dagger b + bb^\dagger)\ln(\cosh(s))\right)$$

$$\times \exp\left(\frac{1}{2}b^2 e^{-i\theta}\tanh(s)\right).$$

From $b|0\rangle = 0|0\rangle$ and $|\zeta\rangle = S(\zeta)|0\rangle$ we find

$$|\zeta\rangle = \sqrt{(\operatorname{sech}(s))}\sum_{n=0}^{\infty}\frac{\sqrt{((2n)!)}}{n!}\left(-\frac{1}{2}e^{i\theta}\tanh(s)\right)^n |2n\rangle.$$

Consequently the single-mode squeezed state is a superposition only of even number states. Thus the single-mode squeezed states are not complete. Note that the coherent states are overcomplete. The coherent squeezed states $D(\beta)S(\zeta)|0\rangle$ also form an overcomplete set. One has

$$\frac{1}{\pi}\int_{\mathbb{C}}d^2\beta|\beta,\zeta\rangle\langle\beta,\zeta| = I$$

where $I$ is the identity operator.

**Problem 4.** Consider the two-mode squeezed state

$$|\psi\rangle = e^{s(b_1^\dagger b_2^\dagger - b_1 b_2)}|00\rangle$$

where $|00\rangle \equiv |0\rangle \otimes |0\rangle$ and $s$ is the squeezing parameter. This state can also be written as

$$|\psi\rangle = \frac{1}{\cosh(s)}\sum_{n=0}^{\infty}(\tanh(s))^n|n\rangle. \otimes |n\rangle.$$

This is the *Schmidt basis* for this state. The density operator $\rho$ is given by $\rho = |\psi\rangle\langle\psi|$. Calculate the partial traces using the number states.

**Solution 4.** We have

$$\rho = \frac{1}{(\cosh(s))^2}\sum_{n=0}^{\infty}(\tanh(s))^n|n\rangle \otimes |n\rangle \sum_{m=0}^{\infty}(\tanh(s))^m\langle m| \otimes \langle m|$$

$$= \frac{1}{(\cosh(s))^2}\sum_{m=0}^{\infty}\sum_{n=0}^{\infty}(\tanh(s))^n(\tanh(s))^m|n\rangle\langle m| \otimes |n\rangle\langle m|.$$

Let $I$ be the identity operator. Using that $\langle k|n \rangle = \delta_{kn}$ and $\langle m|k \rangle = \delta_{mk}$ we have

$$\rho_1 = \sum_{k=0}^{\infty} (I \otimes \langle k|) \rho (I \otimes |k\rangle)$$

$$= \frac{1}{(\cosh(s))^2} \sum_{k=0}^{\infty} (I \otimes \langle k|) \sum_{m,n=0}^{\infty} (\tanh(s))^n (\tanh(s))^m |n\rangle\langle m| \otimes |n\rangle\langle m| (I \otimes |k\rangle)$$

$$= \frac{1}{(\cosh(s))^2} \sum_{k=0}^{\infty} \sum_{m,n=0}^{\infty} (\tanh(s))^n (\tanh(s))^m |n\rangle\langle m| \delta_{kn} \delta_{mk}$$

$$= \frac{1}{(\cosh(s))^2} \sum_{k=0}^{\infty} (\tanh(s))^{2k} |k\rangle\langle k|.$$

We obtain the same result for $\rho_2$.

**Problem 5.** Use the reduced density operators $\rho_1$ and $\rho_2$ from the previous problem and calculate the entanglement

$$E(s) := -\mathrm{tr}(\rho_1 \log_2(\rho_1)) = -\mathrm{tr}(\rho_2 \log_2(\rho_2)).$$

Discuss $E$ as a function of the squeezing parameter $s$.

**Solution 5.** We have

$$E(s) = -\mathrm{tr}\left( \left( \frac{1}{\cosh^2(s)} \sum_{k=0}^{\infty} (\tanh^{2k}(s)) |k\rangle\langle k| \right) \right.$$

$$\left. \times \log_2 \left( \frac{1}{\cosh^2(s)} \sum_{\ell=0}^{\infty} (\tanh^{2\ell}(s)) |\ell\rangle\langle \ell| \right) \right).$$

The two matrices inside the trace are diagonal matrices and thus the product is again a diagonal matrix. Thus

$$E(s) = -\frac{1}{\cosh^2(s)} \sum_{k=0}^{\infty} \tanh^{2k}(s) \log_2 \left( \frac{\tanh^{2k}(s)}{\cosh^2(s)} \right).$$

Using the property of log it follows that

$$E(s) = -\frac{\log_2(\tanh^2(s))}{\cosh^2(s)} \sum_{k=0}^{\infty} k \tanh^{2k}(s) + \frac{\log_2(\cosh^2(s))}{\cosh^2(s)} \sum_{k=0}^{\infty} \tanh^{2k}(s).$$

The identity

$$\sum_{k=0}^{\infty} \tanh^{2k}(s) \equiv \cosh^2(s)$$

follows from a geometric series. The identity

$$\sum_{k=0}^{\infty} k \tanh^{2k}(s) \equiv \sinh^2(s) \cosh^2(s)$$

can be obtained from the first identity by parameter differentiation with respect to $s$. Using these results we obtain

$$\begin{aligned} E(s) &= -\sinh^2(s)(\log_2(\sinh^2(s)) - \log_2(\cosh^2(s)) + \log_2(\cosh^2(s)) \\ &= -\sinh^2(s)\log_2(\sinh^2(s)) + (\sinh^2(s) + 1)\log_2(\cosh^2(s)) \\ &= -\sinh^2(s)\log_2(\sinh^2(s)) + \cosh^2(s)\log_2(\cosh^2(s)). \end{aligned}$$

For $s = 0$ we have $E(s = 0) = 0$.

**Problem 6.**   Consider the product Hilbert space $\mathbb{C}^2 \otimes \mathcal{H}$, where $\mathcal{H}$ denotes an arbitrary Hilbert space. For the one-Bose system one would set $\mathcal{H} = \ell_2(\mathbb{N}_0)$. An arbitrary pure state in this product Hilbert space can be written as

$$|\psi\rangle = |0\rangle \otimes |\phi_0\rangle + |1\rangle \otimes |\phi_1\rangle$$

where $|\phi_0\rangle, \phi_1 \in \mathcal{H}$ and $|0\rangle$, $|1\rangle$ forms an orthonormal basis in the Hilbert space $\mathbb{C}^2$. The condition that the state $|\psi\rangle$ to be normalized, i.e. $\langle\psi|\psi\rangle = 1$ leads to the constraint

$$\langle\phi_0|\phi_0\rangle + \langle\phi_1|\phi_1\rangle = 1.$$

If we assume that $|\phi_0\rangle$ and $|\phi_1\rangle$ have identical norms, then $|\psi\rangle$ takes the form

$$|\psi\rangle = \frac{1}{\sqrt{2}}(|0\rangle \otimes |\varphi_0\rangle + |1\rangle \otimes |\varphi_1\rangle)$$

where $|\phi_0\rangle = \frac{1}{\sqrt{2}}|\varphi_0\rangle$, $|\phi_1\rangle = \frac{1}{\sqrt{2}}|\varphi_1\rangle$ and $|\varphi_0\rangle$, $\varphi_1\rangle$ are normalized. Defining the reduced density operators using the partial trace as

$$\rho_1 = \text{tr}_{\mathbb{C}^2}(|\psi\rangle\langle\psi|), \quad \rho_2 = \text{tr}_{\mathcal{H}}(|\psi\rangle\langle\psi|)$$

the entanglement of $|\psi\rangle$ is given by

$$E(|\psi\rangle) = -\text{tr}(\rho_1 \log_2(\rho_1)) = -\text{tr}(\rho_2 \log_2(\rho_2)).$$

Find $\rho_1$, $\rho_2$, the nonzero eigenvalues of $\rho_1$, $\rho_2$ and $E(|\psi\rangle)$.

**Solution 6.**   Straightforward calculation yields

$$\rho_1 = |\phi_0\rangle\langle\phi_0| + |\phi_1\rangle\langle\phi_1|$$

and

$$\rho_2 = \langle\phi_0|\phi_0\rangle|0\rangle\langle0| + \langle\phi_1|\phi_0\rangle|0\rangle\langle1| + \langle\phi_0|\phi_1\rangle|1\rangle\langle0| + \langle\phi_1|\phi_1\rangle|1\rangle\langle1|.$$

Applying the constraint $\langle\phi_0|\phi_0\rangle + \langle\phi_1|\phi_1\rangle = 1$ we find that the nonzero eigenvalues of $\rho_1$ and $\rho_2$ are given by

$$\lambda = \frac{1}{2}(1 + \sqrt{(1 - 2\langle\phi_0|\phi_0\rangle)^2 + 4|\langle\phi_0|\phi_1\rangle|^2})$$

and $1-\lambda$. Thus $E(|\psi\rangle) = -\lambda \log_2(\lambda) - (1-\lambda)\log_2(1-\lambda)$. The entanglement is described exclusively by $\langle\phi_0|\phi_0\rangle$ and $|\langle\phi_0|\phi_1\rangle|^2$.

## Programming Problem

**Problem 1.**   Give an implementation of number states so that they can utilized to calculate traces.

**Solution 1.**

```
/* numberstates.cpp */

#include <iostream>
#include "symbolicc++.h"

int main(void)
{
// b is the Bose annihilation operator, bd is the creation operator
// N[j] is the number state |j>, DN[j] is the dual state <j|
Symbolic b("b"), bd("bd"), N("N"), DN("DN"), m("m"), n("n"), x("x");
b = ~b; bd = ~bd; N = ~N; DN = ~DN;
BindingEquations rules = (b*N[0]==0,DN[0]*bd==0,
                         (n,b*N[n]==sqrt(n)*N[n-1]),
                         (n,bd*N[n]==sqrt(n+1)*N[n+1]),
                         (n,DN[n]*b==D[n+1]*sqrt(n+1)),
                         (n,DN[n]*bd==DN[n-1]*sqrt(n)),
                         (n,DN[n]*N[n]==1),
                         (m,n,DN[m]*N[n]==0));
Symbolic r1 = b*b*N[n];
cout << "r1 = " << r1.subst_all(rules) << endl;
cout << "r1(n=1) = " << r1[n==1].subst_all(rules) << endl;
cout << "DN[2]*N[3] = " << (DN[2]*N[3]).subst_all(rules) << endl;
cout << "DN[4]*N[4] = " << (DN[4]*N[4]).subst_all(rules) << endl;
Symbolic r2 = b*N[n];
cout << "r2 = " << r2.subst_all(rules) << endl;
Symbolic r3 = DN[n]*r2;
```

```
cout << "r3 = " << r3.subst_all(rules) << endl;
Symbolic r4 = b*N[n+1];
cout << "r4 = " << r4.subst_all(rules) << endl;
Symbolic r5 = DN[n]*r4;
cout << "r5 = " << r5.subst_all(rules) << endl;
Symbolic r6 = b*N[n];
cout << "r6 = " << r6.subst_all(rules) << endl;
Symbolic r7 = DN[n+1]*r6;
cout << "r7 = " << r7.subst_all(rules) << end;
return 0;
}
```

# 21.3   Supplementary Problems

**Problem 1.**   Consider the density operator

$$\rho = \frac{\exp(-\hbar\omega b^\dagger b/k_B T)}{\text{tr}(\exp(-\hbar\omega b^\dagger b/k_B T))}.$$

Show that $\rho$ expressed with number states $|n\rangle$ $(n = 0, 1, \ldots)$ is given by

$$\rho = \sum_{n=0}^{\infty} \frac{\langle n\rangle^n}{(1 + \langle n\rangle)^{n+1}} |n\rangle\langle n|$$

where

$$\langle n\rangle := \text{tr}(\rho b^\dagger b) = (\exp(\hbar\omega/k_B T) - 1)^{-1}.$$

**Problem 2.**   Let $|0\rangle$, $|1\rangle$ be the standard basis in $\mathbb{C}^2$ and $|\beta_1\rangle$, $|\beta_2\rangle$ be coherent states. Consider the normalized state

$$|\psi\rangle = c_0|0\rangle \otimes |\beta_1\rangle + c_1|1\rangle \otimes |\beta_2\rangle$$

where $c_0$ and $c_1$ are the normalization constants. Find $\rho_1 = \text{tr}_{\mathbb{C}^2}(|\psi\rangle\langle\psi|)$.

**Problem 3.**   Can the squeezed state $|\zeta\rangle$ be used to calculate the trace of an operator? Calculate

$$\frac{1}{\pi} \int_{\mathbb{C}} |\zeta\rangle\langle\zeta| d\zeta.$$

Discuss.

# Chapter 22

# Entanglement

## 22.1 Introduction

In the original paper of Einstein, Podolsky and Rosen the spin version of entanglement was not used. They considered measurement of position and momentum observables for two particles in one-dimensional motion. The entangled state

$$|\psi\rangle = \int_{-\infty}^{\infty} |p\rangle \otimes |-p\rangle e^{-i\ell p/\hbar} dp$$

was studied, where the first component in the tensor product refers to particle 1 and the second to particle 2. The state $|\psi\rangle$ is thus a superposition of simultaneous eigenkets of the momentum operators $\hat{P}_1$ and $\hat{P}_2$ of the two particles with associated eigenvalues $p$ and $-p$, respectively. Thus $|\psi\rangle$ is itself an eigenket of

$$\hat{P}_1 \otimes I + I \otimes \hat{P}_2$$

with the eigenvalue 0. The entangled state $|\psi\rangle$ is also an eigenket of the operator

$$\hat{Q}_1 \otimes I + I \otimes \hat{Q}_2$$

where $\hat{Q}_1$ and $\hat{Q}_2$ are the position operators of the two particles. The maximally entangled state of the original EPR pair can also be written as

$$|\psi\rangle = \frac{1}{(2\pi)^{3/2}} \int d\mathbf{k} \exp(i\mathbf{k} \cdot \mathbf{r}_1) \exp(i\mathbf{k} \cdot \mathbf{r}_2) = \delta(\mathbf{r}_1 + \mathbf{r}_2)$$

where $\delta$ denotes the Dirac delta function and $\cdot$ is the scalar product. If the potential energy between the two particles is assumed to be spherically symmetric, we can assume that the vectors $\mathbf{r}_1$ and $\mathbf{r}_2$ lie in the $xy$-plane in the laboratory frame without loss of generality.

Entangled state are the Schrödinger cat states

$$\frac{1}{\sqrt{N_\beta}}(|\beta\rangle \otimes |-\beta\rangle + |-\beta\rangle \otimes |\beta\rangle)$$

where $|\beta\rangle$ denotes a coherent state. For squeezed states $|\zeta\rangle$ we can consider the entangled state

$$\frac{1}{\sqrt{N_\zeta}}(|\zeta\rangle \otimes |-\zeta\rangle + |-\zeta\rangle \otimes |\zeta\rangle).$$

The two-mode state

$$|\psi\rangle = e^{s(b_1^\dagger b_2^\dagger - b_1 b_2)}|0\rangle \otimes |0\rangle$$

$$= \frac{1}{\cosh(s)} \sum_{n=0}^{\infty} (\tanh(s))^n |n\rangle \otimes |n\rangle$$

where $s$ is the squeezing parameter and $|n\rangle$ are the number states is an entangled state. Let $|0\rangle$, $|1\rangle$ be an orthonormal basis in $\mathbb{C}^2$ and $|\beta\rangle$ be a coherent state. Then

$$\frac{1}{\sqrt{N}}(|0\rangle \otimes |\beta\rangle + |1\rangle \otimes |-\beta\rangle)$$

is an entangled state. One can also consider

$$\frac{1}{\sqrt{N}}(|0\rangle \otimes |\zeta\rangle + |1\rangle \otimes |-\zeta\rangle)$$

where $|\zeta\rangle$ is a squeezed state.

## 22.2   Solved Problems

**Problem 1.**   Let

$$f(\mathbf{p},\mathbf{q}) := \frac{2}{N}\left(\sum_{j=1}^{N} q_j\right)^2 + \frac{1}{N}\sum_{j,k=1}^{N}(p_j - p_k)^2$$

and

$$g(\mathbf{p},\mathbf{q}) := \frac{2}{N}\left(\sum_{j=1}^{N} p_j\right)^2 + \frac{1}{N}\sum_{j,k=1}^{N}(q_j - q_k)^2.$$

The *Wigner function* of the pure entangled $N$-mode state is given by

$$W(\mathbf{q},\mathbf{p}) = \left(\frac{2}{\pi}\right)^N \exp\left(-e^{-2s}f(\mathbf{p},\mathbf{q}) - e^{2s}g(\mathbf{p},\mathbf{q})\right) \qquad (1)$$

where $\mathbf{q} = (q_1, q_2, \ldots, q_N)$ and $\mathbf{p} = (p_1, p_2, \ldots, p_N)$ are the positions and momenta of the $N$ modes and $s$ is the squeezing parameter with equal squeezing in all initial modes. Consider the case $N = 2$. What happens if $s \to \infty$?

**Solution 1.**   The state $W(\mathbf{q},\mathbf{p})$ is always positive, symmetric among the $N$ modes and becomes peaked at $q_i - q_j = 0$ $(i, j = 1, 2, \ldots, N)$ and $p_1 + p_2 + \cdots + p_n = 0$ for large squeezing parameter $s$. From (1) we have

$$W(q_1, q_2, p_1, p_2) =$$

$$\frac{4}{\pi^2} \exp\left(-e^{-2s}\left((q_1 + q_2)^2 + (p_1 - p_2)^2\right) - e^{2s}\left((p_1 + p_2)^2 + (q_1 - q_2)^2\right)\right).$$

For $s \to \infty$ we find in the sense of generalized functions

$$C\delta(q_1 - q_2)\delta(p_1 + p_2)$$

where $\delta$ denotes the *Dirac delta function*. This makes a connection to the original *EPR state* of Einstein, Podolsky and Rosen. Thus for large $s$ the function $W$ peaks at $q_1 - q_2 = 0$ and $p_1 + p_2 = 0$.

**Problem 2.**   Consider the operator

$$U(r) := e^{-r(b_1^\dagger b_2^\dagger - b_1 b_2)}$$

where $b_1^\dagger$, $b_2^\dagger$ are Bose creation operators and $b_1, b_2$ are Bose annihilation operators and $r \in \mathbb{R}$. Thus $b_1^\dagger = b^\dagger \otimes I$, $b_2^\dagger = I \otimes b^\dagger$. Let $|0\rangle \otimes |0\rangle$ be the vacuum state, i.e.

$$(b \otimes I)(|0\rangle \otimes |0\rangle) = 0|0\rangle \otimes |0\rangle, \qquad (I \otimes b)(|0\rangle \otimes |0\rangle) = 0|0\rangle \otimes |0\rangle.$$

(i) Calculate $|\psi(r)\rangle = U(r)(|0\rangle \otimes |0\rangle)$.
(ii) Let

$$\hat{X}_1 := b_1 + b_1^\dagger = b \otimes I + b^\dagger \otimes I, \qquad \hat{Y}_1 := -i(b_1 - b_1^\dagger) = -i(b \otimes I - b^\dagger \otimes I),$$

$$\hat{X}_2 := b_2 + b_2^\dagger = I \otimes b + I \otimes b^\dagger, \qquad \hat{Y}_2 := -i(b_2 - b_2^\dagger) = -i(I \otimes b - I \otimes b^\dagger).$$

Find $\mathrm{var}(\hat{X}_1 + \hat{X}_2)$, $\mathrm{var}(\hat{Y}_1 - \hat{Y}_2)$, where

$$\mathrm{var}(A) := \langle A^2 \rangle - \langle A \rangle^2$$

is the *variance*.
(iii) What happens in the limit $r \to \infty$ to the state $|\psi(r)\rangle$?

**Solution 2.**    (i) We find

$$|\psi(r)\rangle = U(r)(|0\rangle \otimes |0\rangle) = \sqrt{(1 - \lambda^2)} \sum_{n=0}^{\infty} \lambda^n |n\rangle \otimes |n\rangle$$

where $\lambda = \tanh(r)$ and therefore $\sqrt{1 - \lambda^2} = 1/\cosh(r)$. The entanglement of this state can be viewed as an entanglement between quadrature phases in the two modes (EPR entanglement) or as an entanglement between number and phase in the two modes.
(ii) We find

$$\mathrm{var}(\hat{X}_1 + \hat{X}_2) = 2e^{-2r}, \qquad \mathrm{var}(\hat{Y}_1 - \hat{Y}_2) = 2e^{-2r}.$$

(iii) The state $|\psi(r)\rangle$ approaches a simultaneous eigenstate of $\hat{X}_1 + \hat{X}_2$ and $\hat{Y}_1 - \hat{Y}_2$.

**Problem 3.**    Consider a quantum-mechanical system governed by the Hamilton operator

$$\hat{H} = \hbar\omega_1 b_1^\dagger b_1 + \hbar\omega_2 b_2^\dagger b_2 + \hbar\chi b_1^\dagger b_1 b_2^\dagger b_2$$

where $b_1$ and $b_2$ are Bose annihilation operators for two distinct harmonic oscillator modes, respectively and $\chi$ is a coupling constant. Such a Hamilton operator for optical systems describes a four-wave mixing process, when the constant $\chi$ is then proportional to the third order susceptibility. It can also be used to describe two distinct modes interaction in Bose condensate. Furthermore, it describes the effective interaction of output pump and probe fields of an optical-cavity mediated by a two-level atom, in the dispersive limit. Let

$$|\psi(t = 0)\rangle := |\beta_1\rangle \otimes |\beta_2\rangle$$

where $|\beta_1\rangle$ and $|\beta_2\rangle$ are coherent states.

(i) Find $U(t)|\psi(t=0)\rangle$, where

$$U(t) = \exp(-i\hat{H}t/\hbar).$$

(ii) Consider the special case $t = \pi/\chi$. Discuss.

(iii) Consider the four cases a) $\omega_1 = 2\chi$, $\omega_2 = 2\chi$, b) $\omega_1 = 2\chi$, $\omega_2 = \chi$, c) $\omega_1 = \chi$, $\omega_2 = 2\chi$, d) $\omega_1 = \chi$, $\omega_2 = \chi$ for $|\psi(\pi/\chi)\rangle$.

**Solution 3.** (i) We find

$$|\psi(t)\rangle = U(t)|\beta_1\rangle \otimes |\beta_2\rangle = e^{-|\beta_1|^2/2} \sum_{m=0}^{\infty} \frac{(\beta_1 e^{-i\omega_1 t})^m}{\sqrt{m}} |m\rangle \otimes |\beta_2 e^{-i\omega_2 t} e^{-i\chi mt}\rangle.$$

(ii) For $t = \pi/\chi$ we have

$$\exp(-i\chi mt) = \exp(-im\pi) = \begin{cases} 1 & m = \text{even} \\ -1 & m = \text{odd} \end{cases}$$

Thus

$$|\psi(\pi/\chi)\rangle = |\beta_{1+}e^{-i\pi\omega_1/\chi}\rangle \otimes |\beta_2 e^{-i\pi\omega_2/\chi}\rangle + |\beta_{1-}e^{-i\pi\omega_1/\chi}\rangle \otimes |-\beta_2 e^{-i\pi\omega_2/\chi}\rangle$$

or

$$|\psi(\pi/\chi)\rangle = |\beta_1 e^{-i\pi\omega_1/\chi}\rangle \otimes |\beta_{2+}e^{-i\pi\omega_2/\chi}\rangle + |-\beta_1 e^{-i\pi\omega_1/\chi}\rangle \otimes |\beta_{2-}e^{-i\pi\omega_2/\chi}\rangle$$

where

$$|\epsilon_{\pm}e^{-i\pi\omega_k\chi}\rangle := \frac{1}{2}(|\epsilon e^{-i\pi\omega_k/\chi}\rangle \pm |-\epsilon e^{-i\pi\omega_k/\chi}\rangle)$$

with $k = 1, 2$ and $\epsilon = \beta_1, \beta_2$. Hence the state is entangled.

(iii) For case a) we find

$$|\Phi_+\rangle = |\beta_1\rangle \otimes |\beta_{2+}\rangle + |-\beta_1\rangle \otimes |\beta_{2-}\rangle.$$

For case b) we find

$$|\Phi_-\rangle = |\beta_1\rangle \otimes |\beta_{2+}\rangle - |-\beta_1\rangle \otimes |\beta_{2-}\rangle.$$

For case c) we find

$$|\Psi_+\rangle = |\beta_1\rangle \otimes |\beta_{2-}\rangle + |-\beta_1\rangle \otimes |\beta_{2+}\rangle.$$

For case d) we find

$$|\Psi_-\rangle = |\beta_1\rangle \otimes |\beta_{2-}\rangle - |-\beta_1\rangle \otimes |\beta_{2+}\rangle.$$

These states may be considered as *Bell states*. However these states are not perfectly orthogonal, but for large-amplitude fields $|\beta_1|, |\beta_2| \ll 1$ this can be achieved approximately. Furthermore there is an asymmetry in these states.

**Problem 4.** Discuss the *entanglement* of the state

$$|\Psi\rangle = \sum_s \sum_i \delta(\omega_s + \omega_i - \omega_p)\delta(\mathbf{k}_s + \mathbf{k}_i - \mathbf{k}_p)b_s^\dagger(\omega(\mathbf{k}_i))b_i^\dagger(\omega(\mathbf{k}_i))|0\rangle$$

which appears with *spontaneous parametric down conversion*. Here $\omega_j$, $\mathbf{k}_j$ ($j = s, i, p$) are the frequencies and wave vectors of the signal (s), idler (i), and pump (p) respectively, $\omega_p$ and $\mathbf{k}_p$ can be considered as constants while $b_s^\dagger$ and $b_i^\dagger$ are the respective Bose creation operators for the signal and idler.

**Solution 4.** The entanglement of this state can be thought of as the superposition of an infinite number of two-photon states, corresponding to the infinite number of ways the spontaneous parametric down conversion signal-idler can satisfy the expression for energy and momentum conservation (owing to the delta functions)

$$\hbar\omega_s + \hbar\omega_i = \hbar\omega_p, \qquad \hbar\mathbf{k}_s + \hbar\mathbf{k}_i = \hbar\mathbf{k}_p.$$

Even if there is no precise knowledge of the momentum for either the signal or the idler, the state does give precise knowledge of the momentum correlation of the pair. In EPR's language, the momentum for neither the signal photon nor the idler photon is determined. However, if measurement on one of the photons yields a certain value, then the momentum of the other photon is determined.

**Problem 5.** Consider the function

$$G(x_1, x_2; r) = \frac{1}{\sqrt{2\pi}}\exp\left(-\frac{1}{4}(x_1 + x_2)^2 e^{2s} - \frac{1}{4}(x_1 - x_2)^2 e^{-2s}\right)$$

where $s > 0$ is the squeezing parameter. Find

$$\lim_{s\to\infty}(G(x_1, x_2; s), \phi(x_1, x_2))$$

in the sense of generalized functions, where $\phi \in S(\mathbb{R}^2)$. Here $S(\mathbb{R}^2)$ is the set of all infinitely-differentiable functions which decrease as $|\mathbf{x}| \to \infty$, together with all their derivatives, faster than any power of $|\mathbf{x}|^{-1}$.

**Solution 5.** We find

$$\lim_{r\to\infty}(G(x_1, x_2; s), \phi(x_1, x_2)) \to \phi(x_1, x_1).$$

Thus
$$\lim_{s \to \infty} G(x_1, x_2; s) \to \delta(x_1 - x_2)$$
in the sense of generalized functions, where $\delta$ is the Dirac delta function.

**Problem 6.** Consider the operator
$$U_\epsilon = \exp(\epsilon(b_1^\dagger b_2 - b_2^\dagger b_1)), \qquad \epsilon \in \mathbb{R}.$$

(i) Find $U_\epsilon^\dagger b_1 U_\epsilon$, $U_\epsilon^\dagger b_2 U_\epsilon$. Consider the special case $\epsilon = \pi/4$.
(ii) Find $D = U_{\pi/4}^\dagger (b_1^\dagger b_1 - b_2^\dagger b_2) U_{\pi/4}$.
(iii) Solve the eigenvalue problem $D|\delta\rangle = d|\delta\rangle$.

**Solution 6.** (i) Using the expansion
$$e^{\hat{A}} \hat{B} e^{-\hat{A}} = \hat{B} + [\hat{A}, \hat{B}] + \frac{1}{2}[\hat{A}, [\hat{A}, \hat{B}]] + \cdots + \frac{1}{n!}[\hat{A}, [\hat{A}, \cdots, [\hat{A}, \hat{B}] \ldots]] + \cdots$$

we find
$$U_\epsilon^\dagger b_1 U_\epsilon = b_1 \cos(\epsilon) + b_2 \sin(\epsilon), \qquad U_\epsilon^\dagger b_2 U_\epsilon = -b_1 \sin(\epsilon) + b_2 \cos(\epsilon).$$

For the special case $\epsilon = \pi/4$ we obtain
$$U_{\pi/4}^\dagger b_1 U_{\pi/4} = \frac{1}{\sqrt{2}}(b_2 + b_1), \qquad U_{\pi/4}^\dagger b_2 U_{\pi/4} = \frac{1}{\sqrt{2}}(b_2 - b_1)$$

since $\sin(\pi/4) = \cos(\pi/4) = 1/\sqrt{2}$.
(ii) From (ii) we find
$$U_\epsilon^\dagger b_1^\dagger U_\epsilon = b_1^\dagger \cos(\epsilon) + b_2^\dagger \sin(\epsilon), \qquad U_\epsilon^\dagger b_2^\dagger U_\epsilon = -b_1^\dagger \sin(\epsilon) + b_1^\dagger \cos(\epsilon).$$

Thus
$$D = b_1^\dagger b_2 + b_1 b_2^\dagger.$$

(iii) The eigenvalue problem $D|\delta\rangle = d|\delta\rangle$ can be rewritten as
$$(b_1^\dagger b_1 - b_2^\dagger b_2)|\nu\rangle = d|\nu\rangle$$

where
$$|\nu\rangle = U_{\pi/4}|\delta\rangle.$$

The eigenvalue problem can easily be solved since $b^\dagger b|n\rangle = n|n\rangle$. We find
$$|\nu^{(n)}\rangle = \begin{cases} |n+d\rangle \otimes |n\rangle & d \in \mathbb{Z}^+ \\ |n\rangle \otimes |n\rangle & d = 0 \\ |n\rangle \otimes |n-d\rangle & d \in \mathbb{Z}^- \end{cases}$$

where $\mathbb{Z}^+$ denotes the positive set of integers and $\mathbb{Z}^-$ denotes the negative set of integers. The eigenvalue $d$ has countable degeneracy corresponding to the one-integer parameter set $|\nu^{(n)}\rangle$ of eigenstates. In order to solve for the original eigenvalues we have to compute their transformation under the action of the operator $U_{\pi/4}$, i.e.

$$|\delta^{(n)}\rangle = U^{\dagger}_{\pi/4}|\nu^{(n)}\rangle.$$

We consider the *Schwinger two-bosons realization* of the $su(2)$ Lie algebra

$$J_+ := b_1 b_2^{\dagger}, \qquad J_- := b_1^{\dagger} b_2, \qquad J_3 := \frac{1}{2}(b_2^{\dagger} b_2 - b_1^{\dagger} b_1)$$

with $[J_+, J_-] = 2J_3$, $[J_3, J_{\pm}] = \pm J_{\pm}$. Thus

$$U_{\pi/4} = \exp\left(\frac{\pi}{4}(J_+ - J_-)\right).$$

Using the Baker-Campbell-Hausdorff formula we find

$$\exp(\xi J_+ - \bar{\xi} J_-) = \exp(\eta J_+) \exp(\beta J_3) \exp(-\bar{\eta} J_-)$$

where

$$\eta = \frac{\xi}{|\xi|} \tan(\xi), \qquad \beta = \ln(1 + |\eta|^2).$$

Thus

$$U_{\pi/4} = \exp(b_1 b_2^{\dagger}) \exp(\ln 2(b_2^{\dagger} b_2 - b_1^{\dagger} b_1)) \exp(-b_1^{\dagger} b_2)$$

where we used that $\tan(\pi/4) = 1$.

**Problem 7.**   Consider the Hamilton operator

$$\hat{H} = \sum_{i,j=0}^{N-1} h_{ij} b_i^{\dagger} b_j + \sum_{i,j,l,m=0}^{N-1} V_{ijlm} b_i^{\dagger} b_j^{\dagger} b_m b_l.$$

The operators $b_j^{\dagger}$ are Bose creation operators and the operators $b_j$ are Bose annihilation operators. Let

$$\hat{n}_j := b_j^{\dagger} b_j$$

be the particle number operator of mode $j$ for an appropriate basis. Show that an eigenstate $|\psi\rangle$ of $\hat{H}$ is entangled or

$$[\hat{H}, \hat{n}_j]|\psi\rangle = 0|\psi\rangle$$

i.e. we have eigenvalue equation with eigenvalue 0.

**Solution 7.** Suppose $|\psi\rangle$ is not entangled. We write $|\psi\rangle$ as

$$|\psi\rangle = |n_0\rangle \otimes \cdots \otimes |n_{N-1}\rangle$$

where $\{\, |0_j\rangle, |1_j\rangle, \ldots \,\}$ is a basis for particles in mode $j$ with $0 \le j < N$. We define the creation, annihilation and number operators, for particles in mode $j$, by

$$b_j^\dagger |n_j\rangle := \sqrt{n_j + 1}\,|n + 1_j\rangle, \qquad b_j |n_j\rangle := \sqrt{n_j}\,|n - 1_j\rangle$$

and $\hat{n}_j := b_j^\dagger b_j$. We have the eigenvalue equation for $|\psi\rangle$, namely $\hat{H}|\psi\rangle = \lambda|\psi\rangle$. Thus

$$\begin{aligned}
[\hat{H}, \hat{n}_i]|\psi\rangle &= \hat{H}\hat{n}_i|\psi\rangle - \hat{n}_i\hat{H}|\psi\rangle = \hat{H}n_i|\psi\rangle - \lambda\hat{n}_i|\psi\rangle \\
&= n_i\hat{H}|\psi\rangle - \lambda\hat{n}_i|\psi\rangle = \lambda n_i|\psi\rangle - \lambda n_i|\psi\rangle \\
&= 0|\psi\rangle.
\end{aligned}$$

**Problem 8.** Let $|\beta\rangle$ be a coherent state. Consider the entangled coherent state

$$|\psi\rangle = C(|\beta_1\rangle \otimes |\beta_2\rangle + e^{i\phi}|-\beta_1\rangle \otimes |-\beta_2\rangle)$$

where $C$ is the normalization factor and $\phi \in \mathbb{R}$.
(i) Find the normalization factor $C$.
(ii) Calculate the *partial trace* using the basis $\{\, |n\rangle \otimes I \,:\, n = 0, 1, 2, \ldots \,\}$ where $\{\, |n\rangle \,:\, n = 0, 1, 2, \ldots \,\}$ are the number states and $I$ is the identity operator.

**Solution 8.** (i) Since

$$\langle\beta|\gamma\rangle = \exp\left(-\frac{1}{2}(|\beta|^2 + |\gamma|^2) + \beta\gamma^*\right)$$

for coherent states $|\beta\rangle$ and $|\gamma\rangle$, we have

$$\langle\beta|\beta\rangle = 1, \qquad \langle\beta|-\beta\rangle = \exp(-2|\beta|^2).$$

We find from the condition $\langle\psi|\psi\rangle = 1$ that

$$1 = |C|^2(2 + 2\cos(\phi)\exp(-2|\beta_1|^2 - 2|\beta_2|^2)).$$

Thus

$$C = \frac{1}{\sqrt{2 + 2\cos(\phi)\exp(-2|\beta_1|^2 - 2|\beta_2|^2)}}.$$

(ii) We have to calculate

$$\mathrm{tr}_1(|\psi\rangle\langle\psi|) = C^2 \sum_{n=0}^{\infty} (((\langle n| \otimes I)(|\beta_1\rangle \otimes |\beta_2\rangle + e^{i\phi}|-\beta_1\rangle \otimes |-\beta_2\rangle))$$
$$\times ((\langle \beta_1| \otimes \langle \beta_2| + e^{-i\phi}\langle \beta_1-| \otimes \langle \beta_2-|)(|n\rangle \otimes I)).$$

Thus

$$\mathrm{tr}_1(|\psi\rangle\langle\psi|) = C^2 \sum_{n=0}^{\infty} (\langle n|\beta_1\rangle\langle\beta_1|n\rangle |\beta_2\rangle\langle\beta_2| + e^{-i\phi}\langle n|\beta_1\rangle\langle\beta_1-|n\rangle |\beta_2\rangle\langle\beta_2-|$$
$$+ e^{i\phi}\langle n|-\beta_1\rangle\langle\beta_1|n\rangle |-\beta_2\rangle\langle\beta_2| + \langle n|-\beta_1\rangle\langle\beta_1-|n\rangle |-\beta_2\rangle\langle\beta_2-|).$$

Using

$$\langle n|\beta_1\rangle\langle\beta_1|n\rangle = \frac{e^{-|\beta_1|^2}(|\beta_1|^2)^n}{n!}$$

$$\langle n|\beta_1\rangle\langle\beta_1-|n\rangle = \frac{e^{-|\beta_1|^2}(-|\beta_1|^2)^n}{n!}$$

$$\langle n|-\beta_1\rangle\langle\beta_1|n\rangle = \frac{e^{-|\beta_1|^2}(-|\beta_1|^2)^n}{n!}$$

$$\langle n|-\beta_1\rangle\langle\beta_1-|n\rangle = \frac{e^{-|\beta_1|^2}(|\beta_1|^2)^n}{n!}$$

and

$$\sum_{n=0}^{\infty} \frac{(|\beta|^2)^n}{n!} = e^{|\beta|^2}, \qquad \sum_{n=0}^{\infty} \frac{(-|\beta|^2)^n}{n!} = e^{-|\beta|^2}$$

we arrive at

$$\mathrm{tr}_1(|\psi\rangle\langle\psi|) = C^2(|\beta_2\rangle\langle\beta_2| + e^{i\phi}e^{-2|\beta_1|^2}|\beta_2\rangle\langle\beta_2-|$$
$$+ e^{-i\phi}e^{-2|\beta_1|^2}|-\beta_2\rangle\langle\beta_2| + |-\beta_2\rangle\langle\beta_2-|).$$

**Problem 9.** A *beam splitter* is a simple device which can act to entangle output optical fields. The input field described by the Bose annihilation operator $b_1$ is superposed on the other input field with Bose annihilation operator $b_2$ by a lossless symmetric beam splitter with amplitude reflection and transmission coefficients $r$ and $t$. The output field annihilation operators are given by

$$\tilde{b}_1 = \hat{B}b_1\hat{B}^{\dagger}, \qquad \tilde{b}_2 = \hat{B}b_2\hat{B}^{\dagger}$$

where the *beam splitter operator* is

$$\hat{B} := \exp\left(\frac{\theta}{2}(b_1^{\dagger}b_2e^{i\phi} - b_1b_2^{\dagger}e^{-i\phi})\right)$$

with the amplitude reflection and transmission coefficients

$$t := \cos(\theta/2), \qquad r := \sin(\theta/2).$$

The beam splitter gives the phase difference $\phi$ between the reflected and transmitted fields.

(i) Assume that the input states are two independent number states $|n_1\rangle \otimes |n_2\rangle$, where $n_1, n_2 = 0, 1, 2, \ldots$. Calculate the state $\hat{B}(|n_1\rangle \otimes |n_2\rangle)$.

(ii) Consider the special case $n_1 = 0$ and $n_2 = N$.

**Solution 9.** (i) We obtain the state

$$\hat{B}(|n_1\rangle \otimes |n_2\rangle) = \sum_{m_1=0}^{\infty} \sum_{m_2=0}^{\infty} ((\langle m_1| \otimes \langle m_2|)\hat{B}(|n_1\rangle \otimes |n_2\rangle))|m_1\rangle \otimes |m_2\rangle$$

$$= \sum_{m_1=0}^{\infty} \sum_{m_2=0}^{\infty} B_{n_1 n_2}^{m_1 m_2} |m_1\rangle \otimes |m_2\rangle$$

where

$$B_{n_1 n_2}^{m_1 m_2} = e^{-i\phi(n_1 - m_1)} \sum_{k=0}^{n_1} \sum_{\ell=0}^{n_2} (-1)^{n_1 - k} r^{n_1 + n_2 - k - \ell} t^{k+\ell}$$

$$\times \frac{\sqrt{n_1! n_2! m_1! m_2!}}{k!(n_1 - k)!\ell!(n_2 - \ell)!} \delta_{m_1, n_2 + k - \ell} \delta_{m_2, n_1 - k + \ell}$$

with $\delta_{m,n}$ is the Kronecker delta. When the total number of input photons is $N = n_1 + n_2$, the output state becomes an $(N+1)$-dimensional entangled state.

(ii) We obtain from the results of (i)

$$B(|0\rangle \otimes |N\rangle) = \sum_{k=0}^{N} c_k^N |k\rangle \otimes |N - k\rangle$$

where the expansion coefficients are given by

$$c_k^N = \binom{N}{k}^{1/2} r^k t^{N-k} e^{ik\phi}.$$

**Problem 10.** The *beam splitter operator* is given by

$$\hat{B} = \exp\left(\frac{\theta}{2}(b_1^\dagger b_2 e^{i\phi} - b_1 b_2^\dagger e^{-i\phi})\right).$$

The input field described by the Bose operator $b_1$ is superposed with another input field with Bose operator $b_2$ by a lossless symmetric beam splitter with amplitude reflection and transmission coefficients $r$ and $t$, i.e.

$$t := \cos(\theta/2), \qquad r := \sin(\theta/2).$$

The output-field Bose annihilation operators are given by

$$\tilde{b}_1 = \hat{B} b_1 \hat{B}^\dagger, \qquad \tilde{b}_2 = \hat{B} b_2 \hat{B}^\dagger.$$

Consider the input state (product state of two independent Fock states)

$$|n_1, n_2\rangle \equiv |n_1\rangle |n_2\rangle \equiv |n_1\rangle \otimes |n_2\rangle$$

where $n_1, n_2 = 0, 1, \ldots$. Calculate $\hat{B}|n_1, n_2\rangle$. Is this state entangled?

**Solution 10.**  We obtain

$$\hat{B}|n_1, n_2\rangle = \sum_{N_1=0}^{\infty} \sum_{N_2=0}^{\infty} |N_1, N_2\rangle\langle N_1, N_2|\hat{B}|n_1, n_2\rangle = \sum_{N_1=0}^{\infty} \sum_{N_2=0}^{\infty} B_{n_1 n_2}^{N_1 N_2} |N_1, N_2\rangle$$

where

$$B_{n_1 n_2}^{N_1 N_2} = e^{-i\phi(n_1 - N_1)} \sum_{k=0}^{n_1} \sum_{l=0}^{n_2} (-1)^{n_1 - k} r^{n_1 + n_2 - k - l} t^{k+l}$$

$$\times \frac{\sqrt{n_1! n_2! N_1! N_2!}}{k!(n_1 - k)! l!(n_2 - l)!} \delta_{N_1, n_2 + k - l} \delta_{N_2, n_1 - k + l}$$

where $\delta$ is the Kronecker delta function. When the total number of input photons is $N = n_1 + n_2$ with $N \geq 1$, the output state is an $(N+1)$ entangled state.

**Problem 11.**  Consider the product Hilbert space $\mathbb{C}^2 \otimes \mathcal{H}$ where $\mathcal{H}$ denotes an arbitrary Hilbert space. For the one-Bose system we would set $\mathcal{H} = l_2(\mathbb{N})$. An arbitrary pure state in this product Hilbert space can be written as

$$|\psi\rangle := |0\rangle \otimes |\phi_0\rangle + |1\rangle \otimes |\phi_1\rangle$$

where $|\phi_0\rangle, |\phi_1\rangle \in \mathcal{H}$ and $\{|0\rangle, |1\rangle\}$ forms an orthonormal basis in $\mathbb{C}^2$. The condition for the state $|\psi\rangle$ to be normalized, i.e. $\langle \psi|\psi\rangle = 1$, leads to the constraint $\langle \phi_0|\phi_0\rangle + \langle \phi_1|\phi_1\rangle = 1$. If we assume that $|\phi_0\rangle$ and $|\phi_1\rangle$ have identical norms, then $|\psi\rangle$ takes the form

$$|\psi\rangle = \frac{1}{\sqrt{2}}(|0\rangle \otimes |\varphi_0\rangle + |1\rangle \otimes |\varphi_1\rangle)$$

where $|\phi_0\rangle = \frac{1}{\sqrt{2}}|\varphi_0\rangle$, $|\phi_1\rangle = \frac{1}{\sqrt{2}}|\varphi_1\rangle$ and $|\varphi_0\rangle$, $|\varphi_1\rangle$ are normalized. Defining the reduced density matrices (using the partial trace)

$$\rho_1 := \mathrm{tr}_{\mathbb{C}^2}(|\psi\rangle\langle\psi|), \qquad \rho_2 := \mathrm{tr}_{\mathcal{H}}(|\psi\rangle\langle\psi|)$$

the entanglement of $|\psi\rangle$ is given by

$$E(|\psi\rangle) := -\mathrm{tr}(\rho_1 \log_2(\rho_1)) = -\mathrm{tr}(\rho_2 \log_2(\rho_2)).$$

Describe the entanglement.

**Solution 11.** Straightforward calculation yields

$$\rho_1 = |\phi_0\rangle\langle\phi_0| + |\phi_1\rangle\langle\phi_1|$$

and

$$\rho_2 = \langle\phi_0|\phi_0\rangle|0\rangle\langle0| + \langle\phi_1|\phi_0\rangle|0\rangle\langle1| + \langle\phi_0|\phi_1\rangle|1\rangle\langle0| + \langle\phi_1|\phi_1\rangle|1\rangle\langle1|.$$

Applying the constraint we find that the non-zero eigenvalues of $\rho_1$ and $\rho_2$ are given by

$$\lambda(\langle\phi_0|\phi_0\rangle, |\langle\phi_0|\phi_1\rangle|^2) := \frac{1}{2}\left(1 + \sqrt{(1 - 2\langle\phi_0|\phi_0\rangle)^2 + 4|\langle\phi_0|\phi_1\rangle|^2}\right)$$

and $1 - \lambda$. Thus

$$E(|\psi\rangle) = -\lambda\log_2(\lambda) - (1 - \lambda)\log_2(1 - \lambda).$$

The entanglement is described exclusively by $\langle\phi_0|\phi_0\rangle$ and $|\langle\phi_0|\phi_1\rangle|^2$. Furthermore we have the inequality

$$|\langle\phi_0|\phi_1\rangle|^2 \le \langle\phi_0|\phi_0\rangle - \langle\phi_0|\phi_0\rangle^2 \le \frac{1}{4}.$$

**Problem 12.** Consider the superposition (*macroscopic quantum superposition states*) $|\phi_0\rangle = c_0(|\alpha\rangle + |-\alpha\rangle)$ and $|\phi_1\rangle = c_1(|\beta\rangle + |-\beta\rangle)$ where $c_0, c_1 \in \mathbb{C}$ and $|\alpha\rangle$, $|-\alpha\rangle$, $|\beta\rangle$, $|-\beta\rangle$, are coherent states, i.e.

$$|\psi\rangle = c_0|0\rangle \otimes (|\alpha\rangle + |-\alpha\rangle) + c_1|1\rangle \otimes (|\beta\rangle + |-\beta\rangle).$$

Discuss the entanglement.

**Solution 12.** The conditions for entanglement from the previous problem can be applied. In this case the normalization condition yield

$$2|c_0|^2(1 + e^{-2|\alpha|^2}) + 2|c_1|^2(1 + e^{-2|\beta|^2}) = 1.$$

Consequently $\langle\phi_0|\phi_0\rangle = 2|c_0|^2(1 + e^{-2|\alpha|^2})$ and

$$|\langle\phi_0|\phi_1\rangle|^2 = \langle\phi_0|\phi_0\rangle(1 - \langle\phi_0|\phi_0\rangle)\frac{(e^{-\frac{1}{2}|\alpha-\beta|^2} + e^{-\frac{1}{2}|\alpha+\beta|^2})^2}{(1 + e^{-2|\alpha|^2})(1 + e^{-2|\beta|^2})}.$$

It is convenient to define the real valued quantities

$$p_{00} := \langle\phi_0|\phi_0\rangle = 2|c_0|^2(1 + e^{-2|\alpha|^2}), \qquad p_{01} := \frac{(e^{-\frac{1}{2}|\alpha-\beta|^2} + e^{-\frac{1}{2}|\alpha+\beta|^2})^2}{(1 + e^{-2|\alpha|^2})(1 + e^{-2|\beta|^2})}.$$

Thus we obtain

$$|\langle\phi_0|\phi_1\rangle|^2 = p_{00}(1 - p_{00})p_{01}.$$

The maximum entanglement occurs when $\langle\phi_0|\phi_1\rangle = 0$ and $\langle\phi_0|\phi_0\rangle = \frac{1}{2}$. Since the above equation implies $\langle\phi_0|\phi_1\rangle \neq 0$ for $\langle\phi_0|\phi_0\rangle = \frac{1}{2}$, the maximum entanglement is approached asymptotically for $\alpha = 0$, $|\beta| \to \infty$ or $\beta = 0$, $|\alpha| \to \infty$. This is due to

$$|\alpha - \beta|^2 = |\alpha|^2 + |\beta|^2 - 2\Re(\alpha\bar{\beta}), \qquad |\alpha + \beta|^2 = |\alpha|^2 + |\beta|^2 + 2\Re(\alpha\bar{\beta}).$$

In other words, for $\alpha, \beta \neq 0$ one term shrinking in the numerator of $p_{01}$ implies the other is growing. To find the entanglement we first determine the eigenvalues of $\rho_1$ and $\rho_2$ which are now given by

$$\lambda = \frac{1}{2}\left(1 + \sqrt{1 - 4(1 - p_{01})p_{00}(1 - p_{00})}\right).$$

**Problem 13.** Consider the case when $|\phi_0\rangle$ is described by a number state and $|\phi_1\rangle$ is described by a coherent state, i.e.

$$|\psi\rangle = c_0|0\rangle \otimes |n\rangle + c_1|1\rangle \otimes |\alpha\rangle.$$

The scalar product between a number state $|n\rangle$ and a coherent state $|\alpha\rangle$ is given by

$$\langle n|\alpha\rangle = e^{-\frac{1}{2}|\alpha|^2}\frac{\alpha^n}{\sqrt{n!}}.$$

Discuss the entanglement.

**Solution 13.** The normalization condition for this case gives

$$|c_0|^2 + |c_1|^2 = 1.$$

Consequently $\langle\phi_0|\phi_0\rangle = |c_0|^2$, and

$$|\langle\phi_0|\phi_1\rangle|^2 = |c_0|^2(1 - |c_0|^2)e^{-|\alpha|^2}\frac{|\alpha|^{2n}}{n!}.$$

It is again convenient to define the quantities $p_{00}$ and $p_{01}$ as

$$p_{00} := \langle \phi_0 | \phi_0 \rangle = |c_0|^2, \qquad p_{01} := e^{-|\alpha|^2} \frac{|\alpha|^{2n}}{n!}.$$

Thus we obtain $|\langle \phi_0 | \phi_1 \rangle|^2 = p_{00}(1 - p_{00})p_{01}$. The entanglement can again be determined from $p_{00}$ and $p_{01}$ and proceeds as described in the previous problem.

**Problem 14.** Let $|\beta\rangle$ be a coherent state. Consider the entangled state

$$|\psi\rangle = \frac{1}{\sqrt{N_\beta}}(|\beta\rangle \otimes |\beta\rangle - |-\beta\rangle \otimes |-\beta\rangle).$$

(i) Find the normalization $N_\beta$.
(ii) Express the entangled state $|\psi\rangle$ using Fock states $|n\rangle \otimes |m\rangle$ with $n, m \in \mathbb{N}_0$.

**Solution 14.** Using $\langle \beta | - \beta \rangle = \exp(-2|\beta|^2)$ we obtain

$$N_\beta = 2 - 2\exp(-4|\beta|^2).$$

(ii) We obtain

$$|\psi\rangle = \frac{2\exp(-|\beta|^2)}{\sqrt{N_\beta}} \sum_{n,m \,|\, n+m \text{ odd}}^{\infty} \frac{\beta^{n+m}}{\sqrt{n!m!}} |n\rangle \otimes |m\rangle.$$

Thus the total number of photons is always odd.

**Problem 15.** Let $|\beta\rangle$, $|\gamma\rangle$ be coherent states. Consider the balanced entangled coherent state

$$|\psi\rangle = \frac{1}{\sqrt{N}}(|\beta\rangle \otimes |\gamma\rangle + e^{i\phi}|-\beta\rangle \otimes |-\gamma\rangle)$$

and $\phi = \pi/2$. Consider the operator

$$U = \exp(i\pi(b^\dagger b \otimes I + I \otimes b^\dagger b)).$$

Find $U|\psi\rangle$ and $U^2|\psi\rangle$. Discuss.

**Solution 15.** Since $e^{i\pi} = -1$ and $e^{i\pi/2} = i$ and

$$\exp(i\phi b^\dagger b)|\beta\rangle = |\beta e^{i\phi}\rangle$$

we find the entangled state

$$U|\psi\rangle = \frac{i}{\sqrt{2}}(|\beta\rangle \otimes |\gamma\rangle - i| - \beta\rangle \otimes | - \gamma\rangle).$$

Applying the operator $U$ again yields $U^2|\psi\rangle = |\psi\rangle$. Note that

$$e^{i\pi b^\dagger b}be^{-i\pi b^\dagger b} = -b.$$

**Problem 16.**   Consider the unitary evolution operator for the *beam split-ter*

$$U_{BS} = \exp(\theta(b_1^\dagger b_2 e^{i\phi} - b_1 b_2^\dagger e^{-i\phi}))$$

where the real angular parameter $\theta$ determines the transmission and reflection coefficients via $T = t^2 = \cos^2(\theta)$ and $R = r^2 = \sin^2(\theta)$. The internal phase shift $\phi$ between the reflected and transmitted modes is given by the beam splitter itself. To control $\phi$ we can place a phase shifter in one of the output channels.

(i) Let $|\psi_{in}\rangle = |0\rangle \otimes |1\rangle$, i.e. the one input is a one-photon state and the other the vacuum state. Calculate $|\psi_{out}\rangle = U_{BS}|\psi_{in}\rangle$.

(ii) To test quantum non locality of the state $|\psi_{out}\rangle$ we apply the *displaced parity operator* based on joint parity measurements

$$\hat{\Pi}_{12}(\beta_1, \beta_2) := D_1(\beta_1)D_2(\beta_2)\exp(i\pi(\hat{n}_1 + \hat{n}_2))D_1^\dagger(\beta_1)D_2^\dagger(\beta_2)$$

where $D_1(\beta_1)$ and $D_2(\beta_2)$ are the unitary displacement operators. Calculate

$$\Pi_{12}(\beta_1, \beta_2) := \langle\psi_{out}|\hat{\Pi}_{12}(\beta_1, \beta_2)|\psi_{out}\rangle.$$

(iii) The *two mode Bell function* $B(\beta_1, \beta_2)$ can be written as

$$B(\beta_1, \beta_2) = \Pi_{12}(0,0) + \Pi_{12}(\beta_1, 0) + \Pi_{12}(0, \beta_2) - \Pi_{12}(\beta_1, \beta_2).$$

For local realistic theory $B(\beta_1, \beta_2)$ should satisfy the *Bell-CHSH inequality*

$$|B(\beta_1, \beta_2)| \leq 2.$$

The violation of this inequality indicates quantum non locality of the single photon entangled state. Calculate $B(\beta_1, \beta_2)$ and discuss the case where $|\beta_1|^2 = |\beta_2|^2$.

**Solution 16.**   (i) We find

$$|\psi_{out}\rangle = t|0\rangle \otimes |1\rangle + re^{i\phi}|1\rangle \otimes |0\rangle.$$

(ii) We obtain

$$\Pi(\beta_1, \beta_2) = \langle \psi_{out} | \hat{\Pi}(\beta_1, \beta_2) | \psi_{out} \rangle$$
$$= (4|re^{-i\phi}\beta_1 + t\beta_2|^2 - 1) \exp(-2(|\beta_1|^2 + |\beta_2|^2)).$$

(iii) Let $|\beta_1|^2 = |\beta_2|^2 = J$ and let $\gamma_{12}$ be an arbitrary phase space difference between the two coherent displacements $\beta_1$ and $\beta_2$. Then we can write $\beta_2 = \beta_1 e^{i\gamma_{12}}$. Thus the two-mode Bell function is given by

$$B(\beta_1, \beta_2) = -1 + (4J - 2)e^{-2J} - (4J - 1)e^{-4J} - 8rtJe^{-4J}\cos(\Delta)$$

where $\Delta := \gamma_{12} + \phi$. When $\gamma_{12} = -\phi$ we obtain the maximal value $|B|_{max}$ of the two mode Bell function

$$|B|_{max} = 1 + (4J - 1)e^{-4J} + 8rtJe^{-4J} - (4J - 2)e^{-2J}.$$

**Problem 17.** There are various ways in which photons can be entangled. The photon is a spin-1 particle. One has a) polarization entanglement, b) momentum (direction) entanglement c) time-energy entanglement d) orbital angular momentum states entanglement. Describe the different types of entanglement for photons. Parametric down conversion can produce photons that are entangled both in polarization and in space.

**Solution 17.** Polarization entanglement. The highest contrast in experiments can be achieved for polarization-entangled states created by parametric down-conversion. Type-II sources can produce polarization entanglement directly. Parametric down-conversion or spontaneous parametric fluorescence is the spontaneous reverse process of second-harmonic generation, or more generally speaking three-wave mixing in nonlinear optical media (for example Beta-Barium Borate crystal). In nonlinear optics the polarization **P** depends nonlinear on the electric field **E**. The nonlinearity is given by a power series expansion of the polarization vector (summation convention is used)

$$P_i = \chi_{ij}^{(1)} E_j + \chi_{ijk}^{(2)} E_j E_k + \chi_{ijkl}^{(3)} E_j E_k E_l + \cdots$$

where $\chi_{ij}^{(1)}$ describes the normal refractive properties of a material including any kind of birefringence. $\chi_{ijk}^{(2)}$ is the coefficient tensor for three-wave, because two $E$ terms can lead to another $P$ term, mixing in strongly nonlinear material. The $\chi_{ijkl}^{(3)}$ term describes effects that occur at even higher intensities, e.g. Kerr-lensing or phase conjugation.

In down-conversion one has a high-frequency pump field and two lower frequency down-converted fields. Let $\omega$ be the frequency. Energy conservation yields

$$\hbar\omega_p = \hbar\omega_1 + \hbar\omega_2$$

whereas phase matching is given by

$$\hbar \mathbf{k}_p = \hbar \mathbf{k}_1 + \hbar \mathbf{k}_2$$

where $\mathbf{k}$ is the wave vector. As most nonlinear media are birefringent the second criterion can be satisfied by choosing an appropriate cut, thus managing the refractive indices and therefore wave velocities such that light can be emitted in specific directions. In type-II spontaneous parametric down-conversion a pump beam is incident on a nonlinear optical crystal in which pump photons spontaneously split with a low probability into two orthogonally polarized photons called signal and idler. From energy and momentum conversation one finds that the wavelength and emission directions of the down conversion photons are correlated. They depend on the pump wavelength as well as on the angle between optical axis of the crystal and the pump beam. In the degenerate case (signal and idler having the same wavelength) the photons leave the crystal symmetrically with respect to the pump beam along two cones. For certain orientations of the optical crystal, the two emission cones intersect and the photons emerging along the intersection directions can not be assigned to one of the two orthogonally polarized cones anymore and thus form a polarization entangled pair. The polarization entangled state (Bell state) is described by

$$\frac{1}{\sqrt{2}}(|h\rangle \otimes |v\rangle + e^{i\phi}|v\rangle \otimes |h\rangle)$$

where $h$ and $v$ denote horizontal and vertical polarizations of light. The state is fully entangled (Bell state). By using only standard optical elements in one of the two output beams, one can transform any one of the Bell states into any of the other.

The energy-entangled states from down-conversion photons are the most universal. They are present for any pair of photons. Since there are many ways to partition the energy of the pump photon, each daughter photon has a broad spectrum, and hence a narrow wave packet in time. The sum of the two daughter photons energies is well-defined. They must add up to the energy of the monochromatic pump laser photon. This correlation is given by the energy entangled state

$$|\psi\rangle = \int_0^{E_p} A(E)|E\rangle_s \otimes |E_p - E\rangle_i \, dE$$

where each ket describes the energy of one of the photons, $s$ and $i$ denotes the signal and idler, respectively, and $A(E)$ is the spectral distribution of the collected down conversion light.

Another entanglement from the parametric down-conversion process is the momentum direction entanglement. From the emission of a parametric down-conversion source two pairs of spatial (momentum direction) modes are extracted by pinholes. Photon pairs are emitted such that whenever a photon is emitted into one of the inner two modes its partner will be found in the opposite outer mode due to the phase matching in the crystal. The superposition of the two inner and the two outer modes on the beam-splitter serves to measure coincidence rates in various superposition of the initial spatial modes. After the beam-splitters we cannot distinguish the upper two modes from the lower two and therefore interference will be observed in the various coincidence rates.

In a discrete version (time-bin entanglement) of energy-time entangled states sources one sends a double pulse through the down-conversion crystal. If the delay between the two pump pulses equals the time difference between the short and the long arms of the Mach-Zehnder interferometer then again there are two indistinguishable ways of obtaining a coincidence detection.

## 22.3   Supplementary Problems

**Problem 1.**   Consider the number states $|n\rangle$ $(n = 0, 1, 2, \ldots)$. Are the states

$$\frac{1}{\sqrt{2}}(|n\rangle \otimes |n\rangle + |n+1\rangle \otimes |n+1\rangle)$$

$$\frac{1}{\sqrt{2}}(|n\rangle \otimes |n+1\rangle + |n+1\rangle \otimes |n\rangle)$$

entangled?

**Problem 2.**   Let $|\zeta\rangle$ be a squeezed state. Is the state

$$\frac{1}{\sqrt{N}}(|\zeta\rangle \otimes |-\zeta\rangle + |-\zeta\rangle \otimes |\zeta\rangle)$$

entangled?

**Problem 3.**   Let $|n\rangle$ be a number state and $|\beta\rangle$ be a coherent state. Is the state

$$\frac{1}{\sqrt{N}}(|n\rangle \otimes |\beta\rangle + |n\rangle \otimes |-\beta\rangle)$$

entangled? Does it depend on $n \in \mathbb{N}_0$ and $\beta \in \mathbb{C}$?

**Problem 4.**    Consider the coherent state $|\beta\rangle$ and the squeezed state $|\zeta\rangle$ state. Is the state

$$\frac{1}{\sqrt{N}}(|\beta\rangle \otimes |\zeta\rangle + |\zeta\rangle \otimes |\beta\rangle)$$

entangled? Here $N$ is a normalization factor.

**Problem 5.**    Let

$$b_0^\dagger = b^\dagger \otimes I \otimes I, \quad b_1^\dagger = I \otimes b^\dagger \otimes I, \quad b_2^\dagger = I \otimes I \otimes b^\dagger$$

be Bose creation operators. Let $|\zeta\rangle$ ($\zeta \in \mathbb{C}$) be a squeezed state with input mode $b_0^\dagger$ and $|\beta_1\rangle$, $|\beta_2\rangle$ ($\beta_1, \beta_2 \in \mathbb{C}$) be coherent states with input modes $b_1^\dagger$, $b_2^\dagger$, respectively. Show that the state

$$|\zeta, \beta_1, \beta_2\rangle =$$

$$e^{-(|\beta_1|^2+|\beta_2|^2)/2} \sum_{n=0,n_1=0,n_2=0}^{\infty} C_n(\zeta) \frac{(b_0^\dagger)^n}{\sqrt{n!}} \frac{(\beta_1 b_1^\dagger)^{n_1}}{n_1!} \frac{(\beta_2 b_2^\dagger)^{n_2}}{n_2!} |0\rangle_0 \otimes |0\rangle_1 \otimes |0\rangle_2$$

is entangled. Here $C_n(\zeta)$ is the coefficient of the squeezed state with squeezing parameter $se^{i\theta}$ and is equal to 0 for all odd values of $n$ and for $n$ even given by

$$C_n(\zeta) = \frac{\sqrt{n!}}{\sqrt{\cosh(s)}((n/2)!)} \left(-\frac{1}{2}e^{i\theta}\tanh(s)\right)^{n/2}, \quad n \text{ even}.$$

# Chapter 23

# Continuous Variable Teleportation

## 23.1 Introduction

Quantum continuous variables provide a new approach to quantum information processing and quantum communication. They describe highly excited quantum systems such as multi-photon fields of light. Continuous variables offer additional advantages over the single-photon system. They involve the use of highly efficient telecommunication photodiodes. The coherent sources of continuous entanglement are also orders of magnitude more efficient than the spontaneous sources of discrete entanglement. Teleportation schemes can be demonstrated involving bright light sources. Entangled states build from coherent states $|\beta\rangle$ are utilized and also Schrödinger cat states

$$\frac{1}{\sqrt{N}}(|\beta\rangle + |-\beta\rangle).$$

## 23.2 Solved Problems

**Problem 1.** Consider the numbers states $|n\rangle$ with $n = 0, 1, 2, \ldots$. Let

$$|\Psi_{in}\rangle = \alpha|0\rangle_1 + \beta|1\rangle_1 + \gamma|2\rangle_1 \tag{1}$$

be an input quantum state in mode 1. An ancilla quantum state

$$|\Psi_{ancilla}\rangle = |0\rangle_2|0\rangle_3 \equiv |0\rangle_2 \otimes |0\rangle_3$$

in the vacuum state is also given. The modes 2 and 3 pass through a first parametric amplifier whose transformation is given by the operator ($\theta_1 \in \mathbb{R}$)

$$U_{23}(\theta_1) = \exp(\theta_1(b_2^\dagger b_3^\dagger - b_2 b_3)).$$

(i) Calculate

$$|\Psi'_{ancilla}\rangle = U_{23}(\theta_1)|\Psi_{ancilla}\rangle = U_{23}(\theta_1)(|0\rangle_2 \otimes |0\rangle_3)$$

and set $\gamma_1 = \tanh^2(\theta_1)$.

(ii) Consider the product state

$$|\Psi_{in}\rangle|\Psi'_{ancilla}\rangle \equiv |\Psi_{in}\rangle \otimes |\Psi'_{ancilla}\rangle.$$

The output mode 2 of the first parametric amplifier and the mode 1 are used as the input modes of the second parametric amplifier ($\theta_2 \in \mathbb{R}$)

$$U_{12}(\theta_2) = \exp(\theta_2(b_1^\dagger b_2^\dagger - b_1 b_2)) \otimes I_3.$$

Calculate

$$|\Psi_{out}\rangle = U_{12}(\theta_2)(|\Psi_{in}\rangle \otimes |\Psi'_{ancilla}\rangle)$$

which contains only the state $|1\rangle_1|1\rangle_2$ and set $\gamma_2 = \tanh^2(\theta_2)$.

(iii) Find the projection

$$(_1\langle 1|_2 \otimes \langle 1| \otimes I)|\Psi_{out}\rangle.$$

(iv) Assume that we want to transform the input state (1) into

$$|\Psi_{out}\rangle = \alpha|0\rangle_3 + \beta|1\rangle_3 - \gamma|2\rangle_3.$$

What is the relation between the coefficients $\gamma_1$ and $\gamma_2$?

**Solution 1.**    (i) Since

$$[b_2^\dagger b_3^\dagger, b_2 b_3] = -b_2^\dagger b_2 - b_3^\dagger b_3 - I$$

and

$$[b_2^\dagger b_2, b_2^\dagger b_3^\dagger] = b_2^\dagger b_3^\dagger, \qquad [b_3^\dagger b_3, b_2^\dagger b_3^\dagger] = b_2^\dagger b_3^\dagger$$

$$[b_2^\dagger b_2, b_2 b_3] = -b_2 b_3, \qquad [b_3^\dagger b_3, b_2 b_3] = -b_2 b_3$$

we have a Lie algebra with the basis $b_2^\dagger b_3^\dagger$, $b_2 b_3$, $b_2^\dagger b_2$, $b_3^\dagger b_3$, $I$. Thus we can *disentangle* the operator $U_{23}(\theta_1)$ as

$$U_{23}(\theta_1) = \exp(\sqrt{\gamma_1}b_2^\dagger b_3^\dagger)\exp(\ln(1-\gamma_1)(b_2^\dagger b_2 + b_3^\dagger b_3 + I)/2)\exp(-\sqrt{\gamma_1}b_2 b_3).$$

Now we have

$$\exp(-\sqrt{\gamma_1}b_2b_3)|0\rangle_2|0\rangle_3 = (|0\rangle_2 \otimes |0\rangle_3)$$
$$\exp(\ln(1-\gamma_1)(b_2^\dagger b_2 + b_3^\dagger b_3 + I)/2)|0\rangle_2|0\rangle_3 = \exp(\ln((1-\gamma_1)^{1/2})(|0\rangle_2 \otimes |0\rangle_3)$$
$$= \sqrt{1-\gamma_1}(|0\rangle_2 \otimes |0\rangle_3)$$

and

$$\exp(\sqrt{\gamma_1}b_2^\dagger b_3^\dagger)|0\rangle_2|0\rangle_3 = \sum_{n=0}^{\infty} \gamma_1^{n/2}|n\rangle_2|n\rangle_3.$$

Thus

$$U_{23}(\theta_1)|\Psi_{ancilla}\rangle = \sqrt{1-\gamma_1}\sum_{n=0}^{\infty} \gamma_1^{n/2}(|n\rangle_2 \otimes |n\rangle_3).$$

(ii) Using the result from (i) since the structure of the operator $U_{12}$ is the same as $U_{23}$ we obtain

$$|\Psi_{out}\rangle = \sqrt{(1-\gamma_1)(1-\gamma_2)}|1\rangle_1|1\rangle_2(\sqrt{\gamma_2}\alpha|0\rangle_3 + \sqrt{\gamma_1}(1-2\gamma_2)\beta|1\rangle_3$$
$$+\gamma_1\sqrt{\gamma_2}(3\gamma_2 - 2)\gamma|2\rangle_3 + \Psi_{others}$$

(iii) The projection yields the state

$$|\Psi_{out}\rangle = \sqrt{(1-\gamma_1)(1-\gamma_2)}(\sqrt{\gamma_2}\alpha|0\rangle_3 + \sqrt{\gamma_1}(1-2\gamma_2)\beta|1\rangle_3$$
$$+\gamma_1\sqrt{\gamma_2}(3\gamma_2 - 2)\gamma|2\rangle_3.$$

(iv) The relation for the parameters is

$$\sqrt{\gamma_2} = \sqrt{\gamma_1}(1-2\gamma_2) = -\gamma_1\sqrt{\gamma_2}(3\gamma_2 - 2)$$

with the solution

$$\gamma_1 = \frac{21-7\sqrt{2}}{9+4\sqrt{2}} \approx 0.757, \qquad \gamma_2 = \frac{3-\sqrt{2}}{7} \approx 0.226.$$

**Problem 2.** Let $|\beta\rangle$ be a coherent state. Let $b$ and $b^\dagger$ be Bose annihilation and creation operators, respectively. Let $D(\mu)$ be the displacement operator ($\mu \in \mathbb{C}$). Consider the product state

$$|\psi\rangle := \frac{1}{\pi}\int_{\mathbb{C}} d^2\beta|\beta\rangle \otimes |\beta^*\rangle.$$

This is a maximally entangled continuous-variable state. The state is not normalized. For teleportation we assume the unknown state $|\phi\rangle$ to be in mode 1, the sender's part of the quantum channel to be in mode 2, and the receiver's part in mode 3. Calculate

$$({}_{12}\langle\psi| \otimes I_3)(D_1^\dagger(\mu) \otimes I_2 \otimes I_3)(|\phi\rangle_1 \otimes |\psi\rangle_{23})$$

where $I_2$ is the identity operator acting on mode 2, $I_3$ the identity operator acting on mode 3, and $D_1^\dagger$ indicates that the operator acts on mode 1.

**Solution 2.** Using the completeness relation of coherent states

$$\frac{1}{\pi} \int_{\mathbb{C}} d^2\beta |\beta\rangle\langle\beta| = I$$

yields

$$|\gamma\rangle = \frac{1}{\pi} \int_{\mathbb{C}} d^2\beta |\beta\rangle\langle\beta|\gamma\rangle$$

where we used $I|\beta\rangle = |\beta\rangle$. Applying this expansion and the identity

$$\langle\gamma|\beta\rangle = \langle\beta^*|\gamma^*\rangle$$

we find

$$({}_{12}\langle\psi| \otimes I_3)(D_1^\dagger(\mu) \otimes I_2 \otimes I_3)(|\phi\rangle_1 \otimes |\psi\rangle_{23})$$

$$= \frac{1}{\pi^2} \int_{\mathbb{C}} \int_{\mathbb{C}} d^2\beta d^2\gamma \langle\gamma|D^\dagger(\mu)|\phi\rangle\langle\gamma^*|\beta\rangle|\beta^*\rangle_3$$

$$= \frac{1}{\pi^2} \int_{\mathbb{C}} \int_{\mathbb{C}} d^2\beta d^2\gamma \langle\beta^*|\gamma\rangle\langle\gamma|D^\dagger(\mu)|\phi\rangle|\beta^*\rangle_3$$

$$= \frac{1}{\pi} \int_{\mathbb{C}} d^2\beta \langle\beta^*|D^\dagger(\mu)|\phi\rangle|\beta^*\rangle_3$$

$$= \frac{1}{\pi} \int_{\mathbb{C}} d^2\beta |\beta^*\rangle\langle\beta^*|D_3^\dagger(\mu)|\phi\rangle_3$$

$$= D_3^\dagger(\mu)|\phi\rangle_3$$

where we used the identity

$$\frac{1}{\pi} \int_{\mathbb{C}} d^2\gamma \langle\beta^*|\gamma\rangle\langle\gamma|D^\dagger(\mu)|\phi\rangle = \langle\beta^*|D^\dagger(\mu)|\phi\rangle.$$

We conclude that after the joint measurement, the sender's state is projected onto the state which is a unitarily transformed unknown state. Upon receiving the measurement outcome $\mu$, the receiver recovers the unknown state by using the appropriate unitary transformation $D(\mu)$.

## 23.3 Supplementary Problem

**Problem 1.** Let

$$|\beta\rangle = e^{-|\beta|^2/2} \sum_{n=0}^{\infty} \frac{\beta^n}{\sqrt{n!}}|n\rangle, \quad \beta \in \mathbb{C}$$

be a coherent state. Consider the "coherent Bell states" given by

$$|C\Phi^+\rangle = \frac{1}{\sqrt{N_+}}(|\beta\rangle \otimes |\beta\rangle + |-\beta\rangle \otimes |-\beta\rangle)$$

$$|C\Phi^-\rangle = \frac{1}{\sqrt{N_-}}(|\beta\rangle \otimes |\beta\rangle - |-\beta\rangle \otimes |-\beta\rangle)$$

$$|C\Psi^+\rangle = \frac{1}{\sqrt{N_+}}(|\beta\rangle \otimes |-\beta\rangle + |-\beta\rangle \otimes |\beta\rangle)$$

$$|C\Psi^-\rangle = \frac{1}{\sqrt{N_-}}(|\beta\rangle \otimes |-\beta\rangle - |-\beta\rangle \otimes |\beta\rangle)$$

in analogy to the Bell states in $\mathbb{C}^4$

$$|\Phi^+\rangle = \frac{1}{\sqrt{2}}(|0\rangle \otimes |0\rangle + |1\rangle \otimes |1\rangle)$$

$$|\Phi^-\rangle = \frac{1}{\sqrt{2}}(|0\rangle \otimes |0\rangle - |1\rangle \otimes |1\rangle)$$

$$|\Psi^+\rangle = \frac{1}{\sqrt{2}}(|0\rangle \otimes |1\rangle + |1\rangle \otimes |0\rangle)$$

$$|\Psi^-\rangle = \frac{1}{\sqrt{2}}(|0\rangle \otimes |1\rangle - |1\rangle \otimes |0\rangle).$$

Note that

$$N_\pm = 2(1 \pm e^{-4\beta\bar{\beta}}).$$

For the Bell states in $\mathbb{C}^4$ we have

$$\langle\Phi^+|\Phi^-\rangle = 0$$

etc.. Do we have

$$\langle C\Phi^+|C\Phi^-\rangle = 0$$

etc.? Consider the normalized qubit state

$$|\psi\rangle = a_0|0\rangle + a_1|1\rangle, \quad |a_0|^2 + |a_1|^2 = 1.$$

We define the linear operators

$$R_1(b_+|\beta\rangle + b_-|-\beta\rangle) = b_-|\beta\rangle + b_+|-\beta\rangle$$
$$R_3(b_+|\beta\rangle + b_-|-\beta\rangle) = b_+|\beta\rangle - b_-|-\beta\rangle$$

in analogy to the Pauli spin matrices $\sigma_1$ and $\sigma_3$. Note that $\sigma_1\sigma_3 = -i\sigma_2$. Consider the normalized state

$$|\gamma\rangle = \frac{1}{\sqrt{N_{+,-}}}(b_+|\beta\rangle + b_-|-\beta\rangle)$$

Describe teleportation utilizing the expression

$$
\begin{aligned}
|\psi\rangle \otimes |C\Phi^+\rangle = \frac{1}{\sqrt{N_+}} \Bigg( & \frac{1}{\sqrt{2}}(|0\rangle \otimes |\beta\rangle + |1\rangle \otimes |-\beta\rangle) \otimes |\gamma\rangle \\
& + \frac{1}{\sqrt{2}}(|0\rangle \otimes |\beta\rangle - |1\rangle \otimes |-\beta\rangle) \otimes R_3|\gamma\rangle \\
& + \frac{1}{\sqrt{2}}(|0\rangle \otimes |-\beta\rangle + |1\rangle \otimes |\beta\rangle) \otimes R_1|\gamma\rangle \\
& + \frac{1}{\sqrt{2}}(|0\rangle \otimes |-\beta\rangle - |1\rangle \otimes |\beta\rangle) \otimes (R_1 R_3)|\gamma\rangle \Bigg).
\end{aligned}
$$

Can this scheme also be applied if we replace the coherent state by squeezed states?

# Chapter 24

# Swapping and Cloning

## 24.1 Introduction

Swapping and cloning need not only be studied for finite-dimensional systems but also for continuous variables. We can therefore investigate whether coherent states $|\beta\rangle$ and squeezed states $|\zeta\rangle$ can be swapped or cloned.

## 24.2 Solved Problems

**Problem 1.** Can two coherent states be swapped, i.e. can we find a unitary transformation (*swap operator*) such that

$$U_{swap}(|\beta_1\rangle \otimes |\beta_2\rangle) = |\beta_2\rangle \otimes |\beta_1\rangle$$

holds? Consider the unitary operator

$$U(z) := e^{z b_1^\dagger b_2 - z^* b_1 b_2^\dagger}, \qquad z \in \mathbb{C}.$$

**Solution 1.** Yes, we can find a swap operator. From the unitary operator given above we find $U(z)(|0\rangle \otimes |0\rangle) = |0\rangle \otimes |0\rangle$. Now we have

$$U(z)(|\beta_1\rangle \otimes |\beta_2\rangle) = U(z)(D(\beta_1) \otimes D(\beta_2))|0\rangle \otimes |0\rangle$$
$$= U(z)D_1(\beta_1)D_2(\beta_2)U^{-1}(z)|0\rangle \otimes |0\rangle$$

495

where $D(\beta)$ is the displacement operator. Thus

$$U(z)D_1(\beta_1)D_2(\beta_2)U^{-1}(z) = U(z)\exp(\beta_1 b_1^\dagger - \beta_1^* b_1 + \beta_2 b_2^\dagger - \beta_2^* b_2)U(z)^{-1}$$

and therefore

$$\begin{aligned}
U(z)D_1(\beta_1)D_2(\beta_2)U^{-1}(z) &= \exp(\beta_1(U(z)b_1 U(z)^{-1})^\dagger - \beta_1^*(U(z)b_1 U(z)^{-1}) \\
&\quad + \beta_2(U(z)b_2 U(z)^{-1})^\dagger - \beta_2^*(U(z)b_2 U(z)^{-1})) \\
&\equiv \exp(X).
\end{aligned}$$

Calculating the unitary transformations in the exponent, we find

$$\begin{aligned}
X = &\left( \cos(|z|)\beta_1 + \frac{z\sin(|z|)}{|z|}\beta_2 \right) b_1^\dagger - \left( \cos(|z|)\beta_1^* + \frac{z^*\sin(|z|)}{|z|}\beta_2^* \right) b_1 \\
&+ \left( \cos(|z|)\beta_2 - \frac{z^*\sin(|z|)}{|z|}\beta_1 \right) b_2^\dagger - \left( \cos(|z|)\beta_2^* - \frac{z\sin(|z|)}{|z|}\beta_1^* \right) b_2.
\end{aligned}$$

Thus

$$\begin{aligned}
\exp(X) &= D_1 \left( \cos(|z|)\beta_1 + \frac{z\sin(|z|)}{|z|}\beta_2 \right) D_2 \left( \cos(|z|)\beta_2 - \frac{z^*\sin(|z|)}{|z|}\beta_1 \right) \\
&= D \left( \cos(|z|)\beta_1 + \frac{z\sin(|z|)}{|z|}\beta_2 \right) \otimes D \left( \cos(|z|)\beta_2 - \frac{z^*\sin(|z|)}{|z|}\beta_1 \right).
\end{aligned}$$

Therefore, we have

$$|\beta_1\rangle \otimes |\beta_2\rangle \to \left| \cos(|z|)\beta_1 + \frac{z\sin(|z|)}{|z|}\beta_2 \right\rangle \otimes \left| \cos(|z|)\beta_2 - \frac{z^*\sin(|z|)}{|z|}\beta_1 \right\rangle.$$

If we write $z = |z|e^{i\delta}$, then we can write

$$|\beta_1\rangle \otimes |\beta_2\rangle \to |\cos(|z|)\beta_1 + e^{i\delta}\sin(|z|)\beta_2\rangle \otimes |\cos(|z|)\beta_2 - e^{-i\delta}\sin(|z|)\beta_1\rangle.$$

Choosing $\sin(|z|) = 1$ yields

$$|\beta_1\rangle \otimes |\beta_2\rangle \to |e^{i\delta}\beta_2\rangle \otimes |-e^{-i\delta}\beta_1\rangle = |e^{i\delta}\beta_2\rangle \otimes |e^{-i(\delta+\pi)}\beta_1\rangle. \tag{1}$$

Applying the unitary operator

$$V = e^{-i\delta b_1^\dagger b_1} e^{i(\delta+\pi)b_2^\dagger b_2} = e^{-i\delta b^\dagger b} \otimes e^{i(\delta+\pi)b^\dagger b}$$

from the left, we find $|\beta_1\rangle \otimes |\beta_2\rangle \to |\beta_2\rangle \otimes |\beta_1\rangle$. If we set $\beta_1 = \beta$ and $\beta_2 = 0$ in (1) we obtain

$$|\beta\rangle \otimes |0\rangle = |\cos(|z|)\beta\rangle \otimes |-e^{-i\delta}\sin(|z|)\beta\rangle = |\cos(|z|)\beta\rangle \otimes |e^{-i(\delta+\pi)}\sin(|z|)\beta\rangle.$$

**Problem 2.**   We cannot *clone* coherent states, i.e., we cannot find a unitary operator which maps

$$|\beta\rangle \otimes |0\rangle \to |\beta\rangle \otimes |\beta\rangle.$$

Use the result from problem 1 (equation (1))

$$|\beta\rangle \otimes |0\rangle \to |\cos(|z|)\beta\rangle \otimes |e^{-i(\delta+\pi)}\sin(|z|)\beta\rangle \tag{1}$$

to find an approximation.

**Solution 2.**   Applying the operator $I \otimes e^{i(\delta+\pi)b^\dagger b}$ to the right-hand side of (1), we obtain

$$|\cos(|z|)\beta\rangle \otimes |\sin(|z|)\beta\rangle.$$

If we set $|z| = \pi/4$ we obtain

$$|\beta\rangle \otimes |0\rangle \to \left|\frac{\beta}{\sqrt{2}}\right\rangle \otimes \left|\frac{\beta}{\sqrt{2}}\right\rangle.$$

This is called *imperfect cloning*.

**Problem 3.**   We consider three infinite-dimensional Hilbert spaces $\mathcal{H}_1$, $\mathcal{H}_2$, $\mathcal{H}_3$ and the product Hilbert space $\mathcal{H}_3 \otimes \mathcal{H}_1 \otimes \mathcal{H}_2$ with $\mathcal{H}_1 = \mathcal{H}_2 = \mathcal{H}_3$. Consider the heterodyne-current operator $Z := b_1 + b_2^\dagger$, where the Bose annihilation operator acts in the Hilbert space $\mathcal{H}_1$ and the Bose creation operator $b_2^\dagger$ acts on the Hilbert space $\mathcal{H}_2$. We have $[Z, Z^\dagger] = 0$ and the eigenvalue equation $Z|z\rangle\rangle_{12} = z|z\rangle\rangle_{12}$ with $z \in \mathbb{C}$. The eigenstates $|z\rangle\rangle$ are given by

$$|z\rangle\rangle_{12} = D_1(z)|0\rangle\rangle_{12} = D_2(z^*)|0\rangle\rangle_{12}$$

where $D_1$ denotes the displacement operator for mode 1, $D_2$ the displacement operator for mode 2 and

$$|0\rangle\rangle_{12} := \frac{1}{\sqrt{\pi}} \sum_{n=0}^{\infty} (-1)^n |n\rangle_1 \otimes |n\rangle_2$$

in the Fock basis (number basis). The expression

$$_{32}\langle\langle z|z'\rangle\rangle_{12} = \frac{1}{\pi} D_1(z') T_{13} D_3^\dagger(z)$$

where

$$T_{13} := \sum_{n=0}^{\infty} |n\rangle_{13}\langle n| \equiv \sum_{n=0}^{\infty} (|n\rangle \otimes I \otimes I)(I \otimes I \otimes \langle n|)$$

denotes the transfer operator which obviously satisfies $T_{13}|\psi\rangle_3 = |\psi\rangle_1$ for any vector $|\psi\rangle$. For a cloning operation consider the input state in the product Hilbert space $\mathcal{H}_3 \otimes \mathcal{H}_1 \otimes \mathcal{H}_2$

$$|\psi\rangle = |\phi\rangle_3 \otimes \int_{\mathbb{C}} d^2z f(z, z^*)|z\rangle\rangle_{12}.$$

where $|\phi\rangle_3$ is the original state in the Hilbert space $\mathcal{H}_3$ to be cloned in $\mathcal{H}_3$ itself and $\mathcal{H}_1$ and $\mathcal{H}_2$ is an ancillary Hilbert space. The cloning transformation is realized by the unitary operator

$$U = \exp\left(\left(\frac{1}{2}(b_3 + b_3^\dagger) + \frac{1}{2}(b_3 - b_3^\dagger)\right)Z^\dagger - \left(\frac{1}{2}(b_3 + b_3^\dagger) - \frac{1}{2}(b_3 - b_3^\dagger)\right)Z\right)$$

where $Z = b_1 + b_2^\dagger = b \otimes I + I \otimes b^\dagger$. Let $|\psi\rangle_{out} = U|\psi\rangle$.
(i) Calculate the commutator $[b_3 b_1^\dagger + b_3 b_2, b_3^\dagger b_1 + b_3^\dagger b_2^\dagger]$ and discuss.
(ii) Evaluate the one-site restricted density matrix $\rho_3$ corresponding to the state $|\psi\rangle_{out}$ for the Hilbert space $\mathcal{H}_3$.
(iii) Evaluate the one-site restricted density matrix $\rho_1$ corresponding to the state $|\psi\rangle_{out}$ for the Hilbert space $\mathcal{H}_1$.
(iv) Compare the two density matrices.

**Solution 3.**    (i) Using $[b_j, b_k^\dagger] = \delta_{jk}I$ we find

$$[b_3 b_1^\dagger + b_3 b_2, b_3^\dagger b_1 + b_3^\dagger b_2^\dagger] = I + b_1^\dagger b_1 + b_2^\dagger b_2 + b_1^\dagger b_2^\dagger + b_1 b_2.$$

The right-hand side does not depend on $b_3$ and $b_3^\dagger$.
(ii) Let $|w\rangle\rangle_{12}$ be an eigenstate of the operator $Z$. We have $\rho = |\psi\rangle\langle\psi|$. Thus for the partial trace we have to calculate

$$\rho_3 = \int_{\mathbb{C}} d^2w \int_{\mathbb{C}} d^2z \int_{\mathbb{C}} d^2z' f(z, z^*) f^*(z', z'^*) A$$

where $A \equiv {}_{12}\langle\langle w|D_3^\dagger(z)|\phi\rangle_{33}\langle\phi|D_3(z') \otimes |z\rangle\rangle_{12}{}_{12}\langle\langle z'|w\rangle\rangle_{12}$. Using the completeness and orthogonality of the eigenstates $|w\rangle\rangle_{12}$ of the operator $Z$ we find

$$\rho_3 = \int_{\mathbb{C}} d^2z |f(z, z^*)|^2 D_3^\dagger(z)|\phi\rangle_{33}\langle\phi|D_3(z).$$

(iii) For $\rho_1$ we have to calculate

$$\rho_1 = \int_{\mathbb{C}} d^2w \int_{\mathbb{C}} \frac{d^2z}{\pi} \int_{\mathbb{C}} \frac{d^2z'}{\pi} f(z, z^*) f^*(z', z'^*)$$
$$\times D_1(z) T_{13}(D_3^\dagger(w) D_3^\dagger(z)|\phi\rangle_{33}\langle\phi|D_3(z') D_3(w)) T_{31} D_3^\dagger(z').$$

Using the completeness and orthogonality of the eigenstates $|w\rangle\rangle$ of the operator $Z$ we find after integration over $z$ and $z'$

$$\rho_1 = \int_C d^2w |\tilde{f}(w, w^*)|^2 D_1^\dagger(w)|\phi\rangle_{11}\langle\phi|D_1(z)$$

where $\tilde{f}(w, w^*)$ denotes the *Fourier transform* over the complex plane

$$\tilde{f}(w, w^*) = \frac{1}{\pi}\int_C d^2z e^{wz^* - w^*z} f(z, z^*).$$

(iv) For

$$f(z, z^*) = \sqrt{\frac{2}{\pi}} e^{-|z|^2}$$

one has two identical clones, i.e., $\rho_3 = \rho_1$ which are given by the original state $|\psi\rangle$ degraded by Gaussian noise.

**Problem 4.** Let $\alpha \in \mathbb{R}$. Find

$$e^{\alpha(b^\dagger \otimes b - b \otimes b^\dagger)}(|1\rangle \otimes |0\rangle).$$

**Solution 4.** We have

$$(b^\dagger \otimes b - b \otimes b^\dagger)(|1\rangle \otimes |0\rangle) = -|0\rangle \otimes |1\rangle$$
$$(b^\dagger \otimes b - b \otimes b^\dagger)(-|0\rangle \otimes |1\rangle) = -|1\rangle \otimes |0\rangle$$
$$(b^\dagger \otimes b - b \otimes b^\dagger)(-|1\rangle \otimes |0\rangle) = |0\rangle \otimes |1\rangle$$
$$(b^\dagger \otimes b - b \otimes b^\dagger)(|0\rangle \otimes |1\rangle) = |1\rangle \otimes |0\rangle.$$

Therefore we find

$$e^{\alpha(b^\dagger \otimes b - b \otimes b^\dagger)} = |1\rangle \otimes |0\rangle \left(1 - \frac{\alpha^2}{2!} + \frac{\alpha^4}{4!} - \cdots\right) + |0\rangle \otimes |1\rangle \left(-\alpha + \frac{\alpha^3}{3!} - \cdots\right)$$
$$= \cos(\alpha)|1\rangle \otimes |0\rangle - \sin(\alpha)|0\rangle \otimes |1\rangle.$$

If $\alpha = \pi/2$, then

$$\exp(\alpha(b^\dagger \otimes b - b \otimes b^\dagger))(|1\rangle \otimes |0\rangle) = -|0\rangle \otimes |1\rangle$$

and if $\alpha = 3\pi/2$, then

$$\exp(\alpha(b^\dagger \otimes b - b \otimes b^\dagger))(|1\rangle \otimes |1\rangle) = |0\rangle \otimes |1\rangle.$$

## 24.3    Supplementary Problems

**Problem 1.**    Consider the four Bell states of entangled coherent states

$$|C\Phi^+\rangle = \frac{1}{\sqrt{N_+}}(|\beta\rangle \otimes |\beta\rangle + |-\beta\rangle \otimes |-\beta\rangle)$$

$$|C\Phi^-\rangle = \frac{1}{\sqrt{N_+}}(|\beta\rangle \otimes |\beta\rangle - |-\beta\rangle \otimes |-\beta\rangle)$$

$$|C\Psi^+\rangle = \frac{1}{\sqrt{N_+}}(|\beta\rangle \otimes |-\beta\rangle + |-\beta\rangle \otimes |\beta\rangle)$$

$$|C\Psi^-\rangle = \frac{1}{\sqrt{N_+}}(|\beta\rangle \otimes |-\beta\rangle - |-\beta\rangle \otimes |\beta\rangle).$$

Let $U_{BS}$ be the unitary operator of the beam splitter

$$U_{BS} = \exp\left(\frac{\pi}{4}(b^\dagger \otimes b - b \otimes b^\dagger)\right).$$

Find the states

$$U_{BS}|C\Phi^+\rangle, \quad U_{BS}|C\Phi^-\rangle, \quad U_{BS}|C\Psi^+\rangle, \quad U_{BS}|C\Psi^-\rangle.$$

Study also the case with the coherent state $|\beta\rangle$ replaced by a squeezed state $|\zeta\rangle$.

**Problem 2.**    (i) Find a unitary operator $U$ such that

$$U|\beta\rangle = |-\beta\rangle.$$

(ii) Find a unitary operator $V$ such that

$$V|\zeta\rangle = |-\zeta\rangle.$$

# Chapter 25

# Homodyne Detection

## 25.1 Introduction

In optical homodyne detection one mixes a local oscillator field (for example a coherent state or squeezed state of light) with a signal field at a balanced (50/50) *beam splitter* (unitary operator)

$$U_{BS} = \exp(i\pi(b_S^\dagger \otimes b_{LO} + b_S \otimes b_{LO}^\dagger)/4)$$

where the signal field mode is represented by the Bose operator $b_S$ and the local oscillator is represented by the Bose operator $b_{LO}$. Then

$$U_{BS} \begin{pmatrix} b_S \otimes I \\ I \otimes b_{LO} \end{pmatrix} U_{BS}^\dagger = \begin{pmatrix} U_{BS}(b_S \otimes I)U_{BS}^\dagger \\ U_{BS}(I \otimes b_{LO}U_{BS}^\dagger) \end{pmatrix} = \frac{1}{\sqrt{2}} \begin{pmatrix} b_S \otimes I - iI \otimes b_{LO} \\ I \otimes b_{LO} - ib_S \otimes I \end{pmatrix}.$$

Thus the signal wave is overlapped on a beam splitter with a relatively strong local oscillator wave in the matching optical mode. The two fields emerging from the beam splitter are incident on two high-efficiency photodiodes whose output photocurrents are subtracted. The photocurrent difference is proportional to the value of the electric field operator. Thus optical homodyne detection corresponds to the difference photon counting of the two output electric fields from the beam splitter.

In heterodyne detection the signal electric field and another electric field, the auxiliary electric field, feed the same port of a beam splitter. The local electric field oscillator enters the other port of the beam splitter. The frequencies of the signal, auxiliary and local oscillator fields are different. One has $\omega_S + \omega_A = 2\omega_L$, $\omega_S - \omega_A = 2\omega$ with $\omega_L \gg \omega$.

## 25.2   Solved Problems

**Problem 1.**   A *homodyne detector* is constructed by placing a photo-counter in each arm after the beam splitter and then considering the difference photocurrent between the two modes

$$\hat{D} := \tilde{b}_1^\dagger \tilde{b}_1 - \tilde{b}_2^\dagger \tilde{b}_2.$$

Express the homodyne photocurrent in terms of the input modes $b_1$, $b_2$.

**Solution 1.**   Straightforward calculation yields

$$\hat{D} = (2\tau - 1)(b_1^\dagger b_1 - b_2^\dagger b_2) + 2\sqrt{\tau(1-\tau)}(b_1^\dagger b_2 + b_1 b_2^\dagger).$$

This expression reduces to

$$\hat{D} = b_1^\dagger b_2 + b_1 b_2^\dagger \equiv b^\dagger \otimes b + b \otimes b^\dagger$$

for a balanced ($\tau = 1/2$) beam splitter.

**Problem 2.**   Let $b_1$, $b_2$ be Bose annihilation operators. Consider the operators

$$d_\pm = \frac{1}{\sqrt{2}}(b_1 \pm b_2 e^{i\theta})$$

where $\theta$ is a phase shift. Let $I := d_+^\dagger d_+ - d_-^\dagger d_-$. Find $I$ in terms of the original operators $b_1$, $b_2$.

**Solution 2.**   We have

$$
\begin{aligned}
I &= d_+^\dagger d_+ - d_-^\dagger d_- \\
&= \frac{1}{2}(b_1^\dagger + b_2^\dagger e^{-i\theta})(b_1 + b_2 e^{i\theta}) - \frac{1}{2}(b_1^\dagger - b_2^\dagger e^{-i\theta})(b_1 - b_2 e^{i\theta}) \\
&= b_2^\dagger b_1 e^{-i\theta} + b_1^\dagger b_2 e^{i\theta}.
\end{aligned}
$$

This plays a role in homodyne measurement, where $b_1$ describes the signal field and $b_2$ describes the local oscillator field.

**Problem 3.** Let $b_S$, $b_{LO}$, $b_P$ be the Bose annihilation operators corresponding to the signal, the local oscillator and the photodetector input field, respectively. Let $0 < \epsilon < 1$. By shifting the $b_P$ field with a constant phase, we have

$$b_P = \sqrt{\epsilon}b_S + i\sqrt{1-\epsilon}b_{LO}.$$

Let

$$b_S = \hat{x}_S + i\hat{y}_S, \quad b_{LO} = \hat{x}_{LO} + i\hat{y}_{LO}, \quad b_P = \hat{x}_P + i\hat{y}_P$$

where $\hat{x}$, $\hat{y}$ are the (hermitian) quadrature operators. Let

$$\hat{n}_S = b_S^\dagger b_S, \quad \hat{n}_{LO} = b_{LO}^\dagger b_{LO}, \quad \hat{n}_P = b_P^\dagger b_P$$

be the photon number operators and let $\langle\,\rangle$ be the average with respect to a quantum state. Find the mean photodetector output $\langle\hat{n}_P\rangle$.

**Solution 3.** We obtain

$$\langle\hat{n}_P\rangle = \epsilon\hat{n} + (1-\epsilon)\hat{n} - 2(\epsilon(1-\epsilon))^{1/2}\langle\hat{x}_S\hat{y}_{LO} - \hat{x}_{LO}\hat{y}_S\rangle.$$

**Problem 4.** Suppose that $b_1^\dagger, b_2^\dagger$ are Bose creation operators and $b_1, b_2$ are Bose annihilation operators and $I$ is the identity operator. Consider the linear operator

$$Z := b \otimes I + I \otimes b^\dagger$$

where $b_1 := b \otimes I$ and $b_2^\dagger := I \otimes b^\dagger$. Thus $Z = b_1 + b_2^\dagger$. The operator is called the *heterodyne-current operator*. One also finds the notation

$$Z = b_S + b_I$$

where the subscripts $S$ and $I$, respectively denote the signal mode at frequency $\omega_0 + \Delta\omega$ and the imaging mode at frequency $\omega_0 - \Delta\omega$ ($\Delta\omega \ll \omega$).
(i) Calculate the commutator $[Z, Z^\dagger]$.
(ii) Find the states $Z(|0\rangle \otimes |0\rangle)$, $Z^\dagger(|0\rangle \otimes |0\rangle)$.
(iii) Find the state $Z^2(|0\rangle \otimes |0\rangle)$.

**Solution 4.** (i) We have

$$Z^\dagger = b_1^\dagger + b_2 \equiv b^\dagger \otimes I + I \otimes b$$

and

$$
\begin{aligned}
[Z, Z^\dagger] &= (b \otimes I + I \otimes b^\dagger)(b^\dagger \otimes I + I \otimes b) \\
&\quad -(b^\dagger \otimes I + I \otimes b)(b \otimes I + I \otimes b^\dagger) \\
&= bb^\dagger \otimes I + b \otimes b + b^\dagger \otimes b^\dagger + I \otimes b^\dagger b \\
&\quad -b^\dagger b \otimes I - b^\dagger \otimes b^\dagger - b \otimes b - I \otimes bb^\dagger \\
&= (bb^\dagger - b^\dagger b) \otimes I + I \otimes (b^\dagger b - bb^\dagger) \\
&= I \otimes I - I \otimes I = 0.
\end{aligned}
$$

(ii) We have

$$Z(|0\rangle \otimes |0\rangle) = (b \otimes I + I \otimes b^\dagger)|0\rangle \otimes |0\rangle$$
$$= (b \otimes I)(|0\rangle \otimes |0\rangle) + (I \otimes b^\dagger)(|0\rangle \otimes |0\rangle)$$
$$= |0\rangle \otimes b^\dagger|0\rangle = |0\rangle \otimes |1\rangle$$

since $b|0\rangle = 0$. Analogously

$$Z^\dagger|0\rangle \otimes |0\rangle = b^\dagger|0\rangle \otimes |0\rangle = |1\rangle \otimes |0\rangle.$$

(iii) We find

$$Z^2(|0\rangle \otimes |0\rangle) = \sqrt{2}(|0\rangle \otimes |2\rangle).$$

# 25.3   Supplementary Problems

**Problem 1.**   (i) Show that the spectrum of the operator

$$\hat{K} = b_1^\dagger b_2 + b_1 b_2^\dagger \equiv b^\dagger \otimes b + b \otimes b^\dagger$$

is discrete and coincides with the set $\mathbb{Z}$ of relative integers. Note that

$$\hat{K}(|0\rangle \otimes |0\rangle) = 0(|0\rangle \otimes |0\rangle)$$

$$\hat{K}(|1\rangle) \otimes |1\rangle) = \sqrt{2}|2\rangle \otimes |0\rangle + \sqrt{2}|0\rangle \otimes |2\rangle.$$

(ii) Find

$$(\langle\beta| \otimes \langle\beta|)\hat{K}(|\beta\rangle \otimes |\beta\rangle), \quad (\langle\zeta| \otimes \langle\zeta|)\hat{K}(|\zeta\rangle \otimes |\zeta\rangle).$$

**Problem 2.**   Let $b_1$, $b_2$ be Bose annihilation operators and

$$\hat{K} = b_1 \cos(\phi) + b_2 \sin(\phi).$$

Show that

$$\hat{K}^\dagger\hat{K} = b_1^\dagger b_1 \cos^2(\phi) + b_2^\dagger b_2 \sin^2(\phi) + (b_1^\dagger b_2 + b_1 b_2^\dagger)\sin(\phi)\cos(\phi).$$

Study the case that $b_1$, $b_2$ are Fermi annihilation operators.

# Chapter 26

# Hamilton Operators

---

## 26.1 Introduction

Most experimental realizations of quantum logic gates (Hadamard gate, quantum phase gate, controlled-NOT gate) involve several qubits and number states. A Hamilton operator $\hat{H}$ must describe the interaction. Thus in quantum computing we are faced with two problems. One is to determine the Hamilton operator $\hat{H}$ for the system such that the time-evolution

$$\exp(-i\hat{H}t/\hbar)$$

represents the execution of the computation. The other one is to build the hardware described by this Hamilton operator.

For example the Hamilton operator that produces squeezed states is given by

$$\hat{H} = \hbar\omega_0 b^\dagger b + \hbar\kappa(b^2 \epsilon e^{i\omega t} + (b^\dagger)^2 \bar{\epsilon} e^{-i\omega t})$$

where $b^\dagger$, $b$ are Bose creation and annihilation operators, $\omega_0$ is the frequency of the degenerate signal/idler mode, $\epsilon$ is the classical pump field of frequency $\omega$ and $\kappa$ is the coupling constant between the pump and signal modes.

## 26.2   Solved Problems

**Problem 1.**   Consider a single continuous variable corresponding to a linear operator $X$. Let $P$ be the operator of the conjugate variable, i.e.

$$[X, P] = iI. \tag{1}$$

Consider the *Kerr-Hamilton operator*

$$K = H^2 = (X^2 + P^2)^2.$$

The *Kerr-Hamilton operator* corresponds to a $\chi^3$ process in nonlinear optics. The linear unbounded operators $X$ and $P$ could correspond to quadrature amplitudes of a mode of the electromagnetic field. The quadrature amplitudes are the real and imaginary parts of the complex electric field. Let

$$S := \frac{1}{2}(XP + PX).$$

Calculate the commutators

$$[K, X], \quad [K, P], \quad [X, [K, S]], \quad [P, [K, S]].$$

Discuss.

**Solution 1.**   Since

$$P^2 X^2 = X^2 P^2 - 4iXP - 2I, \qquad P^2 X = XP^2 - 2iP$$

we find

$$[K, X] = \frac{i}{2}(X^2 P + PX^2 + 2P^3)$$
$$[K, P] = -\frac{i}{2}(P^2 X + XP^2 + 2X^3)$$
$$[X, [K, S]] = P^3$$
$$[P, [K, S]] = X^3.$$

Thus the algebra generated by $X$, $P$, $H$, $S$, $K$ by calculating commutators includes all third order polynomials in $X$ and $P$. We can construct Hamilton operators that are arbitrary hermitian polynomials in any order of $X$ and $P$. We have

$$[P^3, P^m X^n] = iP^{m+2} X^{n-1} + \text{lower order terms}$$

and

$$[X^3, P^m X^n] = iP^{m-1} X^{n+2} + \text{lower order terms}.$$

**Problem 2.** A nonlinear four-wave mixing device can be described by the Hamilton operator

$$\hat{H} = \frac{1}{4}\hbar\omega(b_1^\dagger b_2 + b_1 b_2^\dagger)^2 \equiv \frac{1}{4}\hbar\omega(b^\dagger \otimes b + b \otimes b^\dagger)^2.$$

(i) Let $|n\rangle$ be a number state. Consider the normalized input state $|n\rangle \otimes |0\rangle$. Find the normalized state

$$\exp(-i\hat{H}t/\hbar)(|n\rangle \otimes |0\rangle)$$

for the interaction time $t = \pi/\omega$.

(ii) Consider the normalized input state $|n\rangle \otimes |0\rangle$. Find the state

$$\exp(-i\hat{H}t/\hbar)(|n\rangle \otimes |0\rangle)$$

for the interaction time $t = \pi/(2\omega)$.

**Solution 2.** (i) For the interaction time $t = \pi/\omega$ the output state is $i|0\rangle \otimes |n\rangle$ (for $n$ even) and $\exp(-i\pi/4)|n\rangle \otimes |0\rangle$ for $n$ odd. Thus the device acts as an even-odd filter, switching the even numbers from one mode to the other. Under these operating conditions it can be used as a device to measure parity without counting the photon number. It is sufficient to detect any photons in either of the output channels.

(ii) If the interaction time is $t = \pi/(2\omega)$ the output state will have the form

$$\frac{1}{\sqrt{2}}(|n\rangle \otimes |0\rangle + e^{-i(n+1)\pi/2}|0\rangle \otimes |n\rangle)$$

which is a maximally entangled state for $n$ photons.

**Problem 3.** *Cross phase modulation* is described by the Hamilton operator

$$\hat{H} = -\hbar\omega b_1^\dagger b_1 b_2^\dagger b_2 = -\hbar\omega(b^\dagger b \otimes b^\dagger b)$$

where $\omega$ is a function of the third order susceptibility $\chi^{(3)}$. Consider the two-mode number state $|m\rangle \otimes |n\rangle$, i.e. $m, n = 0, 1, 2, \ldots$. Find the state $\exp(-i\hat{H}t/\hbar)(|m\rangle \otimes |n\rangle)$.

**Solution 3.** Since $b_1 = b \otimes I$ and $b_2 = I \otimes b$ we have

$$b_1^\dagger b_1 |m\rangle \otimes |n\rangle = m|m\rangle \otimes |n\rangle, \quad b_2^\dagger b_2 |m\rangle \otimes |n\rangle = n|m\rangle \otimes |n\rangle.$$

Thus

$$\exp(-i\hat{H}t/\hbar)|m\rangle \otimes |n\rangle = e^{i\omega tmn}|m\rangle \otimes |n\rangle.$$

**Problem 4.**    A *Kerr medium* is nonlinear in the sense that its refractive index $n$ has a component which varies with the intensity of the propagating electric field $\mathbf{E}$, that is

$$n = n_0 + n_2|\mathbf{E}|^2$$

where $n_0$ and $n_2$ are constants. For a single-mode field, described by Bose creation and annihilation operators $b^\dagger$ and $b$, propagating through a low-loss Kerr media, the interaction Hamilton operator can be written as

$$\hat{H} = \chi b^{\dagger 2} b^2.$$

(i) Show that the *number state* $|n\rangle$ is an eigenstate.
(ii) Assume that the initial state is a coherent state $|\beta\rangle$. Find $|\beta(t)\rangle$.
(iii) Let $\chi t = \pi r/s$ where $r$ and $s$ are mutually prime with $r < s$. Write $\exp(-i\pi r n^2/s)$ as a discrete Fourier transform. Express $|\beta(t)\rangle$ using this expansion.

**Solution 4.**    (i) Since $b^{\dagger 2}b^2 \equiv b^\dagger b(b^\dagger b - I)$ and $b^\dagger b|n\rangle = n|n\rangle$, we have

$$\hat{H}|n\rangle = \chi(n^2 - n)|n\rangle.$$

Thus the eigenvalues are $\chi(n^2 - n)$.
(ii) The solution of the time dependent Schrödinger equation ($\hbar = 1$)

$$i\frac{d|\beta\rangle}{dt} = \hat{H}|\beta\rangle$$

is given by

$$|\beta(t)\rangle = \exp(-i\hat{H}t)|\beta\rangle.$$

Using the result from (i) we find

$$|\beta(t)\rangle = \sum_{n=0}^{\infty} c_n e^{-i\chi t(n^2 - n)}|n\rangle$$

where

$$c_n := \exp(-|\beta|^2/2)\frac{\beta^n}{\sqrt{n!}}.$$

Since $n^2 - n$ is always an even number, the system will revive whenever $\chi t$ is a multiple of $\pi$.
(iii) Let $\chi t = \pi r/s$ where $r$, $s$ are mutually prime with $r < s$. Then we can write the quadratic (in $n$) phase in terms of linear phases using the discrete Fourier transform

$$\exp(-i\pi n^2 r/s) = \sum_{p=0}^{\ell-1} a_p^{(r,s)} \exp(-2\pi i p n/\ell)$$

where
$$\ell = \begin{cases} s & \text{if } r \text{ is odd, } s \text{ is even or vice-versa} \\ 2s & \text{if both } r \text{ and } s \text{ are odd.} \end{cases}$$

Thus
$$a_p^{(r,s)} = \frac{1}{\ell} \sum_{k=0}^{\ell-1} \exp(-i\pi r k^2/s + 2\pi i p k/\ell)$$

and
$$|\beta(t)\rangle = \sum_{p=0}^{\ell-1} a_p^{(r,s)} |\beta \exp(i\pi(r/s - 2p/\ell))\rangle.$$

**Problem 5.** The Hamilton operator for the *second harmonic generation* can be written as
$$\hat{H} = i\hbar\frac{\kappa}{2}(b^{\dagger 2}b_{sh} - b^2 b_{sh}^\dagger)$$

where $b$ is the fundamental cavity mode Bose operator, $b_{sh}$ is the second-harmonic mode Bose operator and $\kappa$ is the nonlinear coupling. Using the Heisenberg equation of motion find the time evolution of $b$ and $b_{sh}$.

**Solution 5.** The *Heisenberg equation of motion* of an operator $\hat{A}$ is given by
$$i\hbar\frac{d\hat{A}}{dt} = [\hat{A}, \hat{H}](t).$$

The commutation relations are given by
$$[b, b^\dagger] = I, \quad [b_{sh}, b_{sh}^\dagger] = I$$
$$[b, b] = [b_{sh}, b_{sh}] = [b, b_{sh}] = [b, b_{sh}^\dagger] = 0.$$

Thus we find the operator-valued nonlinear differential equations
$$\frac{db}{dt} = \kappa b^\dagger b_{sh}, \quad \frac{db_{sh}}{dt} = -\frac{\kappa}{2}b^2.$$

In a more realistic model, cavity photon losses must be taken into account.

**Problem 6.** A single spin-$\frac{1}{2}$ particle is placed on a cantilever tip. The tip can oscillate only in the $z$-direction. A ferromagnetic particle, whose magnetic moment points in the positive $z$-direction, produces a non-uniform magnetic field at the spin. A uniform magnetic field, $\mathbf{B}_0$, oriented in the positive $z$-direction, determines the ground state of the spin. A rotating magnetic field, $\mathbf{B}_1(t)$, induces transitions between the ground state and excited states of the spin. It is given by
$$B_x(t) = B_1\cos(\omega t + \phi(t)), \quad B_y(t) = -B_1\sin(\omega t + \phi(t)), \quad B_z(t) = 0$$

where $\phi(t)$ describes a smooth change in phase required for a cyclic adiabatic inversion of the spin

$$|d\phi(t)/dt| \ll \omega.$$

In the reference frame rotating with $\mathbf{B}_1(t)$, the time-dependent Hamilton operator is given by

$$\hat{H}(t) = \frac{P_z^2}{2m_c^*} + \frac{m_c^* \omega_c^2 Z^2}{2} - \hbar \left( \omega_L - \omega - \frac{d\phi}{dt} \right) S_3 - \hbar \omega_1 S_1 - g\mu \frac{\partial B_z}{\partial Z} Z S_3$$

where $Z$ is the coordinate of the oscillator which describes the dynamics of the quasi-classical cantilever tip, $P_z$ is its momentum, $m_c^*$ and $\omega_c$ are the effective mass and the frequency of the cantilever, $S_3$ and $S_1$ are the $z-$ and the $x-$ component of the spin,

$$S_1 = \frac{1}{2} \begin{pmatrix} 0 & 1 \\ 1 & 0 \end{pmatrix}, \qquad S_3 = \frac{1}{2} \begin{pmatrix} 1 & 0 \\ 0 & -1 \end{pmatrix},$$

$\omega_L$ is its Larmor frequency, $\omega_1$ is the *Rabi frequency* (the frequency of the spin precession around the magnetic field $\mathbf{B}_1(t)$ at the resonance condition $\omega = \omega_L$, $d\phi/dt = 0$), $g$ and $\mu$ are the $g$-factors and the magnetic moment of the spin and we defined $m_c^* = m_c/4$ as the effective cantilever mass. The operator acts in the product Hilbert space $L_2(\mathbb{R}) \otimes \mathbb{C}^2$. One sets

$$\omega_c = (k_c/m_c^*)^{1/2}, \quad \omega_L = \gamma B_z, \quad \omega_1 = \gamma B_1$$

where $\gamma = g\mu/\hbar$ is the gyromagnetic ratio of the spin, $m_c$ and $k_c$ are the mass and the force constant of the cantilever, $B_z$ includes the uniform magnetic field $B_0$ and the magnetic field produced by the ferromagnetic particle.

(i) Cast the Hamilton operator in dimensionless form $\hat{H}/\hbar\omega_c \to \hat{K}$ by introducing the quantities

$$E_0 := \hbar\omega_c, \quad F_0 := \sqrt{k_c E_0}, \quad Z_0 := \sqrt{E_0/k_c}, \quad P_0 := \hbar/Z_0$$

with $\omega = \omega_L$ and using the dimensionless time $\tau := \omega_c t$.

(ii) The dimensionless time-dependent Schrödinger equation

$$i\frac{\partial \Psi}{\partial \tau} = \hat{K}\Psi$$

where

$$\Psi(\tau, z) = \begin{pmatrix} \Psi_1(\tau, z) \\ \Psi_2(\tau, z) \end{pmatrix}$$

can be solved using the expansions

$$\Psi_1(\tau, z) = \sum_{n=0}^{\infty} A_n(\tau)|n\rangle, \qquad \Psi_2(\tau, z) = \sum_{n=0}^{\infty} B_n(\tau)|n\rangle$$

$$|n\rangle = \pi^{1/4} 2^{n/2} (n!)^{1/2} e^{-z^2/2} H_n(z)$$

where $\{|n\rangle : n = 0, 1, \dots\}$ are number states. Here $H_n(z)$ are the Hermitian polynomials. Find the time evolution of the complex expansion coefficients $A_n$ and $B_n$.

(iii) What would be an initial state closest to the classical limit?

**Solution 6.** (i) Since $\hat{H}/(\hbar\omega_c) \to \hat{K}$ we find

$$\hat{K} = \frac{\hat{H}}{\hbar\omega_c} = \frac{1}{2}(p_z^2 + z^2) + S_3 \frac{d\phi}{d\tau} - \epsilon S_1 - 2\eta z S_3$$

where we used $\omega_L = \omega$ and

$$p_z := \frac{P_z}{P_0}, \quad z := \frac{Z}{Z_0}, \quad \epsilon := \frac{\omega_1}{\omega_c}, \quad \eta = \frac{g\mu}{2F_c} \frac{\partial B_z}{\partial Z}, \quad \omega_c dt = d\tau.$$

(ii) Inserting the series expansions into the dimensionless Schrödinger equation we find the system of linear differential equations with time-dependent coefficients for the complex amplitudes $A_n(\tau)$ and $B_n(\tau)$

$$i\frac{dA_n}{d\tau} = \left(n + \frac{1}{2} + \frac{1}{2}\frac{d\phi}{d\tau}\right) A_n - \frac{\eta}{\sqrt{2}}\left(\sqrt{n}A_{n-1} + \sqrt{n+1}A_{n+1}\right) - \frac{\epsilon}{2}B_n$$

$$i\frac{dB_n}{d\tau} = \left(n + \frac{1}{2} + \frac{1}{2}\frac{d\phi}{d\tau}\right) B_n + \frac{\eta}{\sqrt{2}}\left(\sqrt{n}B_{n-1} + \sqrt{n+1}B_{n+1}\right) - \frac{\epsilon}{2}A_n$$

where we used the Bose operators $b$ and $b^\dagger$ defined by

$$b|n\rangle = \sqrt{n}|n-1\rangle, \qquad b^\dagger|n\rangle = \sqrt{n+1}|n+1\rangle$$

and

$$\frac{1}{2}\left(p_z^2 + z^2\right)|n\rangle = \left(n + \frac{1}{2}\right)|n\rangle$$

with

$$z := \frac{1}{\sqrt{2}}(b^\dagger + b), \quad p_z := \frac{i}{\sqrt{2}}(b^\dagger - b), \quad [b, b^\dagger] = I.$$

(iii) We can choose the coherent state

$$\Psi_1(z, 0) = \sum_{n=0}^{\infty} A_n(0)|n\rangle, \quad \Psi_2(z, 0) = 0, \quad A_n(0) = \frac{\beta^n}{\sqrt{n!}} \exp(-|\beta|^2/2).$$

**Problem 7.** Consider two *Bose-Einstein condensates* which both occupy the ground-state of their respective traps. They are described by the atom Bose annihilation (creation) operators $b_1$ $(b_1^\dagger)$ and $b_2$ $(b_2^\dagger)$. Atoms are released from each trap with momenta (wave vectors) $\mathbf{k}_1$ and $\mathbf{k}_2$, respectively, producing an interference pattern which enables a relative phase to be measured. The *intensity* $I(\mathbf{x}, t)$ of the atomic field is given by

$$I(\mathbf{x}, t) = I_0 \langle \psi | (b_1^\dagger(t) e^{i\mathbf{k}_1 \cdot \mathbf{x}} + b_2^\dagger(t) e^{i\mathbf{k}_2 \cdot \mathbf{x}})(b_1(t) e^{-i\mathbf{k}_1 \cdot \mathbf{x}} + b_2(t) e^{-i\mathbf{k}_2 \cdot \mathbf{x}}) | \psi \rangle \quad (1)$$

where $I_0$ is the single atom intensity. Atoms within each condensate collide. This can be described using the Hamilton operator

$$\hat{H} = \frac{1}{2} \hbar \chi ((b_1^\dagger b_1)^2 + (b_2^\dagger b_2)^2) \quad (2)$$

where $\chi$ is the collision rate between the atoms within each condensate. Cross-collisions between the two condensates, described by the term $b_1^\dagger b_1 b_2^\dagger b_2$ could also be included. Using the Hamilton operator given by (2) the intensity $I(\mathbf{x}, t)$ is given by

$$I(\mathbf{x}, t) = I_0 (\langle \psi | b_1^\dagger b_1 | \psi \rangle + \langle \psi | b_2^\dagger b_2 | \psi \rangle + \langle \psi | b_1^\dagger \exp(i\chi t (b_1^\dagger b_1 - b_2^\dagger b_2)) b_2 | \psi \rangle e^{-i\phi(\mathbf{x})}$$
$$+ \langle \psi | b_2^\dagger \exp(-i\chi t (b_1^\dagger b_1 - b_2^\dagger b_2)) b_1 | \psi \rangle e^{i\phi(\mathbf{x})})$$

where $\phi(\mathbf{x}) := (\mathbf{k}_2 - \mathbf{k}_1) \cdot \mathbf{x}$. Calculate $I(\mathbf{x}, t)$ for the product state

$$|\psi\rangle = |\beta_1\rangle \otimes |\beta_2\rangle$$

where $|\beta_1\rangle$ and $|\beta_2\rangle$ are coherent states.

**Solution 7.**   Since $b|\beta\rangle = \beta|\beta\rangle$, $\langle \beta | b^\dagger = \langle \beta | \beta^*$, and

$$e^{i\chi t b^\dagger b} |\beta\rangle = \sum_{j=0}^{\infty} \frac{1}{j!} (e^{i\chi t} - 1)(b^\dagger)^j b^j |\beta\rangle = \sum_{j=0}^{\infty} \frac{1}{j!} (e^{i\chi t} - 1)(b^\dagger)^j \beta^j$$

we have

$$\langle \beta | e^{i\chi t b^\dagger b} |\beta\rangle = \exp((e^{i\chi t} - 1)\beta^* \beta).$$

We also have

$$\langle \beta | e^{i\chi t b^\dagger b} b |\beta\rangle = \beta \exp((e^{i\chi t} - 1)\beta^* \beta), \quad \langle \beta | b^\dagger e^{i\chi t b^\dagger b} |\beta\rangle = \beta^* \exp((e^{i\chi t} - 1)\beta^* \beta).$$

Since $b_1^\dagger b_1 = b^\dagger b \otimes I$ and $b_2^\dagger b_2 = I \otimes b^\dagger b$ we have

$$e^{i\chi t (b_1^\dagger b_1 - b_2^\dagger b_2)} = e^{i\chi t b_1^\dagger b_1} e^{-i\chi t b_2^\dagger b_2} = e^{i\chi t (b^\dagger b \otimes I)} e^{-i\chi t (I \otimes b^\dagger b)}$$
$$= (e^{i\chi t b^\dagger b} \otimes I)(I \otimes e^{-i\chi t b^\dagger b}) = e^{i\chi t b^\dagger b} \otimes e^{-i\chi t b^\dagger b}.$$

Thus

$$\langle \psi | b_1^\dagger b_1 | \psi \rangle = \beta_1^* \beta_1, \quad \langle \psi | b_2^\dagger b_2 | \psi \rangle = \beta_2^* \beta_2$$

and

$$\langle\psi|b_1^\dagger e^{ixt(b_1^\dagger b_1 - b_2^\dagger b_2)} b_2|\psi\rangle e^{-i\phi(\mathbf{x})} = \beta_1^*\beta_2 \exp((e^{ixt} - 1)\beta_1^*\beta_1)$$
$$\times \exp((e^{-ixt} - 1)\beta_2^*\beta_2)e^{-i\phi(\mathbf{x})}$$

$$\langle\psi|b_2^\dagger e^{-ixt(b_1^\dagger b_1 - b_2^\dagger b_2)} b_1|\psi\rangle e^{i\phi(\mathbf{x})} = \beta_1\beta_2^* \exp((e^{-ixt} - 1)\beta_1^*\beta_1)$$
$$\times \exp((e^{ixt} - 1)\beta_2^*\beta_2)e^{i\phi(\mathbf{x})}.$$

**Problem 8.** Consider the Hamilton operator $\hat{H}$ of a coupled one Bose one Fermi system

$$\hat{H} = \omega b^\dagger b \otimes I_F + J I_B \otimes c^\dagger c + \alpha(b^\dagger \otimes c + b \otimes c^\dagger)$$

where $\omega, J, \alpha$ are positive quantities, $I_B$ is the identity operator in the space of the Bose operators and $I_F$ is the identity operator in the space of the Fermi operators. Here $b$ and $b^\dagger$ are Bose annihilation and creation operators, respectively, and $c$ and $c^\dagger$ are Fermi annihilation and creation operators, respectively. The commutation and anticommutation relations are

$$[b, b^\dagger] = I_B, \quad [b, b] = [b^\dagger, b^\dagger] = 0$$
$$[c, c^\dagger]_+ = I_F, \quad [c, c]_+ = [c^\dagger, c^\dagger]_+ = 0.$$

Find the eigenvalues of the Hamilton operator $\hat{H}$. Find the eigenvectors. Use the matrix representations for $c^\dagger$, $c$, $b^\dagger$, $b$, i.e.

$$c^\dagger = \begin{pmatrix} 0 & 0 \\ 1 & 0 \end{pmatrix}, \quad c = \begin{pmatrix} 0 & 1 \\ 0 & 0 \end{pmatrix}.$$

For the Fermi states we set

$$|0\rangle_F = \begin{pmatrix} 1 \\ 0 \end{pmatrix}.$$

Thus

$$|1\rangle_F = c^\dagger|0\rangle_F = \begin{pmatrix} 0 \\ 1 \end{pmatrix}.$$

For the Bose operators we have

$$b^\dagger = \begin{pmatrix} 0 & 0 & 0 & 0 & \dots \\ 1 & 0 & 0 & 0 & \dots \\ 0 & \sqrt{2} & 0 & 0 & \dots \\ 0 & 0 & \sqrt{3} & 0 & \dots \\ & & \dots & & \end{pmatrix}, \quad b = \begin{pmatrix} 0 & 1 & 0 & 0 & 0 & \dots \\ 0 & 0 & \sqrt{2} & 0 & 0 & \dots \\ 0 & 0 & 0 & \sqrt{3} & 0 & \dots \\ 0 & 0 & 0 & 0 & \sqrt{4} & \dots \\ & & \dots & & & \end{pmatrix}.$$

For the number states of the Bose operators we have

$$|n\rangle_B = \frac{1}{\sqrt{n!}} b^{\dagger n} |0\rangle_B$$

where $n = 0, 1, 2, \ldots$.

**Solution 8.**    Using the matrix representation for the number operators $c^\dagger c$ and $b^\dagger b$ given above we find

$$c^\dagger c = \begin{pmatrix} 0 & 0 \\ 0 & 1 \end{pmatrix}$$

and

$$b^\dagger b = \mathrm{diag}(0, 1, 2, \ldots).$$

Thus it follows that

$$b^\dagger b \otimes I_F = \mathrm{diag}(0, 1, 2, \ldots) \otimes \begin{pmatrix} 1 & 0 \\ 0 & 1 \end{pmatrix} = \mathrm{diag}(0, 0, 1, 1, 2, 2, \ldots)$$

and

$$I_B \otimes c^\dagger c = \mathrm{diag}(1, 1, 1, \ldots) \otimes \begin{pmatrix} 0 & 0 \\ 0 & 1 \end{pmatrix} = \mathrm{diag}(0, 1, 0, 1, \ldots).$$

For the interacting terms we find

$$b \otimes c^\dagger = (0) \oplus \begin{pmatrix} 0 & 1 \\ 0 & 0 \end{pmatrix} \oplus \begin{pmatrix} 0 & \sqrt{2} \\ 0 & 0 \end{pmatrix} \oplus \begin{pmatrix} 0 & \sqrt{3} \\ 0 & 0 \end{pmatrix} \oplus \cdots$$

$$b^\dagger \otimes c = (0) \oplus \begin{pmatrix} 0 & 0 \\ 1 & 0 \end{pmatrix} \oplus \begin{pmatrix} 0 & 0 \\ \sqrt{2} & 0 \end{pmatrix} \oplus \begin{pmatrix} 0 & 0 \\ \sqrt{3} & 0 \end{pmatrix} \oplus \cdots$$

where $\oplus$ denotes the direct sum of matrices. Adding up the matrices we obtain the matrix representation of the Hamilton operator

$$\hat{H} = (0) \oplus \begin{pmatrix} J & \alpha \\ \alpha & \omega \end{pmatrix} \oplus \begin{pmatrix} \omega + J & \sqrt{2}\alpha \\ \sqrt{2}\alpha & 2\omega \end{pmatrix} \oplus \begin{pmatrix} 2\omega + J & \sqrt{3}\alpha \\ \sqrt{3}\alpha & 3\omega \end{pmatrix} \oplus \cdots$$

$$\oplus \begin{pmatrix} n\omega + J & \sqrt{n+1}\alpha \\ \sqrt{n+1}\alpha & (n+1)\omega \end{pmatrix} \oplus \cdots .$$

Thus one eigenvalue is 0. The eigenvector for the eigenvalue 0 is given by $|0\rangle_B \otimes |0\rangle_F$. Thus this state is not entangled. To find the other eigenvalues we have to find the eigenvalues of the $2 \times 2$ matrices

$$\begin{pmatrix} n\omega + J & \sqrt{n+1}\alpha \\ \sqrt{n+1}\alpha & (n+1)\omega \end{pmatrix}$$

where $n = 0, 1, 2, \ldots$. We find

$$E_\pm(n) = \frac{J}{2} + \frac{\omega}{2} + n\omega \pm \sqrt{\frac{1}{4}(J - \omega)^2 + (n + 1)\alpha^2}.$$

Whether 0 is the lowest eigenvalue depends on the values of $J$, $\omega$ and $\alpha$. Consider the case with $n = 0$. Then

$$E_-(n = 0) = \frac{J}{2} + \frac{\omega}{2} - \sqrt{\frac{1}{4}(J - \omega)^2 + \alpha^2}.$$

The eigenvector for the eigenvalue $E_-(n = 0)$ is given by ($\alpha \neq 0$)

$$|0\rangle_B \otimes |1\rangle_F + \frac{1}{\alpha}\left(-\sqrt{\frac{1}{4}(J - \omega)^2 + \alpha^2} + \frac{\omega}{2} - \frac{J}{2}\right)|1\rangle_B \otimes |0\rangle_F.$$

Thus this state is entangled except if $-\sqrt{(J - \omega)^2/4 + \alpha^2} + \omega/2 - J/2 = 0$. The condition $E_-(n = 0) = 0$ yields $\alpha^2 = J\omega$. Then we find the state

$$|0\rangle_B \otimes |1\rangle_F - \frac{J}{\alpha}|1\rangle_B \otimes |0\rangle_F.$$

This state is also entangled if $J \neq 0$. Thus we have an unentangled state with eigenvalue 0 and for given parameter values of $J$, $\omega$ and $\alpha$ we can have an entangled state with this eigenvalue.

**Problem 9.** Let $b$, $b^\dagger$ be Bose annihilation and creation operators. Let

$$\sigma_+ := \frac{1}{2}(\sigma_1 + i\sigma_2) = \begin{pmatrix} 0 & 1 \\ 0 & 0 \end{pmatrix}, \quad \sigma_- := \frac{1}{2}(\sigma_1 - i\sigma_2) = \begin{pmatrix} 0 & 0 \\ 1 & 0 \end{pmatrix}.$$

Consider the Hamilton operator

$$\hat{H} = b^\dagger \otimes \sigma_- + b \otimes \sigma_+$$

which describes a single atom coupled to a single mode of an electromagnetic field. $\sigma_\pm$ act on the atom and $b$, $b^\dagger$ act on the field.
(i) Calculate $\hat{H}^2$.
(ii) Calculate the commutator $[b^\dagger \otimes \sigma_-, b \otimes \sigma_+]$.
(iii) Let $|n\rangle$ be a number state and $|\beta\rangle$ be a coherent state. Calculate

$$A(n, \beta) := (\langle n| \otimes I_2)e^{i\theta\hat{H}}(|\beta\rangle \otimes I_2)$$

where $\theta \in \mathbb{R}$ and $I_2$ is $2 \times 2$ unit matrix.

**Solution 9.** (i) Since $\sigma_-\sigma_- = 0$, $\sigma_+\sigma_+ = 0$ and

$$\sigma_+\sigma_- = \begin{pmatrix} 1 & 0 \\ 0 & 0 \end{pmatrix}, \quad \sigma_-\sigma_+ = \begin{pmatrix} 0 & 0 \\ 0 & 1 \end{pmatrix}$$

we find

$$(b^\dagger \otimes \sigma_- + b \otimes \sigma_+)(b^\dagger \otimes \sigma_- + b \otimes \sigma_+) = bb^\dagger \otimes \sigma_+\sigma_- + b^\dagger b \otimes \sigma_-\sigma_+.$$

Thus we can write in matrix notation

$$\begin{pmatrix} bb^\dagger & 0 \\ 0 & b^\dagger b \end{pmatrix}.$$

(ii) We have

$$[b^\dagger \otimes \sigma_-, b \otimes \sigma_+] = b^\dagger b \otimes \sigma_-\sigma_+ - bb^\dagger \otimes \sigma_+\sigma_-.$$

(iii) Since $b|\beta\rangle = \beta|\beta\rangle$ and $b|n\rangle = \sqrt{n}|n-1\rangle$ we obtain

$$A(n,\alpha) = \exp(-|\beta|^2) \frac{|\beta|^2}{n!} \begin{pmatrix} \cos(\theta\sqrt{n}) & \frac{i\sqrt{n}}{\beta}\sin(\theta\sqrt{n}) \\ \frac{in}{\sqrt{n+1}}\sin(\theta\sqrt{n+1}) & \cos(\theta\sqrt{n+1}) \end{pmatrix}.$$

**Problem 10.**   Consider the model Hamilton operator for ions trapped inside an optical cavity

$$\hat{H} := \hat{H}_0 + \hat{V}$$

where

$$\hat{H}_0 = \left(\hbar\nu a^\dagger a + \frac{1}{2}I_a\right) \otimes I_b \otimes I_2 + I_a \otimes \hbar\omega_c b^\dagger b \otimes I_2 + I_a \otimes I_b \otimes \frac{\hbar\omega_0}{2}\sigma_3$$

and

$$\hat{V} = \hbar\Omega(\exp(i\eta_L(a^\dagger + a) - i(\omega_L t + \phi)I_a) \otimes I_b \otimes \sigma_+ + h.c.)$$
$$+ \hbar g\sin(\eta_c(a^\dagger + a)) \otimes (b^\dagger + b) \otimes (\sigma_+ + \sigma_-).$$

Here $a^\dagger$ ($a$) and $b^\dagger$ ($b$) are Bose creation (annihilation) operators for the vibrational phonon and the cavity field photon, respectively and $\omega_0$ is the transition frequency of the two-level ion. The ion-phonon and ion-cavity coupling constants are $\Omega$ and $g$, and $\sigma_k$ ($k = z, +, -$) are the Pauli operators describing the internal state of the ion. Thus we consider a two-level ion radiated by the single mode cavity field of frequency $\omega_c$ and an external laser field of frequency $\omega_L$. The operators $I_a$, $I_b$ and $I_2$ are the identity operators in their respective Hilbert spaces, where $I_2$ is the $2 \times 2$ unit matrix. Thus we have a tripartite system. The parameters $\eta_L$ and $\eta_c$ are the *Lamb-Dicke parameters*.
(i) Consider the unitary operator

$$U_0(t) = \exp(-i\hat{H}_0 t/\hbar).$$

Calculate (interaction picture)

$$H_I(t) = U_0^\dagger(t)\hat{V}U_0(t) \equiv \exp(i\hat{H}_0 t/\hbar)\hat{V}\exp(-i\hat{H}_0 t/\hbar).$$

(ii) Discuss how a Hadamard gate can be realized.

**Solution 10.** (i) Straightforward calculation yields

$$\hat{H}_I(t) = \hbar\Omega\left(\hat{O}_0^L \exp(i(\delta_{0L}t - \phi))\right) \otimes I_b \otimes \sigma_+$$

$$+\hbar\Omega\left(\sum_{k=1}^{\infty}(i\eta_L)^k \hat{O}_k^L a^k \exp(i((\delta_{0L} - k\nu)t - \phi))\right) \otimes I_b \otimes \sigma_+$$

$$+\hbar\Omega\left(\sum_{k=1}^{\infty}(i\eta_L)^k a^{\dagger k}\hat{O}_k^L \exp(i((\delta_{0L} + k\nu)t - \phi))\right) \otimes I_b \otimes \sigma_+$$

$$+\hbar g\left(\sum_{k=1,3,\ldots}^{\infty}(i^{k-1}\eta_c^k)a^{\dagger k}\hat{O}_k^c \exp(i(\delta_{0c} + k\nu + 2\omega_c)t)\right) \otimes b^\dagger \otimes \sigma_+$$

$$+\hbar g\left(\sum_{k=1,3,\ldots}^{\infty}(i^{k-1}\eta_c^k)\hat{O}_k^c a^k \exp(i(\delta_{0c} - k\nu + 2\omega_c)t)\right) \otimes b^\dagger \otimes \sigma_+$$

$$+\hbar g\left(\sum_{k=1,3,\ldots}^{\infty}(i^{k-1}\eta_c^k)\hat{O}_k^c a^k \exp(i(\delta_{0c} - k\nu)t)\right) \otimes b \otimes \sigma_+$$

$$+\hbar g\left(\sum_{k=1,3,\ldots}^{\infty}(i^{k-1}\eta_c^k)a^{\dagger k}\hat{O}_k^c \exp(i(\delta_{0c} + k\nu)t)\right) \otimes b \otimes \sigma_+$$

$$+h.c.$$

where

$$\hat{O}_k^L := \exp\left(\frac{-\eta_L^2}{2}\right)\sum_{p=0}^{\infty}\frac{(i\eta_L)^{2p}a^{\dagger p}a^p}{p!(p+k)!}, \qquad \hat{O}_k^c := \exp\left(\frac{-\eta_c^2}{2}\right)\sum_{p=0}^{\infty}\frac{(i\eta_c)^{2p}a^{\dagger p}a^p}{p!(p+k)!}$$

and

$$\delta_{0L} := \omega_0 - \omega_L, \qquad \delta_{0c} := \omega_0 - \omega_c.$$

(ii) A basis is

$$|m\rangle \otimes |n\rangle \otimes |g\rangle, \qquad |m\rangle \otimes |n\rangle \otimes |e\rangle$$

where $m = 0, 1, \ldots, \infty$ denotes the state of ionic vibrational motion, $n = 0, 1, \ldots, \infty$ denotes the state of the quantized cavity field and $|g\rangle$ and $|e\rangle$ denote the ground state and excited state, respectively for the two-level ion. Using appropriate values for the parameters and the time we can find an implementation of the Hadamard gate

$$|m\rangle \otimes |0\rangle \otimes |g\rangle \rightarrow \frac{1}{2}(|m\rangle \otimes |0\rangle \otimes |g\rangle + |m\rangle \otimes |0\rangle \otimes |e\rangle)$$

$$|m\rangle \otimes |0\rangle \otimes |e\rangle \rightarrow \frac{1}{2}(|m\rangle \otimes |0\rangle \otimes |g\rangle - |m\rangle \otimes |0\rangle \otimes |e\rangle).$$

**Problem 11.** Consider a square lattice with lattice constant $a$ and periodic boundary conditions. Let $c_{\mathbf{j}}^\dagger$ ($c_{\mathbf{j}}$) denotes the creation (annihilation) operator for an electron in the Wannier state at the lattice site $\mathbf{j}$. The Hamilton operator $\hat{H}$ of spinless tight-binding electrons in the presence of a magnetic field can be written as

$$\hat{H} = \sum_{\langle \mathbf{j}_1, \mathbf{j}_2 \rangle} t_{\mathbf{j}_1 \mathbf{j}_2} c_{\mathbf{j}_1}^\dagger c_{\mathbf{j}_2}$$

with

$$t_{\mathbf{j}_1 \mathbf{j}_2} = -t \exp\left(-i\frac{e}{\hbar} \int_{\mathbf{j}_1}^{\mathbf{j}_2} \mathbf{A} \cdot d\mathbf{r}\right)$$

The summation $\langle \mathbf{j}_1, \mathbf{j}_2 \rangle$ runs over the nearest neighbour site on the square lattice. The uniform magnetic field $\mathbf{B}$ is applied in $z$-direction. Choosing the *Landau gauge*

$$\mathbf{A} = B(0, x, 0)$$

the line integral in $t_{\mathbf{j}_1, \mathbf{j}_2}$ can be written as

$$\frac{e}{\hbar} \int_{\mathbf{j}_1}^{\mathbf{j}_2} \mathbf{A} \cdot d\mathbf{r} = \begin{cases} 0 & \mathbf{j}_1 = (m, n) \ \mathbf{j}_2 = (m+1, n) \\ -2\pi m\Phi/\Phi_0 & \mathbf{j}_1 = (m, n) \ \mathbf{j}_2 = (m, n+1) \end{cases}$$

where the integers $m$ and $n$ refer to the $x = 1$, $y = 2$ coordinates of the square lattice sites. $\Phi = Ba^2$ is the magnetic flux through a unit plaquette. $\Phi_0$ stands for the magnetic flux quantum $h/e$. Find the Hamilton operator in the *Bloch representation*. The Fourier transform is given by

$$c_{\mathbf{k}} = \frac{1}{\sqrt{N}} \sum_{\mathbf{j}} e^{i(k_1 j_1 + k_2 j_2)} c_{\mathbf{j}}$$

with the inverse

$$c_{\mathbf{j}} = \frac{1}{\sqrt{N}} \sum_{\mathbf{k}} e^{-i(k_1 j_1 + k_2 j_2)} c_{\mathbf{k}}$$

where $\mathbf{k}$ runs over the first Brillouin zone.

**Solution 11.** We obtain the Hamilton operator in Bloch representation

$$\hat{H} = -2t \sum_{\mathbf{k}} \cos(k_1 a) c^\dagger(\mathbf{k}) c(\mathbf{k})$$

$$-t \sum_{\mathbf{k}} \left( \exp(-ik_2 a) c^\dagger \left( k_1 + 2\frac{\pi\Phi}{a\Phi_0}, k_2 \right) c(k_1, k_2) \right.$$

$$\left. + \exp(ik_2 a) c^\dagger \left( k_1 - 2\frac{\pi\Phi}{a\Phi_0}, k_2 \right) c(k_1, k_2) \right)$$

where $c^\dagger(\mathbf{k})$ creates an electron in the Block state with wave vector $\mathbf{k}$ and $c(\mathbf{k})$ annihilates an electron in the Block state with wave vector $\mathbf{k}$. If $\Phi/\Phi_0 = p/q$ is rational, the magnetic Brillouin zone can be reduced to $0 \leq k_1 \leq 2\pi/a$ and $0 \leq k_2 \leq 2\pi/(qa)$. We obtain

$$c \left( k_1 + 2\frac{\pi\Phi}{a\Phi_0}\ell, k_2 \right) = c \left( k_1 + 2\frac{\pi\Phi}{a\Phi_0}(\ell + q), k_2 \right)$$

with $\ell$ are integers.

# 26.3 Supplementary Problems

**Problem 1.** Let $b_1^\dagger$, $b_2^\dagger$ be Bose creation operators. Consider the Hamilton operator

$$\hat{H} = \hbar\omega b_1^\dagger b_1 (b_2^\dagger + b_2).$$

Find the unitary operator

$$U = \exp(-i\hat{H}t/\hbar) = \exp(-i\omega t b_1^\dagger b_1 (b_2^\dagger + b_2)).$$

**Problem 2.** Let $b_1^\dagger$, $b_2^\dagger$ be Bose creation operators. Study the Hamilton operator

$$\hat{H} = \hbar\omega_1 b_1^\dagger b_1 + \hbar\omega_2 b_2^\dagger b_2 + \frac{\kappa_1}{2}(b_1^\dagger)^2 b_1^2 + \frac{\kappa_2}{2}(b_2^\dagger)^2 b_2^2 + \hbar\omega(b_1^\dagger b_2 + b_1 b_2^\dagger) + \kappa_{12} b_1^\dagger b_1 b_2^\dagger b_2.$$

**Problem 3.** Let $|\beta\rangle$ be a coherent state. Then the operator

$$U = I - 2|\beta\rangle\langle\beta|$$

is unitary, i.e. $UU^\dagger = I$. Find a self-adjoint operator $K$ such that

$$U = \exp(-iK).$$

Study also the cases

$$V = I - 2|\zeta\rangle\langle\zeta|$$

where $|\zeta\rangle$ is a squeezed state and

$$W = I - 2|n\rangle\langle n|$$

where $|n\rangle$ is a number state.

# Bibliography

**Books**

Barnett S. M. and Radmore P. M.
*Methods in Theoretical Quantum Optics*, Oxford University Press (1997)

Bell J. S.
*Speakable and unspeakable in quantum mechanics*, Cambridge University Press (1989)

Benenti G., Casati G. and Strini G.
*Principles of Quantum Computing and Information*, Volume I: Basic Concepts, World Scientific (2007)

Blank J., Exner P. and Havlíček M.
*Hilbert Space Operators in Quantum Physics*, second edition, Springer (2008)

Bohm A.
*Quantum Mechanics*, Third edition, Springer (1994)

Bouwmeister D., Ekert A. K. and Zeilinger A.
*The Physics of Quantum Information: Quantum Cryptography. Quantum Teleportation, Quantum Computing*, Springer (2000)

Braunstein S. L. and Pati A. K.
*Quantum Information with Continuous Variables*, Kluwer (2003)

Cerf N. J., Leuchs G. and Polzik E. S.
*Quantum Information with continuous variables of atoms and light*, Imperial College Press (2007)

Constantinescu F. and Magyari E.
*Problems in Quantum Mechanics*, Pergamon Press (1971)

Dirac P. A. M.
*The Principle of Quantum Mechanics*, Clarendon Press (1958)

Flügge S.
*Practical Quantum Mechanics*, Springer (1974)

Gasiorowicz S.
*Quantum Physics*, Wiley (1974)

Gilmore R.
*Lie Groups, Lie Algebras, and Some of Their Applications*, Wiley (1974)

Glimm J. and Jaffe A.
*Quantum Physics*, Springer (1981)

Gruska J.
*Quantum Computing*, McGraw-Hill (1999)

Hardy Y. and Steeb W.-H.
*Classical and Quantum Computing with C++ and Java Simulations*
Birkhauser-Verlag (2002)

Hayashi M.
*Quantum Information: An Introduction*, Springer (2010)

Hirvensalo M.
*Quantum Computing*, second edition, Springer (2004)

Kaye P., Laflamme R. and Mosca M.
*An Introduction to Quantum Computing*, Oxford University Press (2007)

Kim Y. S. and Noz M. E.
*Phase Space Picture of Quantum Mechanics*, World Scientific (1991)

Kitaev A. Yu., Shen A. H. and Vyalyi M. N.,
*Classical and Quantum Computation*, American Mathematical Society (2002)

Klauder J. R. and Skagerstam B.
*Coherent States - Applications in Physics and Mathematical Physics*, World
Scientific (1985)

Kowalski K. and Steeb W.-H.
*Nonlinear Dynamical Systems and Carleman Linearization*, World Scientific, (1991)

Lewand R. E.
*Cryptological Mathematics*, Mathematical Association of America (2000)

Loudon R.
*The Quantum Theory of Light*, (3rd edition), Oxford University Press (2000)

Louisell W. H.
*Quantum Statistical Properties of Radiation*, Wiley (1973)

Mandel L. and Wolf E.
*Optical coherence and Quantum Physics*, Cambridge University Press (1995)

Mermin N. D.
*Quantum Computer Science*, Cambridge University Press (2007)

Merzbacher E.
*Quantum Mechanics*, 2nd edition, Wiley (1970)

Nielsen M. A. and Chuang I. L.
*Quantum Computation and Quantum Information*, Cambridge University Press (2000)

Prugovećki E.
*Quantum Mechanics in Hilbert Space*, second edition, Academic Press (1981)

Rieffel E. and Polak W.
*Quantum Computing: A Gentle Introduction*, MIT Press 2011

Sakurai J. J.
*Modern Quantum Mechanics*, Addison-Wesley (1994)

Scully M. O. and Zubairy M. S.
*Quantum Optics*, Cambridge University Press (1997)

Steeb W.-H. and Hardy Y.
*Matrix Calculus and Kronecker Product: A Practical Approach to Linear and Multilinear Algebra*, second edition, World Scientific (2010)

Steeb W.-H.
*Continuous Symmetries, Lie Algebras, Differential Equations and Computer Algebra*, second edition, World Scientific (2007)

Steeb W.-H.
*Hilbert Spaces, Wavelets, Generalized Functions and Quantum Mechanics*, Kluwer Academic Publishers (1998)

Steeb W.-H. and Hardy Y.
*Quantum Mechanics using Computer Algebra*, second edition, World Scientific (2010)

Steeb W.-H.
*Problems and Solutions in Theoretical and Mathematical Physics, Third Edition, Volume I: Introductory Level*, World Scientific (2010)

Steeb W.-H.
*Problems and Solutions in Theoretical and Mathematical Physics, Third Edition, Volume II: Advanced Level*, World Scientific (2010)

Steeb W.-H. and Hardy Y.
*Bose, Spin and Fermi Systems: Problems and Solutions*, World Scientific (2015)

Stolze J. and Suter D.
*Quantum Computing: A short course from theory to experiment*, Wiley (2008)

von Neumann J.
*Mathematical Foundations of Quantum Mechanics*, Princeton University Press (1955)

Walls D. F. and Milburn G. J.
*Quantum Optics*, Springer (1994)

Wilde M. M.
*Quantum Information Theory*, Cambridge University Press (2017)

**Papers**

**Part I**

Alber G., Delgado A., Gisin N. and Jex I., *Efficient bipartite quantum state purification in arbitrary dimensional Hilbert spaces*, J. Phys. A: Math. Gen. **34**, 8821 (2001)

Ban M., *Information and Entropy in Quantum Measurement Processes*, Int. J. Theor. Phys. **37**, 2491 (1998)

Bennett C. H., DiVincenzo D. P., Smolin J. A. and Wootters W. K., *Mixed State Entanglement and Quantum Error Correction*, Phys. Rev. A **54**, 3824 (1996)

Bennett C. H., Brassard G., Popescu S., Schumacher B., Smolin J. A. and Wootters W.K., *Purification of Noisy Entanglement and Faithful Teleportation via Noisy Channels*, Phys. Rev. Lett. **77**, 722 (1996)

Benioff P., *Models of quantum Turing machines*, Fortsch. Phys. **46**, 423 (1997)

Berman G. P., Borgonovi F., Chapline G., Gurvitz S. A., Hammel P. C., Pelekhov D. V., Suter A. and Tsifrinovich V. I., *Application of Magnetic Resonance Force Microscopy Cyclic Adiabatic Inversion for a Single-Spin Measurement*, J. Phys. A : Math. Gen. **36**, 4417 (2003)

Brandt H. E., *Riemannian Geometry of Quantum Computation*, Nonlinear Analysis: Theory, Methods and Applications **71**, e474 (2009)

Brassard G., Braunstein S. L. and Cleve R. *Teleportation as a quantum computation*, Physica D **120**, 43 (1998)

Brassard G., Cleve R. and Tapp A., *The cost of Exactly Simulating Quantum Entanglement with Classical Communication*, Phys. Rev. Lett. **83**, 1874 (1999)

Braunstein S. L., Caves C. M., Jozsa R., Linden N., Popescu S. and Schack R., *Separability of Very Noisy Mixed States and Implications for NMR Quantum Computing*, Phys. Rev. Lett. **83**, 1054 (1999)

Braunstein S. L., Mann A. and Revzen M., *Maximal Violation of Bell Inequalities for Mixed States*, Phys. Rev. Lett. **68**, 3259 (1992)

Brukner C., Pan J.-W., Simon C., Weihs G. and Zeilinger A., *Probabilistic Instantaneous Quantum Computation*, Phys. Rev. A **67**, 034304 (2003)

Bruß D., *Characterizing Entanglement*, J. Math. Phys. **43**, 4237 (2002)

Buhrman H., Cleve R. and van Dam W., *Quantum Entanglement and Communication Complexity*, SIAM Journ. on Computing **30**, 1829 (2001)

Childs A. M., Leung D., Verstraete F. and Vidal G., *Asymptotic entanglement capacity of the Ising and anisotropic Heisenberg interactions*, Quantum Information and Computation **3**, 97 (2003)

Chuang I. L. and Yamamoto Y., *A Simple Quantum Computer*, Phys. Rev. A **52**, 3489 (1995)

Cleve R. and Buhrman H., *Substituting Quantum Entanglement for Communication*, Phys. Rev. A **56**, 1201 (1997)

Cleve R., Ekert A., Macchiavello C. and Mosca M., *Quantum Algorithms Revisited*, Proc. Roy. Soc. Lond. A **454**, 339 (1998)

Dehaene J. and De Moor B., *Clifford group, stabilizer states, and linear and quadratic operations over GF(2)*, Phys. Rev. A **68**, 042318 (2003)

de Almeida N. G., *All-quantum teleportation*, arXiv:1703.07278

Deutsch D., *Quantum Computation Networks*, Proc. R. Soc. London A **425**, 73 (1989)

Deutsch D., Ekert A., Josza R., Macchiavello C., Popescu S. and Sanpera A., *Quantum Privacy Amplification and the Security of Quantum Cryptography over Noisy Channels*, Phys. Rev. Lett. **77**, 2818 (1996)

Deutsch D. and Jozsa R., *Rapid solutions of problems by quantum computation*, Proc. R. Soc. London **439**, 553 (1992)

Edamatsu K., *Complete and Deterministic Bell State Measurement Using Spin Products*, arXiv:1612.08578

Ekert A., *Quantum Cryptography Based on Bell's Theorem*, Phys. Rev. Lett. **67**, 661 (1991)

Ekert A. and Knight P. L., *Entangled quantum systems and the Schmidt decomposition*, American Journal of Physics, **63**, 415 (1995)

Fan H., Matsumoto K. and Imai H., *Quantify entanglement by concurrence hierarchy*, J. Phys. A: Math. Gen. **36**, 4151 (2003)

Gleason A. M., *Measures on the closed subspaces of a Hilbert space*, Indiana University Mathematics Journal, **6**, 885 (1957)

Gottesman D., *Grand Mathematical Challenge for the Twenty-First Century and the Millennium*, edited by S. J. Lomonaco Jr., (American Mathematical Society, Providence, Rhode Island), 221 (2002)

Gottesman D., *A Class of Quantum Error-Correcting Codes Saturating the Quantum Hamming Bound*, Phys. Rev. A **54**, 1862 (1996)

Gottesman D., *An Introduction to Quantum Error Correction and Fault-Tolerant Quantum Computation*, arXiv:0904.2557v1

Gottesman D., *An Introduction to Quantum Error Corrections and Fault-Tolerant Quantum Computation*, in "Quantum Information Science and Its Contributions to Mathematics", Proceedings of Symposia in Applied Mathematics, **68**, 13 (2010)

Grabowski J., Kuś M. and Marmo G., *Geometry of quantum systems: density states and entanglement*, J. Phys. A: Math. Gen. **38**, 10217 (2005)

Grover L., *Quantum Mechanics helps searching for a needle in a haystack*, Phys. Rev. Lett. **79**, 325 (1997)

Grudka A. and Wojcik A., *How to encode the states of two non-entangled qubits in one qutrit*, Phys. Lett. A **314**, 350 (2003)

Hardy Y., Steeb W.-H. and Stoop R., *Fully Entangled Quantum States in $C^{N^2}$ and Bell Measurement*, Int. J. Theor. Phys. **42**, 2314 (2003)

Hardy Y., Steeb W.-H. and Stoop R., *Entanglement, Disentanglement and Wigner Functions*, Physica Scripta **69**, 166 (2004)

Hardy Y. and Steeb W.-H., *Unitary Operators, Entanglement, and Gram-Schmidt Orthogonalization*, Int. J. of Mod. Phys. C **20**, 891 (2009)

Hardy Y. and Steeb W.-H., *Genetic Algorithms and Optimization Problems in Quantum Computing*, Int. J. Mod. Phys. C **21**, 1359 (2010)

Henderson L. and Vedral V., *Information, Relative Entropy of Entanglement and Irreversibility*, Physical Review Letters, **84**, 2263 (2000)

Horodecki M., P. Horodecki and R. Horodecki, *Separability of Mixed States: necessary and sufficient conditions*, Physics Letters A, **223**, 1 (1996)

Jordan T. F., *Quantum mysteries explored*, Am. J. Phys. **62**, 874 (1994)

Jozsa R., *Quantum Algorithms and the Fourier transform*, Proc. Roy. Soc. London A **454**, 323 (1998)

Kielpinski D., *A small trapped-ion quantum register*, J. Opt. B: Quantum Semiclass. Opt. **5**, R121 (2003)

Kimura G., *The Bloch vector for N-level systems*, Phys. Lett. A **314**, 319 (2003)

Korbelář M. and Tolar J., *Symmetries of the finite Heisenberg group for composite systems*, J. Phys. A: Math. Theor. **43**, 375302 (2010)

Kuzmich A., Walmsley A. and Mandel L., *Violation of a Bell-type inequality in the homodyne measurement of light in an Einstein-Podolsky-Rosen state*, Phys. Rev. A **64**, 063804 (2001)

Laflamme L., Miquel C., Paz J. P. and Zurek W. H., *Perfect Quantum Error Correction Code*, Phys. Rev. Lett. **77**, 198 (1996)

Maassen H. and Uffink J. B. M., *Generalized Entropic Uncertainty Relations*, Phys. Rev. Lett. **60**, 1103 (1988)

Mermin N. D., *From Classical State-Swapping to Quantum Teleportation*, Phys. Rev. A **65**, 012320 (2002)

Nielsen M. A., *Quantum information theory*, Ph.D. thesis, University of New Mexico, (1998)

Nielsen M. A., Dawson C. M., Dodd J. L., Gilchrist A., Mortimer D., Osborne T. J., Bremner M. J., Harrow A. W. and Hines A., *Quantum dynamics as a physical resource*, Physical Review A, **67**, 052301 (2003)

Nielsen M. A. and Chuang L., *Programmable Quantum Gate Arrays*, Phys. Rev. Lett. **79**, 321 (1997)

Pan J.-W., Bouwmeester D., Daniell M., Weinfurter H. and Zeilinger A., *Experimental test of quantum nonlocality in three-photon Greenberger-Horne-Zeilinger entanglement*, Nature **403**, 515 (2000)

Planat M. and Jorrand P., *Group theory for quantum gates and quantum coherence*, J. Phys. A: Math. Theor. **41**, 182001 (2008)

Sharma S. S., *Tripartite GHZ state generation with trapped ion in optical cavity*, Phys. Lett. A **311**, 111 (2003)

Shor P. W., *Scheme of reducing decoherence in quantum computer memory*, Phys. Rev. A **52**, 2493 (1995)

Shor P. W., *Polynomial-time algorithm for prime factorization and discrete logarithms on a quantum computer*, Soc. Ind. Appl. Math. J. Comp. 1484 (1997)

Steeb W.-H. and Hardy Y., *Entangled Quantum States and a C++ Implementation*, Int. J. Mod. Phys. C **11**, 69 (2000)

Steeb W.-H. and Hardy Y., *Quantum Computing and SymbolicC++ Simulations*, Int. J. Mod. Phys. C **11**, 323 (2000)

Steeb W.-H. and Hardy Y., *Energy Eigenvalue Level Motion and a SymbolicC++ Implementation*, Int. J. Mod. Phys. C **11**, 1347 (2000)

Steeb W.-H. and Hardy Y., *Fermi Systems, Hubbard Model and a SymbolicC++ Implementation*, Int. J. Mod. Phys. C **12**, 235 (2001)

Steeb W.-H. and Hardy Y., *Energy Eigenvalue Level Motion with Two Parameters*, Z. Naturforsch. **56 a**, 565 (2001)

Steeb W.-H. and Hardy Y., *Entangled Quantum States and the Kronecker Product*, Z. Naturforsch. **57 a**, 689 (2002)

Steeb W.-H., Hardy Y., Stoop R., *Discrete wavelets and perturbation theory*, J. Phys A: Math. Gen. **36**, 6807 (2003)

Steeb W.-H. and Hardy Y., *A hierarchy of Hamilton operators and entanglement*, Central European Journal of Physics, **7**, 854 (2009)

Steeb W.-H., *Diabolic Points and Entanglement*, Physica Scripta **81**, 025012 (2010)

Steeb W.-H., *Hamilton Operators, Discrete Symmetries, Brute Force and SymbolicC++*, Int. J. Mod. Phys. C **24**, 1250095 (2013)

Sutter D., Scholz V. B., Winter A. and Renner R., *Approximate Degradable Quantum Channels*, arXiv:1412.0980v3

Thomsen M. K., Glück R. and Axelsen H. B., *Reversible arithmetic logic unit for quantum arithmetic*, J. Phys. A: Math. Theor. **43**, 382002 (2010)

Trifonov A., Tsegaye T., Björk G., Söderholm J., Goobar E., Atatüre M. and Sergienko A. V., *Experimental demonstration of the relative phase operator*, J. Opt. B: Quantum Semiclass. Opt., **2**, 105 (2000)

Tyson J., *Operator-Schmidt decomposition of the quantum Fourier transform on $C^{N_1} \otimes C^{N_2}$*, J. Phys. A: Math. Gen. **36** 6813 (2003) van Loock P. and Braunstein S. L., *Greenberger-Horne-Zeilinger nonlocality in phase space*, Phys. Rev. A **63**, 022106 (2001)

Wang X. and Sanders B. C., *Entanglement capability of self-inverse Hamiltonian evolution*, Phys. Rev. A **68**, 014301 (2003)

Werner R. F., *Quantum states with Einstein-Podolsky-Rosen correlations admitting a hidden-variable model*, Phys. Rev. A **40**, 4277 (1989)

Wong A. and Christensen N., *Potential multiparticle entanglement measure*, Phys. Rev. A **63**, 044301

Yang G., Hung W. N. N., Song X. and Perkowski M., *Majority based reversible logic gates*, Theoretical Computer Science **334**, 259 (2005)

Yura F., *Entanglement cost of three-level antisymmetric states*, J. Phys. A: Math. Gen. **36**, L237 (2003)

**Part II**

Ban M., *The phase operator in quantum information processing*, J. Phys. A: Math. Gen. **35**, L193 (2002)

Ban M., *Phase-space approach to continuous variable quantum teleportation*, Phys. Rev. A **69**, 054304 (2004)

Banerji J., *Non-linear wave packet dynamics of coherent states*, PRAMANA J. Phys. **56**, 267 (2001)

Barak R. and Ben-Aryeh Y., *Photon statistics and entanglement in coherent-squeezed linear Mach-Zehnder and Michelson interferometers*, J. Opt. Soc. Am. B **25**, 361 (2008)

Bartlett S. D., Sanders B. C., Braunstein S. L. and Nemoto K., *Efficient Classical Simulations of Continuous Variable Quantum Information Processes*, Phys. Rev. Lett. **88**, 097904 (2002)

Barnett S. M., Jeffers J. and Gatti A., *Quantum optics of lossy beam splitters*, Phys. Rev. A **57**, 2134 (1998)

Braunstein S. L. and Kimble H. J., *Teleportation of Continuous Quantum Variables*, Phys. Rev. Lett. **80**, 869 (1998)

Braunstein S. L., Fuchs C. A., Kimble H. J. and van Loock P., *Quantum versus classical domains for teleportation with continuous variables*, Phys. Rev. A **64**, 022321 (2001)

Buzek V., Vidiella-Barranco and Knight P. L., *Superpositions of coherent states: Squeezing and dissipation*, Phys. Rev. A **45**, 6570 (1992)

Carmichael H. J., Milburn G. J. and Walls D. F., *Squeezing in a detuned parametric amplifier*, J. Phys. A: Math. Gen. **17**, 469 (1984)

Cerf N. J., Adami C. and Kwiat P. G., *Optical Simulation of Quantum Logic*, Phys. Rev. A **57**, 1477 (1998)

Cirac J. I., *Quantum Information: Entanglement, Purification, Error Correction, and Quantum Optical Implementations*, in Fundamentals of Quantum Information, Heiss D. (Ed.), Springer, Berlin (2002)

Chruściński D., *Spectral Properties of the Squeeze Operator*, Phys. Lett. A **327**, 290-295 (2004)

D'Ariano G. M., De Martin F. and Sacchi M. F., *Continuous Variable Cloning via Network of Parametric Gates*, Phys. Rev. Lett. **86**, 914 (2001)

DasGupta A., *Disentanglement formulas: An alternative derivation and some applications to squeezed coherent states*, Am. J. Phys. **64**, 1422 (1996)

de Oliveira M. C. and Milburn G. J., *Discrete teleportation protocol of continuum spectra field states*, Phys. Rev. A **65**, 032304 (2002)

Dell'Anno F., De Siena S. and Illuminati F., *Multiphoton quantum optics and quantum state engineering*, Physics Reports, **428**, 53 (2006)

Dodonov V. V., Man'ko V. I. and Nikonov D. E., *Even and odd coherent states for multimode parametric systems*, Phys. Rev. A **51**, 3328 (1995)

Einstein A., Podolsky B. and Rosen N., *Can Quantum-Mechanical Description of Physical Reality Be Considered Complete?*, Phys. Rev. **47**, 777 (1935)

Eisert J., *Optimizing Linear Optics Quantum Gates*, Phys. Rev. Lett. **95**, 040502 (2005)

Ferraro A., Olivares S. and Paris M. G. A., *Gaussian states in a continuous variable quantum information*, arXiv:quant-ph/0503237v1, 2005

Fujii K., *Basic Properties of Coherent and Generalized Coherent Operators Revisited*, Mod. Phys. Lett. A **16**, 1277 (2001)

Hardy Y. and Steeb W.-H., *Fermi-Bose Systems, Macroscopic Quantum Superposition States and Entanglement*, Int. J. Theor. Phys. **43**, 2207 (2004)

Hong-Yi F. and Klauder J. R., *Canonical coherent state representation of some squeeze operators*, J. Phys. A: Math. Gen. **21**, L725-L730 (1988)

Imoto N., Haus H. A. and Yamamoto Y., *Quantum nondemolition measurement of the photon number via the optical Kerr effect*, Phys. Rev. A **32**, 2287 (1985)

Jeffers J., Barnett S. M. and Pegg D., *Retrodiction as a tool for micromaser field measurement*, J. Mod. Opt. **49**, 925 (2002)

Joo J. and Ginossar E., *Hybrid teleportation via entangled coherent states in circuit quantum electrodynamics*, arXiv:1509.02859

Knill E., Laflamme R. and Milburn G. J., *A scheme for efficient quantum computation with linear optics*, Nature **409**, 46 (2001)

Kok P., Munro W. J., Nemoto K., Ralph T. C., Dowling J. P. and Milburn G. J., *Linear optical quantum computing with photonic qubits*, Rev. Mod. Phys. **79**, 1 (2007)

Koniorczyk M., Buzek V. and Janszky J., *Wigner-function description of quantum teleportation in arbitrary dimensions and a continuous limit*, Phys. Rev. A **64**, 034301 (2001)

Kok P., Munro W. J., Nemoto K., Ralph T. C., Dowling J. P. and Milburn G. J., *Linear Optical Quantum Computing*, arXiv:quant-ph/0512071v2

Kok P. and Braunstein S. L., *Multi-dimensional Hermite polynomials in quantum optics*, J. Phys. A: Math. Gen. **34**, 6185 (2001)

Lee J. and Kim M. S., *Quantum Teleportation Via Mixed Two-Mode Squeezed States in the Coherent-State Representation*, J. Kor. Phys. Soc. **42**, 457 (2003)

Li Shang-Bin and Xu Jing-Bo, *Quantum probabilistic teleportation via entangled coherent states*, Phys. Lett. A **309**, 321 (2003)

Lloyd S. and Braunstein S. L., *Quantum Computing over Continuous Variables*, Phys. Rev. Lett. **82**, 1784 (1999)

Lo C. F., *Eigenfunctions and eigenvalues of squeeze operators*, Phys. Rev. A **42**, 752 (1990)

Milburn G. J. and Braunstein S. L., *Quantum teleportation with squeezed states*, Phys. Rev. A **60**, 937 (1999)

Pan Jian-Wei, Daniell M., Gasparoni S., Weihs G., and Zeilinger A., *Experimental Demonstration of Four-Photon Entanglement and High-Fidelity Teleportation*, Phys. Rev. Lett. **86**, 4435 (2001)

Paris M. G. A., *Generation of mesoscopic quantum superpositions through Kerr-stimulated degenerate downconversion*, J. Opt. B: Quantum Semiclass. Opt. **1**, 662 (1999)

Pegg D. T., Barnett S. M. and Jeffers J., *Quantum theory of preparation and measurement*, J. Mod. Opt. **49**, 913 (2002)

Ralph T. C., Munro W. J. and Milburn G. J., *Quantum Computation with Coherent States, Linear Interactions and Superposed Resources*, arXiv:quant-ph/0110115v2

Schrödinger E., *Die gegenwärtige Situation in der Quantenmechanik*, Naturwissenschaften **23**, pp. 807-812; 823-828; 844-849 (1935)

Serafini A., De Siena S., Illuminati F. and Paris M. G. A., *Minimum decoherence cat-like states in Gaussian noisy channels*, arXiv:quant-ph/0310005v2

Sharma S. S. and Vidiella-Barranco A., *Fast quantum logic gates with trapped ions interacting with external laser and quantized cavity field beyond Lamb-Dicke regime*, Phys. Lett. A **309** 345 (2003)

Slosser J. J. and Milburn G. J., *Creating Metastable Schrödinger Cat States*, Phys. Rev. Lett. **75**, 418 (1995)

Steeb W.-H., *Entanglement, Bose operators, coherent states and classical dynamical systems*, J. Phys. A: Math. Theor. **40**, 9577 (2007)

Steeb W.-H. and Hardy Y., *Quantum Optics Networks, Unitary Operators and Computer Algebra*, Int. J. of Mod. Phys. C **19**, 1069 (2008)

Steeb W.-H., *Ordinary Differential Equations, First Integrals, Conformal Invariants, Bose Operators, Coherent States and Entanglement*, Int. J. Geom. Methods in Modern Physics, **5**, 1057 (2008)

Thamm R., Kolley E. and Kolley W., *Green's Functions for Tight-Binding Electrons on a Square Lattice in a Magnetic Field*, phys. stat. sol. (b) **170**, 163 (1992)

van Enk S. J. and Hirota O., *Entangled coherent states: Teleportation and decoherence*, Phys. Rev. A **64**, 022313 (2001)

van Loock P. and Braunstein S. L., *Multipartite Entanglement for Continuous Variables: A Quantum Teleportation Network*, **84**, 3482 (2000)

Vaziri A., Mair A., Weihs G. and Zeilinger A., *Entanglement of the Angular Orbital Momentum States of the Photons*, Nature **412**, 313 (2001)

Youn S.-H., *Conditional Generation Scheme for Entangled Vacuum Evacuated Coherent States by Mixing Two Coherent Beams with a Squeezed Vaccum State*, arXiv:1511.07956

Yurke B. and Stoler D., *Generating quantum mechanical superpositions of macroscopically distinguishable states via amplitude dispersion*, Phys. Rev. Lett. **57**, 13 (1986)

Zou Xu-Bo, Pahlke K. and Mathis W., *The non-deterministic quantum logic operation and the teleportation of the superposition of the vacuum state and the single-photon state via parametric amplifiers*, Phys. Lett. A **311** 271 (2003)

# Index

Printed in the United States
By Bookmasters